RULES FOR DERIVATIVES (continued)

Constant Times a Function Let k be a real number. Then the derivative of $y = k \cdot f(x)$ is

$$y' = k \cdot f'(x).$$

Sum or Difference Rule If $f(x) = u(x) \pm v(x)$, then $f'(x) = u'(x) \pm v'(x)$.

Product Rule If $f(x) = u(x) \cdot v(x)$, then

$$f'(x) = u(x) \cdot v'(x) + v(x) \cdot u'(x).$$

Quotient Rule If $f(x) = \dfrac{u(x)}{v(x)}$, and $v(x) \neq 0$, then

$$f'(x) = \frac{v(x) \cdot u'(x) - u(x) \cdot v'(x)}{[v(x)]^2}.$$

Chain Rule If y is a function of u, say $y = f(u)$, and if u is a function of x, say $u = g(x)$, then $y = f(u) = f[g(x)]$, and

$$\frac{dy}{dx} = \frac{dy}{du} \cdot \frac{du}{dx}.$$

Chain Rule (Alternate Form) If $y = f[g(x)]$, then $y' = f'[g(x)] \cdot g'(x)$.

Generalized Power Rule Let u be a function of x, and let $y = u^n$, for any real number n. Then $y' = n \cdot u^{n-1} \cdot u'$.

4.3

Logarithmic Function If $f(x) = \ln |g(x)|$, then

$$f'(x) = \frac{d}{dx} \ln |g(x)| = \frac{g'(x)}{g(x)}.$$

4.4

Exponential Function If $f(x) = e^{g(x)}$, then

$$f'(x) = \frac{d}{dx} e^{g(x)} = g'(x)\, e^{g(x)}.$$

3.2 FIRST DERIVATIVE TEST

Let c be a critical number for a function f. Suppose that f is differentiable on (a, b), and that c is the only critical number for f in (a, b).

1. (c) is a relative maximum of f if the derivative $f'(x)$ is positive in the interval (a, c) and negative in the interval (c, b).
2. $f(c)$ is a relative minimum of f if the derivative $f'(x)$ is negative in the interval (a, c) and positive in the interval (c, b).

3.4 SECOND DERIVATIVE TEST

Let f'' exist on some open interval containing c, and let $f'(c) = 0$.

1. If $f''(c) > 0$, then $f(c)$ is a relative minimum.
2. If $f''(c) < 0$, then $f(c)$ is a relative maximum.
3. If $f''(c) = 0$, then the test gives no information about extrema.

Calculus with Applications

Fifth Edition

TO THE STUDENT

If you want further help with this course, you may wish to buy a copy of the *Student's Solution Manual* (ISBN 0-673-46755-4) that accompanies this textbook. This manual provides detailed, step-by-step solutions to the odd-numbered exercises in the textbook and can help you study and understand the course material. Also included for each chapter is a practice test, with answers so you can check your work. Your college bookstore either has this manual or can order it for you.

Ask your instructor for the program *GraphExplorer* that will be of help in approaching some of the exercises marked with [icon]. This program contains a comprehensive graphing utility and is available in IBM and Macintosh formats.

If you would like to find out how spreadsheets can help you learn and explore some of the topics discussed in this text, you might want to obtain a copy of Sam Spero's *The Electronic Spreadsheet and Elementary Calculus* (ISBN 0-673-46595-0) from your college bookstore. This booklet will help you get started using spreadsheets to enhance your problem-solving skills. Knowledge of spreadsheets is not assumed, and the material is adaptable to any spreadsheet program that may be available to you.

TO THE INSTRUCTOR

We will make available as ''Special Topic'' supplements the following titles: *Sequences, Series, and Other Topics*; *Linear Programming: The Graphical Method* and *Linear Programming: The Simplex Method*; and *Introduction to the Graphing Calculator*. Please place your order with your HarperCollins sales representative. These supplements are free to you and your students.

Calculus with Applications

Fifth Edition

Margaret L. Lial
American River College

Charles D. Miller

Raymond N. Greenwell
Hofstra University

HarperCollins*CollegePublishers*

Sponsoring Editor: George Duda
Developmental Editor: Adam Bryer
Project Editor: Janet Tilden
Art Director: Julie Anderson
Text and Cover Designer: Jeanne Calabrese
Photo Researcher: Karen Koblik
Photo Acknowledgments: Unless otherwise acknowledged, all photographs are the property of ScottForesman. *Cover*: Peter Aaron/Esto Photographics; *page xix*: Peter Aaron/Esto Photographics; *page 1*: T. Zimmerman/F.P.G.; *page 68*: Bob Daemmrich/ Stock Boston; *page 201*: David Young-Wolf/PhotoEdit; *page 248*: Bill Pierce/Rainbow; *page 305*: Cecil Brunswick/Tony Stone Worldwide; *page 363*: NASA; *page 477*: Didier Klein/Vandystadt/ALLSPORT; *page 517*: James T. Jones/Frazier Photolibrary; *page 548*: Alpha/F.P.G.
Production Administrator: Brian Branstetter
Manufacturing Manager: Paula Keller
Compositor: Waldman Graphics
Printer and Binder: R.R. Donnelley & Sons Company
Cover Printer: The Lehigh Press, Inc.

Calculus with Applications, Fifth Edition

Library of Congress Cataloging-in-Publication Data
Lial, Margaret L.
Calculus with applications / Margaret L. Lial, Charles D. Miller, Raymond N. Greenwell.—5th ed.
p. cm.
Includes index.
ISBN 0-673-46726-0
1. Calculus I. Miller, Charles David. II. Greenwell, Raymond N. III. Title
QA303.L4818 1992
515—dc20 92-27740

93 94 95 9 8 7 6 5 4 3 2

PREFACE

Calculus with Applications is a solid, application-oriented text for students majoring in business, management, economics, or the life or social sciences. A prerequisite of three to four semesters of high school algebra is assumed. Many new features, including new exercises, new applications, and motivational section-opening questions, make using this edition of the text easier and more enjoyable. Application exercises are designed to be as believable and realistic as possible.

This edition continues to offer the many popular features of the fourth edition: applications grouped by subject matter with subheadings indicating the specific application; extended applications to motivate student interest; careful exposition; fully developed examples with side comments; carefully graded exercises; and an algebra reference, designed to be used either in class or individually as necessary. The index of appications shows the abundant variety of applications included in the text and allows direct reference to particular topics.

NEW AND ENHANCED FEATURES

Conceptual and Writing Exercises To complement the drill and application exercises, several exercises that strengthen conceptual understanding are included in almost every exercise set. Also included are exercises that require the student to respond by writing a few sentences. (Some writing exercises are also conceptual in nature.)

Connections Some exercise sets and many review sections include one or more exercises (labeled ''Connections'') that integrate concepts and skills introduced earlier with those just introduced in the chapter, or that integrate different concepts presented in the chapter.

Summaries A few chapters include a summary of rules or formulas designed to help the student sort out the different ideas introduced in the chapter. These are included in chapters where students traditionally have trouble and become confused about when to use one of several techniques.

Section Openers Most sections open with a thought-provoking question that is answered in an application within the section or in the exercises.

Margin Reviews This feature is designed to help students better understand the ideas being presented. These notes give short explanations or comments reminding students of skills or techniques learned earlier that are needed at this point. Some include a few practice exercises; some include reference to material presented earlier in the text.

Cautions and Notes Common student difficulties and errors are now highlighted graphically and identified with the heading ''Caution.'' Important comments and asides are treated similarly and given the heading ''Notes.''

KEY CONTENT CHANGES

The use of a scientific calculator is assumed in this edition, and references to calculator use occur throughout as appropriate. Although a graphing calculator is not required, its usefulness is mentioned, and some clearly labeled computer or graphing calculator exercises are included in a few sections. These exercises showcase the programming capability of graphing calculators, in addition to their graphing capabilities.

This edition preserves the basic format and topic order of the last edition, with the following changes.

- A new section on the graphing techniques of translations and reflections has been added to Chapter 1 on Functions and Graphs.
- In Chapter 2, the material on continuity has been combined with the introduction of limits in Section 2.1. Limits at infinity are now included in Chapter 3 in a new section discussing curve sketching.
- Applications of the derivative, presented in one long chapter in earlier editions, are now covered in Chapters 3 and 4.
- Some discussion on math of finance has been included in the section on exponential functions.
- Growth and decay applications of exponential and logarithmic functions are presented in Section 5.3, before the derivatives are introduced. These first three sections require no calculus and can be presented earlier with the functions discussed in Chapter 1.
- In Chapter 9, the first section includes elementary and separable differential equations, combining the first two sections of the previous edition.

SUPPLEMENTS

For the Instructor The *Instructor's Guide and Solution Manual* contains solutions to even-numbered exercises; a multiple-choice version and a short-answer version of a final exam, with answers; at least 100 extra test questions per chapter, with answers; and a set of concise teaching tips.

The *Instructor's Answer Manual* provides answers to every exercise in the text.

The *HarperCollins Test Generator* is the foremost product of its kind, enabling the instructor to create multiple versions of tests, to randomly regenerate "variables," and to insert questions of his or her own devising. The Test Generator comes free to adopters.

Accompanying two-color *Overhead Transparencies*, given free to adopters, can help enhance lectures.

The supplements in the series "Special Topics to Accompany *Calculus with Applications*" will be of interest. *Sequences, Series, and Other Topics*; *Linear Programming: The Graphical Method* and *Linear Programming: The Simplex Method*; and *Introduction to the Graphing Calculator* will be offered to you and your students at no charge should you desire to cover these topics.

For the Student The *Student's Solution Manual* (ISBN 0-673-46755-4) provides solutions to odd-numbered exercises and sample chapter tests with answers.

The Electronic Spreadsheet and Elementary Calculus (ISBN 0-673-46595-0), by Sam Spero, Cuyahoga Community College, helps students get started controlling graphing and problem solving by means of the spreadsheet. Knowledge of spreadsheets is not assumed, and the approach is adaptable to all spreadsheet programs.

GraphExplorer provides students and instructors with a comprehensive graphing utility, and is available in IBM and Macintosh formats.

Acknowledgments We wish to thank the following professors for their contributions in reviewing portions of this text.

Dana Dwight Clahane, *El Camino College*
Jerry R. Ehman, *Franklin University*
Madelyn T. Gould, *Dekalb College*
Joseph W. Guest, *Cleveland State Community College*
Vivian Heigl, *University of Wisconsin—Parkside*
Robert Keever, *State University of New York—Plattsburgh*
Jeff Knisley, *East Tennessee State University*
Martin Kotler, *Pace University Pleasantville—Briarcliff Campus*
W. David Laverell, *Calvin College*
Susan S. Lenker, *Central Michigan University*
Billy R. Nail, *Clayton State College*
Charles Rees, *University of New Orleans*
Mary Kay Schippers, *Fort Hays State University*
Arnold L. Schroeder, *Long Beach City College*
Joseph D. Sloan, *Lander College*
C. Donald Smith, *Louisiana State University in Shreveport*
Lyndon C. Weberg, *University of Wisconsin—River Falls*
Earl J. Zwick, *Indiana State University*

Paul Eldersveld, College of DuPage, deserves our gratitude for doing an excellent job coordinating all of the print ancillaries for us, an enormous and time-consuming task. Paul Van Erden, American River College, has created an accurate and complete index for us, and Jim Walker, also of American River College, has compiled the application index and teaching tips. Barbara Bohannon, Peter Grassi, and David Knee, at Hofstra University, have provided many helpful comments. For their invaluable help in maintaining high standards of accuracy in the answer section, we thank C. Donald Smith of Louisiana State University in Shreveport and Stephen Ryan and Margo McCullagh of Hofstra University. We also want to thank Karla Harby for her editorial assistance and the staff at HarperCollins whose contributions have been very important in bringing this project to a successful conclusion: George Duda, Adam Bryer, Linda Youngman, Janet Tilden, and Julie Anderson.

Finally, we want to express our deep appreciation to W. Weston Meyer, Richard E. Klabunde, Allen Blaurock, Alan Opsahl, Jane Gillum, and Henry Robinson, who provided the materials for the "Calculus at Work" essays that give additional perspective on the use of calculus in the workplace, and to Laurie Golson, who edited the essays.

Margaret L. Lial
Raymond N. Greenwell

CONTENTS

SPECIAL TOPICS TO ACCOMPANY CALCULUS WITH APPLICATIONS*

---

*These supplements will be provided free to adopters who request them.

INDEX OF APPLICATIONS

Business and Economics

General Interest

Life Sciences

Physical Sciences

Social Sciences

ALGEBRA REFERENCE

The study of calculus is impossible without a thorough knowledge of elementary algebra. The word *algebra* is derived from the Arabic word *al-jabr*, which appeared in the title of a book by Arab mathematician al-Khowarizmi in the early ninth century A.D. The study of algebra in modern times has grown to include many abstract ideas, but in this text we will be concerned primarily with the topics of elementary algebra found in al-Khowarizmi's book.

This algebra reference is designed for self-study; you can study it all at once or refer to it when needed throughout the course. Since this is a review, answers to all exercises are given in the answer section at the back of the book.

R.1 POLYNOMIALS

A **polynomial** is an expression of the form

$$a_nx^n + a_{n-1}x^{n-1} + \cdots + a_1x + a_0,$$

where $a_0, a_1, a_2, \ldots, a_n$ are real numbers, n is a natural number, and $a_n \neq 0$. Examples of polynomials include

$$5x^4 + 2x^3 + 6x, \qquad 8m^3 + 9m^2 - 6m + 3, \qquad 10p, \qquad \text{and} \qquad -9.$$

Adding and Subtracting Polynomials An expression such as $9p^4$ is a **term;** the number 9 is the **coefficient**, p is the **variable,** and 4 is the **exponent.** The expression p^4 means $p \cdot p \cdot p \cdot p$, while p^2 means $p \cdot p$, and so on. Terms having the same variable and the same exponent, such as $9x^4$ and $-3x^4$, are **like terms.** Terms that do not have both the same variable and the same exponent, such as m^2 and m^4, are **unlike terms.**

Polynomials can be added or subtracted by using the **distributive property,** shown below.

> If a, b, and c are real numbers, then
>
> $a(b + c) = ab + ac$ and $(b + c)a = ba + ca.$

Only like terms may be added or subtracted. For example,

$$12y^4 + 6y^4 = (12 + 6)y^4 = 18y^4,$$

and

$$-2m^2 + 8m^2 = (-2 + 8)m^2 = 6m^2,$$

but the polynomial $8y^4 + 2y^5$ cannot be further simplified. To subtract polynomials, use the fact that $-(a + b) = -a - b$. In the next example, we show how to add and subtract polynomials.

EXAMPLE 1 Add or subtract as indicated.

(a) $(8x^3 - 4x^2 + 6x) + (3x^3 + 5x^2 - 9x + 8)$

Combine like terms.

$$(8x^3 - 4x^2 + 6x) + (3x^3 + 5x^2 - 9x + 8)$$
$$= (8x^3 + 3x^3) + (-4x^2 + 5x^2) + (6x - 9x) + 8$$
$$= 11x^3 + x^2 - 3x + 8$$

(b) $(-4x^4 + 6x^3 - 9x^2 - 12) + (-3x^3 + 8x^2 - 11x + 7)$
$$= -4x^4 + 3x^3 - x^2 - 11x - 5$$

(c) $(2x^2 - 11x + 8) - (7x^2 - 6x + 2)$
$$= (2x^2 - 11x + 8) + (-7x^2 + 6x - 2)$$
$$= -5x^2 - 5x + 6$$ ◀

Multiplying Polynomials The distributive property is used also when multiplying polynomials, as shown in the next example.

Multiply.

(a) $8x(6x - 4)$

$$8x(6x - 4) = 8x(6x) - 8x(4)$$
$$= 48x^2 - 32x$$

(b) $(3p - 2)(p^2 + 5p - 1)$

$$(3p - 2)(p^2 + 5p - 1)$$
$$= 3p(p^2) + 3p(5p) + 3p(-1) - 2(p^2) - 2(5p) - 2(-1)$$
$$= 3p^3 + 15p^2 - 3p - 2p^2 - 10p + 2$$
$$= 3p^3 + 13p^2 - 13p + 2$$ ◀

When two binomials are multiplied, the FOIL method (First, Outer, Inner, Last) is used as a shortcut. This method is shown below.

Find $(2m - 5)(m + 4)$ using the FOIL method.

$$(2m - 5)(m + 4) = \overset{\textbf{F}}{(2m)(m)} + \overset{\textbf{O}}{(2m)(4)} + \overset{\textbf{I}}{(-5)(m)} + \overset{\textbf{L}}{(-5)(4)}$$
$$= 2m^2 + 8m - 5m - 20$$
$$= 2m^2 + 3m - 20$$ ◀

Find $(2k - 5)^2$.
Use FOIL.

$$(2k - 5)^2 = (2k - 5)(2k - 5)$$
$$= 4k^2 - 10k - 10k + 25$$
$$= 4k^2 - 20k + 25$$

Notice that the product of the square of a binomial is the square of the first term, $(2k)^2$, plus twice the product of the two terms, $(2)(2k)(-5)$, plus the square of the last term, $(-5)^2$. ◀

Caution Avoid the common error of writing $(x + y)^2 = x^2 + y^2$. As Example 4 shows, the square of a binomial has three terms, so

$$(x + y)^2 = x^2 + 2xy + y^2.$$

Furthermore, higher powers of a binomial also result in more than two terms. For example, verify by multiplication that

$$(x + y)^3 = x^3 + 3x^2y + 3xy^2 + y^3.$$

R.1 Exercises

Perform the indicated operations.

1. $(2x^2 - 6x + 11) + (-3x^2 + 7x - 2)$
2. $(-4y^2 - 3y + 8) - (2y^2 - 6y - 2)$
3. $-3(4q^2 - 3q + 2) + 2(-q^2 + q - 4)$
4. $2(3r^2 + 4r + 2) - 3(-r^2 + 4r - 5)$
5. $(.613x^2 - 4.215x + .892) - .47(2x^2 - 3x + 5)$
6. $.83(5r^2 - 2r + 7) - (7.12r^2 + 6.423r - 2)$
7. $-9m(2m^2 + 3m - 1)$
8. $(6k - 1)(2k - 3)$
9. $(5r - 3s)(5r + 4s)$
10. $(9k + q)(2k - q)$
11. $\left(\frac{2}{5}y + \frac{1}{8}z\right)\left(\frac{3}{5}y + \frac{1}{2}z\right)$
12. $\left(\frac{3}{4}r - \frac{2}{3}s\right)\left(\frac{5}{4}r + \frac{1}{3}s\right)$
13. $(12x - 1)(12x + 1)$
14. $(6m + 5)(6m - 5)$
15. $(3p - 1)(9p^2 + 3p + 1)$
16. $(2p - 1)(3p^2 - 4p + 5)$
17. $(2m + 1)(4m^2 - 2m + 1)$
18. $(k + 2)(12k^3 - 3k^2 + k + 1)$
19. $(m - n + k)(m + 2n - 3k)$
20. $(r - 3s + t)(2r - s + t)$

R.2 FACTORING

Multiplication of polynomials relies on the distributive property. The reverse process, where a polynomial is written as a product of other polynomials, is called **factoring.** For example, one way to factor the number 18 is to write it as the product $9 \cdot 2$. When 18 is written as $9 \cdot 2$, both 9 and 2 are called **factors** of 18. It is true that $18 = 36 \cdot 1/2$, but 36 and 1/2 are not considered factors of 18; only integers are used as factors. The number 18 also can be written with three integer factors as $2 \cdot 3 \cdot 3$. The integer factors of 18 are ± 1, ± 2, ± 3, ± 6, ± 9, ± 18.

The Greatest Common Factor To factor the algebraic expression $15m + 45$, first note that both $15m$ and 45 can be divided by 15. In fact, $15m = 15 \cdot m$ and $45 = 15 \cdot 3$. Thus, the distributive property can be used to write

$$15m + 45 = 15 \cdot m + 15 \cdot 3 = 15(m + 3).$$

Both 15 and $m + 3$ are factors of $15m + 45$. Since 15 divides into all terms of $15m + 45$ (and is the largest number that will do so), 15 is the **greatest common factor** for the polynomial $15m + 45$. The process of writing $15m + 45$ as $15(m + 3)$ is often called **factoring out** the greatest common factor.

EXAMPLE 1 Factor out the greatest common factor.

(a) $12p - 18q$

Both $12p$ and $18q$ are divisible by 6. Therefore,

$$12p - 18q = 6 \cdot 2p - 6 \cdot 3q = 6(2p - 3q).$$

(b) $8x^3 - 9x^2 + 15x$

Each of these terms is divisible by x.

$$8x^3 - 9x^2 + 15x = (8x^2) \cdot x - (9x) \cdot x + 15 \cdot x$$
$$= x(8x^2 - 9x + 15) \quad \text{or} \quad (8x^2 - 9x + 15)x$$ ◀

Caution When factoring out the greatest common factor in an expression like $2x^2 + x$, be careful to remember the 1 in the second term.

$$2x^2 + x = 2x^2 + 1x = x(2x + 1), \text{ not } x(2x)$$

Factoring Trinomials A polynomial that has no greatest common factor (other than 1) may still be factorable. For example, the polynomial $x^2 + 5x + 6$ can be factored as $(x + 2)(x + 3)$. To see that this is correct, find the product $(x + 2)(x + 3)$; you should get $x^2 + 5x + 6$. To factor a polynomial of three terms such as $x^2 + 5x + 6$, where the coefficient of x^2 is 1, proceed as shown in the following example.

EXAMPLE 2 Factor $y^2 + 8y + 15$.

Since the coefficient of y^2 is 1, factor by finding two numbers whose *product* is 15 and whose *sum* is 8. Since the constant and the middle term are positive, the numbers must both be positive. Begin by listing all pairs of positive integers having a product of 15. As you do this, also form the sum of each pair of numbers.

Products	*Sums*
$15 \cdot 1 = 15$	$15 + 1 = 16$
$5 \cdot 3 = 15$	$5 + 3 = 8$

The numbers 5 and 3 have a product of 15 and a sum of 8. Thus, $y^2 + 8y + 15$ factors as

$$y^2 + 8y + 15 = (y + 5)(y + 3).$$

The answer also can be written as $(y + 3)(y + 5)$. ◀

If the coefficient of the squared term is *not* 1, work as shown below.

EXAMPLE 3 Factor $2x^2 + 9xy - 5y^2$.

The factors of $2x^2$ are $2x$ and x; the possible factors of $-5y^2$ are $-5y$ and y, or $5y$ and $-y$. Try various combinations of these factors until one works (if, indeed, any work). For example, try the product $(2x + 5y)(x - y)$.

$$(2x + 5y)(x - y) = 2x^2 - 2xy + 5xy - 5y^2$$
$$= 2x^2 + 3xy - 5y^2$$

This product is not correct, so try another combination.

$$\begin{aligned}(2x - y)(x + 5y) &= 2x^2 + 10xy - xy - 5y^2 \\ &= 2x^2 + 9xy - 5y^2\end{aligned}$$

Since this combination gives the correct polynomial,

$$2x^2 + 9xy - 5y^2 = (2x - y)(x + 5y).$$

Special Factorizations Four special factorizations occur so often that they are listed here for future reference.

SPECIAL FACTORIZATIONS	$x^2 - y^2 = (x + y)(x - y)$	**Difference of two squares**
	$x^2 + 2xy + y^2 = (x + y)^2$	**Perfect square**
	$x^3 - y^3 = (x - y)(x^2 + xy + y^2)$	**Difference of two cubes**
	$x^3 + y^3 = (x + y)(x^2 - xy + y^2)$	**Sum of two cubes**

A polynomial that cannot be factored is called a **prime polynomial.**

EXAMPLE 4 Factor each of the following.

(a) $64p^2 - 49q^2 = (8p)^2 - (7q)^2 = (8p + 7q)(8p - 7q)$

(b) $x^2 + 36$ is a prime polynomial.

(c) $x^2 + 12x + 36 = (x + 6)^2$

(d) $9y^2 - 24yz + 16z^2 = (3y - 4z)^2$

(e) $y^3 - 8 = y^3 - 2^3 = (y - 2)(y^2 + 2y + 4)$

(f) $m^3 + 125 = m^3 + 5^3 = (m + 5)(m^2 - 5m + 25)$

(g) $8k^3 - 27z^3 = (2k)^3 - (3z)^3 = (2k - 3z)(4k^2 + 6kz + 9z^2)$

Caution In factoring, always look for a common factor first. Since $36x^2 - 4y^2$ has a common factor of 4,

$$36x^2 - 4y^2 = 4(9x^2 - y^2) = 4(3x + y)(3x - y).$$

It would be incomplete to factor it as

$$36x^2 - 4y^2 = (6x + 2y)(6x - 2y)$$

since each factor can be factored still further. To *factor* means to factor completely, so that each polynomial factor is prime.

R.2 Exercises

Factor each of the following. If a polynomial cannot be factored, write prime. *Factor out the greatest common factor as necessary.*

1. $8a^3 - 16a^2 + 24a$

2. $3y^3 + 24y^2 + 9y$

3. $25p^4 - 20p^3q + 100p^2q^2$

4. $60m^4 - 120m^3n + 50m^2n^2$

5. $m^2 + 9m + 14$

6. $x^2 + 4x - 5$

7. $z^2 + 9z + 20$
8. $b^2 - 8b + 7$
9. $a^2 - 6ab + 5b^2$
10. $s^2 + 2st - 35t^2$
11. $y^2 - 4yz - 21z^2$
12. $6a^2 - 48a - 120$
13. $3m^3 + 12m^2 + 9m$
14. $2x^2 - 5x - 3$
15. $3a^2 + 10a + 7$
16. $2a^2 - 17a + 30$
17. $15y^2 + y - 2$
18. $21m^2 + 13mn + 2n^2$
19. $24a^4 + 10a^3b - 4a^2b^2$
20. $32z^5 - 20z^4a - 12z^3a^2$
21. $x^2 - 64$
22. $9m^2 - 25$
23. $121a^2 - 100$
24. $9x^2 + 64$
25. $z^2 + 14zy + 49y^2$
26. $m^2 - 6mn + 9n^2$
27. $9p^2 - 24p + 16$
28. $a^3 - 216$
29. $8r^3 - 27s^3$
30. $64m^3 + 125$

R.3 RATIONAL EXPRESSIONS

Many algebraic fractions are **rational expressions,** which are quotients of polynomials with nonzero denominators. Examples include

$$\frac{8}{x-1}, \quad \frac{3x^2+4x}{5x-6}, \quad \text{and} \quad \frac{2y+1}{y^2}.$$

Methods for working with rational expressions are summarized below.

PROPERTIES OF RATIONAL EXPRESSIONS

For all mathematical expressions P, Q, R, and S, with Q and $S \neq 0$:

$$\frac{P}{Q} = \frac{PS}{QS} \qquad \text{Fundamental property}$$

$$\frac{P}{Q} \cdot \frac{R}{S} = \frac{PR}{QS} \qquad \text{Multiplication}$$

$$\frac{P}{Q} + \frac{R}{Q} = \frac{P+R}{Q} \qquad \text{Addition}$$

$$\frac{P}{Q} - \frac{R}{Q} = \frac{P-R}{Q} \qquad \text{Subtraction}$$

$$\frac{P}{Q} \div \frac{R}{S} = \frac{P}{Q} \cdot \frac{S}{R} \quad (R \neq 0) \qquad \text{Division}$$

Write each rational expression in lowest terms.

(a) $$\frac{8x+16}{4} = \frac{8(x+2)}{4} = \frac{4 \cdot 2(x+2)}{4} = 2(x+2)$$

The first step is to factor both numerator and denominator in order to identify any common factors. The answer also could be written as $2x + 4$, if desired.

(b) $$\frac{k^2+7k+12}{k^2+2k-3} = \frac{(k+4)(k+3)}{(k-1)(k+3)} = \frac{k+4}{k-1}$$

The answer cannot be further reduced.

Caution One of the most common errors in algebra involves incorrect use of the fundamental property of rational expressions. Only common *factors* may be divided or "canceled." It is essential to factor rational expressions before writing them in lowest terms. In Example 1(b), for instance, it is not correct to "cancel" k^2 (or cancel k, or divide 12 by -3) because the additions and subtraction must be performed first. Here they cannot be performed, so it is not possible to divide. Factoring allows the expression to be reduced, however.

EXAMPLE 2 Perform each operation.

(a) $\dfrac{3y + 9}{6} \cdot \dfrac{18}{5y + 15}$

Factor where possible, then multiply numerators and denominators and reduce to lowest terms.

$$\frac{3y + 9}{6} \cdot \frac{18}{5y + 15} = \frac{3(y + 3)}{6} \cdot \frac{18}{5(y + 3)}$$

$$= \frac{3 \cdot 18(y + 3)}{6 \cdot 5(y + 3)}$$

$$= \frac{3 \cdot \mathbf{6} \cdot 3(y + 3)}{\mathbf{6} \cdot 5(y + 3)} = \frac{3 \cdot 3}{5} = \frac{9}{5}$$

(b) $$\frac{m^2 + 5m + 6}{m + 3} \cdot \frac{m}{m^2 + 3m + 2} = \frac{(m + 2)(m + 3)}{m + 3} \cdot \frac{m}{(m + 2)(m + 1)}$$

$$= \frac{m(m + 2)(m + 3)}{(m + 3)(m + 2)(m + 1)} = \frac{m}{m + 1}$$

(c) $\dfrac{9p - 36}{12} \div \dfrac{5(p - 4)}{18}$

$$= \frac{9p - 36}{12} \cdot \frac{18}{5(p - 4)} \qquad \text{Invert and multiply.}$$

$$= \frac{9(p - 4)}{12} \cdot \frac{18}{5(p - 4)} = \frac{27}{10}$$

(d) $\dfrac{4}{5k} - \dfrac{11}{5k}$

As shown in the list of properties above, when two rational expressions have the same denominators, we subtract by subtracting the numerators.

$$\frac{4}{5k} - \frac{11}{5k} = \frac{4 - 11}{5k} = -\frac{7}{5k}$$

(e) $\dfrac{7}{p} + \dfrac{9}{2p} + \dfrac{1}{3p}$

These three fractions cannot be added until their denominators are the same. A common denominator into which p, $2p$, and $3p$ all divide is $6p$. Use the fundamental property to rewrite each rational expression with a denominator of $6p$.

$$\frac{7}{p} + \frac{9}{2p} + \frac{1}{3p} = \frac{6 \cdot 7}{6 \cdot p} + \frac{3 \cdot 9}{3 \cdot 2p} + \frac{2 \cdot 1}{2 \cdot 3p}$$
$$= \frac{42}{6p} + \frac{27}{6p} + \frac{2}{6p}$$
$$= \frac{42 + 27 + 2}{6p}$$
$$= \frac{71}{6p}$$ ◀

R.3 Exercises

Write each rational expression in lowest terms.

1. $\frac{7z^2}{14z}$ **2.** $\frac{25p^3}{10p^2}$ **3.** $\frac{8k + 16}{9k + 18}$ **4.** $\frac{3(t + 5)}{(t + 5)(t - 3)}$

5. $\frac{8x^2 + 16x}{4x^2}$ **6.** $\frac{36y^2 + 72y}{9y}$ **7.** $\frac{m^2 - 4m + 4}{m^2 + m - 6}$ **8.** $\frac{r^2 - r - 6}{r^2 + r - 12}$

9. $\frac{x^2 + 3x - 4}{x^2 - 1}$ **10.** $\frac{z^2 - 5z + 6}{z^2 - 4}$ **11.** $\frac{8m^2 + 6m - 9}{16m^2 - 9}$ **12.** $\frac{6y^2 + 11y + 4}{3y^2 + 7y + 4}$

Perform the indicated operations.

13. $\frac{9k^2}{25} \cdot \frac{5}{3k}$ **14.** $\frac{15p^3}{9p^2} \div \frac{6p}{10p^2}$ **15.** $\frac{a + b}{2p} \cdot \frac{12}{5(a + b)}$

16. $\frac{a - 3}{16} \div \frac{a - 3}{32}$ **17.** $\frac{2k + 8}{6} \div \frac{3k + 12}{2}$ **18.** $\frac{9y - 18}{6y + 12} \cdot \frac{3y + 6}{15y - 30}$

19. $\frac{4a + 12}{2a - 10} \div \frac{a^2 - 9}{a^2 - a - 20}$ **20.** $\frac{6r - 18}{9r^2 + 6r - 24} \cdot \frac{12r - 16}{4r - 12}$

21. $\frac{k^2 - k - 6}{k^2 + k - 12} \cdot \frac{k^2 + 3k - 4}{k^2 + 2k - 3}$ **22.** $\frac{m^2 + 3m + 2}{m^2 + 5m + 4} \div \frac{m^2 + 5m + 6}{m^2 + 10m + 24}$

23. $\frac{2m^2 - 5m - 12}{m^2 - 10m + 24} \div \frac{4m^2 - 9}{m^2 - 9m + 18}$ **24.** $\frac{6n^2 - 5n - 6}{6n^2 + 5n - 6} \cdot \frac{12n^2 - 17n + 6}{12n^2 - n - 6}$

25. $\frac{a + 1}{2} - \frac{a - 1}{2}$ **26.** $\frac{3}{p} + \frac{1}{2}$ **27.** $\frac{2}{y} - \frac{1}{4}$

28. $\frac{1}{6m} + \frac{2}{5m} + \frac{4}{m}$ **29.** $\frac{1}{m - 1} + \frac{2}{m}$ **30.** $\frac{6}{r} - \frac{5}{r - 2}$

31. $\frac{8}{3(a - 1)} + \frac{2}{a - 1}$ **32.** $\frac{2}{5(k - 2)} + \frac{3}{4(k - 2)}$

33. $\frac{2}{x^2 - 2x - 3} + \frac{5}{x^2 - x - 6}$ **34.** $\frac{2y}{y^2 + 7y + 12} - \frac{y}{y^2 + 5y + 6}$

35. $\frac{3k}{2k^2 + 3k - 2} - \frac{2k}{2k^2 - 7k + 3}$ **36.** $\frac{4m}{3m^2 + 7m - 6} - \frac{m}{3m^2 - 14m + 8}$

R.4 EQUATIONS

Linear Equations Equations that can be written in the form $ax + b = 0$, where a and b are real numbers, with $a \neq 0$, are **linear equations.** Examples of linear equations include $5y + 9 = 16$, $8x = 4$, and $-3p + 5 = -8$. Equations that are *not* linear include absolute value equations such as $|x| = 4$. The following properties are used to solve linear equations.

PROPERTIES OF REAL NUMBERS

For all real numbers a, b, and c:

$a(b + c) = ab + ac$ — **Distributive property**

If $a = b$, then $a + c = b + c$. — **Addition property of equality**

(The same number may be added to both sides of an equation.)

If $a = b$, then $ac = bc$. — **Multiplication property of equality**

(The same number may be multiplied on both sides of an equation.)

The following example shows how these properties are used to solve linear equations. Of course, the solutions should always be checked by substitution in the original equation.

Solve $2x - 5 + 8 = 3x + 2(2 - 3x)$.

$2x - 5 + 8 = 3x + 4 - 6x$	Distributive property
$2x + 3 = -3x + 4$	Combine like terms.
$5x + 3 = 4$	Add $3x$ to both sides.
$5x = 1$	Add -3 to both sides.
$x = \frac{1}{5}$	Multiply on both sides by $\frac{1}{5}$.

Check by substituting in the original equation. ◀

Absolute Value Equations Recall the definition of absolute value:

$$|x| = \begin{cases} x \text{ if } x \text{ is greater than or equal to } 0 \\ -x \text{ if } x \text{ is less than } 0. \end{cases}$$

By this definition, $|5| = 5$ and $|-5| = 5$. Think of the absolute value of a number as the distance between 0 and the number. The distance is always positive.

Although equations involving absolute value are not linear equations, sometimes they may be reduced to linear equations by using the following property.

EQUATIONS WITH ABSOLUTE VALUE

If k is greater than or equal to 0, then $|ax + b| = k$ is equivalent to

$$\boldsymbol{ax + b = k} \quad \text{or} \quad \boldsymbol{-(ax + b) = k.}$$

Solve each equation.

(a) $|2x - 5| = 7$

Rewrite $|2x - 5| = 7$ as follows.

$$\begin{aligned} 2x - 5 &= 7 \\ 2x &= 12 \\ x &= 6 \end{aligned} \quad \text{or} \quad \begin{aligned} -(2x - 5) &= 7 \\ -2x + 5 &= 7 \\ -2x &= 2 \\ x &= -1 \end{aligned}$$

Check both answers by substituting them into the original equation.

(b) $|3x + 1| = |x - 1|$

Use a variation of the property given above to rewrite the equation.

$$\begin{aligned} 3x + 1 &= x - 1 \\ 2x &= -2 \\ x &= -1 \end{aligned} \quad \text{or} \quad \begin{aligned} -(3x + 1) &= x - 1 \\ -3x - 1 &= x - 1 \\ -4x &= 0 \\ x &= 0 \end{aligned}$$

Be sure to check both answers. 

Quadratic Equations An equation with 2 as the highest exponent is a *quadratic equation.* A **quadratic equation** has the form $ax^2 + bx + c = 0$, where a, b, and c are real numbers and $a \neq 0$. A quadratic equation written in the form $ax^2 + bx + c = 0$ is said to be in **standard form.**

The simplest way to solve a quadratic equation, but one that is not always applicable, is by factoring. This method depends on the **zero-factor property.**

ZERO-FACTOR PROPERTY

If a and b are real numbers, with $ab = 0$, then

$$a = 0, b = 0, \text{ or both.}$$

Solve $6r^2 + 7r = 3$.

First write the equation in standard form.

$$6r^2 + 7r - 3 = 0$$

Now factor $6r^2 + 7r - 3$ to get

$$(3r - 1)(2r + 3) = 0.$$

By the zero-factor property, the product $(3r - 1)(2r + 3)$ can equal 0 only if

$$3r - 1 = 0 \quad \text{or} \quad 2r + 3 = 0.$$

Solve each of these equations separately to find that the solutions are $1/3$ and $-3/2$. Check these solutions by substituting them in the original equation. ◀

Caution Remember, the zero-factor property requires that the product of two (or more) factors be equal to *zero*, not some other quantity. It would be incorrect to use the zero-factor property with an equation in the form $(x + 3)(x - 1) = 4$, for example.

If a quadratic equation cannot be solved easily by factoring, use the *quadratic formula*. (The derivation of the quadratic formula is given in most algebra books.)

QUADRATIC FORMULA

The solutions of the quadratic equation $ax^2 + bx + c = 0$, where $a \neq 0$, are given by

$$x = \frac{-b \pm \sqrt{b^2 - 4ac}}{2a}.$$

EXAMPLE 4 Solve $x^2 - 4x - 5 = 0$ by the quadratic formula.

The equation is already in standard form (it has 0 alone on one side of the equals sign), so the values of a, b, and c from the quadratic formula are easily identified. The coefficient of the squared term gives the value of a; here, $a = 1$. Also, $b = -4$ and $c = -5$. (Be careful to use the correct signs.) Substitute these values into the quadratic formula.

$$x = \frac{-(-4) \pm \sqrt{(-4)^2 - 4(1)(-5)}}{2(1)} \qquad \text{Let } a = 1,\ b = -4,\ c = -5.$$

$$x = \frac{4 \pm \sqrt{16 + 20}}{2} \qquad (-4)^2 = (-4)(-4) = 16$$

$$x = \frac{4 \pm 6}{2} \qquad \sqrt{16 + 20} = \sqrt{36} = 6$$

The $\pm$ sign represents the two solutions of the equation. To find all of the solutions, first use $+$ and then use $-$.

$$x = \frac{4 + 6}{2} = \frac{10}{2} = 5 \quad \text{or} \quad x = \frac{4 - 6}{2} = \frac{-2}{2} = -1$$

The two solutions are 5 and -1. ◀

Caution Notice in the quadratic formula that the square root is added to or subtracted from the value of $-b$ *before* dividing by $2a$.

EXAMPLE 5 Solve $x^2 + 1 = 4x$.

First, add $-4x$ on both sides of the equals sign in order to get the equation in standard form.

$$x^2 - 4x + 1 = 0$$

Now identify the letters a, b, and c. Here $a = 1$, $b = -4$, and $c = 1$. Substitute these numbers into the quadratic formula.

$$\begin{aligned} x &= \frac{-(-4) \pm \sqrt{(-4)^2 - 4(1)(1)}}{2(1)} \\ &= \frac{4 \pm \sqrt{16 - 4}}{2} \\ &= \frac{4 \pm \sqrt{12}}{2} \end{aligned}$$

Simplify the solutions by writing $\sqrt{12}$ as $\sqrt{4 \cdot 3} = \sqrt{4} \cdot \sqrt{3} = 2\sqrt{3}$. Substituting $2\sqrt{3}$ for $\sqrt{12}$ gives

$$\begin{aligned} x &= \frac{4 \pm 2\sqrt{3}}{2} \\ &= \frac{2(2 \pm \sqrt{3})}{2} \qquad \text{Factor } 4 \pm 2\sqrt{3}. \\ &= 2 \pm \sqrt{3}. \end{aligned}$$

The two solutions are $2 + \sqrt{3}$ and $2 - \sqrt{3}$.

The exact values of the solutions are $2 + \sqrt{3}$ and $2 - \sqrt{3}$. The $\sqrt{\ }$ key on a calculator gives decimal approximations of these solutions (to the nearest thousandth).

$$\begin{aligned} 2 + \sqrt{3} &\approx 2 + 1.732 = 3.732^* \\ 2 - \sqrt{3} &\approx 2 - 1.732 = .268 \end{aligned}$$

Note Sometimes the quadratic formula will give a result with a negative number under the radical sign, such as $2 \pm \sqrt{-3}$. A solution of this type is not a real number. Since this text deals only with real numbers, such solutions cannot be used.

Equations with Fractions When an equation includes fractions, first eliminate all denominators by multiplying both sides of the equation by a **common denominator,** a number that can be divided (with no remainder) by each denominator in the equation. When an equation involves fractions with variable denominators, it is *necessary* to check all solutions in the original equation to be sure that no solution will lead to a zero denominator.

*The symbol $\approx$ means "is approximately equal to."

Solve each equation.

(a) $\frac{r}{10} - \frac{2}{15} = \frac{3r}{20} - \frac{1}{5}$

The denominators are 10, 15, 20, and 5. Each of these numbers can be divided into 60, so 60 is a common denominator. Multiply both sides of the equation by 60 and use the distributive property. (If a common denominator cannot be found easily, all the denominators in the problem can be multiplied together to produce one.)

$$\frac{r}{10} - \frac{2}{15} = \frac{3r}{20} - \frac{1}{5}$$

$$60\left(\frac{r}{10} - \frac{2}{15}\right) = 60\left(\frac{3r}{20} - \frac{1}{5}\right)$$

$$60\left(\frac{r}{10}\right) - 60\left(\frac{2}{15}\right) = 60\left(\frac{3r}{20}\right) - 60\left(\frac{1}{5}\right)$$

$$6r - 8 = 9r - 12$$

Add $-6r$ and 12 to both sides.

$$6r - 8 + (-6r) + 12 = 9r - 12 + (-6r) + 12$$

$$4 = 3r$$

Multiply both sides by $1/3$ to get

$$r = \frac{4}{3}.$$

(b) $\frac{3}{x^2} - 12 = 0$

Begin by multiplying both sides of the equation by x^2 to get $3 - 12x^2 = 0$. This equation could be solved by using the quadratic formula with $a = -12$, $b = 0$, and $c = 3$. Another method, which works well for the type of quadratic equation in which $b = 0$, is shown below.

$$3 - 12x^2 = 0$$

$$3 = 12x^2 \quad \text{Add } 12x^2.$$

$$\frac{1}{4} = x^2 \quad \text{Multiply by } \frac{1}{12}.$$

$$\pm\frac{1}{2} = x \quad \text{Take square roots.}$$

There are two solutions, $-1/2$ and $1/2$.

(c) $\frac{2}{k} - \frac{3k}{k + 2} = \frac{k}{k^2 + 2k}$

Factor $k^2 + 2k$ as $k(k + 2)$. The common denominator for all the fractions is $k(k + 2)$. Multiplying both sides by $k(k + 2)$ gives the following.

$$
\begin{aligned}
2(k + 2) - 3k(k) &= k \\
2k + 4 - 3k^2 &= k \\
-3k^2 + k + 4 &= 0 && \text{Add } -k\text{; rearrange terms.} \\
3k^2 - k - 4 &= 0 && \text{Multiply by } -1. \\
(3k - 4)(k + 1) &= 0 && \text{Factor.}
\end{aligned}
$$

$$
3k - 4 = 0 \quad \text{or} \quad k + 1 = 0
$$

$$
k = \frac{4}{3} \qquad\qquad k = -1
$$

Verify that the solutions are 4/3 and -1. ◀

Caution It is possible to get, as a solution of a rational equation, a number that makes one or more of the denominators in the original equation equal to zero. That number is not a solution, so it is *necessary* to check all potential solutions of rational equations.

EXAMPLE 7 Solve $\dfrac{2}{x-3} + \dfrac{1}{x} = \dfrac{6}{x(x-3)}$.

The common denominator is $x(x-3)$. Multiply both sides by $x(x-3)$ and solve the resulting equation.

$$
\begin{aligned}
\frac{2}{x-3} + \frac{1}{x} &= \frac{6}{x(x-3)} \\
2x + x - 3 &= 6 \\
3x &= 9 \\
x &= 3
\end{aligned}
$$

Checking this potential solution by substitution in the original equation shows that 3 makes two denominators 0. Thus 3 cannot be a solution, so there is no solution for this equation. ◀

R.4 Exercises

Solve each equation.

1. $.2m - .5 = .1m + .7$
2. $\frac{5}{6}k - 2k + \frac{1}{3} = \frac{2}{3}$
3. $3r + 2 - 5(r + 1) = 6r + 4$
4. $2[m - (4 + 2m) + 3] = 2m + 2$
5. $|3x + 2| = 9$
6. $|4 - 7x| = 15$
7. $|2x + 8| = |x - 4|$
8. $|5x + 2| = |8 - 3x|$

Solve each of the following equations by factoring or by using the quadratic formula. If the solutions involve square roots, give both the exact solutions and the approximate solutions to three decimal places.

9. $x^2 + 5x + 6 = 0$
10. $x^2 = 3 + 2x$
11. $m^2 + 16 = 8m$
12. $2k^2 - k = 10$
13. $6x^2 - 5x = 4$
14. $m(m - 7) = -10$

15. $9x^2 - 16 = 0$
16. $z(2z + 7) = 4$
17. $12y^2 - 48y = 0$
18. $3x^2 - 5x + 1 = 0$
19. $2m^2 = m + 4$
20. $p^2 + p - 1 = 0$
21. $k^2 - 10k = -20$
22. $2x^2 + 12x + 5 = 0$
23. $2r^2 - 7r + 5 = 0$
24. $2x^2 - 7x + 30 = 0$
25. $3k^2 + k = 6$
26. $5m^2 + 5m = 0$

Solve each of the following equations.

27. $\frac{3x - 2}{7} = \frac{x + 2}{5}$
28. $\frac{x}{3} - 7 = 6 - \frac{3x}{4}$
29. $\frac{4}{x - 3} - \frac{8}{2x + 5} + \frac{3}{x - 3} = 0$
30. $\frac{5}{2p + 3} - \frac{3}{p - 2} = \frac{4}{2p + 3}$
31. $\frac{2}{m} + \frac{m}{m + 3} = \frac{3m}{m^2 + 3m}$
32. $\frac{2y}{y - 1} = \frac{5}{y} + \frac{10 - 8y}{y^2 - y}$
33. $\frac{1}{x - 2} - \frac{3x}{x - 1} = \frac{2x + 1}{x^2 - 3x + 2}$
34. $\frac{5}{a} + \frac{-7}{a + 1} = \frac{a^2 - 2a + 4}{a^2 + a}$
35. $\frac{2b^2 + 5b - 8}{b^2 + 2b} + \frac{5}{b + 2} = -\frac{3}{b}$

R.5 INEQUALITIES

To write that one number is greater than or less than another number, we use the following symbols.

INEQUALITY SYMBOLS

$<$ means *is less than*
$>$ means *is greater than*
$\leq$ means *is less than or equal to*
$\geq$ means *is greater than or equal to*

Linear Inequalities An equation states that two expressions are equal; an **inequality** states that they are unequal. A **linear inequality** is an inequality that can be simplified to the form $ax < b$. (Properties introduced in this section are given only for $<$, but they are equally valid for $>$, $\leq$, or $\geq$.) Linear equalities are solved with the following properties.

PROPERTIES OF INEQUALITY

For all real numbers a, b, and c:

1. If $a < b$, then $a + c < b + c$.

2. If $a < b$ and if $c > 0$, then $ac < bc$.

3. If $a < b$ and if $c < 0$, then $ac > bc$.

Pay careful attention to property 3; it says that if both sides of an inequality are multiplied by a negative number, the direction of the inequality symbol must be reversed.

Solve $4 - 3y \leq 7 + 2y$.

Use the properties of inequality.

$$4 - 3y + (-4) \leq 7 + 2y + (-4) \quad \text{Add } -4 \text{ to both sides.}$$
$$-3y \leq 3 + 2y$$

Remember that *adding* the same number to both sides never changes the direction of the inequality symbol.

$$-3y + (-2y) \leq 3 + 2y + (-2y) \quad \text{Add } -2y \text{ to both sides.}$$
$$-5y \leq 3$$

Multiply both sides by $-1/5$. Since $-1/5$ is negative, change the direction of the inequality symbol.

$$-\frac{1}{5}(-5y) \geq -\frac{1}{5}(3)$$
$$y \geq -\frac{3}{5}$$

Caution It is a common error to forget to reverse the direction of the inequality sign when multiplying or dividing by a negative number. For example, to solve $-4x \leq 12$, we must multiply by $-1/4$ on both sides *and* reverse the inequality symbol to get $x \geq -3$.

The solution $y \geq -3/5$ in Example 1 represents an interval on the number line. **Interval notation** often is used for writing intervals. With interval notation, $y \geq -3/5$ is written as $[-3/5, \infty)$. This is an example of a **half-open interval,** since one endpoint, $-3/5$, is included. The **open interval** (2, 5) corresponds to $2 < x < 5$, with neither endpoint included. The **closed interval** [2, 5] includes both endpoints and corresponds to $2 \leq x \leq 5$.

The **graph** of an interval shows all points on a number line that correspond to the numbers in the interval. To graph the interval $[-3/5, \infty)$, for example, use a bracket at $-3/5$, since $-3/5$ is part of the solution. To show that the solution includes all real numbers greater than or equal to $-3/5$, draw a heavy arrow pointing to the right (the positive direction). See Figure 1.

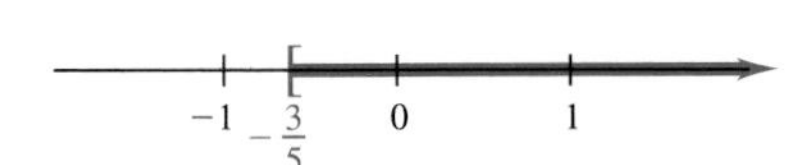

FIGURE 1

EXAMPLE 2 Solve $-2 < 5 + 3m < 20$. Graph the solution.

The inequality $-2 < 5 + 3m < 20$ says that $5 + 3m$ is *between* -2 and 20. Solve this inequality with an extension of the properties given above. Work as follows, first adding -5 to each part.

$$-2 + (-5) < 5 + 3m + (-5) < 20 + (-5)$$
$$-7 < 3m < 15$$

Now multiply each part by $1/3$.

$$-\frac{7}{3} < m < 5$$

A graph of the solution is given in Figure 2; here parentheses are used to show that $-7/3$ and 5 are *not* part of the graph.

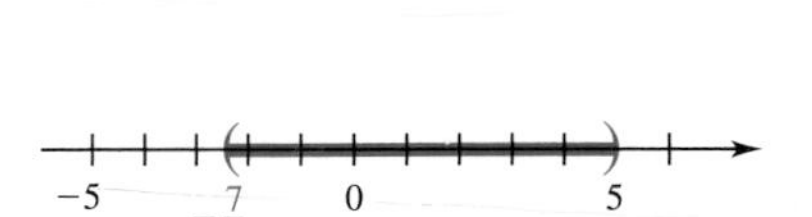

FIGURE 2

Quadratic Inequalities A **quadratic inequality** has the form $ax^2 + bx + c > 0$ (or $<$, or $\leq$, or $\geq$). The highest exponent is 2. The next few examples show how to solve quadratic inequalities.

EXAMPLE 3 Solve the quadratic inequality $x^2 - x - 12 < 0$.

This inequality is solved with values of x that make $x^2 - x - 12$ negative (< 0). The quantity $x^2 - x - 12$ changes from positive to negative or from negative to positive at the point where it equals 0. For this reason, first solve the *equation* $x^2 - x - 12 = 0$.

$$x^2 - x - 12 = 0$$
$$(x - 4)(x + 3) = 0$$
$$x = 4 \quad \text{or} \quad x = -3$$

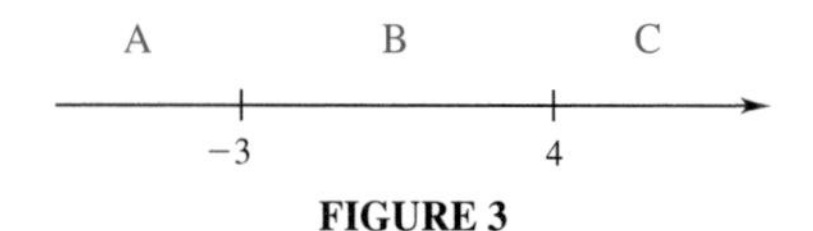

FIGURE 3

Locating -3 and 4 on a number line, as shown in Figure 3, determines three intervals A, B, and C. Decide which intervals include numbers that make $x^2 - x - 12$ negative by substituting a number from each interval in the polynomial. For example,

choose -4 from interval A: $(-4)^2 - (-4) - 12 = 8 > 0$;

choose 0 from interval B: $0^2 - 0 - 12 = -12 < 0$;

choose 5 from interval C: $5^2 - 5 - 12 = 8 > 0$.

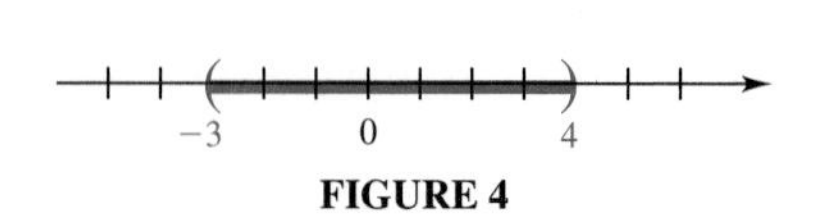

FIGURE 4

Only numbers in interval B satisfy the given inequality, so the solution is $(-3, 4)$. A graph of this solution is shown in Figure 4. ◀

EXAMPLE 4 Solve the quadratic inequality $r^2 + 3r \geq 4$.

First solve the equation $r^2 + 3r = 4$.

$$r^2 + 3r = 4$$
$$r^2 + 3r - 4 = 0$$
$$(r - 1)(r + 4) = 0$$
$$r = 1 \quad \text{or} \quad r = -4$$

These two solutions determine three intervals on the number line: $(-\infty, -4)$, $(-4, 1)$, and $(1, \infty)$. Substitute a number from each interval into the original inequality to determine the solution. The solution includes the numbers less than or equal to -4, together with the numbers greater than or equal to 1. This solution is written in interval notation as

$$(-\infty, -4] \cup [1, \infty).*$$

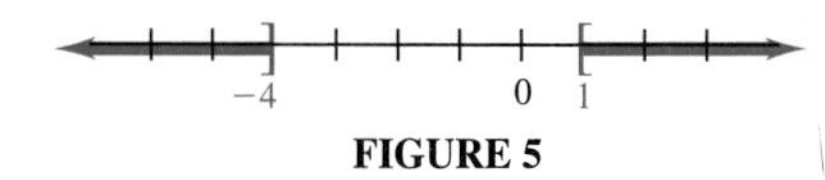

FIGURE 5

A graph of the solution is given in Figure 5. ◀

Inequalities with Fractions Inequalities with fractions can be solved in a manner similar to the way quadratic inequalities are solved.

Solve $\dfrac{2x - 3}{x} \geq 1$.

---

*The symbol ∪ indicates the *union* of two sets, which includes all elements in either set.

As with quadratic inequalities, first solve the corresponding equation.

$$\frac{2x - 3}{x} = 1$$

$$2x - 3 = x$$

$$x = 3$$

The solution, $x = 3$, determines the intervals on the number line where the fraction may change from greater than 1 to less than 1. This change also may occur on either side of a number that makes the denominator equal 0. Here, the x-value that makes the denominator 0 is $x = 0$. Test each of the three intervals determined by the numbers 0 and 3.

For $(-\infty, 0)$, choose -1: $\quad \frac{2(-1) - 3}{-1} = 5 \geq 1.$

For $(0, 3)$, choose 1: $\quad \frac{2(1) - 3}{1} = -1 \not\geq 1.$

For $(3, \infty)$, choose 4: $\quad \frac{2(4) - 3}{4} = \frac{5}{4} \geq 1.$

The symbol $\not\geq$ means "is *not* greater than or equal to." Testing the endpoints 0 and 3 shows that the solution is $(-\infty, 0) \cup [3, \infty)$. ◀

Solve $\frac{x^2 - 3x}{x^2 - 9} < 4.$

Begin by solving the corresponding equation.

$$\frac{x^2 - 3x}{x^2 - 9} = 4$$

$$x^2 - 3x = 4x^2 - 36 \qquad \text{Multiply by } x^2 - 9.$$

$$0 = 3x^2 + 3x - 36 \qquad \text{Get 0 on one side.}$$

$$0 = x^2 + x - 12 \qquad \text{Multiply by } \frac{1}{3}.$$

$$0 = (x + 4)(x - 3) \qquad \text{Factor.}$$

$$x = -4 \quad \text{or} \quad x = 3$$

Now set the denominator equal to 0 and solve that equation.

$$x^2 - 9 = 0$$

$$(x - 3)(x + 3) = 0$$

$$x = 3 \quad \text{or} \quad x = -3$$

The intervals determined by the three (different) solutions are $(-\infty, -4)$, $(-4, -3)$, $(-3, 3)$, and $(3, \infty)$. Testing a number from each interval and the endpoints of the intervals in the given inequality shows that the solution is

$$(-\infty, -4) \cup (-3, 3) \cup (3, \infty).$$ ◀

Caution Don't forget to solve the equation formed by setting the denominator equal to zero. Any number that makes the denominator zero always creates two intervals on the number line. For instance, in Example 5, 0 makes the denominator of the rational inequality equal to 0, so we know that there may be a sign change from one side of 0 to the other (as was indeed the case).

R.5 Exercises

Solve each inequality and graph the solution.

1. $-3p - 2 \geq 1$

2. $6k - 4 < 3k - 1$

3. $m - (4 + 2m) + 3 < 2m + 2$

4. $-2(3y - 8) \geq 5(4y - 2)$

5. $3p - 1 < 6p + 2(p - 1)$

6. $x + 5(x + 1) > 4(2 - x) + x$

7. $-7 < y - 2 < 4$

8. $8 \leq 3r + 1 \leq 13$

9. $-4 \leq \dfrac{2k - 1}{3} \leq 2$

10. $-1 \leq \dfrac{5y + 2}{3} \leq 4$

11. $\dfrac{3}{5}(2p + 3) \geq \dfrac{1}{10}(5p + 1)$

12. $\dfrac{8}{3}(z - 4) \leq \dfrac{2}{9}(3z + 2)$

Solve each of the following quadratic inequalities. Graph each solution.

13. $(m + 2)(m - 4) < 0$

14. $(t + 6)(t - 1) \geq 0$

15. $y^2 - 3y + 2 < 0$

16. $2k^2 + 7k - 4 > 0$

17. $q^2 - 7q + 6 \leq 0$

18. $2k^2 - 7k - 15 \leq 0$

19. $6m^2 + m > 1$

20. $10r^2 + r \leq 2$

21. $2y^2 + 5y \leq 3$

22. $3a^2 + a > 10$

23. $x^2 \leq 25$

24. $p^2 - 16p > 0$

Solve the following inequalities.

25. $\dfrac{m - 3}{m + 5} \leq 0$

26. $\dfrac{r + 1}{r - 1} > 0$

27. $\dfrac{k - 1}{k + 2} > 1$

28. $\dfrac{a - 5}{a + 2} < -1$

29. $\dfrac{2y + 3}{y - 5} \leq 1$

30. $\dfrac{a + 2}{3 + 2a} \leq 5$

31. $\dfrac{7}{k + 2} \geq \dfrac{1}{k + 2}$

32. $\dfrac{5}{p + 1} > \dfrac{12}{p + 1}$

33. $\dfrac{3x}{x^2 - 1} < 2$

34. $\dfrac{8}{p^2 + 2p} > 1$

35. $\dfrac{z^2 + z}{z^2 - 1} \geq 3$

36. $\dfrac{a^2 + 2a}{a^2 - 4} \leq 2$

R.6 EXPONENTS

Integer Exponents Recall that $a^2 = a \cdot a$, while $a^3 = a \cdot a \cdot a$, and so on. In this section a more general meaning is given to the symbol a^n.

DEFINITION OF EXPONENT If n is a natural number, then

$$a^n = a \cdot a \cdot a \cdot \cdots \cdot a,$$

where a appears as a factor n times.

In the expression a^n, n is the **exponent** and a is the **base.** This definition can be extended by defining a^n for zero and negative integer values of n.

ZERO AND NEGATIVE EXPONENTS

If a is any nonzero real number, and if n is a positive integer, then

$$a^0 = 1 \quad \text{and} \quad a^{-n} = \frac{1}{a^n}.$$

(The symbol 0^0 is meaningless.)

EXAMPLE 1

(a) $6^0 = 1$

(b) $(-9)^0 = 1$

(c) $3^{-2} = \frac{1}{3^2} = \frac{1}{9}$

(d) $9^{-1} = \frac{1}{9^1} = \frac{1}{9}$

(e) $\left(\frac{3}{4}\right)^{-1} = \frac{1}{(3/4)^1} = \frac{1}{3/4} = \frac{4}{3}$ ◀

The following properties follow from the definitions of exponents given above.

PROPERTIES OF EXPONENTS

For any integers m and n, and any real numbers a and b for which the following exist:

1. $a^m \cdot a^n = a^{m+n}$
2. $\frac{a^m}{a^n} = a^{m-n}$
3. $(a^m)^n = a^{mn}$
4. $(ab)^m = a^m \cdot b^m$
5. $\left(\frac{a}{b}\right)^m = \frac{a^m}{b^m}$

EXAMPLE 2

Use the properties of exponents to simplify each of the following. Leave answers with positive exponents. Assume that all variables represent positive real numbers.

(a) $7^4 \cdot 7^6 = 7^{4+6} = 7^{10}$ Property 1

(b) $\frac{9^{14}}{9^6} = 9^{14-6} = 9^8$ Property 2

(c) $\frac{r^9}{r^{17}} = r^{9-17} = r^{-8} = \frac{1}{r^8}$ Property 2

(d) $(2m^3)^4 = 2^4 \cdot (m^3)^4 = 16m^{12}$ Properties 3 and 4

(e) $(3x)^4 = 3^4 \cdot x^4$ Property 4

(f) $\left(\dfrac{9}{7}\right)^6 = \dfrac{9^6}{7^6}$ Property 5

(g) $\dfrac{2^{-3} \cdot 2^5}{2^4 \cdot 2^{-7}} = \dfrac{2^2}{2^{-3}} = 2^{2-(-3)} = 2^5$

(h) $2^{-1} + 3^{-1} = \dfrac{1}{2} + \dfrac{1}{3} = \dfrac{5}{6}$ ◀

Caution If Example 2(e) were written $3x^4$, the properties of exponents would not apply. When no parentheses are used, the exponent refers only to the factor closest to it. Also notice in Examples 2(g) and 2(h) that a negative exponent does *not* indicate a negative number.

Roots For *even* values of n, the expression $a^{1/n}$ is defined to be the **positive nth root** of a. For example, $a^{1/2}$ denotes the positive second root, or **square root,** of a, while $a^{1/4}$ is the positive fourth root of a. When n is *odd*, there is only one nth root, which has the same sign as a. For example, $a^{1/3}$, the **cube root** of a, has the same sign as a. By definition, if $b = a^{1/n}$, then $b^n = a$.

EXAMPLE 3 (A calculator will be helpful here.)

(a) $121^{1/2} = 11$, since 11 is positive and $11^2 = 121$.

(b) $625^{1/4} = 5$, since $5^4 = 625$.

(c) $256^{1/4} = 4$

(d) $64^{1/6} = 2$

(e) $27^{1/3} = 3$

(f) $(-32)^{1/5} = -2$

(g) $128^{1/7} = 2$

(h) $(-49)^{1/2}$ is not a real number. ◀

Rational Exponents In the following definition, the domain of an exponent is extended to include all rational numbers.

DEFINITION OF $a^{m/n}$

For all real numbers a for which the indicated roots exist, and for any rational number m/n,

$$a^{m/n} = (a^{1/n})^m.$$

EXAMPLE 4

(a) $27^{2/3} = (27^{1/3})^2 = 3^2 = 9$

(b) $32^{2/5} = (32^{1/5})^2 = 2^2 = 4$

(c) $64^{4/3} = (64^{1/3})^4 = 4^4 = 256$

(d) $25^{3/2} = (25^{1/2})^3 = 5^3 = 125$ ◀

All the properties for integer exponents given in this section also apply to any rational exponent on a nonnegative real-number base.

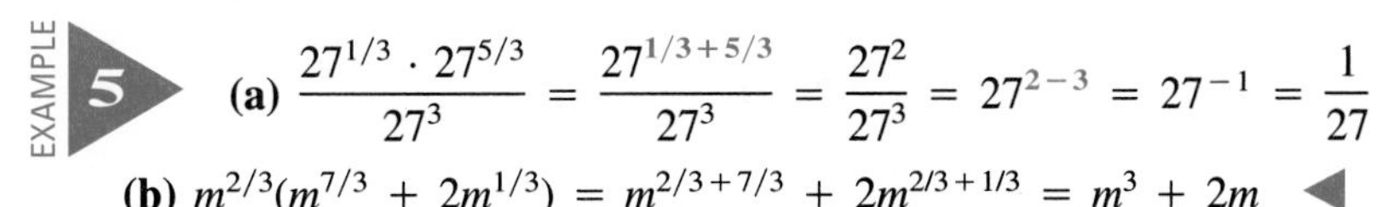

EXAMPLE 5 **(a)** $\dfrac{27^{1/3} \cdot 27^{5/3}}{27^3} = \dfrac{27^{1/3+5/3}}{27^3} = \dfrac{27^2}{27^3} = 27^{2-3} = 27^{-1} = \dfrac{1}{27}$

(b) $m^{2/3}(m^{7/3} + 2m^{1/3}) = m^{2/3+7/3} + 2m^{2/3+1/3} = m^3 + 2m$ ◀

In calculus, it is often necessary to factor expressions involving fractional exponents, as shown in the next example.

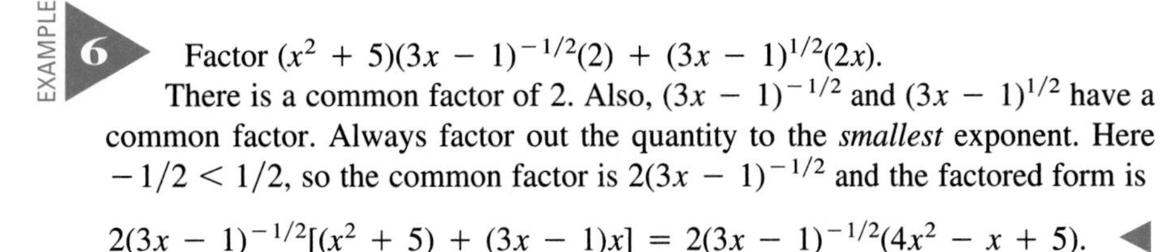

EXAMPLE 6 Factor $(x^2 + 5)(3x - 1)^{-1/2}(2) + (3x - 1)^{1/2}(2x)$.

There is a common factor of 2. Also, $(3x - 1)^{-1/2}$ and $(3x - 1)^{1/2}$ have a common factor. Always factor out the quantity to the *smallest* exponent. Here $-1/2 < 1/2$, so the common factor is $2(3x - 1)^{-1/2}$ and the factored form is

$$2(3x - 1)^{-1/2}[(x^2 + 5) + (3x - 1)x] = 2(3x - 1)^{-1/2}(4x^2 - x + 5).$$ ◀

R.6 Exercises

Evaluate each expression. Write all answers without exponents.

1. 8^{-2} **2.** 3^{-4} **3.** 6^{-3} **4.** 5^0

5. $(-12)^0$ **6.** $2^{-1} + 4^{-1}$ **7.** -2^{-4} **8.** $(-2)^{-4}$

9. $-(-3)^{-2}$ **10.** $-(-3^{-2})$ **11.** $\left(\dfrac{5}{8}\right)^2$ **12.** $\left(\dfrac{6}{7}\right)^3$

13. $\left(\dfrac{1}{2}\right)^{-3}$ **14.** $\left(\dfrac{1}{5}\right)^{-3}$ **15.** $\left(\dfrac{2}{7}\right)^{-2}$ **16.** $\left(\dfrac{4}{3}\right)^{-3}$

Simplify each expression. Assume that all variables represent positive real numbers. Write answers with only positive exponents.

17. $\dfrac{7^5}{7^9}$ **18.** $\dfrac{3^{-4}}{3^2}$ **19.** $\dfrac{2^{-5}}{2^{-2}}$ **20.** $\dfrac{6^{-1}}{6}$

21. $4^{-3} \cdot 4^6$ **22.** $\dfrac{8^9 \cdot 8^{-7}}{8^{-3}}$ **23.** $\dfrac{10^8 \cdot 10^{-10}}{10^4 \cdot 10^2}$ **24.** $\left(\dfrac{5^{-6} \cdot 5^3}{5^{-2}}\right)^{-1}$

25. $\dfrac{x^4 \cdot x^3}{x^5}$ **26.** $\dfrac{y^9 \cdot y^7}{y^{13}}$ **27.** $\dfrac{(4k^{-1})^2}{2k^{-5}}$ **28.** $\dfrac{(3z^2)^{-1}}{z^5}$

29. $\dfrac{2^{-1}x^3y^{-3}}{xy^{-2}}$ **30.** $\dfrac{5^{-2}m^2y^{-2}}{5^2m^{-1}y^{-2}}$ **31.** $\left(\dfrac{a^{-1}}{b^2}\right)^{-3}$ **32.** $\left(\dfrac{2c^2}{d^3}\right)^{-2}$

Evaluate each expression, assuming that $a = 2$ and $b = -3$.

33. $a^{-1} + b^{-1}$ **34.** $b^{-2} - a$ **35.** $\dfrac{2b^{-1} - 3a^{-1}}{a + b^2}$

36. $\dfrac{3a^2 - b^2}{b^{-3} + 2a^{-1}}$ **37.** $\left(\dfrac{a}{3}\right)^{-1} + \left(\dfrac{b}{2}\right)^{-2}$ **38.** $\left(\dfrac{2b}{5}\right)^2 - 3\left(\dfrac{a^{-1}}{4}\right)$

Write each number without exponents.

39. $81^{1/2}$
40. $27^{1/3}$
41. $8^{2/3}$
42. $1000^{2/3}$
43. $32^{2/5}$
44. $-125^{2/3}$
45. $\left(\frac{4}{9}\right)^{1/2}$
46. $\left(\frac{64}{27}\right)^{1/3}$
47. $16^{-5/4}$
48. $625^{-1/4}$
49. $\left(\frac{27}{64}\right)^{-1/3}$
50. $\left(\frac{121}{100}\right)^{-3/2}$

Simplify each expression. Write all answers with only positive exponents. Assume that all variables represent positive real numbers.

51. $2^{1/2} \cdot 2^{3/2}$
52. $27^{2/3} \cdot 27^{-1/3}$
53. $\frac{4^{2/3} \cdot 4^{5/3}}{4^{1/3}}$
54. $\frac{3^{-5/2} \cdot 3^{3/2}}{3^{7/2} \cdot 3^{-9/2}}$
55. $\frac{7^{-1/3} \cdot 7r^{-3}}{7^{2/3} \cdot (r^{-2})^2}$
56. $\frac{12^{3/4} \cdot 12^{5/4} \cdot y^{-2}}{12^{-1} \cdot (y^{-3})^{-2}}$
57. $\frac{6k^{-4} \cdot (3k^{-1})^{-2}}{2^3 \cdot k^{1/2}}$
58. $\frac{8p^{-3} \cdot (4p^2)^{-2}}{p^{-5}}$
59. $\frac{a^{4/3} \cdot b^{1/2}}{a^{2/3} \cdot b^{-3/2}}$
60. $\frac{x^{1/3} \cdot y^{2/3} \cdot z^{1/4}}{x^{5/3} \cdot y^{-1/3} \cdot z^{3/4}}$
61. $\frac{k^{-3/5} \cdot h^{-1/3} \cdot t^{2/5}}{k^{-1/5} \cdot h^{-2/3} \cdot t^{1/5}}$
62. $\frac{m^{7/3} \cdot n^{-2/5} \cdot p^{3/8}}{m^{-2/3} \cdot n^{3/5} \cdot p^{-5/8}}$

Factor each expression.

63. $(x^2 + 2)(x^2 - 1)^{-1/2}(x) + (x^2 - 1)^{1/2}(2x)$
64. $(3x - 1)(5x + 2)^{1/2}(15) + (5x + 2)^{-1/2}(5)$
65. $(2x + 5)^2\left(\frac{1}{2}\right)(x^2 - 4)^{-1/2}(2x) + (x^2 - 4)^{1/2}(2)(2x + 5)$
66. $(4x^2 + 1)^2(2x - 1)^{-1/2} + (2x - 1)^{1/2}(2)(4x^2 + 1)$

R.7 RADICALS

We have defined $a^{1/n}$ as the positive nth root of a for appropriate values of a and n. An alternative notation for $a^{1/n}$ uses radicals.

RADICALS

If n is an even natural number and $a > 0$, or n is an odd natural number, then

$$\mathbf{a^{1/n} = \sqrt[n]{a}.}$$

The symbol $\sqrt[n]{}$ is a **radical sign,** the number a is the **radicand,** and n is the **index** of the radical. The familiar symbol $\sqrt{a}$ is used instead of $\sqrt[2]{a}$.

(a) $\sqrt[4]{16} = 16^{1/4} = 2$

(b) $\sqrt[5]{-32} = -2$

(c) $\sqrt[3]{1000} = 10$

(d) $\sqrt[6]{\frac{64}{729}} = \frac{2}{3}$ ◀

With $a^{1/n}$ written as $\sqrt[n]{a}$, $a^{m/n}$ also can be written using radicals.

$$a^{m/n} = (\sqrt[n]{a})^m \quad \text{or} \quad a^{m/n} = \sqrt[n]{a^m}$$

The following properties of radicals depend on the definitions and properties of exponents.

PROPERTIES OF RADICALS

For all real numbers a and b and natural numbers m and n such that $\sqrt[n]{a}$ and $\sqrt[n]{b}$ are real numbers:

1. $(\sqrt[n]{a})^n = a$

2. $\sqrt[n]{a^n} = \begin{cases} |a| & \text{if } n \text{ is even} \\ a & \text{if } n \text{ is odd} \end{cases}$

3. $\sqrt[n]{a} \cdot \sqrt[n]{b} = \sqrt[n]{ab}$

4. $\dfrac{\sqrt[n]{a}}{\sqrt[n]{b}} = \sqrt[n]{\dfrac{a}{b}} \qquad (b \neq 0)$

5. $\sqrt[m]{\sqrt[n]{a}} = \sqrt[mn]{a}$

Property 3 can be used to simplify certain radicals. For example, since $48 = 16 \cdot 3$,

$$\sqrt{48} = \sqrt{16 \cdot 3} = \sqrt{16} \cdot \sqrt{3} = 4\sqrt{3}.$$

EXAMPLE 2

(a) $\sqrt{1000} = \sqrt{100 \cdot 10} = \sqrt{100} \cdot \sqrt{10} = 10\sqrt{10}$

(b) $\sqrt{128} = \sqrt{64 \cdot 2} = 8\sqrt{2}$

(c) $\sqrt{108} = \sqrt{36 \cdot 3} = 6\sqrt{3}$

(d) $\sqrt[3]{54} = \sqrt[3]{27 \cdot 2} = \sqrt[3]{27} \cdot \sqrt[3]{2} = 3\sqrt[3]{2}$

(e) $\sqrt{288m^5} = \sqrt{144 \cdot m^4 \cdot 2m} = 12m^2\sqrt{2m}$

(f) $$\begin{aligned} 2\sqrt{18} - 5\sqrt{32} &= 2\sqrt{9 \cdot 2} - 5\sqrt{16 \cdot 2} \\ &= 2\sqrt{9} \cdot \sqrt{2} - 5\sqrt{16} \cdot \sqrt{2} \\ &= 2(3)\sqrt{2} - 5(4)\sqrt{2} = -14\sqrt{2} \end{aligned}$$ ◀

Rationalizing Denominators The next example shows how to *rationalize* (remove all radicals from) the denominator in an expression containing radicals.

EXAMPLE 3

Simplify each of the following expressions by rationalizing the denominator.

(a) $\dfrac{4}{\sqrt{3}}$

To rationalize the denominator, multiply by $\sqrt{3}/\sqrt{3}$ (or 1) so that the denominator of the product is a rational number.

$$\frac{4}{\sqrt{3}} \cdot \frac{\sqrt{3}}{\sqrt{3}} = \frac{4\sqrt{3}}{3}$$

(b) $\dfrac{2}{\sqrt[3]{x}}$

Here, we need a perfect cube under the radical sign to rationalize the denominator. Multiplying by $\sqrt[3]{x^2}/\sqrt[3]{x^2}$ gives

$$\frac{2}{\sqrt[3]{x}} \cdot \frac{\sqrt[3]{x^2}}{\sqrt[3]{x^2}} = \frac{2\sqrt[3]{x^2}}{\sqrt[3]{x^3}} = \frac{2\sqrt[3]{x^2}}{x}.$$

(c) $\dfrac{1}{1 - \sqrt{2}}$

The best approach here is to multiply both numerator and denominator by the number $1 + \sqrt{2}$. The expressions $1 + \sqrt{2}$ and $1 - \sqrt{2}$ are conjugates.* Doing so gives

$$\frac{1}{1 - \sqrt{2}} = \frac{1(1 + \sqrt{2})}{(1 - \sqrt{2})(1 + \sqrt{2})} = \frac{1 + \sqrt{2}}{1 - 2} = -1 - \sqrt{2}. \blacktriangleleft$$

Sometimes it is advantageous to rationalize the *numerator* of a rational expression.

EXAMPLE 4 Rationalize the numerator of $\dfrac{\sqrt{2} - \sqrt{3}}{5}$.

Multiply numerator and denominator by the conjugate of the numerator, $\sqrt{2} + \sqrt{3}$.

$$\frac{\sqrt{2} - \sqrt{3}}{5} \cdot \frac{\sqrt{2} + \sqrt{3}}{\sqrt{2} + \sqrt{3}} = \frac{2 - 3}{5(\sqrt{2} + \sqrt{3})} = \frac{-1}{5(\sqrt{2} + \sqrt{3})} \blacktriangleleft$$

EXAMPLE 5 Simplify $\sqrt{m^2 - 4m + 4}$.

Factor the polynomial as $m^2 - 4m + 4 = (m - 2)^2$. Then

$$\sqrt{(m - 2)^2} = m - 2 \quad \text{if} \quad m - 2 \geq 0.$$

If $m - 2 \leq 0$, $\sqrt{(m - 2)^2} = -(m - 2) = 2 - m$. ◀

Caution Avoid the common error of writing $\sqrt{a + b}$ as $\sqrt{a} + \sqrt{b}$. We must add a^2 and b^2 *before* taking the square root. For example $\sqrt{16 + 9} = \sqrt{25} = 5$, *not* $\sqrt{16} + \sqrt{9} = 4 + 3 = 7$. This idea applies as well to higher roots:

$$\sqrt[3]{a^3 + b^3} \neq \sqrt[3]{a^3} + \sqrt[3]{b^3},$$
$$\sqrt[4]{a^4 + b^4} \neq \sqrt[4]{a^4} + \sqrt[4]{b^4},$$

and so on.

*If a and b are real numbers, the *conjugate* of $a + b$ is $a - b$.

R.7 Exercises

Simplify each expression. Assume that all variables represent positive real numbers.

1. $\sqrt[3]{125}$
2. $\sqrt[4]{1296}$
3. $\sqrt[5]{-3125}$
4. $\sqrt{50}$
5. $\sqrt{2000}$
6. $\sqrt{32y^5}$
7. $7\sqrt{2} - 8\sqrt{18} + 4\sqrt{72}$
8. $4\sqrt{3} - 5\sqrt{12} + 3\sqrt{75}$
9. $2\sqrt{5} - 3\sqrt{20} + 2\sqrt{45}$
10. $3\sqrt{28} - 4\sqrt{63} + \sqrt{112}$
11. $\sqrt[3]{2} - \sqrt[3]{16} + 2\sqrt[3]{54}$
12. $2\sqrt[3]{3} + 4\sqrt[3]{24} - \sqrt[3]{81}$
13. $\sqrt[3]{32} - 5\sqrt[3]{4} + 2\sqrt[3]{108}$
14. $\sqrt{2x^3y^2z^4}$
15. $\sqrt{98r^3s^4t^{10}}$
16. $\sqrt[3]{16z^5x^8y^4}$
17. $\sqrt[4]{x^8y^7z^{11}}$
18. $\sqrt{a^3b^5} - 2\sqrt{a^7b^3} + \sqrt{a^3b^9}$
19. $\sqrt{p^7q^3} - \sqrt{p^5q^9} + \sqrt{p^9q}$

Rationalize each denominator. Assume that all radicands represent positive real numbers.

20. $\dfrac{5}{\sqrt{7}}$
21. $\dfrac{-2}{\sqrt{3}}$
22. $\dfrac{-3}{\sqrt{12}}$
23. $\dfrac{4}{\sqrt{8}}$
24. $\dfrac{3}{1 - \sqrt{5}}$
25. $\dfrac{5}{2 - \sqrt{6}}$
26. $\dfrac{-2}{\sqrt{3} - \sqrt{2}}$
27. $\dfrac{1}{\sqrt{10} + \sqrt{3}}$
28. $\dfrac{1}{\sqrt{r} - \sqrt{3}}$
29. $\dfrac{5}{\sqrt{m} - \sqrt{5}}$
30. $\dfrac{y - 5}{\sqrt{y} - \sqrt{5}}$
31. $\dfrac{z - 11}{\sqrt{z} - \sqrt{11}}$
32. $\dfrac{\sqrt{x} + \sqrt{x + 1}}{\sqrt{x} - \sqrt{x + 1}}$
33. $\dfrac{\sqrt{p} + \sqrt{p^2 - 1}}{\sqrt{p} - \sqrt{p^2 - 1}}$

Rationalize each numerator. Assume that all radicands represent positive real numbers.

34. $\dfrac{1 + \sqrt{2}}{2}$
35. $\dfrac{1 - \sqrt{3}}{3}$
36. $\dfrac{\sqrt{x} + \sqrt{x + 1}}{\sqrt{x} - \sqrt{x + 1}}$
37. $\dfrac{\sqrt{p} + \sqrt{p^2 - 1}}{\sqrt{p} - \sqrt{p^2 - 1}}$

Simplify each root, if possible.

38. $\sqrt{16 - 8x + x^2}$
39. $\sqrt{4y^2 + 4y + 1}$
40. $\sqrt{4 - 25z^2}$
41. $\sqrt{9k^2 + h^2}$

Applications at a glance . . .

Torque on a Porsche . . . average global temperature . . . estimating the passage of time . . . postage rates . . .

CHAPTER 1

FUNCTIONS AND GRAPHS

(See page 65.)

Figure 1(a) shows the speed in miles per hour of a Porsche 928 over elapsed time in seconds as the car accelerates from rest.* For example, the figure shows that 15 seconds after starting, the car is going 90 miles per hour. Figure 1(b) shows the variation in blood pressure for a typical person.† (Systolic and diastolic pressures are the upper and lower limits in the periodic changes in pressure that produce the pulse. The length of time between peaks is called the period of the pulse.) After .8 seconds, the blood pressure is the same as its starting value, 80 millimeters. Figures 1(a) and 1(b) illustrate *functions*, which are rules or procedures that yield just one value of one variable from any given value of another variable.

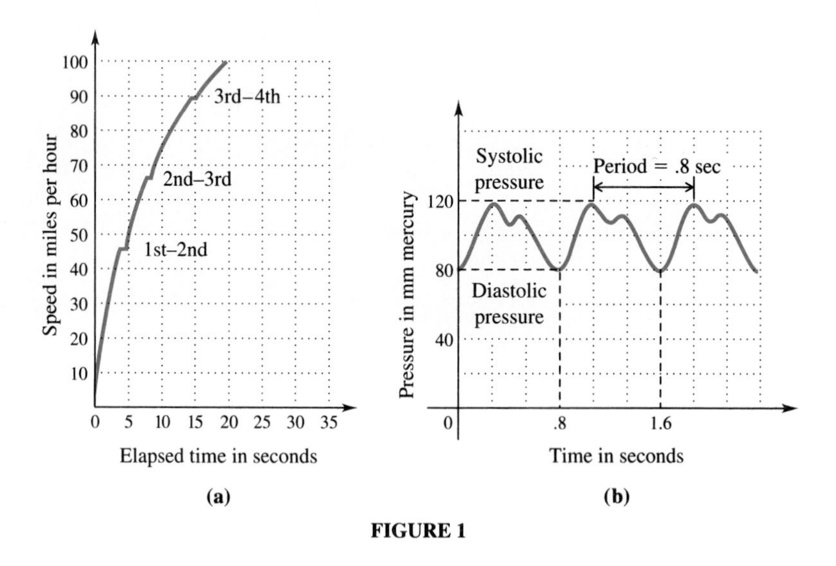

FIGURE 1

Functions are useful in describing many situations involving two variables because each value of one variable corresponds to only one value of the other variable. We can find the exact speed of the Porsche after 10 seconds because the car cannot have two different speeds at the same moment. Similarly, a person can have only one blood pressure at a given instant. A function used in this way, as a mathematical description of a real-world situation, is called a **mathematical model.** Constructing a mathematical model requires a solid understanding of the situation to be modeled, along with a good knowledge of the possible mathematical ideas that can be used to construct the model. In this chapter we discuss a variety of functions useful in modeling.

---

*From *Road & Track*, April and May 1978. Reprinted with permission.

†From *Calculus for the Life Sciences* by Rodolfo De Sapio. Copyright © 1976, 1978 by W. H. Freeman and Company. Reprinted by permission.

1.1 FUNCTIONS

How has the amount of energy used in the United States varied with time, and how much money has been saved as a result of decreased use of energy?

After developing the concept of function, we will answer this question in one of the exercises.

The cost of leasing an office in an office building depends on the number of square feet of space in the office. Typical associations between the area of an office in square feet and the monthly rent in dollars are shown in Figure 2.

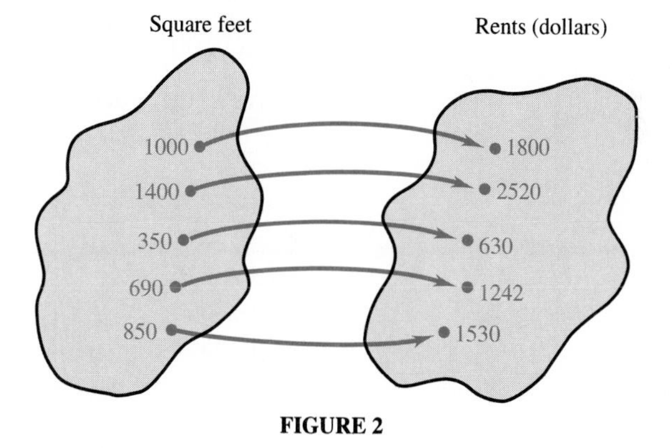

FIGURE 2

Another association might be set up between investments in a mutual fund and the corresponding earnings, assuming an annual return of 6%. Typical associations are shown in Figure 3.

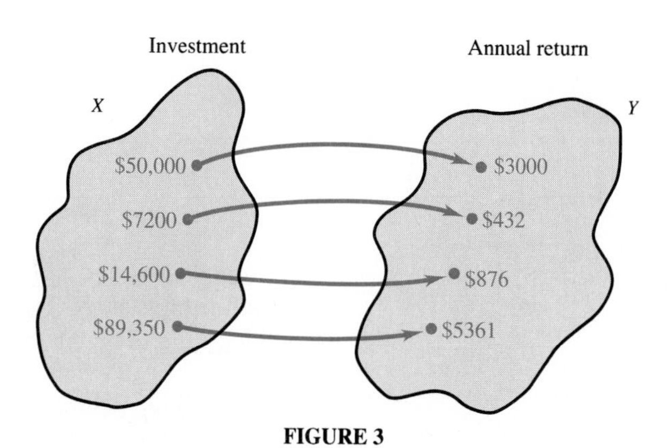

FIGURE 3

In this second example, we could use a formula to show how the numbers in set X are used to obtain the numbers in set Y. If x is a dollar amount from set X, then the corresponding annual return y in dollars from set Y can be found with the formula

$$y = .06 \times x \qquad \text{or} \qquad y = .06x.$$

In this example, x, the investment in the mutual fund, is called the **independent variable.** Because the annual return *depends* on the amount of the investment, y is called the **dependent variable.** When a specific number, say 2000, is substituted for x, then y takes on one specific value—in this case, $.06 \times 2000 = 120$. The variable y is said to be a function of x.

This pair of numbers, one for x and one for y, can be written as an **ordered pair** (x, y), where the order of the numbers is important. Using ordered pairs, the information shown in Figure 3 can be expressed as the set

$$\{(50{,}000,\ 3000),\ (7200,\ 432),\ (14{,}600,\ 876),\ (89{,}350,\ 5361)\}.$$

Alternatively, the ordered pairs can be given in a table, as shown in Figure 4.

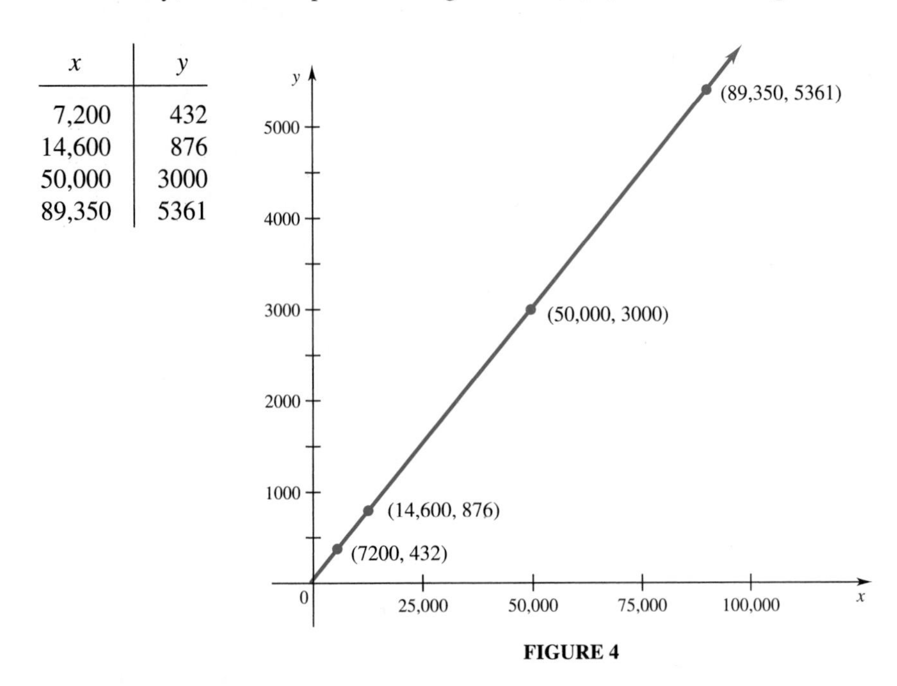

x	y
7,200	432
14,600	876
50,000	3000
89,350	5361

FIGURE 4

The set of points in a plane that correspond to the ordered pairs of a function is called the **graph** of the function. The function defined by $y = .06x$ that contains the ordered pairs shown in the table also can be illustrated with a graph, as shown in Figure 4.

The following definition of a function summarizes the discussion above.

DEFINITION OF FUNCTION A **function** is a rule that assigns to each element of a set X exactly one element from a set Y.

You may wonder precisely what we mean by a rule. An alternative definition of a function is the following:

ALTERNATIVE DEFINITION OF FUNCTION A **function** is a set of ordered pairs in which no two ordered pairs have the same first number.

For instance, in the ordered pairs in the investment example, notice that each one has a different first number: 50,000, 7200, 14,600, or 89,350. A function can have two ordered pairs with the same *second* number, however, as will be seen in Example 1(c).

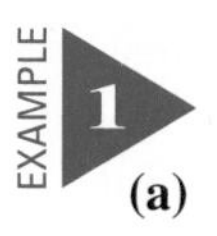

Which of the following are functions?

(a)

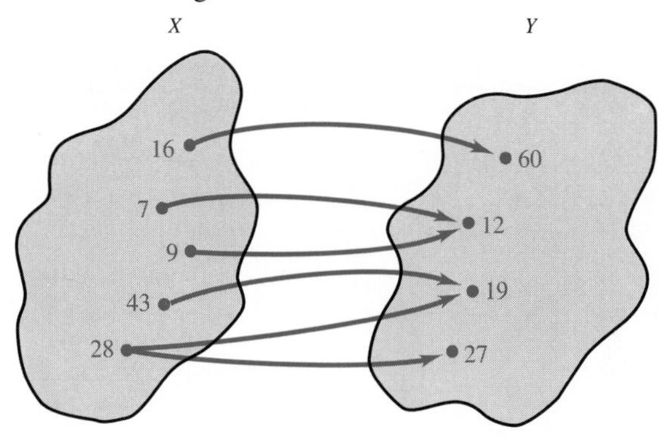

FIGURE 5

Figure 5 shows that an x-value of 28 corresponds to *two* y-values, 19 and 27. In a function, each x must correspond to exactly one y, so this correspondence is not a function.

(b) The optical reader at the checkout counter in many stores that converts codes to prices

For each code, the reader produces exactly one price, so this is a function.

(c) The x^2 key on a calculator

This correspondence between input and output is a function because the calculator produces just one x^2 (one y-value) for each x-value entered. Notice also that two x-values, such as 3 and -3, produce the same y-value of 9, but this does not violate the definition of a function.

(d)

x	1	1	2	2	3	3
y	3	-3	5	-5	8	-8

Since at least one x-value corresponds to more than one y-value, this table does not define a function.

(e) The set of ordered pairs with first elements mothers and second elements their children

Here the mother is the independent variable and the child is the dependent variable. For a given mother, there may be several children, so this correspondence is not a function.

(f) The set of ordered pairs with first elements children and second elements their birth mothers

In this case the child is the independent variable and the mother is the dependent variable. Since each child has only one birth mother, this is a function. ◀

As shown in Example 1, there are many ways to define functions. Almost every function we will use in this book will be defined by an equation, such as the equation $y = .06x$ discussed earlier in this section.

As we have seen, a function is a correspondence between the elements of two sets. These sets are given special names.

DOMAIN AND RANGE

The set of all possible values for the independent variable in a function is called the **domain** of the function; the set of all possible values for the dependent variable is the **range.**

The domain and range may or may not be the same set. For example, the domain of the function in Figure 2 is a set of positive whole numbers representing area, and the range is a set of positive whole numbers representing dollars. For the function in Figure 3, both the domain and range are a set of positive decimal numbers representing dollars and cents.

EXAMPLE 2 Decide whether each of the following equations represents a function. (Assume that x represents the independent variable here, an assumption we shall make throughout this book.) Give the domain and range of any functions.

(a) $y = -4x + 11$

For a given value of x, calculating $-4x + 11$ produces exactly one value of y. (For example, if $x = -7$, then $y = -4(-7) + 11 = 39$.) Since one value of the independent variable leads to exactly one value of the dependent variable, $y = -4x + 11$ is a function. Both x and y may take on any real-number values, so both the domain and range are the set of all real numbers.

(b) $y^2 = x$

Suppose $x = 36$. Then $y^2 = x$ becomes $y^2 = 36$, from which $y = 6$ or $y = -6$. Since one value of the independent variable can lead to two values of the dependent variable, $y^2 = x$ does not represent a function. ◀

The following agreement on domains is customary.

AGREEMENT ON DOMAINS Unless otherwise stated, assume that the domain of all functions defined by an equation is the largest set of real numbers that are meaningful replacements for the independent variable.

For example, suppose

$$y = \frac{-4x}{2x - 3}.$$

Any real number can be used for x except $x = 3/2$, which makes the denominator equal 0. By the agreement on domains, the domain of this function is the set of all real numbers except $3/2$.

Caution When finding the domain of a function, there are two operations to avoid: (1) dividing by zero; and (2) taking the square root (or any even root) of a negative number. Later chapters will present other functions, such as logarithms, which require further restrictions on the domain. For now, just remember these two restrictions on the domain.

From now on, we will write domains and ranges in **interval notation.** With this notation, the set of real numbers less than 4 is written $(-\infty, 4)$. The symbol ∞ (the symbol for infinity) is not a real number; it shows that the interval includes *all* real numbers less than 4.

A square bracket is used to indicate that a number is included in the interval. For example, $2 \le x < 5$ is written $[2, 5)$, indicating that 2 is included but 5 is not. Similarly, $x \ge 10$ is written $[10, \infty)$. Using interval notation, the set of all real numbers is the interval $(-\infty, \infty)$. Parentheses and brackets also are used to graph intervals on the number line.

The Inequalities section of the Algebra Reference discusses the graphs of intervals. For example, the interval $[2, 5)$ is graphed as shown.

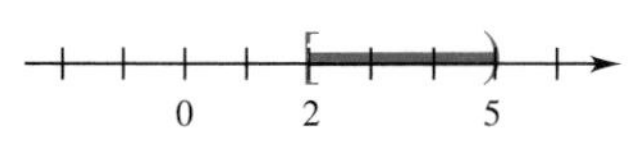

EXAMPLE 3 Find the domain and range for each of the functions defined as follows.

(a) $y = x^2$

Any number may be squared, so the domain is the set of all real numbers, written $(-\infty, \infty)$. Since $x^2 \ge 0$ for every value of x, the range is $[0, \infty)$.

(b) $y = \sqrt{6 - x}$

For y to be a real number, $6 - x$ must be nonnegative. This happens only when $6 - x \ge 0$, or $6 \ge x$, making the domain $(-\infty, 6]$. The range is $[0, \infty)$ because $\sqrt{6 - x}$ is always nonnegative.

(c) $y = \sqrt{2x^2 + 5x - 12}$

The domain includes only those values of x satisfying $2x^2 + 5x - 12 \ge 0$. Using the methods for solving a quadratic inequality produces the domain

$$(-\infty, -4] \cup [3/2, \infty).*$$

As in part (b), the range is $[0, \infty)$.

The Inequalities section of the Algebra Reference demonstrates the method for solving a quadratic inequality. To solve $2x^2 + 5x - 12 \ge 0$, factor the quadratic to get $(2x - 3)(x + 4) \ge 0$. Setting each factor equal to 0 gives $x = 3/2$ or $x = -4$, leading to the intervals $(-\infty, -4]$, $[-4, 3/2]$, and $[3/2, \infty)$. Testing a number from each interval shows that the solution is $(-\infty, -4] \cup [3/2, \infty)$.

---

*The *union* of sets A and B, written $A \cup B$, is defined as the set of all elements in A or B or both.

(d) $y = \dfrac{1}{x + 3}$

Since the denominator cannot be zero, $x \neq -3$ and the domain is

$$(-\infty, -3) \cup (-3, \infty).$$

Because the numerator can never be zero, $y \neq 0$. There are no other restrictions on y, so the range is $(-\infty, 0) \cup (0, \infty)$. ◀

Function Notation The letters f, g, and h frequently are used to represent functions. For example, f might be used to name the function defined by $y = 5 - 3x$. For a given value of x in the domain of a function f, there is exactly one corresponding value of y in the range. To emphasize that y is obtained by applying function f to the element x, replace y with the symbol $f(x)$, read "f of x" or "f at x." Here x is the independent variable; either y or $f(x)$ represents the dependent variable.

Using $f(x)$ to replace y in the equation $y = 5 - 3x$ gives

$$f(x) = 5 - 3x.$$

If 2 is chosen as a value of x, $f(x)$ becomes

$$f(2) = 5 - 3 \cdot 2$$
$$f(2) = -1.$$

In a similar manner,

$$f(-4) = 5 - 3(-4) = 17, \qquad f(0) = 5, \qquad f(-6) = 23,$$

and so on.

To understand how a function works, think of a function f as a machine—for example, a calculator or computer—that takes an input x from the domain and uses it to produce an output $f(x)$ (which represents the y value), as shown in Figure 6.

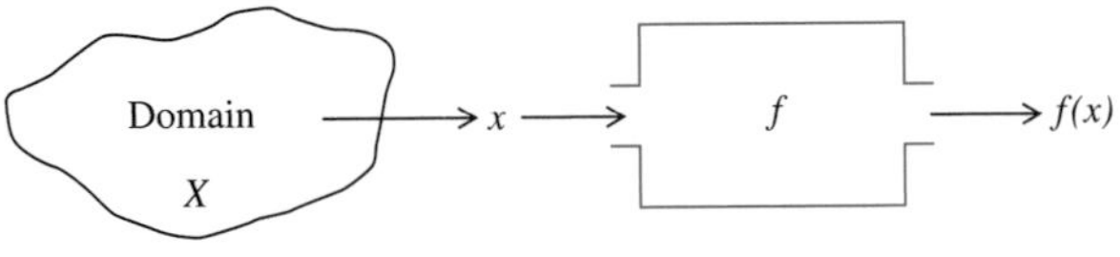

FIGURE 6

EXAMPLE 4 Let $g(x) = -x^2 + 4x - 5$. Find each of the following.

(a) $g(3)$

Replace x with 3.

$$g(3) = -3^2 + 4 \cdot 3 - 5 = -9 + 12 - 5 = -2$$

(b) $g(a)$

Replace x with a to get $g(a) = -a^2 + 4a - 5$.

This replacement of one variable with another is important in later chapters.

(c) $g(x + h) = -(x + h)^2 + 4(x + h) - 5$

$= -(x^2 + 2xh + h^2) + 4(x + h) - 5$

$= -x^2 - 2xh - h^2 + 4x + 4h - 5$

(d) $g\left(\frac{2}{r}\right) = -\left(\frac{2}{r}\right)^2 + 4\left(\frac{2}{r}\right) - 5 = -\frac{4}{r^2} + \frac{8}{r} - 5$ ◀

Caution Notice from part (c) that $g(x + h)$ is *not* the same as $g(x) + h$, which equals $-x^2 + 4x - 5 + h$. There is a significant difference between applying a function to the quantity $x + h$ and applying a function to x and adding h afterwards.

If you tend to get confused when replacing x with $x + h$, as in Example 4(c), you might try replacing the x in the original function with a box, like this:

$$g\left(\boxed{}\right) = -\left(\boxed{}\right)^2 + 4\left(\boxed{}\right) - 5.$$

Then, to compute $g(x + h)$, just enter $x + h$ into the box:

$$g\left(\boxed{x + h}\right) = -\left(\boxed{x + h}\right)^2 + 4\left(\boxed{x + h}\right) - 5$$

and proceed as in Example 4(c).

Graphs As mentioned earlier, it often is useful to draw a graph of the ordered pairs produced by a function. To graph a function, we use a **Cartesian coordinate system,** as shown in Figure 7. The horizontal number line, or ***x*-axis,** represents the elements from the domain of the function, and the vertical or ***y*-axis** represents the elements from the range. The point where the number lines cross is the zero point on both of these number lines; this point is called the **origin.**

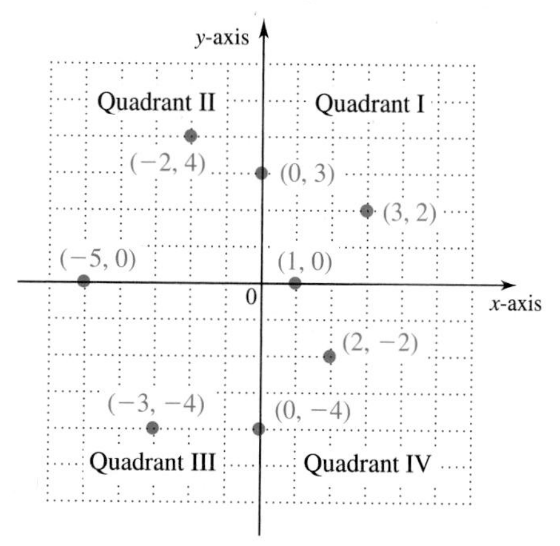

FIGURE 7

The name "Cartesian" honors René Descartes (1596–1650), a brilliant but sickly man. According to legend, Descartes was lying in bed when he noticed an insect crawling on the ceiling and realized that if he could determine the distance from the bug to each of two perpendicular walls, he could describe its position at any given moment. The same idea can be used to locate a point in a plane.

Each point in a Cartesian coordinate system corresponds to an ordered pair of real numbers. Several points and their corresponding ordered pairs are shown in Figure 7. In the point $(-2, 4)$, for example, -2 is the ***x*-coordinate** and 4 is the ***y*-coordinate.** From now on, instead of referring to "the point corresponding to the ordered pair $(-2, 4)$," we will say "the point $(-2, 4)$."

The *x*-axis and *y*-axis divide the plane into four parts or **quadrants.** For example, quadrant I includes points whose *x*- and *y*-coordinates are both positive. The quadrants are numbered as shown in Figure 7. The points of the axes themselves belong to no quadrant.

EXAMPLE 5 Let $f(x) = 3 - 2x$, with domain $\{-2, -1, 0, 1, 2, 3, 4\}$. Graph the ordered pairs produced by this function.

If $x = -2$, then $f(-2) = 3 - 2(-2) = 7$, giving the corresponding ordered pair $(-2, 7)$. Using additional values of x in the equation gives the ordered pairs listed below.

x	-2	-1	0	1	2	3	4
y	7	5	3	1	-1	-3	-5
Ordered Pair	$(-2, 7)$	$(-1, 5)$	$(0, 3)$	$(1, 1)$	$(2, -1)$	$(3, -3)$	$(4, -5)$

These ordered pairs are graphed in Figure 8. ◀

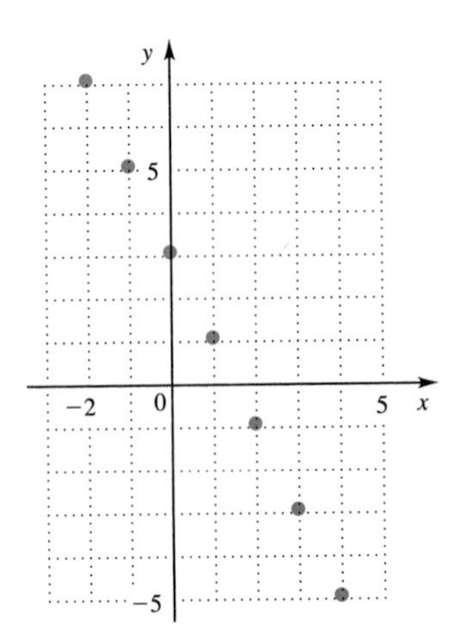

FIGURE 8

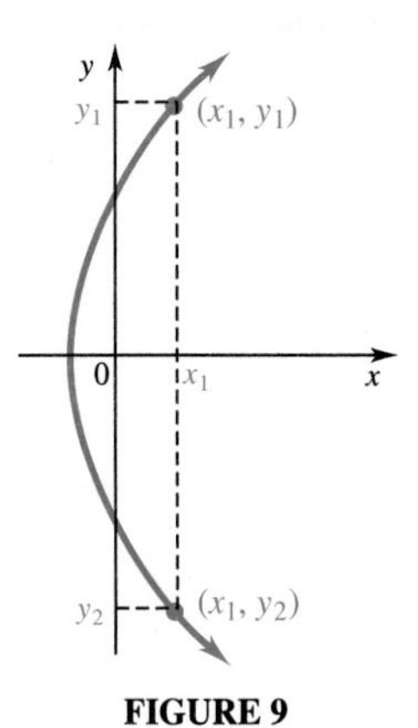

FIGURE 9

If a graph is to represent a function, each value of x from the domain must lead to exactly one value of y. In the graph in Figure 9, the domain value x_1 leads to *two* y-values, y_1 and y_2. Since the given x-value corresponds to two different y-values, this is not the graph of a function. This example suggests the **vertical line test** for the graph of a function.

VERTICAL LINE TEST

If a vertical line intersects a graph in more than one point, the graph is not the graph of a function.

EXAMPLE 6 Use the vertical line test to decide which of the graphs in Figure 10 are graphs of functions.

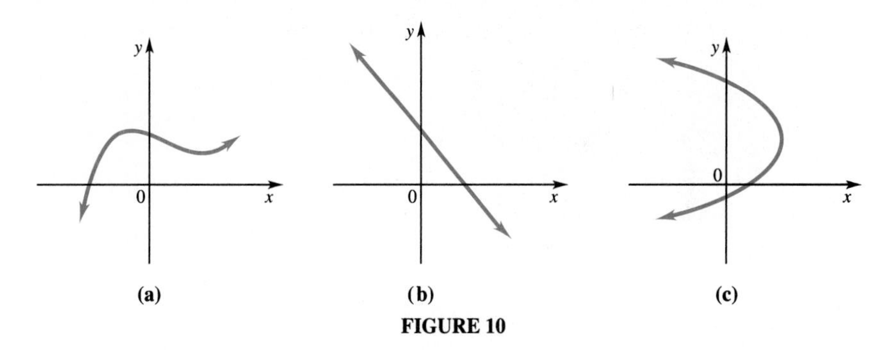

FIGURE 10

(a) Every vertical line intersects this graph in at most one point, so this is the graph of a function.

(b) Again, each vertical line intersects the graph in at most one point, showing that this is the graph of a function.

(c) It is possible for a vertical line to intersect the graph in part (c) twice. This is not the graph of a function. ◀

EXAMPLE 7 An overnight delivery service charges \$25 for a package weighing up to 2 pounds. For each additional pound there is an additional charge of \$3. Let $D(x)$ represent the cost to send a package weighing x pounds. Graph $D(x)$ for x in the interval (0, 6].

For x in the interval (0, 2], $y = 25$. For x in (2, 3], $y = 25 + 3 = 28$. For x in (3, 4], $y = 28 + 3 = 31$, and so on. The graph is shown in Figure 11. ◀

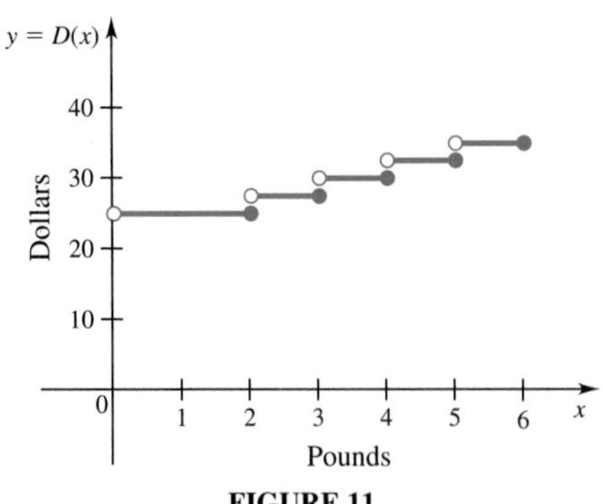

FIGURE 11

The function discussed in Example 7 is called a **step function.** Many real-life situations are best modeled by step functions. Additional examples are given in the exercises.

1.1 Exercises

Which of the following rules define y as a function of x?

1.

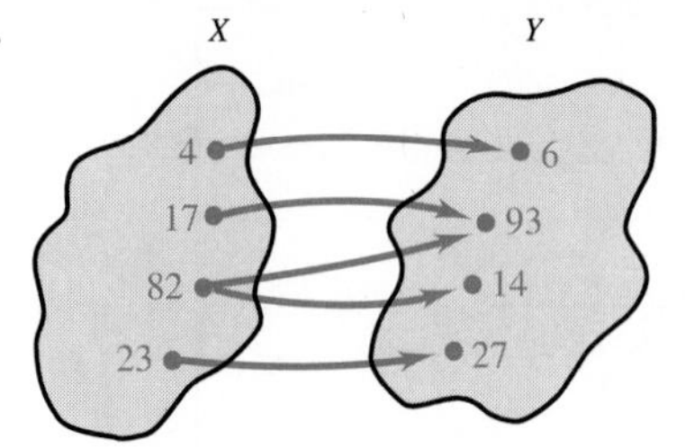

2.

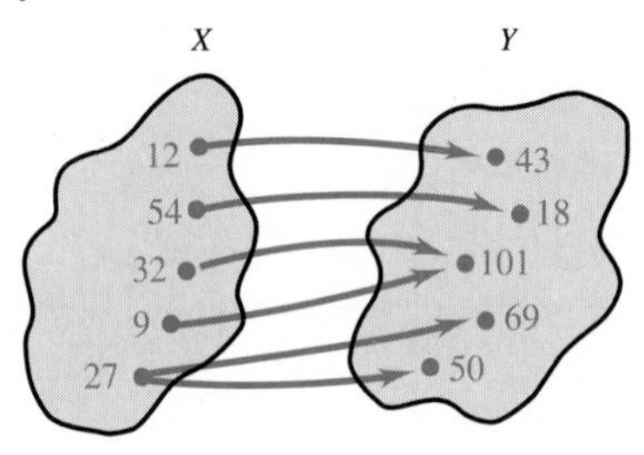

3.

x	y
3	9
2	4
1	1
0	0
−1	1
−2	4
−3	9

4.

x	y
9	3
4	2
1	1
0	0
1	−1
4	−2
9	−3

5. $y = x^3$ **6.** $y = \sqrt{x}$ **7.** $x = |y|$ **8.** $x = y^4 - 1$

List the ordered pairs obtained from each equation, given $\{-2, -1, 0, 1, 2, 3\}$ *as domain. Graph each set of ordered pairs. Give the range.*

9. $y = x - 1$ **10.** $y = 2x + 3$ **11.** $y = -4x + 9$ **12.** $y = -6x + 12$

13. $2x + y = 9$ **14.** $3x + y = 16$ **15.** $2y - x = 5$ **16.** $6x - y = -3$

17. $y = x(x + 1)$ **18.** $y = (x - 2)(x - 3)$ **19.** $y = x^2$ **20.** $y = -2x^2$

21. $y = \dfrac{1}{x + 3}$ **22.** $y = \dfrac{-2}{x + 4}$ **23.** $y = \dfrac{3x - 3}{x + 5}$ **24.** $y = \dfrac{2x + 1}{x + 3}$

Write each expression in interval notation. Graph each interval.

25. $x < 0$ **26.** $x \geq -3$ **27.** $1 \leq x < 2$

28. $-5 < x \leq -4$ **29.** $-9 > x$ **30.** $6 \leq x$

Using the variable x, write each interval as an inequality.

31. $(-4, 3)$ **32.** $[2, 7)$ **33.** $(-\infty, -1]$ **34.** $(3, \infty)$

35.  $-2 \quad 0 \quad 6$

36. $0 \quad 8$

37. 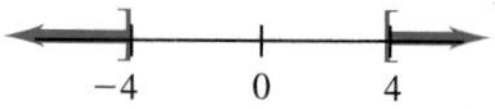 $-4 \quad 0 \quad 4$

38. 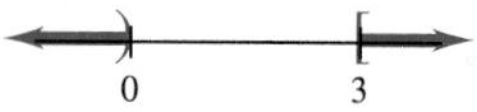 $0 \quad 3$

Give the domain of each function defined as follows.

39. $f(x) = 2x$

40. $f(x) = x + 2$

41. $f(x) = x^4$

42. $f(x) = (x - 2)^2$

43. $f(x) = \sqrt{16 - x^2}$

44. $f(x) = |x - 1|$

45. $f(x) = (x - 3)^{1/2}$

46. $f(x) = (3x + 5)^{1/2}$

47. $f(x) = \dfrac{2}{x^2 - 4}$

48. $f(x) = \dfrac{-8}{x^2 - 36}$

49. $f(x) = -\sqrt{\dfrac{2}{x^2 + 9}}$

50. $f(x) = -\sqrt{\dfrac{5}{x^2 + 36}}$

51. $f(x) = \sqrt{x^2 - 4x - 5}$

52. $f(x) = \sqrt{15x^2 + x - 2}$

53. $f(x) = \dfrac{1}{\sqrt{x^2 - 6x + 8}}$

54. $f(x) = \sqrt{\dfrac{x + 1}{x - 1}}$

Give the domain and the range of each function. Where arrows are drawn, assume the function continues in the indicated direction.

55.

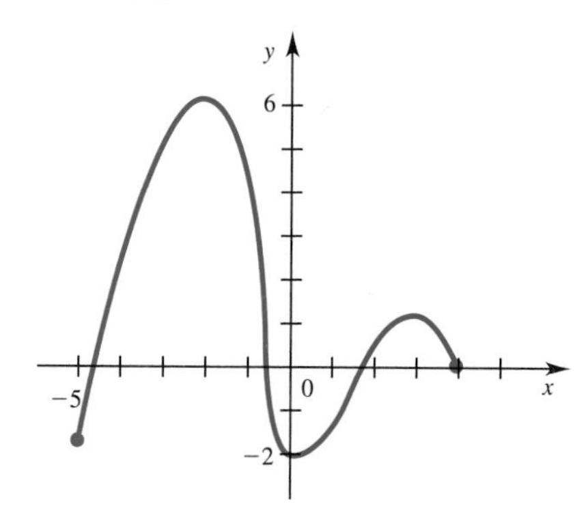

56.

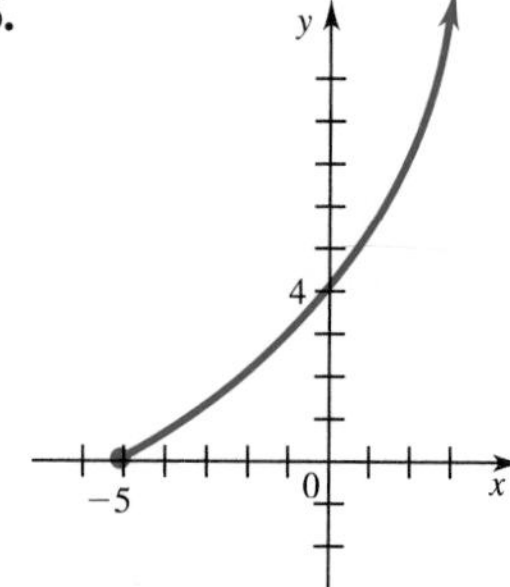

57.

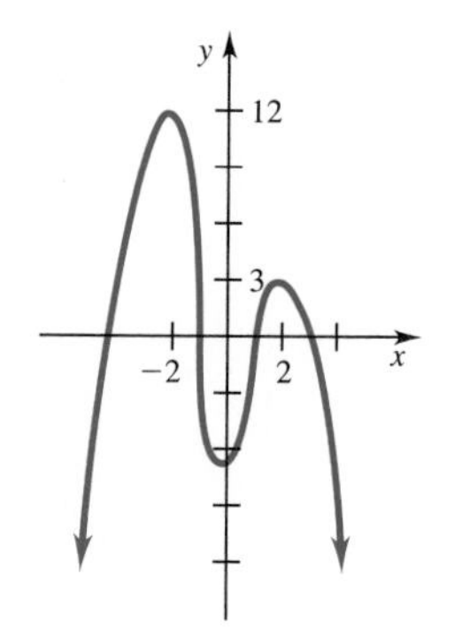

58.

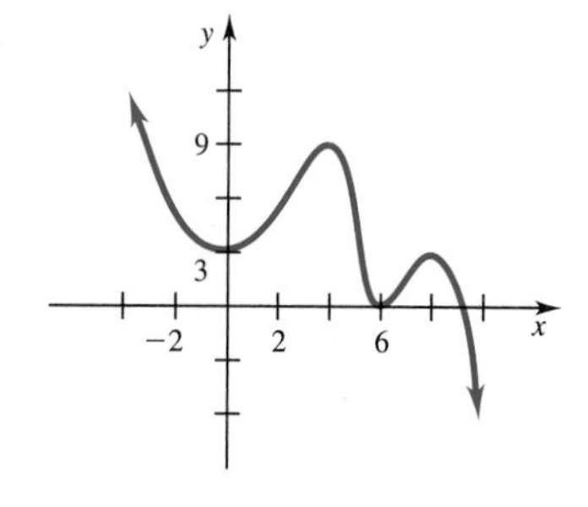

For each function, find **(a)** $f(4)$, **(b)** $f(-3)$, **(c)** $f(-1/2)$, **(d)** $f(a)$, *and* **(e)** $f(2/m)$.

59. $f(x) = 3x + 2$

60. $f(x) = 5x - 6$

61. $f(x) = -x^2 + 5x + 1$

62. $f(x) = (x + 3)(x - 4)$

63. $f(x) = \dfrac{2x + 1}{x - 2}$

64. $f(x) = \dfrac{3x - 5}{2x + 3}$

Use each graph to find **(a)** $f(-2)$, **(b)** $f(0)$, **(c)** $f(1/2)$, *and* **(d)** $f(4)$.

65.

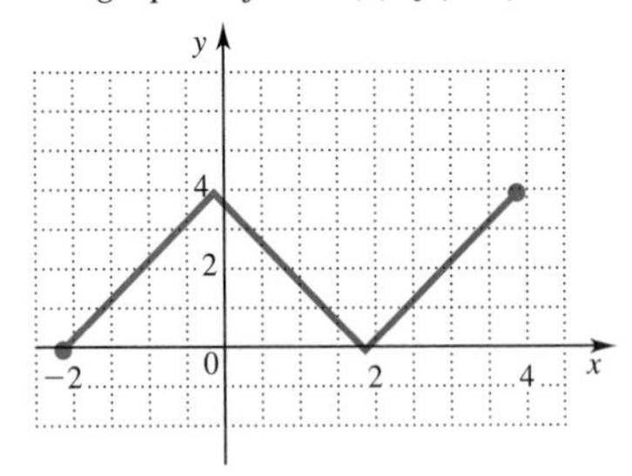

66.

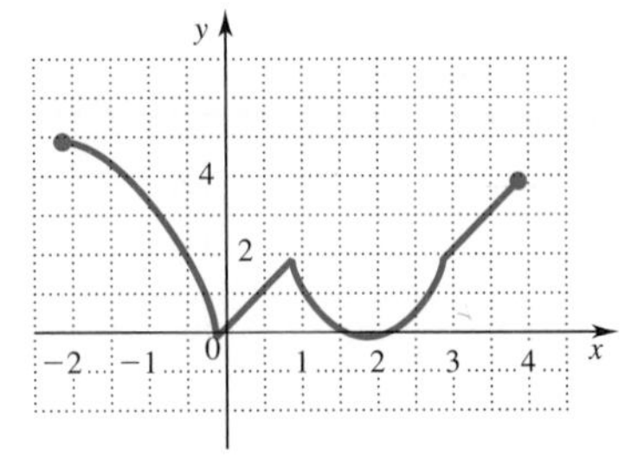

67.

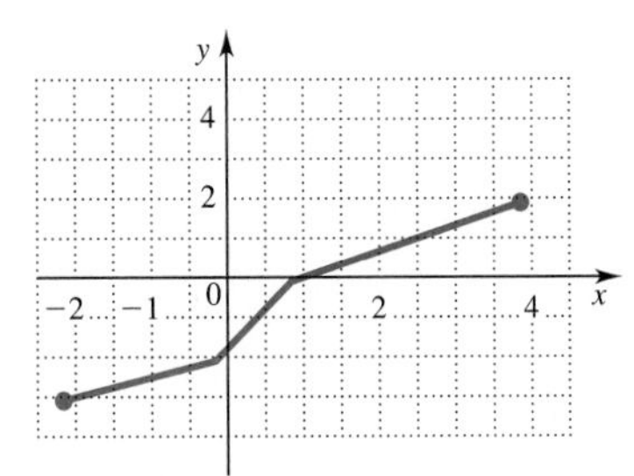

68.

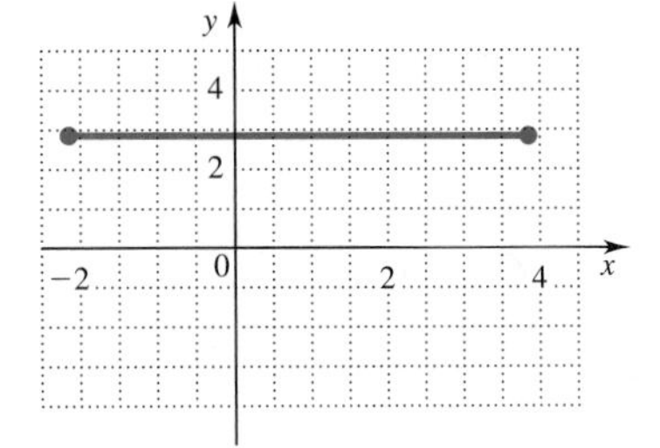

Let $f(x) = 6x - 2$ *and* $g(x) = x^2 - 2x + 5$ *to find the following values.*

69. $f(m - 3)$

70. $f(2r - 1)$

71. $g(r + h)$

72. $g(z - p)$

73. $g\left(\dfrac{3}{q}\right)$

74. $g\left(-\dfrac{5}{z}\right)$

Decide whether each graph represents a function.

75.

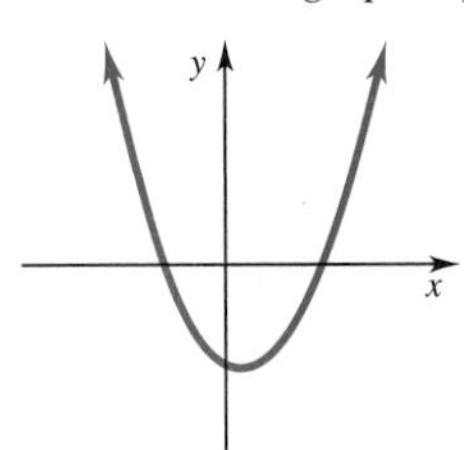

76.

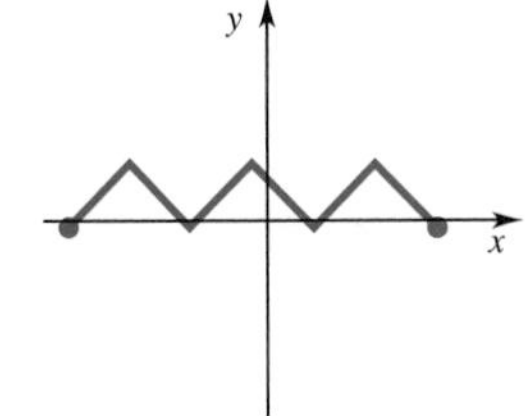

77.

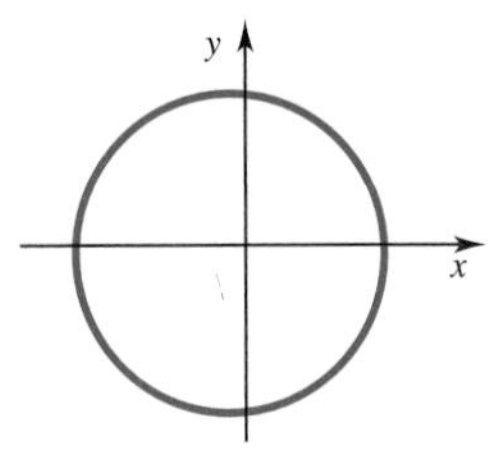

78.

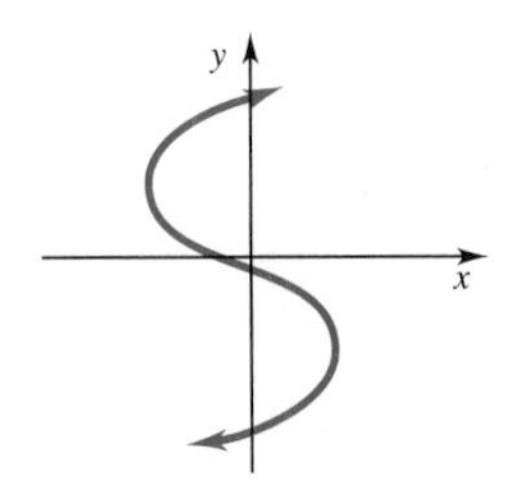

79.

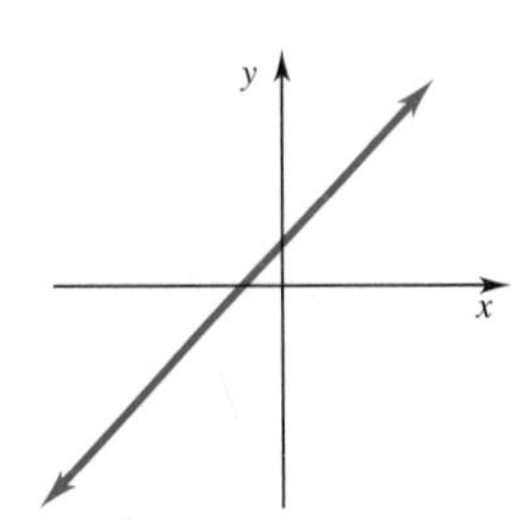

80.

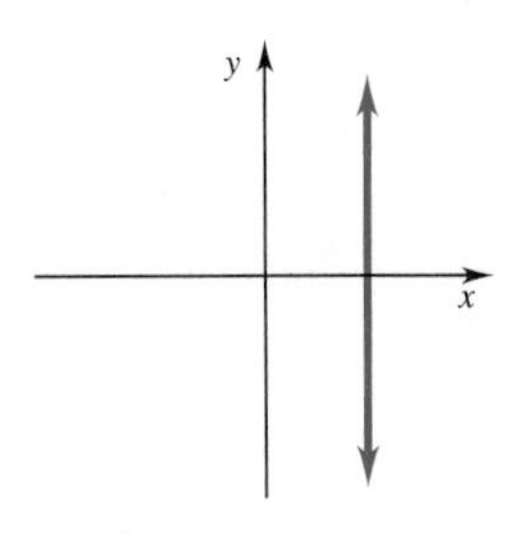

For each function defined as follows, find **(a)** $f(x+h)$, **(b)** $f(x+h)-f(x)$, *and* **(c)** $\dfrac{f(x+h)-f(x)}{h}$.

81. $f(x) = x^2 - 4$

82. $f(x) = 8 - 3x^2$

83. $f(x) = 6x + 2$

84. $f(x) = 4x - 11$

85. $f(x) = \dfrac{1}{x}$

86. $f(x) = -\dfrac{1}{x^2}$

Applications

General Interest

87. Energy Use The graph* shows two functions of time. The first, which we will designate $I(t)$, shows the energy intensity of the United States, which is defined as the thousands of BTUs of energy consumed per dollar of Gross National Product (GNP) in year t (in 1982 dollars). The second, which we will designate $S(t)$, represents the national savings per year (also in 1982 dollars) due to a decreased use of energy.

(a) Find $I(1976)$. **(b)** Find $S(1984)$.

(c) Find $I(t)$ and $S(t)$ at the time when the two graphs cross. In what year does this occur?

(d) What significance, if any, is there to the point at which the two graphs cross?

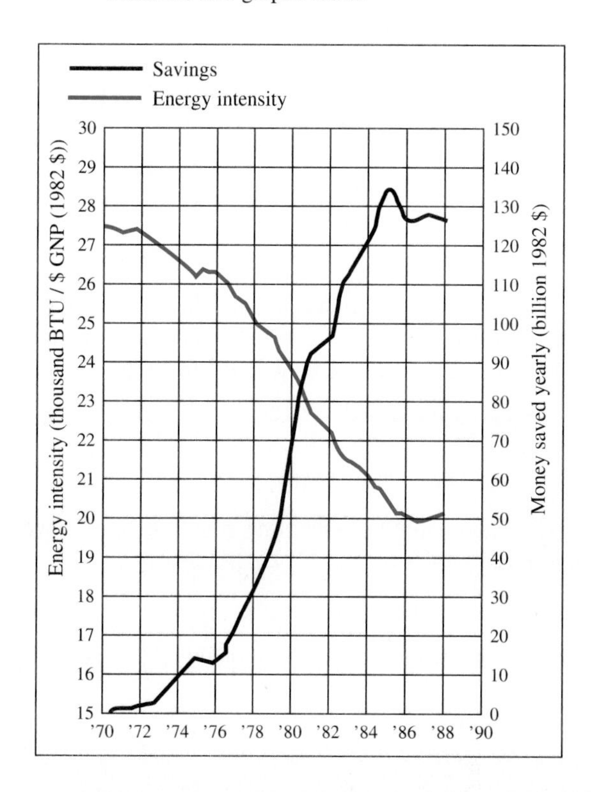

Business and Economics

88. Saw Rental A chain-saw rental firm charges \$7 per day or fraction of a day to rent a saw, plus a fixed fee of \$4 for resharpening the blade. Let $S(x)$ represent the cost of renting a saw for x days.

Find each of the following.

(a) $S\left(\dfrac{1}{2}\right)$ **(b)** $S(1)$ **(c)** $S\left(1\dfrac{1}{4}\right)$

(d) $S\left(3\dfrac{1}{2}\right)$ **(e)** $S(4)$ **(f)** $S\left(4\dfrac{1}{10}\right)$

(g) What does it cost to rent a saw for $4\frac{9}{10}$ days?

(h) A portion of the graph of $y = S(x)$ is shown here. Explain how the graph could be continued.

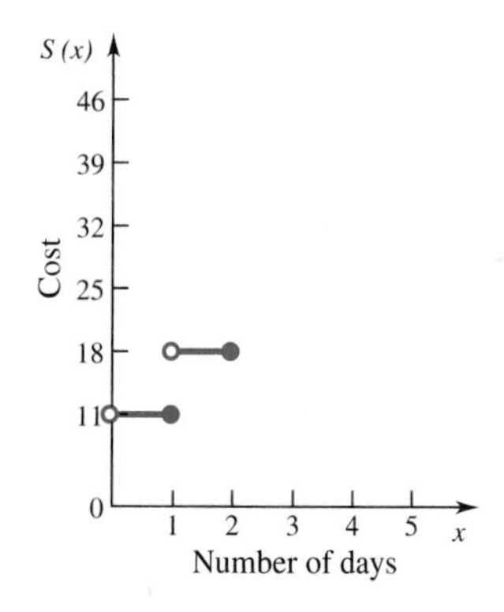

(i) What is the independent variable?

(j) What is the dependent variable?

(k) Write a sentence or two explaining what part (f) and its answer represent.

(l) We have left $x = 0$ out of the graph. Discuss why it should or shouldn't be included. If it were included, how would you define $S(0)$?

---

*Graph entitled "Energy Use Down, Savings Up" from "The Greenhouse Effect" by the Union of Concerned Scientists. Reprinted by permission of the Union of Concerned Scientists.

89. Postage In 1991, it cost 29¢ in the United States to mail a first-class letter weighing up to 1 oz, and 23¢ for each additional ounce or fraction of an ounce. Let $C(x)$ represent the postage for a first-class letter weighing x oz. Find each of the following.

(a) $C\left(\frac{2}{3}\right)$ **(b)** $C\left(1\frac{1}{3}\right)$ **(c)** $C(2)$ **(d)** $C\left(3\frac{1}{8}\right)$

(e) What does it cost to send a letter that weighs $2\frac{5}{8}$ oz?

(f) Write a sentence or two explaining what part (e) and its answer represent.

(g) Graph the function defined by $y = C(x)$ for $0 < x \le 4$.

(h) What is the independent variable?

(i) What is the dependent variable?

(j) We have left $x = 0$ out of the domain. Discuss why it should or shouldn't be included. If it were included, how would you define $S(0)$?

1.2 LINEAR FUNCTIONS

How much vinyl siding should a supplier produce each month, and how much should the supplier charge for it?

In this section, we will learn one method for solving such problems.

Many practical situations can be described (or at least approximated) with a *linear function.* Some examples are the relationships between Fahrenheit and Celsius temperatures, between price and supply of some consumer goods, and between demand and supply of some commodities.

LINEAR FUNCTION

A function f is **linear** if its equation can be written as

$$f(x) = ax + b$$

for real numbers a and b.

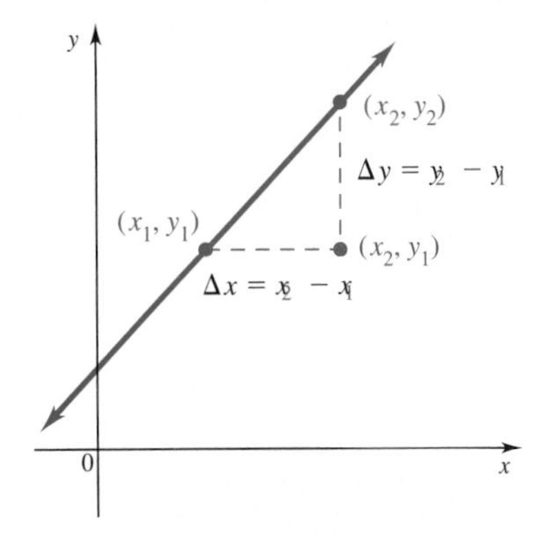

FIGURE 12

As the name implies, every linear function has a graph that is a straight line. An important characteristic of a straight line is its *slope*, a numerical measure of the steepness of the line. To find this measure, start with the line through the two distinct points (x_1, y_1) and (x_2, y_2), as shown in Figure 12. (Assume $x_1 \neq x_2$.) The difference

$$x_2 - x_1$$

is called the *change in x* and written with the symbol Δx (read "delta x"), where Δ is the Greek letter delta. In the same way, the *change in y* is

$$\Delta y = y_2 - y_1.$$

The **slope** of a nonvertical line is defined as the quotient of the change in y and the change in x.

SLOPE OF A LINE

The slope m of the nonvertical line through the distinct points (x_1, y_1) and (x_2, y_2) is

$$m = \frac{\Delta y}{\Delta x} = \frac{y_2 - y_1}{x_2 - x_1}.$$

Similar triangles can be used to show that the slope does not depend on which pair of points on the line is used to calculate slope. That is, the same slope will be obtained for any two points on the line.

The slope of a line can be found only if the line is nonvertical, because $x_2 \neq x_1$ for a nonvertical line, so that the denominator $x_2 - x_1 \neq 0$. The slope of a vertical line is not defined.

EXAMPLE 1 Find the slope of the line through each of the following pairs of points.

(a) $(-4, 8)$ and $(2, -3)$

Choosing $x_1 = -4$, $y_1 = 8$, $x_2 = 2$, and $y_2 = -3$ gives $\Delta y = -3 - 8 = -11$ and $\Delta x = 2 - (-4) = 6$. By definition, the slope is

$$m = \frac{\Delta y}{\Delta x} = -\frac{11}{6}.$$

(b) $(2, 7)$ and $(2, -4)$

A sketch shows that the line through $(2, 7)$ and $(2, -4)$ is vertical. As mentioned above, the slope of a vertical line is undefined. (An attempt to use the definition of slope here would produce a zero denominator.)

(c) $(5, -3)$ and $(-2, -3)$

By the definition of slope,

$$m = \frac{-3 - (-3)}{-2 - 5} = 0.$$ ◀

Drawing a graph of the line in Example 1(c) shows that it is horizontal, which suggests that the slope of a horizontal line is 0.

The slope of a horizontal line is 0.

The slope of a vertical line is not defined.

Caution The phrase "no slope" should be avoided; specify instead whether the slope is zero or undefined.

Figure 13 (on the next page) shows lines of various slopes. As suggested by the figure, a line with a positive slope goes up from left to right, and a line with a negative slope goes down.

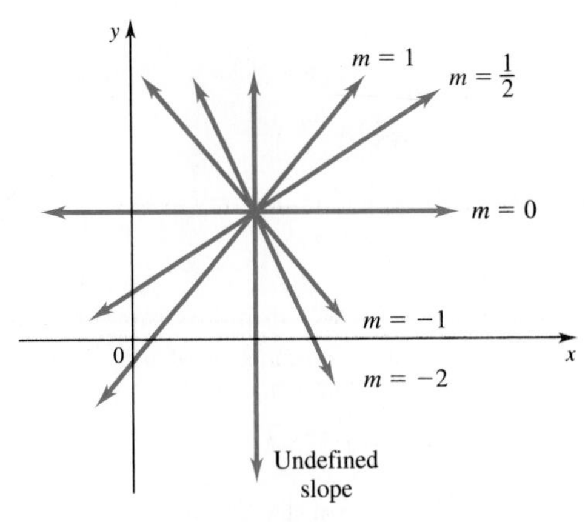

FIGURE 13

EXAMPLE 2 Graph the line that passes through $(-1, 5)$ and has slope $-5/3$.

First, locate the point $(-1, 5)$, as shown in Figure 14. Since the slope of this line is $-5/3$, a change of 3 units horizontally produces a change of -5 units vertically, giving a second point, $(2, 0)$, which is used to complete the graph. ◀

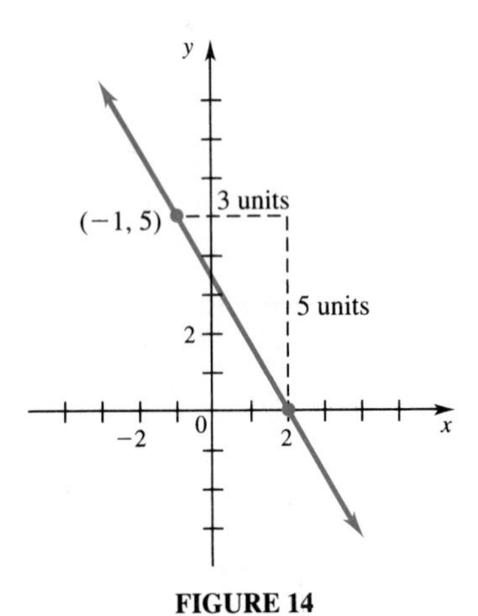

FIGURE 14

Equations of a Line The slope of a line, together with a point on the line, can be used to find an equation of the line. The procedure for finding the equation depends on whether or not the line is vertical. The vertical line through the point $(a, 0)$ passes through all points of the form (a, y), making the equation $x = a$.

VERTICAL LINE

An equation of the vertical line through the point $(a, 0)$ is $x = a$.

Let m be the slope of a nonvertical line. Assume that the fixed point (x_1, y_1) is on the line. Let (x, y) represent any other point on the line. The point (x, y) can be on the line if and only if the slope of the line through (x_1, y_1) and (x, y) is m; that is, if

$$\frac{y - y_1}{x - x_1} = m.$$

Multiplying both sides by $x - x_1$ gives

$$y - y_1 = m(x - x_1).$$

This result is summarized below.

POINT-SLOPE FORM

The line with slope m passing through the point (x_1, y_1) has an equation

$$\mathbf{y - y_1 = m(x - x_1).}$$

This equation is called the **point-slope form** of the equation of a line.

EXAMPLE 3

Write an equation of each line.

(a) Through $(-4, 1)$, with slope -3

Use the point-slope form of the equation of a line, with $x_1 = -4$, $y_1 = 1$, and $m = -3$.

$$\begin{aligned} y - 1 &= -3[x - (-4)] \\ y - 1 &= -3(x + 4) \\ y - 1 &= -3x - 12 \\ y &= -3x - 11 \qquad \text{or} \qquad 3x + y = -11 \end{aligned}$$

(b) Through $(-3, 2)$ and $(2, -4)$

First, find the slope with the definition of slope:

$$m = \frac{-4 - 2}{2 - (-3)} = \frac{-6}{5}.$$

Either $(-3, 2)$ or $(2, -4)$ can be used for (x_1, y_1). Choosing $x_1 = -3$ and $y_1 = 2$ gives

$$\begin{aligned} y - 2 &= \frac{-6}{5}[x - (-3)] \\ 5(y - 2) &= -6(x + 3) \\ 5y - 10 &= -6x - 18 \\ 5y &= -6x - 8 \qquad \text{or} \qquad 6x + 5y = -8. \end{aligned}$$

Verify that using $(2, -4)$ instead of $(-3, 2)$ leads to the same result. ◀

Any value of x where a graph crosses the x-axis is called an **x-intercept** for the graph. Any value of y where the graph crosses the y-axis is called a **y-intercept** for the graph. The graph in Figure 15 has x-intercepts x_1, x_2, and x_3 and y-intercept y_1. As suggested by the graph, x-intercepts can be found by letting $y = 0$ and y-intercepts by letting $x = 0$.

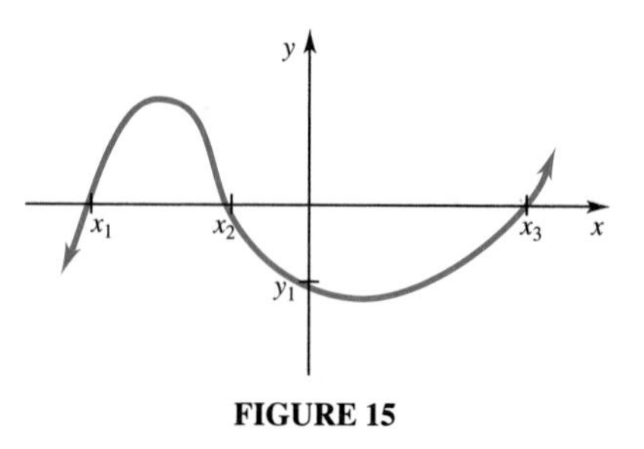

FIGURE 15

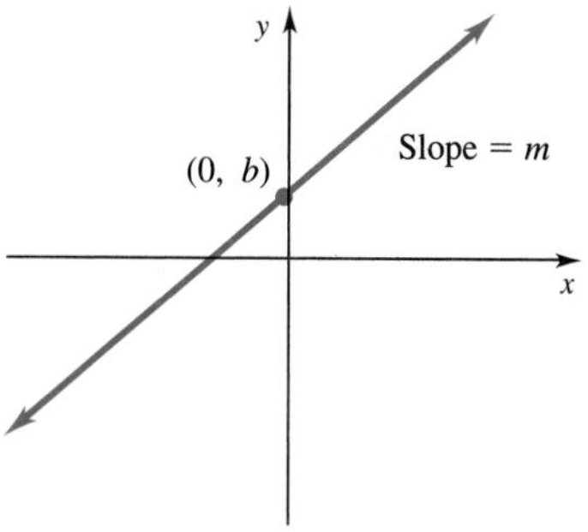

Figure 16 shows a line with y-intercept b; the line goes through $(0, b)$. If the slope of the line is m, then by the point-slope form, an equation of the line is

$$y - y_1 = m(x - x_1)$$
$$y - b = m(x - 0)$$
$$y = mx + b.$$

FIGURE 16

This result, which shows both the slope and the y-intercept, is the **slope-intercept form** of the equation of a line. Reversing these steps shows that any equation of the form $y = mx + b$ has a graph that is a line with slope m that passes through the point $(0, b)$.

SLOPE-INTERCEPT FORM

The line with slope m passing through the point $(0, b)$ has an equation

$$\boldsymbol{y = mx + b.}$$

This equation is called the slope-intercept form of the equation of a line.

This result, together with the fact that vertical lines have equations of the form $x = a$, shows that every line has an equation of the form $ax + by + c = 0$, where a and b are not both 0. Conversely, assuming $b \neq 0$ and solving $ax + by + c = 0$ for y gives $y = (-a/b)x - c/b$. By the result above, this equation is a line with slope $-a/b$ and y-intercept $-c/b$. If $b = 0$, solve for x to get $x = -c/a$, a vertical line. In any case, the equation $ax + by + c = 0$ has a straight line for its graph.

If a and b are not both 0, then the equation $ax + by + c = 0$ has a line for its graph. Also, any line has an equation of the form $ax + by + c = 0$.

EXAMPLE 4 Graph $3x + 2y = 6$.

This equation has a line for its graph because it can be written as $3x + 2y - 6 = 0$. Two distinct points on the line are enough to locate the graph. The intercepts often provide the necessary points. To find the x-intercept, let $y = 0$.

For review on solving a linear equation, see Section R.4 of the Algebra Reference.

$$3x + 2(0) = 6$$
$$3x = 6$$
$$x = 2$$

The x-intercept is 2. Let $x = 0$ to find that the y-intercept is 3. These two intercepts were used to get the graph shown in Figure 17.

Alternatively, solve $3x + 2y = 6$ for y to get

$$3x + 2y = 6$$
$$2y = -3x + 6$$
$$y = -\frac{3}{2}x + 3.$$

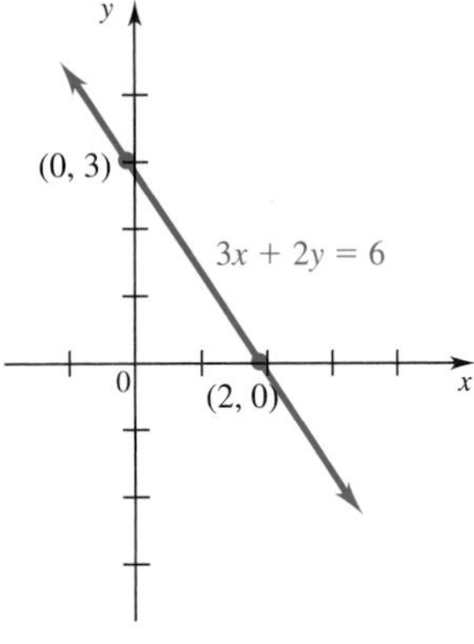

FIGURE 17

By the slope-intercept form, the graph of this equation is the line with y-intercept 3 and slope $-3/2$. (This means that the line goes down 3 units for each 2 units it goes to the right.) ◀

EXAMPLE 5 Graph $y = -3$.

For a line to have an x-intercept, there must be a value of x that makes $y = 0$. Here, however, $y = -3 \neq 0$. This means the line has no x-intercept, a situation that can happen only if the line is parallel to the x-axis (see Figure 18). ◀

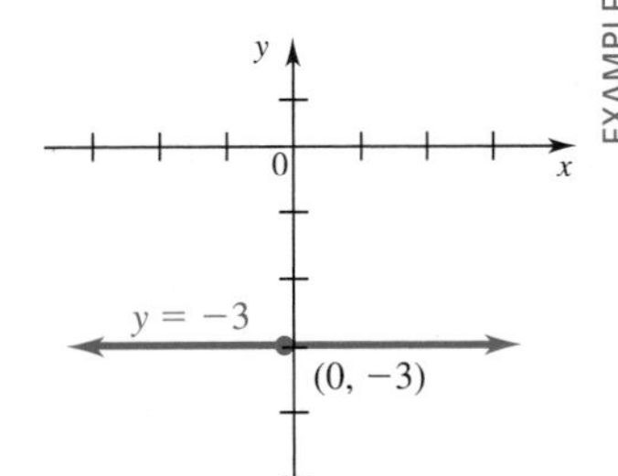

FIGURE 18

Example 5 suggests the following generalization.

HORIZONTAL LINE An equation of the horizontal line through $(0, a)$ is $y = a$.

Parallel and Perpendicular Lines One application of slope involves deciding whether two lines are parallel. Since two parallel lines are equally "steep," they should have the same slope. Also, two lines with the same "steepness" are parallel.

PARALLEL LINES Two nonvertical lines are parallel if and only if they have the same slope.

EXAMPLE 6 Find the equation of the line that passes through the point (3, 5) and is parallel to the line $2x + 5y = 4$.

The slope of $2x + 5y = 4$ can be found by writing the equation in slope-intercept form.

$$\begin{aligned} 2x + 5y &= 4 \\ y &= -\frac{2}{5}x + \frac{4}{5} \end{aligned}$$

This result shows that the slope is $-2/5$. Since the lines are parallel, $-2/5$ is also the slope of the line whose equation is needed. This line passes through (3, 5). Substituting $m = -2/5$, $x_1 = 3$, and $y_1 = 5$ into the point-slope form gives

$$\begin{aligned} y - y_1 &= m(x - x_1) \\ y - 5 &= -\frac{2}{5}(x - 3) \\ 5(y - 5) &= -2(x - 3) \\ 5y - 25 &= -2x + 6 \\ 2x + 5y &= 31. \end{aligned}$$ ◀

As mentioned above, two nonvertical lines are parallel if and only if they have the same slope. Two lines having slopes with a product of -1 are perpendicular. A proof of this fact, which depends on similar triangles from geometry, is given as Exercise 61 in this section.

PERPENDICULAR LINES Two lines, neither of which is vertical, are perpendicular if and only if their slopes have a product of -1.

Find the slope of the line L perpendicular to the line having the equation $5x - y = 4$.

To find the slope, write $5x - y = 4$ in slope-intercept form:

$$y = 5x - 4.$$

The slope is 5. Since the lines are perpendicular, if line L has slope m, then

$$\begin{aligned} 5m &= -1 \\ m &= -\frac{1}{5}. \end{aligned}$$ ◀

Supply and Demand Linear functions are often good choices for **supply and demand curves.** Typically, as the price of an item increases, the demand for the item decreases, while the supply increases. On the other hand, when demand for an item increases, so does its price, causing the supply of the item to decrease.

For example, several years ago the price of gasoline increased rapidly. As the price continued to escalate, most buyers became more and more prudent in their use of gasoline in order to restrict their demand to an affordable amount. Consequently, the overall demand for gasoline decreased and the supply increased, to a point where there was an oversupply of gasoline. This caused prices to fall until supply and demand were approximately balanced. Many other factors were involved in the situation, but the relationship between price, supply, and demand was nonetheless typical. Some commodities, however, such as medical care, college tuition, and certain luxury items, may be exceptions to these typical relationships.

Although economists consider price to be the independent variable, they have the unfortunate habit of plotting price, usually denoted by p, on the vertical axis, while the usual custom is to graph the independent variable on the horizontal axis. This practice originated with the English economist Alfred Marshall (1842–1924). To conform to this custom, we will write p, the price, as a function of q, the quantity produced, and plot p on the vertical axis. But remember that it is really price that determines how much consumers demand and producers supply, not the other way around.

Supply and demand functions are not necessarily linear (the simplest kind of function). Most functions are *approximately* linear if a small enough piece of the graph is taken, however, allowing applied mathematicians to often use linear functions for simplicity. We will take that approach in this section and the next.

EXAMPLE 8

Suppose that Greg Okjakjian, an economist, has studied the supply and demand for vinyl siding and has determined that the price in dollars per square yard, p, and the quantity demanded monthly in thousands of square yards, q, are related by the linear function

$$p = D(q) = 60 - \frac{3}{4}q, \quad \text{Demand}$$

while the price p and the supply q are related by

$$p = S(q) = \frac{3}{4}q. \quad \text{Supply}$$

(a) Find the demand at a price of \$45 and at a price of \$18.

Start with the demand function

$$p = 60 - \frac{3}{4}q,$$

and replace p with 45.

$$45 = 60 - \frac{3}{4}q$$

Solve this equation to find that

$$q = 20.$$

Thus, at a price of \$45 per square yard, 20,000 square yards are demanded per month.

Similarly, replace 18 for p to find the demand when the price is \$18. Verify that this leads to $q = 56$. When the price is lowered from \$45 to \$18, the demand increases from 20,000 square yards to 56,000 square yards.

(b) Find the supply at a price of \$60 and at a price of \$12.

Substitute 60 for p in the supply equation

$$p = \frac{3}{4}q$$

to find that $q = 80$, so the supply is 80,000 square yards. Similarly, replacing p with 12 in the supply equation gives a supply of 16,000 square yards. If the price decreases from \$60 to \$12 per square yard, the supply also decreases, from 80,000 square yards to 16,000 square yards.

(c) Graph both functions on the same axes.

The results of part (a) are written as the ordered pairs (20, 45) and (56, 18). The line through the corresponding points is the graph of $p = 60 - (3/4)q$, shown in black in Figure 19.

Use the ordered pairs (80, 60) and (16, 12) from the work in part (b) to get the supply graph shown in color in Figure 19. ◀

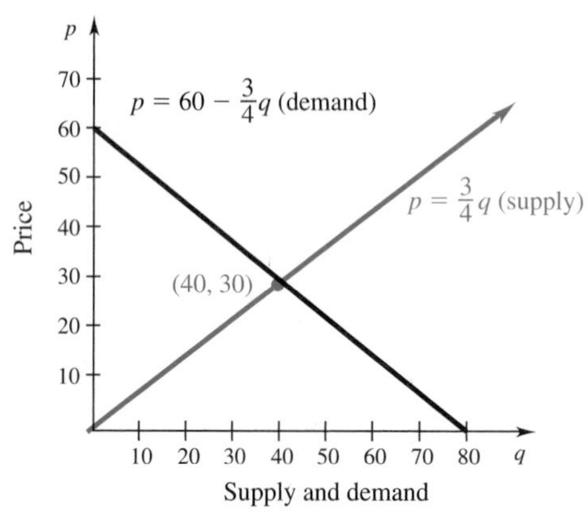

FIGURE 19

In Example 8 supply and demand are determined by

$$p = \frac{3}{4}q \quad \text{and} \quad p = 60 - \frac{3}{4}q,$$

respectively. To find a price where supply and demand are equal, solve the equation

$$\frac{3}{4}q = 60 - \frac{3}{4}q,$$

getting

$$q = 40.$$

Supply and demand will be equal when $q = 40$. This happens at a price of

$$p = \frac{3}{4}q$$

$$p = \frac{3}{4}(40) = 30,$$

or \$30. (Find the same result by using the demand function.) If the price of the item is more than \$30, the supply will exceed the demand. At a price less than \$30, the demand will exceed the supply. Only at a price of \$30 will demand and supply be equal. For this reason, \$30 is called the *equilibrium price*. When the price is \$30, demand and supply both equal 40,000 square yards, the *equilibrium quantity*.

Generalizing, the **equilibrium price** of a commodity is the price at the point where the supply and demand graphs for that commodity cross. The **equilibrium quantity** is the demand and the supply at the same point.

Another important issue is how, in practice, the equations of the supply and demand functions can be found. This issue is important for many problems involving linear functions in this section and the next. Data must be collected, and if the plotted points lie perfectly along a line, the equation can easily be found using any two points. What usually happens, however, is that the data are scattered, and there is no line that goes through all the points. In this case, a line must be found that approximates the linear trend in the data as much as possible (assuming the points lie approximately along a line). This is usually done by the *method of least squares,* also referred to as *linear regression.* A discussion of this method can be found in the chapter on functions of several variables, or in any elementary statistics text.

1.2 Exercises

Find the slope of each line that has a slope.

1. Through $(4, 5)$ and $(-1, 2)$
2. Through $(5, -4)$ and $(1, 3)$
3. Through $(8, 4)$ and $(8, -7)$
4. Through $(1, 5)$ and $(-2, 5)$
5. $y = 2x$
6. $y = 3x - 2$
7. $5x - 9y = 11$
8. $4x + 7y = 1$
9. $x = -6$
10. The x-axis
11. A line parallel to $2y - 4x = 7$
12. A line perpendicular to $6x = y - 3$
13. Through $(-1.978, 4.806)$ and $(3.759, 8.125)$
14. Through $(11.72, 9.811)$ and $(-12.67, -5.009)$

Find the slope of each of the lines shown.

15.

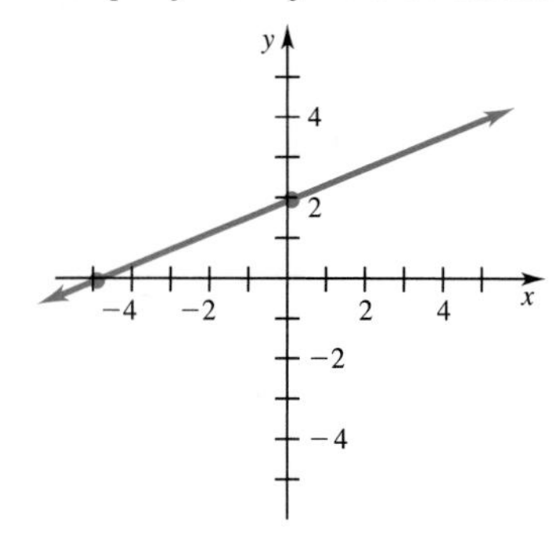

16.

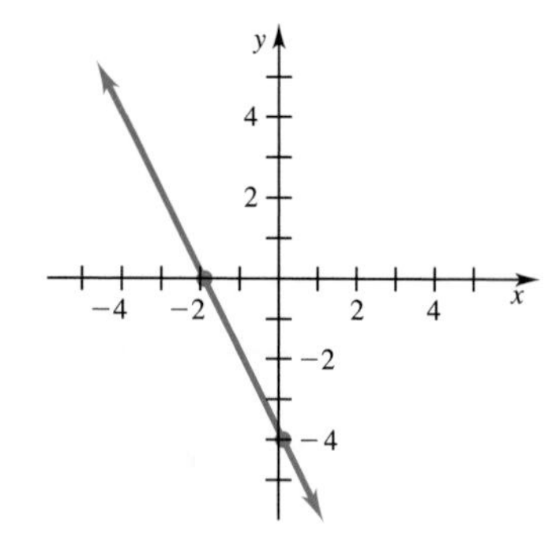

17.

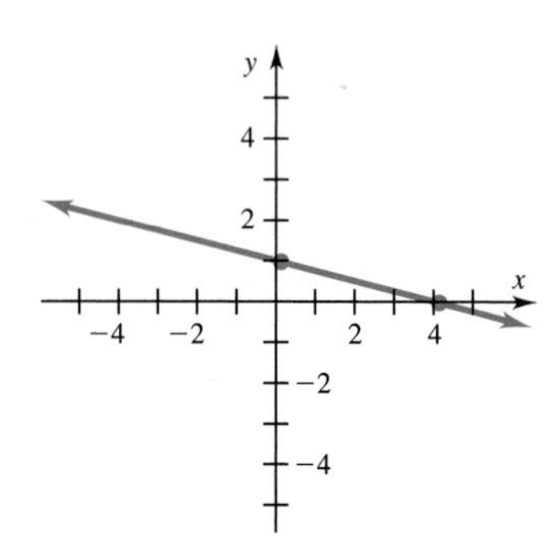

18.

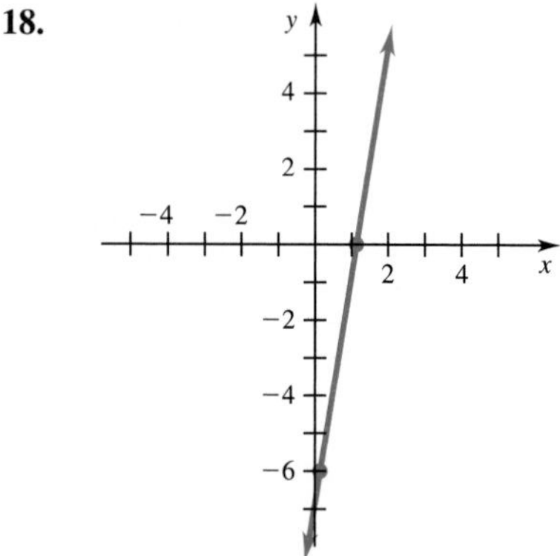

Write an equation in the form $ax + by = c$ for each line.

19. Through (1, 3), $m = -2$

20. Through (2, 4), $m = -1$

21. Through (6, 1), $m = 0$

22. Through (−8, 1), with undefined slope

23. Through (4, 2) and (1, 3)

24. Through (8, −1) and (4, 3)

25. Through (1/2, 5/3) and (3, 1/6)

26. Through (−2, 3/4) and (2/3, 5/2)

27. x-intercept 3, y-intercept −2

28. x-intercept −2, y-intercept 4

29. Vertical, through (−6, 5)

30. Horizontal, through (8, 7)

31. Through (−1.76, 4.25), with slope −5.081

32. Through (5.469, 11.08), with slope 4.723

Graph each line.

33. Through (−1, 3), $m = 3/2$

34. Through (−2, 8), $m = -1$

35. Through (3, −4), $m = -1/3$

36. Through (−2, −3), $m = -3/4$

37. $3x + 5y = 15$

38. $2x - 3y = 12$

39. $4x - y = 8$

40. $x + 3y = 9$

41. $x + 2y = 0$

42. $3x - y = 0$

43. $x = -1$

44. $y + 2 = 0$

45. $y = -3$

46. $x = 5$

Write an equation for each line in Exercises 47–54.

47. Through (−1, 4), parallel to $x + 3y = 5$

48. Through (2, −5), parallel to $y - 4 = 2x$

49. Through (3, −4), perpendicular to $x + y = 4$

50. Through (−2, 6), perpendicular to $2x - 3y = 5$

51. x-intercept −2, parallel to $y = 2x$

52. y-intercept 3, parallel to $x + y = 4$

53. The line with y-intercept 2 and perpendicular to $3x + 2y = 6$

54. The line with x-intercept −2/3 and perpendicular to $2x - y = 4$

55. Do the points (4, 3), (2, 0), and (−18, −12) lie on the same line? (*Hint:* Find the equation of the line through two of the points.)

56. Find k so that the line through (4, −1) and (k, 2) is

(a) parallel to $3y + 2x = 6$,

(b) perpendicular to $2y - 5x = 1$.

57. Use slopes to show that the quadrilateral with vertices at (1, 3), (−5/2, 2), (−7/2, 4), and (2, 1) is a parallelogram.

58. Use slopes to show that the square with vertices at (−2, 3), (4, 3), (4, −1), and (−2, −1) has diagonals that are perpendicular.

59. Use similar triangles from geometry to show that the slope of a line is the same, no matter which two distinct points on the line are chosen to compute it.

60. Suppose that $(0, b)$ and (x_1, y_1) are distinct points on the line $y = mx + b$. Show that $(y_1 - b)/x_1$ is the slope of the line, and that $m = (y_1 - b)/x_1$.

61. To show that two perpendicular lines, neither of which is vertical, have slopes with a product of −1, go through the following steps. Let line L_1 have equation $y = m_1x + b_1$, and let L_2 have equation $y = m_2x + b_2$. Assume that L_1 and L_2 are perpendicular, and use right triangle MPN shown in the figure. Prove each of the following statements.

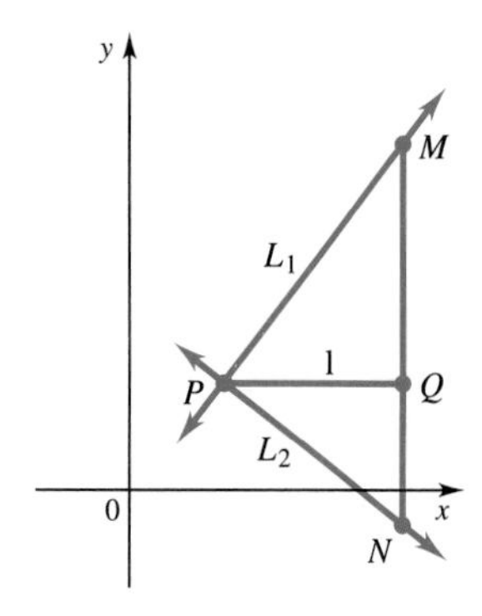

(a) MQ has length m_1.

(b) QN has length $-m_2$.

(c) Triangles MPQ and PNQ are similar.

(d) $m_1/1 = 1/-m_2$ and $m_1m_2 = -1$

Applications

Business and Economics

62. Supply and Demand Let the supply and demand functions for sugar be given in dollars by the following equations.

$$S(q) = 1.4q - .6 \qquad D(q) = -2q + 3.2$$

(a) Graph these equations on the same axes.

(b) Find the equilibrium quantity.

(c) Find the equilibrium price.

63. Supply and Demand Let the supply and demand functions for wool sweaters be given in dollars by the following equations.

$$S(q) = \frac{2}{5}q \qquad D(q) = 100 - \frac{2}{5}q$$

(a) Graph these equations on the same axes.

(b) Find the equilibrium quantity.

(c) Find the equilibrium price.

64. Cost Analysis A company finds that it can make a total of 20 small trailers for \$13,900 and 10 for \$7500. Let y be the total cost to produce x trailers. Assume that the relationship between the number of trailers produced and the cost is linear.

(a) Find the slope and an equation for the line in the form $y = mx + b$.

(b) Use your answer from part (a) to find out how many small trailers the company can make for \$20,000.

(c) Use your answer from part (a) to find out how many small trailers the company can make for \$1000.

65. Rental Car Cost The cost to rent a midsized car is \$27 per day or fraction of a day. If the car is picked up in Lansing and dropped off in West Lafayette, there is a fixed \$25 dropoff charge. Let $C(x)$ represent the cost of renting the car for x days, taking it from Lansing to West Lafayette. Find each of the following.

(a) $C(3/4)$ **(b)** $C(9/10)$ **(c)** $C(1)$ **(d)** $C\left(1\frac{5}{8}\right)$

(e) Find the cost of renting the car for 2.4 days.

(f) Graph $y = C(x)$.

(g) Is C a function?

(h) Is C a linear function?

Life Sciences

66. Pollution When a certain industrial pollutant is introduced into a river, the reproduction of catfish declines. In a given period of time, depositing 3 tons of the pollutant results in a fish population of 37,000. Also, 12 tons of pollutant produce a fish population of 28,000. Let y be the fish population when x tons of pollutant are introduced into the river.

(a) Assuming a linear relationship, find the slope and an equation of the line in the form $y = mx + b$.

(b) Use your answer from part (a) to find out how many tons of the pollutant would result in the elimination of the fish population.

(c) Use your answer from part (a) to find out how much pollution should be allowed to maintain a fish population of 45,000.

67. Human Growth A person's tibia bone goes from ankle to knee. A male with a tibia 40 cm in length has a height of 177 cm, while a tibia 43 cm in length corresponds to a height of 185 cm.

(a) Write a linear equation showing how the height of a male, h, relates to the length of his tibia, t.

(b) Suppose an archaeologist finds a collection of tibias having lengths of 38–45 cm. Assuming these belonged to males, estimate what their heights would have been.

(c) Estimate the length of the tibia for a male whose height is 190 cm.

68. Human Growth The radius bone goes from the wrist to the elbow. A female whose radius bone is 24 cm long is 167 cm tall, while a radius of 26 cm corresponds to a height of 174 cm.

(a) Write a linear equation showing how the height of a female, h, corresponds to the length of her radius bone, r.

(b) Suppose an archaeologist finds a collection of radius bones having lengths of 23–27 cm. Assuming these belonged to females, estimate what their heights would have been.

(c) Estimate the length of a radius bone for a height of 170 cm.

Social Sciences

69. Voting Trends According to research done by the political scientist James March, if the Democrats win 45% of the two-party vote for the House of Representatives, they win 42.5% of the seats. If the Democrats win 55% of the vote, they win 67.5% of the seats.

(a) Let y be the percent of seats won and x the percent of the two-party vote. Assuming a linear relationship, find the slope and an equation of the line in the form $y = mx + b$.

(b) Use your answer from part (a) to find the percent of the vote the Democrats would need to capture 60% of the seats in the House of Representatives.

(c) Could the equation found in part (a) be valid for every percent of the vote between 0% and 100%? Explain.

70. Voting Trends If the Republicans win 45% of the two-party vote, they win 32.5% of the seats (see Exercise 69). If they win 60% of the vote, they get 70% of the seats. Let y represent the percent of the seats and x the percent of the vote.

(a) Assume the relationship is linear and find the slope and an equation of the line in the form $y = mx + b$.

(b) Use your answer from part (a) to find the percent of the vote the Republicans would need to capture 60% of the seats in the House of Representatives.

(c) Could the equation found in part (a) be valid for every possible percent of the vote between 0% and 100%? Explain.

Physical Sciences

71. Global Warming In 1990, the Intergovernmental Panel on Climate Change predicted that the average temperature on the earth would rise .3° C per decade in the absence of international controls on greenhouse emissions.* Let t measure the time in years since 1970, when the average global temperature was 15° C.

(a) Find an equation giving the average global temperature in degrees Celsius as a function of t, the number of years since 1970.

(b) Scientists have estimated that the sea level will rise by 65 cm if the average global temperature rises to 19° C. According to your answer to part (a), when would this occur?

---

*See *Science News*, June 23, 1990, p. 391.

1.3 LINEAR MATHEMATICAL MODELS

What diameter drill bit should a homeowner purchase, if the required length is known?

Later in this section, this question will be answered using a linear function.

Throughout this book, we construct mathematical models that describe real-world situations. In this section we describe some situations that lead to linear functions as mathematical models.

Temperature One of the most common linear relationships found in everyday situations deals with temperature. Recall that water freezes at 32° Fahrenheit and 0° Celsius, while it boils at 212° Fahrenheit and 100° Celsius. The ordered pairs (0, 32) and (100, 212) are graphed in Figure 20 on axes showing Fahrenheit (F) as a function of Celsius (C). The line joining them is the graph of the function.

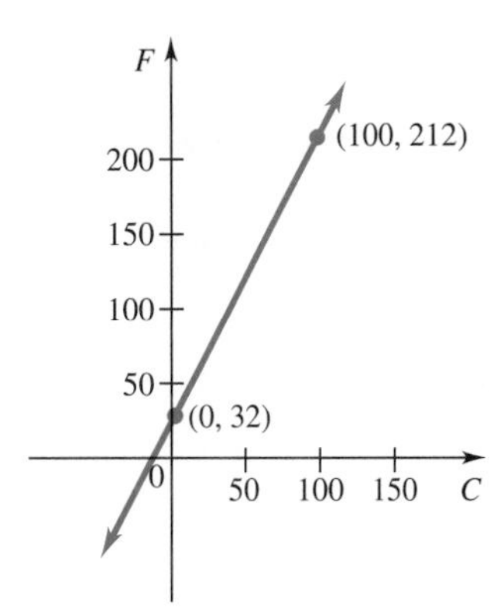

FIGURE 20

EXAMPLE 1 Derive an equation relating F and C.

To derive the required linear equation, first find the slope using the given ordered pairs, (0, 32) and (100, 212).

$$m = \frac{212 - 32}{100 - 0} = \frac{9}{5}$$

The F-intercept of the graph is 32, so by the slope-intercept form, the equation of the line is

$$F = \frac{9}{5}C + 32.$$

With simple algebra this equation can be rewritten to give C in terms of F:

$$C = \frac{5}{9}(F - 32).$$ ◀

Sales Analysis It is common to compare the change in sales of two companies by comparing the rates at which these sales change. If the sales of the two companies can be approximated by linear functions, we can use the work of the last section to find the rate of change of the dependent variable compared to the change in the independent variable.

EXAMPLE 2 The chart below shows sales in two different years for two different companies.

Company	Sales in 1988	Sales in 1991
A	\$10,000	\$16,000
B	5,000	14,000

A study of company records suggests that the sales of both companies have increased linearly (that is, the sales can be closely approximated by a linear function).

(a) Find a linear equation describing the sales for Company A.

To find a linear equation describing the sales, let $x = 0$ represent 1988, so that 1991 corresponds to $x = 3$. Then, by the chart above, the line representing the sales for Company A passes through the points (0, 10,000) and (3, 16,000). The slope of the line through these points is

$$\frac{16{,}000 - 10{,}000}{3 - 0} = 2000.$$

Using the point-slope form of the equation of a line gives

$$\begin{aligned} y - 10{,}000 &= 2000(x - 0) \\ y &= 2000x + 10{,}000 \end{aligned}$$

as an equation describing the sales of Company A.

(b) Find a linear equation describing the sales for Company B.

Since the sales for Company B also have increased linearly, they can be described by a line through (0, 5000) and (3, 14,000). Using the same procedure as in part (a) gives

$$y = 3000x + 5000$$

as the equation describing the sales of Company B. ◀

Average Rate of Change Notice that the sales for Company A in Example 2 increased from \$10,000 to \$16,000 over the period from 1988 to 1991, representing a total increase of \$6000 in three years.

$$\text{Average rate of change in sales} = \frac{\$6000}{3} = \$2000 \text{ per year}$$

This is the same as the slope found in part (a) of the example. Verify that the average annual rate of change in sales for Company B over the three-year period also agrees with the slope found in part (b). Management needs to watch the rate of change in sales closely in order to be aware of any unfavorable trends. If the rate of change is decreasing, then sales growth is slowing down, and this trend may require some response.

AVERAGE RATE OF CHANGE

For $y = f(x)$, as x changes from x to $x + \Delta x$, the **average rate of change in y with respect to x** is given by

$$\frac{\text{Change in } y}{\text{Change in } x} = \frac{f(x + \Delta x) - f(x)}{(x + \Delta x) - x}$$
$$= \frac{f(x + \Delta x) - f(x)}{\Delta x}$$
$$= \frac{\Delta y}{\Delta x}.$$

For data that can be modeled with a linear function, the average rate of change, which is the same as the slope of the line, is constant.

Cost Analysis The cost of manufacturing an item commonly consists of two parts. The first is a **fixed cost** for designing the product, setting up a factory, training workers, and so on. Within broad limits, the fixed cost is constant for a particular product and does not change as more items are made. The second part is a *cost per item* for labor, materials, packing, shipping, and so on. The total value of this second cost *does* depend on the number of items made.

Suppose that the cost of producing clock-radios can be approximated by

$$C(x) = 12x + 100,$$

where $C(x)$ is the cost in dollars to produce x radios. The cost to produce 0 radios is

$$C(0) = 12(0) + 100 = 100,$$

or \$100. This sum, \$100, is the fixed cost.

Once the company has invested the fixed cost into the clock-radio project, what then will be the additional cost per radio? As an example, we first find the cost of a total of 5 radios:

$$C(5) = 12(5) + 100 = 160,$$

or \$160. The cost of 6 radios is

$$C(6) = 12(6) + 100 = 172,$$

or \$172.

The sixth radio itself costs \$172 − \$160 = \$12 to produce. In the same way, the 81st radio costs $C(81) - C(80) = \$1072 - \$1060 = \$12$ to produce. In fact, the $(n + 1)$st radio costs

$$\begin{aligned} C(n + 1) - C(n) &= [12(n + 1) + 100] - (12n + 100) \\ &= 12, \end{aligned}$$

or \$12, to produce. The number 12 is also the slope of the graph of the cost function $C(x) = 12x + 100$. ◀

In economics, **marginal cost** is the rate of change of cost $C(x)$ at a level of production x. The marginal cost is equal to the slope of the cost function. It approximates the cost of producing one additional item. In fact, some books define the marginal cost to be the cost of producing one additional item. With *linear functions*, these two definitions are equivalent, and the marginal cost, which is equal to the slope of the cost function, is constant. For instance, in the clock-radio example, the marginal cost of each radio is \$12. For other types of functions, these two definitions are only approximately equal. Marginal cost is important to management in making decisions in areas such as cost control, pricing, and production planning.

The work in Example 3 can be generalized. Suppose the total cost to make x items is given by the cost function $C(x) = mx + b$. The fixed cost is found by letting $x = 0$:

$$C(0) = m \cdot 0 + b = b;$$

thus, the fixed cost is b dollars. The additional cost of the $(n + 1)$st item is

$$\begin{aligned} C(n + 1) - C(n) &= [m(n + 1) + b] - (mn + b) \\ &= mn + m + b - mn - b \\ &= m. \end{aligned}$$

This is exactly equal to the slope of the line $C(x) = mx + b$, which represents the marginal cost.

COST FUNCTION

In a cost function of the form $C(x) = mx + b$, m represents the marginal cost per item and b the fixed cost. Conversely, if the fixed cost of producing an item is b and the marginal cost is m, then the **cost function** $C(x)$ for producing x items is $C(x) = mx + b$.

Another important concept is the average cost function $\overline{C}(x)$, pronounced "C bar."

AVERAGE COST FUNCTION

If $C(x)$ is the total cost to manufacture x items, then the **average cost** per item is given by

$$\overline{C}(x) = \frac{C(x)}{x}.$$

In Example 3 the average cost per clock-radio is

$$\overline{C}(x) = \frac{C(x)}{x} = \frac{12x + 100}{x} = 12 + \frac{100}{x}.$$

The second term, $100/x$, shows that as more and more items are produced and the fixed cost is spread over a larger number of items, the average cost per item decreases.

EXAMPLE 4 The marginal cost for raising a certain type of frog for laboratory study is \$12 per unit of frogs, while the cost to produce 100 units is \$1500.

(a) Find the cost function $C(x)$, given that it is linear.

Since the cost function is linear, it can be expressed in the form $C(x) = mx + b$. The marginal cost is \$12 per unit, which gives the value for m, leading to $C(x) = 12x + b$. To find b, use the fact that the cost of producing 100 units of frogs is \$1500, or $C(100) = 1500$. Substituting $C(x) = 1500$ and $x = 100$ into $C(x) = 12x + b$ gives

$$\begin{aligned} C(x) &= 12x + b \\ 1500 &= 12 \cdot 100 + b \\ 1500 &= 1200 + b \\ 300 &= b. \end{aligned}$$

The model is given by $C(x) = 12x + 300$, where the fixed cost is \$300.

(b) Find the average cost per item to produce 50 units and 300 units.

The average cost per item is

$$\overline{C}(x) = \frac{C(x)}{x} = \frac{12x + 300}{x} = 12 + \frac{300}{x}.$$

If 50 units of frogs are produced, the average cost is

$$\overline{C}(50) = 12 + \frac{300}{50} = 18,$$

or \$18 per unit. Producing 300 units of frogs will lead to an average cost of

$$\overline{C}(300) = 12 + \frac{300}{300} = 13,$$

or \$13 per unit. ◀

Break-Even Analysis The **revenue** $R(x)$ from selling x units of a product is the product of the price per unit p and the number of units sold (demand) x, so that

$$R(x) = px.$$

The corresponding **profit** $P(x)$ is the difference between revenue $R(x)$ and cost $C(x)$. That is,

$$P(x) = R(x) - C(x).$$

A company can make a profit only if the revenue received from its customers exceeds the cost of producing and selling its goods and services. The number of units at which revenue just equals cost is the **break-even point.**

EXAMPLE 5 A firm producing poultry feed finds that the total cost $C(x)$ of producing and selling x units is given by

$$C(x) = 20x + 100.$$

Management plans to charge $24 per unit for the feed.

(a) How many units must be sold for the firm to break even?

The firm will break even (no profit and no loss) as long as revenue just equals cost, or $R(x) = C(x)$. From the given information, since $R(x) = px$ and $p = \$24$,

$$R(x) = 24x.$$

Substituting for $R(x)$ and $C(x)$ in the equation $R(x) = C(x)$ gives

$$24x = 20x + 100,$$

from which $x = 25$. The firm breaks even by selling 25 units. The graphs of $C(x) = 20x + 100$ and $R(x) = 24x$ are shown in Figure 21. The break-even point (where $x = 25$) is shown on the graph. If the company sells more than 25 units (if $x > 25$), it makes a profit. If it sells less than 25 units, it loses money.

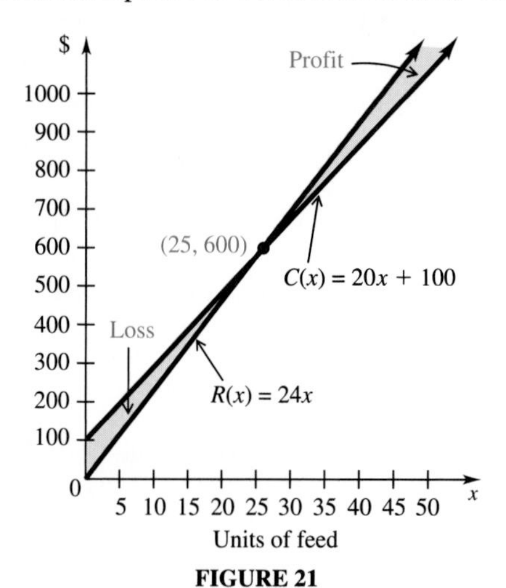

FIGURE 21

(b) What is the profit if 100 units of feed are sold?

Use the formula for profit $P(x)$.

$$\begin{aligned} P(x) &= R(x) - C(x) \\ &= 24x - (20x + 100) \\ &= 4x - 100 \end{aligned}$$

Then $P(100) = 4(100) - 100 = 300$. The firm will make a profit of $300 from the sale of 100 units of feed. ◀

EXAMPLE 6

A do-it-yourself homeowner purchased a set of drill bits. Each drill bit is 1/64 inch larger in diameter than the previous bit, and as the bits increase in diameter, their lengths also increase in a linear way. This means the length of a bit is a linear function of its diameter. The smallest bit has a diameter of 1/16 inch and a length of $1\frac{5}{8}$ inches. The largest bit has a diameter of 1/4 inch and a length of 4 inches. Suppose the homeowner wants a bit with a length of at least $5\frac{1}{4}$ inches. What diameter bit is needed, assuming that bit diameters are available in multiples of 1/64 inch, and that the relationship between a bit's diameter and its length remains constant? (For those who want more drill problems, we hope this example will satisfy them, at least a bit.)

Let x be the diameter of the bit and y its length. The smallest and largest bits in the set then correspond to the ordered pairs $(1/16, 1\frac{5}{8})$ and $(1/4, 4)$. The slope of the line through these two points is

$$m = \frac{4 - 1\frac{5}{8}}{1/4 - 1/16} = \frac{19/8}{3/16} = \frac{38}{3}.$$

Using the point-slope form of a linear equation and the point $(1/4, 4)$, we have

$$y - 4 = \frac{38}{3}\left(x - \frac{1}{4}\right).$$

Verify that this simplifies to

$$y = \frac{38}{3}x + \frac{5}{6}.$$

Now let $y = 5\frac{1}{4}$ in this equation. Go through the details to verify that $x = 53/152$. Unfortunately, the bits are sold only in multiples of 1/64 inch. To find how many 64ths equal 53/152, solve the equation

$$\frac{n}{64} = \frac{53}{152}$$

by multiplying both sides by 64, giving $n = 424/19$, which is approximately 22.3. Since n must be an integer, and a bit *at least* $5\frac{1}{4}$ inches long is needed, let $n = 23$. In other words, the homeowner should request a bit with a diameter of 23/64 inch. ◀

1.3 Exercises

Write a cost function for each situation. Identify all variables used.

1. A chain-saw rental firm charges \$12 plus \$1 per hour.
2. A trailer-hauling service charges \$45 plus \$2 per mile.
3. A parking garage charges 50¢ plus 35¢ per half-hour.
4. For a one-day rental, a car rental firm charges \$44 plus 28¢ per mile.

Assume that each situation can be expressed as a linear cost function, and find the appropriate cost function.

5. Fixed cost, \$100; 50 items cost \$1600 to produce
6. Fixed cost, \$400; 10 items cost \$650 to produce
7. Fixed cost, \$1000; 40 items cost \$2000 to produce
8. Fixed cost, \$8500; 75 items cost \$11,875 to produce
9. Marginal cost, \$50; 80 items cost \$4500 to produce
10. Marginal cost, \$120; 100 items cost \$15,800 to produce
11. Marginal cost, \$90; 150 items cost \$16,000 to produce
12. Marginal cost, \$120; 700 items cost \$96,500 to produce

Applications

Business and Economics

13. **Sales Analysis** Suppose the sales of a particular brand of electric guitar satisfy the relationship

$$S(x) = 300x + 2000,$$

where $S(x)$ represents the number of guitars sold in year x, with $x = 0$ corresponding to 1987.

Find the sales in each of the following years.

(a) 1987 **(b)** 1990 **(c)** 1991

(d) The manufacturer needs sales to reach 4000 guitars by 1996 to pay off a loan. Will sales reach that goal?

(e) Find the annual rate of change of sales.

14. **Sales Analysis** Assume that the sales of a certain appliance dealer are approximated by a linear function. Suppose that sales were \$850,000 in 1982 and \$1,262,500 in 1987. Let $x = 0$ represent 1982.

(a) Find an equation giving the dealer's yearly sales.

(b) Estimate the sales in 1994.

(c) The dealer estimates that a new store will be necessary once sales exceed \$2,170,000. When is this expected to occur?

15. **Sales Analysis** Assume that the sales of a certain automobile parts company are approximated by a linear function. Suppose that sales were \$200,000 in 1981 and \$1,000,000 in 1988. Let $x = 0$ represent 1981 and $x = 7$ represent 1988.

(a) Find the equation giving the company's yearly sales.

(b) Estimate the sales in 1992.

(c) The company wants to negotiate a new contract with the automobile company once sales reach \$2,000,000. When is this expected to occur?

16. **Marginal Cost** In deciding whether to set up a new manufacturing plant, company analysts have established that a reasonable function for the total cost to produce x items is

$$C(x) = 500{,}000 + 4.75x.$$

(a) Find the total cost to produce 100,000 items.

(b) Find the marginal cost of the items to be produced in this plant.

17. **Marginal Cost** The manager of a local restaurant has found that his cost function for producing coffee is $C(x) = .097x$, where $C(x)$ is the total cost in dollars of producing x cups. (He is ignoring the cost of the coffee pot and the cost of labor.)

Find the total cost of producing the following numbers of cups of coffee.

(a) 1000 cups **(b)** 1001 cups

(c) Find the marginal cost of the 1001st cup.

(d) What is the marginal cost for *any* cup?

Average Cost *Let $C(x)$ be the total cost in dollars to manufacture x items. Find the average cost per item in Exercises 18 and 19.*

18. $C(x) = 500{,}000 + 4.75x$

(a) $x = 1000$ **(b)** $x = 5000$ **(c)** $x = 10{,}000$

19. $C(x) = 800 + 20x$

(a) $x = 10$ **(b)** $x = 50$ **(c)** $x = 200$

20. **Break-Even Analysis** The cost to produce x units of squash is $C(x) = 100x + 6000$, while the revenue is $R(x) = 500x$. Find the break-even point.

21. **Break-Even Analysis** The cost to produce x units of wire is $C(x) = 50x + 5000$, while the revenue is $R(x) = 60x$. Find the break-even point and the revenue at the break-even point.

Break-Even Analysis *Suppose that you are the manager of a firm. You are considering the manufacture of a new product, so you ask the accounting department to produce cost estimates and the sales department to produce estimates for revenue and sales. After you receive the data, you must decide whether to go ahead with production of the new product. Analyze the data in Exercises 22–25 (find a break-even point), and then decide what you would do.*

22. $C(x) = 105x + 6000$; $R(x) = 250x$; not more than 400 units can be sold

23. $C(x) = 85x + 900$; $R(x) = 105x$; not more than 38 units can be sold

24. $C(x) = 1000x + 5000$; $R(x) = 900x$ (*Hint:* What does a negative break-even point mean?)

25. $C(x) = 70x + 500$; $R(x) = 60x$

26. **Break-Even Analysis** The graph shows the productivity of U.S. and Japanese workers in appropriate units over a 35-year period. Estimate the break-even point (the point at which workers in the two countries produced the same amounts).*

27. **Break-Even Analysis** The graph gives U.S. imports and exports in billions of dollars over a five-year period. Estimate the break-even point.*

Life Sciences

28. **Medical Expenses** Over a recent three-year period, medical expenses in the U.S. rose in a linear pattern from 10.3% of the gross national product in year 0 to 11.2% in year 3.
    (a) Assuming that this change in medical expenses continues to be linear, write a linear function describing the percent of the gross national product devoted to medical expenses, y, in terms of the year, x.
    (b) Give the average rate of change of medical expenses from year 0 to 3. Compare it with the slope of the line in part (a). What do you find?

29. **Life Span** Some scientists believe there is a limit to how long humans can live.† One supporting argument is that during the last century, life expectancy from age 65 has increased more slowly than life expectancy from birth, so eventually these two will be equal, at which point, according to these scientists, life expectancy should increase no further. In 1900, life expectancy at birth was 46 years, and life expectancy at age 65 was 76. In 1975, these figures had risen to 75 and 80, respectively. In both cases, the increase in life expectancy has been linear. Using these assumptions and the data given, find the maximum life expectancy for humans.

30. **Population** The number of children in the U.S. from 5 to 13 years old decreased from 31.2 million in 1980 to 30.3 million in 1986.
    (a) Write a linear function describing this population y in terms of year x for the given period.
    (b) What was the average rate of change in this population over the period from 1980 to 1986?

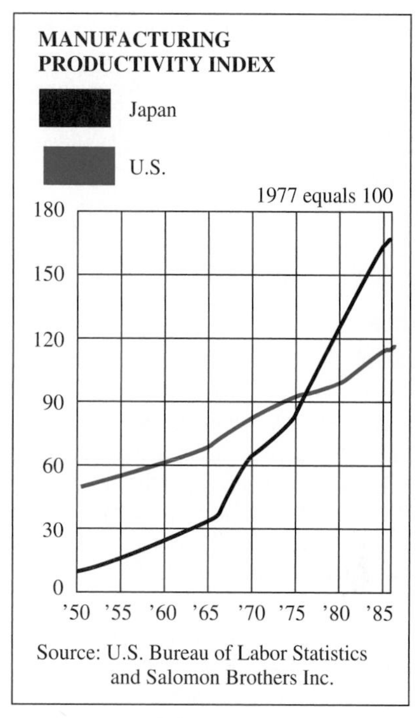

EXERCISE 26

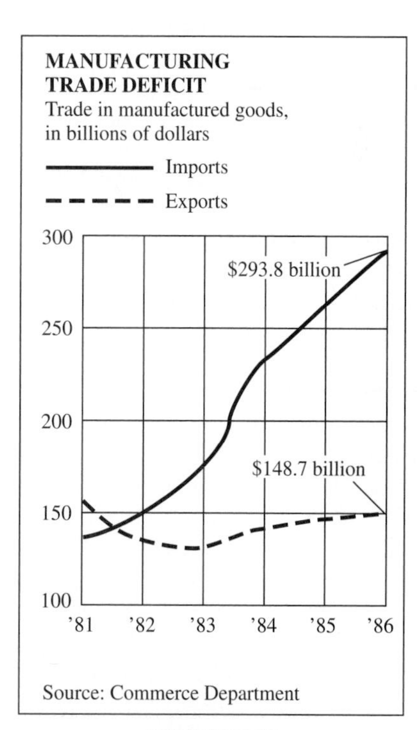

EXERCISE 27

*Figures, "Manufacturing Productivity Index" and "Manufacturing Trade Deficit" as appeared in *The Sacramento Bee*, December 21, 1987. Reprinted by permission of The Associated Press.
†See *Science*, November 15, 1991, pp. 936–38.

31. Ponies Trotting A 1991 study found that the peak vertical force on a trotting Shetland pony increased linearly with the pony's speed, and that when the force reached a critical level, the pony switched from a trot to a gallop.* For one pony, the critical force was 1.16 times its body weight. It experienced a force of .75 times its body weight at a speed of 2 meters per second, and a force of .93 times its body weight at 3 meters per second. At what speed did the pony switch from a trot to a gallop?

Social Sciences

32. Stimulus Effect In psychology, the just-noticeable-difference (JND) for some stimulus is defined as the amount by which the stimulus must be increased so that a person will perceive it as having barely been increased. For example, suppose a research study indicates that a line 40 centimeters in length must be increased to 42 cm before a subject thinks that it is longer. In this case, the JND would be $42 - 40 = 2$ cm. In a particular experiment, the JND is given by

$$y = .03x,$$

where x represents the original length of the line and y the JND.

Find the JND for lines having the following lengths.

(a) 10 cm **(b)** 20 cm **(c)** 50 cm **(d)** 100 cm

(e) Find the rate of change in the JND with respect to the original length of the line.

33. Passage of Time Most people are not very good at estimating the passage of time. Some people's estimations are too fast, and those of others are too slow. One psychologist has constructed a mathematical model for actual time as a function of estimated time: if y represents actual time and x estimated time, then

$$y = mx + b,$$

where m and b are constants that must be determined experimentally for each person. Suppose that for a particular person, $m = 1.25$ and $b = -5$. Find y for each of the following values of x.

(a) 30 min **(b)** 60 min **(c)** 120 min
(d) 180 min

34. Passage of Time In Exercise 33, suppose that for another person, $m = .85$ and $b = 1.2$. Find y for each of the following values of x.

(a) 15 min **(b)** 30 min **(c)** 60 min
(d) 120 min

For this same person, find x for each of the following values of y.

(e) 60 min **(f)** 90 min

Physical Sciences

35. Temperature Use the formulas for conversion between Fahrenheit and Celsius derived in Example 1 to convert each temperature.

(a) 58° F to Celsius

(b) 50° C to Fahrenheit

36. Temperature Use the formulas for conversion between Fahrenheit and Celsius derived in Example 1 to convert each temperature.

(a) 98.6° F to Celsius

(b) 20° C to Fahrenheit

37. Derive a formula giving the Celsius temperature (C) as a linear function of the Fahrenheit temperature (F).

38. Find the temperature at which the Celsius and Fahrenheit temperatures are numerically equal.

General Interest

39. Class Size Let $R(x) = -8x + 240$ represent the number of students present in a large business calculus class, where x represents the number of hours of study required weekly.

Find the number of students present at each of the following levels of required study.

(a) $x = 0$ **(b)** $x = 5$ **(c)** $x = 10$

(d) What is the rate of change of the number of students in the class with respect to the number of hours of study? Interpret the negative sign in the answer.

(e) The professor in charge of the class likes to have exactly 16 students. How many hours of study must he require in order to have exactly 16 students?

40. Drill Bits For the drill bit example in the text, find the diameter of the bit that should be ordered for each of the following minimum lengths. Remember that the diameter must be a multiple of $1/64$ inches.

(a) $4\frac{1}{4}$ inches **(b)** $5\frac{1}{2}$ inches

---

ience, July 19, 1991, pp. 306–8.

1.4 QUADRATIC FUNCTIONS

How much should a company charge for its seminars? When Power and Money, Inc., charges $600 for a seminar on management techniques, it attracts 1000 people. For each $20 decrease in the fee, an additional 100 people will attend the seminar. The managers are wondering how much to charge for the seminar to maximize their revenue.

In this section we will see how knowledge of *quadratic functions* will help provide an answer to the question above.

In this section you will need to know how to solve a quadratic equation by factoring and by the quadratic formula, which are covered in Sections R.2 and R.4 of the Algebra Reference. Factoring is usually easiest; when a polynomial is set equal to zero and factored, then a solution is found by setting any one factor equal to zero. But factoring is not always possible. The quadratic formula will provide the solution to *any* quadratic equation.

A linear function was defined by

$$f(x) = ax + b,$$

for real numbers a and b. In a *quadratic function* the independent variable is squared. A quadratic function is an especially good model for many situations with a maximum or a minimum function value. Quadratic functions also may be used to describe supply and demand curves; cost, revenue, and profit; as well as other quantities. Next to linear functions, they are the simplest type of function, and so well worth studying thoroughly.

QUADRATIC FUNCTION

A **quadratic function** is defined by

$$f(x) = ax^2 + bx + c,$$

where a, b, and c are real numbers, with $a \neq 0$.

The simplest quadratic function has $f(x) = x^2$, with $a = 1$, $b = 0$, and $c = 0$. This function can be graphed by choosing several values of x and then finding the corresponding values of $f(x)$. We then plot the resulting ordered pairs $(x, f(x))$, and draw a smooth curve through them, as in Figure 22. This graph is called a **parabola.** Every quadratic function has a parabola as its graph. The lowest (or highest) point on a parabola is the **vertex** of the parabola. The vertex of the parabola in Figure 22 is (0, 0).

x	$f(x)$
−2	4
−1	1
0	0
1	1
2	4

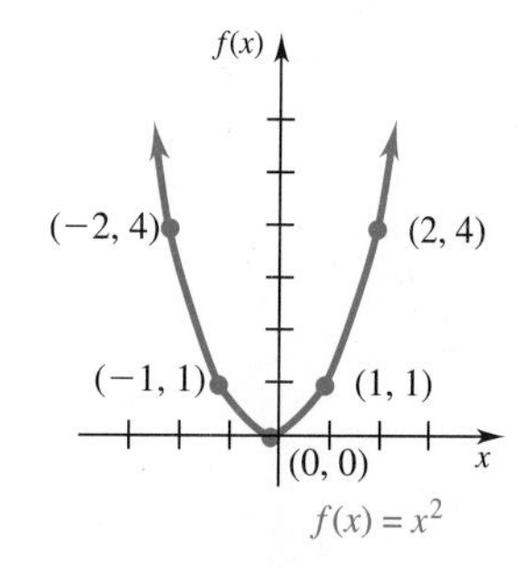

FIGURE 22

If the graph in Figure 22 were folded in half along the y-axis, the two halves of the parabola would match exactly. This means that the graph of a quadratic function is *symmetric* with respect to a vertical line through the vertex; this line is the **axis** of the parabola.

There are many real-world instances of parabolas. For example, cross sections of spotlight reflectors or radar dishes form parabolas. Also, a projectile thrown in the air follows a parabolic path.

EXAMPLE 1 Graph $y = x^2 - 4$.

Each value of y will be 4 less than the corresponding value of y in $y = x^2$. The graph of $y = x^2 - 4$ has the same shape as that of $y = x^2$ but is 4 units lower. See Figure 23. The vertex of the parabola (on this parabola, the *lowest* point) is at $(0, -4)$. The x-intercepts can be found by letting $y = 0$ to get

$$0 = x^2 - 4,$$

from which $x = 2$ and $x = -2$ are the x-intercepts. The axis of the parabola is the vertical line $x = 0$. ◀

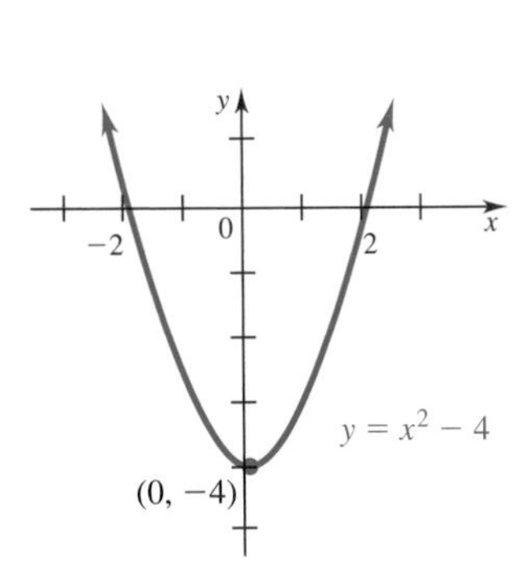

FIGURE 23

Example 1 suggests that the effect of c in $ax^2 + bx + c$ is to lower the graph if c is negative and to raise the graph if c is positive.

EXAMPLE 2 Graph $y = -2x^2$.

By plotting points, we see that since a is negative, the graph opens downward. Since a is larger than 1, the graph is steeper than the original graph. See Figure 24. ◀

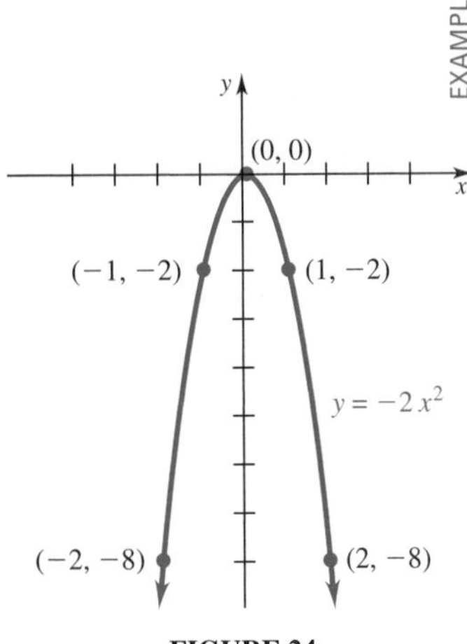

FIGURE 24

Example 2 shows that the sign of a in $ax^2 + bx + c$ determines whether the parabola opens upward or downward, and the magnitude of a determines how steeply the graph increases. The effect of b in $ax^2 + bx + c$ is more complicated; it causes the vertex to shift to the left or right, as well as to be raised or lowered, depending upon the values of a and c. For a quadratic function with $b \neq 0$, the best way to understand what the graph looks like is by first using the quadratic formula to find the x-intercepts, when they exist. Recall from algebra that if $ax^2 + bx + c = 0$, where $a \neq 0$, then

$$x = \frac{-b \pm \sqrt{b^2 - 4ac}}{2a}.$$

Notice that this is the same as

$$x = \frac{-b}{2a} \pm \frac{\sqrt{b^2 - 4ac}}{2a} = \frac{-b}{2a} \pm Q,$$

where $Q = \sqrt{b^2 - 4ac}/2a$. Since a parabola is symmetric with respect to its axis, the vertex is halfway between its two x-intercepts. Halfway between $x = -b/(2a) + Q$ and $x = -b/(2a) - Q$ is $x = -b/(2a)$. Once we have the x-coordinate of the vertex, we can easily get the y-coordinate by substituting the x-coordinate into the original equation. The following example illustrates the complete process.

EXAMPLE 3 Graph $y = -3x^2 - 2x + 1$.

Because the x^2 term has a coefficient of -3, the graph is a parabola opening downward. The x-intercepts are found by taking

$$\begin{aligned} x &= \frac{-b \pm \sqrt{b^2 - 4ac}}{2a} \\ &= \frac{-(-2) \pm \sqrt{(-2)^2 - 4(-3)(1)}}{2(-3)} \\ &= \frac{2 \pm \sqrt{16}}{-6} \\ x &= -1 \quad \text{or} \quad \frac{1}{3}. \end{aligned}$$

(These solutions could also have been found by factoring.) The x-coordinate of the vertex is

$$x = \frac{-b}{2a} = \frac{-(-2)}{2(-3)} = -\frac{1}{3}.$$

We get the y-coordinate by substituting this x-value into the equation.

$$\begin{aligned} y &= -3\left(-\frac{1}{3}\right)^2 - 2\left(-\frac{1}{3}\right) + 1 \\ &= \frac{4}{3}. \end{aligned}$$

Thus, the vertex is $(-1/3, 4/3)$. The axis goes through the vertex; it is the line $x = -1/3$. Finally, the y-intercept, which is c, or 1 in this case, should always be plotted. Plotting the vertex, the two x-intercepts, and the y-intercept yields the graph in Figure 25. ◀

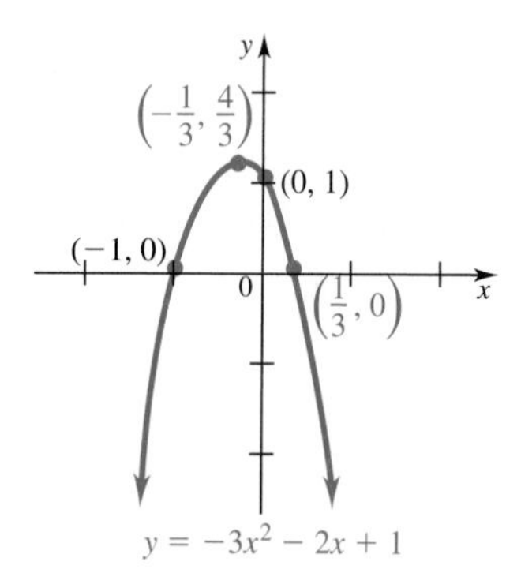

FIGURE 25

In many examples the quadratic will not factor, and the solutions given by the quadratic formula will be irrational numbers. If this happens, you can use a calculator to approximate the solutions. Another situation that may arise is the absence of any x-intercepts, as in the next example.

EXAMPLE 4

Graph $y = x^2 + 4x + 6$.

This does not appear to factor, so we'll try the quadratic formula.

$$x = \frac{-b \pm \sqrt{b^2 - 4ac}}{2a}$$

$$= \frac{-4 \pm \sqrt{4^2 - 4(1)(6)}}{2(1)} = \frac{-4 \pm \sqrt{-8}}{2}$$

As soon as we see the negative under the square root sign, we know there are no x-intercepts. Nevertheless, the vertex is still at

$$x = \frac{-b}{2a} = \frac{-4}{2} = -2.$$

Substituting this into the equation gives

$$y = (-2)^2 + 4(-2) + 6 = 2.$$

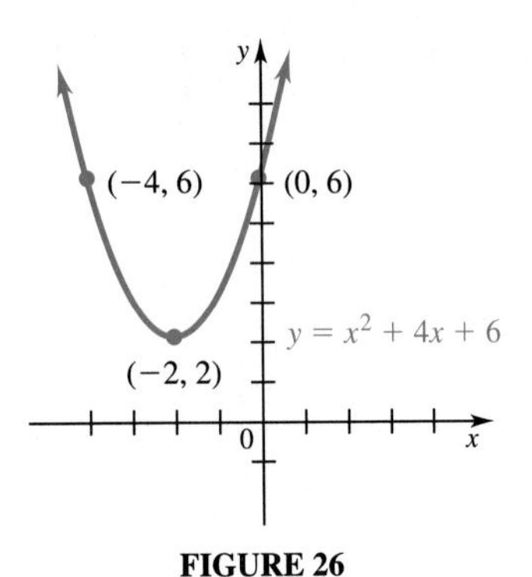

FIGURE 26

The y-intercept is at (0, 6), 2 units to the right of the parabola's axis $x = -2$. We can also plot the mirror image of this point on the opposite side of the parabola's axis: at $x = -4$, 2 units to the left of the axis, y is also equal to 6. Plotting the vertex, the y-intercept, and the point $(-4, 6)$ gives the graph in Figure 26.

We will now return to the example with which we started this section.

EXAMPLE 5

When Power and Money, Inc., charges \$600 for a seminar on management techniques, it attracts 1000 people. For each \$20 decrease in the fee, an additional 100 people will attend the seminar. The managers are wondering how much to charge for the seminar to maximize their revenue.

Let x be the number of \$20 decreases in the price. Then the price charged per person will be

$$\text{Price per person} = 600 - 20x,$$

and the number of people in the seminar will be

$$\text{Number of people} = 1000 + 100x.$$

The total revenue, $R(x)$, is given by the product of the price and the number of people attending, or

$$R(x) = (600 - 20x)(1000 + 100x)$$
$$= 600{,}000 + 40{,}000x - 2000x^2.$$

We see by the negative in the x^2 term that this defines a parabola opening downward, so the maximum revenue is at the vertex. The x-coordinate of the vertex is

$$x = \frac{-b}{2a} = \frac{-40{,}000}{2(-2000)} = 10.$$

The y-coordinate is then

$$\begin{aligned} y &= 600{,}000 + 40{,}000(10) - 2000(10^2) \\ &= 800{,}000. \end{aligned}$$

Therefore, the maximum revenue is \$800,000, which is achieved by charging $600 - 20x = 600 - 20(10) =$ \$400 per person. ◀

Notice in this last example that the maximum revenue was achieved by charging less than the current price of \$600, which was more than made up for by the increase in sales. This is typical of many applications. Mathematics is a powerful tool for solving such problems, since the answer is not what one might have guessed intuitively.

The main difficulty in learning to solve such problems is deriving expressions for the price per person and the number of people. You will find this process easier if you notice that both expressions are linear functions of x. Notice also that you know the constant term in these functions because the original problem told what happens if $x = 0$ (i.e., if there is no \$20 decrease). Finally, notice that the slope of these linear functions is just the amount of change each time x increases by 1. (In this example, an increase in x by 1 is equivalent to a \$20 decrease in the price.) Once you understand these ideas (after getting some practice), you will be able to solve any such maximization problem.

EXAMPLE 6 Ms. Tilden owns and operates Aunt Emma's Pie Shop. She has hired a consultant to analyze her business operations. The consultant tells her that her profit $P(x)$ from the sale of x units of pies is given by

$$P(x) = 120x - x^2.$$

How many units of pies must be sold in order to maximize the profit? What is the maximum possible profit?

Because of the $-x^2$ term, the profit function defines a parabola opening downward with the maximum at the vertex. It can be factored as

$$P(x) = x(120 - x).$$

The two x-intercepts are therefore 0 and 120, so the x-coordinate of the vertex is halfway between those values, at $x = 60$. Since $P(60) = 3600$, the maximum profit of \$3600 is reached when 60 units of pies are sold. In this case, profit increases as more pies are sold up to 60 units and then decreases as more pies are sold past this point. See Figure 27. ◀

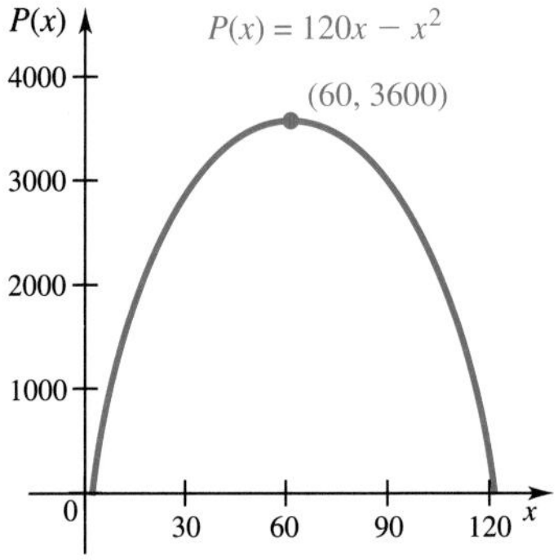

FIGURE 27

Although many examples in this book involve maximization of profit, we do not mean to imply that this should always be the only goal of a company. Far-sighted companies may seek to minimize their impact on the environment or maximize their market share. U.S. companies have been criticized for seeking only to maximize short-term profits, which in the long run may cause them to lose out to foreign competitors who take a longer view.

1.4 Exercises

1. Graph the functions in parts (a)–(d) on the same coordinate system.

   (a) $f(x) = 2x^2$ **(b)** $f(x) = 3x^2$ **(c)** $f(x) = \frac{1}{2}x^2$ **(d)** $f(x) = \frac{1}{3}x^2$

   (e) How does the coefficient affect the shape of the graph?

2. Graph the functions in parts (a)–(d) on the same coordinate system.

   (a) $y = \frac{1}{2}x^2$ **(b)** $y = -\frac{1}{2}x^2$ **(c)** $y = 4x^2$ **(d)** $y = -4x^2$

   (e) What effect does the negative sign have on the graph?

3. Graph the functions in parts (a)–(d) on the same coordinate system.

   (a) $f(x) = x^2 + 2$ **(b)** $f(x) = x^2 - 1$ **(c)** $f(x) = x^2 + 1$ **(d)** $f(x) = x^2 - 2$

   (e) How do these graphs differ from the graph of $f(x) = x^2$?

4. Graph the functions in parts (a)–(d) on the same coordinate system.

   (a) $f(x) = (x - 2)^2$ **(b)** $f(x) = (x + 1)^2$ **(c)** $f(x) = (x + 3)^2$ **(d)** $f(x) = (x - 4)^2$

   (e) How do these graphs differ from the graph of $f(x) = x^2$?

Graph each parabola and give its vertex, axis, x-intercepts, and y-intercept.

5. $y = x^2 + 6x + 5$
6. $y = x^2 - 10x + 21$
7. $y = x^2 - 4x + 4$
8. $y = x^2 + 8x + 16$
9. $y = -2x^2 - 12x - 16$
10. $y = -3x^2 + 12x - 11$
11. $y = 2x^2 + 12x - 16$
12. $y = x^2 - 2x + 3$
13. $y = 3x^2 + 6x + 2$
14. $y = -x^2 - 4x + 2$
15. $y = -x^2 + 6x - 6$
16. $y = 2x^2 - 4x + 5$
17. $y = -3x^2 + 24x - 36$
18. $y = -\frac{1}{3}x^2 + 2x + 4$
19. $y = \frac{5}{2}x^2 + 10x + 8$
20. $y = -\frac{1}{2}x^2 - x - \frac{7}{2}$
21. $y = \frac{2}{3}x^2 - \frac{8}{3}x + \frac{5}{3}$
22. $y = \frac{1}{2}x^2 + 2x + \frac{7}{2}$
23. Let x be in the interval $[0, 1]$. Use a graph to suggest that the product $x(1 - x)$ is always less than or equal to $1/4$. For what values of x does the product equal $1/4$?

Applications

Business and Economics

24. **Minimizing Cost** George Duda runs a sandwich shop. By studying data concerning his past costs, he has found that the cost of operating his shop is given by

$$C(x) = 2x^2 - 20x + 360,$$

where $C(x)$ is the daily cost in dollars to make x batches of sandwiches.

   (a) Find the number of batches George must sell to minimize the cost.

   (b) What is the minimum cost?

   (c) Find a formula for the average daily cost per batch of sandwiches as a function of x.

   (d) Find the average daily cost per batch of sandwiches when 4 batches are produced.

   (e) Find the average daily cost per batch of sandwiches when 7 batches are produced.

25. **Maximizing Revenue** The revenue of a charter bus company depends on the number of unsold seats. If the revenue $R(x)$ is given by

$$R(x) = 5000 + 50x - x^2,$$

where x is the number of unsold seats, find the maximum revenue and the number of unsold seats that corresponds to maximum revenue.

26. Maximizing Revenue A charter flight charges a fare of \$200 per person plus \$4 per person for each unsold seat on the plane. The plane holds 100 passengers. Let x represent the number of unsold seats.

(a) Find an expression for the total revenue received for the flight $R(x)$. (*Hint:* Multiply the number of people flying, $100 - x$, by the price per ticket.)

(b) Graph the expression from part (a).

(c) Find the number of unsold seats that will produce the maximum revenue.

(d) What is the maximum revenue?

(e) Some managers might be concerned about the empty seats, arguing that it doesn't make economic sense to leave any seats empty. Write a few sentences explaining why this is not necessarily so.

27. Maximizing Revenue The demand for a certain type of cosmetic is given by

$$p = 500 - x,$$

where p is the price in dollars when x units are demanded.

(a) Find the revenue $R(x)$ that would be obtained at a price of p. (*Hint:* Revenue = Demand × Price)

(b) Graph the revenue function $R(x)$.

(c) From the graph of the revenue function, estimate the price that will produce maximum revenue.

(d) What is the maximum revenue?

28. Revenue The manager of a peach orchard is trying to decide when to arrange for picking the peaches. If they are picked now, the average yield per tree will be 100 lb, which can be sold for 40¢ per pound. Past experience shows that the yield per tree will increase about 5 lb per week, while the price will decrease about 2¢ per pound per week.

(a) Let x represent the number of weeks that the manager should wait. Find the income per pound.

(b) Find the number of pounds per tree.

(c) Find the total revenue from a tree.

(d) When should the peaches be picked in order to produce maximum revenue?

(e) What is the maximum revenue?

29. Income The manager of an 80-unit apartment complex is trying to decide what rent to charge. Experience has shown that at a rent of \$200, all the units will be full. On the average, one additional unit will remain vacant for each \$20 increase in rent.

(a) Let x represent the number of \$20 increases. Find an expression for the rent for each apartment.

(b) Find an expression for the number of apartments rented.

(c) Find an expression for the total revenue from all rented apartments.

(d) What value of x leads to maximum revenue?

(e) What is the maximum revenue?

Life Sciences

30. Maximizing Population The number of mosquitoes $M(x)$, in millions, in a certain area of Kentucky depends on the June rainfall x, in inches, approximately as follows.

$$M(x) = 10x - x^2$$

Find the rainfall that will produce the maximum number of mosquitoes.

31. Maximizing Chlorophyll Production For the months of June through October, the percent of maximum possible chlorophyll production in a leaf is approximated by $C(x)$, where

$$C(x) = 10x + 50.$$

Here x is time in months with $x = 1$ representing June. From October through December, $C(x)$ is approximated by

$$C(x) = -20(x - 5)^2 + 100,$$

with x as above. Find the percent of maximum possible chlorophyll production in each of the following months.

(a) June **(b)** July **(c)** September
(d) October **(e)** November **(f)** December

32. Maximizing Chlorophyll Production Use your results from Exercise 31 to sketch a graph of $y = C(x)$, from June through December. In what month is chlorophyll production at a maximum?

Physical Sciences

33. Maximizing the Height of an Object If an object is thrown upward with an initial velocity of 32 ft/sec, then its height after t sec is given by

$$h = 32t - 16t^2.$$

Find the maximum height attained by the object. Find the number of seconds it takes the object to hit the ground.

34. Maximizing the Height of an Object If an object is thrown upward with an initial velocity of 64 ft/sec, then its height after t sec is given by

$$h = -16t^2 + 64t$$

Find the maximum height attained by the object. Find the number of seconds it takes the object to hit the ground.

General Interest

35. **Maximizing Area** Glenview Community College wants to construct a rectangular parking lot on land bordered on one side by a highway. It has 320 ft of fencing to use along the other three sides. What should be the dimensions of the lot if the enclosed area is to be a maximum? (*Hint:* Let x represent the width of the lot, and let $320 - 2x$ represent the length.)
36. **Maximizing Area** What would be the maximum area that could be enclosed by the college's 320 ft of fencing if it decided to close the entrance by enclosing all four sides of the lot? (See Exercise 35.)
37. **Number Analysis** Find two numbers whose sum is 20 and whose product is a maximum. (*Hint:* Let x and $20 - x$ be the two numbers, and write an equation for the product.)
38. **Number Analysis** Find two numbers whose sum is 45 and whose product is a maximum.
39. **Parabolic Arch** An arch is shaped like a parabola. It is 30 m wide at the base and 15 m high. How wide is the arch 10 m from the ground?
40. **Parabolic Culvert** A culvert is shaped like a parabola, 18 ft across the top and 12 ft deep. How wide is the culvert 8 ft from the top?

1.5 POLYNOMIAL AND RATIONAL FUNCTIONS

In the previous sections we discussed linear and quadratic functions and their graphs. Both of these functions are special types of *polynomial functions.*

POLYNOMIAL FUNCTION

A **polynomial function** of degree n, where n is a nonnegative integer, is defined by

$$f(x) = a_n x^n + a_{n-1} x^{n-1} + \cdots + a_1 x + a_0,$$

where $a_n, a_{n-1}, \ldots, a_1$, and a_0 are real numbers, with $a_n \neq 0$.

For $n = 1$, a polynomial function takes the form

$$f(x) = a_1 x + a_0,$$

a linear function. A linear function, therefore, is a polynomial function of degree 1. (Note, however, that a linear function of the form $f(x) = a_0$ for a real number a_0 is a polynomial function of degree 0.) A polynomial function of degree 2 is a quadratic function.

Accurate graphs of polynomial functions of degree 3 or higher require methods of calculus to be discussed later. Meanwhile, point plotting can be used to get reasonable sketches of such graphs. A graphing calculator or computer program is helpful for plotting many points rapidly, but such technology requires thought and care, as shall be seen.

The simplest polynomial functions of higher degree are those of the form $f(x) = x^n$. To graph $f(x) = x^3$, for example, find several ordered pairs that satisfy $y = x^3$, then plot the points and connect them with a smooth curve. The graph of $f(x) = x^3$ is shown as a black curve in Figure 28. This same figure also shows the graph of $f(x) = x^5$ in color.

$f(x) = x^3$

x	$f(x)$
-2	-8
-1	-1
0	0
1	1
2	8

$f(x) = x^5$

x	$f(x)$
-1.5	-7.6
-1	-1
0	0
1	1
1.5	7.6

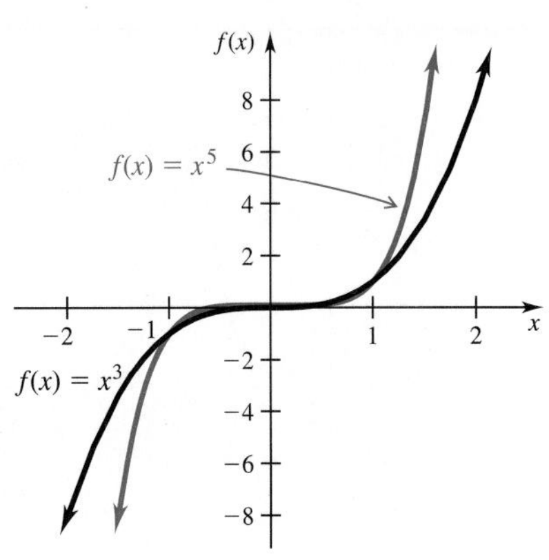

FIGURE 28

The graphs of $f(x) = x^4$ and $f(x) = x^6$ can be sketched in a similar manner. Figure 29 shows $f(x) = x^4$ as a black curve and $f(x) = x^6$ in color. These graphs have symmetry about the y-axis, as does the graph of $f(x) = ax^2$ for a nonzero real number a. As with the graph of $f(x) = ax^2$, the value of a in $f(x) = ax^n$ affects the direction of the graph. When $a > 0$, the graph has the same general appearance as the graph of $f(x) = x^n$. However, if $a < 0$, the graph is rotated 180° about the x-axis.

$f(x) = x^4$

x	$f(x)$
-2	16
-1	1
0	0
1	1
2	16

$f(x) = x^6$

x	$f(x)$
-1.5	11.4
-1	1
0	0
1	1
1.5	11.4

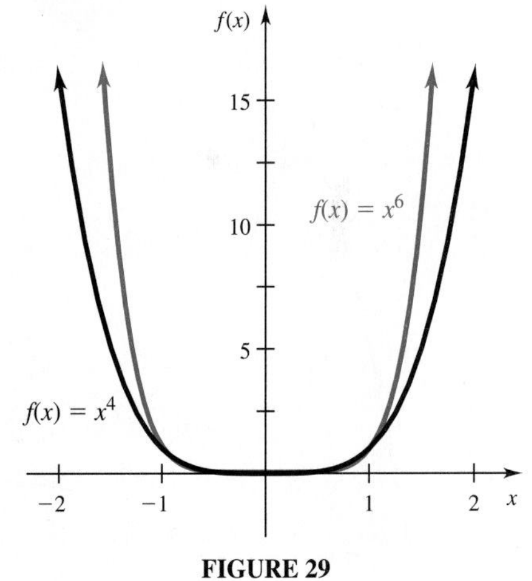

FIGURE 29

EXAMPLE 1 Graph $f(x) = 8x^3 - 12x^2 + 2x + 1$.

Letting x take values from -3 through 3 leads to the values of $f(x)$ given in the following table.

x	-3	-2	-1	0	1	2	3
$f(x)$	-329	-115	-21	1	-1	21	115

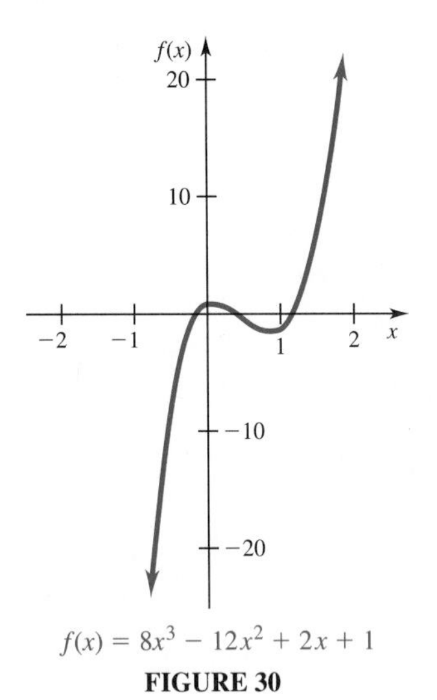

$f(x) = 8x^3 - 12x^2 + 2x + 1$

FIGURE 30

Plot the points with reasonable y-values. You will probably need to calculate additional ordered pairs between these points to get a good idea of how the graph should look. See Figure 30. ◀

We will check our work by graphing this function on a graphing calculator. We press the Graph key, and type in the function, but no graph appears on the screen. Let's check the domain and range over which the calculator is plotting the graph. Here's the problem: the domain is set as $-360 \le x \le 360$ and the range as $-1.6 \le y \le 1.6$, settings left over from a previous graph. They are unsuitable for this function, however. For such large x values, the corresponding y values will be far beyond what the screen can show. If we change the range to $-1{,}000{,}000 \le y \le 1{,}000{,}000$, we find that this also is not a wise choice. The graph fits on the screen, but it looks similar in shape to the graph of $y = x^3$ in Figure 28. The two ''bumps'' around $x = 0$ and $x = 1$ in Figure 30 don't appear. (These ''bumps'' will be referred to as **turning points.** Later another term will be used: *relative extrema.*) And no wonder they don't appear: their heights are minuscule compared with 1,000,000.

Getting a good graph with a computer or calculator requires care in choosing the domain and range. Later we will see how calculus helps us locate the interesting parts of the graph. It will also tell us what is happening between the points we plot by hand, by calculator, or by computer.

Graph $f(x) = 3x^4 - 14x^3 + 54x - 3$.

Complete a table of ordered pairs, and plot the points.

x	−3	−2	−1	0	1	2	3
$f(x)$	456	49	−40	−3	40	41	24

The graph is shown in Figure 31. ◀

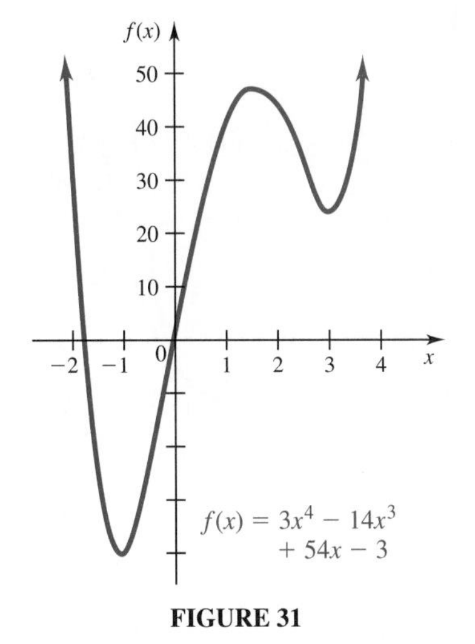

FIGURE 31

As suggested by the graphs above, the domain of a polynomial function is the set of all real numbers. The range of a polynomial function of odd degree is also the set of all real numbers. Some typical graphs of polynomial functions of odd degree are shown in Figure 32. These graphs suggest that for every polynomial function f of odd degree there is at least one real value of x for which $f(x) = 0$. Such a value of x is called a **real zero** of f; these values are also the x-intercepts of the graph.

y

x

Degree 3; three real zeros

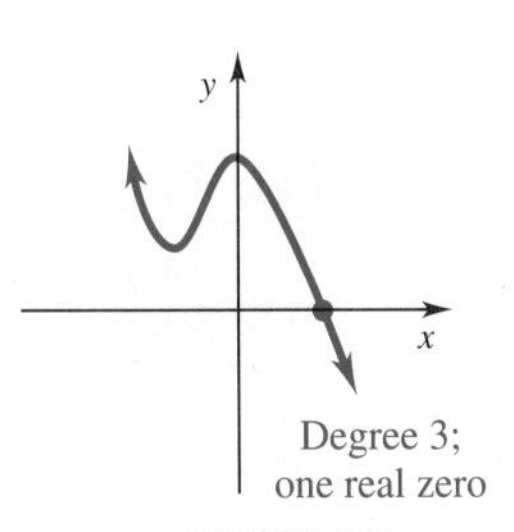

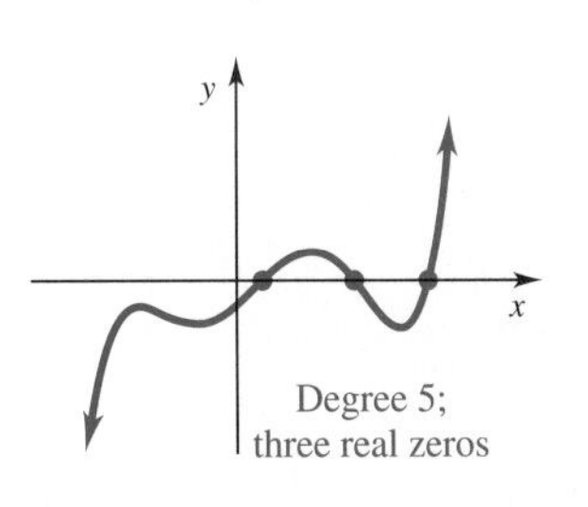

FIGURE 32

Polynomial functions of even degree have a range that takes either the form $(-\infty, k]$ or the form $[k, \infty)$ for some real number k. Figure 33 shows two typical graphs of polynomial functions of even degree.

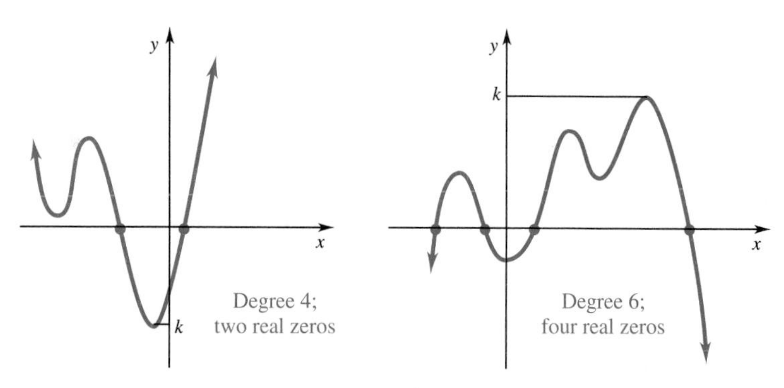

FIGURE 33

A fifth-degree polynomial can have four turning points, as in the last graph of Figure 32, or no turning points, as in Figure 28. By examining the figures in this section, you may notice that the graph of a polynomial of degree n has at most $n - 1$ turning points. In a later chapter we will use calculus to see why this is true. Meanwhile, you can learn much about a polynomial by examining its graph. For example, if you are presented with the second graph in Figure 33 and told it is a polynomial, you know immediately that it is of even degree, because the range is of the form $(-\infty, k]$. It has five turning points, so it must be of degree 6 or higher. It could be of degree 8 or 10 or 12, etc., but you can't be sure from the graph alone.

The shape of a third-degree polynomial is similar to the first graph in Figure 32 if the x^3 term is positive, and to the second graph in Figure 32 if the x^3 term is negative. The turning points may also be flattened out, as they are in Figure 28.

Rational Functions Many situations require mathematical models that are quotients. A common model for such situations is a *rational function.*

RATIONAL FUNCTION

A function defined by

$$f(x) = \frac{p(x)}{q(x)},$$

where $p(x)$ and $q(x)$ are polynomial functions and $q(x) \neq 0$, is called a **rational function.**

Since any values of x such that $q(x) = 0$ are excluded from the domain, a rational function usually has a graph with one or more breaks.

EXAMPLE 3 Graph $y = \frac{1}{x}$.

This function is undefined for $x = 0$, since 0 is not allowed in the denominator of a fraction. For this reason, the graph of this function will not intersect the vertical line $x = 0$, which is the y-axis. Since x can take on any value except 0, the values of x can approach 0 as closely as desired from either side of 0.

x approaches 0 ↓

x	$-.5$	$-.2$	$-.1$	$-.01$	$.01$	$.1$	$.2$	$.5$
$y = \frac{1}{x}$	-2	-5	-10	-100	100	10	5	2

↑ $|y|$ gets larger and larger

The table above suggests that as x gets closer and closer to 0, $|y|$ gets larger and larger. This is true in general: as the denominator gets smaller, the fraction gets larger. Thus, the graph of the function approaches the vertical line $x = 0$ (the y-axis) without ever touching it.

As $|x|$ gets larger and larger, $y = 1/x$ gets closer and closer to 0, as shown in the table below. This is also true in general: as the denominator gets larger, the fraction gets smaller.

x	-100	-10	-4	-1	1	4	10	100
$y = \frac{1}{x}$	$-.001$	$-.01$	$-.25$	-1	1	$.25$	$.01$	$.001$

The graph of the function approaches the horizontal line $y = 0$ (the x-axis). The information from the table supports the graph in Figure 34. ◀

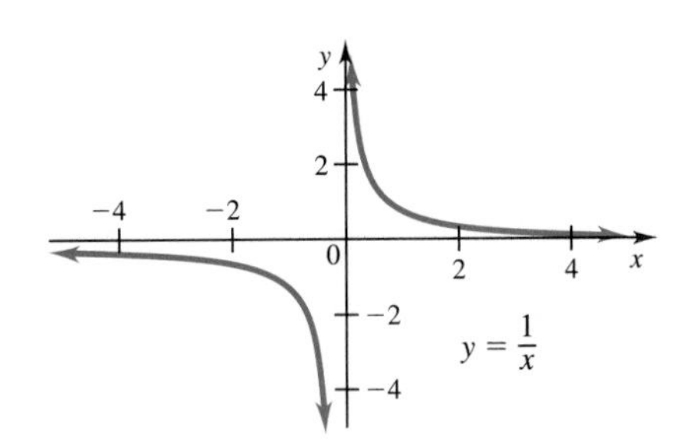

FIGURE 34

In Example 3, the vertical line $x = 0$ and the horizontal line $y = 0$ are *asymptotes,* defined as follows.

ASYMPTOTES

If a number k makes the denominator 0 in a rational function, then the line $x = k$ is a **vertical asymptote.***

If the values of y approach a number k as $|x|$ gets larger and larger, the line $y = k$ is a **horizontal asymptote.**

EXAMPLE 4

Graph $y = \dfrac{3x + 2}{2x + 4}$.

The value $x = -2$ makes the denominator 0, so the line $x = -2$ is a vertical asymptote. To find a horizontal asymptote, find y as x gets larger and larger, as in the following chart.

x	$y = \dfrac{3x + 2}{2x + 4}$	*Ordered Pair*
10	$\dfrac{32}{24} = 1.33$	(10, 1.33)
20	$\dfrac{62}{44} = 1.41$	(20, 1.41)
100	$\dfrac{302}{204} = 1.48$	(100, 1.48)
100,000	$\dfrac{300{,}002}{200{,}004} = 1.49998$	(100,000, 1.49998)

The chart suggests that as x gets larger and larger, $(3x + 2)/(2x + 4)$ gets closer and closer to 1.5, or $3/2$, making the line $y = 3/2$ a horizontal asymptote. Use a calculator to show that as x gets more negative and takes on the values -10, -100, -1000, $-1{,}000{,}000$, and so on, the graph again approaches the line $y = 3/2$. The intercepts should also be noted. When $x = 0$, $y = 2/4 = 1/2$ (the y-intercept). To make a fraction 0, the numerator must be 0; so to make $y = 0$, it is necessary that $3x + 2 = 0$. Solve this for x to get $x = -2/3$ (the x-intercept). These two points, the asymptotes, and several other plotted points (at least one to the left of the vertical asymptote to pin down that side of the graph) produce the graph in Figure 35. ◀

*Actually, we should make sure that $x = k$ does not also make the numerator 0. If both the numerator and denominator are 0, then there may be no vertical asymptote at k.

x	y
3	1.1
2	1
1	5/6
0	1/2
−1	−1/2
−3/2	−5/2
−5/2	11/2
−3	7/2
−4	5/2
−6	2

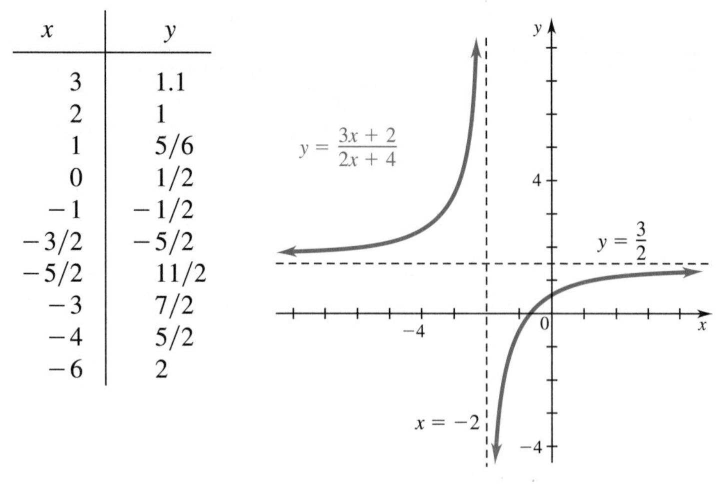

FIGURE 35

In Example 4, $y = 3/2$ was the horizontal asymptote for the rational function $y = (3x + 2)/(2x + 4)$. An equation for the horizontal asymptote can also be found by asking what happens when x gets large. If x is very large, then $3x + 2 \approx 3x$,[†] because the 2 is very small by comparison. In other words, just keep the larger term ($3x$ in this case) and discard the smaller term. Similarly, the denominator is approximately equal to $2x$, so $y \approx 3x/2x = 3/2$. This means that the line $y = 3/2$ is a horizontal asymptote. A more precise way of approaching this idea will be seen later, when limits at infinity are discussed.

Rational functions occur often in practical applications. In many situations involving environmental pollution, much of the pollutant can be removed from the air or water at a fairly reasonable cost, but the last small part of the pollutant can be very expensive to remove. Cost as a function of the percentage of pollutant removed from the environment can be calculated for various percentages of removal, with a curve fitted through the resulting data points. This curve then leads to a mathematical model of the situation. Rational functions are often a good choice for these **cost-benefit models** because they rise rapidly as they approach a vertical asymptote.

Suppose a cost-benefit model is given by

$$y = \frac{18x}{106 - x},$$

where y is the cost (in thousands of dollars) of removing x percent of a certain pollutant. The domain of x is the set of all numbers from 0 to 100 inclusive; any

[†]The symbol $\approx$ means *approximately equal to.*

amount of pollutant from 0% to 100% can be removed. Find the cost to remove the following amounts of the pollutant: 100%, 95%, 90%, and 80%. Graph the function.

Removal of 100% of the pollutant would cost

$$y = \frac{18(100)}{106 - 100} = 300,$$

or \$300,000. Check that 95% of the pollutant can be removed for \$155,000, 90% for \$101,000, and 80% for \$55,000. Using these points, as well as others that could be obtained from the function above, gives the graph shown in Figure 36. ◀

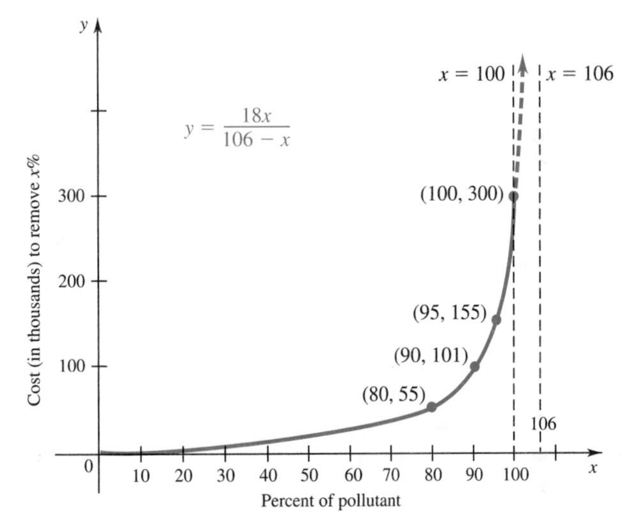

FIGURE 36

1.5 Exercises

Each of the following is the graph of a polynomial function. Give the possible values for the degree of the polynomial, and give the sign (+ or −) for the x^n term.

1.

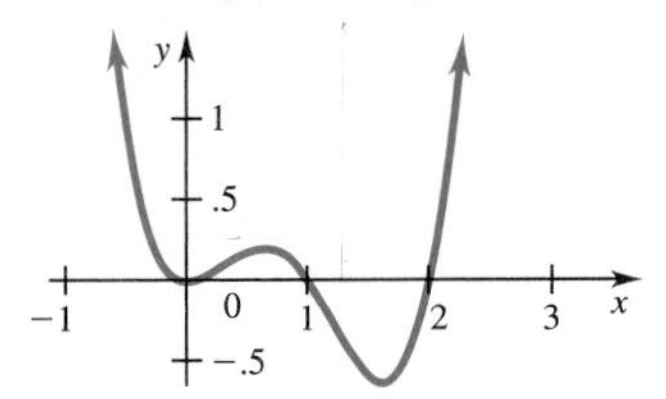

2.

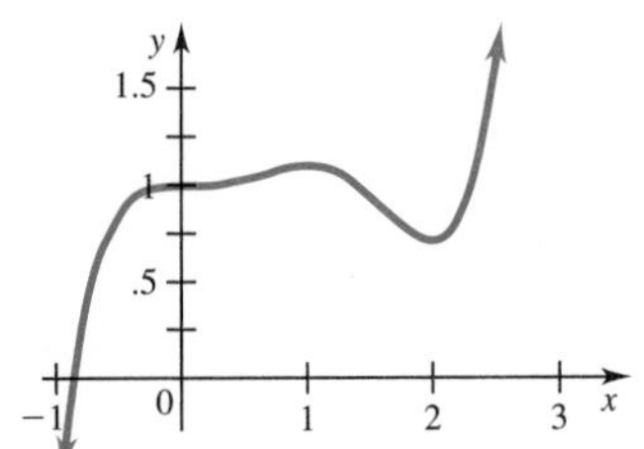

3.

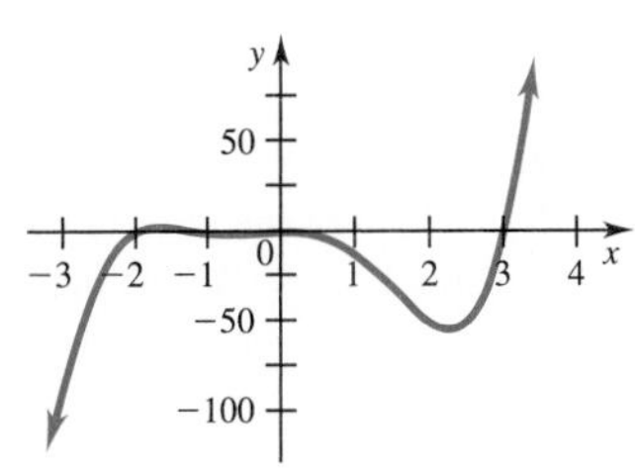

4.

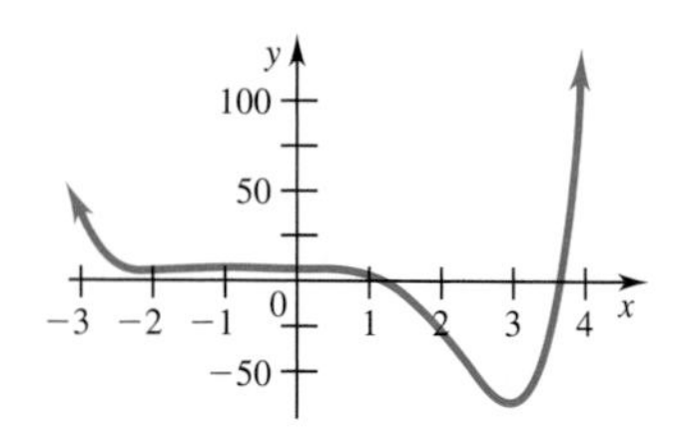

5.

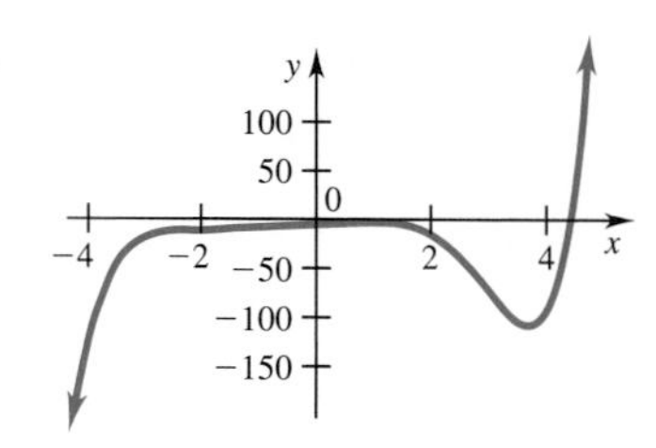

6.

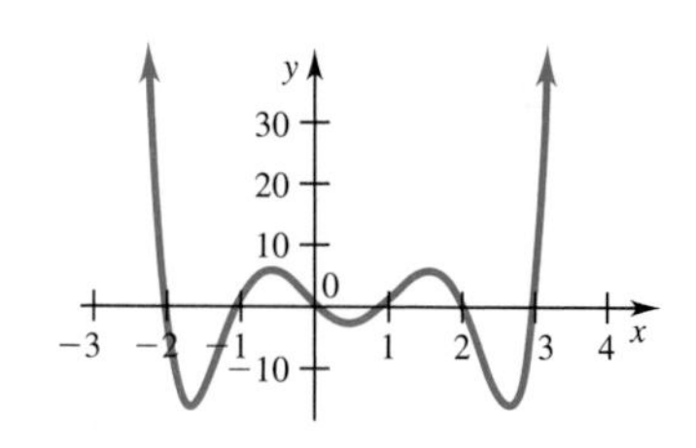

7.

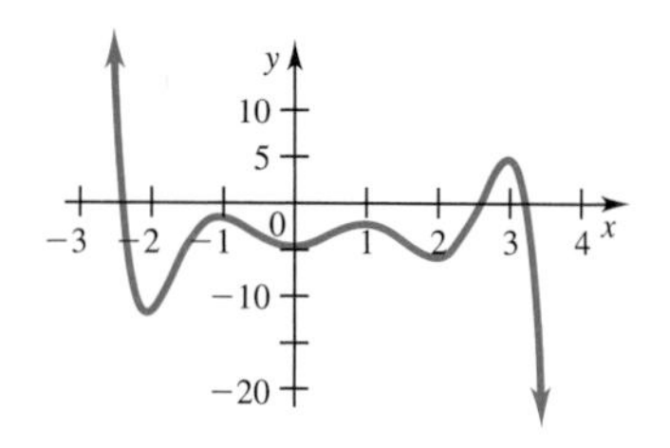

8.

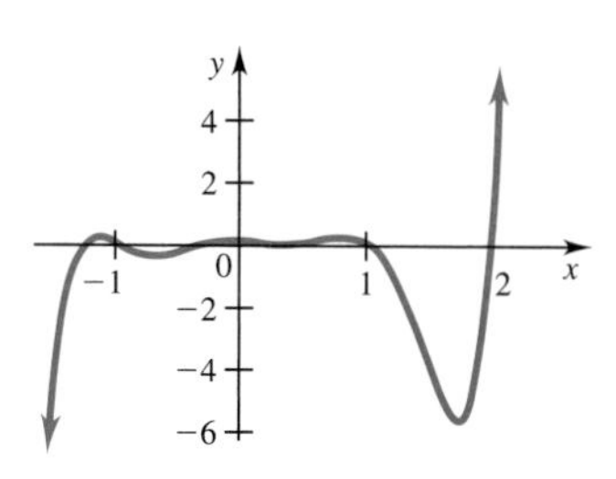

Find the horizontal and vertical asymptotes for each of the following rational functions. Draw the graph of each function, including any x- or y-intercepts.

9. $y = \dfrac{-4}{x - 3}$

10. $y = \dfrac{-1}{x + 3}$

11. $y = \dfrac{2}{3 + 2x}$

12. $y = \dfrac{4}{5 + 3x}$

13. $y = \dfrac{3x}{x - 1}$

14. $y = \dfrac{4x}{3 - 2x}$

15. $y = \dfrac{x + 1}{x - 4}$

16. $y = \dfrac{x - 3}{x + 5}$

17. $y = \dfrac{1 - 2x}{5x + 20}$

18. $y = \dfrac{6 - 3x}{4x + 12}$

19. $y = \dfrac{-x - 4}{3x + 6}$

20. $y = \dfrac{-x + 8}{2x + 5}$

Applications

Business and Economics

21. Average Cost Suppose the average cost per unit $C(x)$, in dollars, to produce x units of margarine is given by

$$C(x) = \frac{500}{x + 30}.$$

(a) Find $C(10)$, $C(20)$, $C(50)$, $C(75)$, and $C(100)$.

(b) Which of the intervals $(0, \infty)$ and $[0, \infty)$ would be a more reasonable domain for C? Why?

(c) Graph $y = C(x)$.

22. Cost Analysis In a recent year, the cost per ton, y, to build an oil tanker of x thousand deadweight tons was approximated by

$$y = \frac{110{,}000}{x + 225}.$$

(a) Find y for $x = 25$, $x = 50$, $x = 100$, $x = 200$, $x = 300$, and $x = 400$.

(b) Graph the function.

Tax Rates *Exercises 23 to 26 refer to the* Laffer curve, *originated by the economist Arthur Laffer. It has been a center of controversy. An idealized version of this curve is shown here. According to this curve, decreasing a tax rate, say from x_2 percent to x_1 percent on the graph, can actually lead to an increase in government revenue. The theory is that people will work harder and earn more money if they are taxed at a lower rate, so the government ends up with more revenue than it would at a higher tax rate. All economists agree on the endpoints—0 revenue at tax rates of both 0% and 100%—but there is much disagreement on the location of the tax rate x_1 that produces the maximum revenue.*

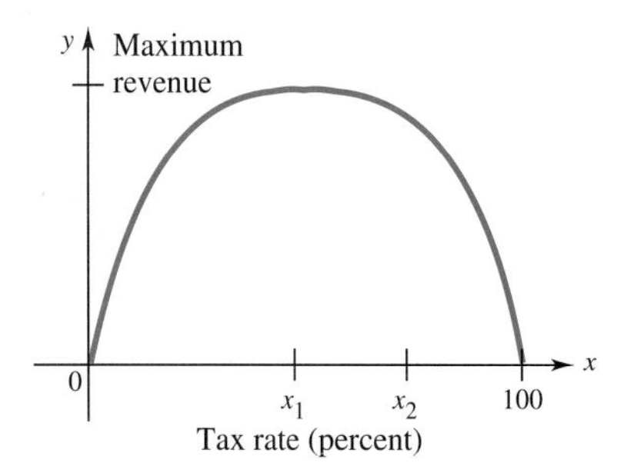

23. A function that might describe the entire Laffer curve is

$$y = x(100 - x)(x^2 + 500),$$

where y is government revenue in hundreds of thousands of dollars from a tax rate of x percent, with the function valid for $0 \le x \le 100$. Find the revenue from the following tax rates.

(a) 10% **(b)** 40% **(c)** 50% **(d)** 80%

(e) Graph the function.

24. Find the equations of two quadratic functions that could describe the Laffer curve by having roots at $x = 0$ and $x = 100$. Give the first a maximum of 100 and the second a maximum of 250, then multiply them together to get a new Laffer curve with a maximum of 25,000. Plot the resulting function.

25. Suppose an economist studying the Laffer curve produced the rational function

$$y = \frac{60x - 6000}{x - 120},$$

where y is government revenue in millions from a tax rate of x percent, with the function valid for $50 \le x \le 100$.

Find the revenue from the following tax rates.

(a) 50% **(b)** 60% **(c)** 80% **(d)** 100%

(e) Graph the function.

26. Suppose the economist in Exercise 25 studies a different tax, this time producing

$$y = \frac{80x - 8000}{x - 110},$$

where y is the government revenue in tens of millions of dollars for a tax rate of x percent, with the function valid for $55 \le x \le 100$.

Find the revenue from the following tax rates.

(a) 55% **(b)** 60% **(c)** 70% **(d)** 90%

(e) 100% **(f)** Graph the function.

27. Cost-Benefit Model Suppose a cost-benefit model is given by

$$y = \frac{6.7x}{100 - x}.$$

where y is the cost in thousands of dollars of removing x percent of a given pollutant.

Find the cost of removing each of the following percents of pollutants.

(a) 50% **(b)** 70% **(c)** 80% **(d)** 90%

(e) 95% **(f)** 98% **(g)** 99%

(h) Is it possible, according to this function, to remove *all* the pollutant?

(i) Graph the function.

28. Cost-Benefit Model Suppose a cost-benefit model is given by

$$y = \frac{6.5x}{102 - x},$$

where y is the cost in thousands of dollars of removing x percent of a certain pollutant.

Find the cost of removing each of the following percents of pollutants.

(a) 0% **(b)** 50% **(c)** 80% **(d)** 90%

(e) 95% **(f)** 99% **(g)** 100%

(h) Graph the function.

Life Sciences

29. Cardiac Output A technique for measuring cardiac output depends on the concentration of a dye after a known amount is injected into a vein near the heart. In a normal heart, the concentration of the dye at time x (in seconds) is given by the function

$$g(x) = -.006x^4 + .140x^3 - .053x^2 + 1.79x.$$

(a) Graph $g(x)$ by point plotting or by using a graphing calculator or computer.

(b) In your graph from part (a), notice that the function initially increases. Considering the form of $g(x)$, do you think it can keep increasing forever? Explain.

(c) Write a short paragraph describing the extent to which the concentration of dye might be described by the function $g(x)$.

30. Population Variation During the early part of the twentieth century, the deer population of the Kaibab Plateau in Arizona experienced a rapid increase, because hunters had reduced the number of natural predators. The increase in population depleted the food resources and eventually caused the population to decline. For the period from 1905 to 1930, the deer population was approximated by

$$D(x) = -.125x^5 + 3.125x^4 + 4000,$$

where x is time in years from 1905.

(a) Use a calculator to find enough points to graph $D(x)$, or use a graphing calculator or computer to graph the function.

(b) From the graph, over what period of time (from 1905 to 1930) was the population increasing? relatively stable? decreasing?

31. Alcohol Concentration The polynomial function

$$A(x) = -.015x^3 + 1.058x$$

gives the approximate alcohol concentration (in tenths of a percent) in an average person's bloodstream x hours after drinking about eight ounces of 100-proof whiskey. The function is approximately valid for x in the interval $[0, 8]$.

(a) Graph $A(x)$.

(b) Using the graph you drew for part (a), estimate the time of maximum alcohol concentration.

(c) In one state, a person is legally drunk if the blood alcohol concentration exceeds .08%. Use the graph from part (a) to estimate the period in which this average person is legally drunk.

32. Drug Dosage To calculate the drug dosage for a child, pharmacists may use the formula

$$d(x) = \frac{Dx}{x + 12},$$

where x is the child's age in years and D is the adult dosage. Let $D = 70$, the adult dosage of the drug Naldecon.

(a) What is the vertical asymptote for this function?

(b) What is the horizontal asymptote for this function?

(c) Graph $d(x)$.

33. Population Biology The function

$$f(x) = \frac{\lambda x}{1 + (ax)^b}$$

is used in population models to give the size of the next generation ($f(x)$) in terms of the current generation (x).*

(a) What is a reasonable domain for this function, considering what x represents?

(b) Graph this function for $\lambda = a = b = 1$.

(c) Graph this function for $\lambda = a = 1$ and $b = 2$.

(d) What is the effect of making b larger?

34. Growth Models The function

$$f(x) = \frac{Kx}{A + x}$$

is used in biology to give the growth rate of a population in the presence of a quantity x of food. This is called Michaelis-Menten kinetics.†

(a) What is a reasonable domain for this function, considering what x represents?

(b) Graph this function for $K = 5$ and $A = 2$.

(c) Show that $y = K$ is a horizontal asymptote.

(d) What do you think K represents?

(e) Show that A represents the quantity of food for which the growth rate is half of its maximum.

Physical Sciences

35. Electronics In electronics, the circuit gain is given by

$$G(R) = \frac{R}{r + R},$$

where R is the resistance of a temperature sensor in the circuit and r is constant. Let $r = 1000$ ohms.

(a) Find any vertical asymptotes of the function.

(b) Find any horizontal asymptotes of the function.

(c) Graph $G(R)$.

36. Oil Pressure The pressure of the oil in a reservoir tends to drop with time. By taking sample pressure readings, petroleum engineers have found that the *change* in pressure in a particular oil reservoir is given by

$$P(t) = t^3 - 25t^2 + 200t,$$

where t is time in years from the date of the first reading.

(a) Graph $P(t)$.

(b) For what time period is the *change* in pressure (the drop in pressure) increasing? decreasing?

---

*See J. Maynard Smith, *Models in Ecology*, Cambridge University Press, 1974.

†See Leah Edelstein-Keshet, *Mathematical Models in Biology*, Random House, 1988.

37. Catastrophe Theory An area of mathematics known as catastrophe theory studies how functions suddenly change their behavior. In catastrophe theory, the family of functions

$$f_t(u) = u^3 + tu$$

arises, where t can take on various values to produce various functions.*

Graph $f_t(u)$ for

(a) $t = 1$ **(b)** $t = -1$

38. Catastrophe theory has been applied to thermodynamics, in which the following family of functions arises:†

$$f(x) = \frac{x^4}{4} + \frac{a}{2}x^2 + bx$$

Graph $f(x)$ for

(a) $a = -5$ and $b = 1$; **(b)** $a = 1$ and $b = -6$.

General Interest

39. Antique Cars Antique car fans often enter their cars in a *concours d'élègance* in which a maximum of 100 points can be awarded to a particular car. Points are awarded for the general attractiveness of the car. Based on a recent article in *Business Week*, we constructed the following mathematical model for the cost, in thousands of dollars, of restoring a car so that it will win x points.

$$C(x) = \frac{10x}{49(101 - x)}$$

Find the cost of restoring a car so that it will win the following numbers of points.

(a) 99 **(b)** 100

Computer/Graphing Calculator

Make an intelligent guess about the shape of each polynomial function. Then sketch the graph of the function by using a graphing calculator or computer program, and compare the graph with what you expected it to look like.

40. $f(x) = x^3 - 7x - 9$

41. $f(x) = -x^3 + 4x^2 + 3x - 8$

42. $f(x) = -6x^3 - 11x^2 + x + 6$

43. $f(x) = 2x^3 + 4x - 1$

44. $f(x) = x^4 - 5x^2 + 7$

45. $f(x) = x^4 + x^3 - 2$

46. $f(x) = -8x^4 + 2x^3 + 47x^2 + 52x + 15$

47. $f(x) = -x^4 - 2x^3 + 3x^2 + 3x + 5$

48. $f(x) = x^5 - 2x^4 - x^3 + 3x^2 + x + 2$

49. $f(x) = -x^5 + 6x^4 - 11x^3 + 6x^2 + 5$

50. Write a paragraph or two describing what you learned about the graphs of polynomial functions from Exercises 40–49.

Find the horizontal and vertical asymptotes for each of the following rational functions. Use this information to make an intelligent guess about the shape of each function. Then sketch the graph of the function by using a graphing calculator or computer program, and compare the graph with what you expected it to look like.

51. $f(x) = \dfrac{-2x^2 + x - 1}{2x + 3}$

52. $f(x) = \dfrac{3x + 2}{x^2 - 4}$

53. $f(x) = \dfrac{2x^2 - 5}{x^2 - 1}$

54. $f(x) = \dfrac{4x^2 - 1}{x^2 + 1}$

55. $f(x) = \dfrac{-2x^2}{x^2 - 10}$

56. $f(x) = \dfrac{5x + 4}{2x^2 - 1}$

57. Write a paragraph or two describing what you learned about the graphs of rational functions from Exercises 51–56.

---

*See T. Poston and I. Stewart, *Catastrophe Theory and Its Applications*, Pitman Publishing, 1978, p. 105.

†Ibid, p. 337.

1.6 TRANSLATIONS AND REFLECTIONS OF FUNCTIONS

We saw in the section on quadratics that the effect of adding a constant to a quadratic function is to raise the function if the constant is positive, and to lower it if the constant is negative. This is true for any function; the movement up or down is referred to as a **vertical translation** of the function.

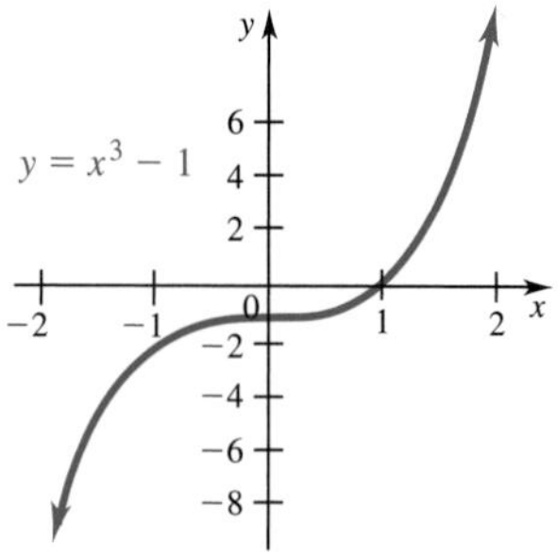

EXAMPLE 1 Graph $f(x) = x^3 - 1$.

The graph of $y = x^3$ is familiar from the previous section. Just lower the graph by 1 to get the graph in Figure 37. ◀

FIGURE 37

The next example illustrates a **horizontal translation** of a function, which is a shift to the left or right.

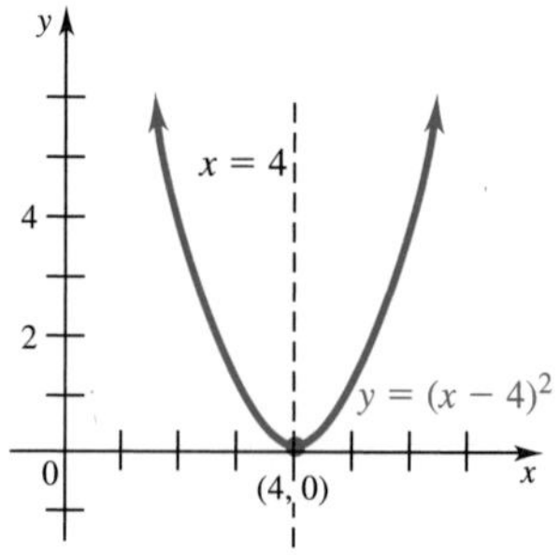

EXAMPLE 2 Graph $f(x) = (x - 4)^2$.

Notice here that 4 is subtracted from the x *before* the squaring occurs. This is not the same as taking the graph of $y = x^2$ and translating it downward by 4. Consider that the vertex of $y = x^2$ is the lowest value of the function, $y = 0$, which occurs when $x = 0$. In the current function, this value of y occurs when $x = 4$. Therefore, the new graph is just the graph of $y = x^2$ shifted 4 units to the right, as shown in Figure 38. ◀

FIGURE 38

It was shown in the section on quadratic functions that multiplying a function by a negative number flips it upside down. This is called a **vertical reflection** of the graph. The following table summarizes what we have seen about translations and reflections, and adds a new feature: **horizontal reflections.**

TRANSLATIONS AND REFLECTIONS OF FUNCTIONS

Let $f(x)$ be any function, and let a, h, and k be positive constants (Figure 39). The graph of $y = f(x) + k$ is the graph of $y = f(x)$ translated upward by an amount k (Figure 40).

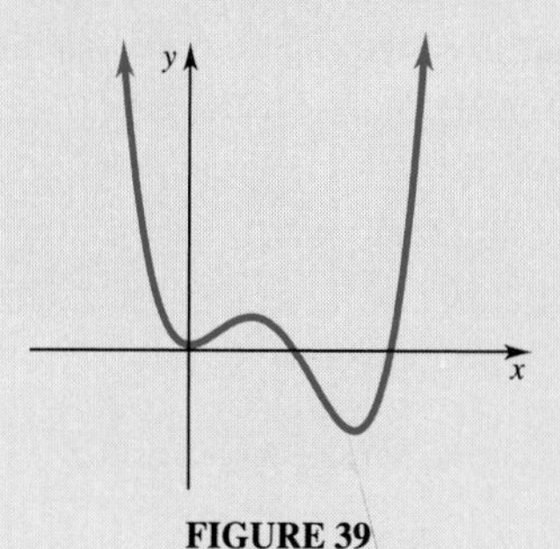

FIGURE 39

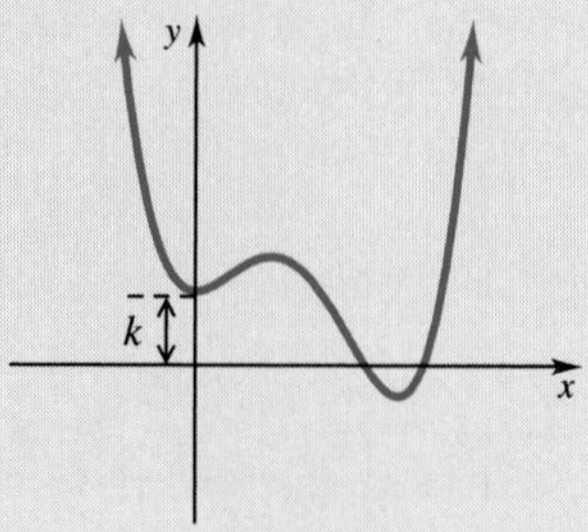

FIGURE 40

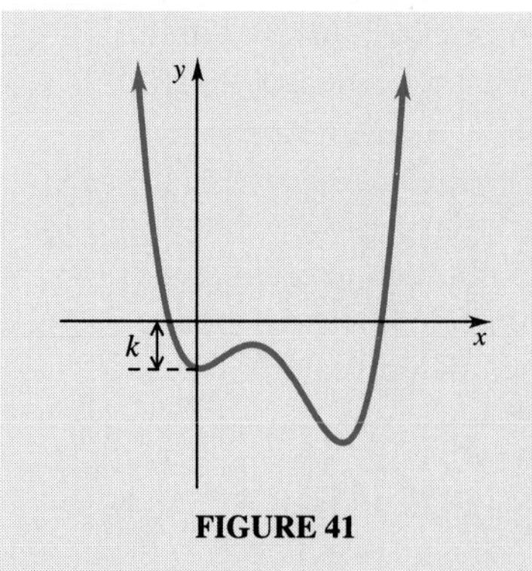

FIGURE 41

The graph of $y = f(x) - k$ is the graph of $y = f(x)$ translated downward by an amount k (Figure 41).

The graph of $y = f(x - h)$ is the graph of $y = f(x)$ translated to the right by an amount h (Figure 42).

The graph of $y = f(x + h)$ is the graph of $y = f(x)$ translated to the left by an amount h (Figure 43).

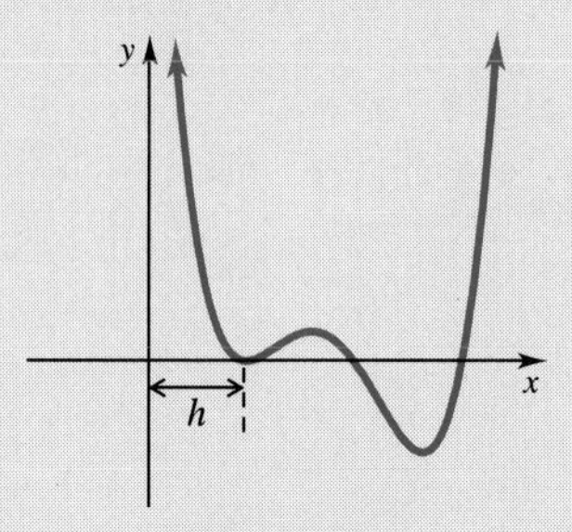

FIGURE 42

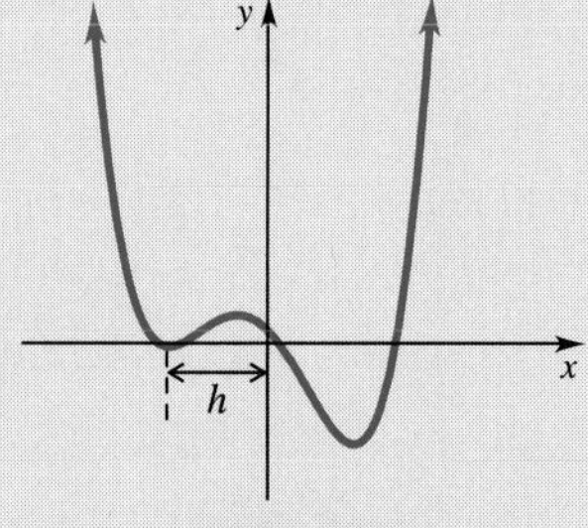

FIGURE 43

The graph of $y = -f(x)$ is the graph of $y = f(x)$ reflected vertically, that is, turned upside down (Figure 44).

The graph of $y = f(-x)$ is the graph of $y = f(x)$ reflected horizontally, that is, its mirror image (Figure 45).

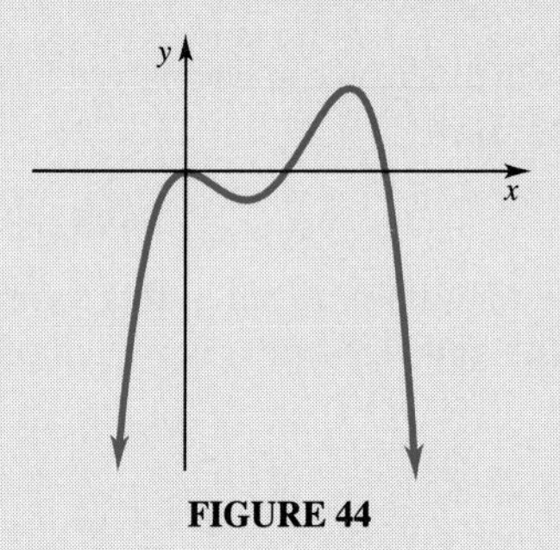

FIGURE 44

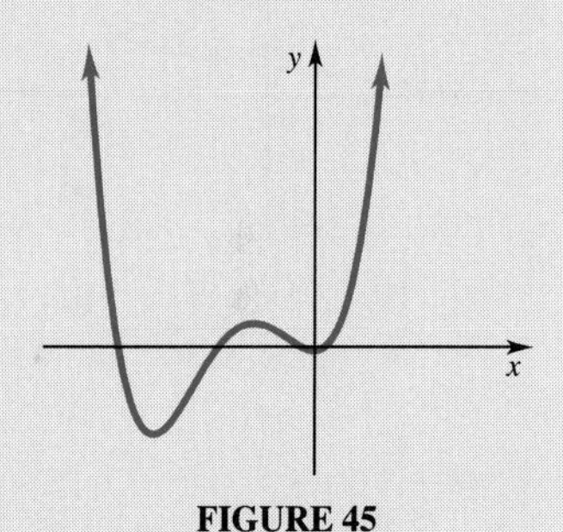

FIGURE 45

It is also possible to consider multiplying x or $f(x)$ by a constant a, as in $f(ax)$ or $af(x)$. When a is positive, this does not change the basic appearance of the graph, except to compress it or stretch it. When a is negative, it also results in a reflection. More details are included in the examples and exercises.

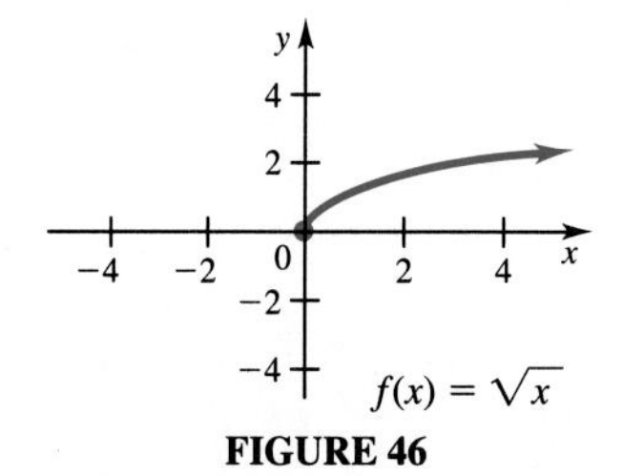

FIGURE 46

EXAMPLE 3 Graph $f(x) = -\sqrt{4 - x} + 3$.

Begin with the simplest possible function, then add each piece in turn. Start with the graph of $f(x) = \sqrt{x}$. As Figure 46 reveals, this is just one-half of the graph of $f(x) = x^2$ lying on its side.

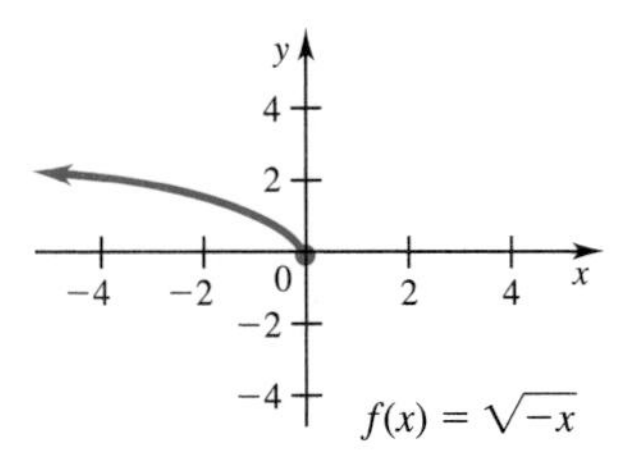

FIGURE 47

Now add another component of the original function, the negative in front of the x, giving $f(x) = \sqrt{-x}$. This is a horizontal reflection of the $f(x) = \sqrt{x}$ graph, as shown in Figure 47.

Next, include the 4 under the square root sign. To get $4 - x$ into the form $f(x - h)$ or $f(x + h)$, we need to factor out the negative: $\sqrt{4 - x} = \sqrt{-(x - 4)}$. Now the 4 is subtracted, so this function is a translation to the right of the function $f(x) = \sqrt{-x}$ by 4 units, as Figure 48 reveals.

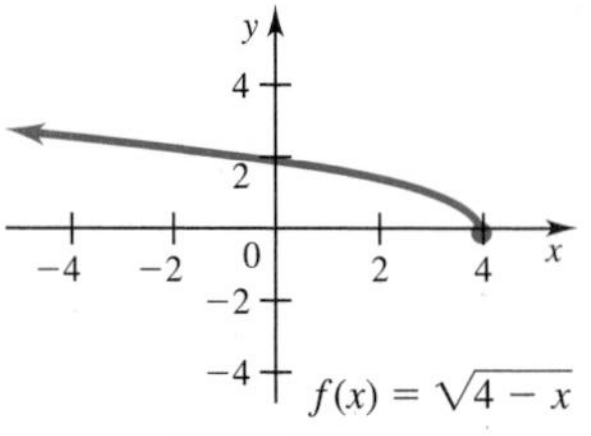

FIGURE 48

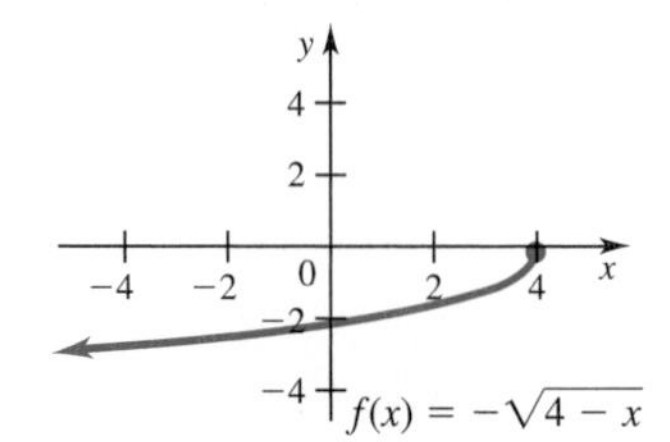

FIGURE 49

The effect of the negative in the front is a vertical reflection, as in Figure 49, which shows the graph of $f(x) = -\sqrt{4 - x}$.

Finally, adding the constant 3 raises the entire graph by 3 units, giving the graph of $f(x) = -\sqrt{4 - x} + 3$ in Figure 50. ◀

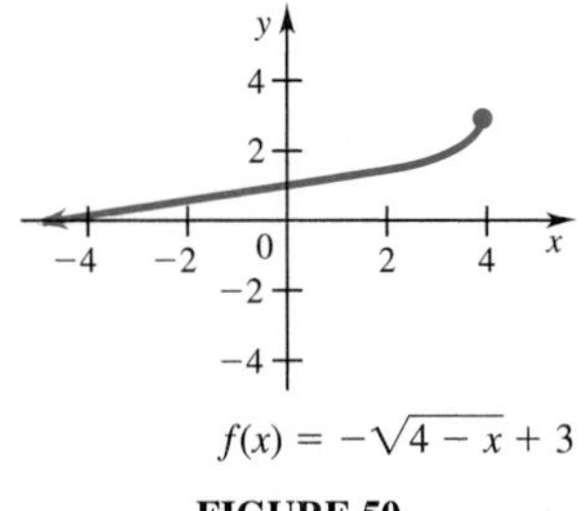

$f(x) = -\sqrt{4 - x} + 3$

FIGURE 50

A quadratic function can be analyzed using the process described in Example 3 if it is in the form

$$y = a(x - h)^2 + k.$$

A quadratic equation not given in this form can be converted by a process called **completing the square**, which is explained next.

EXAMPLE 4 Graph $f(x) = -3x^2 - 2x + 1$.

To rewrite $-3x^2 - 2x + 1$ in the form $a(x - h)^2 + k$, first factor -3 from $-3x^2 - 2x$ to get

$$y = -3\left(x^2 + \frac{2}{3}x\right) + 1.$$

One-half the coefficient of x is $1/3$, and $(1/3)^2 = 1/9$. Add and subtract $1/9$ inside the parentheses as follows:

$$y = -3\left(x^2 + \frac{2}{3}x + \frac{1}{9} - \frac{1}{9}\right) + 1.$$

Using the distributive property and simplifying gives

$$y = -3\left(x^2 + \frac{2}{3}x + \frac{1}{9}\right) - 3\left(-\frac{1}{9}\right) + 1 = -3\left(x^2 + \frac{2}{3}x + \frac{1}{9}\right) + \frac{4}{3}.$$

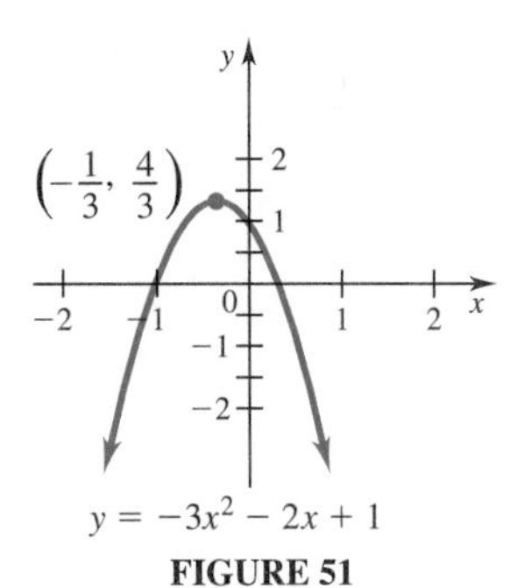

FIGURE 51

Factor to get

$$y = -3\left(x + \frac{1}{3}\right)^2 + \frac{4}{3}.$$

This result shows that the graph is the parabola $y = x^2$ translated 1/3 unit to the left, flipped upside down, and translated 4/3 units upward. This puts the vertex at $(-1/3, 4/3)$. The 3 in front of the squared term causes the parabola to be stretched vertically by a factor of 3. These results are shown in Figure 51. ◀

1.6 Exercises

Use the ideas in this section to graph each of the following functions.

1. $f(x) = (x + 2)^3$ **2.** $f(x) = (x - 3)^3$ **3.** $f(x) = (4 - x)^3 + 10$

4. $f(x) = (1 - x)^3 - 3$ **5.** $f(x) = -2(x + 1)^3 + 5$ **6.** $f(x) = -3(x - 2)^3 - 2$

7. $f(x) = -2x^5 + 5$ **8.** $f(x) = (x - 1)^6 - 4$ **9.** $f(x) = \dfrac{2}{x + 3} - 2$

10. $f(x) = \dfrac{3}{x - 2} + 3$ **11.** $f(x) = \dfrac{-1}{x - 1} + 2$ **12.** $f(x) = \dfrac{-1}{x + 4} - 1$

13. $f(x) = \sqrt{x - 1} + 3$ **14.** $f(x) = \sqrt{x + 1} - 4$ **15.** $f(x) = -\sqrt{-4 - x} - 2$

16. $f(x) = -\sqrt{2 - x} + 2$

Use the method of completing the square to graph each of the following quadratic functions.

17. $f(x) = x^2 - 2x - 2$ **18.** $f(x) = x^2 + 4x + 6$ **19.** $f(x) = x^2 - 4x - 2$

20. $f(x) = x^2 + 6x + 8$ **21.** $f(x) = 2x^2 + 4x - 1$ **22.** $f(x) = 3x^2 - 12x + 10$

23. $f(x) = -3x^2 + 24x - 55$ **24.** $f(x) = -2x^2 - 8x - 13$

Drawing a figure similar to Figures 39–45, show the graph of $f(ax)$ where a satisfies the given condition.

25. $0 < a < 1$ **26.** $1 < a$ **27.** $-1 < a < 0$ **28.** $a < -1$

Drawing a figure similar to Figures 39–45, show the graph of $af(x)$ where a satisfies the given condition.

29. $0 < a < 1$ **30.** $1 < a$ **31.** $-1 < a < 0$ **32.** $a < -1$

33. One observation arising from these methods is that all rational functions in which both the numerator and denominator are linear functions are just translations and reflections of the function $f(x) = c/x$.

(a) Show that if $f(x) = c/x$, then $af(x - h) + k$ is also a rational function.

(b) Going the other way is a bit trickier. Here is an example:

$$f(x) = \frac{2x + 3}{4x + 1} = \frac{2x + 1/2}{4x + 1} + \frac{5/2}{4x + 1} = \frac{1}{2} + \frac{5}{2} \cdot \frac{1}{4x + 1}.$$

How would you know to break up the 3 as $1/2 + 5/2$?

(c) Use the result of part (b) to graph $f(x) = \dfrac{2x + 3}{4x + 1}$.

(d) Use this method to graph $f(x) = \dfrac{2x + 5}{x + 2}$.

Applications

Business and Economics

34. Cost Analysis Ali finds that the cost in dollars to make x chocolate chip cookies in his bakery is approximately equal to

$$C(x) = 10\sqrt{x} + 50.$$

(a) Find the cost to make 100 cookies.

(b) What is the fixed cost (i.e., the cost of making no cookies)?

(c) Sketch the graph of $C(x)$.

(d) Find a formula for $A(x)$, the average cost per cookie.

(e) What is the average cost per cookie when 100 cookies are made?

(f) What is the average cost per cookie when 1600 cookies are made?

35. Advertising A study done by an advertising agency reveals that when x thousands of dollars are spent on advertising, it results in a sales increase in thousands of dollars given by the function

$$S(x) = \frac{1}{25}(x - 10)^3 + 40.$$

(a) Find the increase in sales when no money is spent on advertising.

(b) Find the increase in sales when $10,000 is spent on advertising.

(c) Sketch the graph of $S(x)$.

(d) Find a formula for $A(x)$, the average increase in sales per amount spent on advertising.

(e) What is the average increase in sales per amount spent on advertising when $10,000 is spent on advertising?

(f) What is the average increase in sales per amount spent on advertising when $50,000 is spent on advertising?

Chapter Summary

Key Terms

To understand the concepts presented in this chapter, you should know the meaning and use of the following words. For easy reference, the section in the chapter where a word (or expression) was first used is given with each item.

1.1 **function**
mathematical model
independent variable
dependent variable
ordered pair
domain
range
interval notation
Cartesian coordinate system
***x*-axis**
***y*-axis**
origin
***x*-coordinate**
***y*-coordinate**
quadrant
vertical line test
step function

1.2 **linear function**
slope
point-slope form
slope-intercept form
supply and demand curves
equilibrium price
equilibrium quantity

1.3 **average rate of change**
fixed cost
marginal cost
cost function
break-even point

1.4 **quadratic function**
parabola
vertex
axis

1.5 **polynomial function**
turning point
real zero
rational function
vertical asymptote
horizontal asymptote
cost-benefit model

1.6 **vertical translation**
horizontal translation
vertical reflection
horizontal reflection
completing the square

Chapter 1 Review Exercises

1. What is a function? A linear function? A quadratic function? A rational function?

2. How do you find a vertical asymptote? A horizontal asymptote?

3. What is the marginal cost? The fixed cost?

4. What can you tell about the graph of a polynomial function of degree n before you plot any points?

List the ordered pairs obtained from each of the following if the domain of x for each exercise is $\{-3, -2, -1, 0, 1, 2, 3\}$. Graph each set of ordered pairs. Give the range.

5. $2x - 5y = 10$

6. $3x + 7y = 21$

7. $y = (2x + 1)(x - 1)$

8. $y = (x + 4)(x + 3)$

9. $y = -2 + x^2$

10. $y = 3x^2 - 7$

11. $y = \dfrac{2}{x^2 + 1}$

12. $y = \dfrac{-3 + x}{x + 10}$

13. $y + 1 = 0$

14. $y = 3$

In Exercises 15–18, find **(a)** $f(6)$, **(b)** $f(-2)$, **(c)** $f(-4)$, and **(d)** $f(r + 1)$.

15. $f(x) = 4x - 1$

16. $f(x) = 3 - 4x$

17. $f(x) = -x^2 + 2x - 4$

18. $f(x) = 8 - x - x^2$

19. Let $f(x) = 5x - 3$ and $g(x) = -x^2 + 4x$. Find each of the following.

(a) $f(-2)$ **(b)** $g(3)$ **(c)** $g(-4)$ **(d)** $f(5)$ **(e)** $g(-k)$ **(f)** $g(3m)$ **(g)** $g(k - 5)$ **(h)** $f(3 - p)$

Graph each of the following.

20. $y = 6 - 2x$

21. $y = 4x + 3$

22. $2x + 7y = 14$

23. $3x - 5y = 15$

24. $y = 1$

25. $x + 2 = 0$

26. $x + 3y = 0$

27. $y = 2x$

Find the slope for each line that has a slope.

28. Through $(4, -1)$ and $(3, -3)$

29. Through $(-2, 5)$ and $(4, 7)$

30. Through the origin and $(0, 7)$

31. Through the origin and $(11, -2)$

32. $4x - y = 7$

33. $2x + 3y = 15$

34. $3y - 1 = 14$

35. $x + 4 = 9$

Find an equation in the form $ax + by = c$ for each line.

36. Through $(8, 0)$, with slope $-1/4$

37. Through $(5, -1)$, with slope $2/3$

38. Through $(2, -3)$ and $(-3, 4)$

39. Through $(5, -2)$ and $(1, 3)$

40. Slope 0, through $(-2, 5)$

41. Undefined slope, through $(-1, 4)$

42. Through $(2, -1)$, parallel to $3x - y = 1$

43. Through $(0, 5)$, perpendicular to $8x + 5y = 3$

44. Through $(2, -10)$, perpendicular to a line with undefined slope

45. Through $(3, -5)$, parallel to $y = 4$

46. Through $(-7, 4)$, perpendicular to $y = 8$

Graph each line.

47. Through $(2, -4)$, $m = 3/4$

48. Through $(0, 5)$, $m = -2/3$

49. Through $(-4, 1)$, $m = 3$

50. Through $(-3, -2)$, $m = -1$

Graph each of the following.

51. $y = x^2 - 4$

52. $y = 6 - x^2$

53. $y = 3x^2 + 6x - 2$

54. $y = -\frac{1}{4}x^2 + x + 2$

55. $y = x^2 - 4x + 2$

56. $y = -3x^2 - 12x - 1$

57. $f(x) = x^3 + 5$

58. $f(x) = 1 - x^4$

59. $y = -(x - 1)^3 + 4$

60. $y = -(x + 2)^4 - 2$

61. $y = 2\sqrt{x + 3} + 1$

62. $y = -\sqrt{x - 4} + 2$

63. $f(x) = \frac{8}{x}$

64. $f(x) = \frac{2}{3x - 1}$

65. $f(x) = \frac{4x - 2}{3x + 1}$

66. $f(x) = \frac{6x}{x + 2}$

In Exercises 67–70, complete the square for the indicated exercise, then use the methods of Section 1.6 to graph the function.

67. Exercise 53

68. Exercise 54

69. Exercise 55

70. Exercise 56

Applications

Business and Economics

71. Supply and Demand The supply and demand for crab meat in a local fish store are related by the equations

$$\text{Supply: } p = S(q) = 6q + 3$$

and

$$\text{Demand: } p = D(q) = 19 - 2q,$$

where p represents the price in dollars per pound and q represents the quantity of crab meat in pounds per day. Find the supply and the demand at each of the following prices.

(a) \$10 **(b)** \$15 **(c)** \$18

(d) Graph both the supply and the demand functions on the same axes.

(e) Find the equilibrium price.

(f) Find the equilibrium quantity.

72. Supply For boxes of computer disks, 72 boxes will be supplied at a price of \$6, while 104 boxes will be supplied at a price of \$10. Write a supply function for this product. Assume it is a linear function.

Cost *Find the linear cost functions and average cost functions in Exercises 73–75.*

73. Eight units cost \$300; fixed cost is \$60

74. Twelve units cost \$445; 50 units cost \$1585

75. Thirty units cost \$1500; 120 units cost \$5640

76. Cost The cost of producing x bird cages is $C(x)$, where

$$C(x) = 20x + 100.$$

The product sells for \$40 per cage.

(a) Find the break-even point.

(b) What revenue will the company receive if it sells just that number of cages?

77. Car Rental To rent a midsized car from one agency costs \$40 per day or fraction of a day. If you pick up the car in Boston and drop it off in Utica, there is a fixed \$40 charge. Let $C(x)$ represent the cost of renting the car for x days and taking it from Boston to Utica.

Find each of the following.

(a) $C\left(\frac{3}{4}\right)$ **(b)** $C\left(\frac{9}{10}\right)$ **(c)** $C(1)$

(d) $C\left(1\frac{5}{8}\right)$ **(e)** $C\left(2\frac{1}{9}\right)$

(f) Graph the function defined by $y = C(x)$ for $0 < x \leq 5$.

(g) What is the independent variable?

(h) What is the dependent variable?

Life Sciences

78. Fever A certain viral infection causes a fever that typically lasts 6 days. A model of the fever (in °F) on day x, $1 \leq x \leq 6$, is

$$F(x) = -\frac{2}{3}x^2 + \frac{14}{3}x + 96.$$

According to the model, on what day should the maximum fever occur? What is the maximum fever?

79. AIDS HIV is the virus responsible for AIDS. The following graph shows an estimation of the HIV infection rate in the United States.*

*"Reconstruction and Future Trends of the AIDS Epidemic in the United States" by Dr. Ronald Brookmeyer from *Science*, Volume 253, July 5, 1991, p. 37. Copyright © 1991 by the American Association for the Advancement of Science. Reprinted by permission of the AAAS and Dr. Brookmeyer.

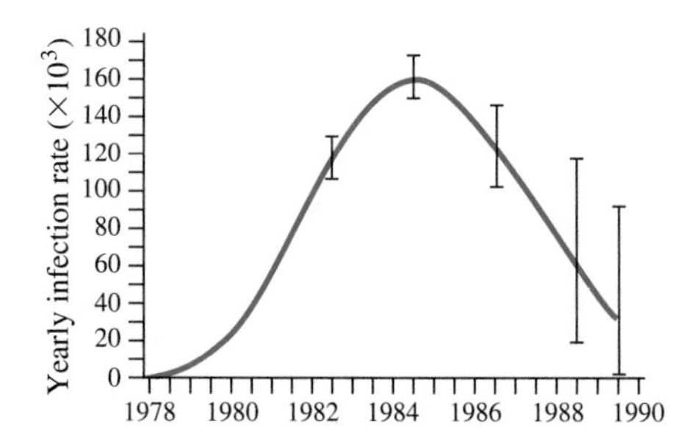

(a) When was the rate the highest?

(b) What was the maximum rate?

(c) What is the range of the function?

80. Pollution The cost to remove x percent of a pollutant is

$$y = \frac{7x}{100 - x},$$

in thousands of dollars.

Find the cost of removing each of the following percents of pollution.

(a) 80% (b) 50% (c) 90%

(d) Graph the function.

(e) Can all of the pollutant be removed?

81. Medicine Dosage Different rules exist for determining medicine dosages for children and adults. Two methods for finding dosages for children are given here. If the child's age in years is A, the adult dosage is d, and the child's dosage is c, then

$$c = \frac{A}{A + 12}d \quad \text{or} \quad c = \frac{A + 1}{24}d.$$

(a) At what age are the two dosages equal?

(b) Carefully graph $c = f(A) = \frac{A}{A + 12}$ and $c = g(A) = \frac{A + 1}{24}$ for the interval $2 \le A \le 13$ on the same axes by plotting points. Give the intervals where $f(A) > g(A)$ and where $g(A) > f(A)$.

(c) From the graphs in part (b), at what ages do the functions f and g appear to differ the most?

General Interest

82. Postage Assume that it costs 30¢ to mail a letter weighing one ounce or less, with each additional ounce, or portion of an ounce, costing 27¢. Let $C(x)$ represent the cost to mail a letter weighing x oz.

Find the costs of mailing letters of the following weights.

(a) 3.4 oz (b) 1.02 oz (c) 5.9 oz (d) 10 oz

(e) Graph C.

(f) Give the domain and range of C.

83. Automobiles The following graph shows the horsepower and torque for a 1991 Porsche 911 Turbo as a function of the engine speed.[†]

(a) Find the engine speed giving the maximum horsepower.

(b) Find the maximum horsepower.

(c) Find the torque when the horsepower is at its maximum.

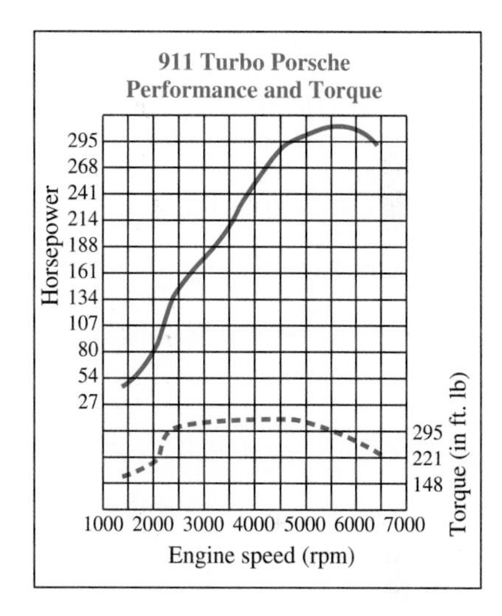

Connections

84. Cost Suppose the cost in dollars to produce x hundreds of nails is given by

$$C(x) = x^2 + 4x + 7.$$

(a) Sketch a graph of $C(x)$.

(b) Find a formula for $C(x + 1) - C(x)$, the cost to produce an additional hundred nails when x hundred are already produced. (This quantity is approximately equal to the marginal cost.)

(c) Find a formula for $A(x)$, the average cost per hundred nails.

(d) Find a formula for $A(x + 1) - A(x)$, the change in the average cost per nail when one additional batch of

[†]Porsche product information kit for the 1991 Turbo Performance and Torque 1991 model year. Reprinted by permission of Porsche Cars North America, Inc.

100 nails is produced. (This quantity is approximately equal to the marginal average cost, discussed in the chapter on the derivative.)

85. Cost Suppose the cost in dollars to produce x posters is given by

$$C(x) = \frac{5x + 3}{x + 1}.$$

(a) Sketch a graph of $C(x)$.

(b) Find a formula for $C(x + 1) - C(x)$, the cost to produce an additional poster when x posters are already produced.

(c) Find a formula for $A(x)$, the average cost per poster.

(d) Find a formula for $A(x + 1) - A(x)$, the change in the average cost per poster when one additional poster is produced. (This quantity is approximately equal to the marginal average cost, discussed in the chapter on the derivative.)

86. Salaries According to Hofstra University's 1991 faculty contract, the increase in faculty salaries in the fourth year of the contract would be the percent increase in the Consumer Price Index (CPI), increased or decreased by 1/4% for each 1% increase or decrease in student enrollment. The minimum increase would be 6%, while the maximum would be 8%.

Suppose the percent change in the CPI that year is 6.5%. Denote the percent increase in faculty salaries by $I(x)$, where x is the percent change in student enrollment.

(a) Find $I(-3)$.

(b) Find $I(-2)$.

(c) Find the percent increase in salaries if enrollment increases by 4%.

(d) Find the percent increase in salaries if enrollment increases by 7%.

(e) Graph $I(x)$ for $-3 \leq x \leq 7$.

(f) What is the range of I?

Extended Application / Marginal Cost—Booz, Allen & Hamilton

Booz, Allen & Hamilton Inc. is a large management consulting firm.* One service it provides to client companies is profitability studies showing ways in which the client can increase profit levels. The client company requesting the analysis presented in this case is a large producer of a staple food. The company buys from farmers and then processes the food in its mills, resulting in a finished product. The company sells some food at retail under its own brands and some in bulk to other companies who use the product in the manufacture of convenience foods.

The client company has been reasonably profitable in recent years, but the management retained Booz, Allen & Hamilton to see whether its consultants could suggest ways of increasing company profits. The management of the company had long operated with the philosophy of trying to process and sell as much of its product as possible, since, they felt, this would lower the average processing cost per unit sold. The consultants found, however, that the client's fixed mill costs were quite low, and that, in fact, processing extra units made the cost per unit start to increase. (There are several reasons for this: the company must run three shifts, machines break down more often, and so on.)

In this application, we shall discuss the marginal cost of two of the company's products. The marginal cost (approximate cost of producing an extra unit) of production for product A was found by the consultants to be approximated by the linear function

$$y = .133x + 10.09,$$

where x is the number of units produced (in millions) and y is the marginal cost. (Here the marginal cost is not a constant, as it was in the examples in the text.)

For example, at a level of production of 3.1 million units, an additional unit of product A would cost about

$$y = .133(3.1) + 10.09 \approx \$10.50.$$

At a level of production of 5.7 million units, an extra unit costs $10.85. Figure 52 shows a graph of the marginal cost function from $x = 3.1$ to $x = 5.7$, the domain over which the function above was found to apply.

The selling price for product A is $10.73 per unit, so that, as shown in Figure 52, the company was losing money on many units of the product that it sold. Since the selling price could

*This case was supplied by John R. Dowdle of the Chicago office of Booz, Allen & Hamilton Inc. Reprinted by permission.

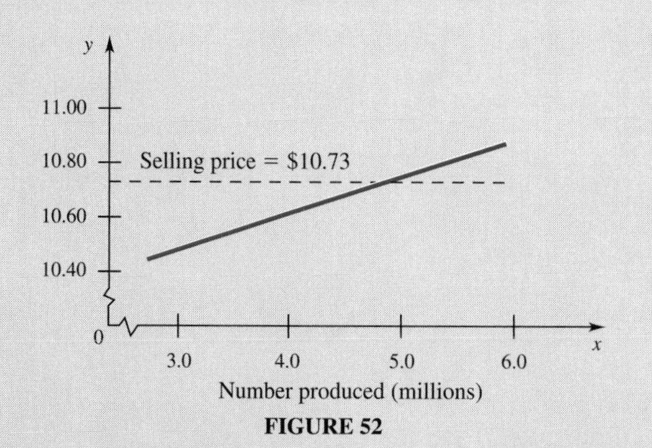

FIGURE 52

not be raised if the company were to remain competitive, the consultants recommended that production of product A be cut.

For product B, the Booz, Allen & Hamilton consultants found a marginal cost function given by

$$y = .0667x + 10.29,$$

with x and y as defined above. Verify that at a production level of 3.1 million units, the marginal cost is about \$10.50, while at a production level of 5.7 million units, the marginal cost is about \$10.67. Since the selling price of this product is \$9.65, the consultants again recommended a cutback in production.

The consultants ran similar cost analyses of other products made by the company and then issued their recommendation: the company should reduce total production by 2.1 million units. The analysts predicted that this would raise profits for the products under discussion from \$8.3 million annually to \$9.6 million—which is very close to what actually happened when the client took the advice.

Exercises

1. At what level of production, x, was the marginal cost of a unit of product A equal to the selling price?

2. Graph the marginal cost function for product B from $x = 3.1$ million units to $x = 5.7$ million units.

3. Find the number of units for which marginal costs equals the selling price for product B.

4. For product C, the marginal cost of production is

$$y = .133x + 9.46.$$

(a) Find the marginal costs at production levels of 3.1 million units and 5.7 million units.

(b) Graph the marginal cost function.

(c) For a selling price of \$9.57, find the production level for which the cost equals the selling price.

Applications at a glance. . .

Penalties for Polluters . . . standard of living in the United States . . . learning curve . . . price of an essential commodity . . .

CHAPTER 2

THE DERIVATIVE

(See page 92.)

In Chapter 1 we were concerned with static situations. We saw how to answer questions like the following:

How much would 100 items cost?

In how many years will the population reach 10,000?

How wide is the arch 10 meters from the ground?

With calculus, we will be able to solve problems that involve changing situations.

How fast is the car moving after 20 seconds?

At what rate is the population growing?

What is the rate of change of profit when sales reach $1000?

Solving each of these problems requires a *derivative*. Since the definition of the derivative involves the idea of a *limit*, we begin with limits.

2.1 LIMITS AND CONTINUITY

Evaluating function notation, such as $f(1) = 3$, was discussed in Section 1.1. Verify that for

$$f(x) = \frac{x^2 - 4}{x - 2},$$

$f(0) = 2$, $f(3) = 5$, and $f(-1) = 1$.

We can find the value of the function defined by

$$f(x) = \frac{x^2 - 4}{x - 2}$$

when $x = 1$ by substitution:

$$f(1) = \frac{1^2 - 4}{1 - 2} = 3.$$

For values of x close to 1, the values of $f(x)$ are close to $f(1) = 3$, as the following table shows.

	x approaches 1 from the left.				↓	x approaches 1 from the right.					
x	.8	.9	.99	.9999	1	1.0000001	1.0001	1.001	1.01	1.05	1.1
$f(x)$	2.8	2.9	2.99	2.9999	3	3.0000001	3.0001	3.001	3.01	3.05	3.1
	$f(x)$ approaches 3.				↑	$f(x)$ approaches 3.					

At $x = 2$, however, $f(2)$ is not defined because the denominator is zero when $x = 2$. For values of x close to (but not equal to) 2, $f(x)$ *is* defined. The following table shows some values of $f(x)$ for x-values closer and closer to 2.

	x approaches 2 from the left.				↓	x approaches 2 from the right.				
x	1.8	1.9	1.99	1.9999	2	2.0000001	2.00001	2.001	2.05	2.1
$f(x)$	3.8	3.9	3.99	3.9999	4	4.0000001	4.00001	4.001	4.05	4.1
	$f(x)$ approaches 4.				↑	$f(x)$ approaches 4.				

The table suggests that, as x gets closer and closer to 2 from either side, $f(x)$ gets closer and closer to 4. In fact, using a calculator, we can make $f(x)$ as close to 4 as we wish by choosing a value of x "close enough" to 2. In such a case, we say "the limit of $f(x)$ as x approaches 2 is 4," which is written as

$$\lim_{x \to 2} f(x) = 4.$$

In the first example, we found that

$$\lim_{x \to 1} f(x) = 3,$$

because the values of $f(x)$ got closer and closer to 3 as x got closer and closer to 1, *from either side* of 1.

The phrase "x approaches 1 from the left" is written $x \to 1^-$. Similarly, "x approaches 1 from the right" is written $x \to 1^+$. These expressions are used to write **one-sided limits.** The **limit from the left** is written

$$\lim_{x \to 1^-} f(x) = 3,$$

and the **limit from the right** is written

$$\lim_{x \to 1^+} f(x) = 3.$$

A **two-sided limit,** such as

$$\lim_{x \to 1} f(x) = 3,$$

exists only if both one-sided limits are the same; that is, if $f(x)$ approaches the same number as x approaches a given number from *either* side.

The examples suggest the following informal definition.

LIMIT OF A FUNCTION

Let f be a function and let a and L be real numbers. If

1. as x takes values closer and closer (but not equal) to a on both sides of a, the corresponding values of $f(x)$ get closer and closer (and perhaps equal) to L; and
2. the value of $f(x)$ can be made as close to L as desired by taking values of x close enough to a;

then L is the **limit** of $f(x)$ as x approaches a, written

$$\lim_{x \to a} f(x) = L.$$

This definition is informal because the expressions "closer and closer to" and "as close as desired" have not been defined. A more formal definition would be needed to prove the rules for limits given later in this section.

Note The definition of a limit describes what happens to $f(x)$ when x is near a. It is not affected by whether or not $f(a)$ is defined. Also, the definition implies that the function values cannot approach two different numbers, so that if a limit exists, it is unique.

Find $\lim_{x\to 2} g(x)$, where $g(x) = \dfrac{x^2 + 4}{x - 2}$.

Make a table of values.

x approaches 2 from the left. → ← *x* approaches 2 from the right.

x	1.8	1.9	1.99	1.999	2	2.001	2.01	2.05
$g(x)$	-36.2	-76.1	-796	-7996	**Limit does not exist.**	8004	804	164

$g(x)$ gets smaller and smaller. → ← $g(x)$ gets larger and larger.

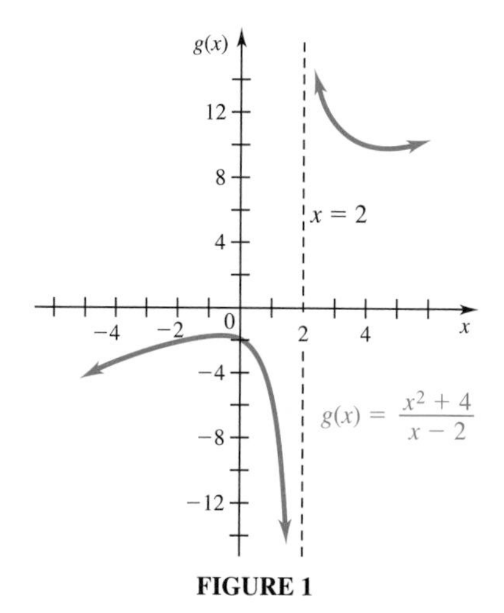

FIGURE 1

Both the table above and the graph in Figure 1 suggest that as $x \to 2$ from the left, $g(x)$ gets more and more negative, hence, smaller and smaller. This is indicated by writing

$$\lim_{x\to 2^-} g(x) = -\infty.$$

The symbol $-\infty$ does not represent a real number; it simply indicates that as $x \to 2^-$, $g(x)$ gets smaller without bound. In the same way, the behavior of the function as $x \to 2$ from the right is indicated by writing

$$\lim_{x\to 2^+} g(x) = \infty.$$

Since there is no real number that $g(x)$ approaches as $x \to 2$ (from either side),

$$\lim_{x\to 2} \frac{x^2 + 4}{x - 2} \text{ does not exist.}$$ ◀

Find $\lim_{x\to 0} \dfrac{|x|}{x}$.

The function $f(x) = |x|/x$ is not defined when $x = 0$. When $x > 0$, the definition of absolute value shows that $f(x) = |x|/x = x/x = 1$. When $x < 0$, then $|x| = -x$ and $f(x) = -x/x = -1$. The graph of f is shown in Figure 2.

As x approaches 0 from the right, x is always positive and the corresponding value of $f(x)$ is 1, so

$$\lim_{x\to 0^+} f(x) = 1.$$

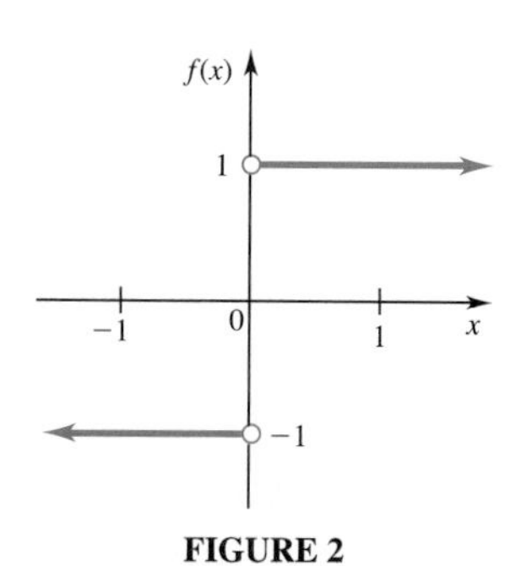

FIGURE 2

But as x approaches 0 from the left, x is always negative and the corresponding value of $f(x)$ is -1, so

$$\lim_{x\to 0^-} f(x) = -1.$$

Thus, as x approaches 0 from either side, the corresponding values of $f(x)$ do not get closer and closer to a *single* real number. Therefore, the limit of $|x|/x$ as x approaches 0 does not exist. ◀

The discussion up to this point can be summarized as follows.

EXISTENCE OF LIMITS

The limit of f as x approaches a may not exist.

1. If $f(x)$ becomes infinitely large or infinitely small as x approaches the number a from either side, we write $\lim_{x \to a} f(x) = \infty$ or $\lim_{x \to a} f(x) = -\infty$. In either case, the limit does not exist.

2. If $\lim_{x \to a^-} f(x) = L$ and $\lim_{x \to a^+} f(x) = M$, and $L \neq M$, then $\lim_{x \to a} f(x)$ does not exist.

Figure 3 illustrates the facts listed above.

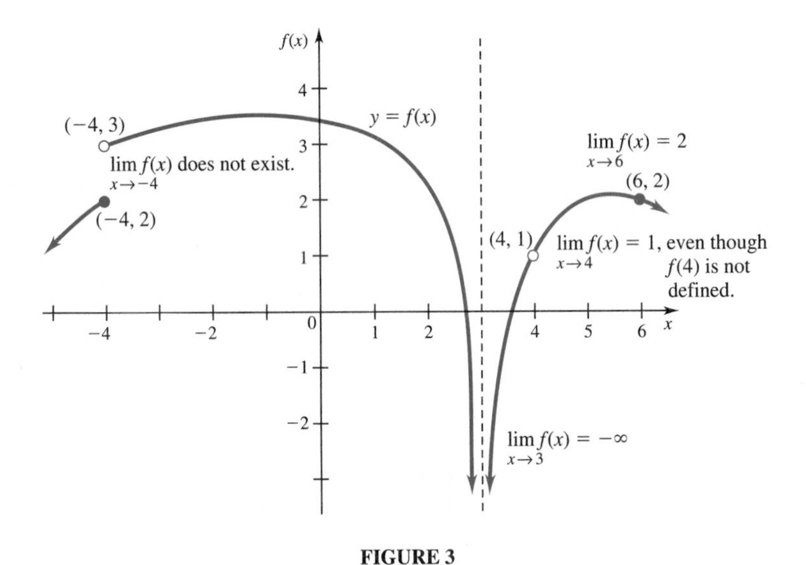

FIGURE 3

Rules for Limits As shown by the examples above, tables and graphs can be used to find limits. However, it is more efficient to use the rules for limits given on the next page. (Proofs of these rules require a formal definition of limit, which we have not given.)

RULES FOR LIMITS

Let a, k, n, A, and B be real numbers, and let f and g be functions such that

$$\lim_{x \to a} f(x) = A \qquad \text{and} \qquad \lim_{x \to a} g(x) = B.$$

1. If k is a constant, then $\lim_{x \to a} k = k$ and $\lim_{x \to a} k \cdot f(x) = k \cdot \lim_{x \to a} f(x)$.
2. $\lim_{x \to a} [f(x) \pm g(x)] = \lim_{x \to a} f(x) \pm \lim_{x \to a} g(x) = A \pm B$

   (The limit of a sum or difference is the sum or difference of the limits.)
3. If $p(x)$ is a polynomial, then $\lim_{x \to a} p(x) = p(a)$.
4. $\lim_{x \to a} [f(x) \cdot g(x)] = [\lim_{x \to a} f(x)] \cdot [\lim_{x \to a} g(x)] = A \cdot B$

   (The limit of a product is the product of the limits.)
5. $\lim_{x \to a} \dfrac{f(x)}{g(x)} = \dfrac{\lim_{x \to a} f(x)}{\lim_{x \to a} g(x)} = \dfrac{A}{B}$ if $B \neq 0$

   (The limit of a quotient is the quotient of the limits, provided the limit of the denominator is not zero.)
6. For any real number n, $\lim_{x \to a} [f(x)]^n = [\lim_{x \to a} f(x)]^n = A^n$.
7. $\lim_{x \to a} f(x) = \lim_{x \to a} g(x)$ if $f(x) = g(x)$ for all $x \neq a$.

This list may seem imposing, but most limit problems have solutions that agree with your common sense. Algebraic techniques and tables of values can also help resolve questions about limits, as some of the following examples will show.

EXAMPLE 3 Find each limit.

(a) $\lim_{x \to 3} [(x^2 + 2x) + (3x^3 - 5x + 1)]$

$$\begin{aligned}
&\lim_{x \to 3} [(x^2 + 2x) + (3x^3 - 5x + 1)] \\
&= \lim_{x \to 3} (x^2 + 2x) + \lim_{x \to 3} (3x^3 - 5x + 1) && \text{Rule 2} \\
&= [3^2 + 2(3)] + [3(3^3) - 5(3) + 1] && \text{Rule 3} \\
&= 15 + 67 \\
&= 82
\end{aligned}$$

(b) $\lim_{x \to -2} (x^2 - 5)(3x + 1)$

$$\begin{aligned}\lim_{x \to -2} (x^2 - 5)(3x + 1) \\ &= [\lim_{x \to -2} (x^2 - 5)][\lim_{x \to -2} (3x + 1)] && \text{Rule 4} \\ &= [(-2)^2 - 5][3(-2) + 1] && \text{Rule 3} \\ &= (-1)(-5) \\ &= 5\end{aligned}$$

(c) $\lim_{x \to -1} 5(3x^2 + 2)$

$$\begin{aligned}\lim_{x \to -1} 5(3x^2 + 2) &= 5 \lim_{x \to -1} (3x^2 + 2) && \text{Rule 1} \\ &= 5\,[3(-1)^2 + 2] && \text{Rule 3} \\ &= 25\end{aligned}$$

(d) $\lim_{x \to 4} \dfrac{x}{x + 2}$

$$\begin{aligned}\lim_{x \to 4} \frac{x}{x + 2} &= \frac{\lim_{x \to 4} x}{\lim_{x \to 4} (x + 2)} && \text{Rule 5} \\ &= \frac{4}{4 + 2} = \frac{2}{3}\end{aligned}$$

(e) $\lim_{x \to 9} \sqrt{4x - 11}$

As $x \to 9$, the expression $4x - 11$ approaches $4 \cdot 9 - 11 = 25$. Using Rule 6, with $n = 1/2$, gives

$$\lim_{x \to 9} (4x - 11)^{1/2} = [\lim_{x \to 9} (4x - 11)]^{1/2} = \sqrt{25} = 5. \blacktriangleleft$$

As Example 3 suggests, the rules for limits actually mean that many limits can be found simply by evaluation. The next examples illustrate some exceptions.

Find $\lim_{x \to 2} \dfrac{x^2 + x - 6}{x - 2}$.

Rule 5 cannot be used here, since

$$\lim_{x \to 2} (x - 2) = 0.$$

For $x \neq 2$, we can, however, simplify the function by rewriting the fraction as

$$\frac{x^2 + x - 6}{x - 2} = \frac{(x + 3)(x - 2)}{x - 2} = x + 3.$$

Now Rule 7 can be used.

$$\lim_{x \to 2} \frac{x^2 + x - 6}{x - 2} = \lim_{x \to 2} (x + 3) = 2 + 3 = 5 \blacktriangleleft$$

Find $\lim_{x \to 4} \frac{\sqrt{x} - 2}{x - 4}$.

As $x \to 4$, the numerator approaches 0 and the denominator also approaches 0, giving the meaningless expression 0/0. Algebra can be used to rationalize the numerator by multiplying both the numerator and the denominator by $\sqrt{x} + 2$. This gives

$$\frac{\sqrt{x} - 2}{x - 4} \cdot \frac{\sqrt{x} + 2}{\sqrt{x} + 2} = \frac{\sqrt{x} \cdot \sqrt{x} - 2\sqrt{x} + 2\sqrt{x} - 4}{(x - 4)(\sqrt{x} + 2)}$$

$$= \frac{x - 4}{(x - 4)(\sqrt{x} + 2)} = \frac{1}{\sqrt{x} + 2}$$

if $x \neq 4$. Now use rules for limits.

$$\lim_{x \to 4} \frac{\sqrt{x} - 2}{x - 4} = \lim_{x \to 4} \frac{1}{\sqrt{x} + 2} = \frac{1}{\sqrt{4} + 2} = \frac{1}{2 + 2} = \frac{1}{4} \blacktriangleleft$$

Find $\lim_{x \to 1} \frac{x + 1}{x^2 - 1}$.

Again, Rule 5 cannot be used since $\lim_{x \to 1} x^2 - 1 = 0$. If $x \neq 1$, the function can be rewritten as

$$\frac{x + 1}{x^2 - 1} = \frac{x + 1}{(x + 1)(x - 1)} = \frac{1}{x - 1}.$$

Then

$$\lim_{x \to 1} \frac{x + 1}{x^2 - 1} = \lim_{x \to 1} \frac{1}{x - 1}$$

by Rule 7. None of the rules can be used to find

$$\lim_{x \to 1} \frac{1}{x - 1},$$

but a table of values would show that the values of $1/(x - 1)$ increase as $x \to 1^+$ and the values of $1/(x - 1)$ decrease as $x \to 1^-$. Therefore,

$$\lim_{x \to 1} \frac{1}{x - 1} \text{ does not exist. } \blacktriangleleft$$

Continuity Sometimes $\lim_{x \to a} f(x)$ and $f(a)$ both exist, but are not equal, as in the next example.

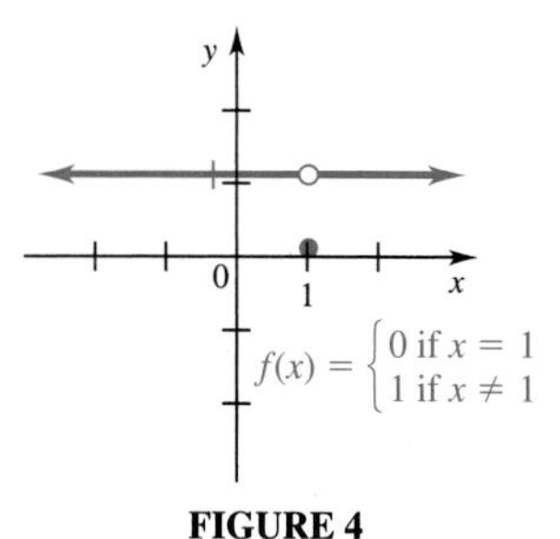

FIGURE 4

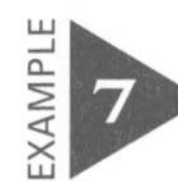

Let f be the function defined by

$$f(x) = \begin{cases} 0 & \text{if } x = 1 \\ 1 & \text{if } x \neq 1 \end{cases}$$

and graphed in Figure 4.

The graph shows that $\lim_{x \to 1} f(x) = 1$, but $f(1) = 0$. For this function,

$$\lim_{x \to 1} f(x) \neq f(1).$$ ◀

The function in Example 7 is said to be *discontinuous at* $x = 1$. A function is discontinuous at some x-value if you must lift your pencil from the paper at that point to sketch the graph. The graph of the function discussed earlier, with

$$f(x) = \frac{x^2 - 4}{x - 2},$$

is shown in Figure 5. The function is discontinuous at $x = 2$, as shown by the "hole" in the graph, because $f(2)$ is undefined. Also, the function g, whose graph is shown in Figure 1, is discontinuous at $x = 2$, because $g(2)$ is undefined. These examples suggest the next definition.

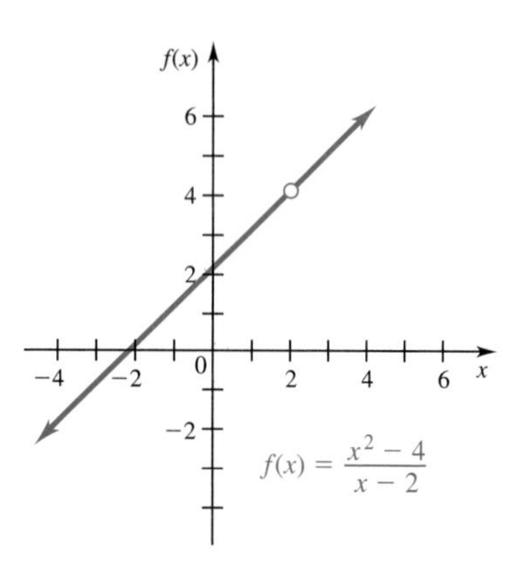

FIGURE 5

DEFINITION OF CONTINUITY AT A POINT

A function f is **continuous** at c if the following three conditions are satisfied:

1. $f(c)$ is defined, **2.** $\lim_{x \to c} f(x)$ exists, and **3.** $\lim_{x \to c} f(x) = f(c)$.

If f is not continuous at c, it is **discontinuous** there.

EXAMPLE 8 Tell why the functions are discontinuous at the indicated points.

(a) $f(x)$ in Figure 6 at $x = 3$

The open circle on the graph of Figure 6 at the point where $x = 3$ means that $f(3)$ is not defined. Because of this, part (1) of the definition fails.

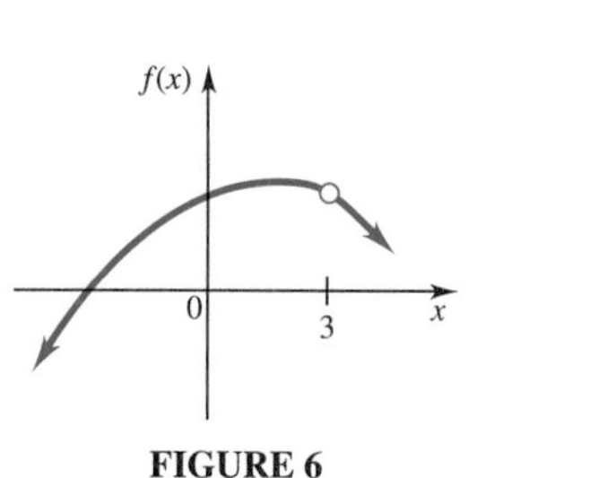

FIGURE 6

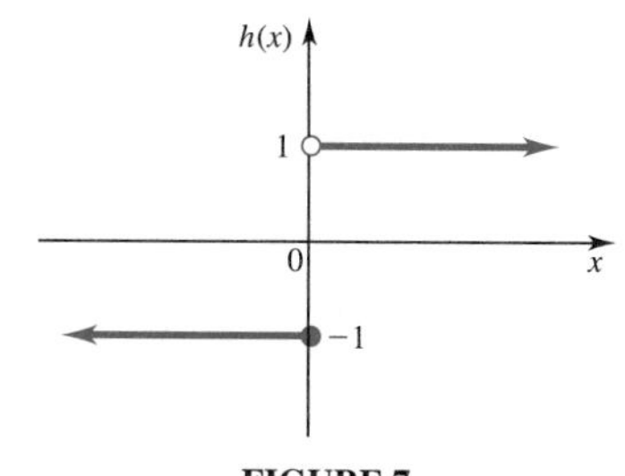

FIGURE 7

(b) $h(x)$ in Figure 7 at $x = 0$

According to the graph of Figure 7, $h(0) = -1$. Also, as x approaches 0 from the left, $h(x)$ is -1. As x approaches 0 from the right, however, $h(x)$ is 1. Since there is no single number approached by the values of $h(x)$ as x approaches 0, $\lim_{x \to 0} h(x)$ does not exist, and part (2) of the definition fails.

(c) $g(x)$ in Figure 8 at $x = 4$

In Figure 8, the heavy dot above 4 shows that $g(4)$ is defined. In fact, $g(4) = 1$. The graph also shows, however, that

$$\lim_{x \to 4} g(x) = -2,$$

so $\lim_{x \to 4} g(x) \neq g(4)$, and part (3) of the definition fails.

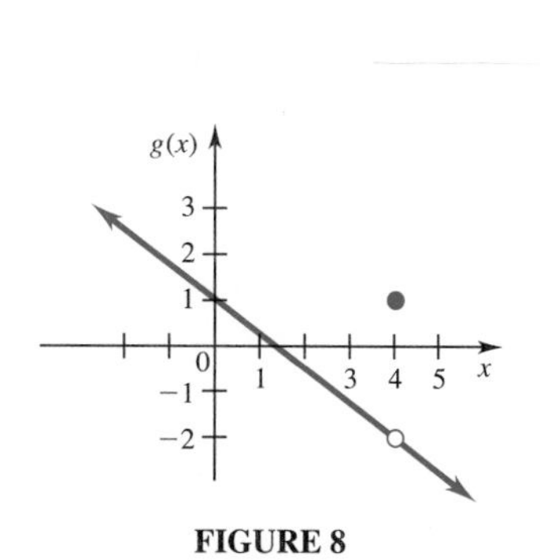

FIGURE 8

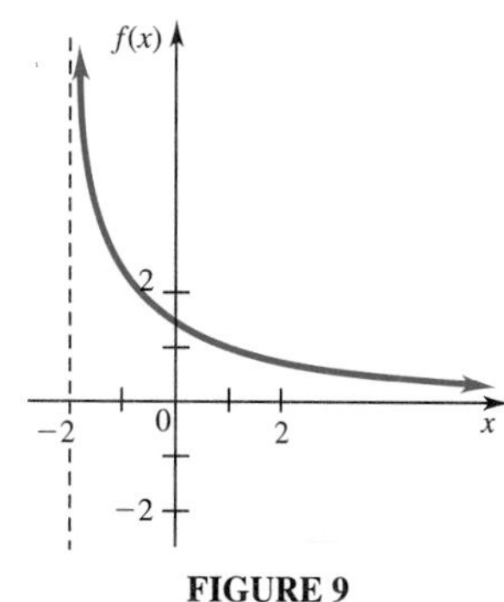

FIGURE 9

(d) $f(x)$ in Figure 9 at $x = -2$

The function f graphed in Figure 9 is not defined at -2, and $\lim_{x \to -2} f(x)$ does not exist. Either of these reasons is sufficient to show that f is not continuous at -2. (Function f *is* continuous at any value of x greater than -2, however.) ◀

The next table lists some key functions and tells where each is continuous.

Continuous Functions

Type of Function	Where It Is Continuous	Graphic Example
Polynomial Function $y = a_nx^n + a_{n-1}x^{n-1} + \cdots + a_1x + a_0$, where $a_n, a_{n-1}, \ldots, a_1, a_0$ are real numbers, not all 0	For all x	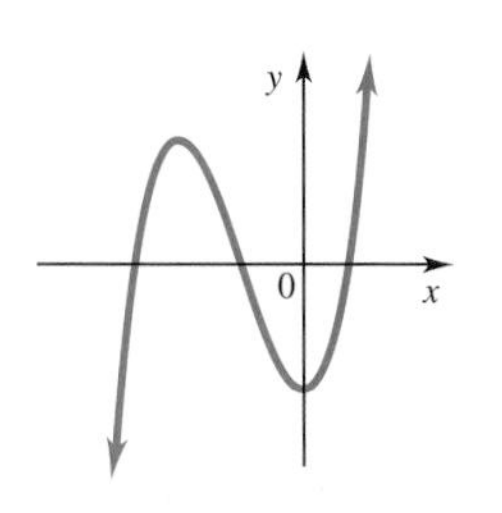
Rational Function $y = \dfrac{p(x)}{q(x)}$, where $p(x)$ and $q(x)$ are polynomials, with $q(x) \neq 0$	For all x where $q(x) \neq 0$	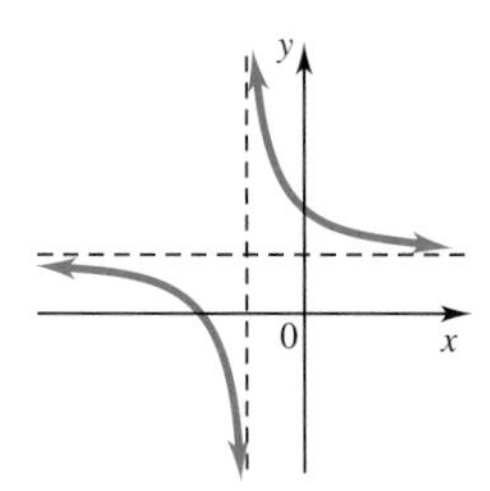
Root Function $y = \sqrt{ax + b}$, where a and b are real numbers, with $a \neq 0$ and $ax + b \geq 0$	For all x where $ax + b > 0$	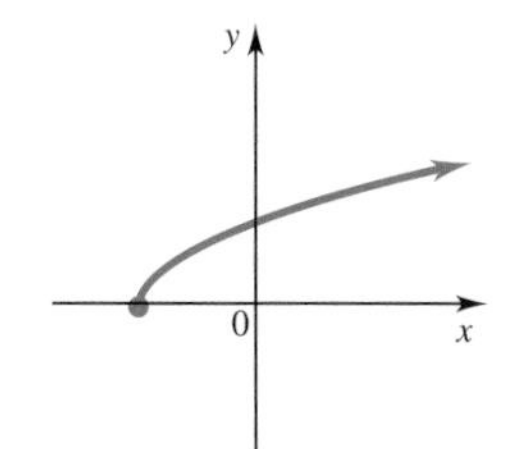

EXAMPLE 9 A trailer rental firm charges a flat \$8 to rent a hitch. The trailer itself is rented for \$22 per day or fraction of a day. Let $C(x)$ represent the cost of renting a hitch and trailer for x days.

(a) Graph C.

The charge for one day is \$8 for the hitch and \$22 for the trailer, or \$30. In fact, if $0 < x \leq 1$, then $C(x) = 30$. To rent the trailer for more than one day, but not more than two days, the charge is $8 + 2 \cdot 22 = 52$ dollars. For any value of x satisfying $1 < x \leq 2$, $C(x) = 52$. Also, if $2 < x \leq 3$, then $C(x) = 74$. These results lead to the graph in Figure 10.

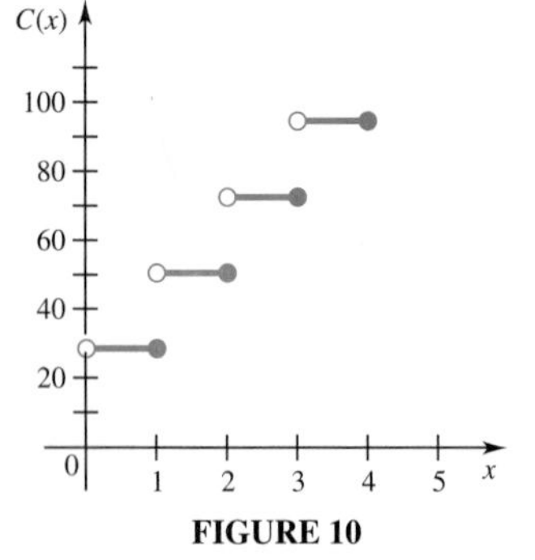

FIGURE 10

(b) Find any points of discontinuity for C.

As the graph suggests, C is discontinuous at $x = 1, 2, 3, 4$, and all other positive integers.

2.1 Exercises

Decide whether each limit exists. If a limit exists, find its value.

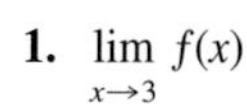

1. $\lim_{x \to 3} f(x)$

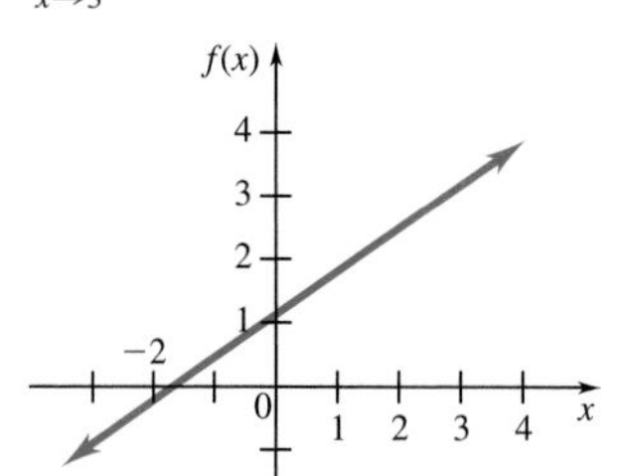

2. $\lim_{x \to 2} F(x)$

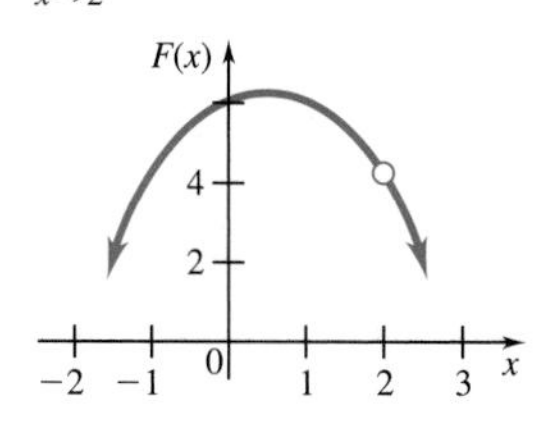

3. $\lim_{x \to -2} f(x)$

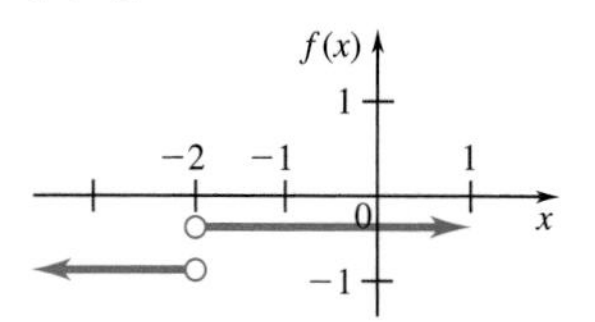

4. $\lim_{x \to 3} g(x)$

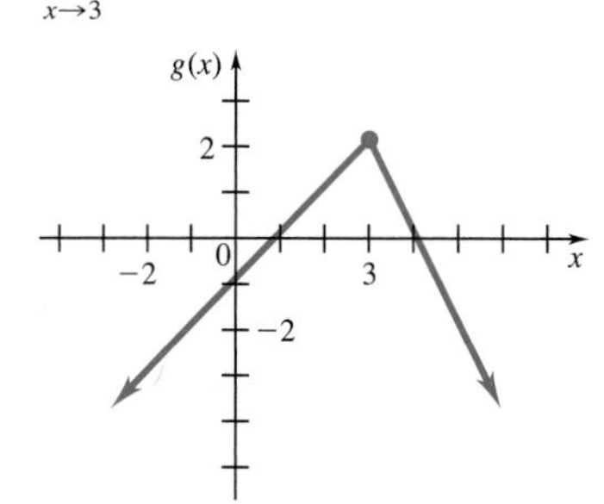

5. $\lim_{x \to 1} g(x)$

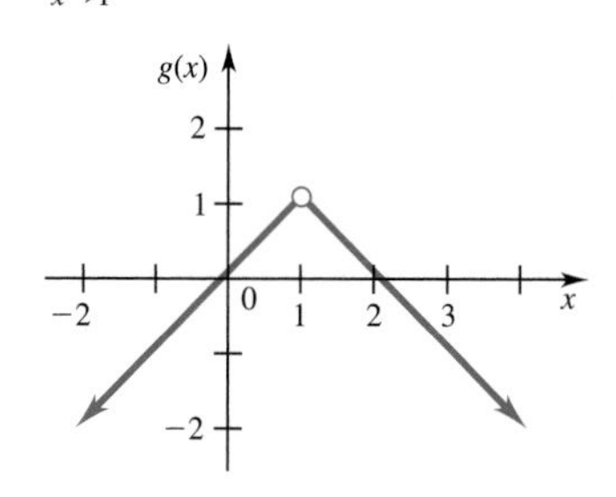

6. $\lim_{x \to 0} f(x)$

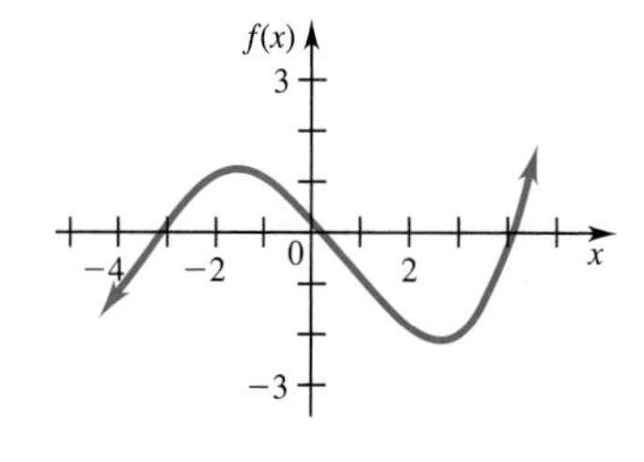

7. Explain why $\lim_{x \to 2} F(x)$ in Exercise 2 exists, but $\lim_{x \to -2} f(x)$ in Exercise 3 does not.

8. In Exercise 5, why does $\lim_{x \to 1} g(x)$ exist, even though $g(1)$ is undefined?

9. Use the table of values below to estimate $\lim_{x \to 1} f(x)$.

x	.9	.99	.999	.9999	1.0001	1.001	1.01	1.1
$f(x)$	1.9	1.99	1.999	1.9999	2.0001	2.001	2.01	2.1

Complete the tables and use the results to find the indicated limits. (You will need a calculator with a $\sqrt{x}$ *key or a computer for Exercises 13 and 14.)*

10. If $f(x) = 2x^2 - 4x + 3$, find $\lim_{x \to 1} f(x)$.

x	.9	.99	.999	1.001	1.01	1.1
$f(x)$			1.000002	1.000002		

11. If $k(x) = \dfrac{x^3 - 2x - 4}{x - 2}$, find $\lim_{x \to 2} k(x)$.

x	1.9	1.99	1.999	2.001	2.01	2.1
$k(x)$						

12. If $f(x) = \dfrac{2x^3 + 3x^2 - 4x - 5}{x + 1}$, find $\lim_{x \to -1} f(x)$.

x	−1.1	−1.01	−1.001	−.999	−.99	−.9
$f(x)$						

13. If $h(x) = \dfrac{\sqrt{x} - 2}{x - 1}$, find $\lim_{x \to 1} h(x)$.

x	.9	.99	.999	1.001	1.01	1.1
$h(x)$						

14. If $f(x) = \dfrac{\sqrt{x} - 3}{x - 3}$, find $\lim_{x \to 3} f(x)$.

x	2.9	2.99	2.999	3.001	3.01	3.1
$f(x)$						

Let $\lim_{x \to 4} f(x) = 16$ *and* $\lim_{x \to 4} g(x) = 8$. *Use the limit rules to find the following limits.*

15. $\lim_{x \to 4} [f(x) - g(x)]$ **16.** $\lim_{x \to 4} [g(x) \cdot f(x)]$ **17.** $\lim_{x \to 4} \dfrac{f(x)}{g(x)}$ **18.** $\lim_{x \to 4} [3 \cdot f(x)]$ **19.** $\lim_{x \to 4} \sqrt{f(x)}$

20. $\lim_{x \to 4} \sqrt[3]{g(x)}$ **21.** $\lim_{x \to 4} [g(x)]^3$ **22.** $\lim_{x \to 4} [1 + f(x)]^2$ **23.** $\lim_{x \to 4} \dfrac{f(x) + g(x)}{2g(x)}$ **24.** $\lim_{x \to 4} \dfrac{5g(x) + 2}{1 - f(x)}$

Use the properties of limits to help decide whether the following limits exist. If a limit exists, find its value.

25. $\lim_{x \to 3} \dfrac{x^2 - 9}{x - 3}$ **26.** $\lim_{x \to -2} \dfrac{x^2 - 4}{x + 2}$ **27.** $\lim_{x \to -2} \dfrac{x^2 - x - 6}{x + 2}$

28. $\lim_{x \to 5} \dfrac{x^2 - 3x - 10}{x - 5}$ **29.** $\lim_{x \to 0} \dfrac{x^3 - 4x^2 + 8x}{2x}$ **30.** $\lim_{x \to 0} \dfrac{-x^5 - 9x^3 + 8x^2}{5x}$

31. $\lim_{x \to 0} \dfrac{[1/(x + 3)] - 1/3}{x}$ **32.** $\lim_{x \to 0} \dfrac{[-1/(x + 2)] + 1/2}{x}$ **33.** $\lim_{x \to 25} \dfrac{\sqrt{x} - 5}{x - 25}$

34. $\lim_{x \to 36} \dfrac{\sqrt{x} - 6}{x - 36}$ **35.** $\lim_{h \to 0} \dfrac{(x + h)^2 - x^2}{h}$ **36.** $\lim_{h \to 0} \dfrac{(x + h)^3 - x^3}{h}$

37. Let $F(x) = \dfrac{3x}{(x + 2)^3}$. **(a)** Find $\lim_{x \to -2} F(x)$. **(b)** Find the vertical asymptote of the graph of $F(x)$.
(c) Compare your answers for parts (a) and (b). What can you conclude?

38. Let $G(x) = \dfrac{-6}{(x - 4)^2}$. **(a)** Find $\lim_{x \to 4} G(x)$. **(b)** Find the vertical asymptote of the graph of $G(x)$.
(c) Compare your answers for parts (a) and (b). Are they related? How?

Find all points $x = a$ *where the function is discontinuous. For each point of discontinuity, give* $f(a)$ *and* $\lim_{x \to a} f(x)$.

39.

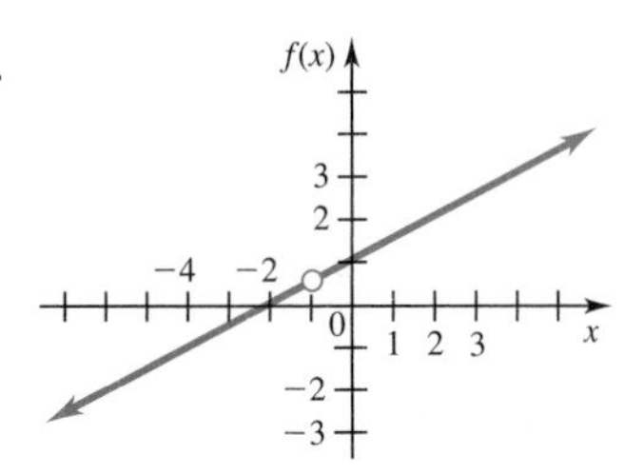

40.

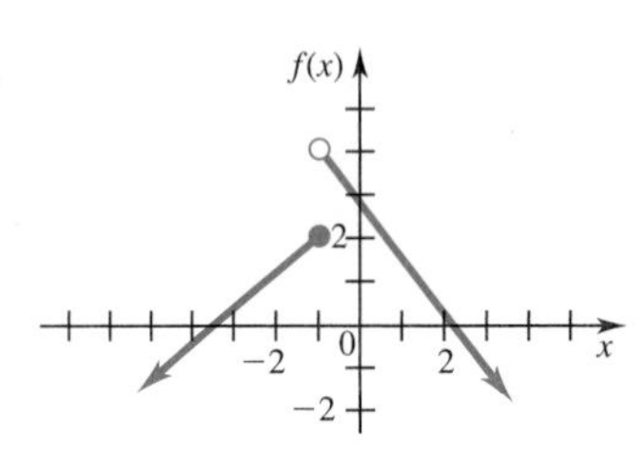

41.

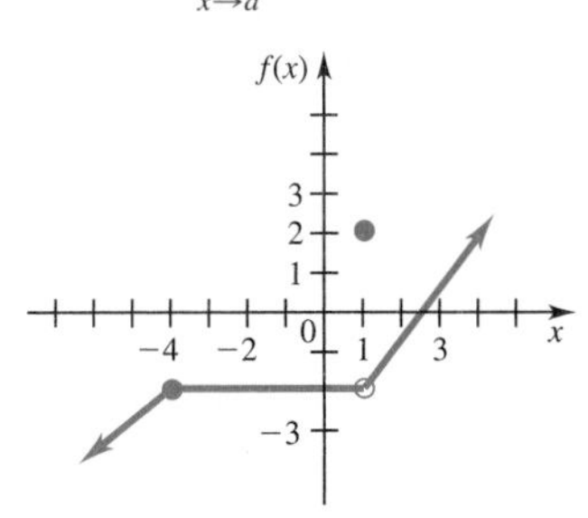

42.

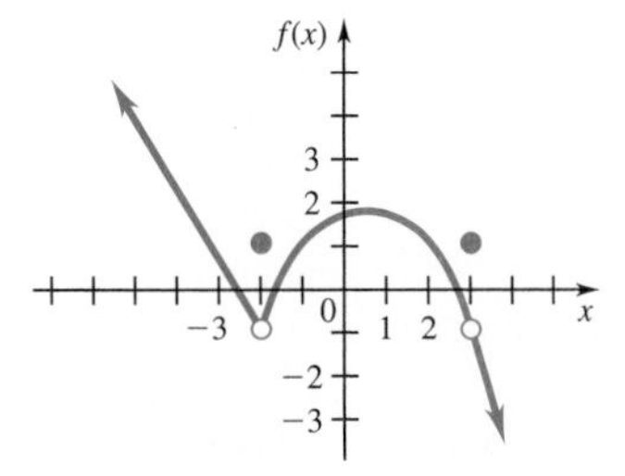

43.

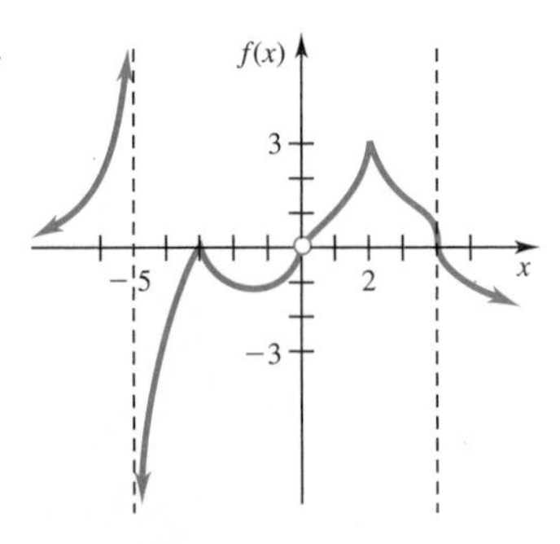

44.

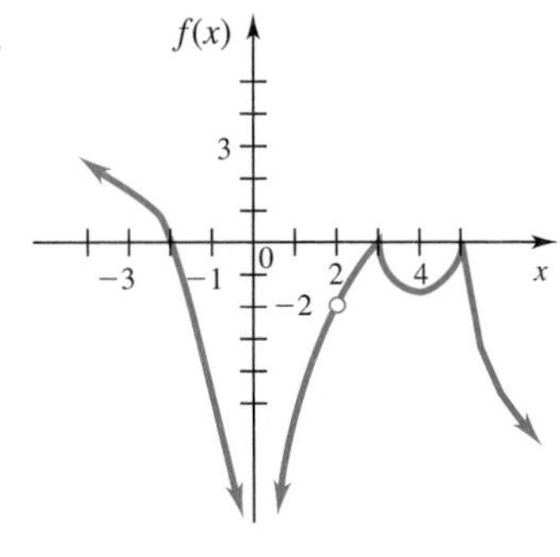

Decide whether each of the following is continuous at the given values of x.

45. $g(x) = \dfrac{1}{x(x-2)}$; $x = 0, 2, 4$

46. $g(x) = \dfrac{-2x}{(2x+1)(3x+6)}$; $x = 0, -1/2, -2$

47. $k(x) = \dfrac{5+x}{2+x}$; $x = 0, -2, -5$

48. $f(x) = \dfrac{4-x}{x-9}$; $x = 0, 4, 9$

49. $g(x) = \dfrac{x^2-4}{x-2}$; $x = 0, 2, -2$

50. $h(x) = \dfrac{x^2-25}{x+5}$; $x = 0, 5, -5$

51. $p(x) = x^2 - 4x + 11$; $x = 0, 2, -1$

52. $q(x) = -3x^3 + 2x^2 - 4x + 1$; $x = -2, 3, 1$

53. $p(x) = \dfrac{|x+2|}{x+2}$; $x = -2, 0, 2$

54. $r(x) = \dfrac{|5-x|}{x-5}$; $x = -5, 0, 5$

Applications

Business and Economics

55. Consumer Demand When the price of an essential commodity (such as gasoline) rises rapidly, consumption drops slowly at first. If the price continues to rise, however, a "tipping" point may be reached, at which consumption takes a sudden, substantial drop. Suppose the accompanying graph shows the consumption of gasoline, $G(t)$, in millions of gallons, in a certain area. We assume that the price is rising rapidly. Here t is time in months after the price began rising. Use the graph to find the following.

(a) $\lim_{t \to 12} G(t)$ **(b)** $\lim_{t \to 16} G(t)$ **(c)** $G(16)$

(d) the tipping point (in months)

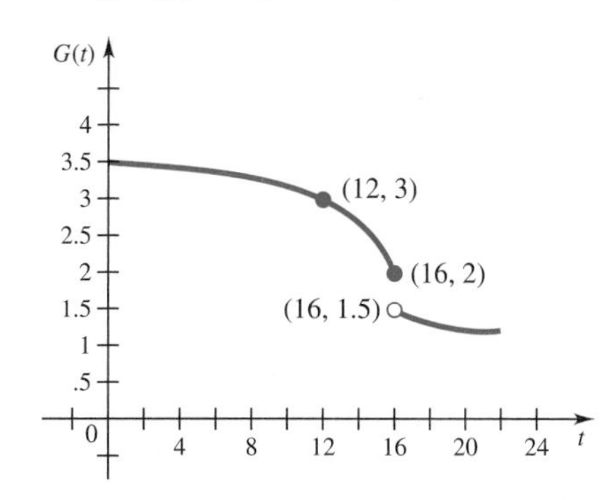

56. Production The graph shows the profit from the daily production of x thousand kilograms of an industrial chemical.
Use the graph to find the following limits.

(a) $\lim_{x \to 6} P(x)$ **(b)** $\lim_{x \to 10^-} P(x)$ **(c)** $\lim_{x \to 10^+} P(x)$

(d) $\lim_{x \to 10} P(x)$

(e) Where is the function discontinuous? What might account for such a discontinuity?

(f) Use the graph to estimate the number of units of the chemical that must be produced before the second shift is as profitable as the first.

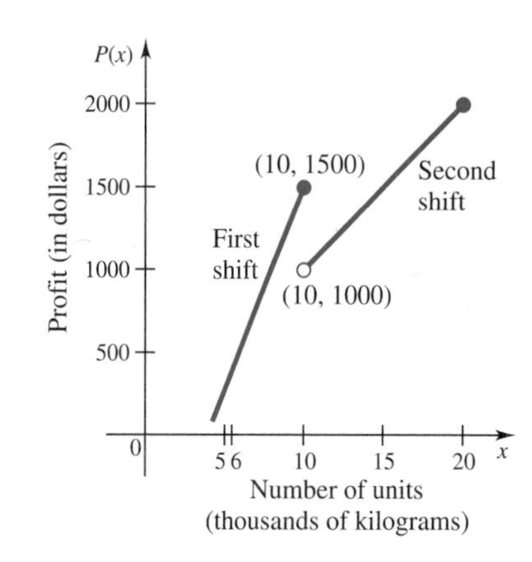

57. Cost Analysis The cost to transport a mobile home depends on the distance, x, in miles that the home is moved. Let $C(x)$ represent the cost to move a mobile home x miles. One firm charges as follows.

Cost per Mile	Distance in Miles
\$4.00	$0 < x \le 150$
\$3.00	$150 < x \le 400$
\$2.50	$400 < x$

Find the cost to move a mobile home the following distances.

(a) 130 mi **(b)** 150 mi **(c)** 210 mi

(d) 400 mi **(e)** 500 mi

(f) Where is C discontinuous?

58. Cost Analysis A company charges \$1.20 per pound for a certain fertilizer on all orders not over 100 lb, and \$1 per pound for orders over 100 lb. Let $F(x)$ represent the cost for buying x lb of the fertilizer. Find the cost of buying

(a) 80 lb; **(b)** 150 lb; **(c)** 100 lb.

(d) Where is F discontinuous?

59. Car Rental Recently, a car rental firm charged \$30 per day or portion of a day to rent a car for a period of 1 to 5 days. Days 6 and 7 were then "free," while the charge for days 8 through 12 was again \$30 per day. Let $C(t)$ represent the total cost to rent the car for t days, where $0 < t \leq 12$. Find the total cost of a rental for the following number of days:

(a) 4 **(b)** 5 **(c)** 6 **(d)** 7 **(e)** 8.

(f) Find $\lim_{t \to 5^-} C(t)$. **(g)** Find $\lim_{t \to 5^+} C(t)$.

(h) Where is C discontinuous on the given interval?

60. Sales Tax Like many states, California suffered a large budget deficit in 1991. As part of the solution, officials raised the sales tax by \$.0125. The graph shows the California state sales tax since it was first established in 1933. Let $T(x)$ represent the sales tax in year x. Find the following.

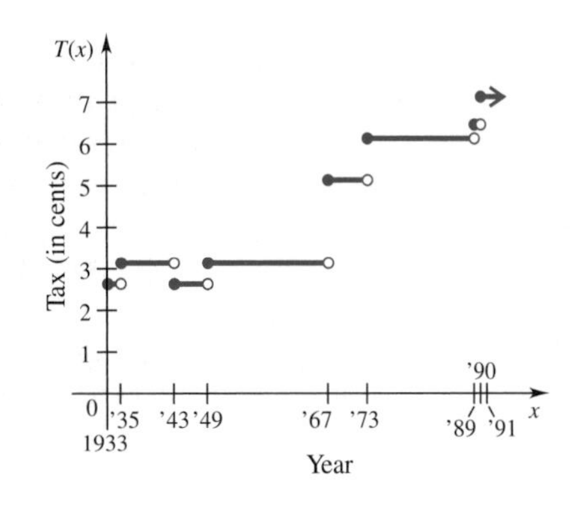

(a) $\lim_{x \to 80} T(x)$ **(b)** $\lim_{x \to 73^-} T(x)$ **(c)** $\lim_{x \to 73^+} T(x)$

(d) $\lim_{x \to 73} T(x)$

(e) List three years for which the graph indicates a discontinuity.

Social Sciences

61. Learning Skills With certain skills (such as musical skills), learning is rapid at first and then levels off, with sudden insights causing learning to take a jump. A typical graph of such learning is shown in the figure. Where is the function discontinuous?

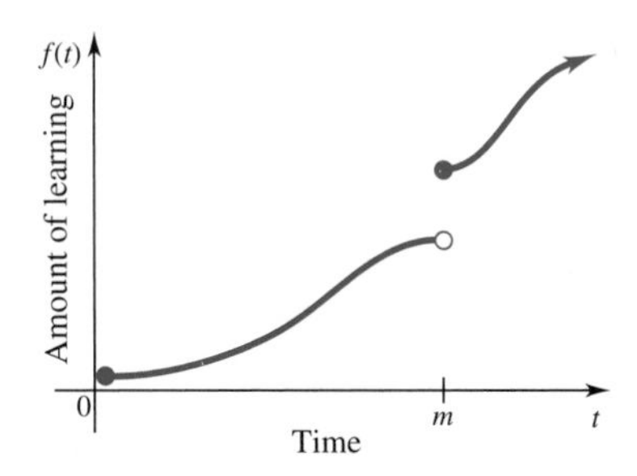

Physical Sciences

62. Temperature Suppose a gram of ice is at a temperature of −100°C. The graph shows the temperature of the ice as increasing numbers of calories of heat are applied. It takes 80 calories to melt one gram of ice at 0°C into water, and 539 calories to boil one gram of water at 100°C into steam. Where is this graph discontinuous?

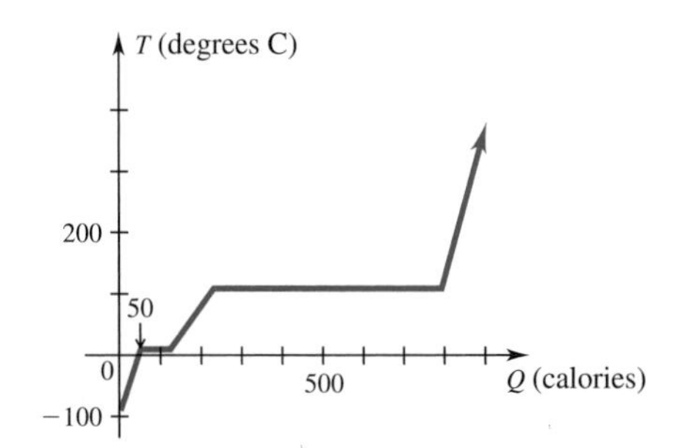

2.2 RATES OF CHANGE

How fast do the blood cells in a capillary move in the seconds after a drug is injected?

How does the manufacturing cost per item change as the number of items manufactured changes?

These questions will be answered in Examples 4 and 5 in this section as we develop a method for finding the rate of change of one variable with respect to a unit change in another variable.

Average Rate of Change Suppose we take a trip from San Francisco. Someone keeps track every half-hour of the distance traveled from San Francisco, with the following results for the first three hours.

Time in Hours	0	.5	1	1.5	2	2.5	3
Distance in Miles	0	20	48	80	104	126	150

Let f be the function giving the distance in miles from San Francisco at time t. Then $f(0) = 0$, $f(1) = 48$, $f(2.5) = 126$, and so on.

Distance equals time multiplied by rate (or speed); so the distance formula is $d = rt$. Solving for rate gives $r = d/t$, or

$$\text{Average speed} = \frac{\text{Distance}}{\text{Time}}.$$

For example, the average speed over the time interval from $t = 0$ to $t = 3$ is

$$\text{Average speed} = \frac{f(3) - f(0)}{3 - 0} = \frac{150 - 0}{3} = 50,$$

or 50 miles per hour. We can use this formula to find the average speed for any interval of time during the trip, as shown below.

Recall the formula for the slope of a line through two points (x_1, y_1) and (x_2, y_2):

$$\frac{y_2 - y_1}{x_2 - x_1}.$$

Find the slope of the line through the following points.

(.5, 20)	and	(1, 48)
(.5, 20)	and	(1.5, 80)
(1, 48)	and	(2, 104)

Compare your answers to the average speeds shown in the chart.

Time Interval	*Average Speed* $= \dfrac{\textit{Distance}}{\textit{Time}}$
$t = .5$ to $t = 1$	$\dfrac{f(1) - f(.5)}{1 - .5} = \dfrac{28}{.5} = 56$
$t = .5$ to $t = 1.5$	$\dfrac{f(1.5) - f(.5)}{1.5 - .5} = \dfrac{60}{1} = 60$
$t = 1$ to $t = 2$	$\dfrac{f(2) - f(1)}{2 - 1} = \dfrac{56}{1} = 56$
$t = 1$ to $t = 3$	$\dfrac{f(3) - f(1)}{3 - 1} = \dfrac{102}{2} = 51$
$t = a$ to $t = b$	$\dfrac{f(b) - f(a)}{b - a}$

The last line in the chart suggests a way to apply this approach to any function f to get the average rate of change of $f(x)$ with respect to x.

AVERAGE RATE OF CHANGE

The average rate of change of $f(x)$ with respect to x for a function f as x changes from a to b, where $a < b$, is

$$\frac{f(b) - f(a)}{b - a}.$$

Note The formula for the average rate of change is the same as the formula for the slope of the line through $(a, f(a))$ and $(b, f(b))$. This connection between slope and rate of change will be examined more closely in the next section.

EXAMPLE 1 A manager is testing two different advertising campaigns in different parts of the country for three months, with the intention of canceling the least effective of the two after that time. The graphs in Figure 11 show sales as a function of time (in months) since the two campaigns began. The graphs suggest that although campaign 1 consistently did better than campaign 2 for the first three months, its *rate of change in sales* is decreasing after that time. The rate of change in sales for campaign 2, however, is increasing at time $t = 3$ months.

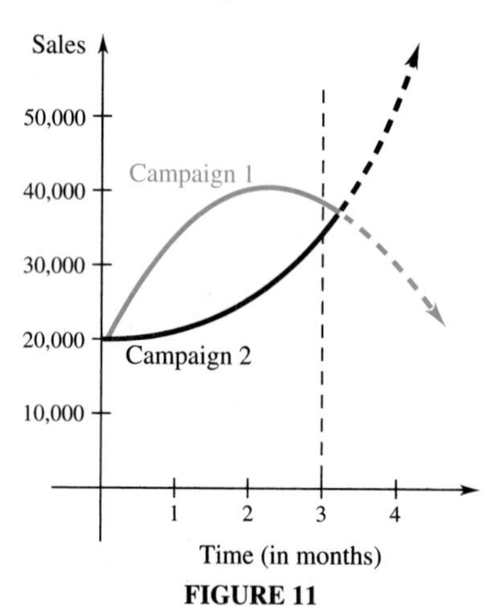

FIGURE 11

On the interval [2, 4], for campaign 1,

$$\text{Average rate of change} = \frac{f(4) - f(2)}{4 - 2}$$

$$= \frac{30{,}000 - 40{,}000}{4 - 2} = -5000.$$

For campaign 2, on [2, 4],

$$\text{Average rate of change} = \frac{50{,}000 - 25{,}000}{4 - 2} = 12{,}500.$$

The negative result for campaign 1 indicates that sales were *decreasing* at the rate of 5000 per month over the two-month period, while the positive result for campaign 2 indicates *increasing* sales at the rate of 12,500 over the same two-month period. These rates show that the second campaign will be the most productive in the future, and campaign 1 should be discontinued. ◀

EXAMPLE 2 The graph in Figure 12 shows the profit $P(x)$, in hundreds of thousands of dollars, from a highly popular new computer program x months after its introduction on the market. The graph shows that profit increases until the program has been on the market for 25 months, after which profit decreases as the popularity of the program decreases. Here the average rate of change of *profit* with respect to *time* on some interval is defined as the change in profit divided by the change in time on the interval.

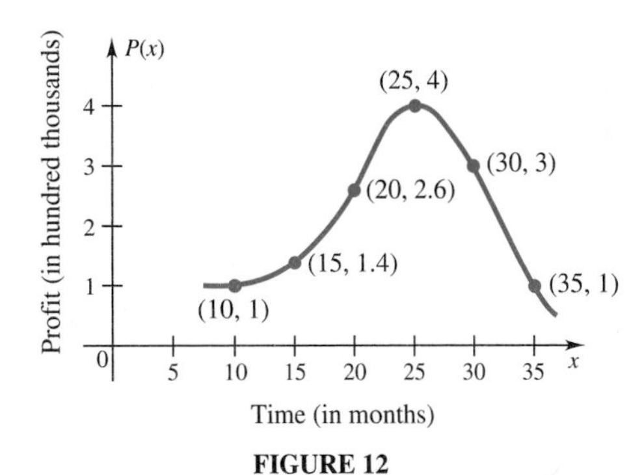

FIGURE 12

(a) On the interval from 15 months to 25 months the average rate of change of profit with respect to time is given by

$$\frac{P(25) - P(15)}{25 - 15} = \frac{4 - 1.4}{25 - 15} = \frac{2.6}{10} = .26.$$

On the average, each month from 15 months to 25 months shows an increase in profit of .26 hundred thousand dollars, or $26,000.

(b) From 25 months to 30 months, the average rate of change is

$$\frac{3 - 4}{30 - 25} = \frac{-1}{5} = -.20,$$

with each month in this interval showing an average *decline* in profits of .20 hundred thousand dollars, or $20,000.

(c) From 10 months to 35 months, the average rate of change of profit is

$$\frac{1 - 1}{35 - 10} = \frac{0}{25} = 0,$$

or $0. ◀

Velocity As Example 2(c) suggests, finding the average rate of change of a function over a large interval can lead to answers that are not very helpful. The results often are more useful if the average rate of change is found over a fairly narrow interval. Finding the exact rate of change at a particular x-value requires a continuous function. For example, suppose a car starts from a stop sign and moves along a straight road. Assume that the distance in feet traveled in t seconds is given by $s(t) = t^2 + 3t$. After 5 seconds, the car has traveled $s(5) = (5)^2 + 3(5) = 40$ feet. The speed of the car at exactly 10 seconds after starting can be estimated by finding the average speed over shorter and shorter time intervals.

Interval	*Average Speed*
$t = 10$ to $t = 10.1$	$\frac{s(10.1) - s(10)}{10.1 - 10} = \frac{132.31 - 130}{.1} = 23.1$
$t = 10$ to $t = 10.01$	$\frac{s(10.01) - s(10)}{10.01 - 10} = \frac{130.2301 - 130}{.01} = 23.01$
$t = 10$ to $t = 10.001$	$\frac{s(10.001) - s(10)}{10.001 - 10} = \frac{130.023001 - 130}{.001} = 23.001$

The results in the chart suggest that the exact speed at $t = 10$ seconds is 23 feet per second.

We can generalize this discussion by finding the average speed from $t = 10$ to $t = 10 + h$, where h represents some nonzero small number. The average speed for this interval is as follows.

$$\begin{aligned}\frac{s(10 + h) - s(10)}{(10 + h) - 10} &= \frac{s(10 + h) - s(10)}{h} \\ &= \frac{[(10 + h)^2 + 3(10 + h)] - 130}{h} \\ &= \frac{100 + 20h + h^2 + 30 + 3h - 130}{h} \\ &= \frac{23h + h^2}{h} \\ &= \frac{h(23 + h)}{h} \\ &= 23 + h\end{aligned}$$

If we take the limit of $23 + h$ as h approaches zero, we get

$$\lim_{h \to 0} \frac{s(10 + h) - s(10)}{h} = \lim_{h \to 0} (23 + h) = 23.$$

We call this limit the *instantaneous velocity at time t.*

To generalize the results of the car example, suppose an object is moving in a straight line, with its position (distance from some fixed point) given by $s(t)$, where t represents time. The quotient

$$\frac{s(t + h) - s(t)}{h}$$

represents the average rate of change of the distance, or the average velocity of the object. The instantaneous velocity at time t (often called just the velocity at time t) is the limit of the quotient above as h approaches 0.

VELOCITY

If $v(t)$ represents the velocity at time t of an object moving in a straight line with position $s(t)$, then

$$\mathbf{v(t) = \lim_{h \to 0} \frac{s(t + h) - s(t)}{h},}$$

provided this limit exists.

Caution Remember, $s(t + h) \neq s(t) + s(h)$. To find $s(t + h)$ replace t with $t + h$ in the expression for $s(t)$. For example, if $s(t) = t^2$,

$$s(t + h) = (t + h)^2 = t^2 + 2th + h^2,$$

but

$$s(t) + s(h) = t^2 + h^2.$$

EXAMPLE 3 The distance in feet of an object from a starting point is given by $s(t) = 2t^2 - 5t + 40$, where t is time in seconds.

(a) Find the average velocity of the object from 2 seconds to 4 seconds.

The average velocity is

$$\frac{s(4) - s(2)}{4 - 2} = \frac{52 - 38}{2} = \frac{14}{2} = 7$$

feet per second.

(b) Find the instantaneous velocity at 4 seconds.

For $t = 4$, the instantaneous velocity is

$$\lim_{h \to 0} \frac{s(4 + h) - s(4)}{h}$$

feet per second.

$$\begin{aligned} s(4 + h) &= 2(4 + h)^2 - 5(4 + h) + 40 \\ &= 2(16 + 8h + h^2) - 20 - 5h + 40 \\ &= 32 + 16h + 2h^2 - 20 - 5h + 40 \\ &= 2h^2 + 11h + 52 \end{aligned}$$

Also,

$$s(4) = 2(4)^2 - 5(4) + 40 = 52.$$

Therefore, the instantaneous velocity at $t = 4$ is

$$\lim_{h\to 0} \frac{2h^2 + 11h}{h} = \lim_{h\to 0} (2h + 11) = 11,$$

or 11 feet per second. ◀

The next examples provide answers to the questions asked at the beginning of this section.

EXAMPLE 4 The velocity of blood cells is of interest to physicians; a slower velocity than normal might indicate a constriction, for example. Suppose the position of a red blood cell in a capillary is given by

$$s(t) = 1.2t + 5,$$

where $s(t)$ gives the position of a cell in millimeters from some initial point and t is time in seconds. Find the velocity of this cell at time t.

Evaluate the limit given above. To find $s(t + h)$, substitute $t + h$ for the variable t in $s(t) = 1.2t + 5$.

$$s(t + h) = 1.2(t + h) + 5$$

Now use the definition of velocity.

$$\begin{aligned} v(t) &= \lim_{h\to 0} \frac{s(t + h) - s(t)}{h} \\ &= \lim_{h\to 0} \frac{1.2(t + h) + 5 - (1.2t + 5)}{h} \\ &= \lim_{h\to 0} \frac{1.2t + 1.2h + 5 - 1.2t - 5}{h} = \lim_{h\to 0} \frac{1.2h}{h} = 1.2 \end{aligned}$$

The velocity of the blood cell is a constant 1.2 millimeters per second. ◀

Note In Example 4, the velocity of 1.2 is the same as the slope of the linear function $s(t) = 1.2t + 5$. As mentioned earlier, we will learn more about this connection in the next section.

Instantaneous Rate of Change We can extend the ideas that led to the definition of velocity to any function f. Earlier, we saw that the average rate of change of $f(x)$ with respect to x as x changes from a to b is

$$\frac{f(b) - f(a)}{b - a}.$$

Replacing b with $x_0 + h$ and a with x_0 and letting h approach 0 gives the following definition.

INSTANTANEOUS RATE OF CHANGE

The **instantaneous rate of change** for a function f when $x = x_0$ is

$$\lim_{h \to 0} \frac{f(x_0 + h) - f(x_0)}{h},$$

provided this limit exists.

EXAMPLE 5 A company determines that the cost in dollars of manufacturing x units of a certain item is

$$C(x) = 100 + 5x - x^2.$$

(a) Find the average rate of change of cost per item for manufacturing between 1 and 5 items.

Use the formula for average rate of change. The cost to manufacture 1 item is

$$C(1) = 100 + 5(1) - (1)^2 = 104,$$

or \$104. The cost to manufacture 5 items is

$$C(5) = 100 + 5(5) - (5)^2 = 100,$$

or \$100. The average rate of change of cost is

$$\frac{C(5) - C(1)}{5 - 1} = \frac{100 - 104}{4} = -1.$$

Thus, on the average, cost is decreasing at the rate of \$1 per item when production is increased from 1 to 5 items.

(b) Find the instantaneous rate of change of cost with respect to the number of items produced when just 1 item is produced.

The instantaneous rate of change for $x = 1$ is given by

$$\lim_{h \to 0} \frac{C(1 + h) - C(1)}{h}$$

$$= \lim_{h \to 0} \frac{[100 + 5(1 + h) - (1 + h)^2] - [100 + 5(1) - 1^2]}{h}$$

$$= \lim_{h \to 0} \frac{[100 + 5 + 5h - 1 - 2h - h^2 - 104]}{h}$$

$$= \lim_{h \to 0} \frac{3h - h^2}{h} \quad \text{Combine terms.}$$

$$= \lim_{h \to 0} (3 - h) \quad \text{Divide by } h.$$

$$= 3. \quad \text{Calculate the limit.}$$

When 1 item is manufactured, the cost is increasing at the rate of \$3 per item. Thus, the instantaneous rate of change of cost represents the approximate cost of manufacturing an additional item. As mentioned in Chapter 1, this rate of change of cost is called the marginal cost.

(c) Find the instantaneous rate of change of cost when 5 items are made.

The instantaneous rate of change for $x = 5$ is given by

$$\lim_{h \to 0} \frac{C(5 + h) - C(5)}{h}$$

$$= \lim_{h \to 0} \frac{[100 + 5(5 + h) - (5 + h)^2] - [100 + 5(5) - 5^2]}{h}$$

$$= \lim_{h \to 0} \frac{[100 + 25 + 5h - 25 - 10h - h^2] - 100}{h}$$

$$= \lim_{h \to 0} \frac{-5h - h^2}{h}$$

$$= \lim_{h \to 0} (-5 - h)$$

$$= -5.$$

When 5 items are manufactured, the cost is *decreasing* at the rate of \$5 per item—that is, the marginal cost when $x = 5$ is $-\$5$. Notice that as the number of items produced goes up, the marginal cost goes down, as might be expected. ◀

2.2 Exercises

Find the average rate of change for each function over the given interval.

1. $y = x^2 + 2x$ between $x = 0$ and $x = 3$

2. $y = -4x^2 - 6$ between $x = 2$ and $x = 5$

3. $y = 2x^3 - 4x^2 + 6x$ between $x = -1$ and $x = 1$

4. $y = -3x^3 + 2x^2 - 4x + 1$ between $x = 0$ and $x = 1$

5. $y = \sqrt{x}$ between $x = 1$ and $x = 4$

6. $y = \sqrt{3x - 2}$ between $x = 1$ and $x = 2$

7. $y = \frac{1}{x - 1}$ between $x = -2$ and $x = 0$

8. $y = \frac{-5}{2x - 3}$ between $x = 2$ and $x = 4$

Use the properties of limits to find the following limits for $s(t) = t^2 + 5t + 2$.

9. $\lim_{h \to 0} \frac{s(6 + h) - s(6)}{h}$

10. $\lim_{h \to 0} \frac{s(1 + h) - s(1)}{h}$

11. $\lim_{h \to 0} \frac{s(10 + h) - s(10)}{h}$

Use the properties of limits to find the following limits for $s(t) = t^3 + 2t + 9$.

12. $\lim_{h \to 0} \frac{s(1 + h) - s(1)}{h}$

13. $\lim_{h \to 0} \frac{s(4 + h) - s(4)}{h}$

Find the instantaneous rate of change for each function at the given value.

14. $f(x) = x^2 + 2x$ at $x = 0$

15. $s(t) = -4t^2 - 6$ at $t = 2$

16. $g(t) = 1 - t^2$ at $t = -1$

17. $F(x) = x^2 + 2$ at $x = 0$

18. Explain the difference between the average rate of change of y as x changes from a to b, and the instantaneous rate of change of y at $x = a$.

19. If the instantaneous rate of change of $f(x)$ with respect to x is positive when $x = 1$, is f increasing or decreasing there?

Applications

Business and Economics

20. Catalog Sales The graph shows the total sales in thousands of dollars from the distribution of x thousand catalogs. Find and interpret the average rate of change of sales with respect to the number of catalogs distributed for the following changes in x.

(a) 10 to 20 **(b)** 20 to 30 **(c)** 30 to 40

(d) What is happening to the average rate of change of sales as the number of catalogs distributed increases?

(e) Explain why (d) might happen.

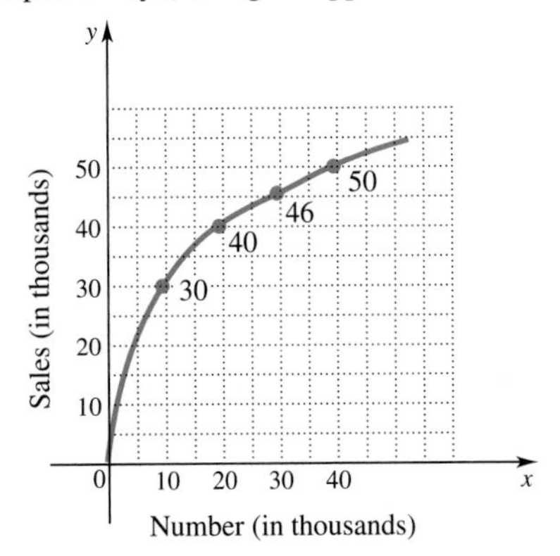

21. Sales The graph shows annual sales (in appropriate units) of a Nintendo game. Find the average annual rate of change in sales for the following changes in years.

(a) 1 to 4 **(b)** 4 to 7 **(c)** 7 to 12

(d) What do your answers for (a) to (c) tell you about the sales of this product?

(e) Give an example of another product that might have such a sales curve.

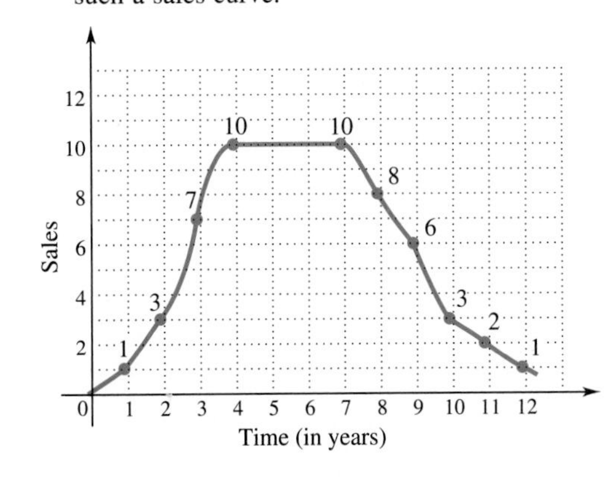

22. Market Share IBM Europe has been steadily losing its share of the market since 1985, as shown in the figure.* Estimate the average drop in market share (in percent) over the following intervals.

(a) 1985 to 1986 **(b)** 1986 to 1988

(c) 1988 to 1989 **(d)** 1989 to 1990

(e) From your answers to parts (a)–(d), does the decline in market share seem to be increasing or tapering off?

(f) In which one-year period was the decline the greatest?

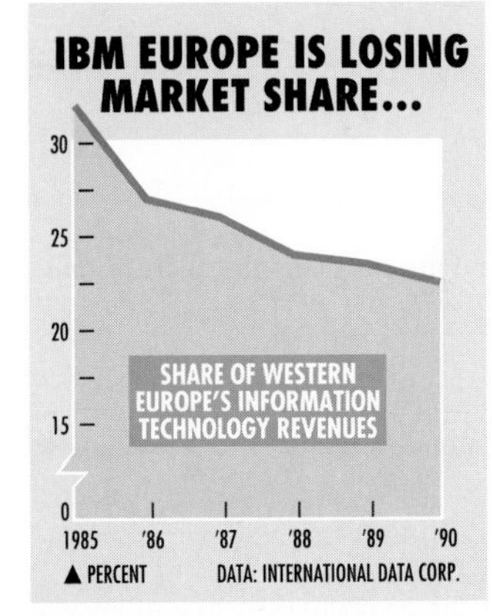

23. Profit Margin The figure* shows the profit margin (in percent) for IBM Europe. Estimate the average rate of change over the following intervals.

(a) 1985 to 1987 **(b)** 1987 to 1988

(c) 1988 to 1989 **(d)** 1989 to 1990

(e) In what time period(s) did the profit margin increase?

(f) Do the results in parts (a)–(d) indicate an improving, stable, or declining profit margin?

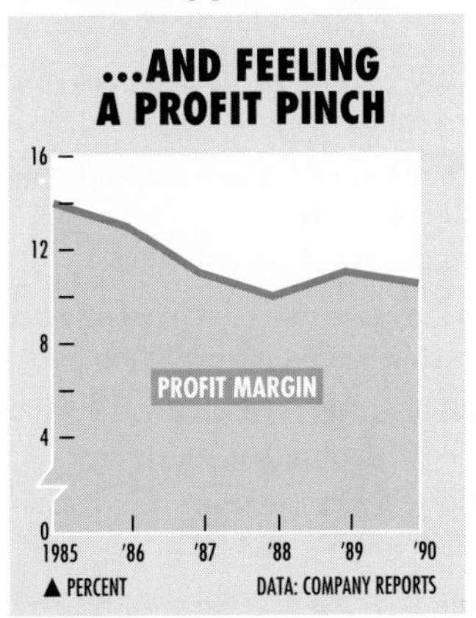

*Graphs entitled "IBM Europe Is Losing Market Share . . . and Feeling a Profit Pinch" reprinted from May 6, 1991 issue of *Business Week* by special permission, copyright © 1991 by McGraw-Hill, Inc.

24. Profit Suppose that the total profit in hundreds of dollars from selling x items is given by

$$P(x) = 2x^2 - 5x + 6.$$

Find the average rate of change of profit for the following changes in x.

(a) 2 to 4 **(b)** 2 to 3

(c) Use $P(x) = 2x^2 - 5x + 6$ to complete the following table.

h	1	.1	.01	.001	.0001
$2 + h$	3	2.1	2.01	2.001	2.0001
$P(2 + h)$	9	4.32	4.0302	4.003002	4.00030002
$P(2)$	4	4	4	4	4
$P(2 + h) - P(2)$	5	.32	.0302	.003002	.00030002
$\dfrac{P(2 + h) - P(2)}{h}$	5	3.2	____	____	____

(d) Use the bottom row of the chart to find $\lim_{h\to 0} \dfrac{P(2 + h) - P(2)}{h}$.

(e) Use the rule for limits from Section 1 of this chapter to find

$$\lim_{h\to 0} \frac{P(2 + h) - P(2)}{h}.$$

(f) What is the instantaneous rate of change of profit with respect to the number of items produced when $x = 2$? (This number, called the *marginal profit* at $x = 2$, is the approximate profit from producing the third item.)

25. Profit Redo the chart in Exercise 24. This time, change the second line to $4 + h$, the third line to $P(4 + h)$, and so on. Then find

$$\lim_{h\to 0} \frac{P(4 + h) - P(4)}{h}.$$

(As in Exercise 24, this result is the marginal profit, the approximate profit from producing the fifth item.)

26. Revenue The revenue (in thousands of dollars) from producing x units of an item is

$$R = 10x - .002x^2.$$

(a) Find the average rate of change of revenue when production is increased from 1000 to 1001 units.

(b) Find the marginal revenue when 1000 units are produced.

(c) Find the additional revenue if production is increased from 1000 to 1001 units.

(d) Compare your answers for parts (a) and (c). What do you find?

27. Demand Suppose customers in a hardware store are willing to buy $N(p)$ boxes of nails at p dollars per box, as given by

$$N(p) = 80 - 5p^2, \quad 1 \le p \le 4.$$

(a) Find the average rate of change of demand for a change in price from \$2 to \$3.

(b) Find the instantaneous rate of change of demand when the price is \$2.

(c) Find the instantaneous rate of change of demand when the price is \$3.

(d) As the price is increased from \$2 to \$3, how is demand changing? Is the change to be expected?

Life Sciences

28. Bacteria Population The graph shows the population in millions of bacteria t minutes after a bactericide is introduced into a culture. Find and interpret the average rate of change of population with respect to time for the following time intervals.

(a) 1 to 2 **(b)** 2 to 3 **(c)** 3 to 4 **(d)** 4 to 5

(e) How long after the bactericide was introduced did the population begin to decrease?

(f) At what time did the rate of decrease of the population slow down?

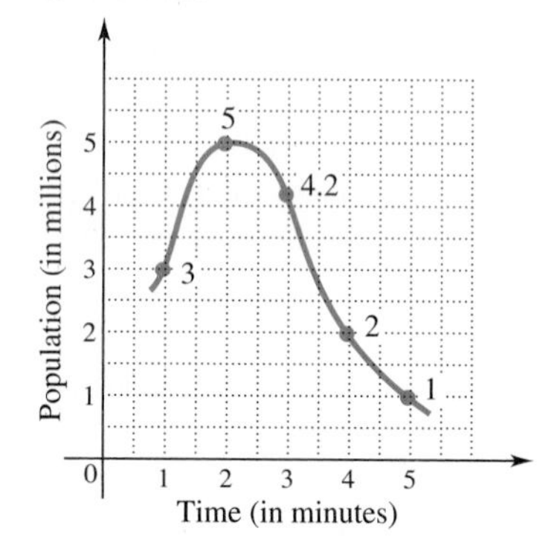

29. Polluter Fines The graph* shows changes since 1981 in the amounts paid to the U.S. Treasury in fines by polluters.

(a) Find the average rate of change for civil penalties from 1987 to 1988, and from 1988 to 1989.

(b) Find the average rate of change of criminal penalties from 1987 to 1988, and from 1988 to 1989.

(c) Compare your results for parts (a) and (b). What do they tell you? What might account for the differences?

---

*Graph entitled "Higher EPA Fines Make Pollution Costly" by Mark Holmes from *National Geographic*, February 1991. Copyright © 1991 by National Geographic Society. Reprinted by permission of National Geographic Society.

(d) Find the average rate of change for civil penalties from 1981 to 1989. What was the general trend over this eight-year period? What might explain this trend?

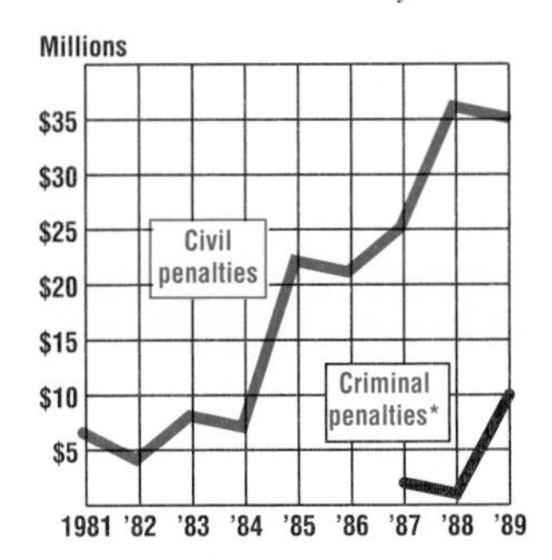

Social Sciences

30. Memory In an experiment, teenagers memorize a list of 15 words and then listen to rock music for certain periods of time. After t minutes of listening to rock music, the typical subject can remember approximately

$$R(t) = -.03t^2 + 15$$

of the words, where $0 \leq t \leq 20$. Find the instantaneous rate at which the typical student remembers the words after the following times have elapsed.

(a) 5 min (b) 15 min

31. Drug Trade The U.S. government estimates that Americans spent more than $40 billion for illicit drugs in 1990. However, as shown in the graph†, Americans have spent less on drugs since 1988. Estimate the average rate of change in the amounts spent on each of the following drugs from 1988 to 1989, and then from 1989 to 1990.

(a) Cocaine (b) Heroin (c) Marijuana

(d) Which drug had the greatest decrease in amount spent?

(e) List some possible reasons for the observed decreases and some reasons why not all drug expenditures decreased at the same rate.

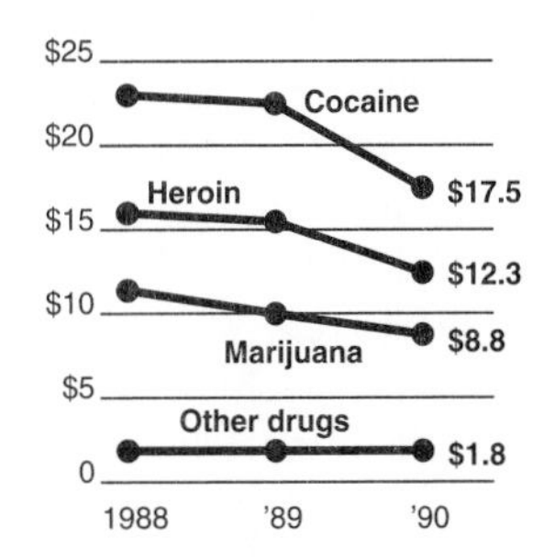

†Graph from "What Americans spend on drugs" as appeared in *The Sacramento Bee*, June 20, 1991. Reprinted by permission of Knight-Ridder/Tribune News Wire.

Physical Sciences

32. Temperature The graph shows the temperature T in degrees Celsius as a function of the altitude h in feet when an inversion layer is over Southern California. (An inversion layer is formed when air at a higher altitude, say 3000 ft, is warmer than air at sea level, even though air normally is cooler with increasing altitude.) Estimate and interpret the average rate of change in temperature for the following changes in altitude.

(a) 1000 to 3000 ft (b) 1000 to 5000 ft

(c) 3000 to 9000 ft (d) 1000 to 9000 ft

(e) At what altitude at or below 7000 ft is the temperature highest? lowest? How would your answer change if 7000 ft is changed to 10,000 ft?

(f) At what altitude is the temperature the same as it is at 1000 ft?

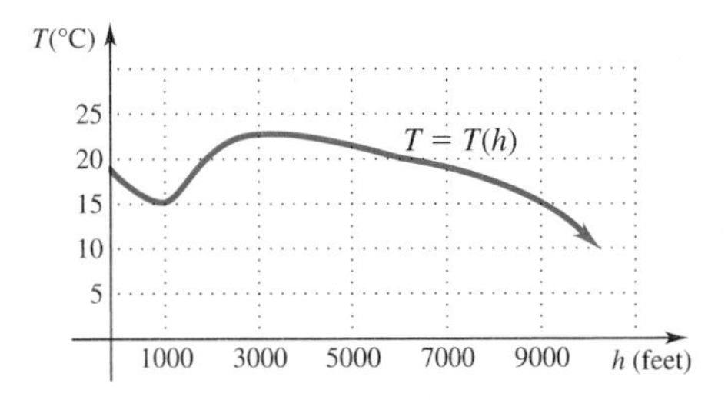

33. Velocity A car is moving along a straight test track. The position in feet of the car, $s(t)$, at various times t is measured, with the following results.

t (seconds)	0	2	6	10	12	18
$s(t)$ (feet)	0	10	14	18	30	36

Find and interpret the average velocities for the following changes in t.

(a) 0 to 2 sec (b) 2 to 6 sec (c) 6 to 10 sec

(d) 10 to 12 sec (e) 12 to 18 sec

(f) Compare the average velocities found in parts (a)–(e), and describe when the car is slowing down, speeding up, or maintaining a constant average velocity.

34. Velocity The distance of a particle from some fixed point is given by

$$s(t) = t^2 + 5t + 2,$$

where t is time measured in seconds.

Find the average velocity of the particle over each of the following intervals.

(a) 4 to 6 sec **(b)** 4 to 5 sec

(c) Complete the following table.

h	1	.1	.01	.001	.0001
$4 + h$	5	4.1	4.01	4.001	4.0001
$s(4 + h)$	52	39.31	38.1301	38.013001	38.00130001
$s(4)$	38	38	38	38	38
$s(4 + h) - s(4)$	14	1.31	.1301	.013001	.00130001
$\frac{s(4 + h) - s(4)}{h}$	14	13.1	______	______	______

(d) Find $\lim_{h\to 0} \frac{s(4 + h) - s(4)}{h}$, and give the instantaneous velocity of the particle when $t = 4$.

2.3 DEFINITION OF THE DERIVATIVE

If the distances of a vehicle from a starting point are known for the corresponding elapsed times, how can the speed of the vehicle be determined?

In this section, we continue our discussion of the relationship among distance, rate, and time. We develop a formula for instantaneous velocity that will be used in Example 7 to answer the question posed above.

In the previous section, the formula

$$\lim_{h\to 0} \frac{f(x_0 + h) - f(x_0)}{h}$$

was used to calculate the instantaneous rate of change of a function f at the point where $x = x_0$. The same formula can be used to find the slope of a line *tangent* to the graph of a function f at the point $x = x_0$. (Remember that the slope of a line is really the rate of change of y with respect to x.)

In geometry, a tangent line to a circle is defined as a line that touches the circle at only one point. The semicircle in Figure 13 is the graph of a function. Comparing the various lines in Figure 14 to the tangent at point P in Figure 13 suggests that the lines in Figure 14 at P_1 and P_3 are tangent to the curve, while

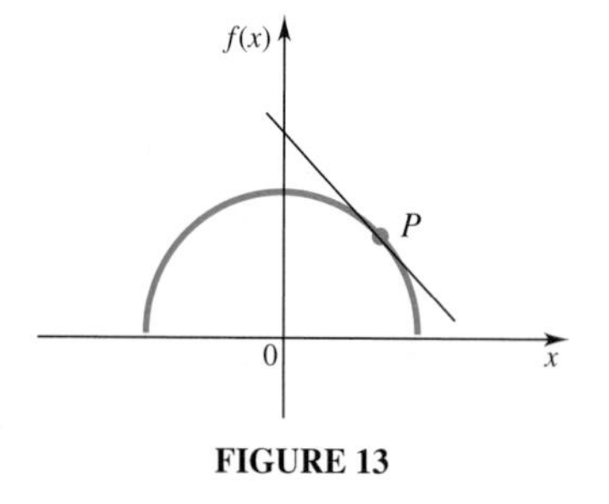

FIGURE 13

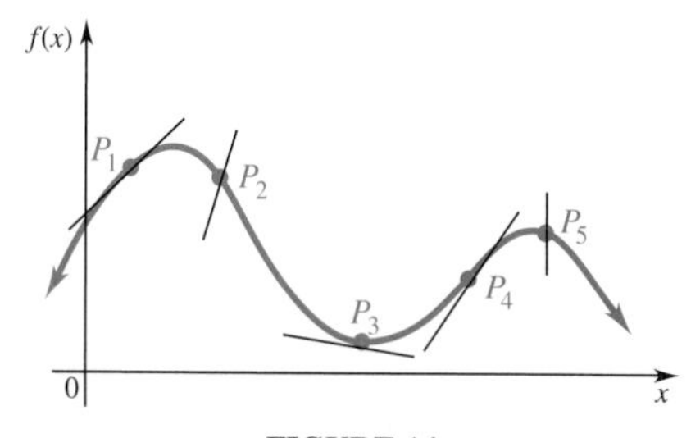

FIGURE 14

the lines at P_2 and P_5 are not. The tangent lines just touch the curve, while the other lines pass through it. To decide about the line at P_4, we need to define the idea of a tangent line to the graph of a function more carefully.

To see how we might define the slope of a line tangent to the graph of a function f at a given point, let R be a fixed point with coordinates $(x_0, f(x_0))$ on the graph of a function $y = f(x)$, as in Figure 15. Choose a different point S on the graph and draw the line through R and S; this line is called a **secant line.** If S has coordinates $(x_0 + h, f(x_0 + h))$, then by the definition of slope, the slope of the secant line RS is given by

$$\text{Slope of secant} = \frac{\Delta y}{\Delta x} = \frac{f(x_0 + h) - f(x_0)}{x_0 + h - x_0} = \frac{f(x_0 + h) - f(x_0)}{h}.$$

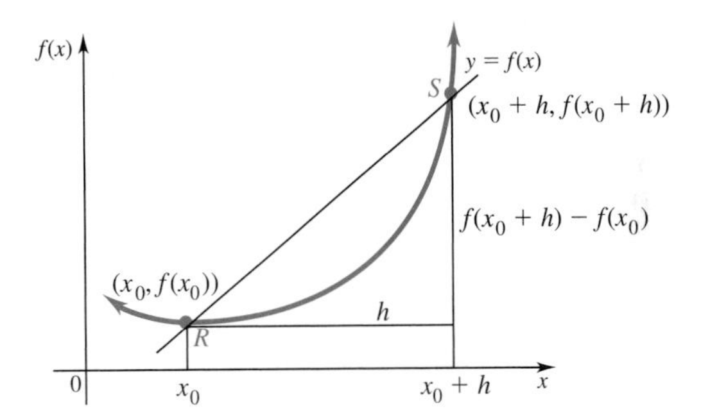

FIGURE 15

This slope corresponds to the average rate of change of y with respect to x over the interval from x_0 to $x_0 + h$. As h approaches 0, point S will slide along the curve, getting closer and closer to the fixed point R. See Figure 16, which shows successive positions S_1, S_2, S_3, and S_4 of the point S. If the slopes of the corresponding secant lines approach a limit as h approaches 0, then this limit is defined to be the slope of the tangent line at point R.

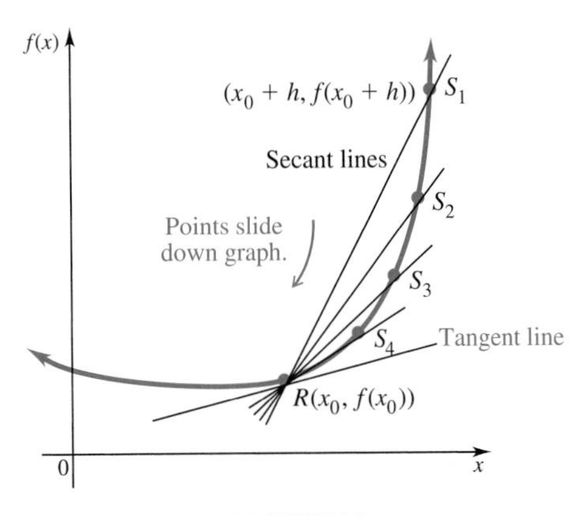

FIGURE 16

TANGENT LINE

The **tangent line** to the graph of $y = f(x)$ at the point $(x_0, f(x_0))$ is the line through this point having slope

$$\lim_{h \to 0} \frac{f(x_0 + h) - f(x_0)}{h},$$

provided this limit exists. If this limit does not exist, then there is no tangent at the point.

The slope of this line at a point is also called the **slope of the curve** at the point and corresponds to the instantaneous rate of change of y with respect to x at the point.

In certain applications of mathematics it is necessary to determine the equation of a line tangent to the graph of a function at a given point, as in the next example.

EXAMPLE 1 Find the slope of the tangent line to the graph of $f(x) = x^2 + 2$ at $x = -1$. Find the equation of the tangent line.

Use the definition given above, with $f(x) = x^2 + 2$ and $x_0 = -1$. The slope of the tangent line is given by

$$\begin{aligned}
\text{Slope of tangent} &= \lim_{h \to 0} \frac{f(x_0 + h) - f(x_0)}{h} \\
&= \lim_{h \to 0} \frac{[(-1 + h)^2 + 2] - [(-1)^2 + 2]}{h} \\
&= \lim_{h \to 0} \frac{[1 - 2h + h^2 + 2] - [1 + 2]}{h} \\
&= \lim_{h \to 0} \frac{-2h + h^2}{h} \\
&= \lim_{h \to 0} (-2 + h) = -2.
\end{aligned}$$

In Section 1.2, we saw that the equation of a line can be found with the point-slope form $y - y_1 = m(x - x_1)$, if the slope m and the coordinates (x_1, y_1) of a point on the line are known. Use the point-slope form to find the equation of the line with slope 3 that goes through the point $(-1, 4)$.

Let $m = 3$, $x_1 = -1$, $y_1 = 4$. Then

$$\begin{aligned}
y - y_1 &= m(x - x_1) \\
y - 4 &= 3(x - (-1)) \\
y - 4 &= 3x + 3 \\
y &= 3x + 7.
\end{aligned}$$

The slope of the tangent line at $(-1, f(-1)) = (-1, 3)$ is -2. The equation of the tangent line can be found with the point-slope form of the equation of a line from Chapter 1.

$$\begin{aligned}
y - y_1 &= m(x - x_1) \\
y - 3 &= -2[x - (-1)] \\
y - 3 &= -2(x + 1) \\
y - 3 &= -2x - 2 \\
y &= -2x + 1
\end{aligned}$$

Figure 17 shows a graph of $f(x) = x^2 + 2$, along with a graph of the tangent line at $x = -1$. ◀

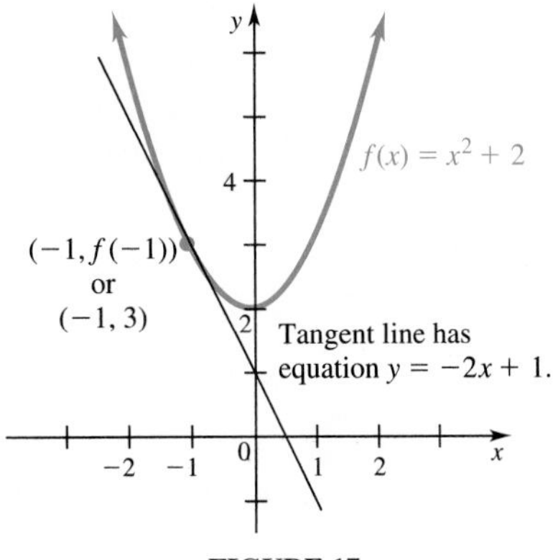

FIGURE 17

The Derivative The special limit

$$\lim_{h \to 0} \frac{f(x_0 + h) - f(x_0)}{h}$$

has now been developed in two different ways—as the instantaneous rate of change of y with respect to x and also as the slope of a tangent line. This particular limit is so important that it is given a special name, the *derivative*.

DERIVATIVE

The **derivative** of the function f at x, written $f'(x)$, is defined as

$$f'(x) = \lim_{h \to 0} \frac{f(x + h) - f(x)}{h},$$

provided this limit exists.

The notation $f'(x)$ is read "f-prime of x". The alternative notation y' (read "y-prime") is sometimes used for the derivative of $y = f(x)$.

Notice that the derivative is a function of x since $f'(x)$ varies as x varies. This differs from the slope of the tangent, given above, in which a specific x-value, x_0, is used. This new function given by $y' = f'(x)$ has as its domain all the points at which the specified limit exists, and its value at x is $f'(x)$. The function f' is called the derivative of f with respect to x. If x is a value in the domain of f, and if $f'(x)$ exists, then f is **differentiable** at x.

In summary, the derivative of a function f is a new function f'. The mathematical process that produces f' is called **differentiation.** This new function has several interpretations. Two of them are given here.

1. The function f' represents the *slope* of the graph of $f(x)$ at any point x. If the derivative is evaluated at the point $x = x_0$, then it represents the slope of the curve at that point.

2. The function f' represents the *instantaneous rate of change* of y with respect to x. This instantaneous rate of change could be interpreted as marginal cost, revenue, or profit (if the original function represented cost, revenue, or profit) or velocity (if the original function described displacement along a line). From now on we will use "rate of change" to mean "instantaneous rate of change."

The next few examples show how to use the definition to find the derivative of a function by means of a four-step procedure.

EXAMPLE 2 Let $f(x) = x^2$.

(a) Find the derivative.

By definition, for all values of x where the following limit exists, the derivative is given by

$$f'(x) = \lim_{h \to 0} \frac{f(x + h) - f(x)}{h}.$$

Use the following sequence of steps to evaluate this limit.

Step 1 Find $f(x + h)$.

Replace x with $x + h$ in the equation for $f(x)$. Simplify the result.

$$\begin{aligned} f(x) &= x^2 \\ f(x + h) &= (x + h)^2 \\ &= x^2 + 2xh + h^2 \end{aligned}$$

(Note that $f(x + h) \neq f(x) + h$, since $f(x) + h = x^2 + h$.)

Step 2 Find $f(x + h) - f(x)$.

Since $f(x) = x^2$,

$$f(x + h) - f(x) = (x^2 + 2xh + h^2) - x^2 = 2xh + h^2.$$

Step 3 Find and simplify the quotient $\dfrac{f(x + h) - f(x)}{h}$.

$$\begin{aligned} \frac{f(x + h) - f(x)}{h} &= \frac{2xh + h^2}{h} \\ &= \frac{h(2x + h)}{h} = 2x + h \end{aligned}$$

Step 4 Finally, find the limit as h approaches 0. In this step, h is the variable and x is fixed.

$$\begin{aligned} f'(x) &= \lim_{h \to 0} \frac{f(x + h) - f(x)}{h} \\ &= \lim_{h \to 0} (2x + h) \\ &= 2x + 0 = 2x \end{aligned}$$

(b) Calculate and interpret $f'(3)$. Use the function defined by $f'(x) = 2x$.

$$f'(3) = 2 \cdot 3 = 6$$

The number 6 is the slope of the tangent line to the graph of $f(x) = x^2$ at the point where $x = 3$, that is, at $(3, f(3)) = (3, 9)$. See Figure 18. ◀

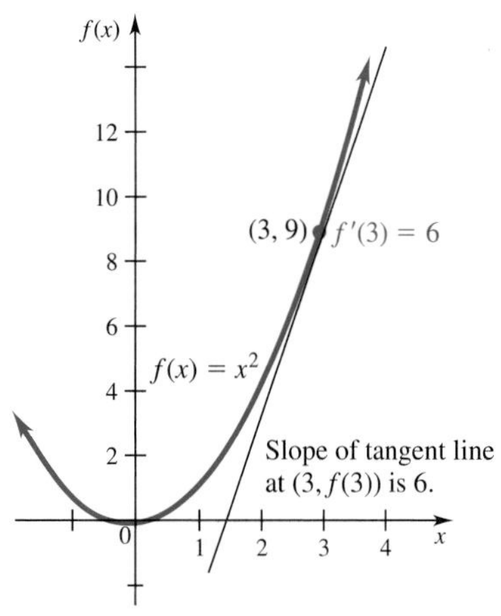

FIGURE 18

Caution In Example 2(b), do not confuse $f(3)$ and $f'(3)$. The value $f(3)$ is the y-value that corresponds to $x = 3$. It is found by substituting 3 for x in $f(x)$; $f(3) = 3^2 = 9$. On the other hand, $f'(3)$ is the slope of the tangent line to the curve at $x = 3$; as Example 2 shows, $f'(3) = 2 \cdot 3 = 6$.

FINDING $f'(x)$ FROM THE DEFINITION OF DERIVATIVE

The four steps used when finding the derivative $f'(x)$ for a function $y = f(x)$ are summarized here.

1. Find $f(x + h)$.
2. Find and simplify $f(x + h) - f(x)$.
3. Divide by h to get $\dfrac{f(x + h) - f(x)}{h}$.
4. Let $h \to 0$; $f'(x) = \lim_{h \to 0} \dfrac{f(x + h) - f(x)}{h}$ if this limit exists.

EXAMPLE 3

Let $f(x) = 2x^3 + 4x$. Find $f'(x)$, $f'(2)$, and $f'(-3)$.

Go through the four steps used above to find $f'(x)$.

Step 1 Find $f(x + h)$ by replacing x with $x + h$.

$$\begin{aligned} f(x + h) &= 2(x + h)^3 + 4(x + h) \\ &= 2(x^3 + 3x^2h + 3xh^2 + h^3) + 4(x + h) \\ &= 2x^3 + 6x^2h + 6xh^2 + 2h^3 + 4x + 4h \end{aligned}$$

Step 2 $f(x + h) - f(x) = 2x^3 + 6x^2h + 6xh^2 + 2h^3 + 4x + 4h - 2x^3 - 4x$
$= 6x^2h + 6xh^2 + 2h^3 + 4h$

Step 3

$$\frac{f(x+h) - f(x)}{h} = \frac{6x^2h + 6xh^2 + 2h^3 + 4h}{h}$$
$$= \frac{h(6x^2 + 6xh + 2h^2 + 4)}{h}$$
$$= 6x^2 + 6xh + 2h^2 + 4$$

Step 4 Now use the rules for limits to get

$$f'(x) = \lim_{h \to 0} \frac{f(x+h) - f(x)}{h}$$
$$= \lim_{h \to 0} (6x^2 + 6xh + 2h^2 + 4)$$
$$= 6x^2 + 6x(0) + 2(0)^2 + 4$$
$$f'(x) = 6x^2 + 4.$$

Use this result to find $f'(2)$ and $f'(-3)$.

$$f'(2) = 6 \cdot 2^2 + 4$$
$$= 28$$
$$f'(-3) = 6 \cdot (-3)^2 + 4 = 58$$

EXAMPLE 4 Let $f(x) = \frac{4}{x}$. Find $f'(x)$.

Step 1 $f(x + h) = \frac{4}{x + h}$

Step 2

$$f(x+h) - f(x) = \frac{4}{x+h} - \frac{4}{x}$$
$$= \frac{4x - 4(x+h)}{x(x+h)} \quad \text{Find a common denominator.}$$
$$= \frac{4x - 4x - 4h}{x(x+h)} \quad \text{Simplify the numerator.}$$
$$= \frac{-4h}{x(x+h)}$$

Step 3

$$\frac{f(x+h) - f(x)}{h} = \frac{\frac{-4h}{x(x+h)}}{h}$$
$$= \frac{-4h}{x(x+h)} \cdot \frac{1}{h} \quad \text{Invert and multiply.}$$
$$= \frac{-4}{x(x+h)}$$

Step 4 $$f'(x) = \lim_{h\to 0} \frac{f(x+h) - f(x)}{h}$$

$$= \lim_{h\to 0} \frac{-4}{x(x+h)}$$

$$= \frac{-4}{x(x+0)}$$

$$f'(x) = \frac{-4}{x(x)} = \frac{-4}{x^2}$$ ◀

Notice that in Example 4 neither $f(x)$ nor $f'(x)$ is defined when $x = 0$.

EXAMPLE 5 Let $f(x) = \sqrt{x}$. Find $f'(x)$.

Step 1 $f(x+h) = \sqrt{x+h}$

Step 2 $f(x+h) - f(x) = \sqrt{x+h} - \sqrt{x}$

Step 3 $$\frac{f(x+h) - f(x)}{h} = \frac{\sqrt{x+h} - \sqrt{x}}{h}$$

At this point, in order to be able to divide by h, multiply both numerator and denominator by $\sqrt{x+h} + \sqrt{x}$; that is, rationalize the *numerator*.

$$\frac{f(x+h) - f(x)}{h} = \frac{\sqrt{x+h} - \sqrt{x}}{h} \cdot \frac{\sqrt{x+h} + \sqrt{x}}{\sqrt{x+h} + \sqrt{x}}$$

$$= \frac{(\sqrt{x+h})^2 - (\sqrt{x})^2}{h(\sqrt{x+h} + \sqrt{x})}$$

$$= \frac{x + h - x}{h(\sqrt{x+h} + \sqrt{x})}$$

$$= \frac{1}{\sqrt{x+h} + \sqrt{x}}$$

Step 4 $$f'(x) = \lim_{h\to 0} \frac{1}{\sqrt{x+h} + \sqrt{x}} = \frac{1}{\sqrt{x} + \sqrt{x}} = \frac{1}{2\sqrt{x}}$$ ◀

EXAMPLE 6 The cost in dollars to manufacture x graphing calculators is given by $C(x) = -.005x^2 + 20x + 150$. Find the rate of change of cost with respect to the number manufactured when 100 calculators are made and when 1000 calculators are made.

The rate of change of cost is given by the derivative of the cost function,

$$C'(x) = \lim_{h\to 0} \frac{C(x+h) - C(x)}{h}.$$

Going through the steps for finding $C'(x)$ gives

$$C'(x) = -.01x + 20.$$

When $x = 100$,

$$C'(100) = -.01(100) + 20 = 19.$$

This rate of change of cost per calculator gives the marginal cost at $x = 100$, which means the approximate cost of producing the 101st calculator is \$19.

When 1000 calculators are made, the marginal cost is

$$C'(1000) = -.01(1000) + 20 = 10,$$

or \$10. ◀

EXAMPLE 7 A sales representative for a textbook publishing company frequently makes a four-hour drive from her home in a large city to a university in another city. During several trips, she periodically jotted down the distance (in miles) she had traveled from home. This data is shown by the points in Figure 19. She was a mathematics major in college, so she decided to find a function that described her distance from home at time t (in hours). The position of the points suggested the curve shown in the figure, with equation $s(t) = -5t^3 + 30t^2$.

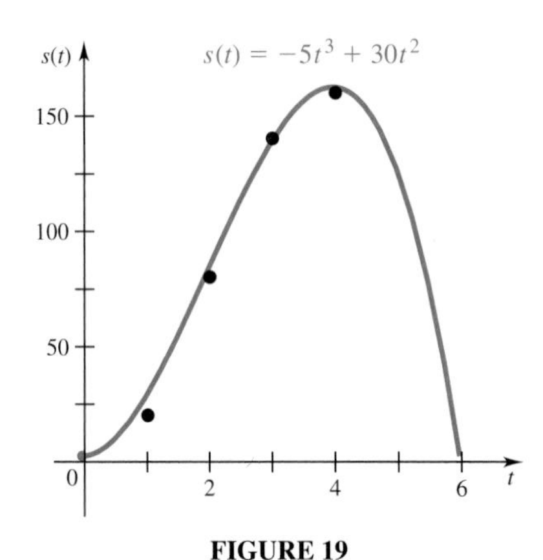

FIGURE 19

Notice that this function is suitable only over the domain [0, 4]. Use the function to answer the following questions.

(a) How far from home will she be after 1 hour? After 1 1/2 hours?

Her distance from home after 1 hour is

$$s(1) = -5(1)^3 + 30(1)^2 = 25,$$

or 25 miles. After 1 1/2 (or 3/2) hours, her distance from home is

$$s\left(\frac{3}{2}\right) = -5\left(\frac{3}{2}\right)^3 + 30\left(\frac{3}{2}\right)^2 = \frac{405}{8} = 50.625,$$

or 50.625 miles.

(b) How far apart are the two cities?

Since the trip takes 4 hours and the distance is given by $s(t)$, the university city is $s(4) = 160$ miles from her home.

(c) How fast is she driving 1 hour into the trip? $1\frac{1}{2}$ hours into the trip?

Velocity (or speed) is the instantaneous rate of change in position with respect to time. We need to find the value of the derivative $s'(t)$ at $t = 1$ and $t = 1\frac{1}{2}$. Going through the four steps for finding the derivative gives $s'(t) = -15t^2 + 60t$ as the velocity of the car at any time t. At $t = 1$, the velocity is

$$s'(1) = -15(1)^2 + 60(1) = 45,$$

or 45 miles per hour. At $t = 1\frac{1}{2}$, the velocity is

$$s'\left(\frac{3}{2}\right) = -15\left(\frac{3}{2}\right)^2 + 60\left(\frac{3}{2}\right) \approx 56,$$

or about 56 miles per hour.

(d) Does she ever exceed the speed limit of 65 miles per hour on the trip?

To find the maximum velocity, notice that the graph of the velocity function $s'(t) = -15t^2 + 60t$ is a parabola opening downward. The maximum velocity will occur at the vertex. Verify that the vertex of this parabola is (2, 60). Thus, her maximum velocity during the trip is 60 miles per hour, so she never exceeds the speed limit. ◀

Existence of the Derivative The definition of the derivative included the phrase "provided this limit exists." If the limit used to define the derivative does not exist, then of course the derivative does not exist. For example, a derivative cannot exist at a point where the function itself is not defined. If there is no function value for a particular value of x, there can be no tangent line for that value. This was the case in Example 4—there was no tangent line (and no derivative) when $x = 0$.

Derivatives also do not exist at "corners" or "sharp points" on a graph. For example, the function graphed in Figure 20 is the **absolute value function,** defined as

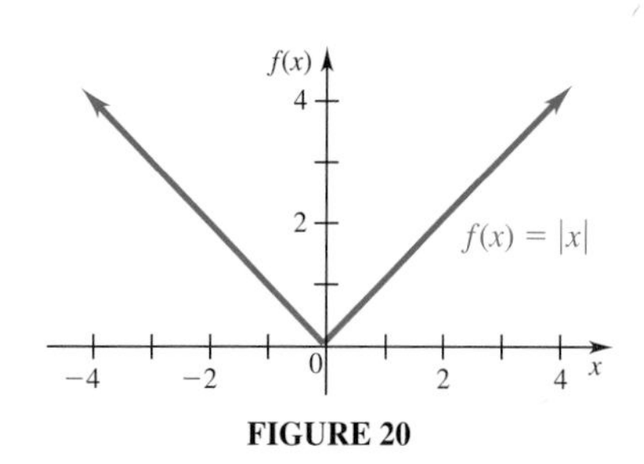

FIGURE 20

$$f(x) = \begin{cases} x & \text{if } x \geq 0 \\ -x & \text{if } x < 0, \end{cases}$$

and written $f(x) = |x|$. By the definition of derivative, the derivative at any value of x is given by

$$\lim_{h \to 0} \frac{f(x + h) - f(x)}{h},$$

provided this limit exists. To find the derivative at 0 for $f(x) = |x|$, replace x with 0 and $f(x)$ with $|0|$ to get

$$\lim_{h \to 0} \frac{|0 + h| - |0|}{h} = \lim_{h \to 0} \frac{|h|}{h}.$$

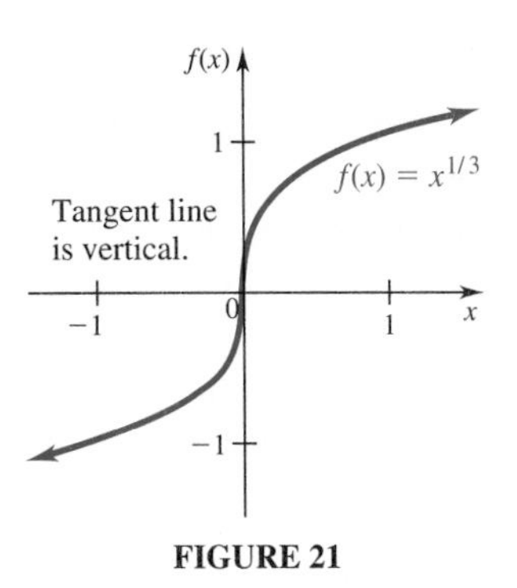

FIGURE 21

In Example 2 in the first section of this chapter, we showed that

$$\lim_{h\to 0} \frac{|h|}{h} \text{ does not exist;}$$

therefore, the derivative does not exist at 0. However, the derivative does exist for all values of x other than 0.

A graph of the function $f(x) = x^{1/3}$ is shown in Figure 21. As the graph suggests, the tangent line is vertical when $x = 0$. Since a vertical line has an undefined slope, the derivative of $f(x) = x^{1/3}$ cannot exist when $x = 0$.

Figure 22 summarizes the various ways that a derivative can fail to exist.

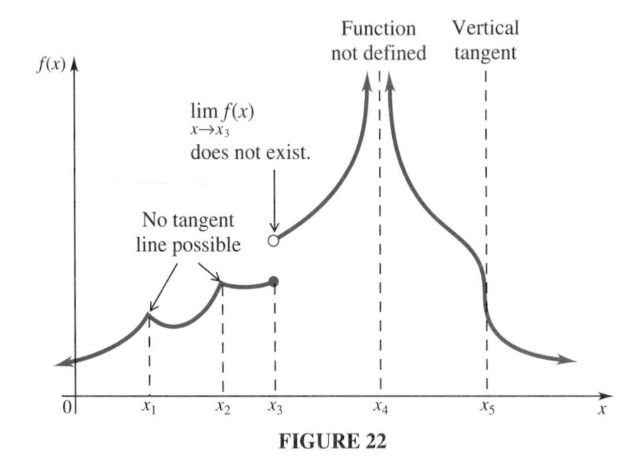

FIGURE 22

EXAMPLE 8 A nova is a star whose brightness suddenly increases and then gradually fades. The cause of the sudden increase in brightness is thought to be an explosion of some kind. The intensity of light emitted by a nova as a function of time is shown in Figure 23.* Notice that although the graph is a continuous curve, it is not differentiable at the point of the explosion. ◀

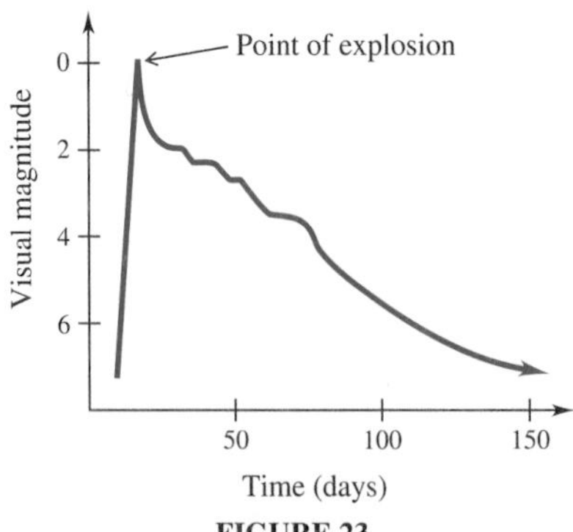

FIGURE 23

*"The Light Curve of a Nova" from *Astronomy: The Structure of the Universe* by William J. Kaufmann, III. Reprinted by permission of William J. Kaufmann, III.

2.3 Exercises

Find the slope of the tangent line to each curve when x has the given value. (Hint for Exercise 5: *In Step 3, multiply numerator and denominator by* $\sqrt{16 + h} + \sqrt{16}$.)

1. $f(x) = -4x^2 + 11x$; $x = -2$
2. $f(x) = 6x^2 - 4x$; $x = -1$
3. $f(x) = -2/x$; $x = 4$
4. $f(x) = 6/x$; $x = -1$
5. $f(x) = \sqrt{x}$; $x = 16$
6. $f(x) = -3\sqrt{x}$; $x = 1$

Find the equation of the tangent line to each curve when x has the given value.

7. $f(x) = x^2 + 2x$; $x = 3$
8. $f(x) = 6 - x^2$; $x = -1$
9. $f(x) = 5/x$; $x = 2$
10. $f(x) = -3/(x + 1)$; $x = 1$
11. $f(x) = 4\sqrt{x}$; $x = 9$
12. $f(x) = \sqrt{x}$; $x = 25$

Estimate the slope of the tangent line to each curve at the given point (x, y).

13.

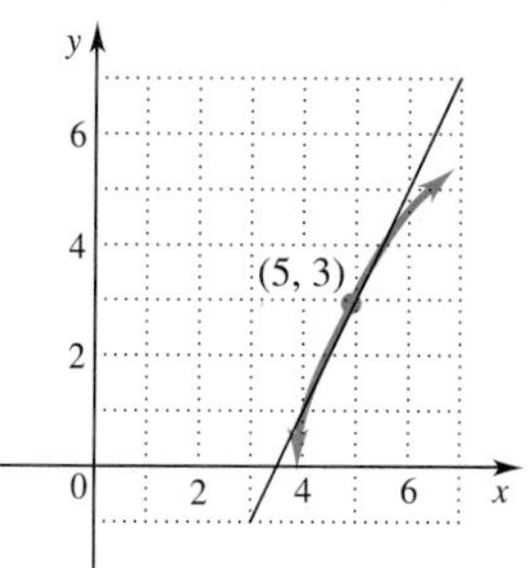

14.

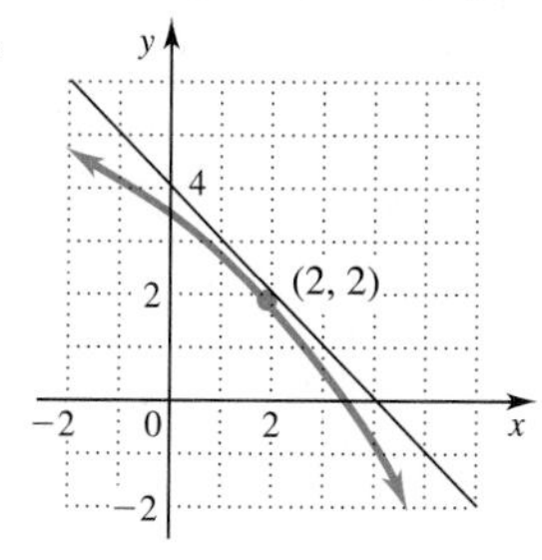

15.

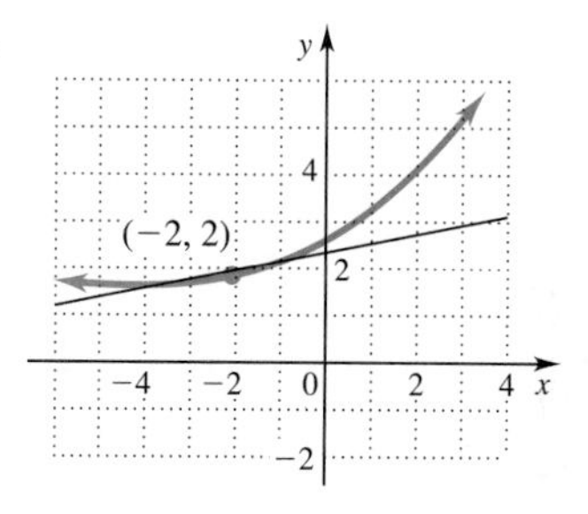

16.

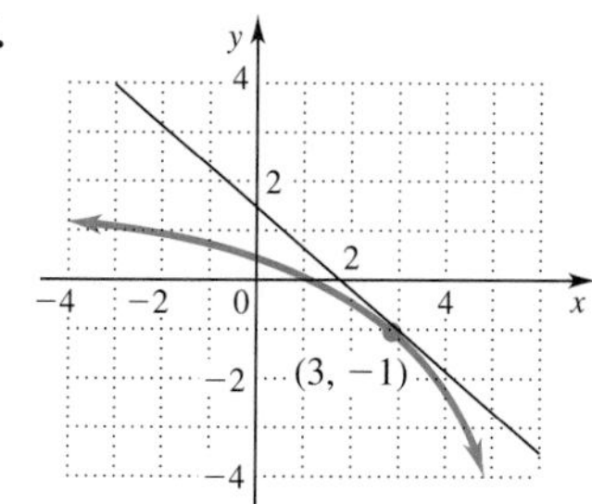

17.

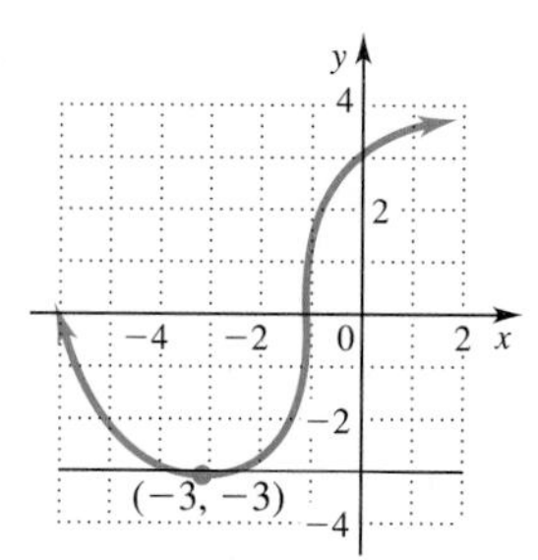

18.

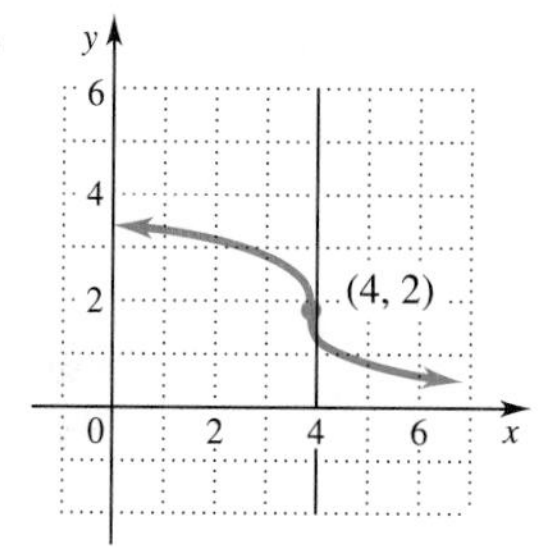

19. By considering, but not calculating, the slope of the tangent line, give the derivative of each of the following.
(a) $f(x) = 5$ **(b)** $f(x) = x$ **(c)** $f(x) = -x$ **(d)** the line $x = 3$ **(e)** the line $y = mx + b$

20. (a) Suppose $g(x) = \sqrt[3]{x}$. Find $g'(0)$
(b) Explain why the derivative of a function does not exist at a point where the tangent line is vertical.

21. If $f(x) = \dfrac{x^2 - 1}{x + 2}$, where is f not differentiable?

Find $f'(x)$, $f'(2)$, $f'(0)$, *and* $f'(-3)$ *for each of the following.*

22. $f(x) = -4x^2 + 11x$
23. $f(x) = 6x^2 - 4x$
24. $f(x) = 8x + 6$
25. $f(x) = -9x - 5$
26. $f(x) = -\dfrac{2}{x}$
27. $f(x) = \dfrac{6}{x}$
28. $f(x) = \sqrt{x}$
29. $f(x) = -3\sqrt{x}$

Find the x-values where the following do not have derivatives.

30.

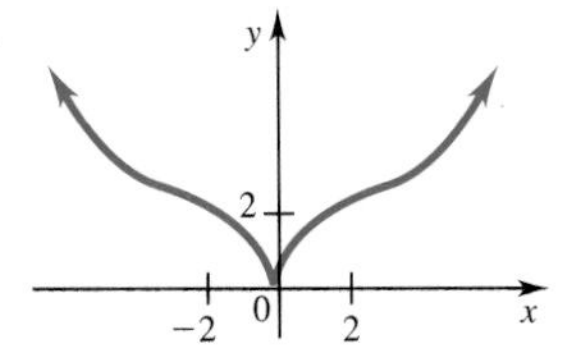

31.

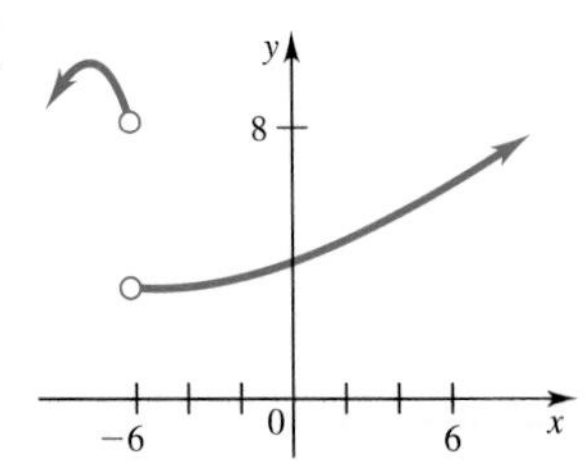

32.

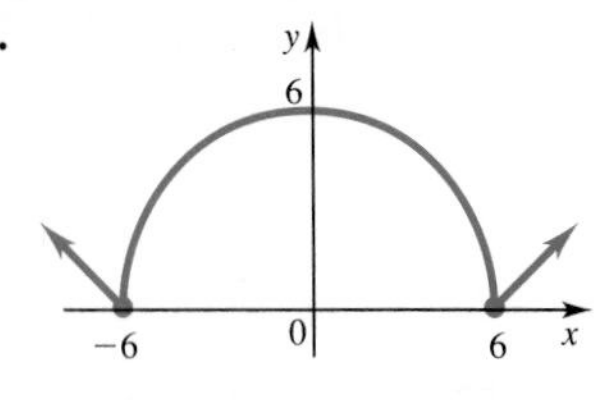

33.

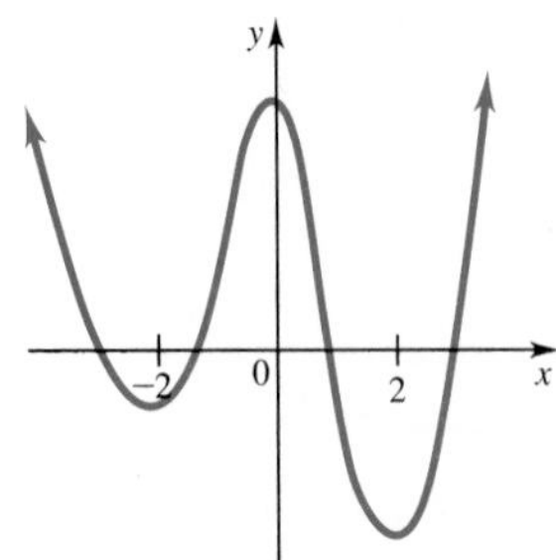

34.

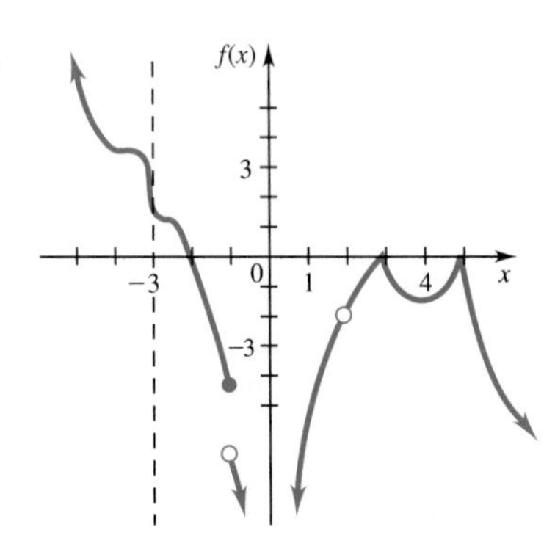

35.

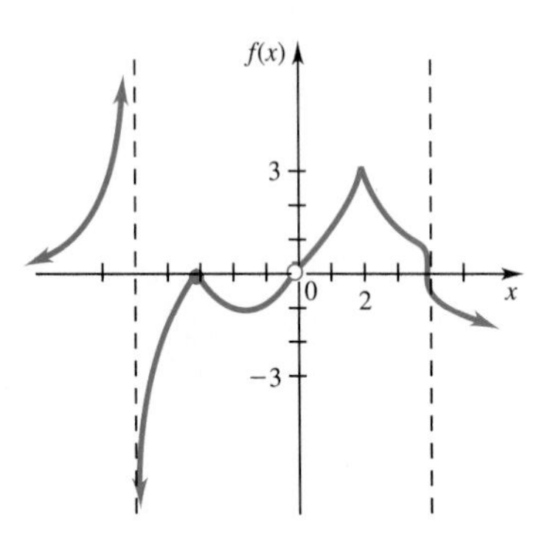

36. If the rate of change of $f(x)$ is zero when $x = a$, what can be said about the tangent line to the graph of $f(x)$ at $x = a$?

37. For the function shown in the sketch, give the intervals or points on the x-axis where the rate of change of $f(x)$ with respect to x is **(a)** positive; **(b)** negative; and **(c)** zero.

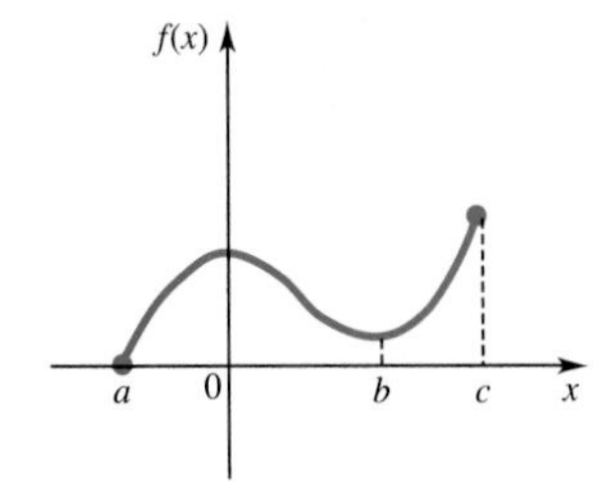

In Exercises 38–39, tell which graph, **(a)** *or* **(b)**, *represents velocity and which represents distance from a starting point. (*Hint: *Consider where the derivative is zero, positive, or negative.)*

38. (a)

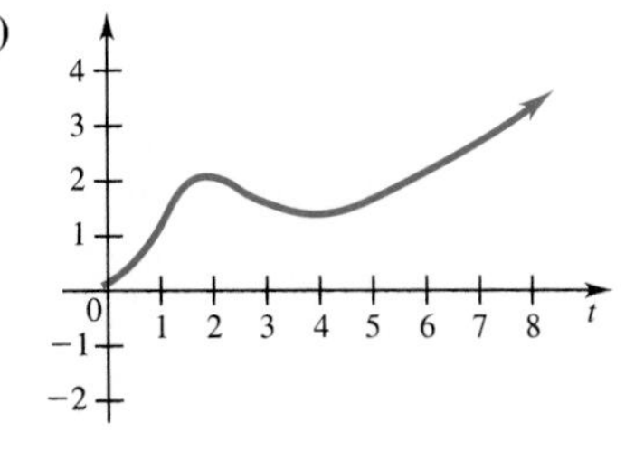

(b)

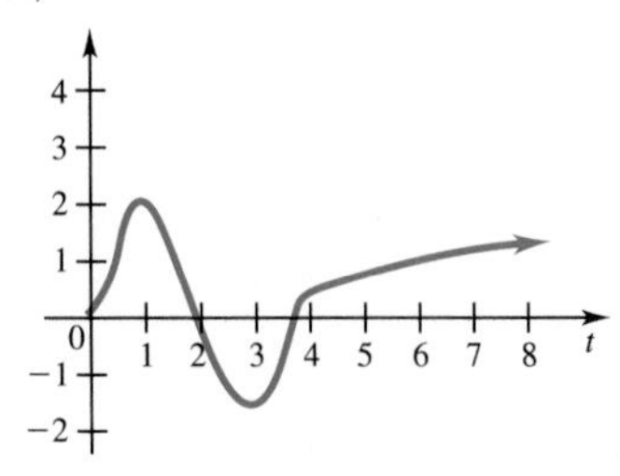

39. (a)

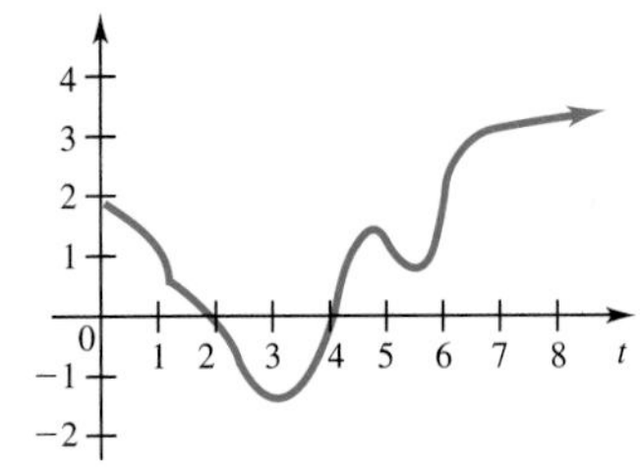

(b)

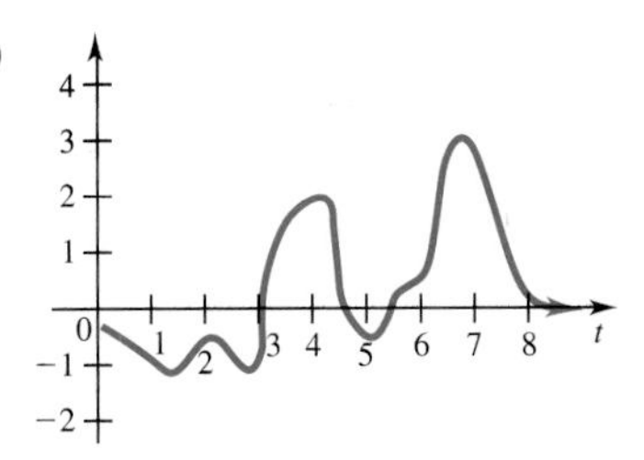

Applications

Business and Economics

40. **Demand** Suppose the demand for a certain item is given by $D(p) = -2p^2 + 4p + 6$, where p represents the price of the item in dollars.

 (a) Given $D'(p) = -4p + 4$, find the rate of change of demand with respect to price.

 (b) Find and interpret the rate of change of demand when the price is \$10.

41. **Debt** The figure gives the percent of outstanding credit card loans that were at least 30 days past due from June 1989 to March 1991.* Assume the function changes smoothly.

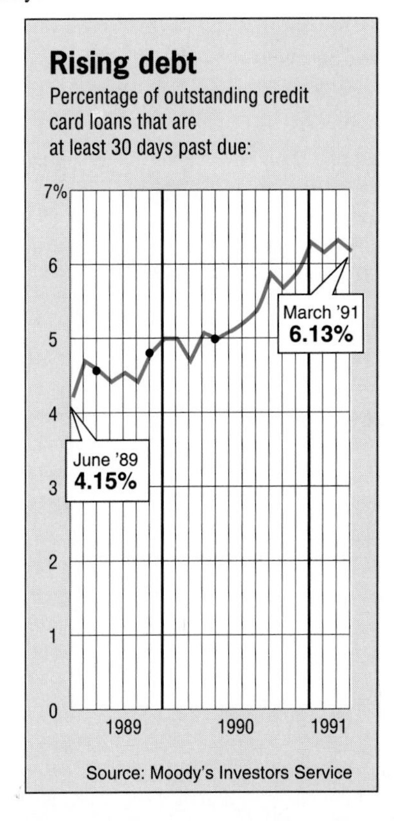

 (a) Estimate and interpret the rate of change in the debt level at the indicated points on the curve.

 (b) During which months in 1990 did the rate of change indicate a decreasing debt level?

 (c) The overall direction of the curve is upward. What does this say about the average rate of change over the time period shown in the figure?

42. **Profit** The profit (in dollars) from the expenditure of x thousand dollars on advertising is given by $P(x) = 1000 + 32x - 2x^2$, with $P'(x) = 32 - 4x$. Find the marginal profit at the following expenditures. In each case, decide whether the firm should increase the expenditure.

 (a) \$8000 (b) \$6000 (c) \$12,000 (d) \$20,000

43. **Revenue** The revenue in dollars generated from the sale of x picnic tables is given by $R(x) = 20x - \frac{x^2}{500}$; therefore, $R'(x) = 20 - \frac{1}{250}x$.

 (a) Find the marginal revenue when 1000 tables are sold.

 (b) Estimate the revenue from the sale of the 1001st table by finding $R'(1000)$.

 (c) Determine the actual revenue from the sale of the 1001st table.

 (d) Compare your answers for (b) and (c). What do you find?

44. **Cost** The cost of producing x tacos is $C(x) = 1000 + .24x^2$, $0 \leq x \leq 30{,}000$; therefore, $C'(x) = .48x$.

 (a) Find the marginal cost.

 (b) Find and interpret the marginal cost at a production level of 100 tacos.

 (c) Find the exact cost to produce the 101st taco.

 (d) Compare the answers to (b) and (c). How are they related?

*Credit Card Defaults Pose Worry for Banks'' as appeared in *The Sacramento Bee*, June 1991. Reprinted by permission of The Associated Press.

Life Sciences

45. Bacteria Population A biologist has estimated that if a bactericide is introduced into a culture of bacteria, the number of bacteria, $B(t)$, present at time t (in hours) is given by $B(t) = 1000 + 50t - 5t^2$ million. If $B'(t) = 50 - 10t$, find the rate of change of the number of bacteria with respect to time after each of the following numbers of hours: **(a)** 3 **(b)** 5 **(c)** 6.

(d) When does the population of bacteria start to decline?

46. Shellfish Population In one research study, the population of a certain shellfish in an area at time t was closely approximated by the graph below. Estimate and interpret the derivative at each of the marked points.

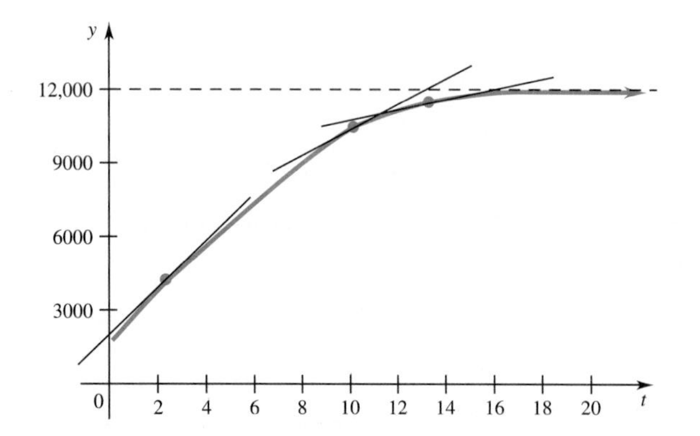

47. Flight Speed The graph shows the relationship between the speed of the arctic tern in flight and the required power expended by its flight muscles.* Several significant flight speeds are indicated on the curve.

(a) The speed V_{mp} minimizes energy costs per unit of time. What is the slope of the line tangent to the curve at the point corresponding to V_{mp}? What is the physical significance of the slope at that point?

(b) The speed V_{mr} minimizes the energy costs per unit of distance covered. Estimate the slope of the curve at the point corresponding to V_{mr}. Give the significance of the slope at that point.

(c) The speed V_{opt} minimizes the total duration of the migratory journey. Estimate the slope of the curve at the point corresponding to V_{opt}. Relate the significance of this slope to the slopes found in (a) and (b).

(d) By looking at the shape of the curve, describe how the power level decreases and increases for various speeds.

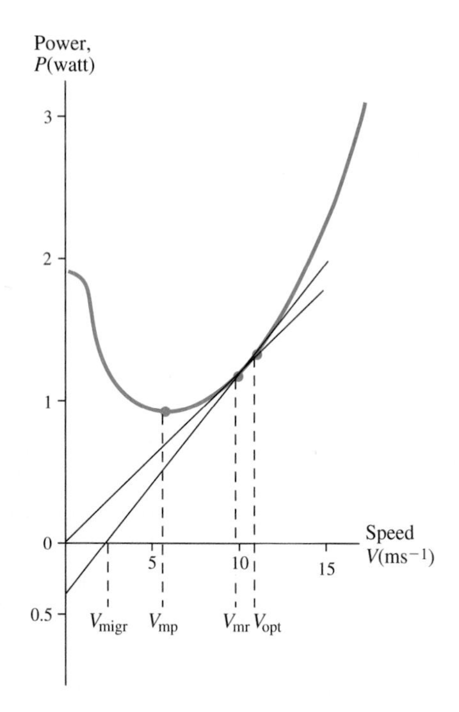

Physical Sciences

48. Temperature The graph shows the temperature in degrees Celsius as a function of the altitude h in feet when an inversion layer is over Southern California. (See Exercise 32 in the previous section.) Estimate and interpret the derivatives of $T(h)$ at the marked points.

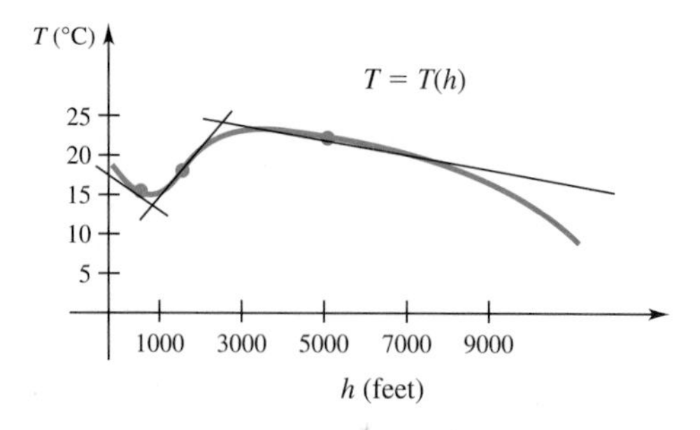

*"Bird Flight and Optimal Migration" by Thomas Alerstam from *Trends in Ecology and Evolution* July 1991. Copyright © 1991 by Elsevier Trends Journals. Reprinted by permission of Elsevier Trends Journals and Thomas Alerstam.

Connections

49. In Exercise 46 a graph was given of a function f that approximated the population of a certain shellfish at time t. In that exercise, the derivative was estimated at each of the marked points.

(a) Fill in the chart below with the estimate of the derivative at each of the points indicated.

t	2	10	13
$f'(t)$			

(b) Plot the points found in (a) and connect them with a smooth curve.

(c) The graph in part (b) is a graph of the derivative of the original function. Discuss what this graph tells of the behavior of the original function.

50. Use the graph given in Exercise 48 to repeat the steps in Exercise 49. Use h-values of 500, 1500, 3500, and 5000. You will need to draw the tangent line at $h = 3500$ yourself.

2.4 TECHNIQUES FOR FINDING DERIVATIVES

How can a manager determine the best production level if the relationship between profit and production is known?

From the shape of a tumor, how can a physician determine how fast the tumor is growing?

These questions can be answered by finding the derivative of an appropriate function. We shall return to them at the end of this section in Examples 7 and 8.

Using the definition to calculate the derivative of a function is a very involved process even for simple functions. In this section we develop rules that make the calculation of derivatives much easier. Keep in mind that even though the process of finding a derivative will be greatly simplified with these rules, *the interpretation of the derivative will not change.* But first, a few words about notation are in order.

Several alternative notations for the derivative are used. In the previous section the symbols y' and $f'(x)$ were used to represent the derivative of $y = f(x)$. Sometimes it is important to show that the derivative is taken with respect to a particular variable; for example, if y is a function of x, the notation

$$\frac{dy}{dx} \quad \text{or} \quad D_x y$$

(both read "the derivative of y with respect to x") can be used for the derivative of y with respect to x. The dy/dx notation for the derivative is sometimes referred to as *Leibniz notation,* named after one of the co-inventors of the calculus, Gottfried Wilhelm von Leibniz (1646–1716). (The other was Sir Isaac Newton, 1642–1727.)

Another notation is used to write a derivative without functional symbols such as f or f'. With this notation, the derivative of $y = 2x^3 + 4x$, for example, which was found in the last section to be $y' = 6x^2 + 4$, would be written

$$\frac{d}{dx}[2x^3 + 4x] = 6x^2 + 4, \quad \text{or} \quad D_x[2x^3 + 4x] = 6x^2 + 4.$$

Either $\frac{d}{dx}[f(x)]$ or $D_x[f(x)]$ represents the derivative of the function f with respect to x.

NOTATIONS FOR THE DERIVATIVE

The derivative of $y = f(x)$ may be written in any of the following ways:

$$\mathbf{f'(x), \quad y', \quad \frac{dy}{dx}, \quad \frac{d}{dx}[f(x)], \quad \text{or} \quad D_x[f(x)].}$$

A variable other than x often may be used as the independent variable. For example, if $y = f(t)$ gives population growth as a function of time, then the derivative of y with respect to t could be written

$$f'(t), \quad \frac{dy}{dt}, \quad \frac{d}{dt}[f(t)], \quad \text{or} \quad D_t[f(t)].$$

Other variables also may be used to name the function, as in $g(x)$ or $h(t)$.

Now we will use the definition

$$f'(x) = \lim_{h \to 0} \frac{f(x + h) - f(x)}{h}$$

to develop some rules for finding derivatives more easily than by the four-step process given in the previous section.

The first rule tells how to find the derivative of a constant function defined by $f(x) = k$, where k is a constant real number. Since $f(x + h)$ is also k, by definition $f'(x)$ is

$$\begin{aligned} f'(x) &= \lim_{h \to 0} \frac{f(x + h) - f(x)}{h} \\ &= \lim_{h \to 0} \frac{k - k}{h} = \lim_{h \to 0} \frac{0}{h} = \lim_{h \to 0} 0 = 0, \end{aligned}$$

establishing the following rule.

CONSTANT RULE

If $f(x) = k$, where k is any real number, then

$$\mathbf{f'(x) = 0.}$$

(The derivative of a constant is 0.)

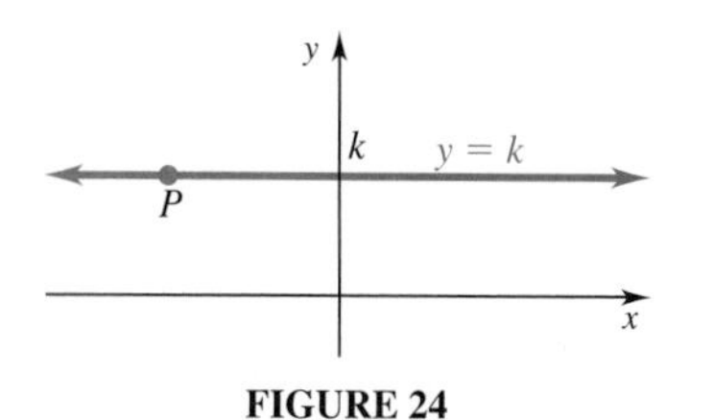

FIGURE 24

Figure 24 illustrates this constant rule geometrically; it shows a graph of the horizontal line $y = k$. At any point P on this line, the tangent line at P is the line itself. Since a horizontal line has a slope of 0, the slope of the tangent line is 0. This agrees with the result above: the derivative of a constant is 0.

EXAMPLE 1

(a) If $f(x) = 9$, then $f'(x) = 0$.

(b) If $y = \pi$, then $y' = 0$.

(c) If $y = 2^3$, then $dy/dx = 0$. ◀

Functions of the form $y = x^n$, where n is a real number, are very common in applications. To get a rule for finding the derivative of such a function, we can use the definition to work out the derivatives for various special values of n. This was done in the previous section in Example 2 to show that for $f(x) = x^2$, $f'(x) = 2x$.

For $f(x) = x^3$, the derivative is found as follows.

$$\begin{aligned} f'(x) &= \lim_{h\to 0} \frac{f(x+h) - f(x)}{h} \\ &= \lim_{h\to 0} \frac{(x+h)^3 - x^3}{h} \\ &= \lim_{h\to 0} \frac{(x^3 + 3x^2h + 3xh^2 + h^3) - x^3}{h} \end{aligned}$$

The binomial theorem (discussed in most intermediate and college algebra texts) was used to expand $(x + h)^3$ in the last step. Now, the limit can be determined.

$$\begin{aligned} f'(x) &= \lim_{h\to 0} \frac{3x^2h + 3xh^2 + h^3}{h} \\ &= \lim_{h\to 0} (3x^2 + 3xh + h^2) \\ &= 3x^2 \end{aligned}$$

The results in the table below were found in a similar way, using the definition of the derivative. (These results are modifications of some of the examples and exercises from the previous section.)

Function	n	*Derivative*
$y = x^2$	2	$y' = 2x = 2x^1$
$y = x^3$	3	$y' = 3x^2$
$y = x^4$	4	$y' = 4x^3$
$y = x^{-1}$	-1	$y' = -1 \cdot x^{-2} = \frac{-1}{x^2}$
$y = x^{1/2}$	1/2	$y' = \frac{1}{2}x^{-1/2} = \frac{1}{2x^{1/2}}$

These results suggest the following rule.

POWER RULE

If $f(x) = x^n$ for any real number n, then

$$f'(x) = nx^{n-1}.$$

(The derivative of $f(x) = x^n$ is found by multiplying by the exponent n and decreasing the exponent on x by 1.)

While the power rule is true for every real-number value of n, a proof is given here only for positive integer values of n. This proof follows the steps used above in finding the derivative of $y = x^3$.

For any real numbers p and q, by the binomial theorem,

$$(p + q)^n = p^n + np^{n-1}q + \frac{n(n-1)}{2}p^{n-2}q^2 + \cdots + npq^{n-1} + q^n.$$

Replacing p with x and q with h gives

$$(x + h)^n = x^n + nx^{n-1}h + \frac{n(n-1)}{2}x^{n-2}h^2 + \cdots + nxh^{n-1} + h^n,$$

from which

$$(x + h)^n - x^n = nx^{n-1}h + \frac{n(n-1)}{2}x^{n-2}h^2 + \cdots + nxh^{n-1} + h^n.$$

Dividing each term by h yields

$$\frac{(x + h)^n - x^n}{h} = nx^{n-1} + \frac{n(n-1)}{2}x^{n-2}h + \cdots + nxh^{n-2} + h^{n-1}.$$

Use the definition of derivative, and the fact that each term except the first contains h as a factor and thus approaches 0 as h approaches 0, to get

$$f'(x) = \lim_{h \to 0} \frac{(x + h)^n - x^n}{h}$$

$$= nx^{n-1} + \frac{n(n-1)}{2}x^{n-2}0 + \cdots + nx0^{n-2} + 0^{n-1}$$

$$f'(x) = nx^{n-1}.$$

This shows that the derivative of $f(x) = x^n$ is $f'(x) = nx^{n-1}$, proving the power rule for positive integer values of n.

(a) If $y = x^6$, find y'.

$$y' = 6x^{6-1} = 6x^5$$

(b) If $y = t = t^1$, find y'.

$$y' = 1t^{1-1} = t^0 = 1$$

At this point you may wish to turn back to the Algebra Reference for a review of negative exponents and rational exponents. The relationship between powers, roots, and rational exponents is explained there.

(c) If $y = 1/x^3$, find dy/dx.

Use a negative exponent to rewrite this equation as $y = x^{-3}$; then

$$\frac{dy}{dx} = -3x^{-3-1} = -3x^{-4} \quad \text{or} \quad \frac{-3}{x^4}.$$

(d) Find $D_x(x^{4/3})$.

$$D_x(x^{4/3}) = \frac{4}{3}x^{4/3-1} = \frac{4}{3}x^{1/3}$$

(e) If $y = \sqrt{z}$, find dy/dz.

Rewrite this as $y = z^{1/2}$; then

$$\frac{dy}{dz} = \frac{1}{2}z^{1/2-1} = \frac{1}{2}z^{-1/2} \quad \text{or} \quad \frac{1}{2z^{1/2}} \quad \text{or} \quad \frac{1}{2\sqrt{z}}.$$

The next rule shows how to find the derivative of the product of a constant and a function.

CONSTANT TIMES A FUNCTION

Let k be a real number. If $f'(x)$ exists, then

$$\boldsymbol{D_x[kf(x)] = kf'(x).}$$

(The derivative of a constant times a function is the constant times the derivative of the function.)

This rule is proved with the definition of the derivative and rules for limits.

$$\begin{aligned} D_x[kf(x)] &= \lim_{h\to 0} \frac{kf(x+h) - kf(x)}{h} \\ &= \lim_{h\to 0} k\frac{[f(x+h) - f(x)]}{h} && \text{Factor out } k. \\ &= k \lim_{h\to 0} \frac{f(x+h) - f(x)}{h} && \text{Limit rule 1} \\ &= kf'(x) && \text{Definition of derivative} \end{aligned}$$

(a) If $y = 8x^4$, find y'.

$$y' = 8(4x^3) = 32x^3$$

(b) If $y = -\frac{3}{4}x^{12}$, find dy/dx.

$$\frac{dy}{dx} = -\frac{3}{4}(12x^{11}) = -9x^{11}$$

(c) Find $D_t(-8t)$.

$$D_t(-8t) = -8(1) = -8$$

(d) Find $D_p(10p^{3/2})$.

$$D_p(10p^{3/2}) = 10\left(\frac{3}{2}p^{1/2}\right) = 15p^{1/2}$$

(e) If $y = \frac{6}{x}$, find $\frac{dy}{dx}$.

Rewrite this as $y = 6x^{-1}$; then

$$\frac{dy}{dx} = 6(-1x^{-2}) = -6x^{-2} \quad \text{or} \quad \frac{-6}{x^2}.$$ ◀

The final rule in this section is for the derivative of a function that is a sum or difference of terms.

SUM OR DIFFERENCE RULE

If $f(x) = u(x) \pm v(x)$, and if $u'(x)$ and $v'(x)$ exist, then

$$\mathbf{f'(x) = u'(x) \pm v'(x).}$$

(The derivative of a sum or difference of functions is the sum or difference of the derivatives.)

The proof of the sum part of this rule is as follows: If $f(x) = u(x) + v(x)$, then

$$\begin{aligned} f'(x) &= \lim_{h\to 0} \frac{[u(x+h) + v(x+h)] - [u(x) + v(x)]}{h} \\ &= \lim_{h\to 0} \frac{[u(x+h) - u(x)] + [v(x+h) - v(x)]}{h} \\ &= \lim_{h\to 0} \left[\frac{u(x+h) - u(x)}{h} + \frac{v(x+h) - v(x)}{h}\right] \\ &= \lim_{h\to 0} \frac{u(x+h) - u(x)}{h} + \lim_{h\to 0} \frac{v(x+h) - v(x)}{h} \\ &= u'(x) + v'(x). \end{aligned}$$

A similar proof can be given for the difference of two functions.

Find the derivative of each function.

(a) $y = 6x^3 + 15x^2$

Let $f(x) = 6x^3$ and $g(x) = 15x^2$; then $y = f(x) + g(x)$. Since $f'(x) = 18x^2$ and $g'(x) = 30x$,

$$\frac{dy}{dx} = 18x^2 + 30x.$$

(b) $p(t) = 12t^4 - 6\sqrt{t} + \dfrac{5}{t}$

Rewrite $p(t)$ as $p(t) = 12t^4 - 6t^{1/2} + 5t^{-1}$; then

$$p'(t) = 48t^3 - 3t^{-1/2} - 5t^{-2}.$$

Also, $p'(t)$ may be written as $p'(t) = 48t^3 - 3/\sqrt{t} - 5/t^2$.

(c) $f(x) = 5\sqrt[3]{x^2} + 4x^{-2} + 7$

Rewrite $f(x)$ as $f(x) = 5x^{2/3} + 4x^{-2} + 7$.

Then
$$D_x f(x) = \frac{10}{3}x^{-1/3} - 8x^{-3} + 0,$$

or
$$D_x f(x) = \frac{10}{3\sqrt[3]{x}} - \frac{8}{x^3}. \blacktriangleleft$$

The rules developed in this section make it possible to find the derivative of a function more directly, so that applications of the derivative can be dealt with more effectively. The following examples illustrate some business applications.

Marginal Analysis In previous sections we discussed the concepts of marginal cost, marginal revenue, and marginal profit. These concepts are summarized here.

In business and economics the rates of change of such variables as cost, revenue, and profit are important considerations. Economists use the word marginal to refer to rates of change: for example, *marginal cost* refers to the rate of change of cost. Since the derivative of a function gives the rate of change of the function, a marginal cost (or revenue, or profit) function is found by taking the derivative of the cost (or revenue, or profit) function. Roughly speaking, the marginal cost at some level of production x is the cost to produce the $(x + 1)$st item. (Similar statements could be made for revenue or profit.)

To see why it is reasonable to say that the marginal cost function is approximately the cost of producing one more unit, look at Figure 25, where $C(x)$ represents the cost of producing x units of some item. Then the cost of producing $x + 1$ units is $C(x + 1)$. The cost of the $x + 1$st unit is therefore $C(x + 1) - C(x)$. This quantity is shown on the graph in Figure 25.

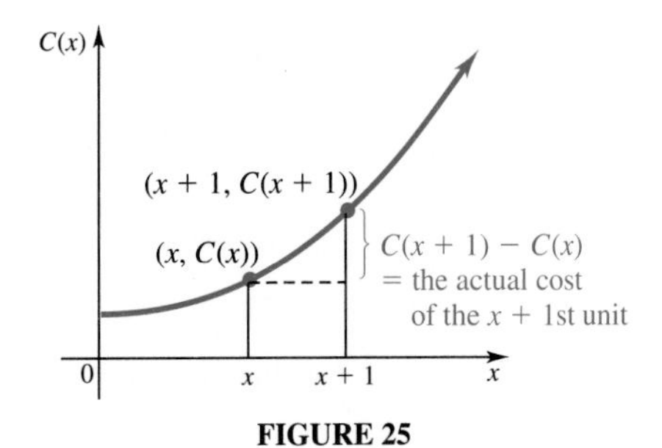

FIGURE 25

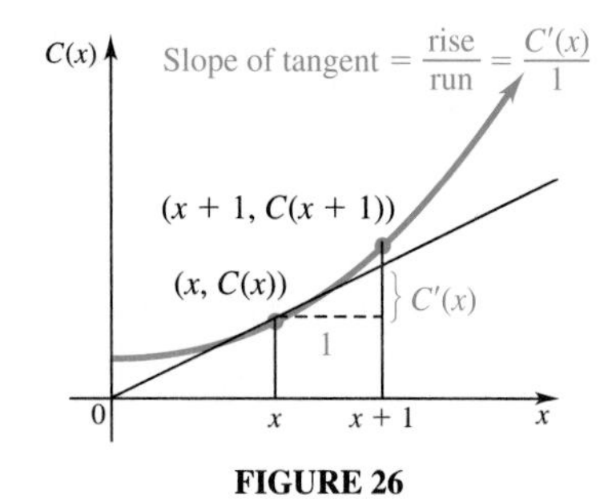

FIGURE 26

Now if $C(x)$ is the cost function, then the marginal cost $C'(x)$ represents the slope of the tangent line at any point $(x, C(x))$. The graph in Figure 26 shows the cost function $C(x)$ and the tangent line at a point $(x, C(x))$. Remember what it means for a line to have a given slope. If the slope of the line is $C'(x)$, then

$$\frac{\Delta y}{\Delta x} = C'(x) = \frac{C'(x)}{1},$$

and beginning at any point on the line and moving 1 unit to the right requires moving $C'(x)$ units up to get back to the line again. The vertical distance from the horizontal line to the tangent line shown in Figure 26 is therefore $C'(x)$.

Superimposing the graphs from Figures 25 and 26 as in Figure 27 shows that $C'(x)$ is indeed very close to $C(x + 1) - C(x)$. The two values are closest when $C'(x)$ is very large, so that 1 unit is relatively small.

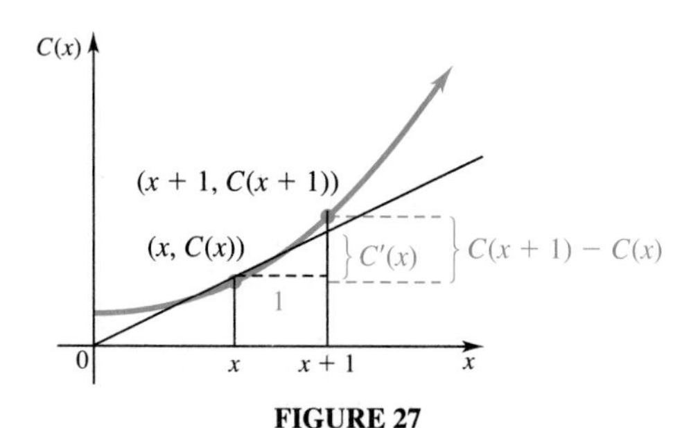

FIGURE 27

EXAMPLE 5

Suppose that the total cost in hundreds of dollars to produce x thousand barrels of a beverage is given by

$$C(x) = 4x^2 + 100x + 500.$$

Find the marginal cost for the following values of x.

(a) $x = 5$

To find the marginal cost, first find $C'(x)$, the derivative of the total cost function.

$$C'(x) = 8x + 100$$

When $x = 5$,

$$C'(5) = 8(5) + 100 = 140.$$

After 5 thousand barrels of the beverage have been produced, the cost to produce one thousand more barrels will be *approximately* 140 hundred dollars, or \$14,000.

The *actual* cost to produce one thousand more barrels is $C(6) - C(5)$:

$$\begin{aligned} C(6) - C(5) &= (4 \cdot 6^2 + 100 \cdot 6 + 500) - (4 \cdot 5^2 + 100 \cdot 5 + 500) \\ &= 1244 - 1100 = 144, \end{aligned}$$

144 hundred dollars, or \$14,400.

(b) $x = 30$

After 30 thousand barrels have been produced, the cost to produce one thousand more barrels will be approximately

$$C'(30) = 8(30) + 100 = 340,$$

or \$34,000. Notice that the cost to produce an additional thousand barrels of beverage has increased by approximately \$20,000 at a production level of 30,000 barrels compared to a production level of 5000 barrels. Management must be careful to keep track of marginal costs. If the marginal cost of producing an extra unit exceeds the revenue received from selling it, then the company will lose money on that unit. ◀

Demand Functions The demand function, defined by $p = f(x)$, relates the number of units x of an item that consumers are willing to purchase to the price p. (Demand functions were also discussed in Chapter 1.) The total revenue $R(x)$ is related to price per unit and the amount demanded (or sold) by the equation

$$R(x) = xp = x \cdot f(x).$$

The demand function for a certain product is given by

$$p = \frac{50{,}000 - x}{25{,}000}.$$

Find the marginal revenue when $x = 10{,}000$ units and p is in dollars.

From the given function for p, the revenue function is given by

$$\begin{aligned} R(x) &= xp \\ &= x\left(\frac{50{,}000 - x}{25{,}000}\right) \\ &= \frac{50{,}000x - x^2}{25{,}000} \\ &= 2x - \frac{1}{25{,}000}x^2. \end{aligned}$$

The marginal revenue is

$$R'(x) = 2 - \frac{2}{25{,}000}x.$$

When $x = 10{,}000$, the marginal revenue is

$$R'(10{,}000) = 2 - \frac{2}{25{,}000}(10{,}000) = 1.2,$$

or \$1.20 per unit. Thus, the next item sold (at sales of 10,000) will produce additional revenue of about \$1.20 per unit. ◀

EXAMPLE 7

Suppose that the cost function for the product in Example 6 is given by

$$C(x) = 2100 + .25x, \quad \text{where } 0 \leq x \leq 30{,}000.$$

Find the marginal profit from the production of the following numbers of units.

(a) 15,000

From Example 6, the revenue from the sale of x units is

$$R(x) = 2x - \frac{1}{25{,}000}x^2.$$

Since profit, P, is given by $P = R - C$,

$$\begin{aligned} P(x) &= R(x) - C(x) \\ &= \left(2x - \frac{1}{25{,}000}x^2\right) - (2100 + .25x) \\ &= 2x - \frac{1}{25{,}000}x^2 - 2100 - .25x \\ &= 1.75x - \frac{1}{25{,}000}x^2 - 2100. \end{aligned}$$

The marginal profit from the sale of x units is

$$P'(x) = 1.75 - \frac{2}{25{,}000}x = 1.75 - \frac{1}{12{,}500}x.$$

At $x = 15{,}000$ the marginal profit is

$$P'(15{,}000) = 1.75 - \frac{1}{12{,}500}(15{,}000) = .55,$$

or \$.55 per unit.

(b) 21,875

When $x = 21{,}875$, the marginal profit is

$$P'(21{,}875) = 1.75 - \frac{1}{12{,}500}(21{,}875) = 0.$$

(c) 25,000

When $x = 25{,}000$, the marginal profit is

$$P'(25{,}000) = 1.75 - \frac{1}{12{,}500}(25{,}000) = -.25,$$

or $-$\$.25 per unit.

As shown by parts (b) and (c), if more than 21,875 units are sold, the marginal profit is negative. This indicates that increasing production beyond that level will *reduce* profit. ◀

The final example shows a medical application of the derivative as the rate of change of a function.

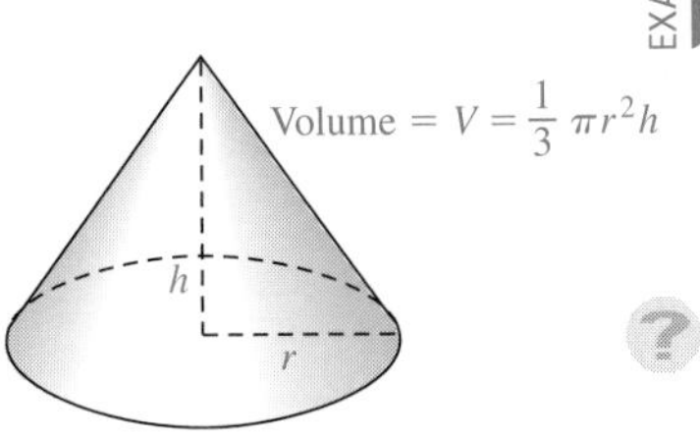

FIGURE 28

EXAMPLE 8

A tumor has the approximate shape of a cone (see Figure 28). The radius of the base of the tumor is fixed by the bone structure at 2 centimeters, but the tumor is growing along the height of the cone (the volume of the tumor therefore being a function of its height). The formula for the volume of a cone is $V = (1/3)\pi r^2 h$, where r is the radius of the base and h is the height of the cone. Find the rate of change in the volume of the tumor with respect to the height.

The symbol dV/dh (instead of V') emphasizes that the rate of change of volume is to be found with respect to the height. For this tumor, r is fixed at 2 centimeters. By substituting 2 for r,

$$V = \frac{1}{3}\pi r^2 h$$

becomes

$$V = \frac{1}{3}\pi 2^2 \cdot h \quad \text{or} \quad V = \frac{4}{3}\pi h.$$

Since $4\pi/3$ is a constant,

$$\frac{dV}{dh} = \frac{4\pi}{3} \approx 4.2 \text{ cubic centimeters per centimeter.}$$

For each additional centimeter that the tumor grows in height, its volume will increase approximately 4.2 cubic centimeters. ◀

2.4 Exercises

Find the derivative of each function defined as follows.

1. $y = 10x^3 - 9x^2 + 6x$
2. $y = 3x^3 - x^2 - \frac{x}{12}$
3. $y = x^4 - 5x^3 + \frac{x^2}{9} + 5$
4. $y = 3x^4 + 11x^3 + 2x^2 - 4x$
5. $f(x) = 6x^{1.5} - 4x^{.5}$
6. $f(x) = -2x^{2.5} + 8x^{.5}$
7. $y = -15x^{3.2} + 2x^{1.9}$
8. $y = -24t^{5/2} - 6t^{1/2}$
9. $y = 8\sqrt{x} + 6x^{3/4}$
10. $y = -100\sqrt{x} - 11x^{2/3}$
11. $g(x) = 6x^{-5} - x^{-1}$
12. $y = 10x^{-2} + 3x^{-4} - 6x$
13. $y = x^{-5} - x^{-2} + 5x^{-1}$
14. $f(t) = \frac{6}{t} - \frac{8}{t^2}$
15. $f(t) = \frac{4}{t} + \frac{2}{t^3}$
16. $y = \frac{9}{x^4} - \frac{8}{x^3} + \frac{2}{x}$
17. $y = \frac{3}{x^6} + \frac{1}{x^5} - \frac{7}{x^2}$
18. $p(x) = -10x^{-1/2} + 8x^{-3/2}$
19. $h(x) = x^{-1/2} - 14x^{-3/2}$
20. $y = \frac{6}{\sqrt[4]{x}}$
21. $y = \frac{-2}{\sqrt[3]{x}}$
22. Which of the following describes the derivative function f' of a quadratic function f?
    (a) quadratic **(b)** linear **(c)** constant **(d)** cubic (third degree)

Find each of the following.

23. $\frac{dy}{dx}$ if $y = 8x^{-5} - 9x^{-4}$
24. $\frac{dy}{dx}$ if $y = -3x^{-2} - 4x^{-5}$
25. $D_x\left[9x^{-1/2} + \frac{2}{x^{3/2}}\right]$

26. $D_x\left[\frac{8}{\sqrt[4]{x}} - \frac{3}{\sqrt{x^3}}\right]$

27. $f'(-2)$ if $f(x) = \frac{x^2}{6} - 4x$

28. $f'(3)$ if $f(x) = \frac{x^3}{9} - 8x^2$

29. Which of the following expressions equals $D_x\left(\frac{6}{x}\right)$?

(a) $\frac{1}{6x^2}$ **(b)** $-\frac{1}{6x^2}$ **(c)** $-\frac{6}{x^2}$ **(d)** $6x^{-2}$

30. Which of the following does *not* equal $\frac{d}{dx}(4x^3 - 6x^{-2})$?

(a) $\frac{12x^2 + 12}{x^3}$ **(b)** $\frac{12x^5 + 12}{x^3}$ **(c)** $12x^2 + \frac{12}{x^3}$ **(d)** $12x^3 + 12x^{-3}$

31. Explain the relationship between the slope and the derivative of $f(x)$ at $x = a$.

In Exercises 32–35 find the slope of the tangent line to the graph of the given function at the given value of x. Find the equation of the tangent line in Exercises 32 and 33.

32. $y = x^4 - 5x^3 + 2;\quad x = 2$

33. $y = -2x^5 - 7x^3 + 8x^2;\quad x = 1$

34. $y = -2x^{1/2} + x^{3/2};\quad x = 4$

35. $y = -x^{-3} + x^{-2};\quad x = 1$

36. Find all points on the graph of $f(x) = 9x^2 - 8x + 4$ where the slope of the tangent line is 0.

37. At what points on the graph of $f(x) = 6x^2 + 4x - 9$ is the slope of the tangent line -2?

38. Use the information given in the figure to find the following values.

(a) $f(1)$ **(b)** $f'(1)$ **(c)** the domain of f **(d)** the range of f

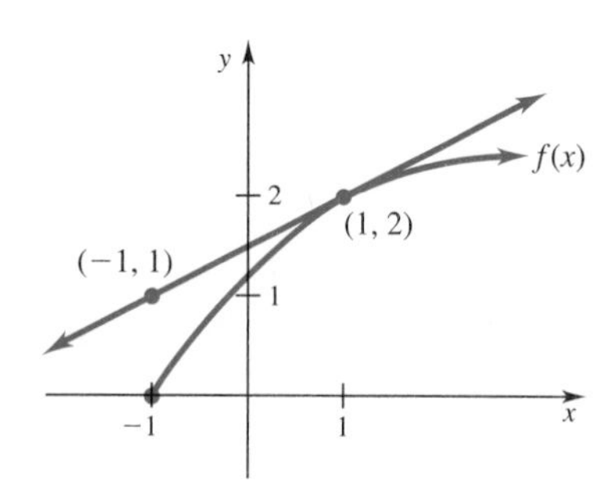

39. Explain the concept of marginal cost. How does it relate to cost? How is it found?

Applications

Business and Economics

40. Revenue Assume that a demand equation is given by $x = 5000 - 100p$. Find the marginal revenue for the following production levels (values of x). (*Hint:* Solve the demand equation for p and use $R(x) = xp$.)

(a) 1000 units **(b)** 2500 units **(c)** 3000 units

41. Profit Suppose that for the situation in Exercise 40 the cost of producing x units is given by $C(x) = 3000 - 20x + .03x^2$. Find the marginal profit for each of the following production levels.

(a) 500 units **(b)** 815 units **(c)** 1000 units

42. Demand If the price of a product is given by

$$p(x) = \frac{1000}{x} + 1000,$$

where x represents the demand for the product, find the rate of change of price when the demand is 10.

43. Sales Often sales of a new product grow rapidly at first and then level off with time. This is the case with the sales represented by the function

$$S(t) = 100 - 100t^{-1},$$

where t represents time in years. Find the rate of change of sales for the following numbers of years.

(a) 1 **(b)** 10

44. Profit The profit in dollars from the sale of x thousand expensive cassette recorders is

$$P(x) = x^3 - 5x^2 + 7x.$$

Find the marginal profit for the following numbers of recorders.

(a) 1000 **(b)** 2000 **(c)** 5000 **(d)** 10,000

(e) Interpret the results from parts (a)–(d).

45. Profit An analyst has found that a company's costs and revenues in dollars for one product are given by

$$C(x) = 2x \quad \text{and} \quad R(x) = 6x - \frac{x^2}{1000},$$

respectively, where x is the number of items produced.

(a) Find the marginal cost function.

(b) Find the marginal revenue function.

(c) Using the fact that profit is the difference between revenue and costs, find the marginal profit function.

(d) What value of x makes marginal profit equal 0?

(e) Find the profit when the marginal profit is 0.

(As we shall see in the next chapter, this process is used to find *maximum* profit.)

Life Sciences

46. Blood Sugar Level Insulin affects the glucose, or blood sugar, level of some diabetics according to the function

$$G(x) = -.2x^2 + 450,$$

where $G(x)$ is the blood sugar level one hour after x units of insulin are injected. (This mathematical model is only approximate, and it is valid only for values of x less than about 40.)

Find the blood sugar level after the following numbers of units of insulin are injected.

(a) 0 **(b)** 25

Find the rate of change of blood sugar level after injection of the following numbers of units of insulin.

(c) 10 **(d)** 25

47. Blood Vessel Volume A short length of blood vessel has a cylindrical shape. The volume of a cylinder is given by $V = \pi r^2 h$. Suppose an experimental device is set up to measure the volume of blood in a blood vessel having a fixed length of 80 mm.

(a) Find a function that gives the rate of change of the volume of blood with respect to the radius of the blood vessel.

Suppose a drug is administered that causes the blood vessel to expand. Find the rate of change of volume for each of the following radii.

(b) 4 mm **(c)** 6 mm **(d)** 8 mm

(e) What is true of the volume as the radius increases?

48. Insect Mating Patterns In an experiment testing methods of sexually attracting male insects to sterile females, equal numbers of males and females of a certain species are permitted to intermingle. Assume that

$$M(t) = 4t^{3/2} + 2t^{1/2}$$

approximates the number of matings observed among the insects in an hour, where t is the temperature in degrees Celsius. (This formula is valid only for certain temperature ranges.) Find the number of matings at the following temperatures.

(a) 16°C **(b)** 25°C

(c) Find the rate of change in the number of matings when the temperature is 16°C.

Physical Sciences

49. Acid Concentration Suppose $P(t) = 100/t$ represents the percent of acid in a chemical solution after t days of exposure to an ultraviolet light source.

Find the percent of acid in the solution after the following numbers of days.

(a) 1 **(b)** 100

(c) Find and interpret $P'(100)$.

Velocity *We saw earlier in this chapter that the velocity of a particle moving in a straight line is given by*

$$\lim_{h \to 0} \frac{s(t+h) - s(t)}{h},$$

where $s(t)$ gives the position of the particle at time t. This limit is actually the derivative of $s(t)$, so the velocity of the particle is given by $s'(t)$. If $v(t)$ represents velocity at time t, then $v(t) = s'(t)$. For each of the position functions in Exercises 50–53, find **(a)** *$v(t)$ and* **(b)** *the velocity when $t = 0$, $t = 5$, and $t = 10$.*

50. $s(t) = 11t^2 + 4t + 2$

51. $s(t) = 25t^2 - 9t + 8$

52. $s(t) = 4t^3 + 8t^2$

53. $s(t) = -2t^3 + 4t^2 - 1$

54. Velocity If a rock is dropped from a 144-foot building, its position (in feet above the ground) is given by $s(t) = -16t^2 + 144$, where t is the time in seconds since it was dropped.

(a) What is its velocity 1 second after being dropped? 2 seconds after being dropped?

(b) When will it hit the ground?

(c) What is its velocity upon impact?

55. Velocity A ball is thrown vertically upward from the ground at a velocity of 64 feet per second. Its distance from the ground at t seconds is given by $s(t) = -16t^2 + 64t$.

(a) How fast is the ball moving 2 seconds after being thrown? 3 seconds after being thrown?

(b) How long after the ball is thrown does it reach its maximum height?

(c) How high will it go?

General Interest

56. Living Standards Living standards are defined by the total output of goods and services divided by the total population. In the United States during the 1980s, living standards were closely approximated by

$$f(x) = -.023x^3 + .3x^2 - .4x + 11.6,$$

where $x = 0$ corresponds to 1981. Find the derivative of f. Use the derivative to find the rate of change in living standards in the following years.

(a) 1981 **(b)** 1983 **(c)** 1988 **(d)** 1989 **(e)** 1990

(f) What do your answers to parts (a)–(e) tell you about living standards in those years?

2.5 DERIVATIVES OF PRODUCTS AND QUOTIENTS

A manufacturer of small motors wants to make the average cost per motor as small as possible. How can this be done?

We show how the derivative is used to solve a problem like this in Example 5, later in this section. In the previous section we developed several rules for finding derivatives. We develop two additional rules in this section, again using the definition of the derivative.

The derivative of a sum of two functions is found from the sum of the derivatives. What about products? Is the derivative of a product equal to the product of the derivatives? For example, if

$$u(x) = 2x + 3 \quad \text{and} \quad v(x) = 3x^2,$$

then

$$u'(x) = 2 \quad \text{and} \quad v'(x) = 6x.$$

Let $f(x)$ be the product of u and v; that is, $f(x) = (2x + 3)(3x^2) = 6x^3 + 9x^2$. By the rules of the preceding section, $f'(x) = 18x^2 + 18x$. On the other hand, $u'(x) = 2$ and $v'(x) = 6x$, with the product $u'(x) \cdot v'(x) = 2(6x) = 12x \neq f'(x)$. In this example, the derivative of a product is *not* equal to the product of the derivatives, nor is this usually the case.

The rule for finding derivatives of products is given below.

PRODUCT RULE

If $f(x) = u(x) \cdot v(x)$, and if $u'(x)$ and $v'(x)$ both exist, then

$$\mathbf{f'(x) = u(x) \cdot v'(x) + v(x) \cdot u'(x).}$$

(The derivative of a product of two functions is the first function times the derivative of the second, plus the second function times the derivative of the first.)

To sketch the method used to prove the product rule, let

$$f(x) = u(x) \cdot v(x).$$

Then $f(x + h) = u(x + h) \cdot v(x + h)$, and, by definition, $f'(x)$ is given by

This proof uses several of the rules for limits given in the first section of this chapter. You may want to review them at this time.

$$\begin{aligned} f'(x) &= \lim_{h \to 0} \frac{f(x + h) - f(x)}{h} \\ &= \lim_{h \to 0} \frac{u(x + h) \cdot v(x + h) - u(x) \cdot v(x)}{h}. \end{aligned}$$

Now subtract and add $u(x + h) \cdot v(x)$ in the numerator, giving

$$\begin{aligned} f'(x) &= \lim_{h \to 0} \frac{u(x + h) \cdot v(x + h) - u(x + h) \cdot v(x) + u(x + h) \cdot v(x) - u(x) \cdot v(x)}{h} \\ &= \lim_{h \to 0} \frac{u(x + h)[v(x + h) - v(x)] + v(x)[u(x + h) - u(x)]}{h} \\ &= \lim_{h \to 0} u(x + h)\left[\frac{v(x + h) - v(x)}{h}\right] + \lim_{h \to 0} v(x)\left[\frac{u(x + h) - u(x)}{h}\right] \\ &= \lim_{h \to 0} u(x + h) \cdot \lim_{h \to 0} \frac{v(x + h) - v(x)}{h} + \lim_{h \to 0} v(x) \cdot \lim_{h \to 0} \frac{u(x + h) - u(x)}{h}. \qquad \textbf{(1)} \end{aligned}$$

If u' and v' both exist, then

$$\lim_{h \to 0} \frac{u(x + h) - u(x)}{h} = u'(x) \quad \text{and} \quad \lim_{h \to 0} \frac{v(x + h) - v(x)}{h} = v'(x).$$

The fact that u' exists can be used to prove

$$\lim_{h \to 0} u(x + h) = u(x),$$

and since no h is involved in $v(x)$,

$$\lim_{h \to 0} v(x) = v(x).$$

Substituting these results into equation (1) gives

$$f'(x) = u(x) \cdot v'(x) + v(x) \cdot u'(x),$$

the desired result.

EXAMPLE 1

Let $f(x) = (2x + 3)(3x^2)$. Use the product rule to find $f'(x)$.

Here f is given as the product of $u(x) = 2x + 3$ and $v(x) = 3x^2$. By the product rule and the fact that $u'(x) = 2$ and $v'(x) = 6x$,

$$\begin{aligned} f'(x) &= u(x) \cdot v'(x) + v(x) \cdot u'(x) \\ &= (2x + 3)(6x) + (3x^2)(2) \\ &= 12x^2 + 18x + 6x^2 = 18x^2 + 18x. \end{aligned}$$

This result is the same as that found at the beginning of the section. ◀

EXAMPLE 2

Find the derivative of $y = (\sqrt{x} + 3)(x^2 - 5x)$.

Let $u(x) = \sqrt{x} + 3 = x^{1/2} + 3$, and $v(x) = x^2 - 5x$. Then

$$\begin{aligned} y' &= u(x) \cdot v'(x) + v(x) \cdot u'(x) \\ &= (x^{1/2} + 3)(2x - 5) + (x^2 - 5x)\left(\frac{1}{2}x^{-1/2}\right). \end{aligned}$$

Simplify by multiplying and combining terms.

$$\begin{aligned} y' &= (2x)(x^{1/2}) + 6x - 5x^{1/2} - 15 + (x^2)\left(\frac{1}{2}x^{-1/2}\right) - (5x)\left(\frac{1}{2}x^{-1/2}\right) \\ &= 2x^{3/2} + 6x - 5x^{1/2} - 15 + \frac{1}{2}x^{3/2} - \frac{5}{2}x^{1/2} \\ &= \frac{5}{2}x^{3/2} + 6x - \frac{15}{2}x^{1/2} - 15 \end{aligned}$$ ◀

We could have found the derivatives above by multiplying out the original functions. The product rule then would not have been needed. In the next section, however, we shall see products of functions where the product rule is essential.

What about *quotients* of functions? To find the derivative of the quotient of two functions, use the next result.

QUOTIENT RULE

If $f(x) = \dfrac{u(x)}{v(x)}$, if all indicated derivatives exist, and if $v(x) \neq 0$, then

$$f'(x) = \frac{v(x) \cdot u'(x) - u(x) \cdot v'(x)}{[v(x)]^2}$$

(The derivative of a quotient is the denominator times the derivative of the numerator, minus the numerator times the derivative of the denominator, all divided by the square of the denominator.)

The proof of the quotient rule is similar to that of the product rule and is left for the exercises.

EXAMPLE 3

Find $f'(x)$ if $f(x) = \dfrac{2x - 1}{4x + 3}$.

Let $u(x) = 2x - 1$, with $u'(x) = 2$. Also, let $v(x) = 4x + 3$, with $v'(x) = 4$. Then, by the quotient rule,

You may want to consult the Rational Expressions section of the Algebra Reference to help you work with the fractions in this section.

$$\begin{aligned} f'(x) &= \frac{v(x) \cdot u'(x) - u(x) \cdot v'(x)}{[v(x)]^2} \\ &= \frac{(4x + 3)(2) - (2x - 1)(4)}{(4x + 3)^2} \\ &= \frac{8x + 6 - 8x + 4}{(4x + 3)^2} \\ &= \frac{10}{(4x + 3)^2}. \end{aligned}$$

Caution In the second step of Example 3, we had the expression

$$\frac{(4x + 3)(2) - (2x - 1)(4)}{(4x + 3)^2}.$$

Students often incorrectly "cancel" the $4x + 3$ in the numerator with one factor of the denominator. Because the numerator is a *difference* of two products, however, you must multiply and combine terms *before* looking for common factors in the numerator and denominator.

EXAMPLE 4

Find $D_x\left(\dfrac{(3 - 4x)(5x + 1)}{7x - 9}\right)$.

This function has a product within a quotient. Instead of multiplying the factors in the numerator first (which is an option), we can use the quotient rule together with the product rule, as follows. Use the quotient rule first to get

$$D_x\left(\frac{(3 - 4x)(5x + 1)}{7x - 9}\right) = \frac{(7x - 9)[D_x(3 - 4x)(5x + 1)] - [(3 - 4x)(5x + 1)D_x(7x - 9)]}{(7x - 9)^2}.$$

Now use the product rule to find $D_x(3 - 4x)(5x + 1)$ in the numerator.

$$= \frac{(7x - 9)[(3 - 4x)5 + (5x + 1)(-4)] - (3 + 11x - 20x^2)(7)}{(7x - 9)^2}$$

$$= \frac{(7x - 9)(15 - 20x - 20x - 4) - (21 + 77x - 140x^2)}{(7x - 9)^2}$$

$$= \frac{(7x - 9)(11 - 40x) - 21 - 77x + 140x^2}{(7x - 9)^2}$$

$$= \frac{-280x^2 + 437x - 99 - 21 - 77x + 140x^2}{(7x - 9)^2}$$

$$= \frac{-140x^2 + 360x - 120}{(7x - 9)^2}$$ ◀

Average Cost Suppose $y = C(x)$ gives the total cost to manufacture x items. As mentioned earlier, the average cost per item is found by dividing the total cost by the number of items. The rate of change of average cost, called the *marginal average cost*, is the derivative of the average cost.

AVERAGE COST

If the total cost to manufacture x items is given by $C(x)$, then the **average cost per item** is

$$\overline{C}(x) = \frac{C(x)}{x}.$$

The **marginal average cost** is the derivative of the average cost function, $\overline{C}'(x)$.

A company naturally would be interested in making the average cost as small as possible. The next chapter will show that this can be done by using the derivative of $C(x)/x$. This derivative often can be found by means of the quotient rule, as in the next example.

Suppose the cost in dollars of manufacturing x hundred small motors is given by

$$C(x) = \frac{3x^2 + 120}{2x + 1}.$$

(a) Find the average cost per hundred motors.

The average cost is defined by

$$\overline{C}(x) = \frac{C(x)}{x} = \frac{3x^2 + 120}{2x + 1} \cdot \frac{1}{x} = \frac{3x^2 + 120}{2x^2 + x}.$$

(b) Find the marginal average cost.

The marginal average cost is given by

$$\frac{d}{dx}(\overline{C}(x)) = \frac{(2x^2 + x)(6x) - (3x^2 + 120)(4x + 1)}{(2x^2 + x)^2}$$

$$= \frac{12x^3 + 6x^2 - 12x^3 - 480x - 3x^2 - 120}{(2x^2 + x)^2}$$

$$= \frac{3x^2 - 480x - 120}{(2x^2 + x)^2}.$$

(c) Average cost is generally minimized when the marginal average cost is zero. Find the level of production that minimizes average cost.

Set the derivative $\overline{C}'(x) = 0$ and solve for x.

$$\frac{3x^2 - 480x - 120}{(2x^2 + x)^2} = 0$$

$$3x^2 - 480x - 120 = 0$$

$$3(x^2 - 160x - 40) = 0$$

Use the quadratic formula to solve this quadratic equation. Discarding the negative solution leaves $x \approx 160$ as the solution. Since x is in hundreds, production of 160 hundred or 16,000 motors will minimize average cost. ◀

2.5 Exercises

Use the product rule to find the derivative of each of the following. (Hint for Exercises 8–11: *Write the quantity as a product.*)

1. $y = (2x - 5)(x + 4)$
2. $y = (3x + 7)(x - 1)$
3. $y = (3x^2 + 2)(2x - 1)$
4. $y = (5x^2 - 1)(4x + 3)$
5. $y = (2t^2 - 6t)(t + 2)$
6. $y = (9x^2 + 7x)(x^2 - 1)$
7. $y = (2x^2 - 4x)(5x^2 + 4)$
8. $y = (2x - 5)^2$
9. $y = (7x - 6)^2$
10. $k(t) = (t^2 - 1)^2$
11. $g(t) = (3t^2 + 2)^2$
12. $y = (x + 1)(\sqrt{x} + 2)$
13. $y = (2x - 3)(\sqrt{x} - 1)$
14. $g(x) = (5\sqrt{x} - 1)(2\sqrt{x} + 1)$
15. $g(x) = (-3\sqrt{x} + 6)(4\sqrt{x} - 2)$

Use the quotient rule to find the derivative of each of the following.

16. $f(x) = \frac{7x + 1}{3x + 8}$
17. $f(x) = \frac{6x - 11}{8x + 1}$
18. $y = \frac{2}{3x - 5}$
19. $y = \frac{-4}{2x - 11}$
20. $y = \frac{5 - 3t}{4 + t}$
21. $y = \frac{9 - 7t}{1 - t}$
22. $y = \frac{x^2 + x}{x - 1}$
23. $y = \frac{x^2 - 4x}{x + 3}$
24. $f(t) = \frac{4t + 11}{t^2 - 3}$
25. $y = \frac{-x^2 + 6x}{4x^2 + 1}$
26. $g(x) = \frac{x^2 - 4x + 2}{x + 3}$
27. $k(x) = \frac{x^2 + 7x - 2}{x - 2}$
28. $p(t) = \frac{\sqrt{t}}{t - 1}$
29. $r(t) = \frac{\sqrt{t}}{2t + 3}$
30. $y = \frac{5x + 6}{\sqrt{x}}$

31. Find the error in the following work.

$$D_x\left(\frac{2x+5}{x^2-1}\right) = \frac{(2x+5)(2x)-(x^2-1)2}{(x^2-1)^2} = \frac{4x^2+10x-2x^2+2}{(x^2-1)^2} = \frac{2x^2+10x+2}{(x^2-1)^2}$$

32. Find the error in the following work.

$$D_x\left(\frac{x^2-4}{x^3}\right) = x^3(2x)-(x^2-4)(3x^2) = 2x^4-3x^4+12x^2 = -x^4+12x^2$$

33. Find an equation of the line tangent to the graph of $f(x) = x/(x-2)$ at $(3, 3)$.

34. Following the steps used to prove the product rule for derivatives, prove the quotient rule for derivatives.

Applications

Business and Economics

35. Average Cost The total cost (in hundreds of dollars) to produce x units of perfume is

$$C(x) = \frac{3x+2}{x+4}.$$

Find the average cost for each of the following production levels.

(a) 10 units **(b)** 20 units **(c)** x units

(d) Find the marginal average cost function.

36. Average Profit The total profit (in tens of dollars) from selling x self-help books is

$$P(x) = \frac{5x-6}{2x+3}.$$

Find the average profit from each of the following sales levels.

(a) 8 books **(b)** 15 books **(c)** x books

(d) Find the marginal average profit function.

(e) Is this a reasonable function for profit? Why?

37. Fuel Efficiency Suppose you are the manager of a trucking firm, and one of your drivers reports that she has calculated that her truck burns fuel at the rate of

$$G(x) = \frac{1}{200}\left(\frac{800+x^2}{x}\right)$$

gallons per mile when traveling at x miles per hour on a smooth, dry road.

(a) If the driver tells you that she wants to travel 20 miles per hour, what should you tell her? (*Hint:* Take the derivative of G and evaluate it for $x = 20$. Then interpret your results.)

(b) If the driver wants to go 40 miles per hour, what should you say? (*Hint:* Find $G'(40)$.)

38. Employee Training A company that manufactures bicycles has determined that a new employee can assemble $M(d)$ bicycles per day after d days of on-the-job training, where

$$M(d) = \frac{200d}{3d+10}.$$

(a) Find the rate of change function for the number of bicycles assembled with respect to time.

(b) Find and interpret $M'(2)$ and $M'(5)$.

Life Sciences

39. Muscle Reaction When a certain drug is injected into a muscle, the muscle responds by contracting. The amount of contraction, s, in millimeters, is related to the concentration of the drug, x, in milliliters, by

$$s(x) = \frac{x}{m+nx},$$

where m and n are constants.

(a) Find $s'(x)$.

(b) Find the rate of contraction when the concentration of the drug is 50 ml, $m = 10$, and $n = 3$.

40. Bacteria Population Assume that the total number (in millions) of bacteria present in a culture at a certain time t (in hours) is given by

$$N(t) = (t-10)^2(2t) + 50.$$

(a) Find $N'(t)$.

Find the rate at which the population of bacteria is changing at each of the following times.

(b) 8 hr **(c)** 11 hr

(d) The answer in part (b) is negative, and the answer in part (c) is positive. What does this mean in terms of the population of bacteria?

Social Sciences

41. Memory Retention Some psychologists contend that the number of facts of a certain type that are remembered after t hr is given by

$$f(t) = \frac{90t}{99t - 90}.$$

Find the rate at which the number of facts remembered is changing after the following numbers of hours.

(a) 1 **(b)** 10

2.6 THE CHAIN RULE

Suppose we know how fast the radius of a circular oil slick is growing, and we know how much the area of the oil slick is growing per unit of change in the radius. How *fast* is the area growing?

The answer to this question involves the chain rule for derivatives. Before discussing the chain rule, we consider the composition of functions. Many of the most useful functions for modeling are created by combining simpler functions. Viewing complex functions as combinations of simpler functions often makes them easier to understand and use.

Composition of Functions Suppose a function f assigns to each element x in set X some element $y = f(x)$ in set Y. Suppose also that a function g takes each element in set Y and assigns to it a value $z = g[f(x)]$ in set Z. By using both f and g, an element x in X is assigned to an element z in Z, as illustrated in Figure 29. The result of this process is a new function called the *composition* of functions g and f and defined as follows.

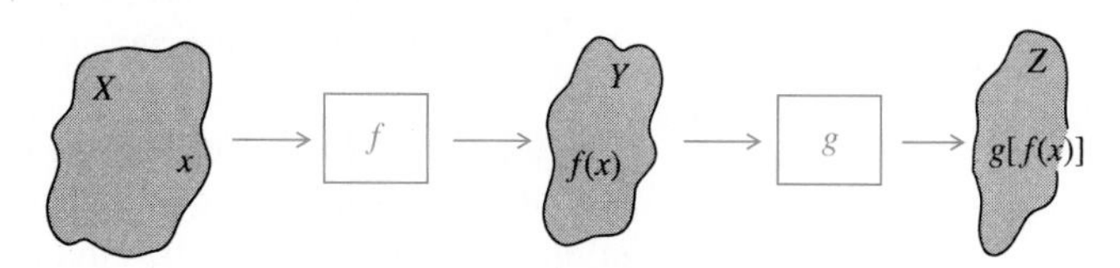

FIGURE 29

COMPOSITE FUNCTION Let f and g be functions. The **composite function,** or **composition,** of g and f is the function whose values are given by $g[f(x)]$ for all x in the domain of f such that $f(x)$ is in the domain of g. (Read $g[f(x)]$ as "g of f of x".)

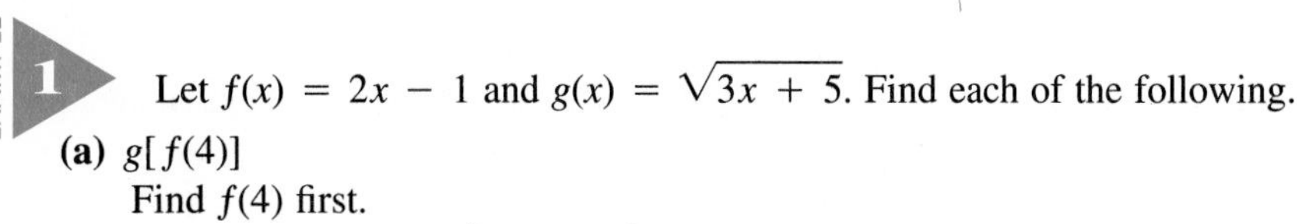

EXAMPLE 1 Let $f(x) = 2x - 1$ and $g(x) = \sqrt{3x + 5}$. Find each of the following.

(a) $g[f(4)]$

Find $f(4)$ first.

$$f(4) = 2 \cdot 4 - 1 = 8 - 1 = 7$$

Then

$$g[f(4)] = g[7] = \sqrt{3 \cdot 7 + 5} = \sqrt{26}.$$

(b) $f[g(4)]$

Since $g(4) = \sqrt{3 \cdot 4 + 5} = \sqrt{17}$,

$$f[g(4)] = 2 \cdot \sqrt{17} - 1 = 2\sqrt{17} - 1.$$

You may want to review how to find the domain of a function. Domain was discussed in Section 1.1.

(c) $f[g(-2)]$ does not exist since -2 is not in the domain of g. ◀

Let $f(x) = 4x + 1$ and $g(x) = 2x^2 + 5x$. Find each of the following.

(a) $g[f(x)]$

Using the given functions, we have

$$\begin{aligned} g[f(x)] &= g[4x + 1] \\ &= 2(4x + 1)^2 + 5(4x + 1) \\ &= 2(16x^2 + 8x + 1) + 20x + 5 \\ &= 32x^2 + 16x + 2 + 20x + 5 \\ &= 32x^2 + 36x + 7. \end{aligned}$$

(b) $f[g(x)]$

By the definition above, with f and g interchanged,

$$\begin{aligned} f[g(x)] &= f[2x^2 + 5x] \\ &= 4(2x^2 + 5x) + 1 \\ &= 8x^2 + 20x + 1. \end{aligned}$$ ◀

As Example 2 shows, it is not always true that $f[g(x)] = g[f(x)]$. (In fact, it is rare to find two functions f and g such that $f[g(x)] = g[f(x)]$. The domain of both composite functions given in Example 2 is the set of all real numbers.

We now return to the question at the beginning of this section.

The Chain Rule A leaking oil well off the Gulf Coast is spreading a circular film of oil over the water surface. At any time t (in minutes) after the beginning of the leak, the radius of the circular oil slick is given by

$$r(t) = 4t, \quad \text{with} \quad \frac{dr}{dt} = 4,$$

where dr/dt is the rate of change in radius over time. The area of the oil slick is given by

$$A(r) = \pi r^2, \quad \text{with} \quad \frac{dA}{dr} = 2\pi r,$$

where dA/dr is the rate of change in area per unit change in radius.

As these derivatives show, the radius is increasing 4 times as fast as the time t, and the area is increasing $2\pi r$ times as fast as the radius r. It seems reasonable, then, that the area is increasing $2\pi r \cdot 4 = 8\pi r$ times as fast as time. That is,

$$\frac{dA}{dt} = \frac{dA}{dr} \cdot \frac{dr}{dt} = 2\pi r \cdot 4 = 8\pi r.$$

Since $r = 4t$,

$$\frac{dA}{dt} = 8\pi r = 8\pi(4t) = 32\pi t.$$

To check this, use the fact that $r = 4t$ and $A = \pi r^2$ to get the same result:

$$A = \pi(4t)^2 = 16\pi t^2, \qquad \text{with} \qquad \frac{dA}{dt} = 32\pi t.$$

The product used above,

$$\frac{dA}{dt} = \frac{dA}{dr} \cdot \frac{dr}{dt},$$

is an example of the **chain rule,** which is used to find the derivative of a composite function.

CHAIN RULE

If y is a function of u, say $y = f(u)$, and if u is a function of x, say $u = g(x)$, then $y = f(u) = f[g(x)]$, and

$$\frac{dy}{dx} = \frac{dy}{du} \cdot \frac{du}{dx}.$$

One way to remember the chain rule is to pretend that dy/du and du/dx are fractions, with du ''canceling out.'' The proof of the chain rule requires advanced concepts and therefore is not given here.

EXAMPLE 3

Find dy/dx if $y = (3x^2 - 5x)^{1/2}$.

Let $y = u^{1/2}$, and $u = 3x^2 - 5x$. Then

$$\frac{dy}{dx} = \frac{dy}{du} \cdot \frac{du}{dx}$$

$$= \frac{1}{2}u^{-1/2} \cdot (6x - 5).$$

Replacing u with $3x^2 - 5x$ gives

$$\frac{dy}{dx} = \frac{1}{2}(3x^2 - 5x)^{-1/2}(6x - 5) = \frac{6x - 5}{2(3x^2 - 5x)^{1/2}}.$$

The following alternative version of the chain rule is stated in terms of composite functions.

CHAIN RULE (ALTERNATIVE FORM)

If $y = f[g(x)]$, then

$$y' = f'[g(x)] \cdot g'(x).$$

(To find the derivative of $f[g(x)]$, find the derivative of $f(x)$, replace each x with $g(x)$, and then multiply the result by the derivative of $g(x)$.)

Use the chain rule to find $D_x(x^2 + 5x)^8$.

Let $f(x) = x^8$ and $g(x) = x^2 + 5x$. Then $(x^2 + 5x)^8 = f[g(x)]$ and

$$D_x(x^2 + 5x)^8 = f'[g(x)]g'(x)$$

Here $f'(x) = 8x^7$, with $f'[g(x)] = 8[g(x)]^7 = 8(x^2 + 5x)^7$ and $g'(x) = 2x + 5$.

$$\begin{aligned} D_x(x^2 + 5x)^8 &= f'[g(x)]g'(x) \\ &= 8[g(x)]^7 g'(x) \\ &= 8(x^2 + 5x)^7(2x + 5) \end{aligned}$$

While the chain rule is essential for finding the derivatives of some of the functions discussed later, the derivatives of the algebraic functions discussed so far can be found by the following *generalized power rule,* a special case of the chain rule.

GENERALIZED POWER RULE

Let u be a function of x, and let $y = u^n$, for any real number n. Then

$$y' = n \cdot u^{n-1} \cdot u'.$$

(The derivative of $y = u^n$ is found by decreasing the exponent on u by 1 and multiplying the result by the exponent n and by the derivative of u with respect to x.)

(a) Use the generalized power rule to find the derivative of $y = (3 + 5x)^2$.

Let $u = 3 + 5x$, and $n = 2$. Then $u' = 5$. By the generalized power rule,

$$\begin{aligned} y' = \frac{dy}{dx} &= n \cdot u^{n-1} \cdot u' \\ &= 2 \cdot (3 + 5x)^{2-1} \cdot \frac{d}{dx}(3 + 5x) \\ &= 2(3 + 5x)^{2-1} \cdot 5 = 10(3 + 5x) = 30 + 50x. \end{aligned}$$

(Labels on the arrows: n, u, $n - 1$, u')

(b) Find y' if $y = (3 + 5x)^{-3/4}$.

Use the generalized power rule, with $u = 3 + 5x$, $n = -3/4$, and $u' = 5$.

$$y' = -\frac{3}{4}(3 + 5x)^{-3/4-1} \cdot 5 = -\frac{15}{4}(3 + 5x)^{-7/4}$$

This result could not have been found by any of the rules given in previous sections. ◀

Caution A common error is to forget to multiply by u' when using the generalized power rule. Remember, the generalized power rule is an example of the chain rule, and so the derivative must involve a "chain," or product, of derivatives.

Sometimes both the generalized power rule and either the product or quotient rule are needed to find a derivative, as the next examples show.

Find the derivative of $y = 4x(3x + 5)^5$.

Write $4x(3x + 5)^5$ as the product

$$(4x) \cdot (3x + 5)^5.$$

To find the derivative of $(3x + 5)^5$, let $u = 3x + 5$, with $u' = 3$. Now use the product rule and the generalized power rule.

$$y' = 4x\overbrace{[5(3x + 5)^4 \cdot 3]}^{\text{Derivative of } (3x+5)^5} + (3x + 5)^5\overbrace{(4)}^{\text{Derivative of } 4x}$$

$$= 60x(3x + 5)^4 + 4(3x + 5)^5$$

$$= 4(3x + 5)^4[15x + (3x + 5)^1] \quad \text{Factor out the greatest common factor, } 4(3x + 5)^4$$

$$= 4(3x + 5)^4(18x + 5)$$ ◀

Find $D_x \dfrac{(3x + 2)^7}{x - 1}$.

Use the quotient rule and the generalized power rule.

$$D_x \frac{(3x + 2)^7}{x - 1} = \frac{(x - 1)[7(3x + 2)^6 \cdot 3] - (3x + 2)^7(1)}{(x - 1)^2}$$

$$= \frac{21(x - 1)(3x + 2)^6 - (3x + 2)^7}{(x - 1)^2}$$

$$= \frac{(3x + 2)^6[21(x - 1) - (3x + 2)]}{(x - 1)^2} \quad \text{Factor out the greatest common factor, } (3x + 2)^6.$$

$$= \frac{(3x + 2)^6[21x - 21 - 3x - 2]}{(x - 1)^2} \quad \text{Simplify inside brackets.}$$

$$= \frac{(3x + 2)^6(18x - 23)}{(x - 1)^2}$$ ◀

Some applications requiring the use of the chain rule or the generalized power rule are illustrated in the next examples.

EXAMPLE 8 The revenue realized by a small city from the collection of fines from parking tickets is given by

$$R(n) = \frac{8000n}{n + 2},$$

where n is the number of work hours each day that can be devoted to parking patrol. At the outbreak of a flu epidemic, 30 work hours are used daily in parking patrol, but during the epidemic that number is decreasing at the rate of 6 work hours per day. How fast is revenue from parking fines decreasing during the epidemic?

We want to find dR/dt, the change in revenue with respect to time. By the chain rule,

$$\frac{dR}{dt} = \frac{dR}{dn} \cdot \frac{dn}{dt}.$$

First find dR/dn, as follows.

$$\frac{dR}{dn} = \frac{(n + 2)(8000) - 8000n(1)}{(n + 2)^2} = \frac{16000}{(n + 2)^2}$$

Since $n = 30$, $dR/dn = 15.625$. Also, $dn/dt = -6$. Thus,

$$\frac{dR}{dt} = (15.625)(-6) = -93.75.$$

Revenue is being lost at the rate of about \$94 per day. ◀

EXAMPLE 9 Suppose a sum of \$500 is deposited in an account with an interest rate of r percent per year compounded monthly. At the end of 10 years, the balance in the account is given by

$$A = 500\left(1 + \frac{r}{1200}\right)^{120}.$$

Find the rate of change of A with respect to r if $r = 5$ or 7.

First find dA/dr using the generalized power rule.

$$\frac{dA}{dr} = (120)(500)\left(1 + \frac{r}{1200}\right)^{119}\left(\frac{1}{1200}\right)$$

$$= 50\left(1 + \frac{r}{1200}\right)^{119}$$

If $r = 5$,

$$\frac{dA}{dr} = 50\left(1 + \frac{5}{1200}\right)^{119}$$
$$\approx 82.01,$$

or \$82.01 per percentage point. If $r = 7$,

$$\frac{dA}{dr} = 50\left(1 + \frac{7}{1200}\right)^{119}$$
$$\approx 99.90,$$

or \$99.90 per percentage point. ◀

The chain rule can be used to develop the formula for **marginal revenue product,** an economic concept that approximates the change in revenue when a manufacturer hires an additional employee. Start with $R = px$, where R is total revenue from the daily production of x units and p is the price per unit. The demand function is $p = f(x)$, as before. Also, x can be considered a function of the number of employees, n. Since $R = px$, and x (and therefore p) depends on n, R can also be considered a function of n. To find an expression for dR/dn, use the product rule for derivatives on the function $R = px$ to get

$$\frac{dR}{dn} = p \cdot \frac{dx}{dn} + x \cdot \frac{dp}{dn}. \qquad \textbf{(1)}$$

By the chain rule,

$$\frac{dp}{dn} = \frac{dp}{dx} \cdot \frac{dx}{dn}.$$

Substituting for dp/dn in equation (1) gives

$$\frac{dR}{dn} = p \cdot \frac{dx}{dn} + x\left(\frac{dp}{dx} \cdot \frac{dx}{dn}\right)$$
$$= \left(p + x \cdot \frac{dp}{dx}\right)\frac{dx}{dn}. \qquad \text{Factor out } \frac{dx}{dn}.$$

The equation for dR/dn gives the marginal revenue product.

EXAMPLE **10** Find the marginal revenue product dR/dn (in dollars) when $n = 20$ if the demand function is $p = 600/\sqrt{x}$ and $x = 5n$.

As shown above,

$$\frac{dR}{dn} = \left(p + x \cdot \frac{dp}{dx}\right)\frac{dx}{dn}.$$

Find dp/dx and dx/dn. From

$$p = \frac{600}{\sqrt{x}} = 600x^{-1/2},$$

we have the derivative

$$\frac{dp}{dx} = -300x^{-3/2}.$$

Also, from $x = 5n$,

$$\frac{dx}{dn} = 5.$$

Then, by substitution,

$$\frac{dR}{dn} = \left[\frac{600}{\sqrt{x}} + x(-300x^{-3/2})\right]5 = \frac{1500}{\sqrt{x}}.$$

If $n = 20$, then $x = 100$ and

$$\frac{dR}{dn} = \frac{1500}{\sqrt{100}} = 150.$$

This means that hiring an additional employee when there are 20 employees will produce an increase in revenue of $150. ◀

For easy reference, a list of the rules for derivatives developed in this chapter is given after the exercises for this section.

2.6 Exercises

Let $f(x) = 4x^2 - 2x$ and $g(x) = 8x + 1$. Find each of the following.

1. $f[g(2)]$ **2.** $f[g(-5)]$ **3.** $g[f(2)]$ **4.** $g[f(-5)]$ **5.** $f[g(k)]$ **6.** $g[f(5z)]$

Find $f[g(x)]$ and $g[f(x)]$ in each of the following.

7. $f(x) = \frac{x}{8} + 12;\quad g(x) = 3x - 1$

8. $f(x) = -6x + 9;\quad g(x) = \frac{x}{5} + 7$

9. $f(x) = \frac{1}{x};\quad g(x) = x^2$

10. $f(x) = \frac{2}{x^4};\quad g(x) = 2 - x$

11. $f(x) = \sqrt{x + 2};\quad g(x) = 8x^2 - 6$

12. $f(x) = 9x^2 - 11x;\quad g(x) = 2\sqrt{x + 2}$

13. $f(x) = \sqrt{x + 1};\quad g(x) = \frac{-1}{x}$

14. $f(x) = \frac{8}{x};\quad g(x) = \sqrt{3 - x}$

15. In your own words, explain how to form the composition of two functions.

Write each function as the composition of two functions. (There may be more than one way to do this.)

16. $y = (3x - 7)^{1/3}$

17. $y = (5 - x)^{2/5}$

18. $y = \sqrt{9 - 4x}$

19. $y = -\sqrt{13 + 7x}$

20. $y = (x^{1/2} - 3)^2 + (x^{1/2} - 3) + 5$

21. $y = (x^2 + 5x)^{1/3} - 2(x^2 + 5x)^{2/3} + 7$

22. What is the distinction between the chain rule and the generalized power rule?

Find the derivative of each function defined as follows.

23. $y = (2x^3 + 9x)^5$

24. $y = (8x^4 - 3x^2)^3$

25. $f(x) = -8(3x^4 + 2)^3$

26. $k(x) = -2(12x^2 + 5)^6$

27. $s(t) = 12(2t^4 + 5)^{3/2}$

28. $s(t) = 45(3t^3 - 8)^{3/2}$

29. $f(t) = 8\sqrt{4t^2 + 7}$

30. $g(t) = -3\sqrt{7t^3 - 1}$

31. $r(t) = 4t(2t^5 + 3)^2$

32. $m(t) = -6t(5t^4 - 1)^2$

33. $y = (x^3 + 2)(x^2 - 1)^2$

34. $y = (3x^4 + 1)^2(x^3 + 4)$

35. $y = (5x^6 + x)^2\sqrt{2x}$

36. $y = 2(3x^4 + 5)^2\sqrt{x}$

37. $y = \dfrac{1}{(3x^2 - 4)^5}$

38. $y = \dfrac{-5}{(2x^3 + 1)^2}$

39. $p(t) = \dfrac{(2t + 3)^3}{4t^2 - 1}$

40. $r(t) = \dfrac{(5t - 6)^4}{3t^2 + 4}$

41. $y = \dfrac{x^2 + 4x}{(3x^3 + 2)^4}$

42. $y = \dfrac{3x^2 - x}{(2x - 1)^5}$

Consider the following table of values of the functions f and g and their derivatives at various points:

x	1	2	3	4
$f(x)$	2	4	1	3
$f'(x)$	−6	−7	−8	−9
$g(x)$	2	3	4	1
$g'(x)$	2/7	3/7	4/7	5/7

Find each of the following.

43. **(a)** $D_x(f[g(x)])$ at $x = 1$ **(b)** $D_x(f[g(x)])$ at $x = 2$

44. **(a)** $D_x(g[f(x)])$ at $x = 1$ **(b)** $D_x(g[f(x)])$ at $x = 2$

Applications

Business and Economics

45. Demand Suppose the demand for a certain brand of vacuum cleaner is given by

$$D(p) = \frac{-p^2}{100} + 500,$$

where p is the price in dollars. If the price, in terms of the cost c, is expressed as

$$p(c) = 2c - 10,$$

find the demand in terms of the cost.

46. Revenue Assume that the total revenue from the sale of x television sets is given by

$$R(x) = 1000\left(1 - \frac{x}{500}\right)^2.$$

Find the marginal revenue when the following numbers of sets are sold.

(a) 400 **(b)** 500 **(c)** 600

(d) Find the average revenue from the sale of x sets.

(e) Find the marginal average revenue.

(f) Write a paragraph covering the following questions. When does the revenue begin to decrease? What sales produce zero marginal average revenue? How should a manager use this information?

47. Interest A sum of \$1500 is deposited in an account with an interest rate of r percent per year, compounded daily. At the end of 5 yr, the balance in the account is given by

$$A = 1500\left(1 + \frac{r}{36500}\right)^{1825}.$$

Find the rate of change of A with respect to r for the following interest rates.

(a) 6% **(b)** 8% **(c)** 9%

48. Demand Suppose a demand function is given by

$$x = 30\left(5 - \frac{p}{\sqrt{p^2 + 1}}\right),$$

where x is the demand for a product and p is the price per item in dollars. Find the rate of change in the demand for the product per unit change in price (i.e., find dx/dp).

49. **Depreciation** A certain truck depreciates according to the formula

$$V = \frac{6000}{1 + .3t + .1t^2},$$

where t is time measured in years and $t = 0$ represents the time of purchase (in years). Find the rate at which the value of the truck is changing at the following times.

(a) 2 yr **(b)** 4 yr

50. **Cost** Suppose the cost in dollars of manufacturing x items is given by

$$C = 2000x + 3500,$$

and the demand equation is given by

$$x = \sqrt{15{,}000 - 1.5p}.$$

(a) Find an expression for the revenue R.

(b) Find an expression for the profit P.

(c) Find an expression for the marginal profit.

(d) Determine the value of the marginal profit when the price is $25.

51. **Marginal Revenue Product** Find the marginal revenue product for a manufacturer with 8 workers if the demand function is $p = 300/x^{1/3}$ and the number of items produced is $8n$, where n is the number of employees.

52. **Marginal Revenue Product** Suppose the demand function for a product is $p = 200/x^{1/2}$ and x is the number of items produced. Find the marginal revenue product if there are 25 employees and each employee produces 15 items.

Life Sciences

53. **Fish Population** Suppose the population P of a certain species of fish depends on the number x (in hundreds) of a smaller fish that serves as its food supply, so that

$$P(x) = 2x^2 + 1.$$

Suppose, also, that the number x (in hundreds) of the smaller species of fish depends upon the amount a (in appropriate units) of its food supply, a kind of plankton. Suppose

$$x = f(a) = 3a + 2.$$

A biologist wants to find the relationship between the population P of the large fish and the amount a of plankton available, that is, $P[f(a)]$. What is the relationship?

54. **Oil Pollution** An oil well off the Gulf Coast is leaking, with the leak spreading oil over the surface as a circle. At any time t (in minutes) after the beginning of the leak, the radius of the circular oil slick on the surface is $r(t) = t^2$ feet. Let $A(r) = \pi r^2$ represent the area of a circle of radius r. Find and interpret $A[r(t)]$.

55. **Thermal Inversion** When there is a thermal inversion layer over a city (as happens often in Los Angeles), pollutants cannot rise vertically but are trapped below the layer and must disperse horizontally. Assume that a factory smokestack begins emitting a pollutant at 8 a.m. Assume that the pollutant disperses horizontally, forming a circle. If t represents the time (in hours) since the factory began emitting pollutants ($t = 0$ represents 8 a.m.), assume that the radius of the circle of pollution is $r(t) = 2t$ miles. Let $A(r) = \pi r^2$ represent the area of a circle of radius r. Find and interpret $A[r(t)]$.

56. **Bacteria Population** The total number of bacteria (in millions) present in a culture is given by

$$N(t) = 2t(5t + 9)^{1/2} + 12,$$

where t represents time (in hours) after the beginning of an experiment. Find the rate of change of the population of bacteria with respect to time for each of the following numbers of hours.

(a) 0 **(b)** 7/5 **(c)** 8

57. **Calcium Usage** To test an individual's use of calcium, a researcher injects a small amount of radioactive calcium into the person's bloodstream. The calcium remaining in the bloodstream is measured each day for several days. Suppose the amount of the calcium remaining in the bloodstream (in milligrams per cubic centimeter) t days after the initial injection is approximated by

$$C(t) = \frac{1}{2}(2t + 1)^{-1/2}.$$

Find the rate of change of the calcium level with respect to time for each of the following numbers of days.

(a) 0 **(b)** 4 **(c)** 7.5

(d) Is C always increasing or always decreasing? How can you tell?

58. **Drug Reaction** The strength of a person's reaction to a certain drug is given by

$$R(Q) = Q\left(C - \frac{Q}{3}\right)^{1/2},$$

where Q represents the quantity of the drug given to the patient and C is a constant.

(a) The derivative $R'(Q)$ is called the *sensitivity* to the drug. Find $R'(Q)$.

(b) Find the sensitivity to the drug if $C = 59$ and a patient is given 87 units of the drug.

(c) Is the patient's sensitivity to the drug increasing or decreasing when $Q = 87$?

RULES FOR DERIVATIVES SUMMARY

Assume all indicated derivatives exist.

Constant Function If $f(x) = k$, where k is any real number, then $\mathbf{f'(x) = 0}$.

Power Rule If $f(x) = x^n$, for any real number n, then $\mathbf{f'(x) = n \cdot x^{n-1}}$.

Constant Times a Function Let k be a real number. Then the derivative of $y = k \cdot f(x)$ is $\mathbf{y' = k \cdot f'(x)}$.

Sum or Difference Rule If $y = f(x) \pm g(x)$, then $\mathbf{y' = f'(x) \pm g'(x)}$.

Product Rule If $f(x) = g(x) \cdot k(x)$, then

$$\mathbf{f'(x) = g(x) \cdot k'(x) + k(x) \cdot g'(x).}$$

Quotient Rule If $f(x) = \frac{g(x)}{k(x)}$, and $k(x) \neq 0$, then

$$\mathbf{f'(x) = \frac{k(x) \cdot g'(x) - g(x) \cdot k'(x)}{[k(x)]^2}.}$$

Chain Rule If y is a function of u, say $y = f(u)$, and if u is a function of x, say $u = g(x)$, then $y = f(u) = f[g(x)]$, and

$$\mathbf{\frac{dy}{dx} = \frac{dy}{du} \cdot \frac{du}{dx}.}$$

Chain Rule (Alternative Form) Let $y = f[g(x)]$. Then $\mathbf{y' = f'[g(x)] \cdot g'(x)}$.

Generalized Power Rule Let u be a function of x, and let $y = u^n$ for any real number n. Then

$$\mathbf{y' = n \cdot u^{n-1} \cdot u'.}$$

Chapter Summary Key Terms

To understand the concepts presented in this chapter, you should know the meaning and use of the following words. For easy reference, the section in the chapter where a word (or expression) was first used is given with each item.

2.1 **limit**
continuous
discontinuous

2.2 **average rate of change**
velocity
instantaneous rate of change

2.3 **secant line**
tangent line
slope of a curve
derivative
differentiable
differentiation

2.4 **marginal analysis**

2.5 **average cost**
marginal average cost

2.6 **marginal revenue product**

Chapter 2 Review Exercises

1. Is a derivative always a limit? Is a limit always a derivative? Explain.
2. Is every continuous function differentiable? Is every differentiable function continuous? Explain.

Decide whether the limits in Exercises 3–16 exist. If a limit exists, find its value.

3. $\lim_{x \to -3} f(x)$

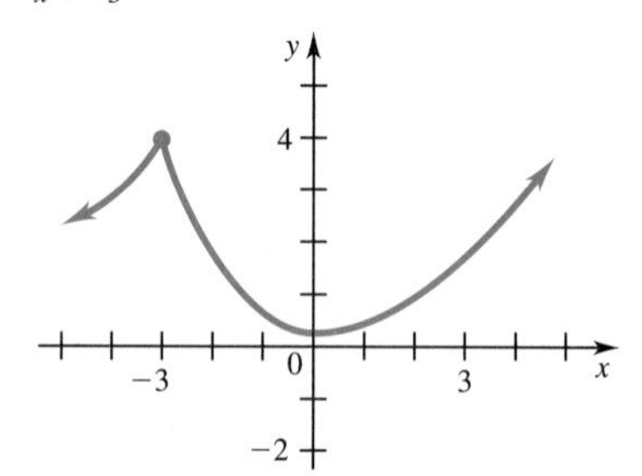

4. $\lim_{x \to -1} g(x)$

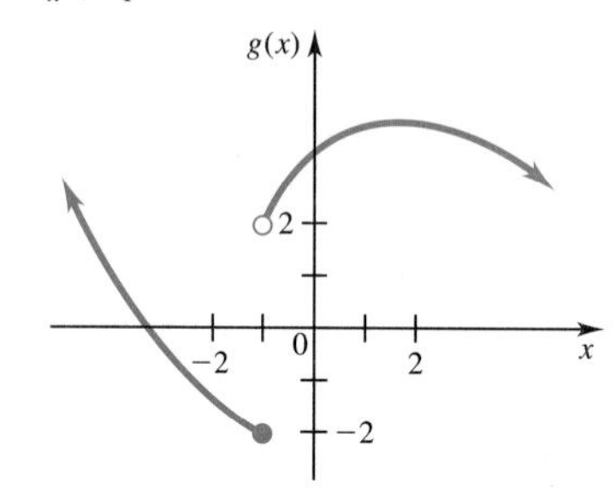

5. $\lim_{x \to 4} f(x)$

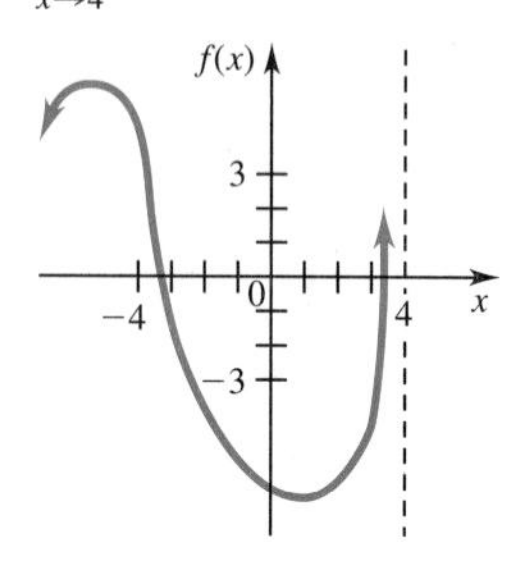

6. $\lim_{x \to 0} \frac{x^2 - 5}{2x}$

7. $\lim_{x \to -1} (2x^2 + 3x + 5)$

8. $\lim_{x \to 2} (-x^2 + 4x + 1)$

9. $\lim_{x \to 6} \frac{2x + 5}{x - 3}$

10. $\lim_{x \to 3} \frac{2x + 5}{x - 3}$

11. $\lim_{x \to 4} \frac{x^2 - 16}{x - 4}$

12. $\lim_{x \to 2} \frac{x^2 + 3x - 10}{x - 2}$

13. $\lim_{x \to -4} \frac{2x^2 + 3x - 20}{x + 4}$

14. $\lim_{x \to 3} \frac{3x^2 - 2x - 21}{x - 3}$

15. $\lim_{x \to 9} \frac{\sqrt{x} - 3}{x - 9}$

16. $\lim_{x \to 16} \frac{\sqrt{x} - 4}{x - 16}$

Identify the x-values where f is discontinuous.

17.

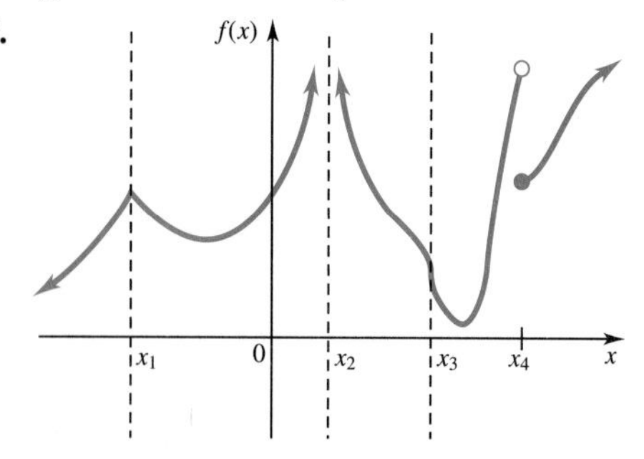

18.

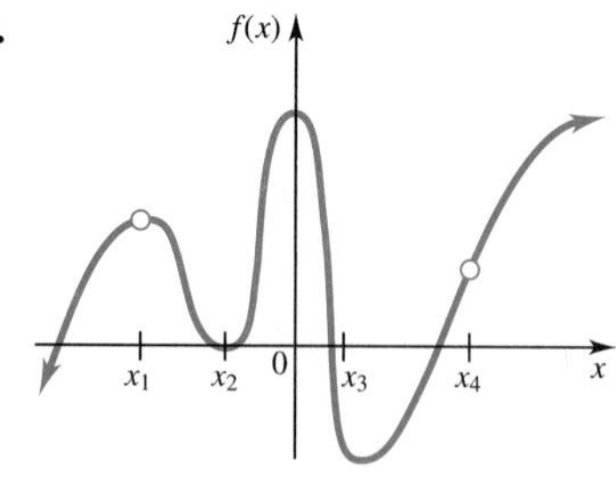

Decide whether each of the following is continuous at the given values of x.

19. $f(x) = \frac{-5}{3x(2x - 1)}$; $x = -5, 0, -1/3, 1/2$

20. $f(x) = \frac{2 - 3x}{(1 + x)(2 - x)}$; $x = 2/3, -1, 2, 0$

21. $f(x) = \frac{x - 6}{x + 5}$; $x = 6, -5, 0$

22. $f(x) = \frac{x^2 - 9}{x + 3}$; $x = 3, -3, 0$

23. $f(x) = x^2 + 3x - 4$; $x = 1, -4, 0$

24. $f(x) = 2x^2 - 5x - 3$; $x = -1/2, 3, 0$

25. Use the graph to find the average rate of change of f on the following intervals.

(a) $x = 0$ to $x = 4$
(b) $x = 2$ to $x = 8$
(c) $x = 2$ to $x = 4$

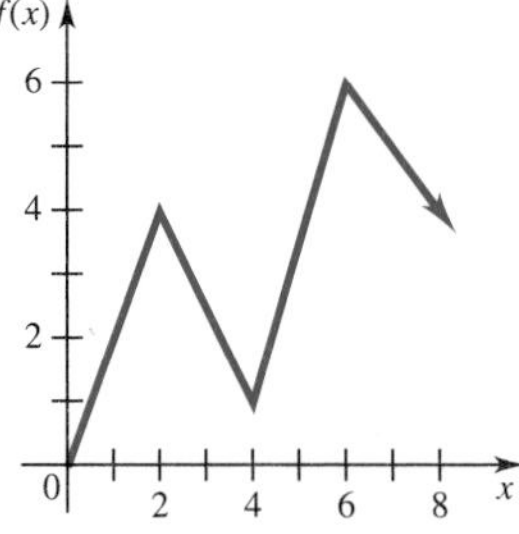

Find the average rate of change for each of the following on the given interval. Then find the instantaneous rate of change at the first x-value.

26. $y = 6x^2 + 2;$ from $x = 1$ to $x = 4$

27. $y = -2x^3 - x^2 + 5;$ from $x = -2$ to $x = 6$

28. $y = \dfrac{-6}{3x - 5};$ from $x = 4$ to $x = 9$

29. $y = \dfrac{x + 4}{x - 1};$ from $x = 2$ to $x = 5$

Use the definition *of the derivative to find the derivative of each of the following.*

30. $y = 4x + 3$

31. $y = 5x^2 + 6x$

32. $y = -x^3 + 7x$

*Find the slope of the tangent line to the given curve at the given value of x. Find the equation of each tangent line (*Hint: *Use the rules for derivatives.)*

33. $y = x^2 - 6x; \quad x = 2$

34. $y = 8 - x^2; \quad x = 1$

35. $y = \dfrac{3}{x - 1}; \quad x = -1$

36. $y = (3x^2 - 5x)(2x); \quad x = -1$

37. $y = \dfrac{3}{x^2 - 1}; \quad x = 2$

38. $y = \sqrt{6x - 2}; \quad x = 3$

39. $y = -\sqrt{8x + 1}; \quad x = 3$

40. Two students are working on taking the derivative of

$$f(x) = \frac{2x}{3x + 4}.$$

The first one uses the quotient rule to get

$$f'(x) = \frac{(3x + 4)2 - 2x(3)}{(3x + 4)^2} = \frac{8}{(3x + 4)^2}.$$

The second converts it into a product and uses the product rule:

$$\begin{aligned} f(x) &= 2x(3x + 4)^{-1} \\ f'(x) &= 2x(-1)(3x + 4)^{-2}(3) + 2(3x + 4)^{-1} \\ &= 2(3x + 4)^{-1} - 6x(3x + 4)^{-2}. \end{aligned}$$

Explain the discrepancies between the two answers. Which procedure do you think is preferable?

41. Two students are working on taking the derivative of

$$f(x) = \frac{2}{(3x + 1)^4}.$$

The first one uses the quotient rule as follows:

$$\begin{aligned} f'(x) &= \frac{(3x + 1)^4 \cdot 0 - 2 \cdot 4(3x + 1)^3 \cdot 3}{(3x + 1)^8} \\ &= \frac{-24(3x + 1)^3}{(3x + 1)^8} = \frac{-24}{(3x + 1)^5}. \end{aligned}$$

The second rewrites the function and uses the generalized power rule as follows:

$$\begin{aligned} f(x) &= 2(3x + 1)^{-4} \\ f'(x) &= (-4)2(3x + 1)^{-5} \cdot 3 = \frac{-24}{(3x + 1)^5}. \end{aligned}$$

Compare the two procedures. Which procedure do you think is preferable?

Use the rules for derivatives to find the derivative of each function defined as follows.

42. $y = 5x^2 - 7x - 9$

43. $y = x^3 - 4x^2$

44. $y = 6x^{7/3}$

45. $y = -3x^{-2}$

46. $f(x) = x^{-3} + \sqrt{x}$

47. $f(x) = 6x^{-1} - 2\sqrt{x}$

48. $y = (3t^2 + 7)(t^3 - t)$

49. $y = (-5t + 4)(t^3 - 2t^2)$

50. $g(t) = -3t^{-1/3}(5t + 7)$

51. $p(t) = 8t^{3/4}(7t - 2)$

52. $y = 12x^{-3/4}\left(\frac{x}{3} + 5\right)$

53. $y = 15x^{-3/5}\left(6 - \frac{x}{3}\right)$

54. $k(x) = \frac{3x}{x + 5}$

55. $r(x) = \frac{-8}{2x + 1}$

56. $y = \frac{x^2 - x + 1}{x - 1}$

57. $y = \frac{2x^3 - 5x^2}{x + 2}$

58. $f(x) = (3x - 2)^4$

59. $k(x) = (5x - 1)^6$

60. $y = \sqrt{2t - 5}$

61. $y = -3\sqrt{8t - 1}$

62. $y = 3x(2x + 1)^3$

63. $y = 4x^2(3x - 2)^5$

64. $r(t) = \frac{5t^2 - 7t}{(3t + 1)^3}$

65. $s(t) = \frac{t^3 - 2t}{(4t - 3)^4}$

Find each of the following.

66. $D_x\left[\frac{\sqrt{x} + 1}{\sqrt{x} - 1}\right]$

67. $D_x\left[\frac{2x + \sqrt{x}}{1 - x}\right]$

68. $\frac{dy}{dt}$ if $y = \sqrt{t^{1/2} + t}$

69. $\frac{dy}{dx}$ if $y = \frac{\sqrt{x - 1}}{x}$

70. $f'(1)$ if $f(x) = \frac{\sqrt{8 + x}}{x + 1}$

71. $f'(-2)$ if $f(t) = \frac{2 - 3t}{\sqrt{2 + t}}$

72. Describe how to tell when a function is discontinuous at $x = a$, where a is a real number.

73. Which of the following does *not* require the product rule or the quotient rule to differentiate?

(a) $f(x) = \frac{x^2}{5x - 1}$ **(b)** $g(x) = 6x\sqrt{x + 2}$ **(c)** $h(x) = \frac{x}{a - x}$ **(d)** $F(x) = \frac{x^2}{1500}$

Find the marginal average cost function of each function defined as follows.

74. $C(x) = \sqrt{x + 1}$

75. $C(x) = \sqrt{3x + 2}$

76. $C(x) = (x^2 + 3)^3$

77. $C(x) = (4x + 3)^4$

Applications

Business and Economics

78. Sales The sales of a company are related to its expenditures on research by

$$S(x) = 1000 + 50\sqrt{x} + 10x,$$

where $S(x)$ gives sales in millions when x thousand dollars is spent on research. Find and interpret dS/dx if the following amounts are spent on research.

(a) \$9000 **(b)** \$16,000 **(c)** \$25,000

(d) As the amount spent on research increases, what happens to sales?

79. Profit Suppose that the profit (in hundreds of dollars) from selling x units of a product is given by

$$P(x) = \frac{x^2}{x - 1}, \text{ where } x > 1.$$

Find and interpret the marginal profit when the following numbers of units are sold.

(a) 4 **(b)** 12 **(c)** 20

(d) What is happening to the marginal profit as the number sold increases?

80. Costs A company finds that its costs are related to the amount spent on training programs by

$$T(x) = \frac{1000 + 50x}{x + 1},$$

where $T(x)$ is costs in thousands of dollars when x hundred dollars are spent on training. Find and interpret $T'(x)$ if the following amounts are spent on training.

(a) \$900 **(b)** \$1900

(c) Are costs per dollar spent on training always increasing or decreasing?

81. Revenue Waverly Products has found that its revenue is related to advertising expenditures by the function

$$R(x) = 5000 + 16x - 3x^2,$$

where $R(x)$ is the revenue in dollars when x hundred dollars are spent on advertising.

(a) Find the marginal revenue function.

(b) Find and interpret the marginal revenue when \$1000 is spent on advertising.

Connections

Business and Economics

82. Cost Analysis A company charges $1.50 per pound when a certain chemical is bought in lots of 125 lb or less, with a price per pound of $1.35 if more than 125 lb are purchased. Let $C(x)$ represent the cost of x lb. Find the cost for the following pounds.

(a) 100 **(b)** 125 **(c)** 140

(d) Graph $y = C(x)$.

(e) Where is C discontinuous?

Find the average cost per pound if the following number of pounds are bought.

(f) 100 **(g)** 125 **(h)** 140

Find and interpret the marginal cost for the following numbers of pounds.

(i) 100 **(j)** 140

83. Marginal Analysis Suppose the profit (in cents) from selling x lb of potatoes is given by

$$P(x) = 15x + 25x^2.$$

Find the average rate of change in profit from selling each of the following amounts.

(a) 6 lb to 7 lb **(b)** 6 lb to 6.5 lb **(c)** 6 lb to 6.1 lb

Find the marginal profit from selling the following amounts.

(d) 6 lb **(e)** 20 lb **(f)** 30 lb

(g) What is the domain of x?

(h) Is it possible for the marginal profit to be negative here? What does this mean?

(i) Find the average profit function.

(j) Find the marginal average profit function.

(k) Is it possible for the marginal average profit to vary here? What does this mean?

(l) Discuss whether this function describes a realistic situation.

Life Science

84. Spread of a Virus The spread of a virus is modeled by

$$V(t) = -t^2 + 6t - 4,$$

where $V(t)$ is the number of people (in hundreds) with the virus and t is the number of weeks since the first case was observed.

(a) Graph $V(t)$.

(b) What is a reasonable domain of t for this problem?

(c) When does the number of cases reach a maximum? What is the maximum number of cases?

(d) Find the rate of change function.

(e) What is the rate of change in the number of cases at the maximum?

(f) Give the sign (+ or −) of the rate of change up to the maximum and after the maximum.

Computer/Graphing Calculator

For each of the following, use a computer or a graphing calculator to graph each function and its derivative on the same axes. Determine the values of x where the derivative is **(a)** *positive,* **(b)** *zero, and* **(c)** *negative.* **(d)** *What is true of the graph of the function in each case?*

85. $g(x) = 6 - 4x + 3x^2 - x^3$

86. $k(x) = 2x^4 - 3x^3 + x$

87. $G(x) = \dfrac{2x}{(x - 1)^2}$

88. $K(x) = \sqrt[3]{(2x - 1)^2}$

Applications at a glance . . .

Timing a Sales Pitch . . . changing listeners' attitudes . . . probability of legislator voting "pro" . . .

CHAPTER 3

CURVE SKETCHING

(See page 156.)

The graph in Figure 1 shows the population of deer on the Kaibab Plateau on the North Rim of the Grand Canyon between 1905 and 1930. In the first two decades of the twentieth century, hunters were effective in reducing the population of predators. With fewer predators, the deer increased in numbers and depleted available food resources. As shown by the graph, the deer population peaked in about 1925 and then declined rapidly.

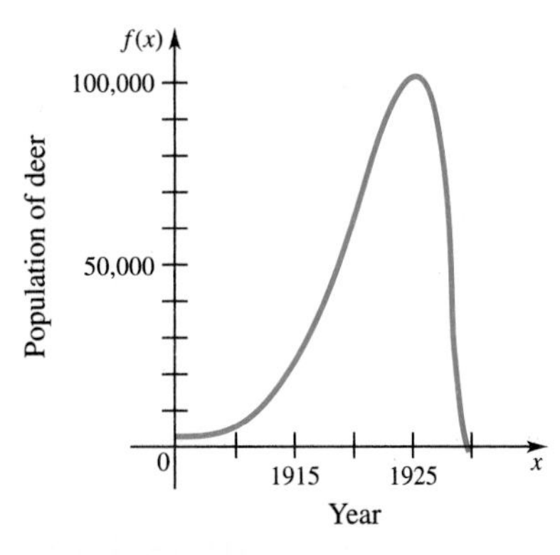

FIGURE 1

Given a graph like the one in Figure 1, often we can locate maximum and minimum values simply by looking at the graph. It is difficult to get *exact* values or *exact* locations of maxima and minima from a graph, however, and many functions are difficult to graph. In Chapter 1 we saw how to find exact maximum and minimum values for quadratic functions by identifying the vertex. A more general approach is to use the derivative of a function to determine precise maximum and minimum values of the function. The procedure for doing this is described in this chapter, which begins with a discussion of increasing and decreasing functions.

3.1 INCREASING AND DECREASING FUNCTIONS

How high is the highest point on a roller coaster?

Suppose that an amusement park has hired an engineering firm to develop a new roller coaster. The ride will include one very steep drop that passes through a tunnel. An engineer has developed a design for this section. The design gives the height h (in feet) of the track above the ground by the polynomial function

$$h = f(x) = -3x^4 + 28x^3 - 84x^2 + 96x,$$

where x is the distance along the ground measured in 10-foot intervals. (In other words, $x = 1$ corresponds to a distance of 10 feet along the ground. This arrangement keeps the values of x smaller, and thus easier to work with.) Meanwhile, a new safety regulation has been passed requiring that such rides be no more than 65 feet high. The engineer is particularly worried about whether this stretch of track, shown in Figure 2, will be too high. The engineer has used a graphing calculator to quickly sketch the curve, and this stretch of track comes very close to the limit. How can he find the *exact value* of the highest point?

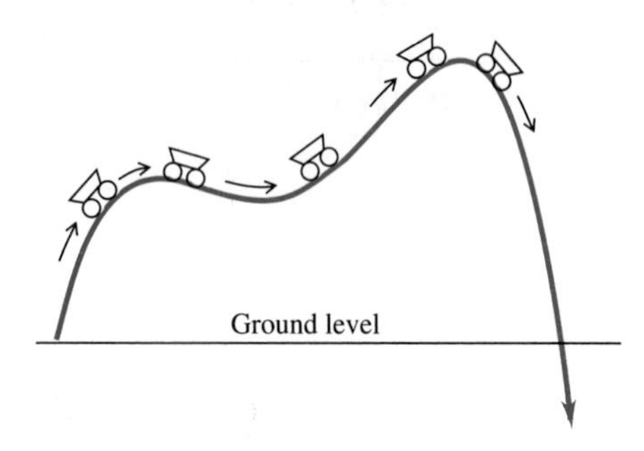

FIGURE 2

The derivative can be used to answer this question. Remember that the derivative of a function at a point gives the slope of the line tangent to the function at that point. Recall also that a line with a positive slope rises from left to right and a line with a negative slope falls from left to right.

Now, picture one of the cars on the roller coaster. When the car is on level ground or parallel to level ground, its floor is horizontal, but as the car moves up the slope, its floor tilts upward. When the car reaches a peak, its floor is again horizontal, but it then begins to tilt downward (very steeply) as the car rolls downhill.

If we think of the roller coaster track as a graph, the floor of the car as it moves from left to right along the track represents the tangent line at each point. Using this analogy, we can see that the slope of the tangent will be *positive* when the car travels uphill and h is *increasing*, and the slope of the tangent will be *negative* when the car travels downhill and h is *decreasing*. (In this case it is also true that the slope will be zero at "peaks" and "valleys.") See Figure 3.

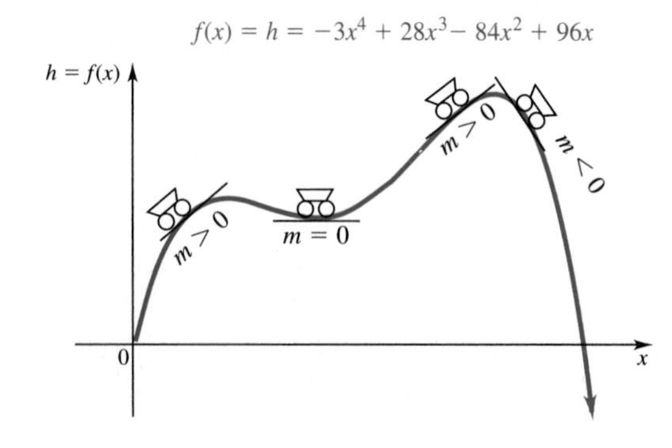

FIGURE 3

In this chapter you will need all of the rules for derivatives you learned in the previous chapter. If any of these are still unclear, go over the Derivative Summary at the end of that chapter and practice some of the Review Exercises before proceeding.

Thus, on intervals where $f'(x) > 0$, $f(x) = h$ will increase, and on intervals where $f'(x) < 0$, $f(x) = h$ will decrease. We can determine where $f(x)$ peaks by finding the intervals on which it increases and decreases. The function

$$f(x) = -3x^4 + 28x^3 - 84x^2 + 96x$$

is differentiated as

$$f'(x) = -12x^3 + 84x^2 - 168x + 96.$$

To determine where $f'(x) > 0$ and where $f'(x) < 0$, first find where $f'(x) = 0$.

$$-12x^3 + 84x^2 - 168x + 96 = 0$$

$$x^3 - 7x^2 + 14x - 8 = 0 \quad \text{Divide each side by } -12.$$

The left side of the equation can be factored as

$$(x - 1)(x - 2)(x - 4) = 0.$$

Therefore, $f'(x)$ will equal 0 when $x = 1, 2,$ or 4. (These values of x correspond to the two "peaks" and one "valley" on the graph of $f(x)$ in Figure 3.)

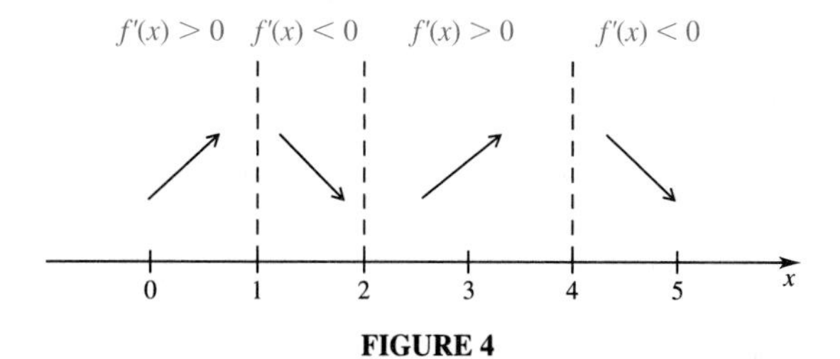

FIGURE 4

The three zeros of $f'(x)$ divide the real number line into four regions, as shown in Figure 4. Evaluating $f'(x)$ at a point in each region shows that $f'(x)$ is positive between 0 and 1, negative between 1 and 2, positive between 2 and 4, and negative for x greater than 4. Thus, $h = f(x)$ increases from 0 to 1, decreases from 1 to 2, increases from 2 to 4, and decreases thereafter. The graph in Figure 5 shows that the highest point occurs at $x = 4$, after the roller coaster has traveled a horizontal distance of 40 feet. The track there is $f(4) = 64$ feet above the ground, within the 65-foot limit.

x	$h = f(x)$
0	0
1	37
2	32
3	45
4	64
5	5

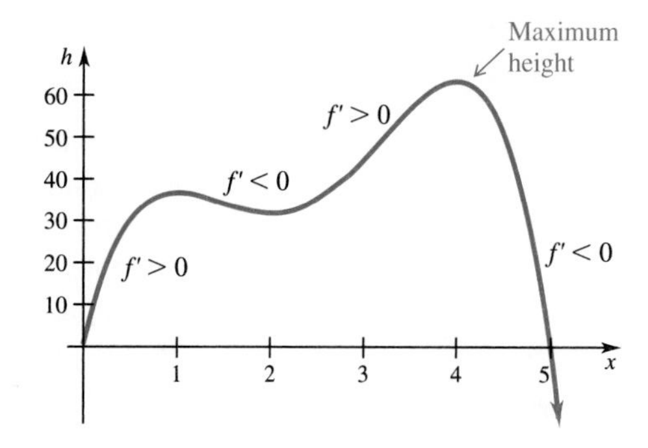

FIGURE 5

As shown in the preceding example, a function is *increasing* if the graph goes *up* from left to right and *decreasing* if its graph goes *down* from left to right. Examples of increasing functions are shown in Figure 6 and examples of decreasing functions in Figure 7.

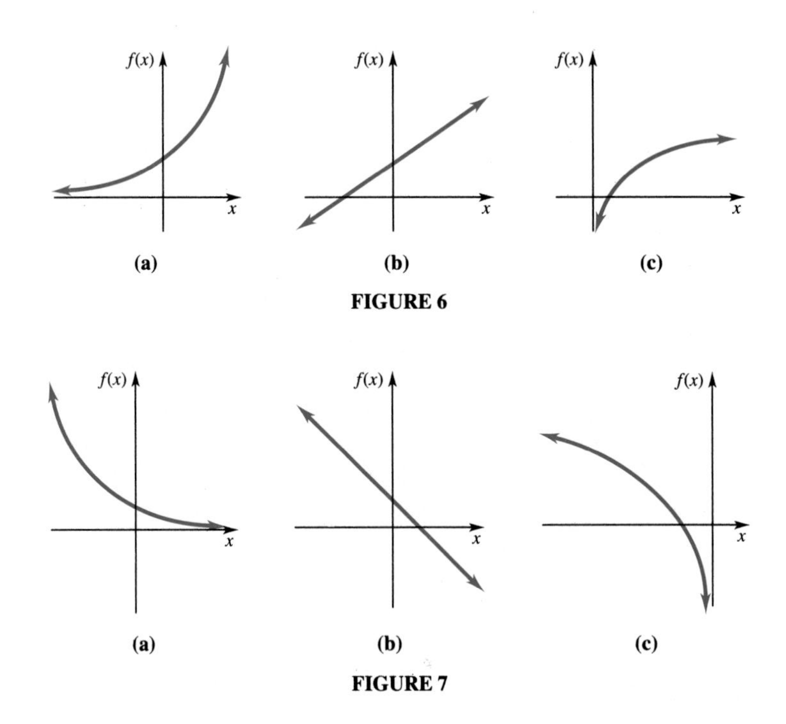

FIGURE 6

FIGURE 7

INCREASING AND DECREASING FUNCTIONS

Let f be a function defined on some interval. Then for any two numbers x_1 and x_2 in the interval, f is **increasing** on the interval if

$$f(x_1) < f(x_2) \quad \text{whenever} \quad x_1 < x_2,$$

and f is **decreasing** on the interval if

$$f(x_1) > f(x_2) \quad \text{whenever} \quad x_1 < x_2.$$

EXAMPLE 1 Where is the function graphed in Figure 8 increasing? Where is it decreasing?

Moving from left to right, the function is increasing up to -4, then decreasing from -4 to 0, constant (neither increasing nor decreasing) from 0 to 4, increasing from 4 to 6, and decreasing from 6 onward. In interval notation, the function is increasing on $(-\infty, -4)$ and $(4, 6)$, decreasing on $(-4, 0)$ and $(6, \infty)$, and constant on $(0, 4)$. ◀

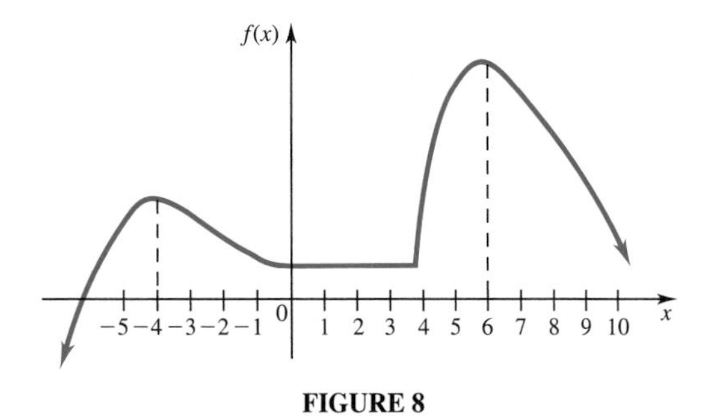

FIGURE 8

The following test summarizes the discussion in the opening example.

TEST FOR INTERVALS WHERE $f(x)$ IS INCREASING AND DECREASING

Suppose a function f has a derivative at each point in an open interval; then
if $f'(x) > 0$ for each x in the interval, f is *increasing* on the interval;
if $f'(x) < 0$ for each x in the interval, f is *decreasing* on the interval;
if $f'(x) = 0$ for each x in the interval, f is *constant* on the interval.

The derivative $f'(x)$ can change signs from positive to negative (or negative to positive) at points where $f'(x) = 0$, and also at points where $f'(x)$ does not exist. The values of x where this occurs are called *critical numbers*. By definition, the **critical numbers** for a function f are those numbers c in the domain of f for which $f'(c) = 0$ or $f'(c)$ does not exist. A **critical point** is a point whose x-coordinate is the critical number c, and whose y-coordinate is $f(c)$.

It is shown in more advanced classes that if the critical numbers of a polynomial function are used to determine open intervals on a number line, then the sign of the derivative at any point in the interval will be the same as the sign of the derivative at any other point in the interval. This suggests that the test for increasing and decreasing functions be applied as follows (assuming that no open intervals exist where the function is constant).

APPLYING THE TEST

1. Locate on a number line those values of x for which $f'(x) = 0$ or $f'(x)$ does not exist. These points determine several open intervals.
2. Choose a value of x in each of the intervals determined in Step 1. Use these values to decide whether $f'(x) > 0$ or $f'(x) < 0$ in that interval.
3. Use the test above to decide whether f is increasing or decreasing on the interval.

EXAMPLE 2 Find the intervals in which the following functions are increasing or decreasing. Locate all points where the tangent line is horizontal. Graph the function.

(a) $f(x) = x^3 + 3x^2 - 9x + 4$

Here $f'(x) = 3x^2 + 6x - 9$. To find the critical numbers, set this derivative equal to 0 and solve the resulting equation by factoring.

$$3x^2 + 6x - 9 = 0$$
$$3(x^2 + 2x - 3) = 0$$
$$3(x + 3)(x - 1) = 0$$
$$x = -3 \quad \text{or} \quad x = 1$$

The tangent line is horizontal at $x = -3$ or $x = 1$. Since there are no values of x where $f'(x)$ fails to exist, the only critical numbers are -3 and 1. To determine where the function is increasing or decreasing, locate -3 and 1 on a number line, as in Figure 9. (Be sure to place the values on the number line in numerical order.) These points determine three intervals, $(-\infty, -3)$, $(-3, 1)$, and $(1, \infty)$.

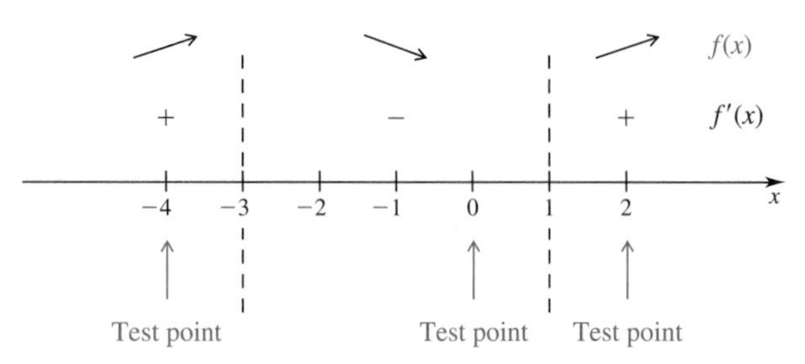

FIGURE 9

Now choose any value of x in the interval $(-\infty, -3)$. Choosing $x = -4$ gives

$$f'(-4) = 3(-4)^2 + 6(-4) - 9 = 15,$$

which is positive. Since one value of x in this interval makes $f'(x) > 0$, all values will do so, and therefore f is increasing on $(-\infty, -3)$. Selecting 0 from the middle interval gives $f'(0) = -9$, so f is decreasing on $(-3, 1)$. Finally, choosing 2 in the right-hand region gives $f'(2) = 15$, with f increasing on $(1, \infty)$. The arrows in each interval in Figure 9 indicate where f is increasing or decreasing.

Up to now our only method of graphing most functions has been by plotting points that lie on the graph, either by hand or using a graphing calculator or computer. Now an additional tool is available: the test for determining where a function is increasing or decreasing. (Other tools are discussed in the next few sections.) To graph the function, plot a point at each of the critical numbers by finding $f(-3) = 31$ and $f(1) = -1$. Also plot points for $x = -4$, 0, and 2, the test values of each interval. Use these points along with the information about where the function is increasing and decreasing to get the graph in Figure 10.

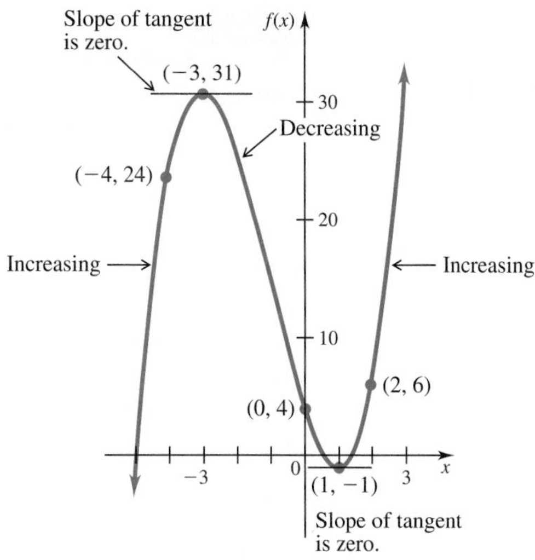

FIGURE 10

Caution Be sure to use $f(x)$, not $f'(x)$, to find the y-values of the points to plot.

(b) $f(x) = \dfrac{x - 1}{x + 1}$

Use the quotient rule to find $f'(x)$.

$$f'(x) = \frac{(x + 1)(1) - (x - 1)(1)}{(x + 1)^2}$$

$$= \frac{x + 1 - x + 1}{(x + 1)^2} = \frac{2}{(x + 1)^2}$$

This derivative is never 0, but it fails to exist when $x = -1$. Since $f(x)$ also does not exist at $x = -1$, however, -1 is not a critical number. There are no critical numbers, and the tangent line is never horizontal. Since the line $x = -1$ is a vertical asymptote for the graph of f, it separates the domain into the two intervals $(-\infty, -1)$ and $(-1, \infty)$. A function can change direction from one side of a vertical asymptote to the other. Use a test point in each of these intervals to find that $f'(x) > 0$ for all x except -1. This means that the function f is increasing on both $(-\infty, -1)$ and $(-1, \infty)$.

When x is large,

$$\frac{x - 1}{x + 1} \approx \frac{x}{x} = 1,$$

so the graph has the line $y = 1$ as a horizontal asymptote. Using all this information and plotting the intercepts and a few points on either side of the vertical asymptote gives the graph in Figure 11. ◀

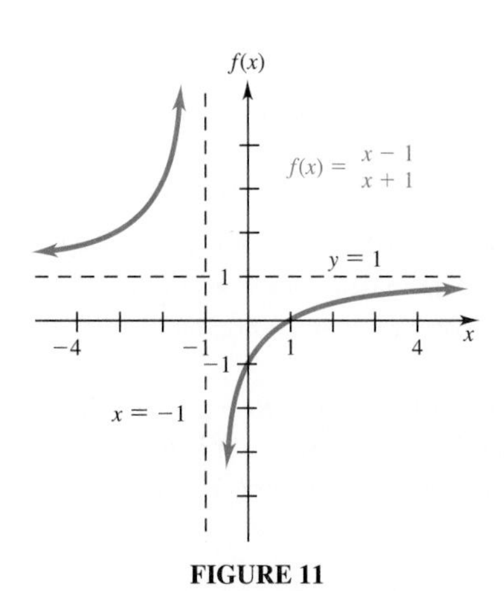

FIGURE 11

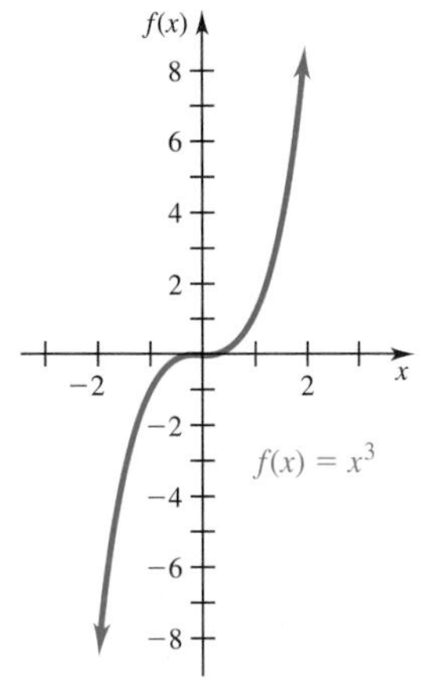

FIGURE 12

Caution It is important to note that the reverse of the test for increasing and decreasing functions is not true—it is possible for a function to be increasing on an interval even though the derivative is not positive at every point in the interval. A good example is given by $f(x) = x^3$, which is increasing on every interval, even though $f'(x) = 0$ when $x = 0$. See Figure 12.

Similarly, it is incorrect to assume that the sign of the derivative in regions separated by critical numbers must alternate between + and −. If this were always so, it would lead to a simple rule for finding the sign of the derivative: just check one test point, and then make the other regions alternate in sign. But this is not true if one of the factors in the derivative is raised to an even power. In the function $f(x) = x^3$ just considered, $f'(x) = 3x^2$, which is positive on both sides of the critical number $x = 0$.

Knowing the intervals where a function is increasing or decreasing can be important in applications, as shown by the next examples.

A company sells computers with the following cost and revenue functions:

$$C(x) = 2000x^2 - x^3, \qquad 0 \le x \le 1000$$
$$R(x) = 2{,}000{,}000x - 1000x^2, \qquad 0 \le x \le 1000$$

where x is the number of computers sold monthly. Determine any intervals on which the profit function is increasing.

First find the profit function $P(x)$.

$$\begin{aligned} P(x) &= R(x) - C(x) \\ &= (2{,}000{,}000x - 1000x^2) - (2000x^2 - x^3) \\ &= 2{,}000{,}000x - 3000x^2 + x^3 \end{aligned}$$

To find any intervals where this function is increasing, set $P'(x) = 0$.

$$P'(x) = 2{,}000{,}000 - 6000x + 3x^2 = 0$$

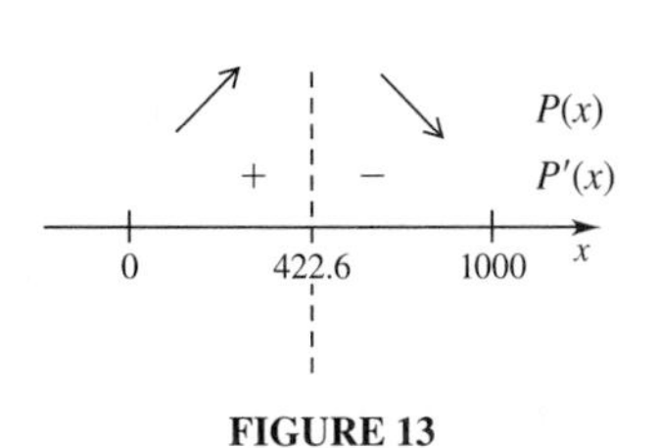

FIGURE 13

Solving this with the quadratic formula gives the approximate solutions $x = 422.6$ and $x = 1577.4$. The latter number is outside of the domain. Use $x = 422.6$ to determine two intervals on a number line, as shown in Figure 13. Choose $x = 0$ and $x = 1000$ as test points.

$$P'(0) = 2{,}000{,}000 - 6000(0) + 3(0^2) = 2{,}000{,}000$$
$$P'(1000) = 2{,}000{,}000 - 6000(1000) + 3(1000^2) = -1{,}000{,}000$$

The test points show that the function increases on (0,422.6) and decreases on (422.6, 1000). See Figure 13. Thus, the profit is increasing when 422 computers or fewer are sold, and decreasing when 423 or more are sold, as shown in Figure 14.

The graph in Figure 14 also shows that the profit will increase as long as the revenue function increases faster than the cost function. That is, increasing production will produce more profit as long as the marginal revenue is greater than the marginal cost. ◀

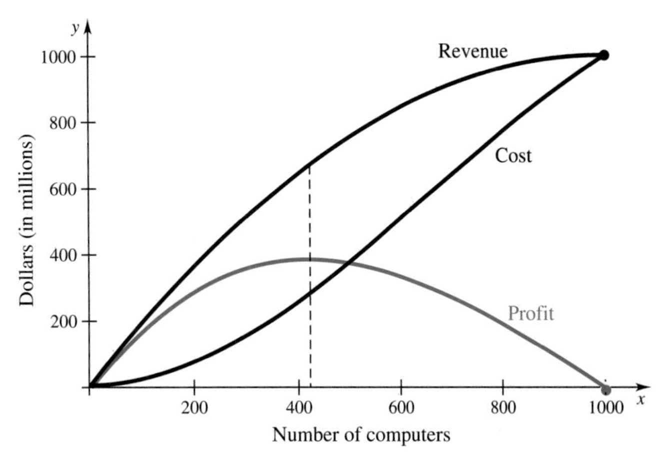

FIGURE 14

EXAMPLE 4 In the exercises in the previous chapter, the function

$$f(t) = \frac{90t}{99t - 90}$$

gave the number of facts recalled after t hours for $t > 10/11$. Find the intervals in which $f(t)$ is increasing or decreasing.

First find the derivative, $f'(t)$.

$$f'(t) = \frac{(99t - 90)(90) - 90t(99)}{(99t - 90)^2}$$

$$= \frac{8910t - 8100 - 8910t}{(99t - 90)^2} = \frac{-8100}{(99t - 90)^2}$$

Since $(99t - 90)^2$ is positive everywhere in the domain of the function and since the numerator is a negative constant, $f'(t) < 0$ for all t in the domain of $f(t)$. Thus $f(t)$ always decreases and, as expected, the number of words recalled decreases steadily over time. ◀

3.1 Exercises

Find the largest open intervals where the functions graphed as follows are **(a)** *increasing, or* **(b)** *decreasing.*

1.

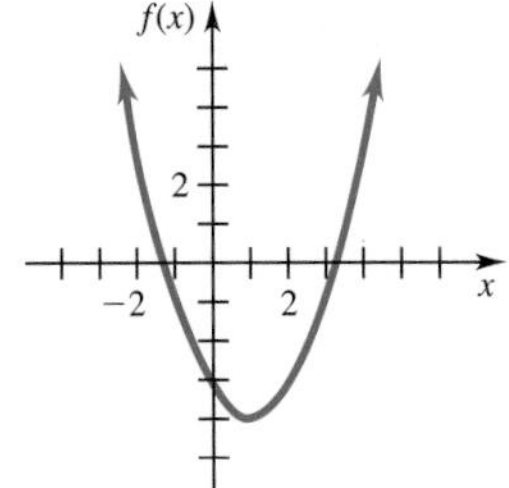

2.

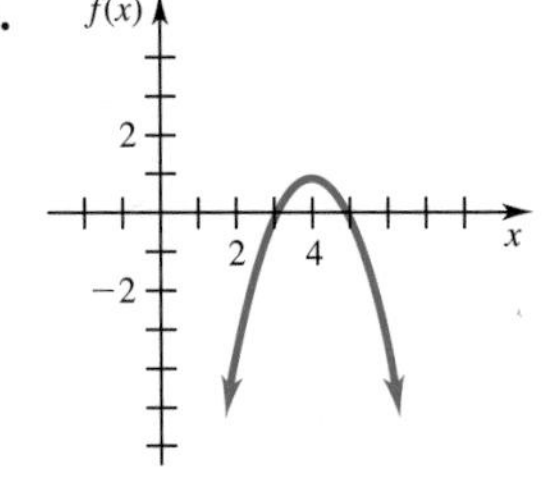

3.

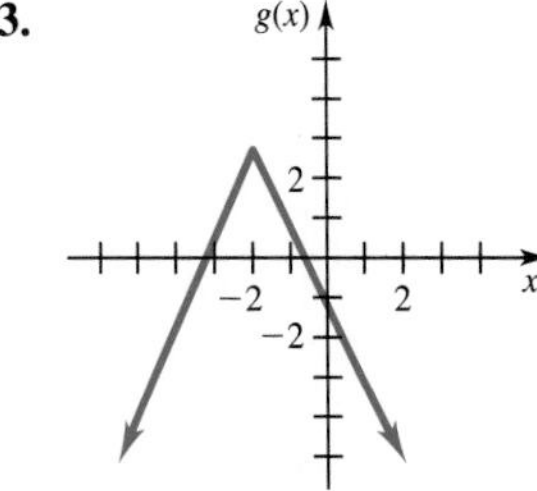

4.

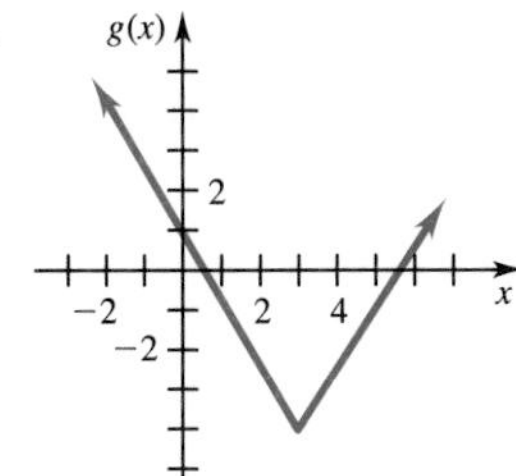

5.

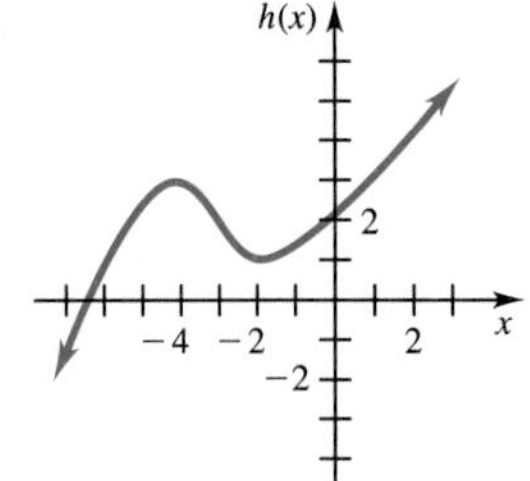

6.

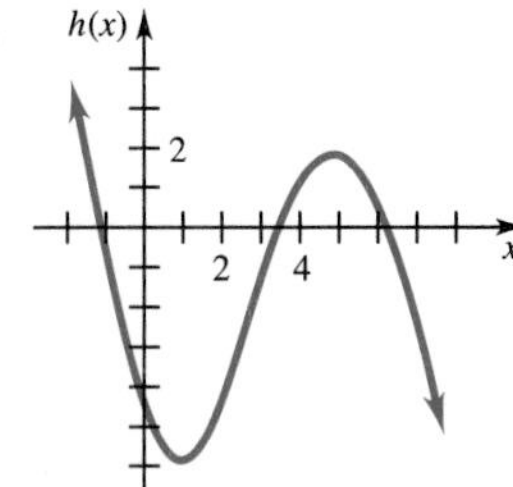

7.

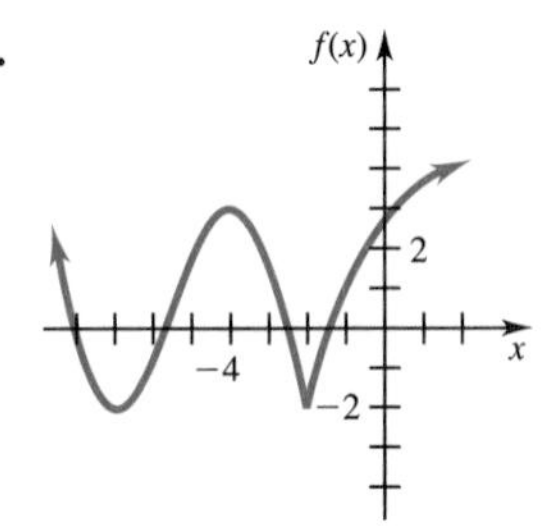

8.

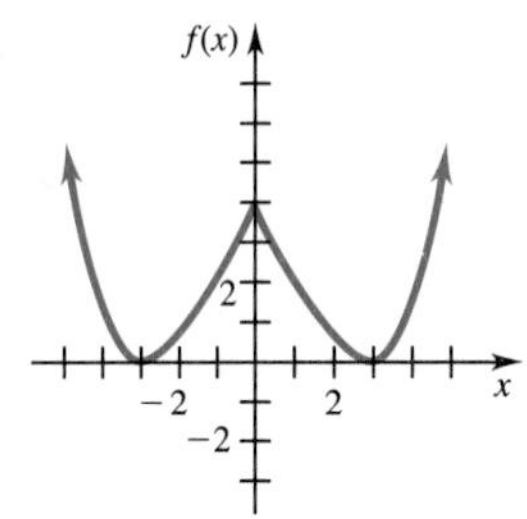

Find the largest open intervals where the following are **(a)** *increasing;* **(b)** *decreasing.*

9. $f(x) = x^2 + 12x - 6$

10. $f(x) = x^2 - 9x + 4$

11. $y = 2 + 3.6x - 1.2x^2$

12. $y = .3 + .4x - .2x^2$

13. $f(x) = \frac{2}{3}x^3 - x^2 - 24x - 4$

14. $f(x) = \frac{2}{3}x^3 - x^2 - 4x + 2$

15. $f(x) = 4x^3 - 15x^2 - 72x + 5$

16. $f(x) = 4x^3 - 9x^2 - 30x + 6$

17. $y = -3x + 6$

18. $y = 6x - 9$

19. $f(x) = \dfrac{x+2}{x+1}$

20. $f(x) = \dfrac{x+3}{x-4}$

21. $y = |x + 4|$

22. $y = -|x - 3|$

23. $f(x) = -\sqrt{x - 1}$

24. $f(x) = \sqrt{5 - x}$

25. $y = \sqrt{x^2 + 1}$

26. $y = x\sqrt{9 - x^2}$

27. $f(x) = x^{2/3}$

28. $f(x) = (x + 1)^{4/5}$

29. Use the techniques of this chapter to find the vertex and intervals where f is increasing and decreasing, given

$$f(x) = ax^2 + bx + c,$$

where we assume $a > 0$. Verify that this agrees with what we found in Chapter 1.

30. Repeat Exercise 29 under the assumption $a < 0$.

Applications

Business and Economics

31. Cost Suppose the total cost $C(x)$ (in dollars) to manufacture a quantity x of weed killer (in hundreds of liters) is given by

$$C(x) = x^3 - 2x^2 + 8x + 50.$$

(a) Where is $C(x)$ decreasing?

(b) Where is $C(x)$ increasing?

32. Housing Starts A county realty group estimates that the number of housing starts per year over the next three years will be

$$H(r) = \frac{300}{1 + .03r^2},$$

where r is the mortgage rate (in percent).

(a) Where is $H(r)$ increasing?

(b) Where is $H(r)$ decreasing?

33. Profit A manufacturer sells video games with the following cost and revenue functions, where x is the number of games sold.

$$C(x) = 4.8x - .0004x^2, \qquad 0 \le x \le 2250$$
$$R(x) = 8.4x - .002x^2, \qquad 0 \le x \le 2250$$

Determine the interval(s) on which the profit function is increasing.

34. Profit A manufacturer of compact disk players has determined that the profit $P(x)$ (in thousands of dollars) is related to the quantity x of players produced (in hundreds) per month by

$$P(x) = \frac{1}{3}x^3 - \frac{7}{2}x^2 + 10x - 2,$$

as long as the number of players produced is fewer than 800 per month.

(a) At what production levels is the profit increasing?

(b) At what levels is it decreasing?

Life Sciences

35. Spread of Infection The number of people $P(t)$ (in hundreds) infected t days after an epidemic begins is approximated by

$$P(t) = 2 + 50t - \frac{5}{2}t^2.$$

When will the number of people infected start to decline?

36. Alcohol Concentration In Chapter 1 we gave the function defined by

$$A(x) = -.015x^3 + 1.058x$$

as the approximate alcohol concentration (in tenths of a percent) in an average person's bloodstream x hr after drinking 8 oz of 100-proof whiskey. The function applies only for the interval [0, 8].

(a) On what time intervals is the alcohol concentration increasing?

(b) On what intervals is it decreasing?

37. Drug Concentration The percent of concentration of a drug in the bloodstream x hr after the drug is administered is given by

$$K(x) = \frac{4x}{3x^2 + 27}.$$

(a) On what time intervals is the concentration of the drug increasing?

(b) On what intervals is it decreasing?

38. Drug Concentration Suppose a certain drug is administered to a patient, with the percent of concentration of the drug in the bloodstream t hr later given by

$$K(t) = \frac{5t}{t^2 + 1}.$$

(a) On what time intervals is the concentration of the drug increasing?

(b) On what intervals is it decreasing?

General Interest

39. Sports Cars The following graph shows the horsepower and torque as a function of the engine speed for a 1991 Porsche 928 GT.

(a) On what intervals is the horsepower increasing with engine speed?

(b) On what intervals is the horsepower decreasing with engine speed?

(c) On what intervals is the torque increasing with engine speed?

(d) On what intervals is the torque decreasing with engine speed?

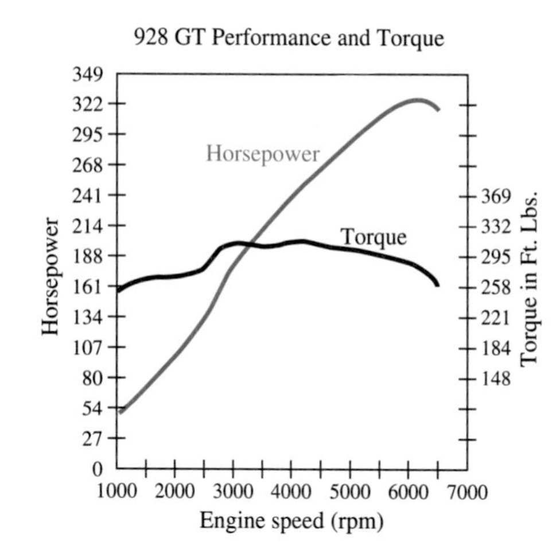

3.2 RELATIVE EXTREMA

In a 30-second commercial, when is the best time to present the sales message?

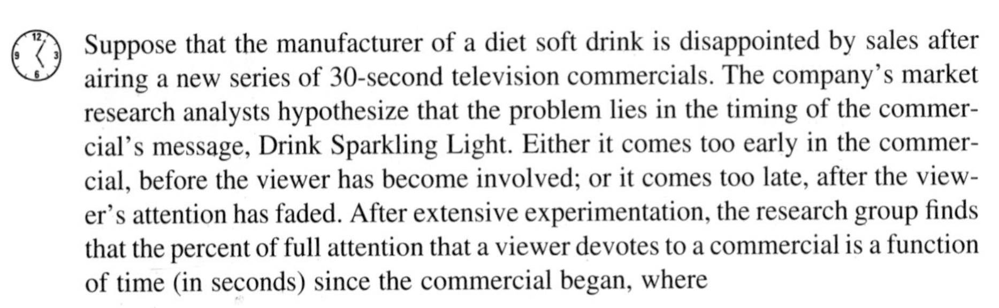

Suppose that the manufacturer of a diet soft drink is disappointed by sales after airing a new series of 30-second television commercials. The company's market research analysts hypothesize that the problem lies in the timing of the commercial's message, Drink Sparkling Light. Either it comes too early in the commercial, before the viewer has become involved; or it comes too late, after the viewer's attention has faded. After extensive experimentation, the research group finds that the percent of full attention that a viewer devotes to a commercial is a function of time (in seconds) since the commercial began, where

$$\text{Viewer's attention} = f(t) = -\frac{3}{20}t^2 + 6t + 20, \qquad 0 \le t \le 30.$$

When is the best time to present the commercial's sales message?

Clearly, the message should be delivered when the viewer's attention is at a maximum. To find this time, find $f'(t)$.

$$f'(t) = -\frac{3}{10}t + 6 = -.3t + 6$$

The derivative $f'(t)$ is greater than 0 when $-.3t + 6 > 0$, $-3t > -60$, or $t < 20$. Similarly, $f'(t) < 0$ when $-.3t + 6 < 0$, $-3t < -60$, or $t > 20$. Thus, attention increases for the first 20 seconds and decreases for the last 10 seconds. The message should appear about 20 seconds into the commercial. At that time the viewer will devote $f(20) = 80\%$ of his attention to the commercial.

The maximum level of viewer attention (80%) in the example above is a *relative maximum*, defined as follows.

RELATIVE MAXIMUM OR MINIMUM

Let c be a number in the domain of a function f. Then $f(c)$ is a **relative maximum** for f if there exists an open interval (a, b) containing c such that

$$f(x) \le f(c)$$

for all x in (a, b), and $f(c)$ is a **relative minimum** for f if there exists an open interval (a, b) containing c such that

$$f(x) \ge f(c)$$

for all x in (a, b).

If c is an endpoint of the domain of f, we only consider x in the half-open interval that is in the domain.

A function has a **relative extremum** (plural: **extrema**) at c if it has either a relative maximum or a relative minimum there.

EXAMPLE 1 Identify the x-values of all points where the graph in Figure 15 has relative extrema.

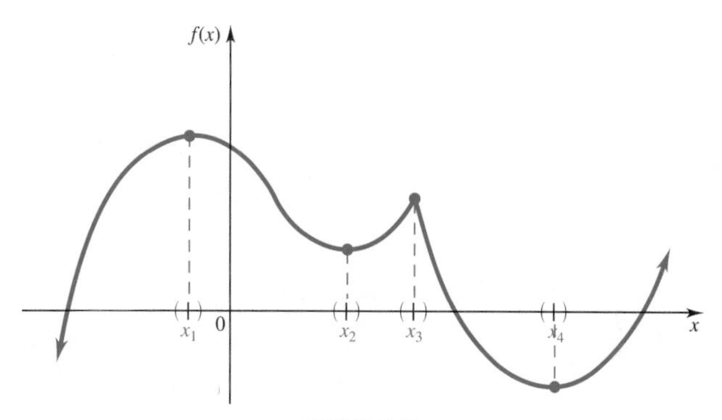

FIGURE 15

The parentheses around x_1 show an open interval containing x_1 such that $f(x) \leq f(x_1)$, so there is a relative maximum of $f(x_1)$ at $x = x_1$. Notice that many other open intervals would work just as well. Similar intervals around x_2, x_3, and x_4 can be used to find a relative maximum of $f(x_3)$ at $x = x_3$ and relative minima of $f(x_2)$ at $x = x_2$ and $f(x_4)$ at $x = x_4$.

The function graphed in Figure 16 has relative maxima when $x = x_2$ or $x = x_4$ and relative minima when $x = x_1$ or $x = x_3$. The tangents at the points having x-values x_1, x_2, and x_3 are shown in the figure. All three tangents are horizontal and have slope 0. There is no single tangent line at the point where $x = x_4$.

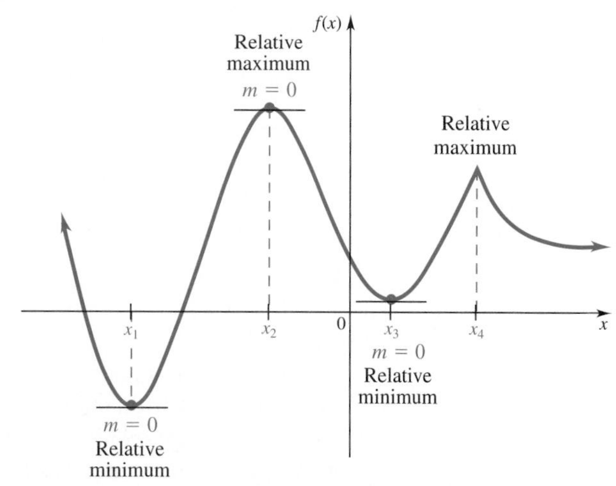

FIGURE 16

Since the derivative of a function gives the slope of a line tangent to the graph of the function, to find relative extrema we first identify all values of x where the derivative of the function (the slope of the tangent line) is 0. Next, we identify all x-values where the derivative *does not* exist but the function *does* exist. Recall from the preceding section that *critical numbers* are these x-values where the derivative is 0 or where the derivative does not exist but the function does exist. A relative extremum *may* exist at a critical number. (A rough sketch of the graph of the function near a critical number often is enough to tell whether an extremum has been found.) These facts about extrema are summarized below.

RELATIVE EXTREMA

If a function f has a relative extremum at a critical number c, then either $f'(c) = 0$, or $f'(c)$ does not exist but $f(c)$ does exist.

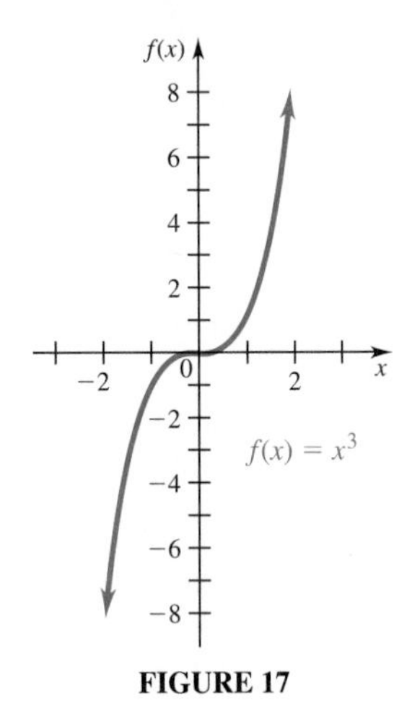

FIGURE 17

Caution Be very careful not to get this result backward. It does *not* say that a function has relative extrema at all critical numbers of the function. For example, Figure 17 shows the graph of $f(x) = x^3$. The derivative, $f'(x) = 3x^2$, is 0 when $x = 0$, so that 0 is a critical number for that function. However, as suggested by the graph of Figure 17, $f(x) = x^3$ has neither a relative maximum nor a relative minimum at $x = 0$ (or anywhere else, for that matter).

First Derivative Test Suppose all critical numbers have been found for some function f. How is it possible to tell from the equation of the function whether these critical numbers produce relative maxima, relative minima, or neither? One way is suggested by the graph in Figure 18.

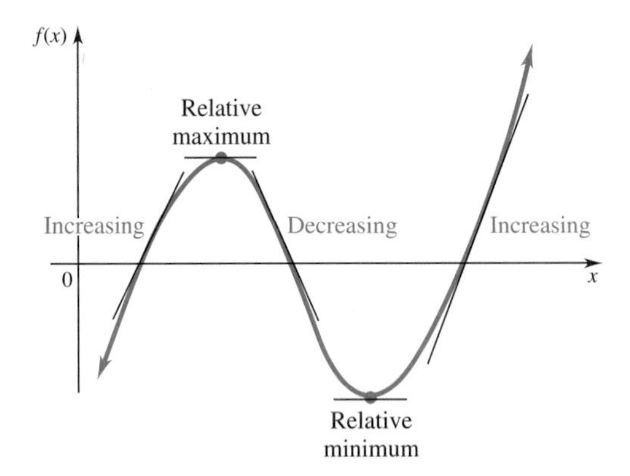

FIGURE 18

As shown in Figure 18, on the left of a relative maximum the tangent lines to the graph of a function have positive slope, indicating that the function is increasing. At the relative maximum, the tangent line is horizontal. On the right of the relative maximum the tangent lines have negative slopes, indicating that the function is decreasing. Around a relative minimum the opposite occurs. As shown by the tangent lines in Figure 18, the function is decreasing on the left of the relative minimum, has a horizontal tangent at the minimum, and is increasing on the right of the minimum.

Putting this together with the methods from Section 1 for identifying intervals where a function is increasing or decreasing gives the following *first derivative test* for locating relative extrema.

FIRST DERIVATIVE TEST

Let c be a critical number for a function f. Suppose that f is differentiable on (a, b), and that c is the only critical number for f in (a, b).

1. $f(c)$ is a relative maximum of f if the derivative $f'(x)$ is positive in the interval (a, c) and negative in the interval (c, b).
2. $f(c)$ is a relative minimum of f if the derivative $f'(x)$ is negative in the interval (a, c) and positive in the interval (c, b).

The sketches in the following table show how the first derivative test works. Assume the same conditions on a, b, and c for the table as those given for the first derivative test.

$f(x)$ has:	Sign of f' in (a, c)	Sign of f' in (c, b)	Sketches
Relative maximum 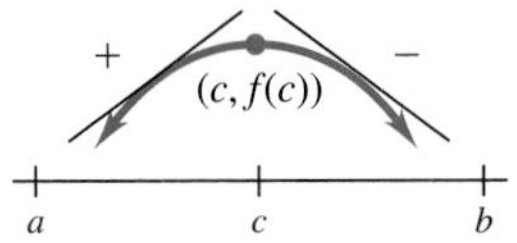 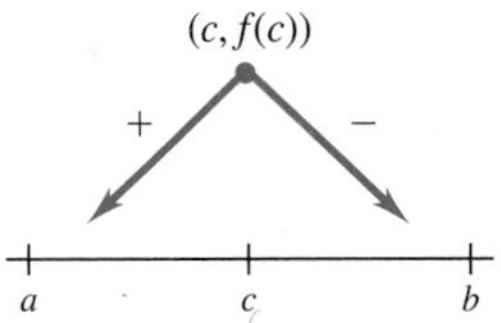	+	–	
Relative minimum 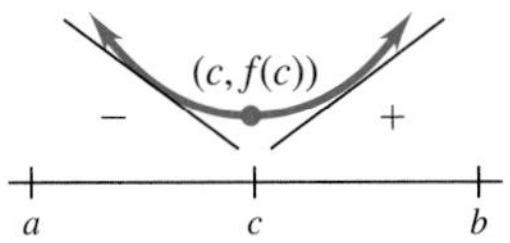 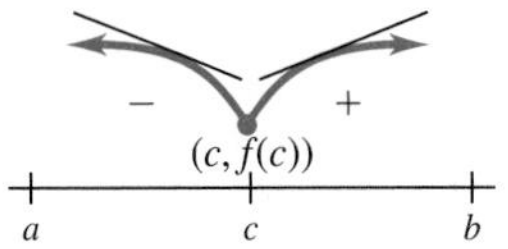	–	+	
No relative extrema 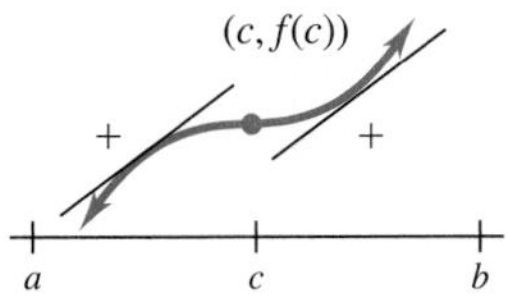 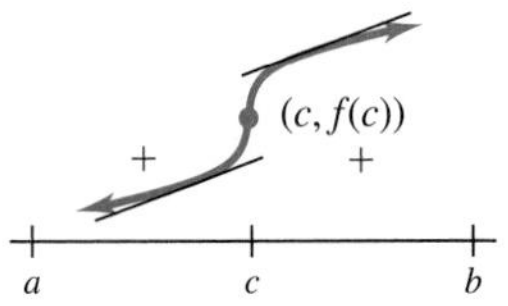	+	+	
No relative extrema 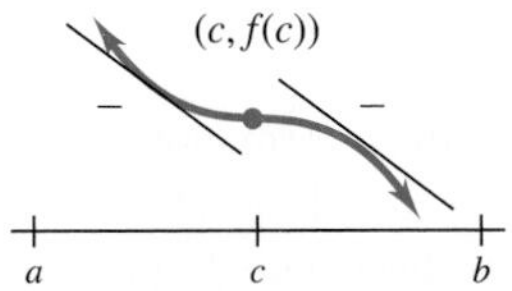 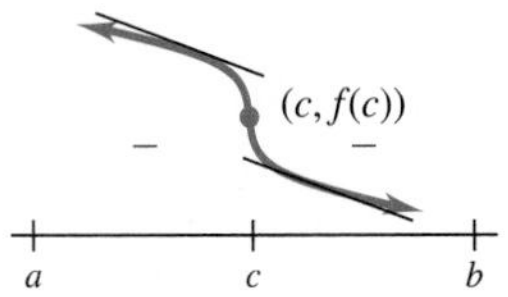	–	–	

EXAMPLE 2 Find all relative extrema for the following functions. Graph each function.

(a) $f(x) = 2x^3 - 3x^2 - 72x + 15$

The derivative is $f'(x) = 6x^2 - 6x - 72$. There are no points where $f'(x)$ fails to exist, so the only critical numbers will be found where the derivative equals 0. Setting the derivative equal to 0 gives

$$\begin{aligned} 6x^2 - 6x - 72 &= 0 \\ 6(x^2 - x - 12) &= 0 \\ 6(x - 4)(x + 3) &= 0 \\ x - 4 = 0 \quad \text{or} \quad x + 3 &= 0 \\ x = 4 \quad \text{or} \quad x &= -3. \end{aligned}$$

As in the previous section, the critical numbers 4 and -3 are used to determine the three intervals $(-\infty, -3)$, $(-3, 4)$, and $(4, \infty)$ shown on the number line in Figure 19. Any number from each of the three intervals can be used as a test point to find the sign of f' in each interval. Using -4, 0, and 5 gives the following information.

$$\begin{aligned} f'(-4) &= 6(-8)(-1) > 0 \\ f'(0) &= 6(-4)(3) < 0 \\ f'(5) &= 6(1)(8) > 0 \end{aligned}$$

Thus, the derivative is positive on $(-\infty, -3)$, negative on $(-3, 4)$, and positive on $(4, \infty)$. By Part 1 of the first derivative test, this means that the function has a relative maximum of $f(-3) = 150$ when $x = -3$; by Part 2, f has a relative minimum of $f(4) = -193$ when $x = 4$.

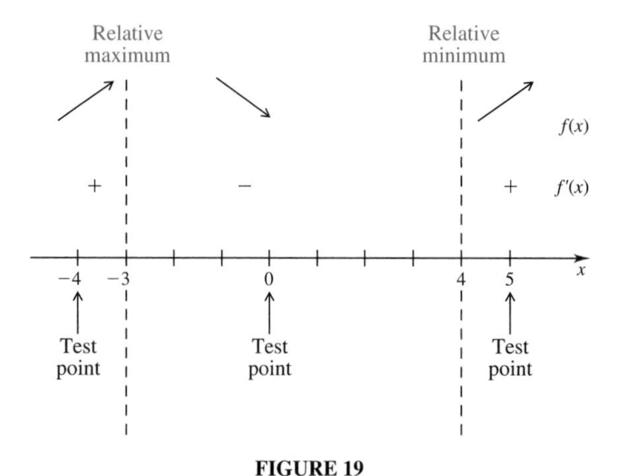

FIGURE 19

Using the information found above and plotting the points from the following chart gives the graph in Figure 20. (Be sure to use $f(x)$, not $f'(x)$, to find the y-values of the points to plot.)

x	-4	-3	0	4	5
$f(x)$	127	150	15	-193	-170

It is not necessary to plot additional points with x-values less than -4 or greater than 5, since any maximum or minimum must occur at a critical number. Thus, the graph will not have other turning points.

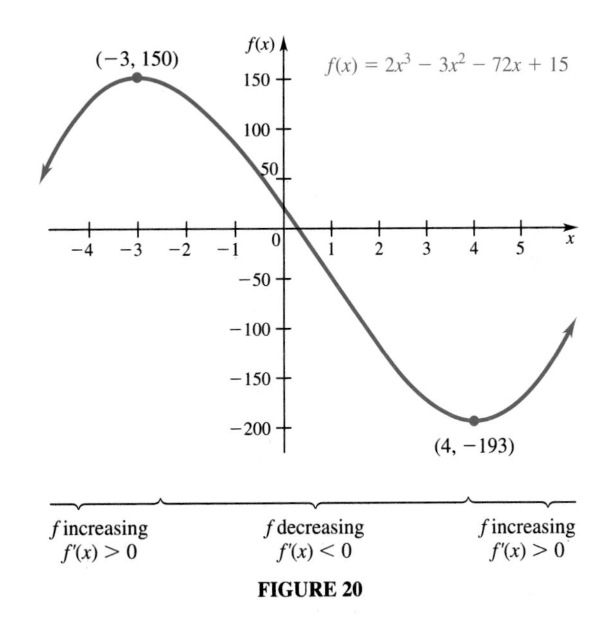

FIGURE 20

(b) $f(x) = 6x^{2/3} - 4x$

Find $f'(x)$:

$$f'(x) = 4x^{-1/3} - 4 = \frac{4}{x^{1/3}} - 4.$$

The derivative fails to exist when $x = 0$, but the function itself is defined when $x = 0$, making 0 a critical number for f. To find other critical numbers, set $f'(x) = 0$.

$$\begin{aligned} f'(x) &= 0 \\ \frac{4}{x^{1/3}} - 4 &= 0 \\ \frac{4}{x^{1/3}} &= 4 \\ 4 &= 4x^{1/3} \\ 1 &= x^{1/3} \\ 1 &= x \end{aligned}$$

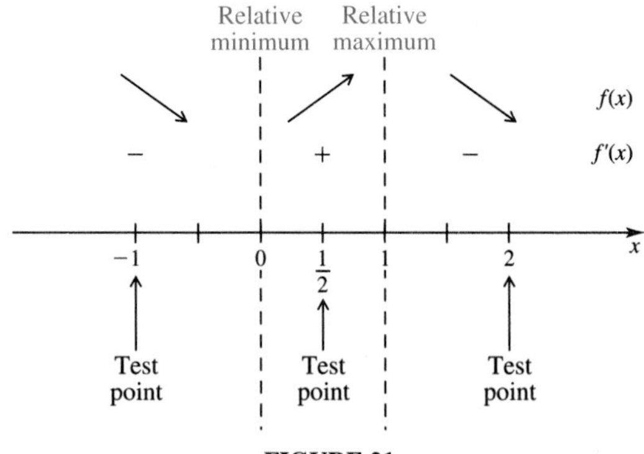

FIGURE 21

The critical numbers 0 and 1 are used to locate the intervals $(-\infty, 0)$, $(0, 1)$, and $(1, \infty)$ on a number line as in Figure 21. Evaluating $f'(x)$ at the test points -1, $1/2$, and 2 and using the first derivative test shows that f has a relative maximum at $x = 1$; the value of this relative maximum is $f(1) = 2$. Also, f has a relative minimum at $x = 0$; this relative minimum is $f(0) = 0$. Using this information and plotting the points from the following table leads to the graph shown in Figure 22.

x	-1	0	1/2	1	2
$f(x)$	10	0	1.8	2	1.5

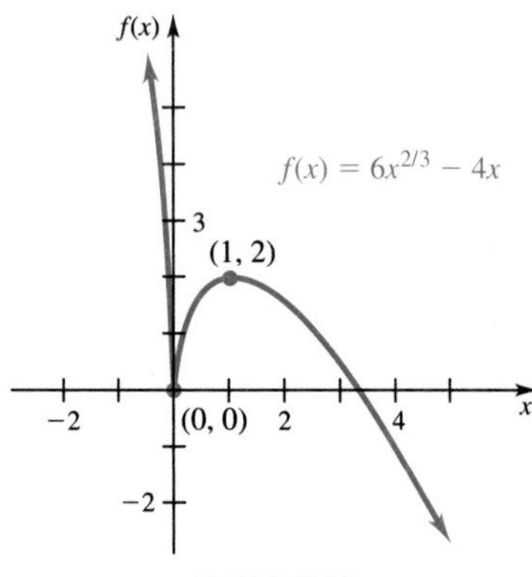

FIGURE 22

To complete the graph as shown, it would be helpful to also plot the points $(-1/4, 3.4)$ and $(4, -.9)$. Note that the graph has a sharp point at the critical number where the derivative does not exist.

(c) $f(x) = x^{2/3}$

The derivative is

$$f'(x) = \frac{2}{3}x^{-1/3} = \frac{2}{3x^{1/3}}.$$

The derivative is never 0, but it does fail to exist when $x = 0$. Since $f(0)$ *does* exist but $f'(0)$ does not, 0 is a critical point. Testing a point on either side of 0 gives the results shown in Figure 23. (Use 8 and -8 as test points because they

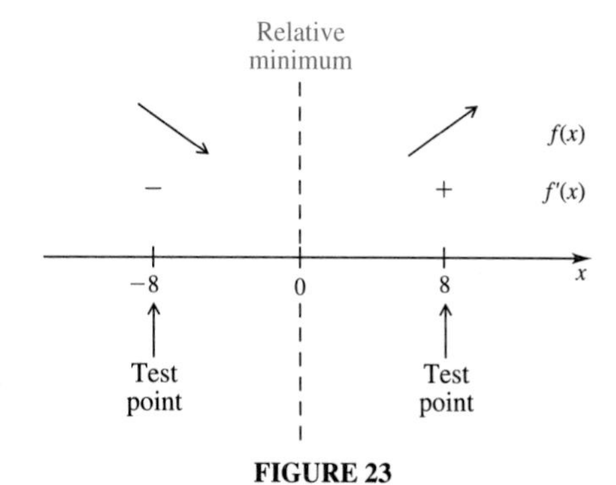

FIGURE 23

are perfect cubes and therefore easy to work with in the equations for $f(x)$ and $f'(x)$.) The function has a relative minimum at 0 with $f(0) = 0^{2/3} = 0$. The function is decreasing on the interval $(-\infty, 0)$ and increasing on $(0, \infty)$. Using this information and plotting a few points gives the graph in Figure 24. Again, there is a sharp point at the critical number because the derivative does not exist there. ◀

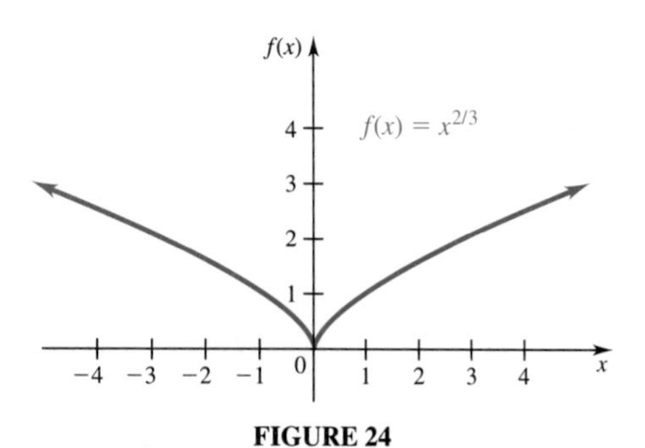

FIGURE 24

Caution A critical number must be in the domain of the function. For example, the derivative of $f(x) = x/(x - 4)$ is $f'(x) = -4/(x - 4)^2$, which fails to exist when $x = 4$. But $f(4)$ does not exist, so 4 is not a critical number, and the function has no relative extrema.

As mentioned at the beginning of this section, finding the maximum or minimum value of a quantity is important in applications of mathematics. The final example gives specific illustrations.

EXAMPLE 3 A small company manufactures and sells bicycles. The production manager has determined that the cost and demand functions for x $(x \geq 0)$ bicycles per week are

$$C(x) = 10 + 5x + \frac{1}{60}x^3 \quad \text{and} \quad p = 90 - x,$$

where p is the price per bicycle.

(a) Find the maximum weekly revenue.

The revenue each week is given by

$$R(x) = xp = x(90 - x) = 90x - x^2.$$

To maximize $R(x) = 90x - x^2$, find $R'(x)$. Then find the critical numbers.

$$\begin{aligned} R'(x) = 90 - 2x &= 0 \\ 90 &= 2x \\ x &= 45 \end{aligned}$$

Since $R'(x)$ exists for all x, 45 is the only critical number. To verify that $x = 45$ will produce a *maximum*, evaluate the derivative on either side of $x = 45$.

$$R'(40) = 10 \quad \text{and} \quad R'(50) = -10$$

This shows that $R(x)$ is increasing up to $x = 45$, then decreasing, so there is a maximum value at $x = 45$ of $R(45) = 2025$. The maximum revenue will be \$2025 and will occur when 45 bicycles are produced and sold each week.

(b) Find the maximum weekly profit.

Since profit equals revenue minus cost, the profit is given by

$$\begin{aligned} P(x) &= R(x) - C(x) \\ &= (90x - x^2) - \left(10 + 5x + \frac{1}{60}x^3\right) \\ &= -\frac{1}{60}x^3 - x^2 + 85x - 10. \end{aligned}$$

Find the derivative and set it equal to 0 to find the critical numbers. (The derivative exists for all x.)

$$P'(x) = -\frac{1}{20}x^2 - 2x + 85 = 0$$

Solving this equation by the quadratic formula gives the solutions $x \approx 25.8$ and $x \approx -65.8$. Since x cannot be negative, the only critical number of concern is 25.8. Determine whether $x = 25.8$ produces a maximum by testing a value on either side of 25.8 in $P'(x)$.

$$P'(0) = 85 \quad \text{and} \quad P'(40) = -75$$

These results show that $P(x)$ increases to $x = 25.8$ and then decreases. Since x must be an integer, verify that $x = 26$ produces a maximum value of $P(26) = 1231.07$. (We should also check $P(25) = 1229.58$.) Thus, the maximum profit of \$1231.07 occurs when 26 bicycles are produced and sold each week. Notice that this is not the same as the number that should be produced to yield maximum revenue.

(c) Find the price the company should charge to realize maximum profit.

As shown in part (b), 26 bicycles per week should be produced and sold to get the maximum profit of \$1231.07 per week. Since the price is given by

$$p = 90 - x,$$

if $x = 26$, then $p = 64$. The manager should charge \$64 per bicycle and produce and sell 26 bicycles per week to get the maximum profit of \$1231.07 per week. Figure 25 shows the graphs of the functions used in this example. Notice that the slopes of the revenue and cost functions are the same at the point where the maximum profit occurs. Why is this true? ◀

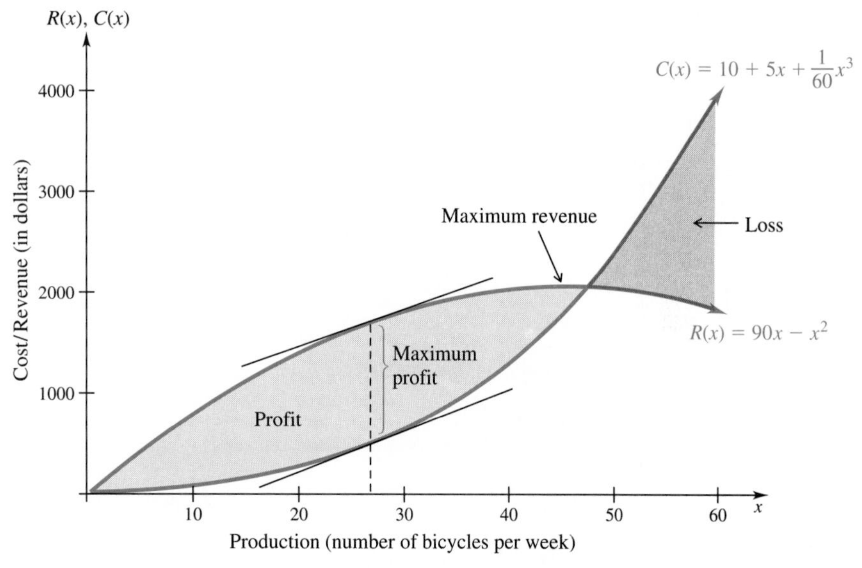

FIGURE 25

The examples in this section involving the maximization of a quadratic function, such as the opening example and the bicycle revenue example, could be solved by the methods described in Chapter 1. But those involving more complicated functions, such as the bicycle profit example, are difficult to analyze without the tools of calculus.

3.2 Exercises

Find the locations and values of all relative extrema for the functions with graphs as follows. Compare with Exercises 1–8 in the preceding section.

1.

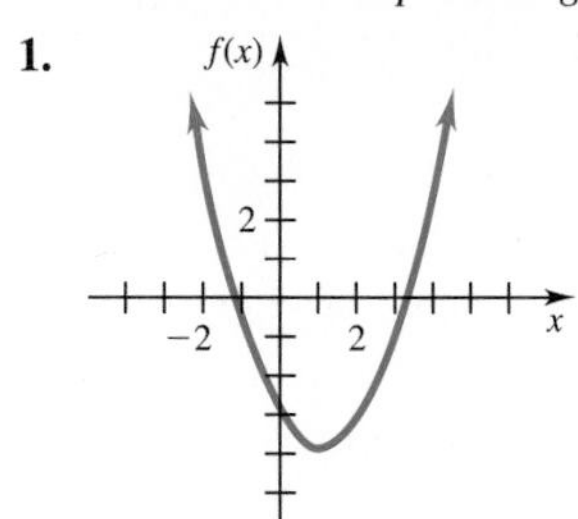

2.

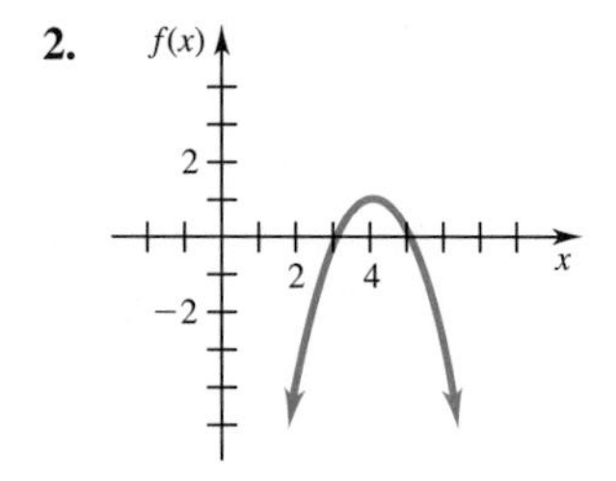

3.

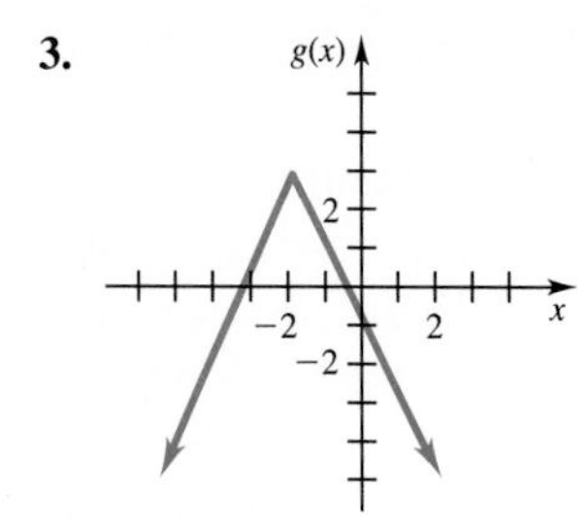

4. 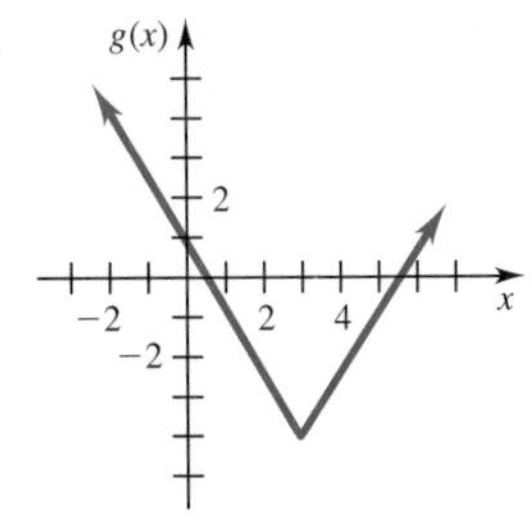

5. 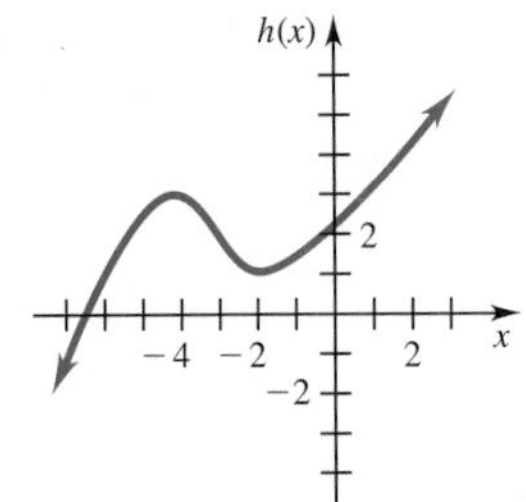

6. 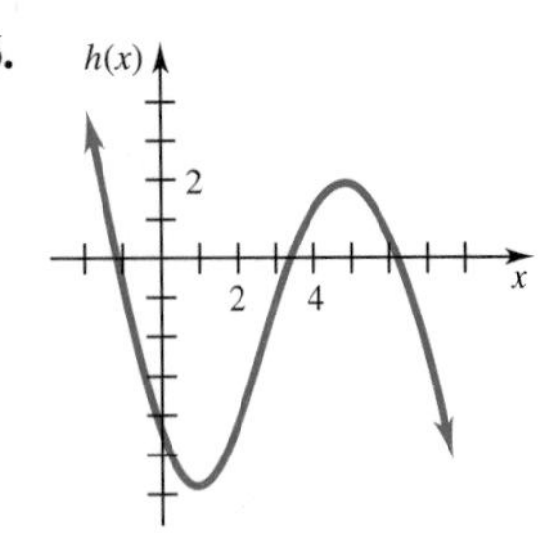

7. 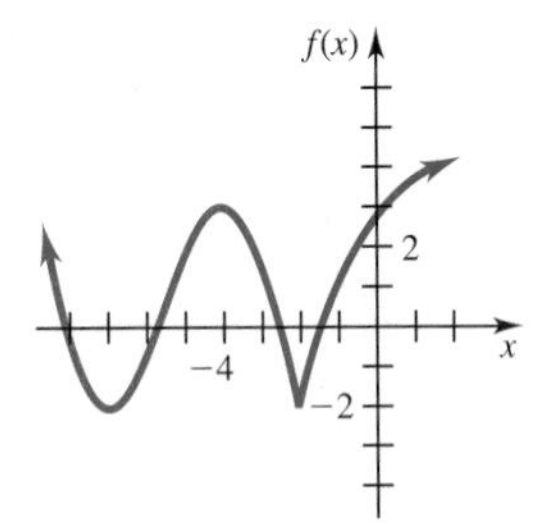

8. 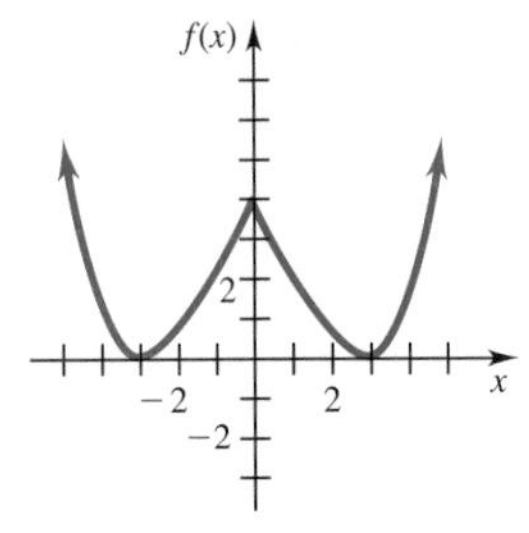

Find the x-values of all points where the functions defined as follows have any relative extrema. Find the value(s) of any relative extrema.

9. $f(x) = x^2 + 12x - 8$

10. $f(x) = x^2 - 4x + 6$

11. $f(x) = 4 - 3x - .5x^2$

12. $f(x) = 3 - .4x - .2x^2$

13. $f(x) = x^3 + 6x^2 + 9x - 8$

14. $f(x) = x^3 + 3x^2 - 24x + 2$

15. $f(x) = -\frac{4}{3}x^3 - \frac{21}{2}x^2 - 5x + 8$

16. $f(x) = -\frac{2}{3}x^3 - \frac{1}{2}x^2 + 3x - 4$

17. $f(x) = 2x^3 - 21x^2 + 60x + 5$

18. $f(x) = 2x^3 + 15x^2 + 36x - 4$

19. $f(x) = x^4 - 18x^2 - 4$

20. $f(x) = x^4 - 8x^2 + 9$

21. $f(x) = -(8 - 5x)^{2/3}$

22. $f(x) = (2 - 9x)^{2/3}$

23. $f(x) = 2x + 3x^{2/3}$

24. $f(x) = 3x^{5/3} - 15x^{2/3}$

25. $f(x) = x - \frac{1}{x}$

26. $f(x) = x^2 + \frac{1}{x}$

27. $f(x) = \frac{x^2}{x^2 + 1}$

28. $f(x) = \frac{x^2}{x - 3}$

29. $f(x) = \frac{x^2 - 2x + 1}{x - 3}$

30. $f(x) = \frac{x^2 - 6x + 9}{x + 2}$

Use the derivative to find the vertex of each parabola.

31. $y = -2x^2 + 8x - 1$

32. $y = 3x^2 - 12x + 2$

33. $y = 2x^2 - 5x + 2$

34. $y = ax^2 + bx + c$

Applications

Business and Economics

Profit *In Exercises 35–37, find* **(a)** *the price p per unit that produces maximum profit;* **(b)** *the number x of units that produces maximum profit; and* **(c)** *the maximum profit P ($P = R - C$).*

35. $C(x) = 75 + 10x; \quad p = 70 - 2x$

36. $C(x) = 25x + 5000; \quad p = 80 - .01x$

37. $C(x) = -13x^2 + 300x; \quad p = -x^2 - x + 360$

38. **Profit** The total profit $P(x)$ (in dollars) from the sale of x units of a certain prescription drug is given by

$$P(x) = -x^3 + 3x^2 + 72x.$$

(a) Find the number of units that should be sold in order to maximize the total profit.

(b) What is the maximum profit?

39. **Revenue** The demand equation for telephones at one store is

$$p = D(q) = \frac{1}{3}q^2 - \frac{25}{2}q + 100, \qquad 0 \le q \le 16,$$

where p is the price (in dollars) and q is the quantity of telephones sold per week. Find the values of q and p that maximize revenue.

40. **Cost** Suppose that the cost function for a product is given by $C(x) = .002x^3 - 9x + 4000$. Find the production level (i.e., value of x) that will produce the minimum average cost per unit $\overline{C}(x)$.

Life Sciences

41. **Activity Level** In the summer the activity level of a certain type of lizard varies according to the time of day. A biologist has determined that the activity level is given by the function

$$a(t) = .008t^3 - .27t^2 + 2.02t + 7,$$

where t is the number of hours after 12 noon. When is the activity level highest? When is it lowest?

Social Sciences

42. **Attitude Change** Social psychologists have found that as the discrepancy between the views of a speaker and those of an audience increases, the attitude change in the audience also increases to a point, but decreases when the discrepancy becomes too large, particularly if the communicator is viewed by the audience as having low credibility.* Suppose that the degree of change, y, can be approximated by the function

$$D(x) = -x^4 + 8x^3 + 80x^2,$$

where x is the discrepancy between the views of the speaker and those of the audience, as measured by scores on a questionnaire. Find the amount of discrepancy the speaker should aim for to maximize the attitude change in the audience.

43. **Film Length** A group of researchers found that people prefer training films of moderate length; shorter films contain too little information, while longer films are boring. For a training film on the care of exotic birds, the researchers determined that the ratings people gave for the film could be approximated by

$$R(t) = \frac{20t}{t^2 + 100},$$

where t is the length of the film (in minutes). Find the film length that received the highest rating.

Physical Sciences

44. **Height** After a great deal of experimentation, two Atlantic Institute of Technology senior physics majors determined that when a bottle of French champagne is shaken several times, held upright, and uncorked, its cork travels according to

$$s(t) = -16t^2 + 64t + 3,$$

where s is its height (in feet) above the ground t seconds after being released. How high will it go?

*See A. H. Eagly and K. Telaak, "Width of the Latitude of Acceptance as a Determinant of Attitude Change" in *Journal of Personality and Social Psychology*, vol. 23, 1972, pp. 388–97.

Computer/Graphing Calculator

Use a computer or graphing calculator to find the approximate x-values (to two significant digits) of all points where the functions defined as follows have any relative extrema. Find the value(s) of any relative extrema. One method is to compute the derivative, then use a graph of the function or an equation solver to find where the derivative is equal to zero.

45. $f(x) = x^5 - x^4 + 4x^3 - 30x^2 + 5x + 6$

46. $f(x) = -x^5 - x^4 + 2x^3 - 25x^2 + 9x + 12$

47. $f(x) = .001x^6 - .02x^5 - x^4 - 4x^3 + 12x^2 - 9x - 17$

48. $f(x) = -.001x^6 - .005x^5 + .5x^4 + 7x^3 - 8x^2 + 15x + 43$

3.3 ABSOLUTE EXTREMA

How do you invest money for the highest rate of return?

If you are in charge of an investment portfolio, you may be happy to have found a rate of return that is a relative maximum. Your boss may not be happy if he or she discovers another rate of return that is even higher. In this situation, the relative maximum is not as important as the *absolute maximum.*

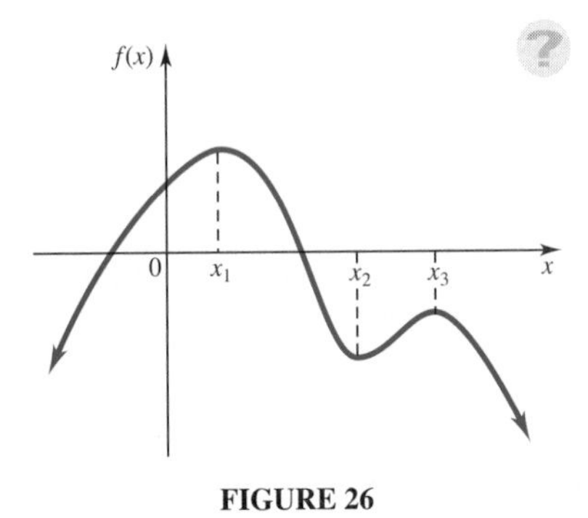

FIGURE 26

The function graphed in Figure 26 has a relative minimum at x_2 and relative maxima at x_1 and x_3. In practical situations it is often desirable to know if a function has one function value that is larger than any other or one function value that is smaller than any other. In Figure 26, $f(x_1) \geq f(x)$ for all x in the domain. There is no function value that is smaller than all others, however, because $f(x) \to -\infty$ as $x \to \infty$ or as $x \to -\infty$.

The largest possible value of a function is called the *absolute maximum* and the smallest possible value of a function is called the *absolute minimum.* As Figure 26 shows, one or both of these may not exist on the domain of the function, $(-\infty, \infty)$ here. Absolute extrema often coincide with relative extrema, as with $f(x_1)$ in Figure 26. Although a function may have several relative maxima or relative minima, it never has more than one *absolute maximum* or *absolute minimum.*

ABSOLUTE MAXIMUM OR MINIMUM

Let f be a function defined on some interval. Let c be a number in the interval. Then $f(c)$ is the **absolute maximum** of f on the interval if

$$f(x) \leq f(c)$$

for every x in the interval, and $f(c)$ is the **absolute minimum** of f on the interval if

$$f(x) \geq f(c)$$

for every x in the interval.

Now look at Figure 27, which shows three functions defined on closed intervals. In each case there is an absolute maximum value and an absolute minimum value. These absolute extrema may occur at the endpoints or at relative extrema. As the graphs in Figure 27 show, an absolute extremum is either the largest or the smallest function value occurring on a closed interval, while a relative extremum is the largest or smallest function value in some (perhaps small) open interval.

Although a function can have only one absolute minimum value and only one absolute maximum value, it can have many points where these values occur.

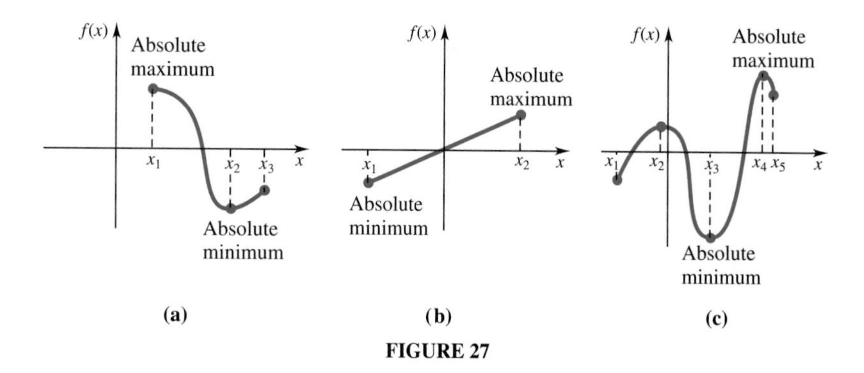

FIGURE 27

(Note that the absolute maximum value and absolute minimum value are numbers, not points.) As an extreme example, consider the function $f(x) = 2$. The absolute minimum value of this function is clearly 2, as is the absolute maximum value. Both the absolute minimum and the absolute maximum occur at every real number x.

One of the main reasons for the importance of absolute extrema is given by the following theorem (which is proved in more advanced courses).

EXTREME VALUE THEOREM

A function f that is continuous on a closed interval $[a, b]$ will have both an absolute maximum and an absolute minimum on the interval.

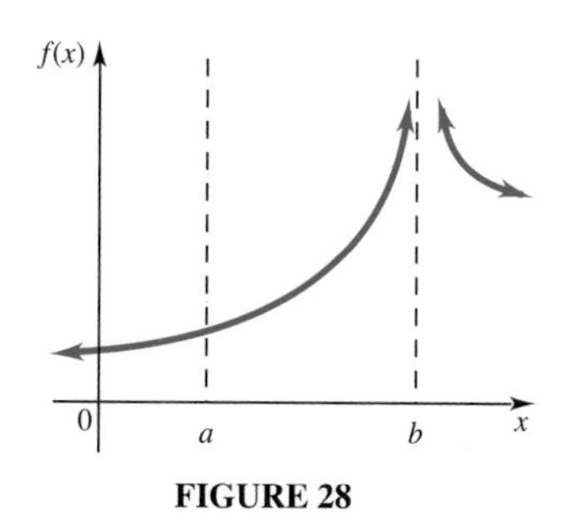

FIGURE 28

As Figure 27 shows, an absolute extremum *may* occur on an open interval. See x_2 in part (a) and x_3 in part (c), for example. The extreme value theorem says that a function *must* have both an absolute maximum and an absolute minimum on a closed interval. The conditions that f be *continuous* and on a *closed* interval are very important. In Figure 28, f is discontinuous at $x = b$. Since $f(x)$ gets larger and larger as $x \to b$, there is no absolute (or relative) maximum on (a, b). Also, as $x \to a^+$, the values of $f(x)$ get smaller and smaller, but there is no smallest value of $f(x)$ on the *open* interval (a, b), since there is no endpoint.

The extreme value theorem guarantees the existence of absolute extrema. To find these extrema, use the following steps.

FINDING ABSOLUTE EXTREMA

To find absolute extrema for a function f continuous on a closed interval $[a, b]$:

1. Find all critical numbers for f in (a, b).
2. Evaluate f for all critical numbers in (a, b).
3. Evaluate f for the endpoints a and b of the interval $[a, b]$.
4. The largest value found in Step 2 or 3 is the absolute maximum for f on $[a, b]$, and the smallest value found is the absolute minimum for f on $[a, b]$.

EXAMPLE 1 Find the absolute extrema of the function

$$f(x) = 2x^3 - x^2 - 20x - 10$$

on the interval $[-2, 4]$.

First look for critical points in the interval $(-2, 4)$. Set the derivative $f'(x) = 6x^2 - 2x - 20$ equal to 0.

$$6x^2 - 2x - 20 = 0$$
$$2(3x^2 - x - 10) = 0$$
$$2(3x + 5)(x - 2) = 0$$
$$3x + 5 = 0 \quad \text{or} \quad x - 2 = 0$$
$$x = -5/3 \qquad\qquad x = 2$$

Since there are no values of x where $f'(x)$ does not exist, the only critical values are $-5/3$ and 2. Both of these critical values are in the interval $(-2, 4)$. Now evaluate the function at $-5/3$ and 2, and at the endpoints of its domain, -2 and 4.

x-value	Value of Function	
−2	10	
−5/3	11.296	
2	−38	←Absolute minimum
4	22	←Absolute maximum

The absolute maximum, 22, occurs when $x = 4$, and the absolute minimum, -38, occurs when $x = 2$. A graph of f on $[-2, 4]$ is shown in Figure 29.

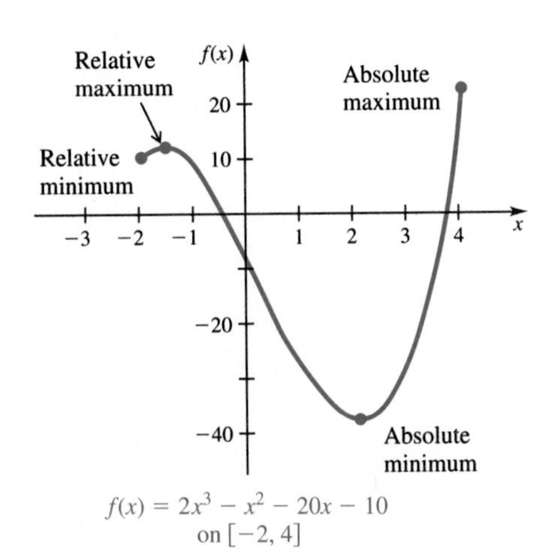

$f(x) = 2x^3 - x^2 - 20x - 10$ on $[-2, 4]$

FIGURE 29

EXAMPLE 2 Find the absolute extrema of the function given by

$$f(x) = 4x + \frac{36}{x}$$

on the interval [1, 6].

Note that f is discontinuous at $x = 0$, but continuous on the closed interval [1, 6], so absolute extrema are guaranteed to exist. First determine any critical points in (1, 6).

$$f'(x) = 4 - \frac{36}{x^2} = 0$$

$$\frac{4x^2 - 36}{x^2} = 0$$

$$4x^2 - 36 = 0$$

$$4x^2 = 36$$

$$x^2 = 9$$

$$x = -3 \quad \text{or} \quad x = 3$$

Although $f'(x)$ does not exist at $x = 0$, f is not defined at $x = 0$, so the critical numbers are -3 and 3. Since -3 is not in the interval [1, 6], 3 is the only critical number. To find the absolute extrema, evaluate $f(x)$ at $x = 3$ and at the endpoints, where $x = 1$ and $x = 6$.

x-value	Value of Function	
1	40	←Absolute maximum
3	24	←Absolute minimum
6	30	

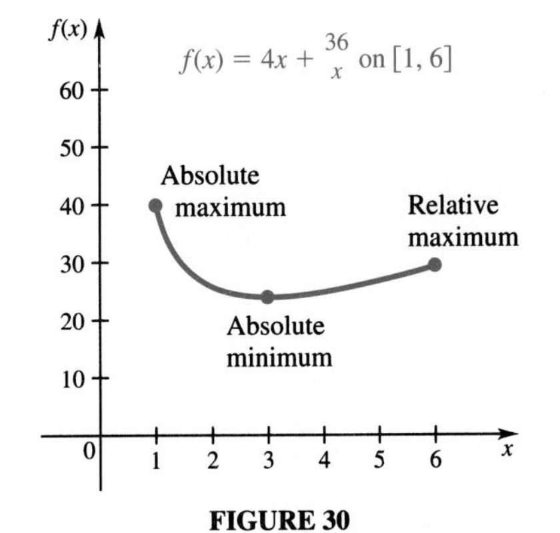

FIGURE 30

Figure 30 shows a graph of f on [1, 6]. ◀

Absolute extrema are particularly important in applications of mathematics. In most applications the domain is limited in some natural way that ensures the occurrence of absolute extrema.

EXAMPLE 3 A company has found that its weekly profit from the sale of x units of an auto part is given by

$$P(x) = -.02x^3 + 600x - 20{,}000.$$

Production bottlenecks limit the number of units that can be made per week to no more than 150, while a long-term contract requires that at least 50 units be made each week. Find the maximum possible weekly profit that the firm can make.

Because of the restrictions, the profit function is defined only for the domain [50, 150]. Look first for critical numbers of the function in the open interval (50, 150). Here $P'(x) = -.06x^2 + 600$. Set this derivative equal to 0 and solve for x.

$$
\begin{aligned}
-.06x^2 + 600 &= 0 \\
-.06x^2 &= -600 \\
x^2 &= 10{,}000
\end{aligned}
$$

$$x = 100 \qquad \text{or} \qquad x = -100$$

Since $x = -100$ is not in the interval [50, 150], disregard it. Now evaluate the function at the remaining critical number, 100, and at the endpoints of the domain, 50 and 150.

***x*-value**	**Value of Function**	
50	7500	
100	20,000	←Absolute maximum
150	2500	

Maximum profit of $20,000 occurs when 100 units are made per week. ◀

3.3 Exercises

Find the locations of any absolute extrema for the functions with graphs as follows.

1.

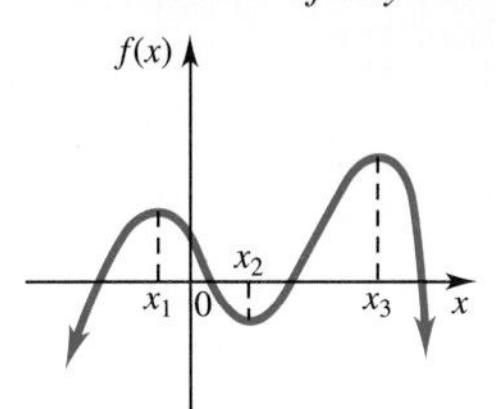

2.

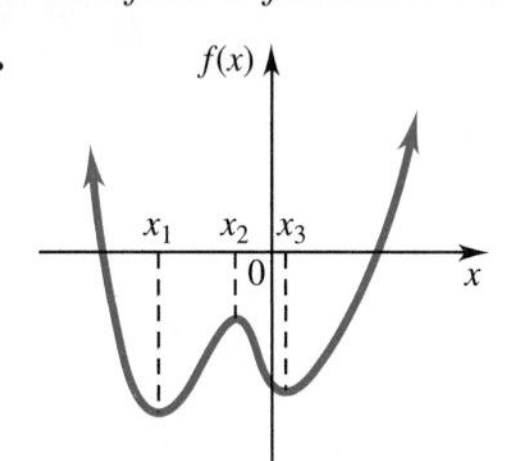

3.

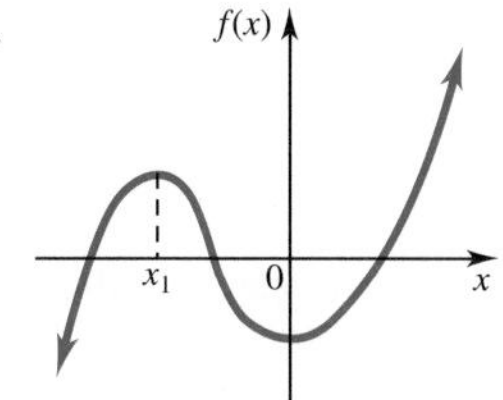

4.

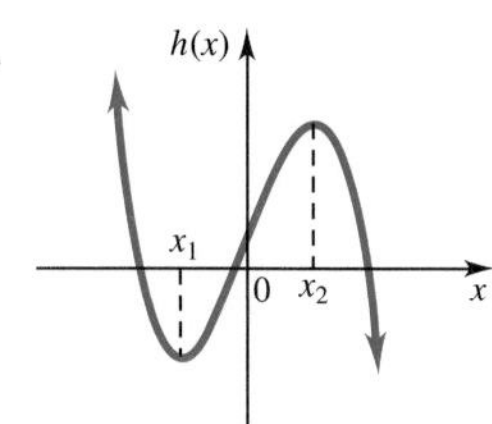

5.

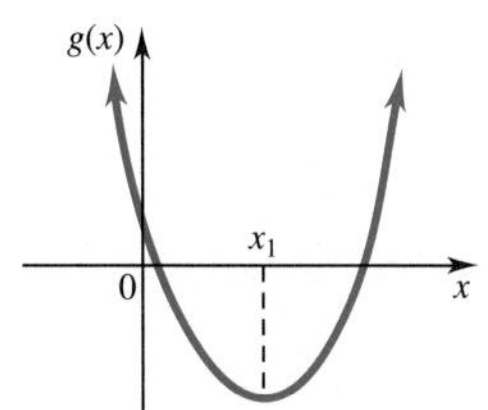

6.

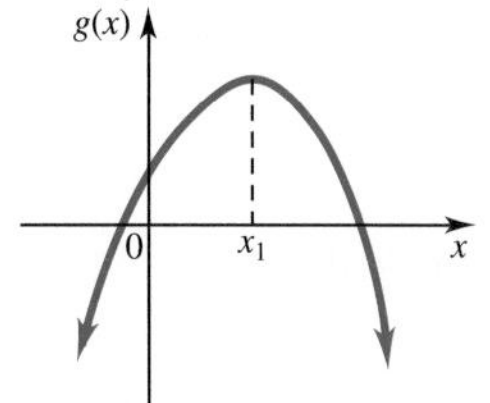

7.

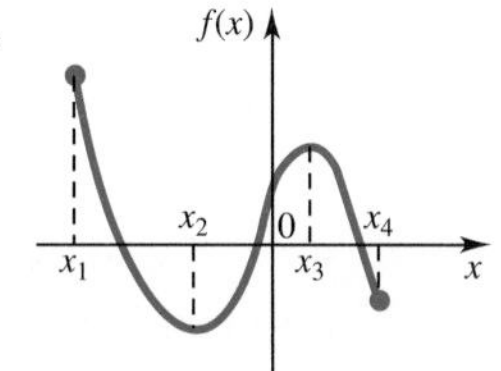

8.

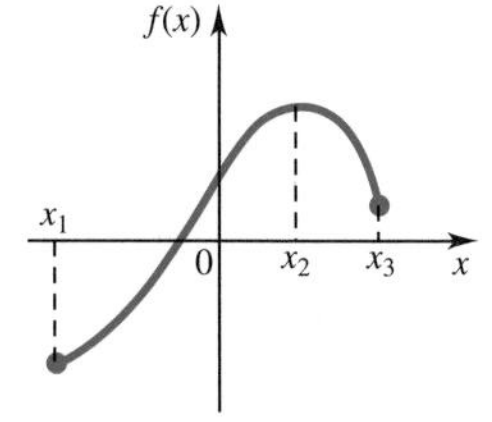

Find the locations of all absolute extrema for the functions defined as follows, with the specified domains.

9. $f(x) = x^2 + 6x + 2$; $[-4, 0]$

10. $f(x) = x^2 - 4x + 1$; $[-6, 5]$

11. $f(x) = 5 - 8x - 4x^2$; $[-5, 1]$

12. $f(x) = 9 - 6x - 3x^2$; $[-4, 3]$

13. $f(x) = x^3 - 3x^2 - 24x + 5$; $[-3, 6]$

14. $f(x) = x^3 - 6x^2 + 9x - 8$; $[0, 5]$

15. $f(x) = \frac{1}{3}x^3 - \frac{1}{2}x^2 - 6x + 3;\ [-4, 4]$

16. $f(x) = \frac{1}{3}x^3 + \frac{3}{2}x^2 - 4x + 1;\ [-5, 2]$

17. $f(x) = x^4 - 32x^2 - 7;\ [-5, 6]$

18. $f(x) = x^4 - 18x^2 + 1;\ [-4, 4]$

19. $f(x) = \frac{1}{1 + x};\ [0, 2]$

20. $f(x) = \frac{-2}{x + 3};\ [1, 4]$

21. $f(x) = \frac{8 + x}{8 - x};\ [4, 6]$

22. $f(x) = \frac{1 - x}{3 + x};\ [0, 3]$

23. $f(x) = \frac{x}{x^2 + 2};\ [0, 4]$

24. $f(x) = \frac{x - 1}{x^2 + 1};\ [1, 5]$

25. $f(x) = (x^2 + 18)^{2/3};\ [-3, 3]$

26. $f(x) = (x^2 + 4)^{1/3};\ [-2, 2]$

27. $f(x) = (x + 1)(x + 2)^2;\ [-4, 0]$

28. $f(x) = (x - 3)(x - 1)^3;\ [-2, 3]$

29. $f(x) = \frac{1}{\sqrt{x^2 + 1}};\ [-1, 1]$

30. $f(x) = \frac{3}{\sqrt{x^2 + 4}};\ [-2, 2]$

Applications

Business and Economics

31. Average Cost An enterprising (although unscrupulous) business student has managed to get his hands on a copy of the out-of-print solutions manual for an applied calculus text. He plans to duplicate it and sell copies in the dormitories. He figures that demand will be between 100 and 1200 copies, and he wants to minimize his average cost of production. After checking into the cost of paper, duplicating, and the rental of a small van, he estimates that the cost in dollars to produce x hundred manuals is given by $C(x) = x^2 + 200x + 100$. How many should he produce in order to make the *average cost* per unit as small as possible? How much will he have to charge to make a profit?

32. Profit The total profit $P(x)$ (in thousands of dollars) from the sale of x hundred thousand automobile tires is approximated by

$$P(x) = -x^3 + 9x^2 + 120x - 400, \quad x \geq 5.$$

Find the number of hundred thousands of tires that must be sold to maximize profit. Find the maximum profit.

Average Cost *In Exercises 33 and 34, find the minimum value of the average cost for the given cost function on the given intervals.*

33. $C(x) = 81x^2 + 17x + 324$ on the following intervals.

(a) $1 \leq x \leq 10$ **(b)** $10 \leq x \leq 20$

34. $C(x) = x^3 + 37x + 250$ on the following intervals.

(a) $1 \leq x \leq 10$ **(b)** $10 \leq x \leq 20$

Life Sciences

35. Pollution A marshy region used for agricultural drainage has become contaminated with selenium. It has been determined that flushing the area with clean water will reduce the selenium for a while, but it will then begin to build up again. A biologist has found that the percent of selenium in the soil x months after the flushing begins is given by

$$f(x) = \frac{x^2 + 36}{2x}, \quad 1 \leq x \leq 12.$$

When will the selenium be reduced to a minimum? What is the minimum percent?

36 Salmon Spawning The number of salmon swimming upstream to spawn is approximated by

$$S(x) = -x^3 + 3x^2 + 360x + 5000, \quad 6 \leq x \leq 20,$$

where x represents the temperature of the water in degrees Celsius. Find the water temperature that produces the maximum number of salmon swimming upstream.

Physical Sciences

37. Gasoline Mileage From information given in a recent business publication, we constructed the mathematical model

$$M(x) = -\frac{1}{45}x^2 + 2x - 20, \quad 30 \leq x \leq 65,$$

to represent the miles per gallon used by a certain car at a speed of x mph. Find the absolute maximum miles per gallon and the absolute minimum.

38. Gasoline Mileage For a certain compact car,

$$M(x) = -.018x^2 + 1.24x + 6.2, \quad 30 \leq x \leq 60,$$

represents the miles per gallon obtained at a speed of x mph. Find the absolute maximum miles per gallon and the absolute minimum.

General Interest

Area *A piece of wire 12 ft long is cut into two pieces. (See the figure.) One piece is made into a circle and the other piece is made into a square. Let the piece of length x be formed into a circle.*

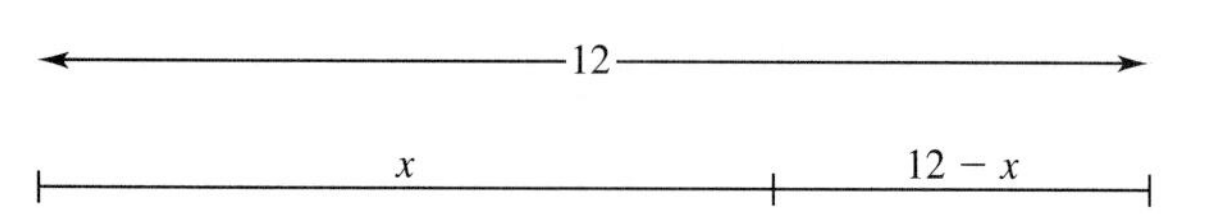

Radius of circle $= \dfrac{x}{2\pi}$ Area of circle $= \pi\left(\dfrac{x}{2\pi}\right)^2$ Side of square $= \dfrac{12 - x}{4}$ Area of square $= \left(\dfrac{12 - x}{4}\right)^2$

39. Where should the cut be made in order to minimize the sum of the areas enclosed by both figures?

40. Where should the cut be made in order to make the sum of the areas maximum? (*Hint:* Remember to use the endpoints of a domain when looking for absolute maxima and minima.)

Computer/Graphing Calculator

Use a computer or a graphing calculator to find the approximate locations (to two significant digits) of all absolute extrema for the functions defined as follows, with the specified domains. One way this may be done is by graphing each function over the indicated domain and focusing on where the function is largest and smallest. Another way is to compute the derivative and then use a graph of the function or an equation solver to find where the derivative is equal to zero.

41. $f(x) = \dfrac{x^3 + 2x + 5}{x^4 + 3x^3 + 10}$; $[-3, 0]$

42. $f(x) = \dfrac{-5x^4 + 2x^3 + 3x^2 + 9}{x^4 - x^3 + x^2 + 7}$; $[-1, 1]$

43. $f(x) = x^{4/5} + x^2 - 4\sqrt{x}$; $[0, 4]$

44. $f(x) = x^{10/3} - x^{4/3} - 4x^2 - 8x$; $[0, 4]$

3.4 HIGHER DERIVATIVES, CONCAVITY, AND THE SECOND DERIVATIVE TEST

Just because the price of a stock is increasing, does that alone make it a good investment?

The following discussion addresses this question.

To understand the behavior of a function on an interval, it is important to know the *rate* at which the function is increasing or decreasing. For example, suppose that your friend, a finance major, has made a study of a young company and is trying to get you to invest in its stock. He shows you the following function, which represents the price $P(t)$ of the company's stock since it became available in January two years ago:

$$P(t) = 17 + t^{1/2},$$

where t is the number of months since the stock became available. He points out that the derivative of the function is always positive, so the price of the stock is always increasing. He claims that you cannot help but make a fortune on it. Should you take his advice and invest?

It is true that the price function increases for all t. The derivative is

$$P'(t) = \frac{1}{2}t^{-1/2} = \frac{1}{2\sqrt{t}},$$

which is always positive because $\sqrt{t}$ is positive for $t > 0$. The catch lies in *how fast* the function is increasing. The derivative $P'(t) = 1/(2\sqrt{t})$ tells how fast the price is increasing at any number of months, t, since the stock became available. For example, when $t = 1$, $P'(t) = 1/2$, and the price is increasing at the rate of 1/2 dollar, or 50 cents, per month. When $t = 4$, $P'(t) = 1/4$; the stock is increasing at 25 cents per month. At $t = 9$ months, $P'(t) = 1/6$, or about 17 cents per month. By the time you could buy in at $t = 24$ months, the price is increasing at 10 cents per month, and the *rate of increase* looks as though it will continue to decrease.

The rate of increase in P' is given by the derivative of $P'(t)$, called the **second derivative** of P and denoted by $P''(t)$. Since $P'(t) = (1/2)t^{-1/2}$,

$$P''(t) = -\frac{1}{4}t^{-3/2} = -\frac{1}{4\sqrt{t^3}}.$$

$P''(t)$ is negative for all positive values of t and therefore confirms the suspicion that the *rate* of increase in price does indeed decrease for all $t \geq 0$. The price of the company's stock will not drop, but the amount of return will certainly not be the fortune that your friend predicts. For example, at $t = 24$ months, when you would buy, the price would be \$21.90. A year later, it would be \$23.00 a share. If you were rich enough to buy 100 shares for \$2190, they would be worth \$2300 in a year. The increase of \$110 is about 5% of the investment—similar to the return that you could get in most savings accounts. The only investors to make a lot of money on this stock would be those who bought early, when the rate of increase was much greater.

In Figure 31* the (generally) decreasing function shows that "things are getting worse," but the overall shape of the graph indicates that the *rate* of decrease is slowing down. In other words, the second derivative is negative.

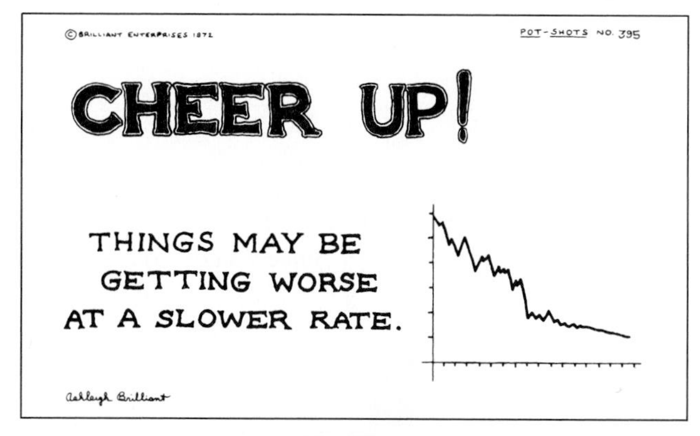

FIGURE 31

*Ashleigh Brilliant Epigrams, Pot-Shots, and Brilliant Thoughts from the book *I May Not Be Totally Perfect, But Parts of Me Are Excellent, and Other Brilliant Thoughts* used by permission of the author, Ashleigh Brilliant, 117 West Valerio St., Santa Barbara, CA 93101.

As mentioned earlier, the second derivative of a function f, written f'', gives the rate of change of the *derivative* of f. Before continuing to discuss applications of the second derivative, we need to introduce some additional terminology and notation.

Higher Derivatives If a function f has a derivative f', then the derivative of f', if it exists, is the second derivative of f, written f''. The derivative of f'', if it exists, is called the **third derivative** of f, and so on. By continuing this process, we can find **fourth derivatives** and other higher derivatives. For example, if $f(x) = x^4 + 2x^3 + 3x^2 - 5x + 7$, then

$$f'(x) = 4x^3 + 6x^2 + 6x - 5, \quad \text{First derivative of } f$$
$$f''(x) = 12x^2 + 12x + 6, \quad \text{Second derivative of } f$$
$$f'''(x) = 24x + 12, \quad \text{Third derivative of } f$$

and

$$f^{(4)}(x) = 24. \quad \text{Fourth derivative of } f$$

NOTATION FOR HIGHER DERIVATIVES

The second derivative of $y = f(x)$ can be written using any of the following notations:

$$f''(x), \quad y'', \quad \frac{d^2y}{dx^2}, \quad \text{or} \quad D_x^2[f(x)].$$

The third derivative can be written in a similar way. For $n \geq 4$, the nth derivative is written $f^{(n)}(x)$.

Let $f(x) = x^3 + 6x^2 - 9x + 8$.

(a) Find $f''(x)$.

To find the second derivative of $f(x)$, find the first derivative, and then take its derivative.

$$f'(x) = 3x^2 + 12x - 9$$
$$f''(x) = 6x + 12$$

(b) Find $f''(0)$.

Since $f''(x) = 6x + 12$,

$$f''(0) = 6(0) + 12 = 12. \blacktriangleleft$$

Find $f''(x)$ for the functions defined as follows.

(a) $f(x) = \sqrt{x}$

Let $f(x) = \sqrt{x} = x^{1/2}$. Then

$$f'(x) = \frac{1}{2}x^{-1/2},$$

and $$f''(x) = -\frac{1}{4}x^{-3/2} = -\frac{1}{4} \cdot \frac{1}{x^{3/2}} = \frac{-1}{4x^{3/2}}.$$

(b) $f(x) = (x^2 - 1)^2$

Here

$$f'(x) = 2(x^2 - 1)(2x) = 4x(x^2 - 1).$$

Use the product rule to find $f''(x)$.

$$\begin{aligned} f''(x) &= 4x(2x) + (x^2 - 1)(4) \\ &= 8x^2 + 4x^2 - 4 \\ &= 12x^2 - 4 \end{aligned}$$ ◀

In the previous chapter we saw that the first derivative of a function represents the rate of change of the function. The second derivative, then, represents the rate of change of the first derivative. If a function describes the position of a particle (along a straight line) at time t, then the first derivative gives the velocity of the particle. That is, if $y = s(t)$ describes the position (along a straight line) of the particle at time t, then $v(t) = s'(t)$ gives the velocity at time t.

The rate of change of velocity is called **acceleration.** Since the second derivative gives the rate of change of the first derivative, the acceleration is the derivative of the velocity. Thus, if $a(t)$ represents the acceleration at time t, then

$$a(t) = \frac{d}{dt}v(t) = s''(t).$$

3 Suppose that a particle is moving along a straight line, with its position at time t given by

$$s(t) = t^3 - 2t^2 - 7t + 9.$$

Find the following.

(a) The velocity at any time t

The velocity is given by

$$v(t) = s'(t) = 3t^2 - 4t - 7.$$

(b) The acceleration at any time t

Acceleration is given by

$$a(t) = v'(t) = s''(t) = 6t - 4.$$

(c) The time intervals (for $t \geq 0$) when the particle is speeding up or slowing down

When the acceleration is positive, the velocity is increasing, so the particle is speeding up. This happens when $6t - 4 > 0$, or $t > 2/3$. A negative acceleration indicates that the particle is slowing down. This happens when $0 < t < 2/3$. ◀

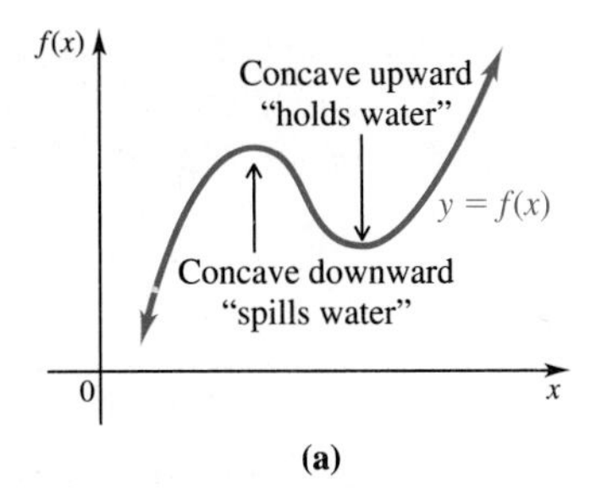

(a)

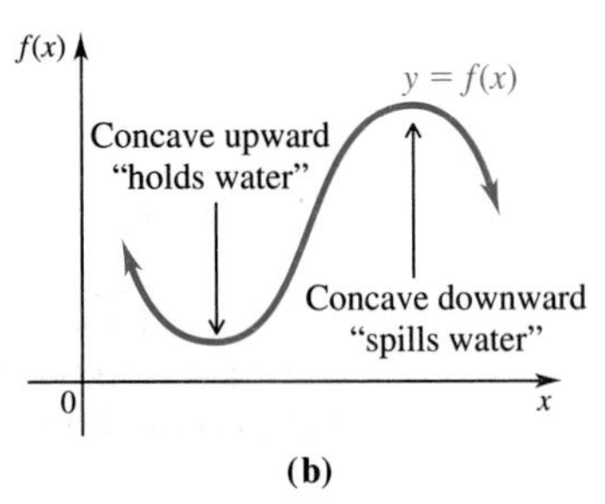

(b)

FIGURE 32

Concavity of a Graph The first derivative has been used to show where a function is increasing or decreasing and where the extrema occur. The second derivative gives the rate of change of the first derivative; it indicates *how fast* the function is increasing or decreasing. The rate of change of the derivative (the second derivative) affects the *shape* of the graph. Intuitively, we say that a graph is *concave upward* on an interval if it ''holds water'' and *concave downward* if it ''spills water.'' See Figure 32.

More precisely, a function is **concave upward** on an interval (a, b) if the graph of the function lies above its tangent line at each point of (a, b). A function is **concave downward** on (a, b) if the graph of the function lies below its tangent line at each point of (a, b). A point where a graph changes concavity is called a **point of inflection.** See Figure 33.

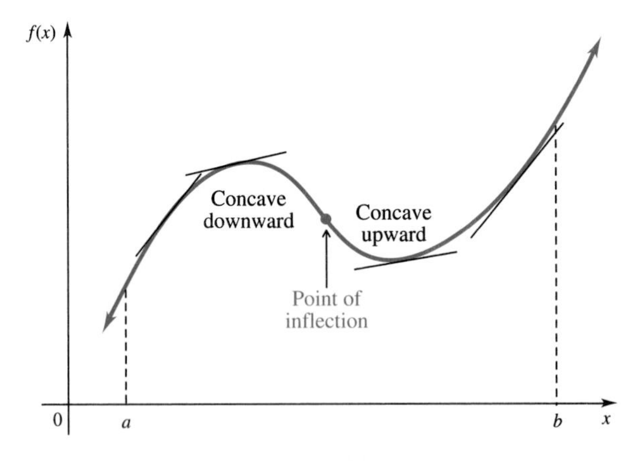

FIGURE 33

Just as a function can be either increasing or decreasing on an interval, it can be either concave upward or concave downward on an interval. Examples of various combinations are shown in Figure 34.

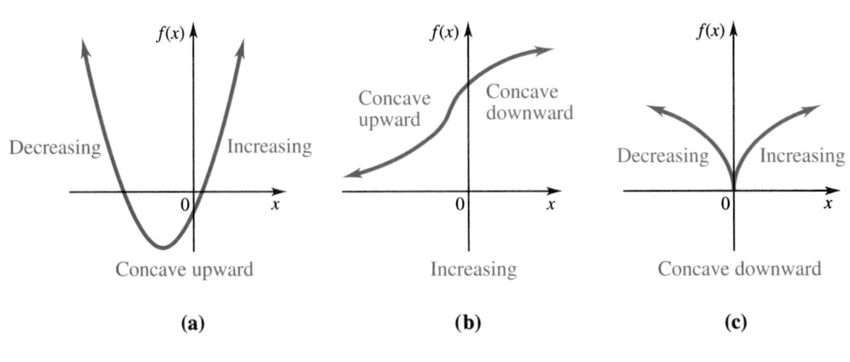

FIGURE 34

Figure 35 shows two functions that are concave upward on an interval (a, b). Several tangent lines are also shown. In Figure 35(a), the slopes of the tangent lines (moving from left to right) are first negative, then 0, and then positive. In Figure 35(b), the slopes are all positive, but they get larger.

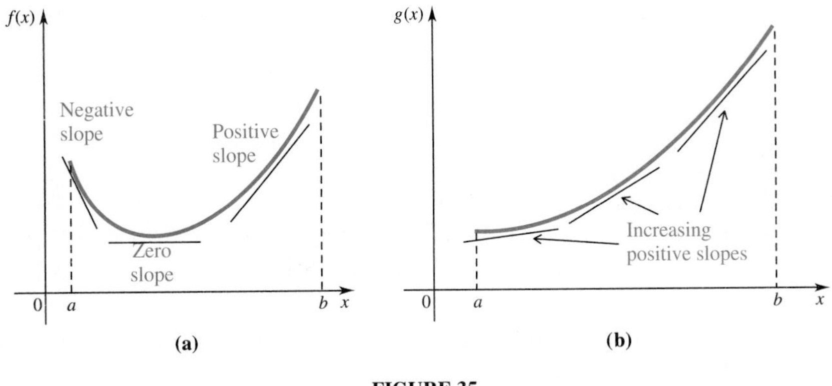

FIGURE 35

In both cases, the slopes are *increasing*. The slope at a point on a curve is given by the derivative. Since a function is increasing if its derivative is positive, its slope is increasing if the derivative of the slope function is positive. Since the derivative of a derivative is the second derivative, a function is concave upward on an interval if its second derivative is positive at each point of the interval.

A similar result is suggested by Figure 36 for functions whose graphs are concave downward. In both graphs, the slopes of the tangent lines are *decreasing* as we move from left to right. Since a function is decreasing if its derivative is negative, a function is concave downward on an interval if its second derivative is negative at each point of the interval. These observations suggest the following test.

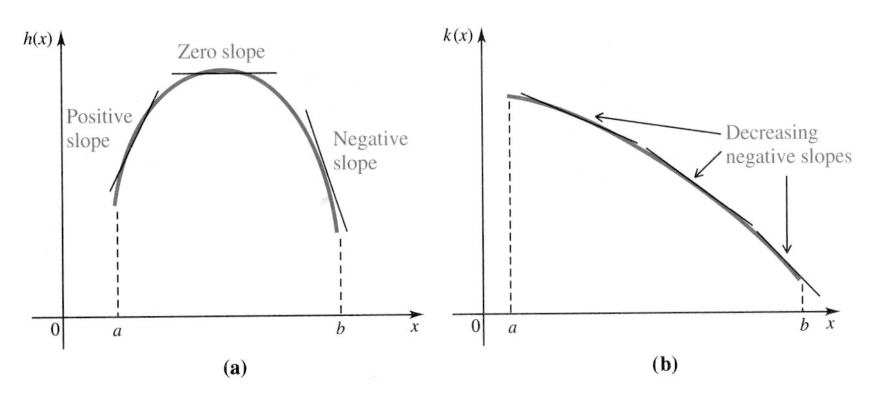

FIGURE 36

TEST FOR CONCAVITY

Let f be a function with derivatives f' and f'' existing at all points in an interval (a, b). Then f is concave upward on (a, b) if $f''(x) > 0$ for all x in (a, b), and concave downward on (a, b) if $f''(x) < 0$ for all x in (a, b).

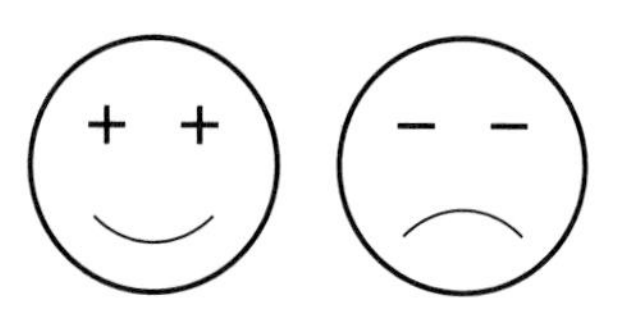

FIGURE 37

An easy way to remember this test is by the faces shown in Figure 37:

When the second derivative is positive at a point (+ +), the graph is concave upward (⌣). When the second derivative is negative at a point (− −), the graph is concave downward (⌢).

EXAMPLE 4 Find all intervals where $f(x) = x^3 - 3x^2 + 5x - 4$ is concave upward or downward.

The first derivative is $f'(x) = 3x^2 - 6x + 5$, and the second derivative is $f''(x) = 6x - 6$. The function f is concave upward whenever $f''(x) > 0$, or

$$\begin{aligned} 6x - 6 &> 0 \\ 6x &> 6 \\ x &> 1. \end{aligned}$$

Also, f is concave downward if $f''(x) < 0$, or $x < 1$. In interval notation, f is concave upward on $(1, \infty)$ and concave downward on $(-\infty, 1)$. A graph of f is shown in Figure 38. ◀

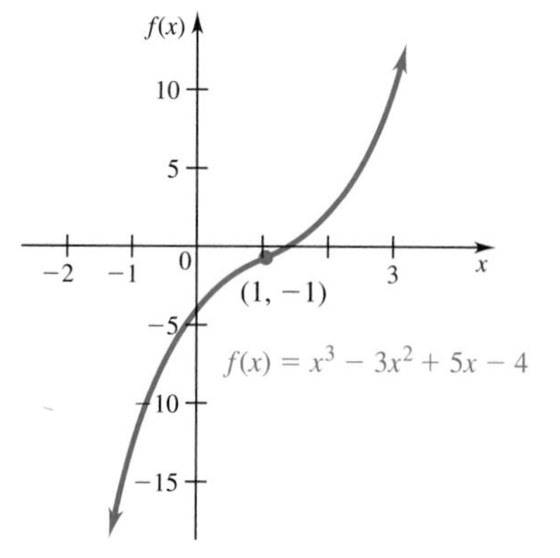

FIGURE 38

The graph of the function f in Figure 38 changes from concave downward to concave upward at $x = 1$. As mentioned earlier, a point where the direction of concavity changes is called a point of inflection. This means that the point $(1, f(1))$ or $(1, -1)$ in Figure 38 is a point of inflection. We can locate this point by finding values of x where the second derivative changes from negative to positive, that is, where the second derivative is 0. Setting the second derivative in Example 4 equal to 0 gives

$$\begin{aligned} 6x - 6 &= 0 \\ x &= 1. \end{aligned}$$

As before, the point of inflection is $(1, f(1))$ or $(1, -1)$.

Example 4 suggests the following result.

At a point of inflection for a function f, the second derivative is 0 or does not exist.

Caution Be careful with this statement. Finding a value $f''(x_0) = 0$ does not mean that a point of inflection has been located. For example, if $f(x) = (x - 1)^4$, then $f''(x) = 12(x - 1)^2$, which is 0 at $x = 1$. The graph of $f(x) = (x - 1)^4$ is always concave upward, however, so it has no point of inflection. See Figure 39.

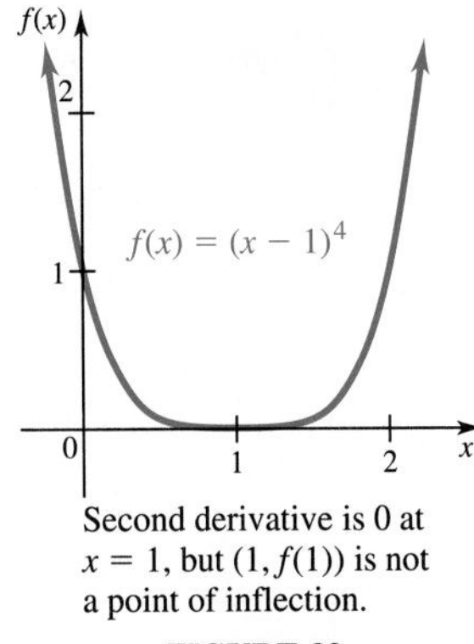

Second derivative is 0 at $x = 1$, but $(1, f(1))$ is not a point of inflection.

FIGURE 39

Second Derivative Test The idea of concavity can be used to decide whether a given critical number produces a relative maximum or a relative minimum. This test, an alternative to the first derivative test, is based on the fact that a curve that has a horizontal tangent at a point c and is concave downward on an open interval containing c also has a relative maximum at c. A relative minimum occurs when a graph has a horizontal tangent at a point d and is concave upward on an open interval containing d. See Figure 40.

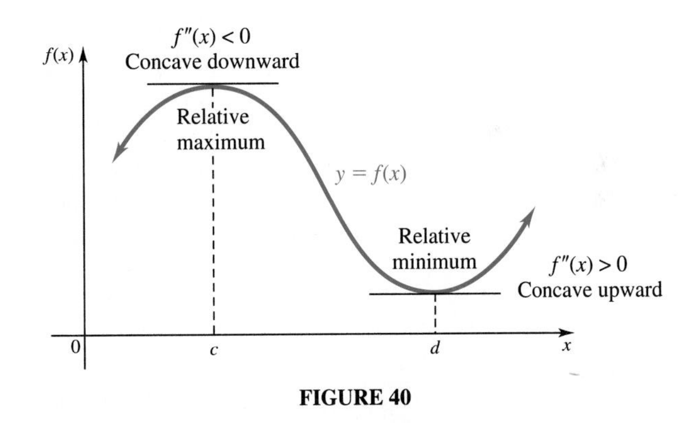

FIGURE 40

A function f is concave upward on an interval if $f''(x) > 0$ for all x in the interval, while f is concave downward on an interval if $f''(x) < 0$ for all x in the interval. These ideas lead to the **second derivative test** for relative extrema.

SECOND DERIVATIVE TEST Let f'' exist on some open interval containing c, and let $f'(c) = 0$.

1. If $f''(c) > 0$, then $f(c)$ is a relative minimum.
2. If $f''(c) < 0$, then $f(c)$ is a relative maximum.
3. If $f''(c) = 0$, then the test gives no information about extrema.

EXAMPLE 5 Find all relative extrema for

$$f(x) = 4x^3 + 7x^2 - 10x + 8.$$

First, find the points where the derivative is 0. Here $f'(x) = 12x^2 + 14x - 10$. Solve the equation $f'(x) = 0$ to get

$$12x^2 + 14x - 10 = 0$$
$$2(6x^2 + 7x - 5) = 0$$
$$2(3x + 5)(2x - 1) = 0$$
$$3x + 5 = 0 \quad \text{or} \quad 2x - 1 = 0$$
$$3x = -5 \qquad 2x = 1$$
$$x = -\frac{5}{3} \qquad x = \frac{1}{2}.$$

Now use the second derivative test. The second derivative is $f''(x) = 24x + 14$. Evaluate $f''(x)$ first at $-5/3$, getting

$$f''\left(-\frac{5}{3}\right) = 24\left(-\frac{5}{3}\right) + 14 = -40 + 14 = -26 < 0,$$

so that by Part 2 of the second derivative test, $-5/3$ leads to a relative maximum of $f(-5/3) = 691/27$. Also, when $x = 1/2$,

$$f''\left(\frac{1}{2}\right) = 24\left(\frac{1}{2}\right) + 14 = 12 + 14 = 26 > 0,$$

with $1/2$ leading to a relative minimum of $f(1/2) = 21/4$. ◀

The second derivative test works only for those critical numbers c that make $f'(c) = 0$. This test does not work for critical numbers c for which $f'(c)$ does not exist (since $f''(c)$ would not exist either). Also, the second derivative test does not work for critical numbers c that make $f''(c) = 0$. In both of these cases, use the first derivative test.

EXAMPLE 6 The graph in Figure 41 shows the population of catfish in a commercial catfish farm as a function of time. As the graph shows, the population increases rapidly up to a point and then increases at a slower rate. The horizontal dashed line shows that the population will approach some upper limit determined by the capacity of the farm. The point at which the rate of population growth starts to slow is the point of inflection for the graph.

To produce the maximum yield of catfish, harvesting should take place at the point of fastest possible growth of the population; here, this is at the point of inflection. The rate of change of the population, given by the first derivative, is increasing up to the inflection point (on the interval where the second derivative is positive) and decreasing past the inflection point (on the interval where the second derivative is negative). ◀

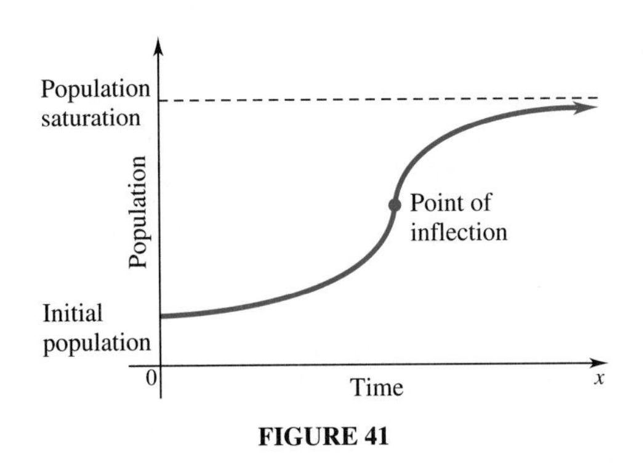

FIGURE 41

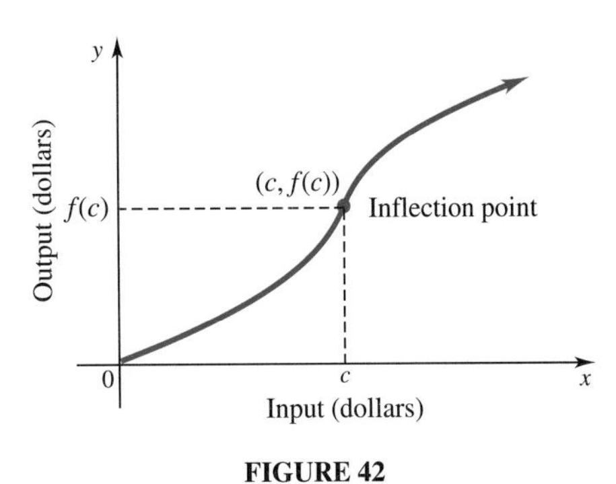

FIGURE 42

The *law of diminishing returns* in economics is related to the idea of concavity. The function graphed in Figure 42 gives the output y from a given input x. If the input were advertising costs for some product, for example, the output might be the corresponding revenue from sales.

The graph in Figure 42 shows an inflection point at $(c, f(c))$. For $x < c$, the graph is concave upward, so the rate of change of the slope is increasing. This indicates that the output y is increasing at a faster rate with each additional dollar spent. When $x > c$, however, the graph is concave downward, the rate of change of the slope is decreasing, and the increase in y is smaller with each additional dollar spent. Thus, further input beyond c dollars produces diminishing returns. The point of inflection at $(c, f(c))$ is called the **point of diminishing returns.** Beyond this point there is a smaller and smaller return for each dollar invested.

EXAMPLE 7 The revenue $R(x)$ generated from sales of a certain product is related to the amount x spent on advertising by

$$R(x) = \frac{1}{15{,}000}(600x^2 - x^3), \quad 0 \le x \le 600,$$

where x and $R(x)$ are in thousands of dollars. Is there a point of diminishing returns for this function? If so, what is it?

Since a point of diminishing returns occurs at an inflection point, look for an x-value that makes $R''(x) = 0$. Write the function as

$$R(x) = \frac{600}{15{,}000}x^2 - \frac{1}{15{,}000}x^3 = \frac{1}{25}x^2 - \frac{1}{15{,}000}x^3.$$

Now find $R'(x)$ and then $R''(x)$.

$$R'(x) = \frac{2x}{25} - \frac{3x^2}{15{,}000} = \frac{2}{25}x - \frac{1}{5000}x^2$$

$$R''(x) = \frac{2}{25} - \frac{1}{2500}x$$

Set $R''(x)$ equal to 0 and solve for x.

$$\frac{2}{25} - \frac{1}{2500}x = 0$$

$$-\frac{1}{2500}x = -\frac{2}{25}$$

$$x = \frac{5000}{25} = 200$$

Test a number in the interval (0, 200) to see that $R''(x)$ is positive there. Then test a number in the interval (200, 600) to find $R''(x)$ negative in that interval. Since the sign of $R''(x)$ changes from positive to negative at $x = 200$, the graph changes from concave upward to concave downward at that point, and there is a point of diminishing returns at the inflection point (200, 1066⅔). Investments in advertising beyond $200,000 return less and less for each dollar invested. Verify that $R'(200) = 8$. This means that when $200,000 is invested, another $1000 invested returns approximately $8000 in additional revenue. Thus it may still be economically sound to invest in advertising beyond the point of diminishing returns. ◀

3.4 Exercises

Find f'' for each of the following. Then find $f''(0)$, $f''(2)$, and $f''(-3)$.

1. $f(x) = 3x^3 - 4x + 5$

2. $f(x) = x^3 + 4x^2 + 2$

3. $f(x) = 3x^4 - 5x^3 + 2x^2$

4. $f(x) = -x^4 + 2x^3 - x^2$

5. $f(x) = 3x^2 - 4x + 8$

6. $f(x) = 8x^2 + 6x + 5$

7. $f(x) = (x + 4)^3$

8. $f(x) = (x - 2)^3$

9. $f(x) = \dfrac{2x + 1}{x - 2}$

10. $f(x) = \dfrac{x + 1}{x - 1}$

11. $f(x) = \dfrac{x^2}{1 + x}$

12. $f(x) = \dfrac{-x}{1 - x^2}$

13. $f(x) = \sqrt{x + 4}$

14. $f(x) = \sqrt{2x + 9}$

15. $f(x) = 5x^{3/5}$

16. $f(x) = -2x^{2/3}$

Find $f'''(x)$, the third derivative of f, and $f^{(4)}(x)$, the fourth derivative of f, for each of the following.

17. $f(x) = -x^4 + 2x^2 + 8$

18. $f(x) = 2x^4 - 3x^3 + x^2$

19. $f(x) = 4x^5 + 6x^4 - x^2 + 2$

20. $f(x) = 3x^5 - x^4 + 2x^3 - 7x$

21. $f(x) = \dfrac{x - 1}{x + 2}$

22. $f(x) = \dfrac{x + 1}{x}$

23. $f(x) = \dfrac{3x}{x - 2}$

24. $f(x) = \dfrac{x}{2x + 1}$

In Exercises 25–40, find the largest open intervals where the functions are concave upward or concave downward. Find any points of inflection.

25.

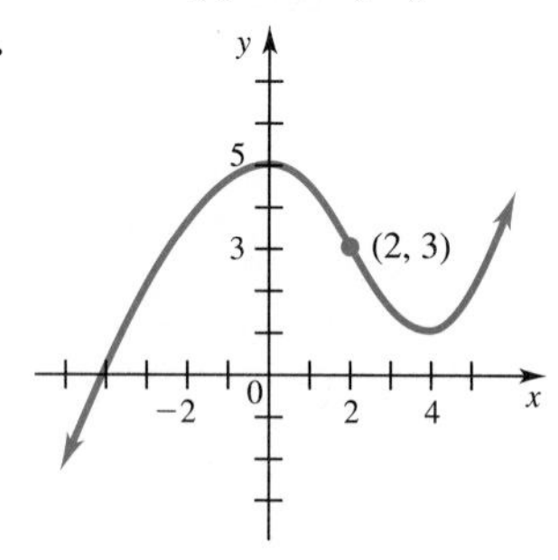

26.

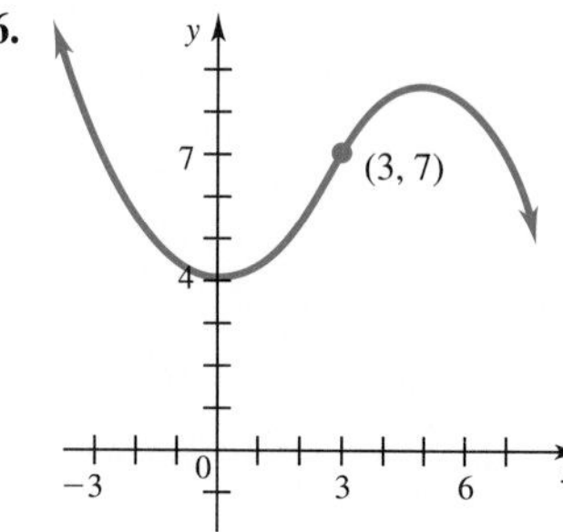

27.

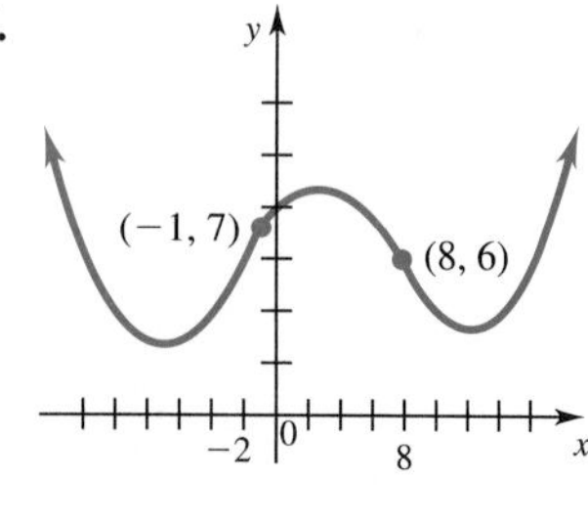

28.

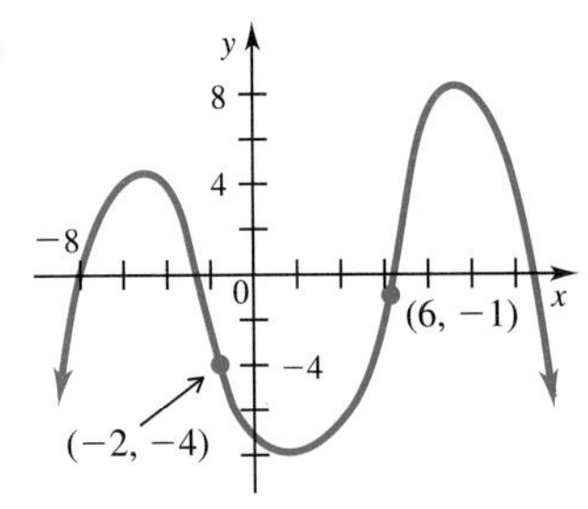

29.

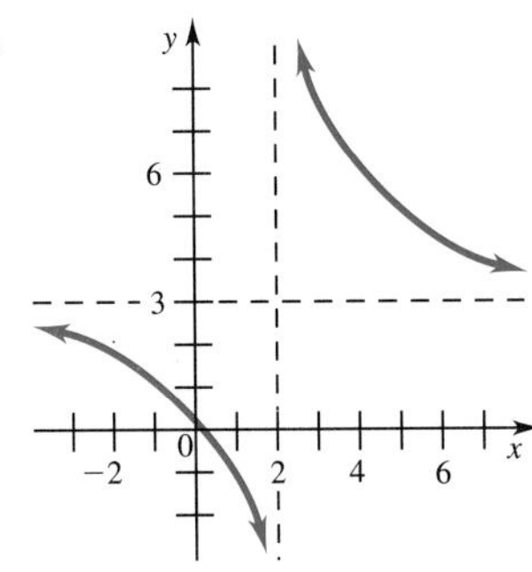

30.

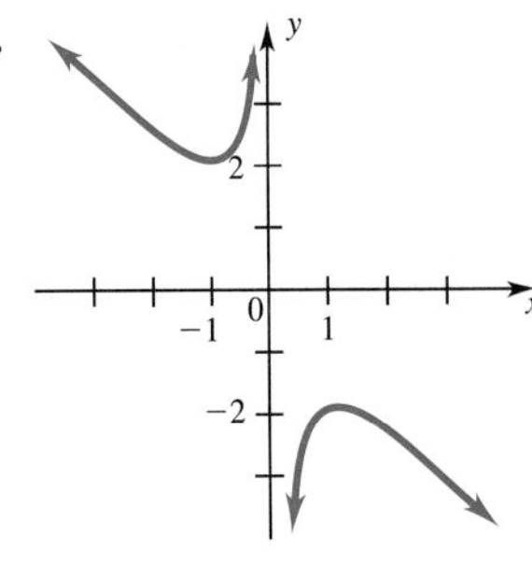

31. $f(x) = x^2 + 10x - 9$

32. $f(x) = 8 - 6x - x^2$

33. $f(x) = x^3 + 3x^2 - 45x - 3$

34. $f(x) = 2x^3 - 3x^2 - 12x + 1$

35. $f(x) = -2x^3 + 9x^2 + 168x - 3$

36. $f(x) = -x^3 - 12x^2 - 45x + 2$

37. $f(x) = \dfrac{3}{x - 5}$

38. $f(x) = \dfrac{-2}{x + 1}$

39. $f(x) = x(x + 5)^2$

40. $f(x) = -x(x - 3)^2$

Find any critical numbers for f in Exercises 41–46 and then use the second derivative test to decide whether the critical numbers lead to relative maxima or relative minima. If $f''(c) = 0$ for a critical number c, then the second derivative test gives no information. In this case, use the first derivative test instead.

41. $f(x) = -x^2 - 10x - 25$

42. $f(x) = x^2 - 12x + 36$

43. $f(x) = 3x^3 - 3x^2 + 1$

44. $f(x) = 2x^3 - 4x^2 + 2$

45. $f(x) = (x + 3)^4$

46. $f(x) = x^3$

Applications

Business and Economics

47. Product Life Cycle The accompanying figure shows the *product life cycle* graph, with typical products marked on it. It illustrates the fact that a new product is often purchased at a faster and faster rate as people become familiar with it. In time, saturation is reached and the purchase rate stays constant until the product is made obsolete by newer products, after which it is purchased less and less.*

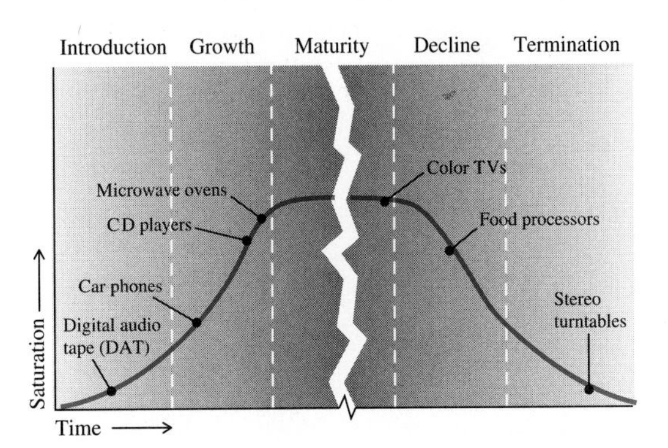

(a) Which products on the left side of the graph are closest to the left-hand point of inflection? What does the point of inflection mean here?

(b) Which product on the right side of the graph is closest to the right-hand point of inflection? What does the point of inflection mean here?

(c) Discuss where home computers, fax machines, and dot-matrix printers should be placed on the graph.

*Based on "The Product Life Cycle: A Key to Strategic Marketing Planning" in *MSU Business Topics* (Winter 1973), p. 30. Reprinted by permission of the publisher, Graduate School of Business Administration, Michigan State University.

Law of Diminishing Returns *In Exercises 48 and 49, find the point of diminishing returns for the given functions, where $R(x)$ represents revenue in thousands of dollars and x represents the amount spent on advertising in thousands of dollars.*

48. $R(x) = 10{,}000 - x^3 + 42x^2 + 800x, \quad 0 \le x \le 20$

49. $R(x) = \frac{4}{27}(-x^3 + 66x^2 + 1050x - 400), \quad 0 \le x \le 25$

50. Risk Aversion In economics, an index of *absolute risk aversion* is defined as

$$I(M) = \frac{-U''(M)}{U'(M)},$$

where M measures how much of a commodity is owned and $U(M)$ is a *utility function,* which measures the ability of quantity M of a commodity to satisfy a consumer's wants. Find $I(M)$ for $U(M) = \sqrt{M}$ and for $U(M) = M^{2/3}$, and determine which indicates a greater aversion to risk.

51. Demand Function The authors of an article† in an economics journal state that if $D(q)$ is the demand function, then the inequality

$$qD''(q) + D'(q) < 0$$

is equivalent to saying that the marginal revenue declines more quickly than does the price. Prove that this equivalence is true.

Life Sciences

52. Population Growth When a hardy new species is introduced into an area, the population often increases as shown.

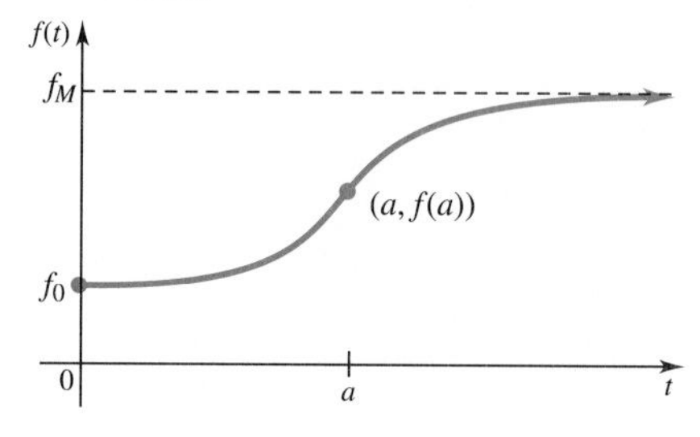

†Drew Fudenberg and Jean Tirole, "Learning by Doing and Market Performance," *Bell Journal of Economics* 14 (1983).

Explain the significance of the following function values on the graph.

(a) f_0 **(b)** $f(a)$ **(c)** f_M

53. Bacteria Population Assume that the number of bacteria $R(t)$ (in millions) present in a certain culture at time t (in hours) is given by

$$R(t) = t^2(t - 18) + 96t + 1000.$$

(a) At what time before 8 hr will the population be maximized?

(b) Find the maximum population.

54. Drug Concentration The percent of concentration of a certain drug in the bloodstream x hr after the drug is administered is given by

$$K(x) = \frac{3x}{x^2 + 4}.$$

For example, after 1 hr the concentration is given by

$$K(1) = \frac{3(1)}{1^2 + 4} = \frac{3}{5}\% = .6\% = .006.$$

(a) Find the time at which concentration is a maximum.

(b) Find the maximum concentration.

55. Drug Concentration The percent of concentration of a drug in the bloodstream x hours after the drug is administered is given by

$$K(x) = \frac{4x}{3x^2 + 27}.$$

(a) Find the time at which the concentration is a maximum.

(b) Find the maximum concentration.

Physical Sciences

56. Chemical Reaction An autocatalytic chemical reaction is one in which the product being formed causes the rate of formation to increase. The rate of a certain autocatalytic reaction is given by

$$V(x) = 12x(100 - x),$$

where x is the quantity of the product present and 100 represents the quantity of chemical present initially. For what value of x is the rate of the reaction a maximum?

Velocity and Acceleration *Each of the functions defined in Exercises 57–62 gives the displacement at time t of a particle moving along a line. Find the velocity and acceleration functions. Then find the velocity and acceleration at $t = 0$ and $t = 4$. Assume that time is measured in seconds and distance is measured in centimeters. Velocity will be in centimeters per second (cm/sec) and acceleration in centimeters per second per second (cm/sec^2).*

57. $s(t) = 8t^2 + 4t$

58. $s(t) = -3t^2 - 6t + 2$

59. $s(t) = -5t^3 - 8t^2 + 6t - 3$

60. $s(t) = 3t^3 - 4t^2 + 8t - 9$

61. $s(t) = \frac{-2}{3t + 4}$

62. $s(t) = \frac{1}{t + 3}$

63. Velocity and Acceleration When an object is dropped straight down, the distance in feet that it travels in t sec is given by

$$s(t) = -16t^2.$$

Find the velocity at each of the following times.

(a) After 3 sec **(b)** After 5 sec **(c)** After 8 sec

(d) Find the acceleration. (The answer here is a constant—the acceleration due to the influence of gravity alone.)

64. Projectile Height If an object is thrown directly upward with a velocity of 256 ft/sec, its height above the ground after t sec is given by $s(t) = 256t - 16t^2$. Find the velocity and the acceleration after t sec. What is the maximum height the object reaches? When does it hit the ground?

Computer/Graphing Calculator

In Exercises 65–68, get a list of values for $f'(x)$ and $f''(x)$ for each function on the given interval as indicated, or sketch the graph of $f'(x)$ and $f''(x)$ on a graphing calculator or computer.

(a) *By looking at the sign of $f'(x)$, give the (approximate) intervals where $f(x)$ is increasing and any intervals where $f(x)$ is decreasing.*

(b) *Give the (approximate) x-values where any maximums or minimums occur.*

(c) *By looking at the sign of $f''(x)$, give the intervals where $f(x)$ is concave upward and where it is concave downward.*

(d) *Give the (approximate) x-values of any inflection points.*

65. $f(x) = .25x^4 - 2x^3 + 3.5x^2 + 4x - 1$; $(-5, 5)$ in steps of .5

66. $f(x) = 10x^3(x - 1)^2$; $(-2, 2)$ in steps of .3

67. $f(x) = 3.1x^4 - 4.3x^3 + 5.82$; $(-1, 2)$ in steps of .2

68. $f(x) = \dfrac{x}{x^2 + 1}$; $(-3, 3)$ in steps of .4

69. Use a graphing calculator or a computer to sketch the graphs of the functions in Exercises 65–68. Compare these graphs with your answers from those exercises. Discuss briefly how well the graphs agree with these answers.

3.5 LIMITS AT INFINITY AND CURVE SKETCHING

What happens to the oxygen concentration in a pond over the long run?

Sometimes it is useful to examine the behavior of the values of $f(x)$ as x gets larger and larger (or smaller and smaller). For example, suppose a small pond normally contains 12 units of dissolved oxygen in a fixed volume of water. Suppose also that at time $t = 0$ a quantity of organic waste is introduced into the pond, with the oxygen concentration t weeks later given by

$$f(t) = \frac{12t^2 - 15t + 12}{t^2 + 1}.$$

As time goes on, what will be the ultimate concentration of oxygen? Will it return to 12 units?

After 2 weeks, the pond contains

$$f(2) = \frac{12 \cdot 2^2 - 15 \cdot 2 + 12}{2^2 + 1} = \frac{30}{5} = 6$$

units of oxygen, and after 4 weeks, it contains

$$f(4) = \frac{12 \cdot 4^2 - 15 \cdot 4 + 12}{4^2 + 1} \approx 8.5$$

units. Choosing several values of t and finding the corresponding values of $f(t)$, or using a graphing calculator or computer, leads to the graph in Figure 43.

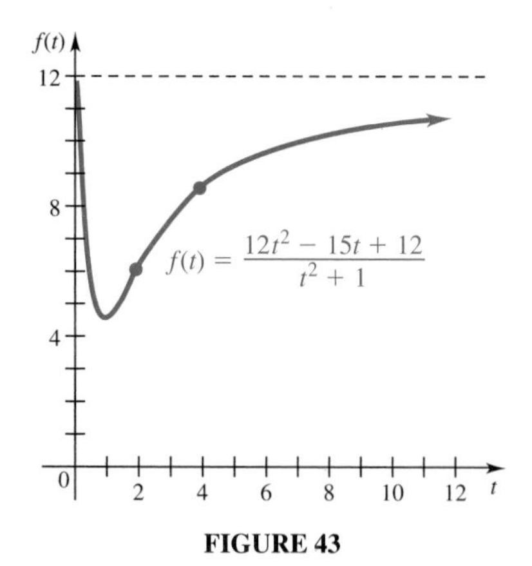

FIGURE 43

The graph suggests that as time goes on, the oxygen level gets closer and closer to the original 12 units. Consider a table of values as t gets larger and larger.

t	10	100	1000	10,000
$f(t)$	10.5	11.85	11.985	11.9985

The table suggests that

$$\lim_{t \to \infty} f(t) = 12,$$

where $t \to \infty$ means that t increases without bound. (Similarly, $x \to -\infty$ means that x *decreases* without bound; that is, x becomes more and more negative.) Thus, the oxygen concentration will approach 12, but it will never be *exactly* 12.

The graphs of $f(x) = 1/x$ (in black) and $g(x) = 1/x^2$ (in color) shown in Figure 44 lead to examples of such **limits at infinity.** The graphs and table of values indicate that $\lim_{x \to \infty} (1/x) = 0$, $\lim_{x \to -\infty} (1/x) = 0$, $\lim_{x \to \infty} (1/x^2) = 0$, and $\lim_{x \to -\infty} (1/x^2) = 0$, suggesting the following rule.

x	$\frac{1}{x}$	$\frac{1}{x^2}$
-100	$-.01$	.0001
-10	$-.1$	.01
-1	-1	1
1	1	1
10	.1	.01
100	.01	.0001

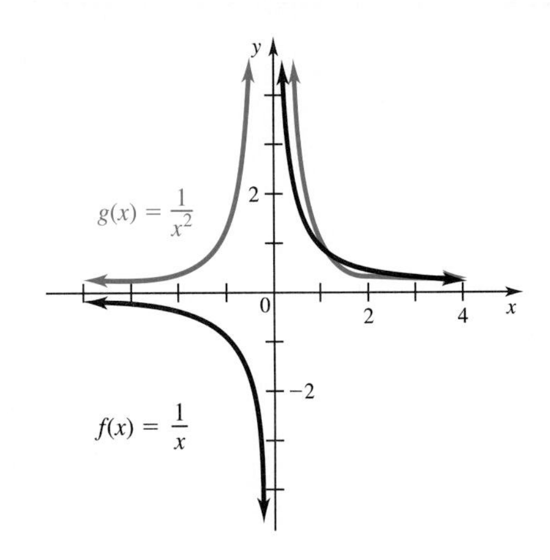

FIGURE 44

LIMITS AT INFINITY

For any positive real number n,

$$\lim_{x\to\infty} \frac{1}{x^n} = 0 \quad \text{and} \quad \lim_{x\to-\infty} \frac{1}{x^n} = 0.^*$$

The rules for limits given in the preceding chapter remain unchanged when a is replaced with ∞ or $-\infty$.

This section requires use of the rules for limits discussed in the section on limits and continuity, with a replaced by ∞ or $-\infty$. One such rule is the following:

If $\lim_{x\to a} f(x) = A$ and $\lim_{x\to a} g(x) = B$, then

$$\lim_{x\to a} \frac{f(x)}{g(x)} = \frac{A}{B} \quad \text{if } B \neq 0.$$

Find where this rule is used in Example 1, and then look up the other rules and find where they are used in this example.

Find each limit.

(a) $\lim_{x\to\infty} \dfrac{8x + 6}{3x - 1}$

We can use the rule $\lim_{x\to\infty} 1/x^n = 0$ to find this limit by first dividing numerator and denominator by x, as follows.

$$\lim_{x\to\infty} \frac{8x + 6}{3x - 1} = \lim_{x\to\infty} \frac{\frac{8x}{x} + \frac{6}{x}}{\frac{3x}{x} - \frac{1}{x}} = \lim_{x\to\infty} \frac{8 + 6 \cdot \frac{1}{x}}{3 - \frac{1}{x}} = \frac{8 + 0}{3 - 0} = \frac{8}{3}$$

(b) $\lim_{x\to-\infty} \dfrac{4x^2 - 6x + 3}{2x^2 - x + 4}$

Divide each term of the numerator and denominator by x^2, the highest power of x.

*If x is negative, x^n does not exist for certain values of n, so the second limit is undefined.

$$\lim_{x\to-\infty}\frac{4x^2-6x+3}{2x^2-x+4}=\lim_{x\to-\infty}\frac{4-6\cdot\frac{1}{x}+3\cdot\frac{1}{x^2}}{2-\frac{1}{x}+4\cdot\frac{1}{x^2}}$$

$$=\frac{4-0+0}{2-0+0}=\frac{4}{2}=2$$

(c) $$\lim_{x\to\infty}\frac{3x+2}{4x^3-1}=\lim_{x\to\infty}\frac{3\cdot\frac{1}{x^2}+2\cdot\frac{1}{x^3}}{4-\frac{1}{x^3}}=\frac{0+0}{4-0}=\frac{0}{4}=0$$

Here, the highest power of x is x^3, which is used to divide each term in the numerator and denominator.

(d) $$\lim_{x\to\infty}\frac{3x^2+2}{4x-3}=\lim_{x\to\infty}\frac{3+\frac{2}{x^2}}{\frac{4}{x}-\frac{3}{x^2}}=\frac{3+0}{0-0}=\frac{3}{0}$$

Division by 0 is undefined, so this limit does not exist. We can actually say more: by examining what happens as larger and larger values of x are put into the function, verify that

$$\lim_{x\to\infty}\frac{3x^2+2}{4x-3}=\infty. \blacktriangleleft$$

The method used in Example 1 is a useful way to rewrite expressions with fractions so that the rules for limits at infinity can be used.

FINDING LIMITS AT INFINITY

If $f(x)=\frac{p(x)}{q(x)}$, for polynomials $p(x)$ and $q(x)$, $q(x)\neq 0$, $\lim_{x\to-\infty} f(x)$ and $\lim_{x\to\infty} f(x)$ can be found as follows.

1. Divide $p(x)$ *and* $q(x)$ by the highest power of x in either polynomial.
2. Use the rules for limits, including the rules for limits at infinity,

$$\lim_{x\to\infty}\frac{1}{x^n}=0 \quad\text{and}\quad \lim_{x\to-\infty}\frac{1}{x^n}=0,$$

to find the limit of the result from Step 1.

Look back at the method for finding horizontal asymptotes in Chapter 1. There we were essentially taking limits at infinity, although we did not use that terminology. Limits at infinity, therefore, give us a method for finding horizontal asymptotes.

Curve Sketching The test for concavity, the test for increasing and decreasing functions, and the concept of limits at infinity help us sketch the graphs of a variety of functions. This process, called *curve sketching*, uses the following steps.

CURVE SKETCHING

To sketch the graph of a function f:

1. Find the y-intercept (if it exists) by substituting $x = 0$ into $f(x)$. Find any x-intercepts by solving $f(x) = 0$ if this is not too difficult.
2. If f is a rational function, find any vertical asymptotes by investigating where the denominator is 0, and find any horizontal asymptotes by finding the limits as $x \to \infty$ and $x \to -\infty$.
3. Find f'. Locate any critical points by solving the equation $f'(x) = 0$ and determining where f' does not exist, but f does. Find any relative extrema and determine where f is increasing or decreasing.
4. Find f''. Locate potential points of inflection by solving the equation $f''(x) = 0$ and determining where $f''(x)$ does not exist. Determine where f is concave upward or concave downward.
5. Plot the intercepts, the critical points, the inflection points, the asymptotes, and other points as needed.

EXAMPLE 2

Graph $f(x) = 2x^3 - 3x^2 - 12x + 1$.

The y-intercept is located at $y = f(0) = 1$. Finding the x-intercepts requires solving the equation $f(x) = 0$. But this is a third-degree equation; since we have not covered a procedure for solving such equations, we will skip this step. This is not a rational function, so we also skip Step 2.

To find the intervals where the function is increasing or decreasing, find the first derivative.

$$f'(x) = 6x^2 - 6x - 12$$

This derivative is 0 when

$$6(x^2 - x - 2) = 0$$
$$6(x - 2)(x + 1) = 0$$
$$x = 2 \quad \text{or} \quad x = -1.$$

These critical numbers divide the number line in Figure 45 into three regions.

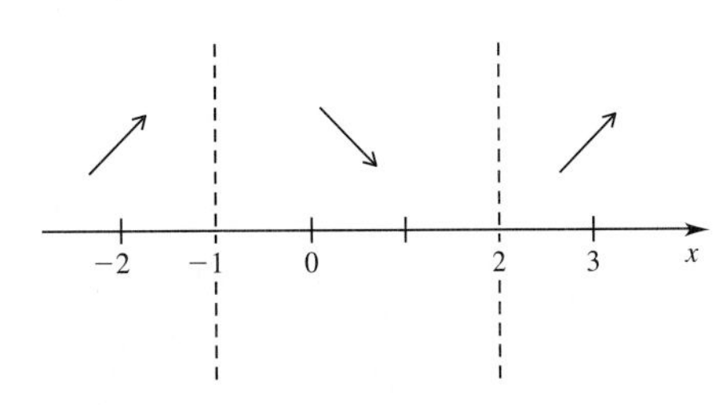

FIGURE 45

Testing a number from each region in $f'(x)$ shows that f is increasing on $(-\infty, -1)$ and $(2, \infty)$, and decreasing on $(-1, 2)$. This is shown with the arrows in Figure 45. By the first derivative test, f has a relative maximum when $x = -1$ and a relative minimum when $x = 2$. The relative maximum is $f(-1) = 8$, while the relative minimum is $f(2) = -19$.

Now use the second derivative to find the intervals where the function is concave upward or downward. Here

$$f''(x) = 12x - 6,$$

which is 0 when $x = 1/2$. Testing a point with x less than $1/2$, and one with x greater than $1/2$, shows that f is concave downward on $(-\infty, 1/2)$ and concave upward on $(1/2, \infty)$. The graph has an inflection point at $(1/2, f(1/2))$, or $(1/2, -11/2)$. This information is summarized in the chart below.

Interval	$(-\infty, -1)$	$(-1, 1/2)$	$(1/2, 2)$	$(2, \infty)$
Sign of f'	+	−	−	+
Sign of f''	−	−	+	+
f Increasing or Decreasing	Increasing	Decreasing	Decreasing	Increasing
Concavity of f	Downward	Downward	Upward	Upward
Shape of Graph				

Use this information and the critical points to get the graph shown in Figure 46.

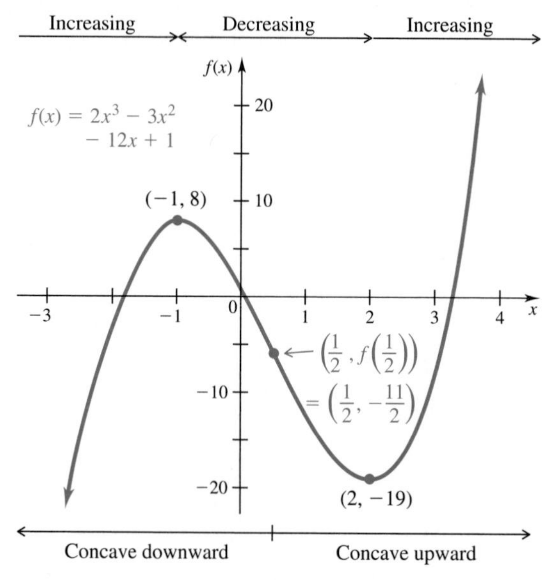

FIGURE 46

EXAMPLE 3

Graph $f(x) = x + 1/x$.

Notice that $x = 0$ is not in the domain of the function, so there is no y-intercept. To find the x-intercept, solve $f(x) = 0$.

$$x + \frac{1}{x} = 0$$

$$x = -\frac{1}{x}$$

$$x^2 = -1$$

Since x^2 is always positive, there is also no x-intercept.

The function is a rational function, but it is not written in the usual form of one polynomial over another. It can be rewritten in that form:

$$f(x) = x + \frac{1}{x} = \frac{x^2 + 1}{x}.$$

Because $x = 0$ makes the denominator 0, the line $x = 0$ is a vertical asymptote. To find any horizontal asymptotes, we investigate

$$\lim_{x\to\infty} \frac{x^2 + 1}{x} = \lim_{x\to\infty} \frac{1 + \frac{1}{x^2}}{\frac{1}{x}} = \frac{1 + 0}{0} = \frac{1}{0}.$$

Division by 0 is undefined, so the limit does not exist. Verify that $\lim_{x\to-\infty} f(x)$ also does not exist, so there are no horizontal asymptotes.

Observe that as x gets very large, the second term $(1/x)$ in $f(x)$ gets very small, so $f(x) = x + (1/x) \approx x$. The graph gets closer and closer to the straight line $y = x$ as x becomes larger and larger. This is what is known as an **oblique asymptote.**

Here $f'(x) = 1 - (1/x^2)$, which is 0 when

$$\frac{1}{x^2} = 1$$

$$x^2 = 1$$

$$x = 1 \quad \text{or} \quad x = -1.$$

The derivative fails to exist at 0, where the vertical asymptote is located. Evaluating $f'(x)$ in each of the regions determined by the critical numbers and the asymptote shows that f is increasing on $(-\infty, -1)$ and $(1, \infty)$ and decreasing on $(-1, 0)$ and $(0, 1)$. By the first derivative test, f has a relative maximum of $y = f(-1) = -2$, when $x = -1$, and a relative minimum of $y = f(1) = 2$ when $x = 1$.

The second derivative is

$$f''(x) = \frac{2}{x^3},$$

which is never equal to 0 and does not exist when $x = 0$. (The function itself also does not exist at 0.) Because of this, there may be a change of concavity, but not an inflection point, when $x = 0$. The second derivative is negative when x is negative, making f concave downward on $(-\infty, 0)$. Also, $f''(x) > 0$ when $x > 0$, making f concave upward on $(0, \infty)$.

Interval	$(-\infty, -1)$	$(-1, 0)$	$(0, 1)$	$(1, \infty)$
Sign of f'	+	−	−	+
Sign of f''	−	−	+	+
f Increasing or Decreasing	Increasing	Decreasing	Decreasing	Increasing
Concavity of f	Downward	Downward	Upward	Upward
Shape of Graph	⌒	⌒	◡	◡

Use this information, the asymptotes, and the critical points to get the graph shown in Figure 47. ◀

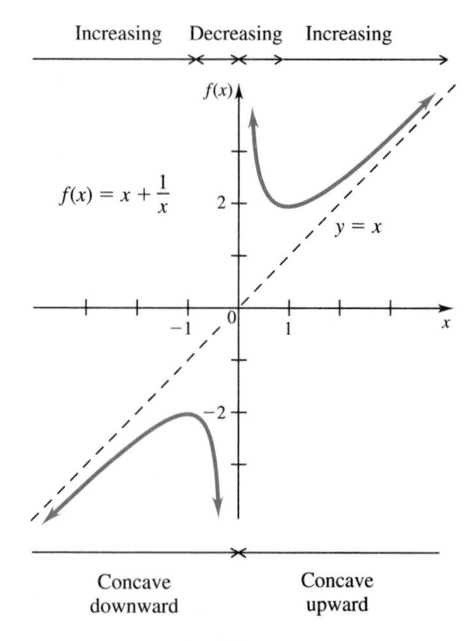

FIGURE 47

EXAMPLE 4

Graph $f(x) = \dfrac{3x^2}{x^2 + 5}$.

The y-intercept is located at $y = f(0) = 0$. Verify that this is also the only x-intercept. There is no vertical asymptote, because $x^2 + 5 \neq 0$ for any value of x. Find any horizontal asymptote by calculating $\lim_{x\to\infty} f(x)$ and $\lim_{x\to-\infty} f(x)$. First, divide both the numerator and the denominator of $f(x)$ by x^2.

$$\lim_{x\to\infty} \frac{3x^2}{x^2 + 5} = \lim_{x\to\infty} \frac{\dfrac{3x^2}{x^2}}{\dfrac{x^2}{x^2} + \dfrac{5}{x^2}} = \frac{3}{1 + 0} = 3$$

Verify that the limit of $f(x)$ as $x \to -\infty$ is also 3. Thus, the horizontal asymptote is $y = 3$.

We now compute $f'(x)$:

$$f'(x) = \frac{(x^2 + 5)(6x) - (3x^2)(2x)}{(x^2 + 5)^2}.$$

Notice that $6x$ can be factored out of each term in the numerator:

$$f'(x) = \frac{(6x)[(x^2 + 5) - x^2]}{(x^2 + 5)^2}$$

$$= \frac{(6x)(5)}{(x^2 + 5)^2} = \frac{30x}{(x^2 + 5)^2}.$$

From the numerator, $x = 0$ is a critical number. The denominator is always positive. (Why?) Evaluating $f'(x)$ in each of the regions determined by $x = 0$ shows that f is decreasing on $(-\infty, 0)$ and increasing on $(0, \infty)$. By the first derivative test, f has a relative minimum when $x = 0$.

The second derivative is

$$f''(x) = \frac{(x^2 + 5)^2(30) - (30x)(2)(x^2 + 5)(2x)}{(x^2 + 5)^4}.$$

Factor $30(x^2 + 5)$ out of the numerator:

$$f''(x) = \frac{30(x^2 + 5)[(x^2 + 5) - (x)(2)(2x)]}{(x^2 + 5)^4}.$$

Divide a factor of $(x^2 + 5)$ out of the numerator and denominator, and simplify the numerator:

$$f''(x) = \frac{30[(x^2 + 5) - (x)(2)(2x)]}{(x^2 + 5)^3}$$

$$= \frac{30[(x^2 + 5) - (4x^2)]}{(x^2 + 5)^3}$$

$$= \frac{30(5 - 3x^2)}{(x^2 + 5)^3}.$$

The numerator of $f''(x)$ is 0 when $x = \pm\sqrt{5/3} \approx \pm 1.29$. Testing a point in each of the three intervals defined by these points shows that f is concave downward on $(-\infty, -1.29)$ and $(1.29, \infty)$, and concave upward on $(-1.29, 1.29)$. The graph has inflection points at $(\pm\sqrt{5/3}, f(\pm\sqrt{5/3})) \approx (\pm 1.29, \pm .75)$.

Interval	**$(-\infty, -1.29)$**	**$(-1.29, 0)$**	**$(0, 1.29)$**	**$(1.29, \infty)$**
Sign of f'	$-$	$-$	$+$	$+$
Sign of f''	$-$	$+$	$+$	$-$
f Increasing or Decreasing	Decreasing	Decreasing	Increasing	Increasing
Concavity of f	Downward	Upward	Upward	Downward
Shape of Graph				

Use this information, the asymptote, the critical point, and the inflection points to get the graph shown in Figure 48. ◀

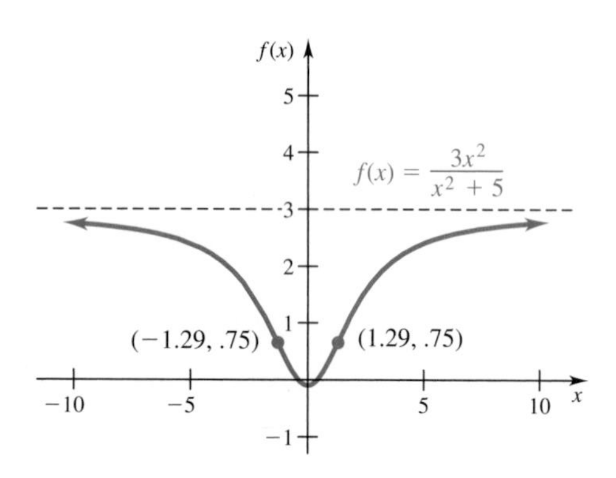

FIGURE 48

As we saw in Chapter 1, a graphing calculator or a computer graphing program, when used with care, can be helpful in studying the behavior of functions. This section has illustrated that calculus is also of great help. The techniques of calculus show where the important points of a function, such as the relative extrema and the inflection points, are located. Furthermore, they tell how the function behaves between and beyond the points that are graphed, something a computer graphing program cannot do.

3.5 Exercises

Decide whether the following limits exist. If a limit exists, find its value.

1. $\lim_{x\to\infty} f(x)$

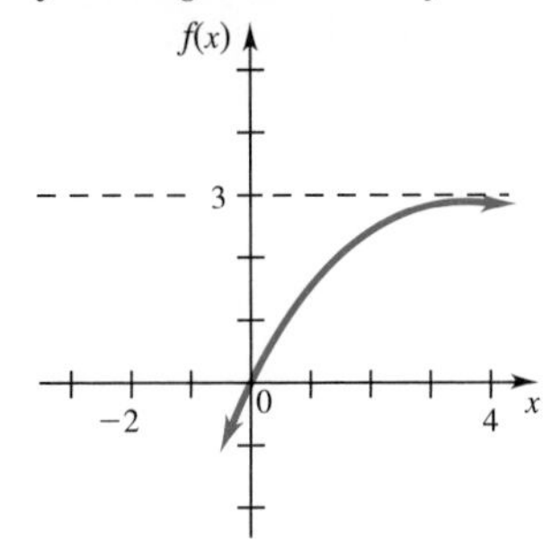

2. $\lim_{x\to-\infty} g(x)$

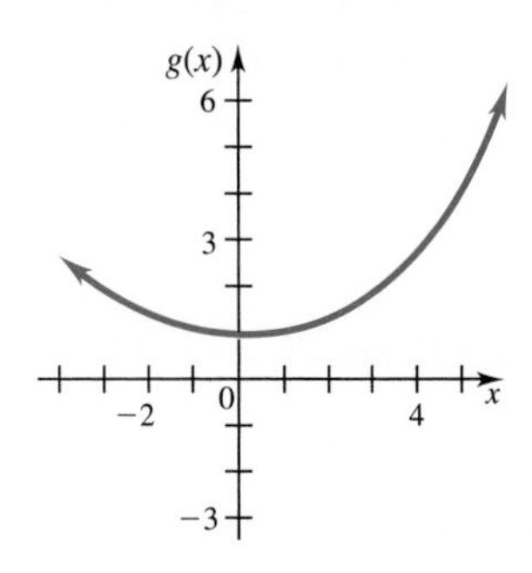

3. $\lim_{x\to\infty} \frac{3x}{5x - 1}$

4. $\lim_{x\to\infty} \frac{5x}{3x - 1}$

5. $\lim_{x\to-\infty} \frac{2x + 3}{4x - 7}$

6. $\lim_{x\to-\infty} \frac{8x + 2}{2x - 5}$

7. $\lim_{x\to\infty} \frac{x^2 + 2x}{2x^2 - 2x + 1}$

8. $\lim_{x\to\infty} \frac{x^2 + 2x - 5}{3x^2 + 2}$

9. $\lim_{x\to\infty} \frac{3x^3 + 2x - 1}{2x^4 - 3x^3 - 2}$

10. $\lim_{x\to\infty} \frac{2x^2 - 1}{3x^4 + 2}$

11. $\lim_{x\to\infty} (\sqrt{x^2 + 4} - x)$

12. $\lim_{x\to\infty} (x - \sqrt{x^2 - 9})$

Graph each of the following functions, including critical points, regions where the function is increasing or decreasing, points of inflection, regions where the function is concave up or concave down, intercepts where possible, and asymptotes where applicable.

13. $f(x) = -2x^3 - 9x^2 + 108x - 10$

14. $f(x) = -2x^3 - 9x^2 + 60x - 8$

15. $f(x) = 2x^3 + \frac{7}{2}x^2 - 5x + 3$

16. $f(x) = x^3 - \frac{15}{2}x^2 - 18x - 1$

17. $f(x) = x^4 - 18x^2 + 5$

18. $f(x) = x^4 - 8x^2$

19. $f(x) = x^4 - 2x^3$

20. $f(x) = x^5 - 15x^3$

21. $f(x) = x + \frac{2}{x}$

22. $f(x) = 2x + \frac{8}{x}$

23. $f(x) = \frac{x^2 + 25}{x}$

24. $f(x) = \frac{x^2 + 4}{x}$

25. $f(x) = \frac{x - 1}{x + 1}$

26. $f(x) = \frac{x}{1 + x}$

27. $f(x) = \frac{x}{x^2 + 1}$

28. $f(x) = \frac{1}{x^2 + 1}$

29. $f(x) = \frac{1}{x^2 - 4}$

30. $f(x) = \frac{x}{x^2 - 1}$

In Exercises 31–34, sketch the graph of a single function that has all of the properties listed.

31.
- **(a)** Continuous and differentiable for all real numbers
- **(b)** Increasing on $(-\infty, -3)$ and $(1, 4)$
- **(c)** Decreasing on $(-3, 1)$ and $(4, \infty)$
- **(d)** Concave downward on $(-\infty, -1)$ and $(2, \infty)$
- **(e)** Concave upward on $(-1, 2)$
- **(f)** $f'(-3) = f'(4) = 0$
- **(g)** Inflection points at $(-1, 3)$ and $(2, 4)$

32.
- **(a)** Continuous for all real numbers
- **(b)** Increasing on $(-\infty, -2)$ and $(0, 3)$
- **(c)** Decreasing on $(-2, 0)$ and $(3, \infty)$
- **(d)** Concave downward on $(-\infty, 0)$ and $(0, 5)$
- **(e)** Concave upward on $(5, \infty)$
- **(f)** $f'(-2) = f'(3) = 0$
- **(g)** $f'(0)$ does not exist
- **(h)** Differentiable everywhere except at $x = 0$
- **(i)** An inflection point at $(5, 1)$

33. **(a)** Continuous for all real numbers
(b) Decreasing on $(-\infty, -6)$ and $(1, 3)$
(c) Increasing on $(-6, 1)$ and $(3, \infty)$
(d) Concave upward on $(-\infty, -6)$ and $(3, \infty)$
(e) Concave downward on $(-6, 3)$
(f) A y-intercept at $(0, 2)$

34. **(a)** Continuous and differentiable everywhere except at $x = 1$, where it has a vertical asymptote
(b) Decreasing everywhere it is defined
(c) A horizontal asymptote at $y = 2$
(d) Concave downward on $(-\infty, 1)$ and $(2, 4)$
(e) Concave upward on $(1, 2)$ and $(4, \infty)$

Applications

Business and Economics

35. Average Cost The cost for manufacturing a particular videotape is

$$c(x) = 15{,}000 + 6x,$$

where x is the number of tapes produced. The average cost per tape, denoted by $\overline{c}(x)$, is found by dividing $c(x)$ by x. Find and interpret $\lim_{x\to\infty} \overline{c}(x)$.

36. Employee Productivity A company training program has determined that, on the average, a new employee can do $P(s)$ pieces of work per day after s days of on-the-job training, where

$$P(s) = \frac{75s}{s + 8}.$$

Find and interpret $\lim_{s\to\infty} P(s)$.

Life Sciences

37. Drug Concentration The concentration of a drug in a patient's bloodstream h hours after it was injected is given by

$$A(h) = \frac{.17h}{h^2 + 2}.$$

Find and interpret $\lim_{h\to\infty} A(h)$.

Social Sciences

38. Legislative Voting Members of a legislature often must vote repeatedly on the same bill. As time goes on, members may change their votes. Suppose that p_0 is the probability that an individual legislator favors an issue before the first roll call vote, and suppose that p is the probability of a change in position from one vote to the next. Then the probability that the legislator will vote "yes" on the nth roll call is given by

$$p_n = \frac{1}{2} + \left(p_0 - \frac{1}{2}\right)(1 - 2p)^n.*$$

For example, the chance of a "yes" on the third roll call vote is

$$p_3 = \frac{1}{2} + \left(p_0 - \frac{1}{2}\right)(1 - 2p)^3.$$

Suppose that there is a chance of $p_0 = .7$ that Congressman Stephens will favor the budget appropriation bill before the first roll call, but only a probability of $p = .2$ that he will change his mind on the subsequent vote. Find and interpret the following.

(a) p_2 **(b)** p_4 **(c)** p_8 **(d)** $\lim_{n\to\infty} p_n$

*See John W. Bishir and Donald W. Drewes, *Mathematics in the Behavioral and Social Sciences* (New York: Harcourt Brace Jovanovich, 1970), p. 538.

Computer/Graphing Calculator

(a) *Use a calculator or a computer to find the following limits by entering larger and larger values.*
(b) *Explain how your answer could have been predicted in advance.*

39. $\lim_{x\to\infty} \frac{\sqrt{9x^2 + 5}}{2x}$

40. $\lim_{x\to-\infty} \frac{\sqrt{9x^2 + 5}}{2x}$

41. $\lim_{x\to-\infty} \frac{\sqrt{36x^2 + 2x + 7}}{3x}$

42. $\lim_{x\to\infty} \frac{\sqrt{36x^2 + 2x + 7}}{3x}$

43. $\lim_{x\to\infty} \frac{(1 + 5x^{1/3} + 2x^{5/3})^3}{x^5}$

44. $\lim_{x\to-\infty} \frac{(1 + 5x^{1/3} + 2x^{5/3})^3}{x^5}$

Chapter Summary — Key Terms

3.1 **increasing function**
decreasing function
critical number
critical point

3.2 **relative maximum**
relative minimum
first derivative test

3.3 **absolute maximum**
absolute minimum
extreme value theorem

3.4 **second derivative**
acceleration
concavity
concave upward and downward
point of inflection
second derivative test

3.5 **limit at infinity**
oblique asymptote

Chapter 3 Review Exercises

1. When given the equation for a function, how can you determine where it is increasing and where it is decreasing?

2. When given the equation for a function, how can you determine where the relative extrema are located? Give two ways to test whether a relative extremum is a minimum or a maximum.

3. What is the difference between a relative extremum and an absolute extremum? Can a relative extremum be an absolute extremum? Is a relative extremum necessarily an absolute extremum?

4. What information about a graph can be found from the second derivative?

Find the largest open intervals where f is increasing or decreasing.

5. $f(x) = x^2 - 5x + 3$

6. $f(x) = -2x^2 - 3x + 4$

7. $f(x) = -x^3 - 5x^2 + 8x - 6$

8. $f(x) = 4x^3 + 3x^2 - 18x + 1$

9. $f(x) = \dfrac{6}{x - 4}$

10. $f(x) = \dfrac{5}{2x + 1}$

Find the locations and values of all relative maxima and minima.

11. $f(x) = -x^2 + 4x - 8$

12. $f(x) = x^2 - 6x + 4$

13. $f(x) = 2x^2 - 8x + 1$

14. $f(x) = -3x^2 + 2x - 5$

15. $f(x) = 2x^3 + 3x^2 - 36x + 20$

16. $f(x) = 2x^3 + 3x^2 - 12x + 5$

Find the second derivative of each of the following, and then find $f''(1)$ and $f''(-3)$.

17. $f(x) = 3x^4 - 5x^2 - 11x$

18. $f(x) = 9x^3 + \dfrac{1}{x}$

19. $f(x) = \dfrac{5x - 1}{2x + 3}$

20. $f(x) = \dfrac{4 - 3x}{x + 1}$

21. $f(t) = \sqrt{t^2 + 1}$

22. $f(t) = -\sqrt{5 - t^2}$

Find the locations and values of all absolute maxima and minima on the given intervals.

23. $f(x) = -x^2 + 5x + 1; \quad [1, 4]$

24. $f(x) = 4x^2 - 8x - 3; \quad [-1, 2]$

25. $f(x) = x^3 + 2x^2 - 15x + 3; \quad [-4, 2]$

26. $f(x) = -2x^3 - 2x^2 + 2x - 1; \quad [-3, 1]$

Decide whether the limits in Exercises 27–32 exist. If a limit exists, find its value.

27. $\lim_{x \to -\infty} g(x)$

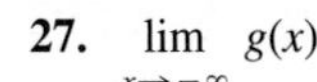

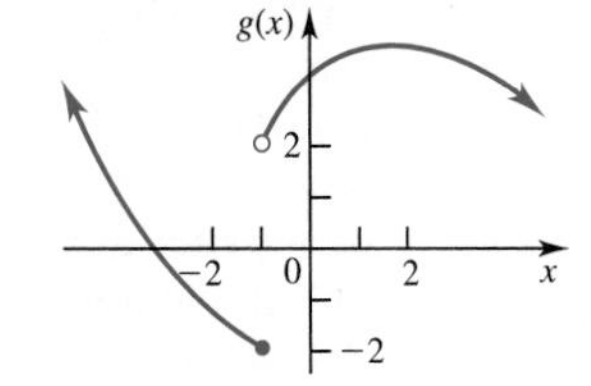

28. $\lim_{x \to \infty} f(x)$

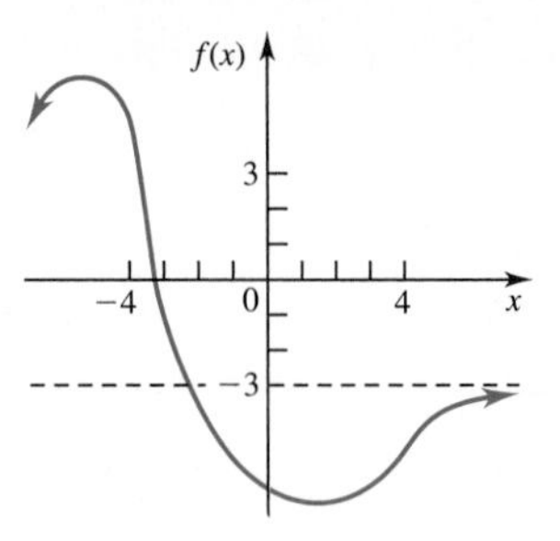

29. $\lim_{x \to \infty} \dfrac{x^2 + 5}{5x^2 - 1}$

30. $\lim_{x \to \infty} \dfrac{x^2 + 6x + 8}{x^3 + 2x + 1}$

31. $\lim_{x \to -\infty} \left(\dfrac{3}{4} + \dfrac{2}{x} - \dfrac{5}{x^2}\right)$

32. $\lim_{x \to -\infty} \left(\dfrac{9}{x^4} + \dfrac{1}{x^2} - 3\right)$

Graph each of the following functions, including critical points, regions where the function is increasing or decreasing, points of inflection, regions where the function is concave up or concave down, intercepts where possible, and asymptotes where applicable.

33. $f(x) = -2x^3 - \dfrac{1}{2}x^2 + x - 3$

34. $f(x) = -\dfrac{4}{3}x^3 + x^2 + 30x - 7$

35. $f(x) = x^4 - \dfrac{4}{3}x^3 - 4x^2 + 1$

36. $f(x) = -\dfrac{2}{3}x^3 + \dfrac{9}{2}x^2 + 5x + 1$

37. $f(x) = \dfrac{x - 1}{2x + 1}$

38. $f(x) = \dfrac{2x - 5}{x + 3}$

39. $f(x) = -4x^3 - x^2 + 4x + 5$

40. $f(x) = x^3 + \dfrac{5}{2}x^2 - 2x - 3$

41. $f(x) = x^4 + 2x^2$

42. $f(x) = 6x^3 - x^4$

43. $f(x) = \dfrac{x^2 + 4}{x}$

44. $f(x) = x + \dfrac{8}{x}$

45. $f(x) = \dfrac{2x}{3 - x}$

46. $f(x) = \dfrac{-4x}{1 + 2x}$

Applications at a glance . . .

Demand for Enrollment in College . . . prediction of crime rate . . . adjusting a profit forecast . . .

CHAPTER 4

APPLICATIONS OF THE DERIVATIVE

(See page 223.)

What do aluminum cans, shipments of antibiotics, elasticity of demand, and a melting icicle have in common? All are involved in applications of the derivative. The previous chapter included examples in which we used the derivative to find the maximum or minimum value of a function. This problem is ubiquitous; consider the efforts people expend trying to maximize their income, or to minimize their costs or the time required to complete a task. In this chapter we will treat the topic of optimization in greater depth.

The derivative is applicable in far wider circumstances, however. In roughly 500 B.C., Heraclitus said, "Nothing endures but change," and his observation has relevance here. If change is continuous, rather than in sudden jumps, the derivative can be used to describe the rate of change. This explains why calculus has been applied to so many fields.

4.1 APPLICATIONS OF EXTREMA

How should boxes and cans be designed to minimize the material needed?

In Examples 3 and 4 we will use the techniques of calculus to find an answer to this question.

In this section, we give several examples showing applications of calculus to maximum and minimum problems. To solve these examples, go through the following steps.

SOLVING APPLIED EXTREMA PROBLEMS

1. Read the problem carefully. Make sure you understand what is given and what is unknown.
2. If possible, sketch a diagram. Label the various parts.
3. Decide on the variable that must be maximized or minimized. Express that variable as a function of *one* other variable. Be sure to find the domain of the function.
4. Find the critical points for the function from Step 3.
5. If the domain has endpoints, evaluate the function at the endpoints and the critical points to see which yields the maximum or minimum. If there are no endpoints, test each critical point for relative maxima or minima using the first or second derivative test.

Caution Do not skip Step 5 in the box above. If a problem asks you to maximize a quantity and you find a critical point at Step 4, do not automatically assume the maximum occurs there, for it may occur at an endpoint or may not exist at all.

An infamous case of such an error occurred in a 1945 study of "flying wing" aircraft designs similar to the Stealth bomber. In seeking to maximize the range of the aircraft (how far it can fly on a tank of fuel), the study's authors found that a critical point occurred when almost all of the volume of the plane was in the wing. They claimed that this critical point was a maximum. But another engineer later found that this critical point, in fact, *minimized* the range of the aircraft!*

EXAMPLE 1 Find two nonnegative numbers x and y for which $2x + y = 30$, such that xy^2 is maximized.

First we must decide what is to be maximized and give it a variable. Here, xy^2 is to be maximized, so let

$$M = xy^2.$$

Now, express M in terms of just *one* variable. Use the equation $2x + y = 30$ to do that. Solve $2x + y = 30$ for either x or y. Solving for y gives

$$\begin{aligned} 2x + y &= 30 \\ y &= 30 - 2x. \end{aligned}$$

Substitute for y in the expression for M to get

$$\begin{aligned} M &= x(30 - 2x)^2 \\ &= x(900 - 120x + 4x^2) \\ &= 900x - 120x^2 + 4x^3. \end{aligned}$$

Note that x must be at least 0. Since y must also be at least 0, $30 - 2x \geq 0$, so $x \leq 15$. Thus x is confined to the interval $[0, 15]$.

Find the critical points for M by finding M', then solving the equation $M' = 0$ for x.

$$\begin{aligned} M' = 900 - 240x + 12x^2 &= 0 \\ 12(75 - 20x + x^2) &= 0 \\ (5 - x)(15 - x) &= 0 \end{aligned}$$

$$x = 5 \quad \text{or} \quad x = 15$$

Find M for the critical numbers $x = 5$ and $x = 15$ as well as for $x = 0$, one endpoint of the domain. (The other endpoint has already been included as a critical number.)

x	M
0	0
5	2000 ← Maximum
15	0

*Wayne Biddle, "Skeleton Alleged in the Stealth Bomber's Closet," *Science*, vol. 244, May 12, 1989.

We see in the chart that the maximum value of the function occurs when $x = 5$. Since $y = 30 - 2(5) = 20$, the values that maximize xy^2 are $x = 5$ and $y = 20$. ◀

We will now return to a problem we first encountered in Chapter 1.

EXAMPLE 2 When Power and Money, Inc., charges $600 for a seminar on management techniques, it attracts 1000 people. For each $20 decrease in the charge, an additional 100 people will attend the seminar.

(a) Find an expression for the total revenue if there are x $20 decreases in price.

The price charged will be

$$\text{Price per person} = 600 - 20x,$$

and the number of people in the seminar will be

$$\text{Number of people} = 1000 + 100x.$$

The total revenue, $R(x)$, is given by the product of the price and the number of people attending, or

$$\begin{aligned} R(x) &= (600 - 20x)(1000 + 100x) \\ &= 600{,}000 + 40{,}000x - 2000x^2. \end{aligned}$$

Notice that x must be at least 0. Furthermore, since we cannot charge a negative amount to attend the seminar, $600 - 20x \geq 0$, so $x \leq 30$. Thus the domain of the function R is $[0, 30]$.

(b) Find the value of x that maximizes revenue.

Here $R'(x) = 40{,}000 - 4000x$. Set this derivative equal to 0.

$$\begin{aligned} 40{,}000 - 4000x &= 0 \\ -4000x &= -40{,}000 \\ x &= 10 \end{aligned}$$

Find $R(x)$ for x equal to 0, 10, and 30 to find which yields the maximum $R(x)$.

x	$R(x)$	
0	600,000	
10	800,000	← Maximum
30	0	

From the table we see that $R(x)$ is maximized when $x = 10$.

(c) Find the amount that Power and Money, Inc., should charge for the seminar to maximize the revenue, and find the maximum revenue.

The price per person to maximize the revenue is $600 - 20(10) = \$400$. From the table in part (b), the revenue is then $R(10) = \$800{,}000$. ◀

EXAMPLE 3

An open box is to be made by cutting a square from each corner of a 12-inch by 12-inch piece of metal and then folding up the sides. What size square should be cut from each corner in order to produce a box of maximum volume?

Let x represent the length of a side of the square that is cut from each corner, as shown in Figure 1(a). The width of the box is $12 - 2x$, with the length also $12 - 2x$. As shown in Figure 1(b), the depth of the box will be x inches. The volume of the box is given by the product of the length, width, and height. In this example, the volume, $V(x)$, depends on x:

$$V(x) = x(12 - 2x)(12 - 2x) = 144x - 48x^2 + 4x^3.$$

Clearly, $0 \leq x$, and since neither the length nor the width can be negative, $0 \leq 12 - 2x$, so $x \leq 6$. Thus the domain of V is the interval $[0, 6]$.

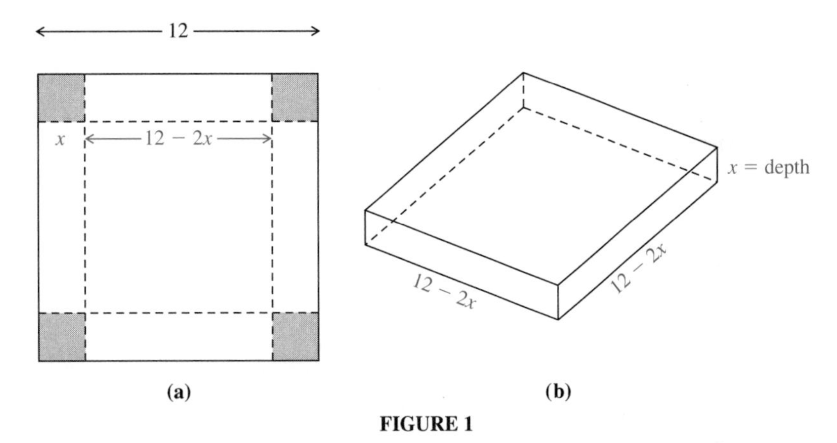

FIGURE 1

The derivative is $V'(x) = 144 - 96x + 12x^2$. Set this derivative equal to 0.

$$12x^2 - 96x + 144 = 0$$
$$12(x^2 - 8x + 12) = 0$$
$$12(x - 2)(x - 6) = 0$$
$$x - 2 = 0 \quad \text{or} \quad x - 6 = 0$$
$$x = 2 \qquad\qquad x = 6$$

Find $V(x)$ for x equal to 0, 2, and 6 to find the depth that will maximize the volume.

x	$V(x)$
0	0
2	128 ← Maximum
6	0

The chart indicates that the box will have maximum volume when $x = 2$ and that the maximum volume will be 128 cubic inches. ◀

EXAMPLE 4

A company wants to manufacture cylindrical aluminum cans with a volume of 1000 cubic centimeters (one liter). What should the radius and height of the can be to minimize the amount of aluminum used?

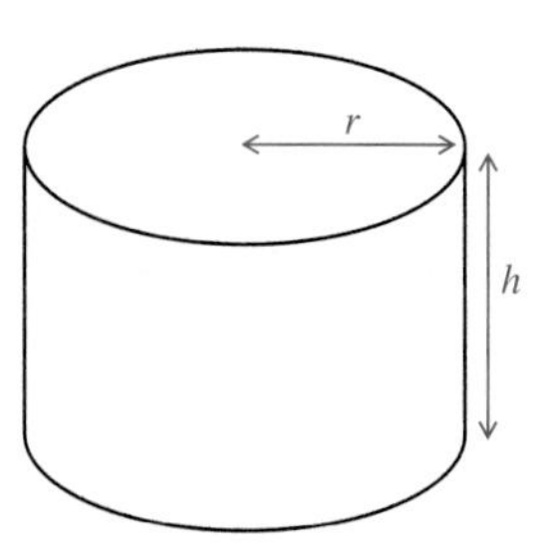

FIGURE 2

The two variables in this problem are the radius and the height of the can, which we shall label r and h, as in Figure 2. Minimizing the amount of aluminum used requires minimizing the surface area of the can, which we will designate S. The surface area consists of a top and a bottom, each of which is a circle with an area πr^2, plus the side. If the side was sliced vertically and unrolled, it would form a rectangle with height h and width equal to the circumference of the can, which is $2\pi r$. Thus the surface area is given by

$$S = 2\pi r^2 + 2\pi rh.$$

The right side of the equation involves two variables. We need to get a function of a single variable. We can do this by using the information about the volume of the can:

$$V = \pi r^2 h = 1000.$$

(Here we have used the formula for the volume of a cylinder.) Solve this for h:

$$h = \frac{1000}{\pi r^2}.$$

(Solving for r would have involved a square root and a more complicated function.)

We now substitute this expression for h into the equation for S to get

$$S = 2\pi r^2 + 2\pi r \frac{1000}{\pi r^2} = 2\pi r^2 + \frac{2000}{r}.$$

There are no restrictions on r other than that it be a positive number, so the domain of S is $(0, \infty)$.

Find the critical points for S by finding S', then solving the equation $S' = 0$ for r.

$$S' = 4\pi r - \frac{2000}{r^2} = 0$$

$$4\pi r^3 = 2000$$

$$r^3 = \frac{500}{\pi}$$

Take the cube root of both sides to get

$$r = \left(\frac{500}{\pi}\right)^{1/3} \approx 5.419$$

centimeters. Substitute this expression into the equation for h to get

$$h = \frac{1000}{\pi 5.419^2} \approx 10.84$$

centimeters. Notice that the height of the can is twice its radius. But we are not yet assured that the critical number we have found is indeed the minimum. Since the domain has no endpoints, we will instead use the second derivative test:

$$S'' = 4\pi + \frac{4000}{r^3}.$$

S'' is positive for any positive value of r, so by the second derivative test, we have found the minimum surface area. ◀

For most living things, reproduction is *seasonal*—it can take place only at selected times of the year.* Large whales, for example, reproduce every two years during a relatively short time span of about two months. Shown on the time axis in Figure 3 are the reproductive periods. Let S = number of adults present during the reproductive period and let R = number of adults that return the next season to reproduce.

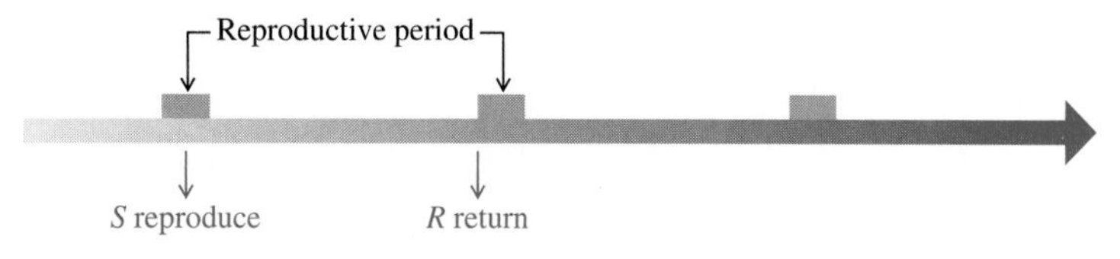

FIGURE 3

If we find a relationship between R and S, $R = f(S)$, then we have formed a **spawner-recruit** function or **parent-progeny** function. These functions are notoriously hard to develop because of the difficulty of obtaining accurate counts and because of the many hypotheses that can be made about the life stages. We will simply suppose that the function f takes various forms.

If $R > S$, we can presumably harvest

$$H = R - S = f(S) - S$$

individuals, leaving S to reproduce. Next season, $R = f(S)$ will return and the harvesting process can be repeated, as shown in Figure 4.

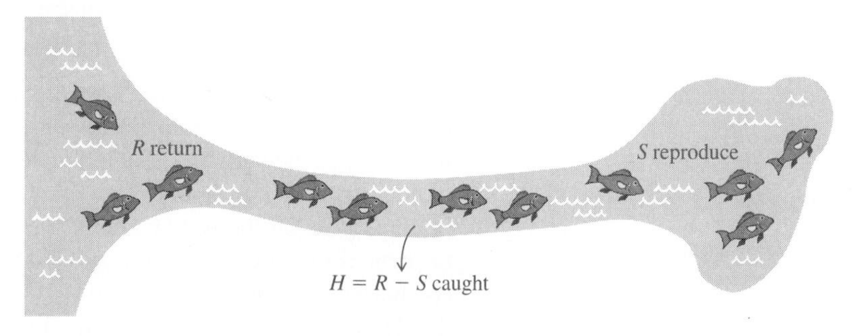

FIGURE 4

---

*From *Mathematics for the Biosciences* by Michael R. Cullen. Copyright © 1983 PWS Publishers. Reprinted by permission.

Let S_0 be the number of spawners that will allow as large a harvest as possible without threatening the population with extinction. Then $H(S_0)$ is called the **maximum sustainable harvest.**

EXAMPLE 5 Suppose the spawner-recruit function for Idaho rabbits is $f(S) = -.025S^2 + 4S$, where S is measured in thousands of rabbits. Find S_0 and the maximum sustainable harvest, $H(S_0)$.

S_0 is the value of S that maximizes H. Since

$$\begin{aligned} H(S) &= f(S) - S \\ &= -.025S^2 + 4S - S \\ &= -.025S^2 + 3S, \end{aligned}$$

the derivative $H'(S) = -.05S + 3$. Set this derivative equal to 0 and solve for S.

$$\begin{aligned} 0 &= -.05S + 3 \\ .05S &= 3 \\ S &= 60 \end{aligned}$$

The number of rabbits needed to sustain the population is $S_0 = 60$ thousand. Note that

$$H''(S) = -.05 < 0,$$

showing that $S = 60$ does indeed produce a *maximum* value of H. At $S_0 = 60$, the harvest is

$$H(60) = -.025(60)^2 + 3(60) = 90,$$

or 90 thousand rabbits. These results indicate that after one reproductive season, a population of 60 thousand rabbits will have increased to 150 thousand. Of these, 90 thousand may be harvested, leaving 60 thousand to regenerate the population. Any harvest larger than 90 thousand will threaten the future of the rabbit population, while a harvest smaller than 90 thousand will allow the population to grow larger each season. Thus, 90 thousand is the maximum sustainable harvest for this population. ◀

Finding extrema for realistic problems requires an accurate mathematical model of the problem. In particular, it is important to be aware of restrictions on the values of the variables. For example, if $T(x)$ closely approximates the number of items that can be manufactured daily on a production line when x is the number of employees on the line, x must certainly be restricted to the positive integers, or perhaps to a few common fractional values. (We can imagine half-time workers, but not $1/49$-time workers.)

On the other hand, to apply the tools of calculus to obtain an extremum for some function, the function must be defined and be meaningful at every real number in some interval. Because of this, the answer obtained from a mathematical model might be a number that is not feasible in the actual problem.

Usually, the requirement that a continuous function be used, rather than one that can take on only certain selected values, is of theoretical interest only. In most cases, the methods of calculus give acceptable results so long as the assumptions of continuity and differentiability are not totally unreasonable. If they lead to the conclusion, say, that $80\sqrt{2}$ workers should be hired, it is usually only necessary to investigate acceptable values close to $80\sqrt{2}$. This was done in Example 3 in the preceding chapter, in the section on Relative Extrema.

When the function to be maximized or minimized is a quadratic function, as in Example 2 or Example 5, we can solve the problem by finding the vertex of the parabola, as we did in Chapter 1. For other problems that would otherwise be very difficult, the calculus techniques of this chapter provide a powerful method of solution.

4.1 Exercises

Applications

General Interest

Number Analysis *Exercises 1–8 involve maximizing and minimizing products and sums of numbers. Solve each problem using steps such as those shown in Exercise 1.*

1. Find two nonnegative numbers x and y such that $x + y = 100$ and the product $P = xy$ is as large as possible, and give the maximum product. Use the following steps.

(a) Solve $x + y = 100$ for y.

(b) Substitute the result from part (a) into $P = xy$, the equation for the variable that is to be maximized.

(c) Find the domain of the function P found in part (b).

(d) Find P'. Solve the equation $P' = 0$.

(e) Evaluate P at any solutions found in part (d), as well as at the endpoints of the domain found in part (c).

(f) Give the maximum value of P, as well as the two numbers x and y whose product is that value.

2. Find two nonnegative numbers x and y whose sum is 250 and whose product is as large as possible, and give the maximum product.

3. Find two nonnegative numbers x and y whose sum is 200, such that the sum of the squares of the two numbers is minimized, and give the minimum sum.

4. Find two nonnegative numbers x and y whose sum is 30, such that the sum of the squares of the two numbers is minimized, and give the minimum sum.

In Exercises 5–8, use steps (a)–(f) from Exercise 1 to find nonnegative numbers x and y satisfying the given requirements and to find the maximum or minimum value of the indicated expression.

5. $x + y = 150$ and x^2y is maximized.

6. $x + y = 45$ and xy^2 is maximized.

7. $x - y = 10$ and xy is minimized.

8. $x - y = 3$ and xy is minimized.

9. **Area** A farmer has 1200 m of fencing. He wants to enclose a rectangular field bordering a river, with no fencing needed along the river. (See the sketch.) Let x represent the width of the field.

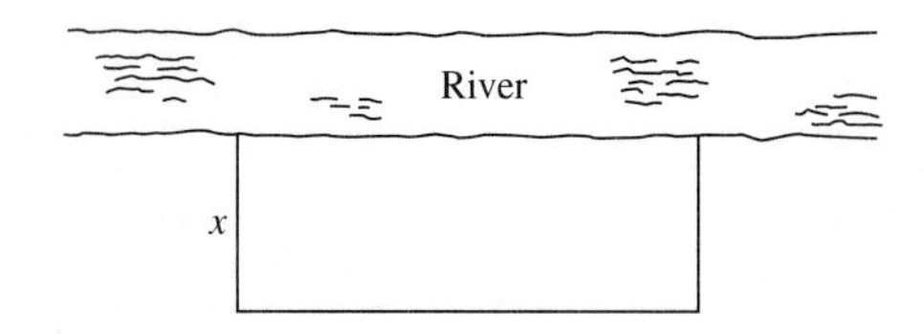

(a) Write an expression for the length of the field.

(b) Find the area of the field (area = length × width).

(c) Find the value of x leading to the maximum area.

(d) Find the maximum area.

10. **Area** Find the dimensions of the rectangular field of maximum area that can be made from 200 m of fencing material. (This fence has four sides.)

11. **Area** An ecologist is conducting a research project on breeding pheasants in captivity. She first must construct suitable pens. She wants a rectangular area with two additional fences across its width, as shown in the sketch. Find the maximum area she can enclose with 3600 m of fencing.

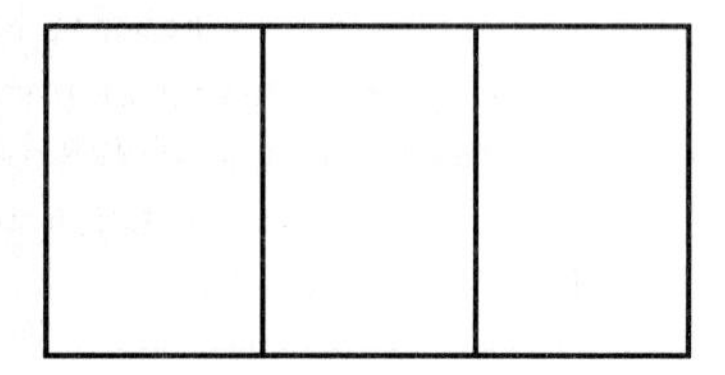

12. **Travel Time** A hunter is at a point on a river bank. He wants to get to his cabin, located 3 mi north and 8 mi west. (See the figure.) He can travel 5 mph on the river but only 2 mph on this very rocky land. How far upriver should he go in order to reach the cabin in minimum time? (*Hint:* distance = rate × time.)

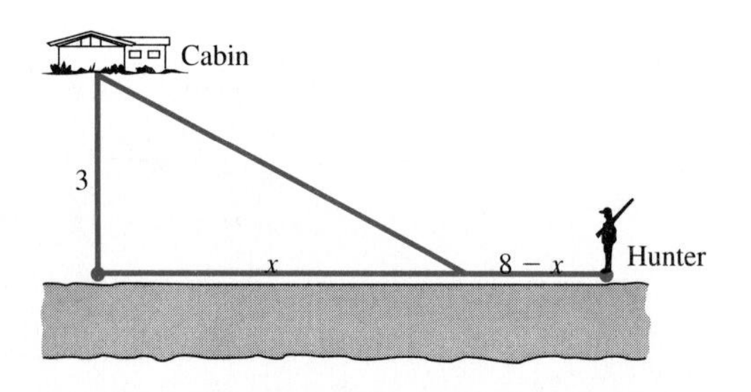

13. **Travel Time** Repeat Exercise 12, but assume the cabin is 19 mi north and 8 mi west.

14. **Postal Regulations** The U.S. Postal Service stipulates that any boxes sent through the mail must have a length plus girth totaling no more than 108 inches. (See the figure.) Find the dimensions of the box with maximum volume that can be sent through the U.S. mail, assuming that the width and the height of the box are equal.

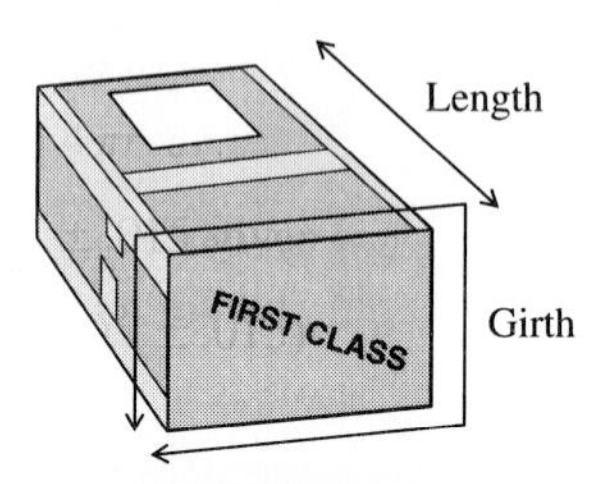

Business and Economics

15. **Revenue** If the price charged for a candy bar is $p(x)$ cents, then x thousand candy bars will be sold in a certain city, where

$$p(x) = 100 - \frac{x}{10}.$$

(a) Find an expression for the total revenue from the sale of x thousand candy bars.

(b) Find the value of x that leads to maximum revenue.

(c) Find the maximum revenue.

16. **Revenue** The sale of cassette tapes of "lesser" performers is very sensitive to price. If a tape manufacturer charges $p(x)$ dollars per tape, where

$$p(x) = 6 - \frac{x}{8},$$

then x thousand tapes will be sold.

(a) Find an expression for the total revenue from the sale of x thousand tapes.

(b) Find the value of x that leads to maximum revenue.

(c) Find the maximum revenue.

17. **Cost** A truck burns fuel at the rate of

$$G(x) = \frac{1}{32}\left(\frac{64}{x} + \frac{x}{50}\right)$$

gallons per mile while traveling at x mph.

(a) If fuel costs \$1.60 per gallon, find the speed that will produce minimum total cost for a 400-mile trip.

(b) Find the minimum total cost.

18. **Cost** A rock-and-roll band travels from engagement to engagement in a large bus. This bus burns fuel at the rate of

$$G(x) = \frac{1}{50}\left(\frac{200}{x} + \frac{x}{15}\right)$$

gallons per mile while traveling at x mph.

(a) If fuel costs \$2 per gallon, find the speed that will produce minimum total cost for a 250-mile trip.

(b) Find the minimum total cost.

19. **Cost with Fixed Area** A fence must be built to enclose a rectangular area of 20,000 ft^2. Fencing material costs \$3 per foot for the two sides facing north and south, and \$6 per foot for the other two sides. Find the cost of the least expensive fence.

20. **Cost with Fixed Area** A fence must be built in a large field to enclose a rectangular area of 15,625 m^2. One side of the area is bounded by an existing fence; no fence is needed there. Material for the fence costs \$2 per meter for the two ends, and \$4 per meter for the side opposite the existing fence. Find the cost of the least expensive fence.

21. **Profit** In planning a small restaurant, it is estimated that a profit of \$5 per seat will be made if the number of seats is between 60 and 80, inclusive. On the other hand, the profit on each seat will decrease by 5¢ for each seat above 80.

(a) Find the number of seats that will produce the maximum profit.

(b) What is the maximum profit?

22. **Timing Income** A local group of scouts has been collecting old aluminum cans for recycling. The group has already collected 12,000 lb of cans, for which they could currently receive \$4 per hundred pounds. The group can continue to collect cans at the rate of 400 lb per day. However, a glut in the old-can market has caused the recycling company to announce that it will lower its price, starting immediately, by \$.10 per hundred pounds per day. The scouts can make only one trip to the recycling center. Find the best time for the trip. What total income will be received?

23. **Packaging Design** A television manufacturing firm needs to design an open-topped box with a square base. The box must hold 32 in^3. Find the dimensions of the box that can be built with the minimum amount of materials. (See the figure.)

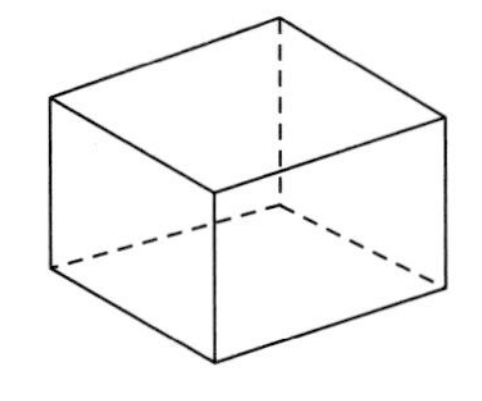

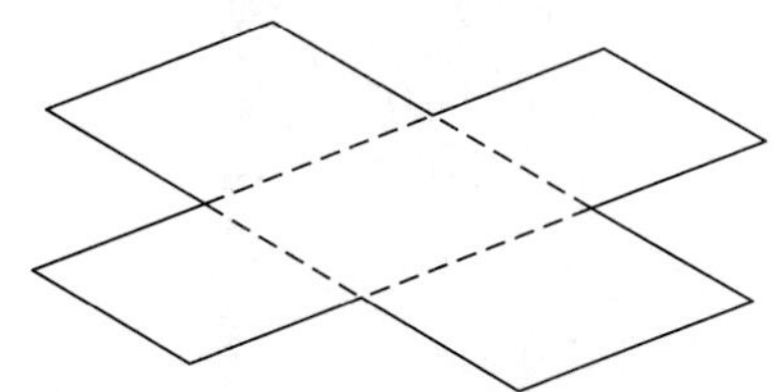

24. **Revenue** A local club is arranging a charter flight to Hawaii. The cost of the trip is \$425 each for 75 passengers, with a refund of \$5 per passenger for each passenger in excess of 75.

(a) Find the number of passengers that will maximize the revenue received from the flight.

(b) Find the maximum revenue.

25. **Packaging Design** A company wishes to manufacture a box with a volume of 36 ft^3 that is open on top and is twice as long as it is wide. Find the dimensions of the box produced from the minimum amount of material.

26. **Packaging Cost** A closed box with a square base is to have a volume of 16,000 cm^3. The material for the top and bottom of the box costs \$3 per square centimeter, while the material for the sides costs \$1.50 per square centimeter. Find the dimensions of the box that will lead to minimum total cost. What is the minimum total cost?

27. **Packaging Design** A cylindrical box will be tied up with ribbon as shown in the figure. The longest piece of ribbon available is 130 cm long, and 10 cm of that are required for the bow. Find the radius and height of the box with the largest possible volume.

28. **Can Design** **(a)** For the can problem in Example 4, the minimum surface area required that the height be twice the radius. Show that this is true for a can of arbitrary volume V.

(b) Do many cans in grocery stores have a height that is twice the radius? If not, discuss why this may be so.

29. **Container Design** Your company needs to design cylindrical metal containers with a volume of 16 ft^3. The top and bottom will be made of a sturdy material that costs \$2 per square foot, while the material for the sides costs \$1 per square foot. Find the radius, height, and cost of the least expensive container.

30. **Container Design** An open box will be made by cutting a square from each corner of a 3-ft by 8-ft piece of cardboard and then folding up the sides. What size square should be cut from each corner in order to produce a box of maximum volume?

31. **Use of Materials** A mathematics book is to contain 36 in^2 of printed matter per page, with margins of 1 in. along the sides and $1\frac{1}{2}$ in. along the top and bottom. Find the dimensions of the page that will require the minimum amount of paper. (See the figure.)

1″ 1″
1½″
36 sq. in.
1½″

32. **Cost** A company wishes to run a utility cable from point A on the shore (see the figure) to an installation at point B on the island. The island is 6 mi from the shore. It costs \$400 per mile to run the cable on land and \$500 per mile underwater. Assume that the cable starts at A and runs along the shoreline, then angles and runs underwater to the island. Find the point at which the line should begin to angle in order to yield the minimum total cost.

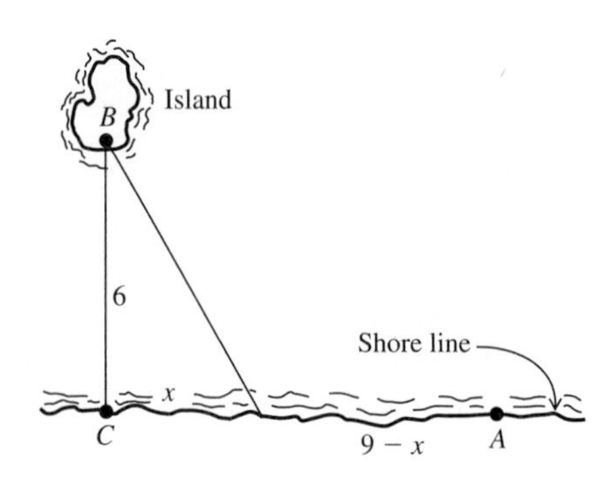

33. **Cost** Repeat Exercise 32, but make point A 7 mi from point C.

34. **Pricing** Decide what you would do if your assistant presented the following contract for your signature:

> Your firm offers to deliver 300 tables to a dealer, at \$90 per table, and to reduce the price per table on the entire order by 25¢ for each additional table over 300.

Find the dollar total involved in the largest possible transaction between the manufacturer and the dealer; then find the smallest possible dollar amount.

Life Sciences

Maximum Sustainable Harvest *Find the maximum sustainable harvest in Exercises 35–40. See Example 5.*

35. $f(S) = -.1S^2 + 11S$

36. $f(S) = -S^2 + 2.2S$

37. $f(S) = 15\sqrt{S}$

38. $f(S) = 12S^{.25}$

39. $f(S) = \dfrac{25S}{S + 2}$

40. $f(S) = .999S$

41. **Pigeon Flight** Homing pigeons avoid flying over large bodies of water, preferring to fly around them instead. (One possible explanation is the fact that extra energy is required to fly over water because air pressure drops over water in the daytime.) Assume that a pigeon released from a boat 1 mi from the shore of a lake (point B in the figure) flies first to point P on the shore and then along the straight edge of the lake to reach its home at L. If L is 2 mi from point A, the point on the shore closest to the boat, and if a pigeon needs $4/3$ as much energy to fly over water as over land, find the location of point P.

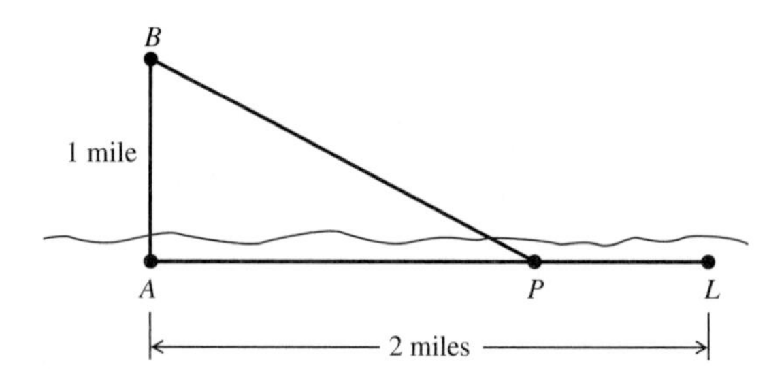

42. Pigeon Flight Repeat Exercise 41, but assume a pigeon needs 10/9 as much energy to fly over water as over land.

43. Salmon Spawning When salmon struggle upstream to their spawning grounds, it is essential that they conserve energy, for they no longer feed once they have left the ocean. Let v_0 = speed (in miles per hour) of the current, d = distance the salmon must travel, and v = speed of the salmon. Hence $(v - v_0)$ is the net velocity of the salmon upstream. (See the figure.)*

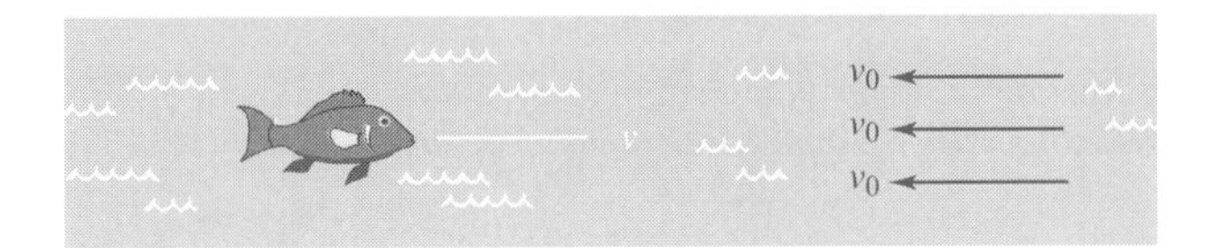

(a) Show that if the journey takes t hr, then

$$t = \frac{d}{v - v_0}.$$

We next will assume that the amount of energy expended per hour when the speed of the salmon is v is directly proportional to v^{α} for some $\alpha > 1$. (Empirical data suggest this.)

(b) Show that the total energy T expended over the journey is given by

$$T = k\frac{v^{\alpha}}{v - v_0} \quad \text{for } v > v_0.$$

(c) Show that T is minimized by selecting velocity

$$v = \frac{\alpha v_0}{\alpha - 1}.$$

(d) If $v_0 = 2$ mph and the salmon make the 20-mi journey by swimming for 40 hr, estimate α. What must you assume in order to do the computations?

44. Lung Function It is well known that during coughing, the diameters of the trachea and bronchi decrease. Let r_0 be the normal radius of an airway at atmospheric pressure P_0. For the trachea, $r_0 \approx .5$ inch. The glottis is the opening at the entrance of the trachea through which air must pass as it either enters or leaves the lungs, as depicted in the illustration.† Assume that, after a deep inspiration of air, the glottis is closed. Then pressure develops in the airways and the radius of the airway decreases. We will assume a simple linear relation between r and P:

$$r - r_0 = a(P - P_0)$$

$$\text{or} \quad \Delta r = a\Delta P,$$

where $a < 0$ and $r_0/2 \le r \le r_0$.

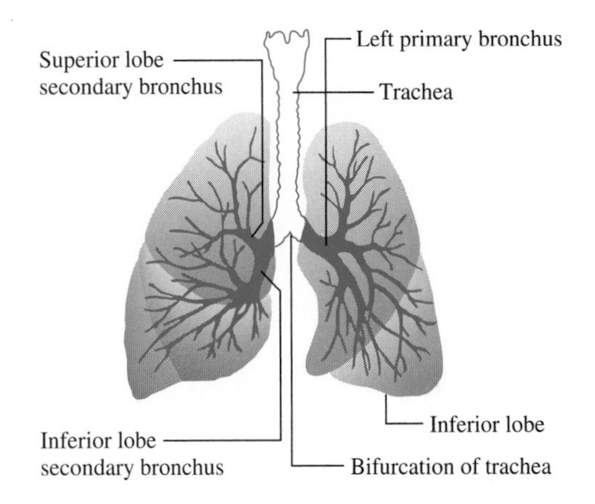

When the glottis is opened, how does the air flow through these passages? We will assume that the flow is governed by **Poiseuille's laws:**

$$\textbf{(i)}\ v = \frac{P - P_0}{k}(r^2 - x^2), \quad 0 \le x \le r;$$

$$\textbf{(ii)}\ F = \frac{dV}{dt} = \frac{\pi(P - P_0)}{2k}r^4.$$

Here, v is the velocity at a distance x from the center of the airway (in cm/sec, e.g.) and F is the flow rate (in cm^3/sec, e.g.). The average velocity $\bar{v}$ over the circular cross section is given by

$$\textbf{(iii)}\ \bar{v} = \frac{F}{(\pi r^2)} = \frac{P - P_0}{2k}r^2.$$

(a) Write the flow rate as a function of r only. Find the value of r that maximizes the rate of flow.

(b) Write both the average velocity $\bar{v}$ and $v(0)$, the velocity in the center of the airway, as functions of r alone. Find the value of r that maximizes these two velocities.

---

*Exercises 43–44 from *Mathematics for the Biosciences* by Michael R. Cullen. Copyright © 1983 PWS Publishers. Reprinted by permission.

†The figure for Exercise 44 was adapted from Barbara R. Landau, *Essential Anatomy and Physiology*, 2nd ed. Glenview, Ill.: Scott, Foresman and Co., Copyright 1980.

Computer/Graphing Calculator

In Exercises 45–47, use a graphing program or an equation solver to determine where the derivative is equal to zero.

45. Can Design Modify the can problem in Example 4 so the cost must be minimized. Assume that aluminum costs 3¢ per square centimeter, and that there is an additional cost of 2¢ per centimeter times the perimeter of the top, and a similar cost for the bottom, to seal the top and bottom of the can to the side.

46. Can Design In this modification of the can problem in Example 4, the cost must be minimized. Assume that aluminum costs 3¢ per square centimeter, and that there is an additional cost of 1¢ per centimeter times the height of the can to make a vertical seam on the side.

47. Can Design This problem is a combination of Exercises 45 and 46. We will again minimize the cost of the can, assuming that aluminum costs 3¢ per square centimeter. In addition, there is a cost of 2¢ per centimeter to seal the top and bottom of the can to the side, plus 1¢ per centimeter to make a vertical seam.

Average Cost *For each of the following cost functions, calculate the average cost function and find where the average cost is smallest. One way to do this is to compute the derivative, then use a graph of the function or an equation solver to find where the derivative is equal to zero.*

48. $C(x) = \frac{1}{2}x^3 + 2x^2 - 3x + 35$

49. $C(x) = .01x^3 + .05x^2 + .2x + 28$

50. $C(x) = 10 + 20x^{1/2} + 16x^{3/2}$

51. $C(x) = 30 + 42x^{1/2} + .2x^{3/2} + .03x^{5/2}$

4.2 FURTHER BUSINESS APPLICATIONS: ECONOMIC LOT SIZE; ECONOMIC ORDER QUANTITY; ELASTICITY OF DEMAND

How many batches of primer should a paint company produce per year to minimize its costs while meeting its customers' demand?

We will answer this question in Example 1 using the concept of *economic lot size*.

In this section we introduce three common business applications of calculus. The first two, *economic lot size* and *economic order quantity*, are related. A manufacturer must determine the production lot (or batch) size that will result in minimum production and storage costs, while a purchaser must decide what quantity of an item to order in an effort to minimize reordering and storage costs. The third application, *elasticity of demand*, deals with the sensitivity of demand for a product to changes in the price of the product.

Economic Lot Size Suppose that a company manufactures a constant number of units of a product per year and that the product can be manufactured in several batches of equal size during the year. If the company were to manufacture the item only once per year, it would minimize setup costs but incur high warehouse costs. On the other hand, if it were to make many small batches, this would increase setup costs. Calculus can be used to find the number of batches per year that should be manufactured in order to minimize total cost. This number is called the **economic lot size.**

Figure 5 shows several of the possibilities for a product having an annual demand of 12,000 units. The top graph shows the results if only one batch of the product is made annually: in this case an average of 6000 items will be held in a warehouse. If four batches (of 3000 each) are made at equal time intervals during a year, the average number of units in the warehouse falls to only 1500. If twelve batches are made, an average of 500 items will be in the warehouse.

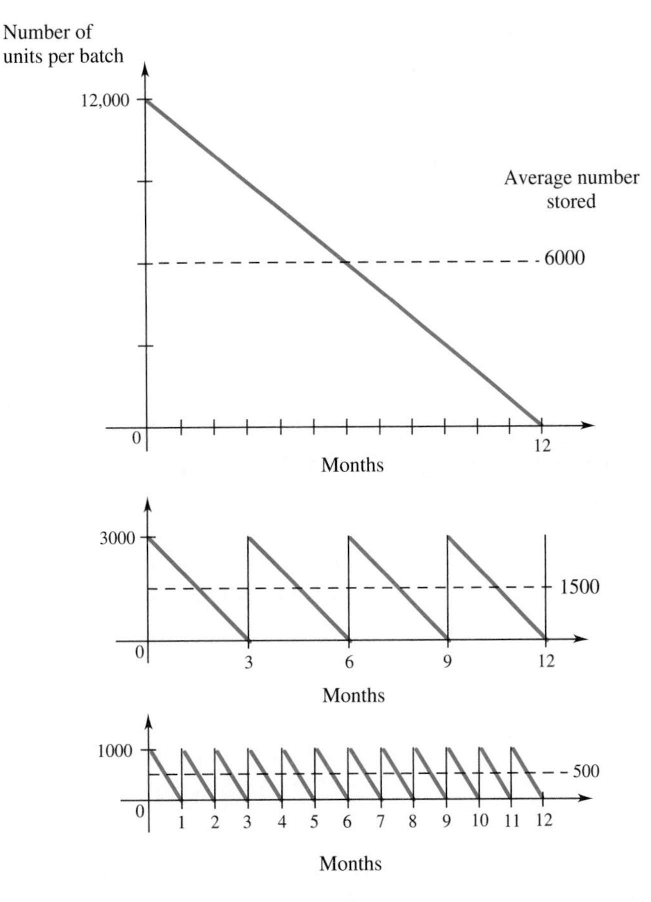

FIGURE 5

The variable in our discussion of economic lot size will be

x = number of batches to be manufactured annually.

In addition, we have the following constants:

k = cost of storing one unit of the product for one year;

f = fixed setup cost to manufacture the product;

g = cost of manufacturing a single unit of the product;

M = total number of units produced annually.

The company has two types of costs associated with the production of its product: a cost associated with manufacturing the item and a cost associated with storing the finished product. During a year the company will produce x batches of the product, with M/x units of the product produced per batch. Each batch has a fixed cost f and a variable cost g per unit, so that the manufacturing cost per batch is

$$f + \frac{gM}{x}.$$

There are x batches per year, so the total annual manufacturing cost is

$$\left(f + \frac{gM}{x}\right)x. \tag{1}$$

Since each batch consists of M/x units, and since demand is constant, it is common to assume an average inventory of

$$\frac{1}{2}\left(\frac{M}{x}\right) = \frac{M}{2x}$$

units per year. The cost of storing one unit of the product for a year is k, so the total storage cost is

$$k\left(\frac{M}{2x}\right) = \frac{kM}{2x}. \tag{2}$$

The total production cost is the sum of the manufacturing and storage costs, or the sum of expressions (1) and (2). If $T(x)$ is the total cost of producing x batches,

$$T(x) = \left(f + \frac{gM}{x}\right)x + \frac{kM}{2x} = fx + gM + \frac{kM}{2x}.$$

Since the only constraint on x is that it be a positive number, the domain of $T(x)$ is $(0, \infty)$. To find the value of x that will minimize $T(x)$, remember that f, g, k, and M are constants and find $T'(x)$.

$$T'(x) = f - \frac{kM}{2}x^{-2}$$

Set this derivative equal to 0.

$$\begin{aligned} f - \frac{kM}{2}x^{-2} &= 0 \\ f &= \frac{kM}{2x^2} \\ 2fx^2 &= kM \\ x^2 &= \frac{kM}{2f} \\ x &= \sqrt{\frac{kM}{2f}} \end{aligned} \tag{3}$$

The second derivative test can be used to show that $\sqrt{kM/(2f)}$ is the annual number of batches that minimizes total production costs. (See Exercise 1.)

This application is referred to as the *inventory problem* and is treated in more detail in management science courses. Please note that equation (3) was derived under very specific assumptions. If the assumptions are changed slightly, a different conclusion might be reached, and it would not necessarily be valid to use equation (3).

In some examples equation (3) may not give an integer value. In this text we will require the number of batches manufactured in a year to be an integer, which is reasonable in many situations. If equation (3) does not give an integer, we must investigate the next integer smaller than x and the next integer larger to see which gives the minimum cost. This same assumption will be held in the *economic order quantity problem*, which will be discussed after the following example.

EXAMPLE 1 A paint company has a steady annual demand for 24,500 cans of automobile primer. The cost accountant for the company says that it costs \$2 to hold one can of paint for one year and \$500 to set up the plant for the production of the primer. Find the number of batches of primer that should be produced in order to minimize total production costs.

Use equation (3) above, with $k = 2$, $M = 24{,}500$, and $f = 500$.

$$x = \sqrt{\frac{kM}{2f}} = \sqrt{\frac{2(24{,}500)}{2(500)}} = \sqrt{49} = 7$$

Seven batches of primer per year will lead to minimum production costs. Each batch will consist of $M/x = 24{,}500/7 = 3500$ cans of primer. ◀

Economic Order Quantity We can extend our previous discussion to the problem of reordering an item that is used at a constant rate throughout the year. Here, the company using a product must decide how often to order and how many units to request each time an order is placed; that is, it must identify the **economic order quantity.** In this case, the variable is

$$x = \text{number of orders to be placed annually.}$$

We also have the following constants.

$$k = \text{cost of storing one unit for one year}$$
$$f = \text{fixed cost to place an order}$$
$$M = \text{total units needed per year}$$

The goal is to minimize the total cost of ordering over a year's time, where

$$\text{Total cost} = \text{Storage cost} + \text{Reorder cost.}$$

Again assume an average inventory of $M/2x$, so the yearly storage cost is

$$\frac{kM}{2x}.$$

The reorder cost is the product of the number of batches ordered each year, x, and the cost per order, f. Thus, the reorder cost is fx, and the total cost is

$$T(x) = fx + \frac{kM}{2x}.$$

This is almost the same formula we derived for the inventory problem, which also had a constant term gM. Since a constant does not affect the derivative, equation (3) is also valid for the economic order quantity problem. As before, the number of units per order is M/x. This illustrates how two different applications might have the same mathematical structure, so a solution to one applies to both.

EXAMPLE 2 A large pharmacy has an annual need for 200 units of a certain antibiotic. It costs $10 to store one unit for one year. The fixed cost of placing an order (clerical time, mailing, and so on) amounts to $40. Find the number of times a year the antibiotic should be ordered, and the number of units to order each time.

Here $k = 10$, $M = 200$, and $f = 40$. We have

$$x = \sqrt{\frac{10(200)}{2(40)}} = \sqrt{25} = 5.$$

The drug should be ordered 5 times a year. Each time,

$$\frac{M}{x} = \frac{200}{5} = 40$$

units should be ordered. ◀

Elasticity of Demand Anyone who sells a product or service is concerned with how a change in price affects demand. The sensitivity of demand to price changes varies with different items. For items such as soft drinks, pepper, and light bulbs, relatively small percentage changes in price will not change the demand for the item much. For cars, home loans, furniture, and computer equipment, however, small percentage changes in price have significant effects on demand.

One way to measure the sensitivity of demand to changes in price is by the ratio of percent change in demand to percent change in price. If q represents the quantity demanded and p the price, this ratio can be written as

Recall from Chapter 1 that the Greek letter Δ, pronounced *delta*, is used in mathematics to mean "change."

$$\frac{\Delta q/q}{\Delta p/p},$$

where Δq represents the change in q and Δp represents the change in p. This ratio is always negative, because q and p are positive, while Δq and Δp have opposite signs. (An *increase* in price causes a *decrease* in demand.) If the absolute value of this ratio is large, it suggests that a relatively small increase in price causes a relatively large drop (decrease) in demand.

This ratio can be rewritten as

$$\frac{\Delta q/q}{\Delta p/p} = \frac{\Delta q}{q} \cdot \frac{p}{\Delta p} = \frac{p}{q} \cdot \frac{\Delta q}{\Delta p}.$$

Suppose $q = f(p)$. (Note that this is the inverse of the way our demand functions have been expressed so far; previously we had $p = D(q)$.) Then $\Delta q = f(p + \Delta p) - f(p)$, and

$$\frac{\Delta q}{\Delta p} = \frac{f(p + \Delta p) - f(p)}{\Delta p}.$$

As $\Delta p \to 0$, this quotient becomes

$$\lim_{\Delta p \to 0} \frac{\Delta q}{\Delta p} = \lim_{\Delta p \to 0} \frac{f(p + \Delta p) - f(p)}{\Delta p} = \frac{dq}{dp},$$

and

$$\lim_{\Delta p \to 0} \frac{p}{q} \cdot \frac{\Delta q}{\Delta p} = \frac{p}{q} \cdot \frac{dq}{dp}.$$

The quantity

$$E = -\frac{p}{q} \cdot \frac{dq}{dp}$$

is positive because dq/dp is negative. E is called the **elasticity of demand** and measures the instantaneous responsiveness of demand to price. For example, E may be 0.2 for medical services, but may be 1.2 for stereo equipment. The demand for essential medical services is much less responsive to price changes than is the demand for nonessential commodities, such as stereo equipment.

If $E < 1$, the relative change in demand is less than the relative change in price, and the demand is called **inelastic.** If $E > 1$, the relative change in demand is greater than the relative change in price, and the demand is called **elastic.** When $E = 1$, the percentage changes in price and demand are relatively equal and the demand is said to have **unit elasticity.**

Given the demand function $q = 300 - 3p, 0 \le p \le 100$, find the following.

(a) Calculate and interpret the elasticity of demand when $p = 25$ and when $p = 75$.

Since $q = 300 - 3p$, $dq/dp = -3$, and

$$\begin{aligned} E &= -\frac{p}{q} \cdot \frac{dq}{dp} \\ &= -\frac{p}{300 - 3p}(-3) \\ &= \frac{3p}{300 - 3p} \\ &= \frac{p}{100 - p}. \end{aligned}$$

Let $p = 25$ to get

$$E = \frac{25}{100 - 25} = \frac{25}{75} = \frac{1}{3} \approx .33.$$

Since $.33 < 1$, the demand is inelastic, and a percentage change in price will result in a smaller percentage change in demand. For example, a 10% increase in price will cause a 3.3% decrease in demand.

If $p = 75$, then

$$E = \frac{75}{100 - 75} = \frac{75}{25} = 3,$$

and since $3 > 1$, demand is elastic. At this point a percentage increase in price will result in a *greater* percentage decrease in demand. Here, a 10% increase in price will cause a 30% decrease in demand.

(b) Determine the price where demand has unit elasticity. What is the significance of this price?

Demand will have unit elasticity at the price p that makes $E = 1$. Here

$$E = \frac{p}{100 - p} = 1$$

$$p = 100 - p$$

$$2p = 100$$

$$p = 50.$$

Demand has unit elasticity at a price of 50. This means that at this price the percentage changes in price and demand are about the same. ◀

The definitions from this discussion can be summarized as follows.

ELASTICITY OF DEMAND

Let $q = f(p)$, where q is demand at a price p. The elasticity of demand is

$$E = -\frac{p}{q} \cdot \frac{dq}{dp}.$$

Demand is inelastic if $E < 1$.

Demand is elastic if $E > 1$.

Demand has unit elasticity if $E = 1$.

Elasticity can be related to the total revenue, R, by considering the derivative of R. Since revenue is given by price times sales (demand),

$$R = pq.$$

Differentiate with respect to p using the product rule.

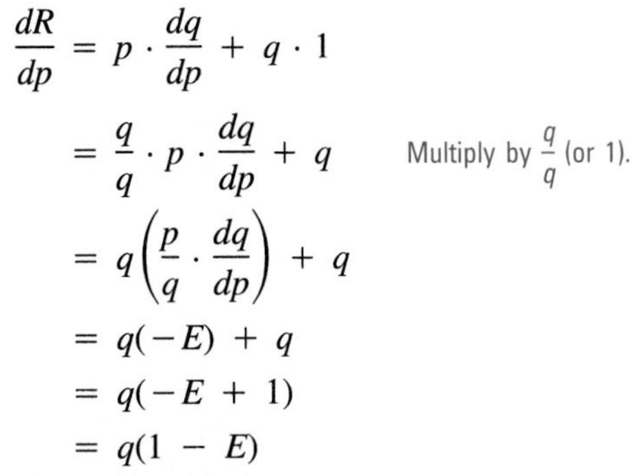
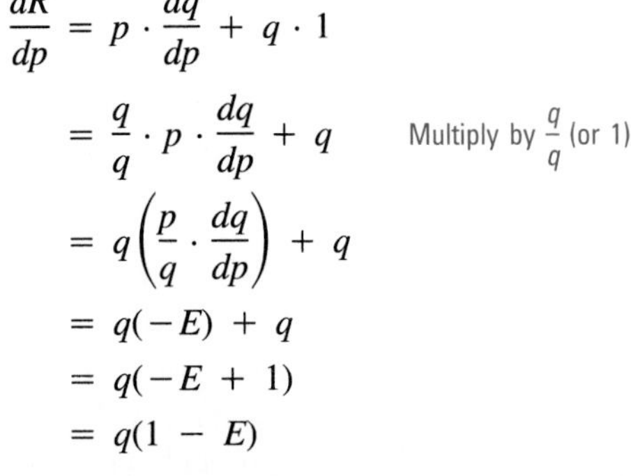

$$\begin{aligned}
\frac{dR}{dp} &= p \cdot \frac{dq}{dp} + q \cdot 1 \\
&= \frac{q}{q} \cdot p \cdot \frac{dq}{dp} + q \qquad \text{Multiply by } \frac{q}{q} \text{ (or 1).} \\
&= q\left(\frac{p}{q} \cdot \frac{dq}{dp}\right) + q \\
&= q(-E) + q \\
&= q(-E + 1) \\
&= q(1 - E)
\end{aligned}$$

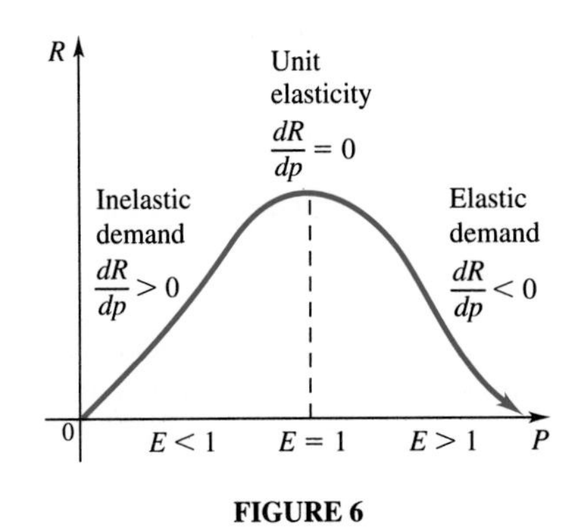

FIGURE 6

Total revenue R is increasing, optimized, or decreasing depending on whether $dR/dp > 0$, $dR/dp = 0$, or $dR/dp < 0$. These three situations correspond to $E < 1$, $E = 1$, or $E > 1$. See Figure 6.

In summary, total revenue is related to elasticity as follows.

REVENUE AND ELASTICITY

1. If the demand is inelastic, total revenue increases as price increases.
2. If the demand is elastic, total revenue decreases as price increases.
3. Total revenue is maximized at the price where demand has unit elasticity.

EXAMPLE 4 Assume that the demand for a product is $q = 216 - 2p^2$, where p is the price.

(a) Find the price intervals where demand is elastic and where demand is inelastic.

Since $q = 216 - 2p^2$, $dq/dp = -4p$, and

$$\begin{aligned}
E &= -\frac{p}{q} \cdot \frac{dq}{dp} \\
&= -\frac{p}{216 - 2p^2}(-4p) \\
&= \frac{4p^2}{216 - 2p^2}.
\end{aligned}$$

To decide where $E < 1$ or $E > 1$, solve the corresponding *equation*.

$$\begin{aligned}
E &= 1 \\
\frac{4p^2}{216 - 2p^2} &= 1 \\
4p^2 &= 216 - 2p^2 \\
6p^2 &= 216 \\
p^2 &= 36 \\
p &= 6
\end{aligned}$$

Substitute a test number on either side of 6 in the expression for E to see which values make $E < 1$ and which make $E > 1$.

$$\text{Let } p = 1: E = \frac{4(1)^2}{216 - 2(1)^2} = \frac{4}{214} < 1.$$

$$\text{Let } p = 10: E = \frac{4(10)^2}{216 - 2(10)^2} = \frac{400}{216 - 200} > 1.$$

Demand is inelastic when $E < 1$. This occurs when $p < 6$. Demand is elastic when $E > 1$; that is, when $p > 6$.

(b) What prices will cause revenue to increase or decrease?

Revenue will increase when demand is inelastic, so keeping prices below \$6 per item will keep demand high enough to continue to increase revenue. When demand is elastic, revenue is decreasing, so any price over \$6 per item will cause demand to decrease to the point where revenue also decreases. ◀

4.2 Exercises

1. In the discussion of economic lot size, use the second derivative to show that the value of x obtained in the text [equation (3)] really leads to the minimum cost.

2. Why do you think that the cost g does not appear in the equation for x [equation (3)]?

Applications

Business and Economics

3. **Lot Size** Find the approximate number of batches of an item that should be produced annually if 100,000 units are to be manufactured, it costs \$1 to store a unit for one year, and it costs \$500 to set up the factory to produce each batch.

4. **Lot Size** A manufacturer has a steady annual demand for 16,800 cases of sugar. It costs \$3 to store one case for one year and it costs \$7 to produce each batch. Find the number of batches of sugar that should be produced each year.

5. **Lot Size** Find the number of units per batch that will be manufactured in Exercise 3.

6. **Lot Size** Find the number of cases per batch in Exercise 4.

7. **Order Quantity** A bookstore has an annual demand for 100,000 copies of a best-selling book. It costs \$.50 to store one copy for one year, and it costs \$60 to place an order. Find the optimum number of copies per order.

8. **Order Quantity** A restaurant has an annual demand for 900 bottles of a California wine. It costs \$1 to store one bottle for one year, and it costs \$5 to place a reorder. Find the optimum number of bottles per order.

9. **Lot Size** Suppose that in the inventory problem, the storage cost depends on the maximum inventory size, rather than the average. This would be more realistic if, for example, the company had to build a warehouse large enough to hold the maximum inventory, and the cost of storage was the same no matter how full or empty the warehouse was. Show that in this case the number of units that should be ordered to minimize the total cost is

$$x = \sqrt{\frac{kM}{f}}.$$

10. **Lot Size** A book publisher wants to know how many times a year a print run should be scheduled. Suppose it costs \$1000 to set up the printing process, and the subsequent cost per book is so low it can be ignored. Suppose further that the annual warehouse cost is \$6 times the maximum number of books stored. Assuming 5000 copies of the book are needed per year, how many times a year

should a print run be scheduled to minimize the total cost, and how many books should be printed in each print run? (See Exercise 9.)

11. **Lot Size** Suppose that in the inventory problem, the storage cost is a combination of the cost described in the text and the cost described in Exercise 9. In other words, suppose there is an annual cost, k_1, for storing a single unit, plus an annual cost per unit, k_2, that must be paid for each unit up to the maximum number of units stored. Show that the number of units that should be ordered to minimize the total cost in this case is

$$x = \sqrt{\frac{(k_1 + 2k_2)M}{2f}}.$$

12. **Lot Size** Every year, Theresa DePalo sells 30,000 cases of her Famous Spaghetti Sauce. It costs her \$1 a year in electricity to store a case, plus she must pay annual warehouse fees of \$2 per case for the maximum number of cases she will store. If it costs her \$750 to set up a production run, plus \$8 per case to manufacture a single case, how many production runs should she have each year to minimize her total costs? (See Exercise 11.)

Elasticity *For each of the following demand functions, find* **(a)** *E, and* **(b)** *values of q (if any) at which total revenue is maximized.*

13. $q = 25{,}000 - 50p$

14. $q = 50 - \dfrac{p}{4}$

15. $q = \dfrac{3000}{p}$

16. $q = \sqrt[3]{\dfrac{2000}{p}}$

Elasticity *In Exercises 17–18, find the elasticity of demand (E) for the given demand function at the indicated values of p. Is the demand elastic, inelastic, or neither at the indicated values? Interpret your results.*

17. $q = 300 - 2p$

(a) $p = \$100$ **(b)** $p = \$50$

18. $q = 400 - .2p^2$

(a) $p = \$20$ **(b)** $p = \$40$

19. **Elasticity** The demand function for q units of a commodity is given by

$$p = (25 - q)^{1/2}, \quad 0 < q < 25.$$

(a) Find E when $q = 16$.

(b) Find the values of q and p that maximize total revenue.

(c) Find the value of E for the q-value found in part (b).

20. **Elasticity** Suppose that a demand function is linear—that is, $q = m - np$ for $0 \le p \le m/n$, where m and n are positive constants. Show that $E = 1$ at the midpoint of the demand curve on the interval $0 \le p \le m/n$; that is, at $p = m/(2n)$.

21. **Elasticity** What must be true about the demand function if $E = 0$?

22. **Elasticity** For some products, the demand function is actually increasing. For example, the most expensive colleges in the United States also tend to have the greatest number of applicants for each student accepted. What is true about the elasticity in this case?

4.3 IMPLICIT DIFFERENTIATION

In almost all of the examples and applications so far, any necessary functions have been defined as

$$y = f(x),$$

with y given **explicitly** in terms of x, or as an **explicit function** of x. For example,

$$y = 3x - 2, \qquad y = x^2 + x + 6, \qquad \text{and} \qquad y = -x^3 + 2$$

are all explicit functions of x. The equation $4xy - 3x = 6$ can be expressed as an explicit function of x by solving for y. This gives

$$\begin{aligned} 4xy - 3x &= 6 \\ 4xy &= 3x + 6 \\ y &= \frac{3x + 6}{4x}. \end{aligned}$$

On the other hand, some equations in x and y cannot be readily solved for y, and some equations cannot be solved for y at all. For example, while it would be possible (but tedious) to use the quadratic formula to solve for y in the equation $y^2 + 2yx + 4x^2 = 0$, it is not possible to solve for y in the equation $y^5 + 8y^3 + 6y^2x^2 + 2yx^3 + 6 = 0$. In equations such as these last two, y is said to be given **implicitly** in terms of x.

In such cases, it may still be possible to find the derivative dy/dx by a process called **implicit differentiation.** In doing so, we assume that there exists some function or functions f, which we may or may not be able to find, such that $y = f(x)$ and dy/dx exists. It is useful to use dy/dx here rather than $f'(x)$ to make it clear which variable is independent and which is dependent.

Find $\dfrac{dy}{dx}$ if $3xy + 4y^2 = 10$.

Differentiate with respect to x on both sides of the equation.

$$3xy + 4y^2 = 10$$

$$\frac{d}{dx}(3xy + 4y^2) = \frac{d}{dx}(10) \tag{1}$$

In Section 1.1, we pointed out that when y is given as a function of x, x is the independent variable and y is the dependent variable. We later defined the derivative dy/dx when y is a function of x. In an equation such as $3xy + 4y^2 = 10$, either variable can be considered the independent variable. If a problem asks for dy/dx, consider x the independent variable; if it asks for dx/dy, consider y the independent variable. A similar rule holds when other variables are used.

Now differentiate each term on the left side of the equation. Think of $3xy$ as the product $(3x)(y)$ and use the product rule. Since

$$\frac{d}{dx}(3x) = 3 \qquad \text{and} \qquad \frac{d}{dx}(y) = \frac{dy}{dx},$$

the derivative of $(3x)(y)$ is

$$(3x)\frac{dy}{dx} + (y)3 = 3x\frac{dy}{dx} + 3y.$$

To differentiate the second term, $4y^2$, use the generalized power rule, since y is assumed to be some function of x.

$$\frac{d}{dx}(4y^2) = 4\overbrace{(2y^1)\frac{dy}{dx}}^{\text{Derivative of } y^2} = 8y\frac{dy}{dx}$$

On the right side of equation (1), the derivative of 10 is 0. Taking the indicated derivatives in equation (1) term by term gives

$$3x\frac{dy}{dx} + 3y + 8y\frac{dy}{dx} = 0.$$

Now solve this result for dy/dx.

$$(3x + 8y)\frac{dy}{dx} = -3y$$

$$\frac{dy}{dx} = \frac{-3y}{3x + 8y}$$

EXAMPLE 2 Find dy/dx for $x^2 + 2xy^2 + 3x^2y = 0$.

Again, differentiate on each side with respect to x.

$$\frac{d}{dx}(x^2 + 2xy^2 + 3x^2y) = \frac{d}{dx}(0)$$

Use the product rule to find the derivatives of $2xy^2$ and $3x^2y$.

$$\underbrace{2x}_{\text{Derivative of } x^2} + \underbrace{2x\left(2y\frac{dy}{dx}\right) + 2y^2}_{\text{Derivative of } 2xy^2} + \underbrace{3x^2\left(\frac{dy}{dx}\right) + 6xy}_{\text{Derivative of } 3x^2y} = \underbrace{0}_{\text{Derivative of } 0}$$

Simplify to get

$$2x + 4xy\left(\frac{dy}{dx}\right) + 2y^2 + 3x^2\left(\frac{dy}{dx}\right) + 6xy = 0,$$

from which

$$(4xy + 3x^2)\frac{dy}{dx} = -2x - 2y^2 - 6xy,$$

or, finally,

$$\frac{dy}{dx} = \frac{-2x - 2y^2 - 6xy}{4xy + 3x^2}.$$

EXAMPLE 3 Find dy/dx for $x + \sqrt{xy} = y^2$.

Take the derivative on each side.

$$\frac{d}{dx}(x + \sqrt{xy}) = \frac{d}{dx}(y^2)$$

Since $\sqrt{xy} = \sqrt{x} \cdot \sqrt{y} = x^{1/2} \cdot y^{1/2}$ for nonnegative values of x and y, taking derivatives gives

$$1 + x^{1/2}\left(\frac{1}{2}y^{-1/2} \cdot \frac{dy}{dx}\right) + y^{1/2}\left(\frac{1}{2}x^{-1/2}\right) = 2y\frac{dy}{dx}$$

$$1 + \frac{x^{1/2}}{2y^{1/2}} \cdot \frac{dy}{dx} + \frac{y^{1/2}}{2x^{1/2}} = 2y\frac{dy}{dx}.$$

Multiply both sides by $2x^{1/2} \cdot y^{1/2}$.

$$2x^{1/2} \cdot y^{1/2} + x\frac{dy}{dx} + y = 4x^{1/2} \cdot y^{3/2} \cdot \frac{dy}{dx}$$

Combine terms and solve for dy/dx.

$$2x^{1/2} \cdot y^{1/2} + y = (4x^{1/2} \cdot y^{3/2} - x)\frac{dy}{dx}$$

$$\frac{dy}{dx} = \frac{2x^{1/2} \cdot y^{1/2} + y}{4x^{1/2} \cdot y^{3/2} - x}$$

EXAMPLE 4 The graph of $x^2 + 5y^2 = 36$ is the ellipse shown in Figure 7. Find the equation of the tangent line at the point (4, 2).

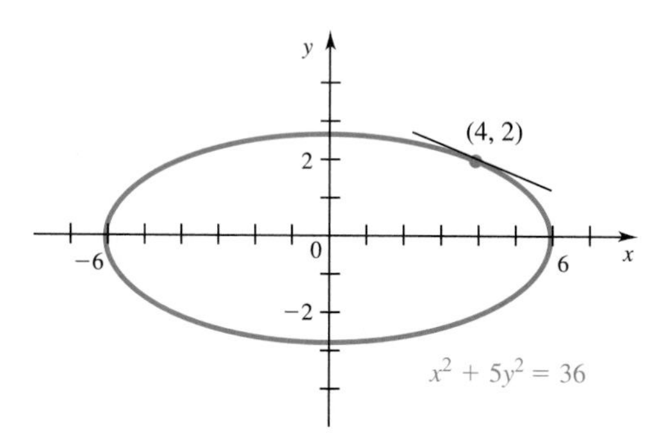

FIGURE 7

A vertical line can cut the graph of the ellipse of Figure 7 in more than one point, so the graph is not the graph of a function. This means that y *cannot* be written as a function of x, so dy/dx would not exist. It seems likely, however, that a tangent line could be drawn to the graph at the point (4, 2). To get around this difficulty, we try to decide if a function f does exist whose graph is exactly the same as our graph in the vicinity of the point (4, 2). Then the tangent line to f at (4, 2) is also the tangent line to the ellipse. See Figure 8.

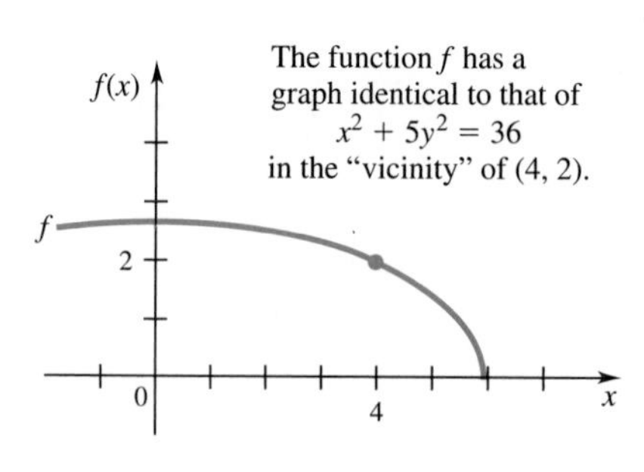

FIGURE 8

Assuming that such a function f exists, dy/dx can be found by implicit differentiation.

$$2x + 10y \cdot \frac{dy}{dx} = 0$$

$$\frac{dy}{dx} = \frac{-2x}{10y} = -\frac{x}{5y}$$

To find the slope of the tangent line at the point (4, 2), let $x = 4$ and $y = 2$. The slope, m, is

$$m = -\frac{x}{5y} = -\frac{4}{5 \cdot 2} = -\frac{2}{5}.$$

The equation of the tangent line is then found by using the point-slope form of the equation of a line.

$$y - y_1 = m(x - x_1)$$

$$y - 2 = -\frac{2}{5}(x - 4)$$

$$5y - 10 = -2x + 8$$

$$2x + 5y = 18$$

This tangent line is graphed in Figure 7. ◀

The steps used in implicit differentiation can be summarized as follows.

IMPLICIT DIFFERENTIATION

To find dy/dx for an equation containing x and y:

1. Differentiate on both sides with respect to x.
2. Place all terms with dy/dx on one side of the equals sign, and all terms without dy/dx on the other side.
3. Factor out dy/dx, and then solve for dy/dx.

When an applied problem involves an equation that is not given in explicit form, implicit differentiation can be used to locate maxima and minima or to find rates of change.

The demand function for a certain commodity is given by

$$p = \frac{500{,}000}{2q^3 + 400q + 5000},$$

where p is the price in dollars and q is the demand in hundreds of units. Find the rate of change of demand with respect to price when $q = 100$ (that is, find dq/dp).

Rewrite the equation as

$$2q^3 + 400q + 5000 = 500{,}000p^{-1}.$$

Use implicit differentiation to find dq/dp.

$$6q^2\frac{dq}{dp} + 400\frac{dq}{dp} = -500{,}000p^{-2}$$

$$(6q^2 + 400)\frac{dq}{dp} = \frac{-500{,}000}{p^2}$$

$$\frac{dq}{dp} = \frac{-500{,}000}{p^2(6q^2 + 400)}$$

When $q = 100$,

$$p = \frac{500{,}000}{2(100)^3 + 400(100) + 5000} \approx .244,$$

and

$$\frac{dq}{dp} = \frac{-500{,}000}{(.244)^2[6(100)^2 + 400]} \approx -139.$$

This means that when demand (q) is 100 hundreds, or 10,000, demand is decreasing at the rate of 139 hundred, or 13,900, units per dollar change in price. ◀

4.3 Exercises

Find dy/dx by implicit differentiation for each of the following.

1. $4x^2 + 3y^2 = 6$

2. $2x^2 - 5y^2 = 4$

3. $2xy + y^2 = 8$

4. $-3xy - 4y^2 = 2$

5. $6xy^2 - 8y + 1 = 0$

6. $-4y^2x^2 - 3x + 2 = 0$

7. $6x^2 + 8xy + y^2 = 6$

8. $8x^2 = 6y^2 + 2xy$

9. $x^3 = y^2 + 4$

10. $x^3 - 6y^2 = 10$

11. $\frac{1}{x} - \frac{1}{y} = 2$

12. $\frac{3}{2x} + \frac{1}{y} = y$

13. $3x^2 = \frac{2 - y}{2 + y}$

14. $2y^2 = \frac{5 + x}{5 - x}$

15. $x^2y + y^3 = 4$

16. $2xy^2 + 2y^3 + 5x = 0$

17. $\sqrt{x} + \sqrt{y} = 4$

18. $2\sqrt{x} - \sqrt{y} = 1$

19. $\sqrt{xy} + y = 1$

20. $\sqrt{2xy} - 1 = 3y^2$

21. $x^4y^3 + 4x^{3/2} = 6y^{3/2} + 5$

22. $(xy)^{4/3} + x^{1/3} = y^6 + 1$

23. $(x^2 + y^3)^4 = x + 2y + 4$

24. $\sqrt{x^2 + y^2} = x^{3/2} - y - 2$

Find the equation of the tangent line at the given point on each curve in Exercises 25–32.

25. $x^2 + y^2 = 25$; $(-3, 4)$

26. $x^2 + y^2 = 100$; $(8, -6)$

27. $x^2y^2 = 1$; $(-1, 1)$

28. $x^2y^3 = 8$; $(-1, 2)$

29. $x^2 + \sqrt{y} = 7$; $(2, 9)$

30. $2y^2 - \sqrt{x} = 4$; $(16, 2)$

31. $y + \frac{\sqrt{x}}{y} = 3$; $(4, 2)$

32. $x + \frac{\sqrt{y}}{3x} = 2$; $(1, 9)$

33. The graph of $x^2 + y^2 = 100$ is a circle having center at the origin and radius 10.

(a) Write the equations of the tangent lines at the points where $x = 6$.

(b) Graph the circle and the tangent lines.

34. The graph of $xy = 1$, shown in the figure, is a hyperbola.

(a) Write the equations of the tangent lines at $x = -1$ and $x = 1$.

(b) Graph the hyperbola and the tangent lines.

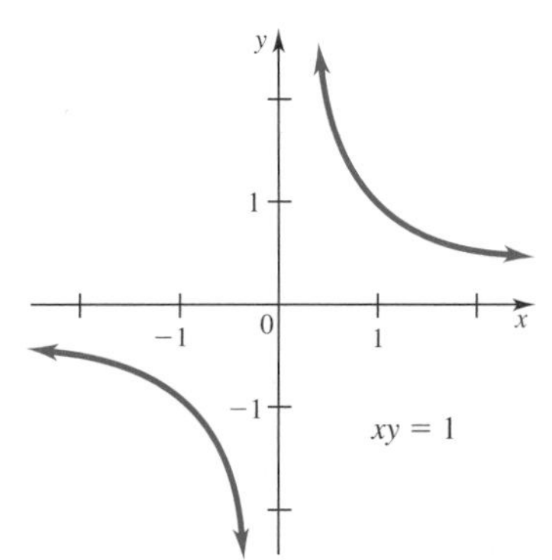

35. The graph of $y^5 - y - x^2 = -1$ is shown in the figure. Find the equation of the line tangent to the graph at the point (1, 1).

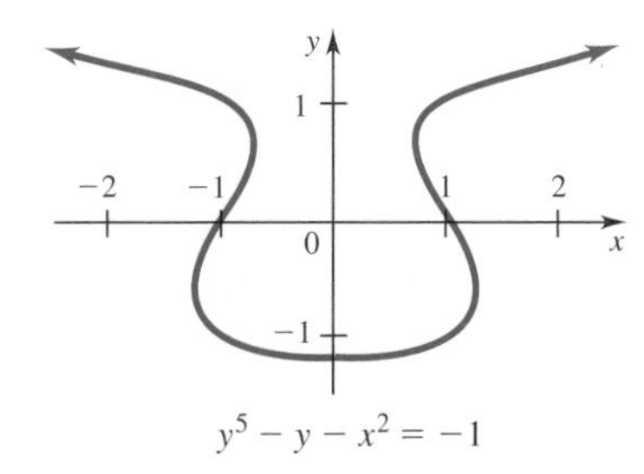

36. The graph of $3(x^2 + y^2)^2 = 25(x^2 - y^2)$ is shown in the figure. Find the equation of the tangent line at the point (2, 1).

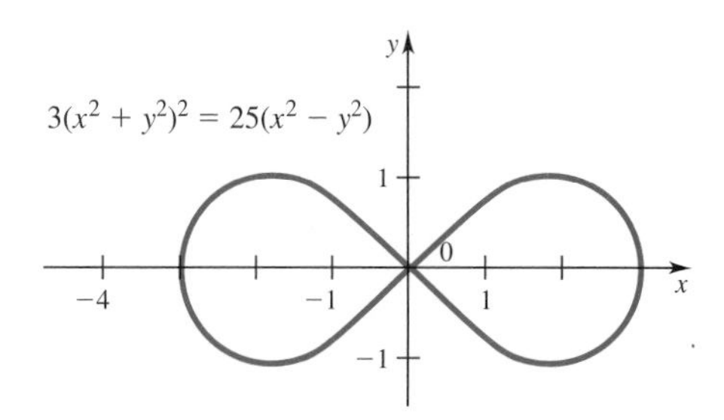

37. Suppose $x^2 + y^2 + 1 = 0$. Use implicit differentiation to find dy/dx. Then explain why the result you got is meaningless. (*Hint:* Can $x^2 + y^2 + 1$ equal 0?)

Let $\sqrt{u} + \sqrt{2v + 1} = 5$. *Find the derivatives in Exercises 38–39.*

38. du/dv

39. dv/du

Applications

Business and Economics

40. Demand The demand equation for a certain product is $2p^2 + q^2 = 1600$, where p is the price per unit in dollars and q is the number of units demanded.

(a) Find and interpret dq/dp.

(b) Find and interpret dp/dq.

41. Cost and Revenue For a certain product, cost C and revenue R are given as follows, where x is the number of units sold (in hundreds).

$$\text{Cost: } C^2 = x^2 + 100\sqrt{x} + 50$$

$$\text{Revenue: } 900(x - 5)^2 + 25R^2 = 22{,}500$$

(a) Find and interpret the marginal cost dC/dx at $x = 5$.

(b) Find and interpret the marginal revenue dR/dx at $x = 5$.

Physical Sciences

Velocity *The position of a particle at time t is given by s. Find the velocity ds/dt.*

42. $s^3 - 4st + 2t^3 - 5t = 0$

43. $2s^2 + \sqrt{st} - 4 = 3t$

4.4 RELATED RATES

If a ladder that is leaning against a wall begins to slide down, how fast is the top of the ladder descending?

Example 2 in this section answers this question using the concept of related rates.

It is common for variables to be functions of time; for example, sales of an item may depend on the season of the year, or a population of animals may be increasing at a certain rate several months after being introduced into an area. Time is often present implicitly in a mathematical model, meaning that derivatives with respect to time must be found by the method of implicit differentiation discussed in the previous section. For example, if a particular mathematical model leads to the equation

$$x^2 + 5y - 3x + 1 = 0,$$

differentiating on both sides with respect to t gives

$$\frac{d}{dt}(x^2 + 5y - 3x + 1) = \frac{d}{dt}(0)$$

$$2x \cdot \frac{dx}{dt} + 5 \cdot \frac{dy}{dt} - 3 \cdot \frac{dx}{dt} = 0$$

$$(2x - 3)\frac{dx}{dt} + 5\frac{dy}{dt} = 0.$$

The derivatives (or rates of change) dx/dt and dy/dt are related by this last equation; for this reason they are called **related rates.**

EXAMPLE 1 A small rock is dropped into a lake. Circular ripples spread over the surface of the water, with the radius of each circle increasing at the rate of 3/2 feet per second. Find the rate of change of the area inside the circle formed by a ripple at the instant that the radius is 4 feet.

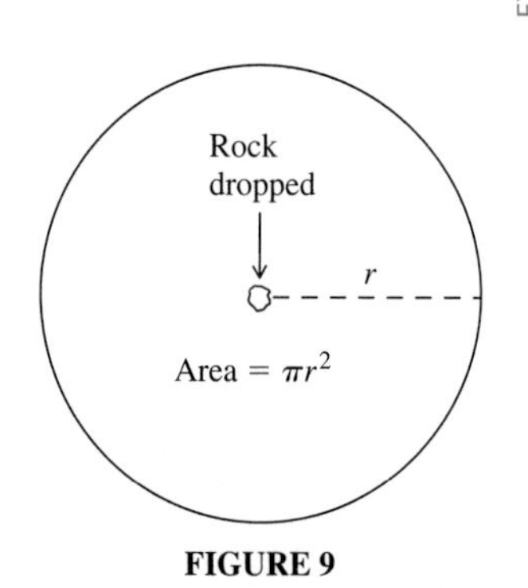

FIGURE 9

As shown in Figure 9, the area A and the radius r are related by

$$A = \pi r^2.$$

Take the derivative of each side with respect to time.

$$\frac{d}{dt}(A) = \frac{d}{dt}(\pi r^2)$$

$$\frac{dA}{dt} = 2\pi r \cdot \frac{dr}{dt} \quad \textbf{(1)}$$

Since the radius is increasing at the rate of 3/2 feet per second,

$$\frac{dr}{dt} = \frac{3}{2}.$$

The rate of change of area at the instant $r = 4$ is given by dA/dt evaluated at $r = 4$. Substituting into equation (1) gives

$$\frac{dA}{dt} = 2\pi \cdot 4 \cdot \frac{3}{2}$$

$$\frac{dA}{dt} = 12\pi \approx 37.7 \text{ square feet per second.}$$

As suggested by Example 1, four basic steps are involved in solving problems about related rates.

SOLVING RELATED RATE PROBLEMS

1. Identify all given quantities, as well as the quantities to be found. Draw a sketch when possible.
2. Write an equation relating the variables of the problem.
3. Use implicit differentiation to find the derivative of both sides of the equation in Step 2 with respect to time.
4. Solve for the derivative giving the unknown rate of change and substitute the given values.

Caution Differentiate *first*, and *then* substitute values for the variables. If the substitutions were performed first, differentiating would not lead to useful results.

EXAMPLE 2 A 50-foot ladder is placed against a large building. The base of the ladder is resting on an oil spill, and it slips (to the right in Figure 10) at the rate of 3 feet per minute. Find the rate of change of the height of the top of the ladder above the ground at the instant when the base of the ladder is 30 feet from the base of the building.

Let y be the height of the top of the ladder above the ground, and let x be the distance of the base of the ladder from the base of the building. By the Pythagorean theorem,

$$x^2 + y^2 = 50^2. \tag{2}$$

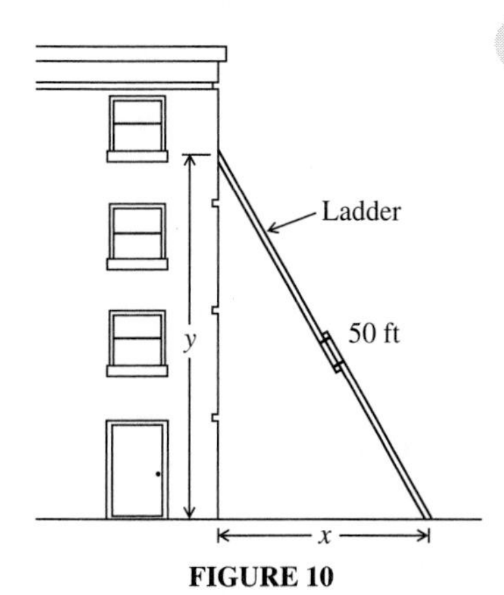

FIGURE 10

Both x and y are functions of time t in minutes after the moment that the ladder starts slipping. Take the derivative of both sides of equation (2) with respect to time, getting

$$\frac{d}{dt}(x^2 + y^2) = \frac{d}{dt}(50^2)$$

$$2x\frac{dx}{dt} + 2y\frac{dy}{dt} = 0. \tag{3}$$

Since the base is sliding at the rate of 3 feet per minute,

$$\frac{dx}{dt} = 3.$$

Also, the base of the ladder is 30 feet from the base of the building. Use this to find y.

$$\begin{aligned} 50^2 &= 30^2 + y^2 \\ 2500 &= 900 + y^2 \\ 1600 &= y^2 \\ y &= 40 \end{aligned}$$

In summary, $y = 40$ when $x = 30$. Also, the rate of change of x over time t is $dx/dt = 3$. Substituting these values into equation (3) to find the rate of change of y over time gives

$$\begin{aligned} 2(30)(3) + 2(40)\frac{dy}{dt} &= 0 \\ 180 + 80\frac{dy}{dt} &= 0 \\ 80\frac{dy}{dt} &= -180 \\ \frac{dy}{dt} &= \frac{-180}{80} = \frac{-9}{4} = -2.25. \end{aligned}$$

At the instant when the base of the ladder is 30 feet from the base of the building, the top of the ladder is sliding down the building at the rate of 2.25 feet per minute. (The minus sign shows that the ladder is sliding *down*, so the distance y is *decreasing*.) ◀

EXAMPLE 3 A cone-shaped icicle is dripping from the roof. The radius of the icicle is decreasing at a rate of .2 centimeters per hour, while the length is increasing at a rate of .8 centimeters per hour. If the icicle is currently 4 centimeters in radius and 20 centimeters long, is the volume of the icicle increasing or decreasing, and at what rate?

For this problem we need the formula for the volume of a cone:

$$V = \frac{1}{3}\pi r^2 h, \tag{4}$$

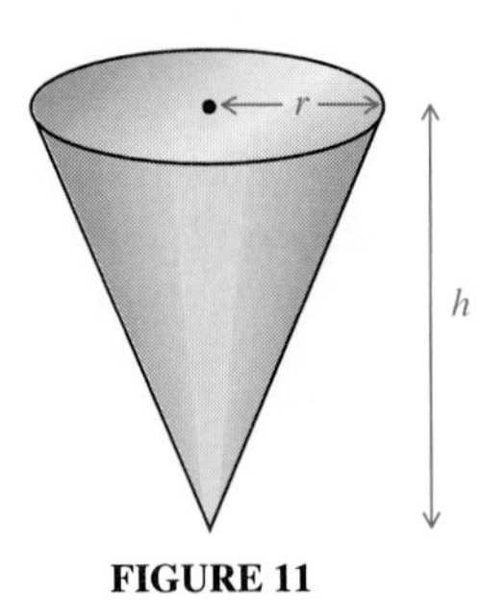

FIGURE 11

where r is the radius of the cone and h is the height of the cone, which in this case is the length of the icicle, as in Figure 11.

In this problem, both r and h are functions of the time t in hours. Taking the derivative of both sides of equation (4) with respect to time yields

$$\frac{dV}{dt} = \frac{1}{3}\pi\left[r^2\frac{dh}{dt} + (h)(2r)\frac{dr}{dt}\right].$$

Since the radius is decreasing at a rate of .2 centimeters per hour and the length is increasing at a rate of .8 centimeters per hour,

$$\frac{dr}{dt} = -.2 \qquad \text{and} \qquad \frac{dh}{dt} = .8.$$

Substituting these, as well as $r = 4$ and $h = 20$, into equation (4) yields

$$\frac{dV}{dt} = \frac{1}{3}\pi[4^2(.8) + (20)(8)(-.2)]$$

$$= \frac{1}{3}\pi(-19.2) \approx -20.$$

Because the sign of dV/dt is negative, the volume of the icicle is decreasing at a rate of 20 cubic centimeters per hour.

A company is increasing production of peanuts at the rate of 50 cases per day. All cases produced can be sold. The daily demand function is given by

$$p = 50 - \frac{q}{200},$$

where q is the number of units produced (and sold) and p is price in dollars. Find the rate of change of revenue with respect to time (in days) when the daily production is 200 units.

The revenue function,

$$R = qp = q\left(50 - \frac{q}{200}\right) = 50q - \frac{q^2}{200},$$

relates R and q. The rate of change of q over time (in days) is $dq/dt = 50$. The rate of change of revenue over time, dR/dt, is to be found when $q = 200$. Differentiate both sides of the equation

$$R = 50q - \frac{q^2}{200}$$

with respect to t.

$$\frac{dR}{dt} = 50\frac{dq}{dt} - \frac{1}{100}q\frac{dq}{dt} = \left(50 - \frac{1}{100}q\right)\frac{dq}{dt}$$

Now substitute the known values for q and dq/dt.

$$\frac{dR}{dt} = \left[50 - \frac{1}{100}(200)\right](50) = 2400$$

Thus revenue is increasing at the rate of \$2400 per day.

EXAMPLE 5 Blood flows faster the closer it is to the center of a blood vessel. According to Poiseuille's laws (see Exercise 44 in the first section of this chapter), the velocity V of blood is given by

$$V = k(R^2 - r^2),$$

where R is the radius of the blood vessel, r is the distance of a layer of blood flow from the center of the vessel, and k is a constant, assumed here to equal 375. See Figure 12.

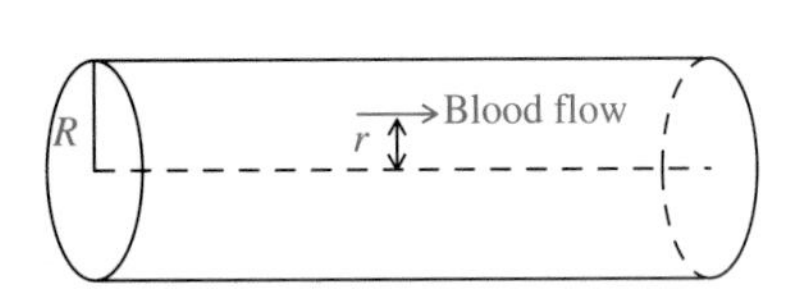

FIGURE 12

(a) Find dV/dt. Treat r as a constant. Assume the given units are compatible.

$$V = 375(R^2 - r^2)$$

$$\frac{dV}{dt} = 375\left(2R\frac{dR}{dt} - 0\right) \qquad r \text{ is a constant.}$$

$$\frac{dV}{dt} = 750R\frac{dR}{dt}$$

(b) Suppose a skier's blood vessel has radius $R = .008$ millimeters and that cold weather is causing the vessel to contract at a rate of $dR/dt = -.001$ millimeters per minute. How fast is the velocity of blood changing?

Find dV/dt. From part (a),

$$\frac{dV}{dt} = 750R\frac{dR}{dt}.$$

Here $R = .008$ and $dR/dt = -.001$, so

$$\frac{dV}{dt} = 750(.008)(-.001) = -.006.$$

That is, the velocity of the blood is decreasing at a rate of $-.006$ millimeters per minute. The units indicate that this is a deceleration (negative acceleration), since it gives the rate of change of velocity. ◀

4.4 Exercises

Evaluate dy/dt for each of the following.

1. $y = 9x^2 + 2x$; $dx/dt = 4, x = 6$

2. $y = \dfrac{3x + 2}{1 - x}$; $dx/dt = -1, x = 3$

3. $y^2 - 5x^2 = -1$; $dx/dt = -3, x = 1, y = 2$

4. $8y^3 + x^2 = 1$; $dx/dt = 2, x = 3, y = -1$

5. $xy - 5x + 2y^3 = -70$; $dx/dt = -5, x = 2, y = -3$

6. $4x^3 - 9xy^2 + y = -80$; $dx/dt = 4, x = -3, y = 1$

7. $\dfrac{x^2 + y}{x - y} = 9$; $dx/dt = 2, x = 4, y = 2$

8. $\dfrac{y^3 - x^2}{x + 2y} = \dfrac{17}{7}$; $dx/dt = 1, x = -3, y = -2$

Applications

Business and Economics

9. Cost A manufacturer of handcrafted wine racks has determined that the cost to produce x units per month is given by $C = .1x^2 + 10{,}000$. How fast is cost per month changing when production is changing at the rate of 10 units per month and the production level is 100 units?

10. Cost/Revenue The manufacturer in Exercise 9 has found that the cost C and revenue R (in dollars) from the production and sale of x units are related by the equation

$$C = \frac{R^2}{400{,}000} + 10{,}000.$$

Find the rate of change of revenue per unit when the cost per unit is changing by \$10 and the revenue is \$20,000.

11. Revenue/Cost/Profit Given the revenue and cost functions $R = 50x - .4x^2$ and $C = 5x + 15$, where x is the daily production (and sales), find the following when 40 units are produced daily and the rate of change of production is 10 units per day.

(a) The rate of change of revenue with respect to time

(b) The rate of change of cost with respect to time

(c) The rate of change of profit with respect to time

12. Revenue/Cost/Profit Repeat Exercise 11, given that 200 units are produced daily and the rate of change of production is 50 units per day.

13. Demand A product sells for \$3.50 currently and has a demand function of

$$p = \frac{8000}{q}.$$

Suppose manufacturing costs are increasing over time at a rate of 15% and the company plans to increase the price p at this rate as well. Find the rate of change of demand over time.

14. Revenue A company is increasing production at the rate of 25 units per day. The daily demand function is given by

$$p = 70 - \frac{q^2}{120}.$$

Find the rate of change of revenue with respect to time (in days) when the daily production (and sales) is 20.

Life Sciences

Blood Velocity *Exercises 15–17 refer to Example 5 in this section.*

15. Find dV/dt if $r = 2$ mm, $k = 3$, $dr/dt = .02$ mm/min, and R is fixed.

16. Find dV/dt if $r = 1$ mm, $k = 4$, $dr/dt = .004$ mm/min, and R is fixed.

17. A cross-country skier has a history of heart problems. She takes nitroglycerin to dilate blood vessels, thus avoiding angina (chest pain) due to blood vessel contraction. Use Poiseuille's law with $k = 555.6$ to find the rate of change of the blood velocity when $R = .02$ mm and R is changing at .003 mm per minute. Assume r is constant.

Social Sciences

18. Crime Rate Sociologists have found that crime rates are influenced by temperature. In a midwestern town of 100,000 people, the crime rate has been approximated as

$$C = \frac{1}{10}(T - 60)^2 + 100,$$

where C is the number of crimes per month and T is the average monthly temperature. The average temperature for May was 76°, and by the end of May the temperature was rising at the rate of 8° per month. How fast is the crime rate rising at the end of May?

19. Learning Skills It is estimated that a person learning a certain assembly-line task takes

$$T(x) = \frac{2 + x}{2 + x^2}$$

minutes to perform the task after x repetitions. Find dT/dt if dx/dt is 4, and 4 repetitions of the task have been performed.

20. Memorization Skills Under certain conditions, a person can memorize W words in t minutes, where

$$W(t) = \frac{-.02t^2 + t}{t + 1}.$$

Find dW/dt when $t = 5$.

Physical Sciences

21. Sliding Ladder A 25-ft ladder is placed against a building. The base of the ladder is slipping away from the building at the rate of 4 ft/min. Find the rate at which the top of the ladder is sliding down the building at the instant when the bottom of the ladder is 7 ft from the base of the building.

22. Distance One car leaves a given point and travels north at 30 mph. Another car leaves the same point at the same time and travels west at 40 mph. At what rate is the distance between the two cars changing at the instant when the cars have traveled 2 hr?

23. **Area** A rock is thrown into a still pond. The circular ripples move outward from the point of impact of the rock so that the radius of the circle formed by a ripple increases at the rate of 2 ft per minute. Find the rate at which the area is changing at the instant the radius is 4 ft.

24. **Volume** A spherical snowball is placed in the sun. The sun melts the snowball so that its radius decreases $1/4$ inch per hour. Find the rate of change of the volume with respect to time at the instant the radius is 4 inches.

25. **Volume** A sand storage tank used by the highway department for winter storms is leaking. As the sand leaks out, it forms a conical pile. The radius of the base of the pile increases at the rate of 1 inch per minute. The height of the pile is always twice the radius of the base. Find the rate at which the volume of the pile is increasing at the instant the radius of the base is 5 inches.

26. **Shadow Length** A man 6 ft tall is walking away from a lamp post at the rate of 50 ft per minute. When the man is 8 ft from the lamp post, his shadow is 10 ft long. Find the rate at which the length of the shadow is increasing when he is 25 ft from the lamp post. (See the figure.)

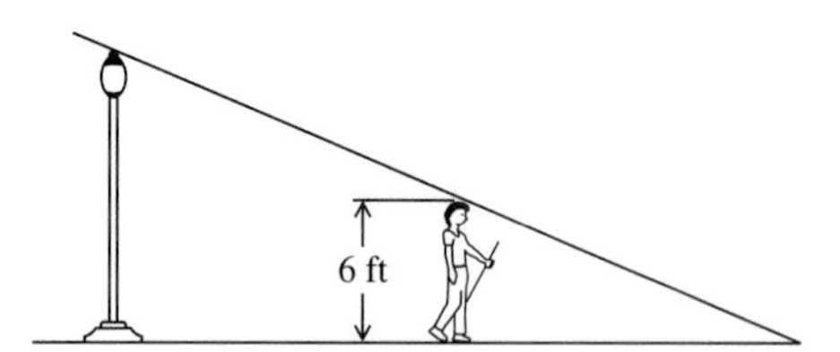

27. **Water Level** A trough has a triangular cross section. The trough is 6 ft across the top, 6 ft deep, and 16 ft long. Water is being pumped into the trough at the rate of 4 cubic feet per minute. Find the rate at which the height of water is increasing at the instant that the height is 4 ft.

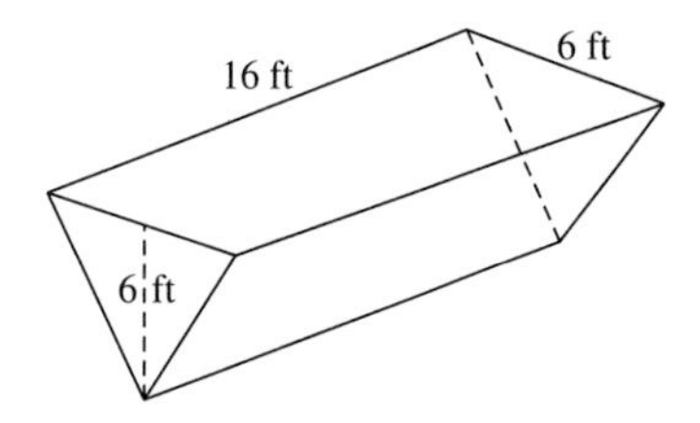

28. **Velocity** A pulley is on the edge of a dock, 8 ft above the water level. A rope is being used to pull in a boat. The rope is attached to the boat at water level. The rope is being pulled in at the rate of 1 ft per second. Find the rate at which the boat is approaching the dock at the instant the boat is 8 ft from the dock.

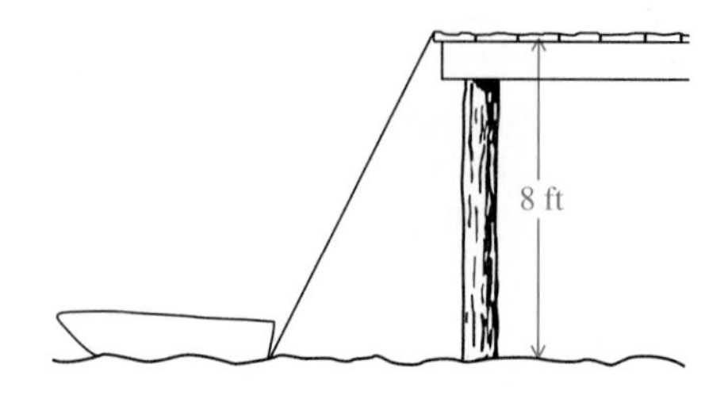

29. **Kite Flying** Karen LaBonte is flying her kite in a wind that is blowing it east at a rate of 50 ft/min. She has already let out 200 ft of string, and the kite is flying 100 ft above her hand. How fast must she let out string at this moment to keep the kite flying with the same speed and altitude?

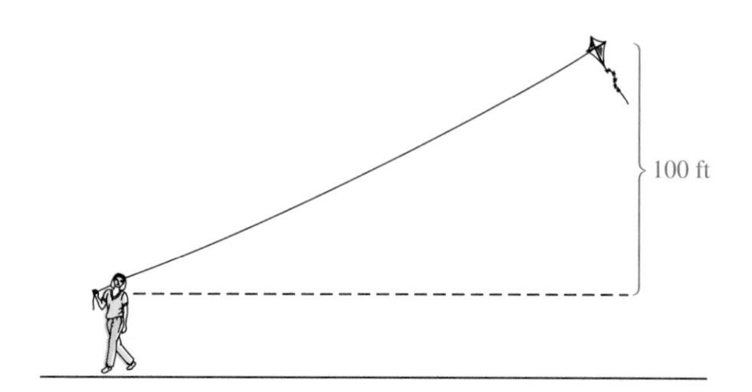

4.5 DIFFERENTIALS

If the estimated sales of microwave ovens turns out to be inaccurate, approximately how much are profits affected?

Using differentials, we will answer this question in Example 4.

As mentioned earlier, the symbol Δx represents a change in the variable x. Similarly, Δy represents a change in y. An important problem that arises in many applications is to determine Δy given specific values of x and Δx. This quantity is often difficult to evaluate. In this section we show a method of approximating Δy that uses the derivative dy/dx.

For values x_1 and x_2,

$$\Delta x = x_2 - x_1.$$

Solving for x_2 gives

$$x_2 = x_1 + \Delta x.$$

For a function $y = f(x)$, the symbol Δy represents a change in y:

$$\Delta y = f(x_2) - f(x_1).$$

Replacing x_2 with $x_1 + \Delta x$ gives

$$\Delta y = f(x_1 + \Delta x) - f(x_1).$$

If Δx is used instead of h, the derivative of a function f at x_1 could be defined as

$$\frac{dy}{dx} = \lim_{\Delta x \to 0} \frac{\Delta y}{\Delta x}.$$

If the derivative exists, then

$$\frac{dy}{dx} \approx \frac{\Delta y}{\Delta x}$$

as long as Δx is close to 0. Multiplying both sides by Δx (assume $\Delta x \neq 0$) gives

$$\Delta y \approx \frac{dy}{dx} \cdot \Delta x.$$

Until now, dy/dx has been used as a single symbol representing the derivative of y with respect to x. In this section, separate meanings for dy and dx are introduced in such a way that their quotient, when $dx \neq 0$, is the derivative of y with respect to x. These meanings of dy and of dx are then used to find an approximate value of Δy.

To define dy and dx, look at Figure 13, which shows the graph of a function $y = f(x)$. The tangent line to the graph has been drawn at the point P. Let Δx be any nonzero real number (in practical problems, Δx is a small number) and locate the point $x + \Delta x$ on the x-axis. Draw a vertical line through $x + \Delta x$. Let this vertical line cut the tangent line at M and the graph of the function at Q.

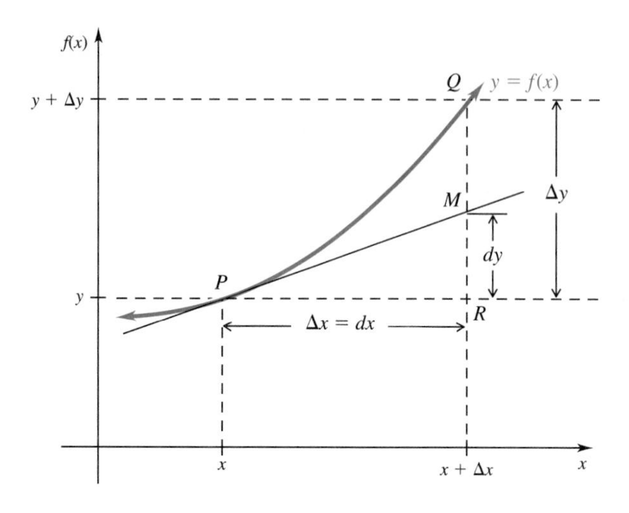

FIGURE 13

Define the new symbol dx to be the same as Δx. Define the new symbol dy to equal the length MR. The slope of PM is $f'(x)$. By the definition of slope, the slope of PM is also dy/dx, so that

$$f'(x) = \frac{dy}{dx},$$

or

$$dy = f'(x)dx.$$

In summary, the definitions of the symbols dy and dx are as follows.

DIFFERENTIALS

For a function $y = f(x)$ whose derivative exists, the **differential** of x, written dx, is an arbitrary real number (usually small); the **differential** of y, written dy, is the product of $f'(x)$ and dx, or

$$\mathbf{dy = f'(x)dx.}$$

The usefulness of the differential is suggested by Figure 13. As dx approaches 0, the value of dy gets closer and closer to that of Δy, so that for small nonzero values of dx

$$dy \approx \Delta y,$$

or

$$\Delta y \approx f'(x)dx.$$

EXAMPLE 1 Find dy for the following functions.

(a) $y = 6x^2$

The derivative is $y' = 12x$, so that

$$dy = 12xdx.$$

(b) If $y = 8x^{-3/4}$, then $dy = -6x^{-7/4}\,dx$.

(c) If $f(x) = \dfrac{2 + x}{2 - x}$, then $dy = \dfrac{4}{(2 - x)^2}dx$. ◀

EXAMPLE 2 Suppose $y = 5x - x^2 + 4$. Find dy for the following values of x and dx.

(a) $x = -3$, $dx = .05$

First, by definition

$$dy = (5 - 2x)dx.$$

If $x = -3$ and $dx = .05$, then

$$dy = [5 - 2(-3)](.05) = 11(.05) = .55.$$

(b) $x = 0$, $dx = -.004$

$$\begin{aligned} dy &= (5 - 2 \cdot 0)(-.004) \\ &= 5(-.004) \\ &= -.02 \end{aligned}$$ ◀

As discussed above,

$$\Delta y = f(x + \Delta x) - f(x).$$

For small nonzero values of Δx, $\Delta y \approx dy$, so that

$$dy \approx f(x + \Delta x) - f(x),$$

or

$$f(x) + dy \approx f(x + \Delta x). \quad \textbf{(1)}$$

Replacing dy with $f'(x)dx$ gives the following result.

DIFFERENTIAL APPROXIMATION

Let f be a function whose derivative exists. For small nonzero values of Δx,

$$\boldsymbol{dy \approx \Delta y,}$$

and

$$\boldsymbol{f(x + \Delta x) \approx f(x) + dy = f(x) + f'(x)dx.}$$

Differentials are used to find an approximate value of the change in a value of the dependent variable corresponding to a given change in the independent variable. When the concept of marginal cost (or profit or revenue) was used to approximate the change in cost for nonlinear functions, the same idea was developed. Thus the differential dy approximates Δy in much the same way as the marginal quantities approximate changes in functions.

For example, for a cost function $C(x)$,

$$dC = C'(x)dx = C'(x)\Delta x.$$

Since $\Delta C \approx dC$,

$$\Delta C \approx C'(x)\Delta x.$$

If the change in production, Δx, is equal to 1, then

$$\begin{aligned} C(x+1) - C(x) &= \Delta C \\ &\approx C'(x)\Delta x \\ &= C'(x), \end{aligned}$$

which snows that marginal cost $C'(x)$ approximates the cost of the next unit produced, as mentioned earlier.

EXAMPLE 3 Let $C(x) = 2x^3 + 300$.

(a) Find ΔC and $C'(x)$ when $\Delta x = 1$ and $x = 3$.

$$\begin{aligned} \Delta C = C(4) - C(3) &= 428 - 354 = 74 \\ C'(x) &= 6x^2 \\ C'(3) &= 54 \end{aligned}$$

Here, the approximation of $C'(3)$ for ΔC is poor, since $\Delta x = 1$ is large relative to $x = 3$.

(b) Find ΔC and $C'(x)$ when $\Delta x = 1$ and $x = 50$.

$$\begin{aligned} \Delta C = C(51) - C(50) &= 265{,}602 - 250{,}300 = 15{,}302 \\ C'(50) &= 6(2500) = 15{,}000 \end{aligned}$$

This approximation is quite good since $\Delta x = 1$ is small compared to $x = 50$. ◀

EXAMPLE 4 An analyst for a manufacturer of small appliances estimates that the total profit (in hundreds of dollars) from the sale of x hundred microwave ovens is given by

$$P(x) = -2x^3 + 51x^2 + 88x.$$

In a report to management, the analyst projected sales for the coming month to be 1000 ovens, for a total profit of $389,000. He now realizes that his sales estimate may have been as much as 100 ovens too high. Approximately how far off is his profit estimate?

Differentials can be used to find the approximate change in P resulting from decreasing x by 1 (= 100 ovens). This change can be approximated by $dP = P'(x)dx$ where $x = 10$ (= 1000 ovens) and $dx = -1$ (= -100 ovens). Since $P'(x) = -6x^2 + 102x + 88$,

$$\begin{aligned}\Delta P \approx dP &= (-6x^2 + 102x + 88)dx \\ &= [-6(100) + 102(10) + 88](-1) \\ &= (-600 + 1020 + 88)(-1) \\ &= (508)(-1) \\ &= -508\end{aligned}$$

hundred dollars. Thus, the profit estimate may have been as much as $50,800 too high. ◀

The mathematical model

$$y = A(x) = \frac{-7}{480}x^3 + \frac{127}{120}x, \qquad 0 \le x < 9,$$

gives the approximate alcohol concentration (in tenths of a percent) in an average person's bloodstream x hours after drinking 8 ounces of 100-proof whiskey.

(a) Approximate the change in y as x changes from 3 to 3.5. Use dy as an approximation of Δy.

Here,

$$dy = \left(\frac{-7}{160}x^2 + \frac{127}{120}\right)dx.$$

In this example, $x = 3$ and $dx = \Delta x = 3.5 - 3 = .5$. Substitution gives

$$\Delta y \approx dy = \left(\frac{-7}{160} \cdot 3^2 + \frac{127}{120}\right)(.5) \approx .33,$$

so $\Delta y \approx .33$. This means that the alcohol concentration increases by about .33 tenths of a percent as x changes from 3 hours to 3.5 hours (The *exact* increase, found by calculating $A(3.5) - A(3)$, is .30 tenths of a percent.)

(b) Approximate the change in y as x changes from 6 to 6.25 hours.

Let $x = 6$ and $dx = 6.25 - 6 = .25$.

$$\Delta y \approx dy = \left(\frac{-7}{160} \cdot 6^2 + \frac{127}{120}\right)(.25) \approx -.13$$

The minus sign shows that alcohol concentration decreases by .13 tenths of a percent as x changes from 6 to 6.25 hours. ◀

The final example in this section shows how differentials are used to estimate errors that might enter into measurements of a physical quantity.

EXAMPLE 6

In a precision manufacturing process, ball bearings must be made with a radius of .6 millimeters, with a maximum error in the radius of $\pm$ .015 millimeters. Estimate the maximum error in the volume of the ball bearing.

The formula for the volume of a sphere is

$$V = \frac{4}{3}\pi r^3.$$

If an error of Δr is made in measuring the radius of the sphere, the maximum error in the volume is

$$\Delta V = \frac{4}{3}\pi(r + \Delta r)^3 - \frac{4}{3}\pi r^3.$$

Rather than calculating ΔV, approximate ΔV with dV, where

$$dV = 4\pi r^2 dr.$$

Replacing r with .6 and $dr = \Delta r$ with $\pm .015$ gives

$$dV = 4\pi(.6)^2(\pm .015)$$
$$\approx \pm .0679.$$

The maximum error in the volume is about .07 cubic millimeters. ◀

4.5 Exercises

Find dy for each of the following.

1. $y = 6x^2$

2. $y = -8x^4$

3. $y = 7x^2 - 9x + 6$

4. $y = -3x^3 + 2x^2$

5. $y = 2\sqrt{x}$

6. $y = 8\sqrt{2x - 1}$

7. $y = \dfrac{8x - 2}{x - 3}$

8. $y = \dfrac{-4x + 7}{3x - 1}$

9. $y = x^2\left(x - \dfrac{1}{x} + 2\right)$

10. $y = -x^3\left(2 + \dfrac{3}{x^2} - \dfrac{5}{x}\right)$

11. $y = \left(2 - \dfrac{3}{x}\right)\left(1 + \dfrac{1}{x}\right)$

12. $y = \left(9 - \dfrac{2}{x^2}\right)\left(3 + \dfrac{1}{x}\right)$

For Exercises 13–24, find dy for the given values of x and Δx.

13. $y = 2x^2 - 5x$; $x = -2, \Delta x = .2$

14. $y = x^2 - 3x$; $x = 3, \Delta x = .1$

15. $y = x^3 - 2x^2 + 3$; $x = 1, \Delta x = -.1$

16. $y = 2x^3 + x^2 - 4x$; $x = 2, \Delta x = -.2$

17. $y = \sqrt{3x}$; $x = 1, \Delta x = .15$

18. $y = \sqrt{4x - 1}$; $x = 5, \Delta x = .08$

19. $y = \dfrac{2x - 5}{x + 1}$; $x = 2, \Delta x = -.03$

20. $y = \dfrac{6x - 3}{2x + 1}$; $x = 3, \Delta x = -.04$

21. $y = -6\left(2 - \dfrac{1}{x^2}\right)$; $x = -1, \Delta x = .02$

22. $y = 9\left(3 + \dfrac{1}{x^4}\right)$; $x = -2, \Delta x = -.015$

23. $y = \dfrac{1 + x}{\sqrt{x}}$; $x = 9, \Delta x = -.03$

24. $y = \dfrac{2 - 5x}{\sqrt{x + 1}}$; $x = 3, \Delta x = .02$

Applications

Business and Economics

25. Demand The demand for grass seed (in thousands of pounds) at a price of x dollars is

$$d(x) = -5x^3 - 2x^2 + 1500.$$

Use the differential to approximate the changes in demand for the following changes in x.

(a) \$2 to \$2.50 **(b)** \$6 to \$6.30

26. Average Cost The average cost to manufacture x dozen marking pencils is

$$A(x) = .04x^3 + .1x^2 + .5x + 6.$$

Use the differential to approximate the changes in the average cost for the following changes in x.

(a) 3 to 4 **(b)** 5 to 6

27. Mortgage Rates A county realty group estimates that the number of housing starts per year is

$$H(r) = \frac{300}{1 + .03r^2},$$

where r is the mortgage rate in percent ($r = 8$ means 8%). Use the differential to approximate the change in housing starts when $r = 10$, if mortgage rates change by .5%.

28. Revenue A company estimates that the revenue (in dollars) from the sale of x units of doghouses is given by

$$R(x) = 625 + .03x + .0001x^2.$$

Use the differential to approximate the change in revenue from the sale of one more doghouse when 1000 doghouses are sold.

29. Profit The profit function for the company in Exercise 28 is

$$P(x) = -390 + 24x + 5x^2 - \frac{1}{3}x^3,$$

where x represents the demand for the product. Find the approximate change in profit for a one-unit change in demand when demand is at a level of 1000 doghouses. Use the differential.

30. Material Requirement A cube 4 inches on an edge is given a protective coating .1 inch thick. About how much coating should a production manager order for 1000 such cubes?

31. Material Requirement Beach balls 1 ft in diameter have a thickness of .03 inch. How much material would be needed to make 5000 beach balls?

Life Sciences

32. Drug Concentration The concentration of a certain drug in the bloodstream x hr after being administered is approximately

$$C(x) = \frac{5x}{9 + x^2}.$$

Use the differential to approximate the changes in concentration for the following changes in x.

(a) 1 to 1.5 **(b)** 2 to 2.25

33. Bacteria Population The population of bacteria (in millions) in a certain culture x hr after an experimental nutrient is introduced into the culture is

$$P(x) = \frac{25x}{8 + x^2}.$$

Use the differential to approximate the changes in population for the following changes in x.

(a) 2 to 2.5 **(b)** 3 to 3.25

34. Area of a Blood Vessel The radius of a blood vessel is 17 mm. A drug causes the radius to change to 16 mm. Find the approximate change in the area of a cross section of the vessel.

35. Volume of a Tumor A tumor is approximately spherical in shape. If the radius of the tumor changes from 14 mm to 16 mm, find the approximate change in volume.

36. Area of an Oil Slick An oil slick is in the shape of a circle. Find the approximate increase in the area of the slick if its radius increases from 1.2 mi to 1.4 mi.

37. Area of a Bacteria Colony The shape of a colony of bacteria on a Petri dish is circular. Find the approximate increase in its area if the radius increases from 20 mm to 22 mm.

Physical Sciences

38. Volume A spherical beach ball is being inflated. Find the approximate change in volume if the radius increases from 4 cm to 4.2 cm.

39. Volume A spherical snowball is melting. Find the approximate change in volume if the radius decreases from 3 cm to 2.8 cm.

General Interest

40. Measurement Error The edge of a square is measured as 3.45 inches, with a possible error of $\pm .002$ inches. Estimate the maximum error in the area of the square.

41. Measurement Error The radius of a circle is measured as 4.87 inches, with a possible error of $\pm .040$ inches. Estimate the maximum error in the area of the circle.

42. Measurement Error A sphere has a radius of 5.81 inches, with a possible error of $\pm .003$ inches. Estimate the maximum error in the volume of the sphere.

43. Measurement Error A cone has a known height of 7.284 inches. The radius of the base is measured as 1.09 inches, with a possible error of $\pm .007$ inch. Estimate the maximum error in the volume of the cone. (The volume of a cone is given by $V = \frac{1}{3}\pi r^2 h$.)

Chapter Summary — Key Terms

4.1 **limits at infinity**
oblique asymptote
4.2 **economic lot size**
economic order quantity
elasticity of demand
4.3 **explicit function**
implicit differentiation
4.4 **related rates**
4.5 **differential**

Chapter 4 Review Exercises

1. When is it necessary to use implicit differentiation?
2. When a term involving y is differentiated in implicit differentiation, it is multiplied by dy/dx. Why? Why aren't terms involving x multiplied by dx/dx?
3. What is the difference between a related rate problem and an applied extremum problem?
4. Why is implicit differentiation used in related rate problems?

Find dy/dx in Exercises 5–12.

5. $x^2y^3 + 4xy = 2$
6. $\dfrac{x}{y} - 4y = 3x$
7. $9\sqrt{x} + 4y^3 = \dfrac{2}{x}$
8. $2\sqrt{y - 1} = 8x^{2/3}$
9. $\dfrac{x + 2y}{x - 3y} = y^{1/2}$
10. $\dfrac{6 + 5x}{2 - 3y} = \dfrac{1}{5x}$
11. $(4y^2 - 3x)^{2/3} = 6x$
12. $(8x + y^{1/2})^3 = 9y^2$
13. Find the equation of the line tangent to the graph of $\sqrt{2x} - 4yx = -22$ at the point (2, 3).
14. The graph of $x^2 + y^2 = 25$ is a circle of radius 5 with center at the origin. Find the equation of the tangent lines when $x = -3$. Graph the circle and the tangent lines.

Find dy/dt in Exercises 15–18.

15. $y = 8x^3 - 7x^2;\quad dx/dt = 4, x = 2$
16. $y = \dfrac{9 - 4x}{3 + 2x};\quad dx/dt = -1, x = -3$
17. $y = \dfrac{1 + \sqrt{x}}{1 - \sqrt{x}};\quad dx/dt = -4, x = 4$
18. $\dfrac{x^2 + 5y}{x - 2y} = 2;\quad dx/dt = 1, x = 2, y = 0$
19. What is a differential? What is it used for?

Find dy.

20. $y = 8x^3 - 2x^2$
21. $y = 4(x^2 - 1)^3$
22. $y = \dfrac{6 - 5x}{2 + x}$
23. $y = \sqrt{9 + x^3}$

Evaluate dy in Exercises 24 and 25.

24. $y = 8 - x^2 + x^3; \quad x = -1, \Delta x = .02$

25. $y = \dfrac{3x - 7}{2x + 1}; \quad x = 2, \Delta x = .003$

26. What is elasticity of demand (in words; no mathematical symbols allowed)? Why is the derivative used to describe elasticity?

27. When solving applied extrema problems, why is it necessary to check the endpoints of the domain?

28. Find two numbers whose sum is 25 and whose product is a maximum.

29. Find nonnegative numbers x and y such that $x = 2 + y$, and xy^2 is minimized.

Applications

Business and Economics

30. Profit Suppose the profit from a product is $P(x) = 40x - x^2$, where x is the price in hundreds of dollars.

(a) At what price will the maximum profit occur?

(b) What is the maximum profit?

31. Profit The total profit (in tens of dollars) from the sale of x hundred boxes of candy is given by

$$P(x) = -x^3 + 10x^2 - 12x.$$

(a) Find the number of boxes of candy that should be sold in order to produce maximum profit.

(b) Find the maximum profit.

32. Packaging Design The packaging department of a corporation is designing a box with a square base and no top. The volume is to be 32 m^3. To reduce cost, the box is to have minimum surface area. What dimensions (height, length, and width) should the box have?

33. Packaging Design Fruit juice will be packaged in cylindrical cans with a volume of 40 cubic inches each. The top and bottom of the can cost 4¢ per square inch, while the sides cost 3¢ per square inch. Find the radius and height of the can of minimum cost.

34. Packaging Design A company plans to package its product in a cylinder that is open at one end. The cylinder is to have a volume of 27π cubic inches. What radius should the circular bottom of the cylinder have to minimize the cost of the material?

35. Order Quantity A store sells 980,000 cases of a product annually. It costs \$2 to store one case for a year and \$20 to place a reorder. Find the number of cases that should be ordered each time.

36. Order Quantity A very large camera store sells 320,000 rolls of film annually. It costs 10¢ to store one roll for one year and \$10 to place a reorder. Find the number of rolls that should be ordered each time.

37. Lot Size In one year, a health food manufacturer produces and sells 240,000 cases of vitamins. It costs \$2 to store a case for one year and \$15 to produce each batch. Find the number of batches that should be produced annually.

38. Lot Size A company produces 128,000 cases of soft drink annually. It costs \$1 to store a case for one year and \$10 to produce one lot. Find the number of lots that should be produced annually.

39. Elasticity Suppose the demand function for a product is given by $q = A/p^k$, where A and k are positive constants. For what values of k is the demand elastic? inelastic?

Life Sciences

40. Pollution A circle of pollution is spreading from a broken underwater waste disposal pipe, with the radius increasing at the rate of 4 ft/min. Find the rate of change of the area of the circle when the radius is 7 ft.

Physical Sciences

41. Sliding Ladder A 50-ft ladder is placed against a building. The top of the ladder is sliding down the building at the rate of 2 ft/min. Find the rate at which the base of the ladder is slipping away from the building at the instant that the base is 30 ft from the building.

42. Spherical Radius A large weather balloon is being inflated with air at the rate of 1.2 ft^3/min. Find the rate of change of the radius when the radius is 1.2 ft.

43. **Water Level** A water trough 2 ft across, 4 ft long, and 1 ft deep has ends in the shape of isosceles triangles. (See the figure.) It is being filled with 3.5 ft^3 of water per minute. Find the rate at which the depth of water in the tank is changing when the water is 1/3 ft deep.

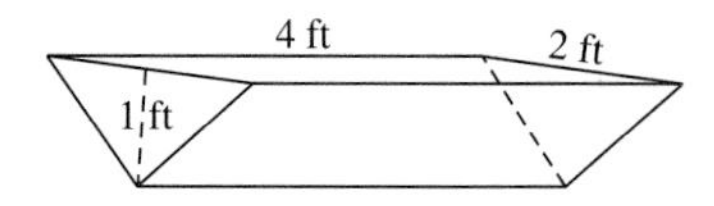

General Interest

44. **Volume** Approximate the volume of coating on a sphere of radius 4 inches if the coating is .02 inch thick.

45. **Area** A square has an edge of 9.2 inches, with a possible error in the measurement of $\pm .04$ inch. Estimate the possible error in the area of the square.

46. **Playground Area** The city park department is planning an enclosed play area in a new park. One side of the area will be against an existing building, with no fence needed there. Find the dimensions of the maximum rectangular area that can be made with 900 m of fence.

Connections

47. (a) Suppose x and y are related by the equation
$$-12x + x^3 + y + y^2 = 4.$$
Find all critical points on the curve.
(b) Determine whether the critical points found in part (a) are relative maxima or relative minima by taking values of x nearby and solving for the corresponding values of y.
(c) Is there an absolute maximum or minimum for x and y in the relationship given in part (a)? Why or why not?

48. In Exercise 47, implicit differentiation was used to find the relative extrema. The exercise was contrived to avoid various difficulties that could have arisen. Discuss some of the difficulties that might be encountered in such problems, and how these difficulties might be resolved.

Extended Application / A Total Cost Model for a Training Program*

In this application, we set up a mathematical model for determining the total costs in setting up a training program. Then we use calculus to find the time interval between training programs that produces the minimum total cost. The model assumes that the demand for trainees is constant and that the fixed cost of training a batch of trainees is known. Also, it is assumed that people who are trained, but for whom no job is readily available, will be paid a fixed amount per month while waiting for a job to open up.

The model uses the following variables.

D = demand for trainees per month
N = number of trainees per batch
C_1 = fixed cost of training a batch of trainees
C_2 = variable cost of training per trainee per month
C_3 = salary paid monthly to a trainee who has not yet been given a job after training
m = time interval in months between successive batches of trainees
t = length of training program in months
$Z(m)$ = total monthly cost of program

The total cost of training a batch of trainees is given by $C_1 + NtC_2$. However $N = mD$, so that the total cost per batch is $C_1 + mDtC_2$.

After training, personnel are given jobs at the rate of D per month. Thus, $N - D$ of the trainees will not get a job the first month, $N - 2D$ will not get a job the second month, and so on. The $N - D$ trainees who do not get a job the first month produce total costs of $(N - D)C_3$, those not getting jobs during the second month produce costs of $(N - 2D)C_3$, and so on. Since $N = mD$, the costs during the first month can be written as

$$(N - D)C_3 = (mD - D)C_3 = (m - 1)DC_3,$$

*Based on "A Total Cost Model for a Training Program" by P. L. Goyal and S. K. Goyal, Faculty of Commerce and Administration, Concordia University. Used with permission.

while the costs during the second month are $(m - 2)DC_3$, and so on. The total cost for keeping the trainees without a job is thus

$$(m - 1)DC_3 + (m - 2)DC_3 + (m - 3)DC_3 + \cdots + 2DC_3 + DC_3,$$

which can be factored to give

$$DC_3[(m - 1) + (m - 2) + (m - 3) + \cdots + 2 + 1].$$

The expression in brackets is the sum of the terms of an arithmetic sequence, discussed in most algebra texts. Using formulas for arithmetic sequences, the expression in brackets can be shown to equal $m(m - 1)/2$, so that we have

$$DC_3\left[\frac{m(m - 1)}{2}\right] \qquad \textbf{(1)}$$

as the total cost for keeping jobless trainees.

The total cost per batch is the sum of the training cost per batch, $C_1 + mDtC_2$, and the cost of keeping trainees without a proper job, given by equation (1). Since we assume that a batch of trainees is trained every m months, the total cost per month, $Z(m)$, is given by

$$Z(m) = \frac{C_1 + mDtC_2}{m} + \frac{DC_3\left[\frac{m(m - 1)}{2}\right]}{m}$$

$$= \frac{C_1}{m} + DtC_2 + DC_3\left(\frac{m - 1}{2}\right).$$

Exercises

1. Find $Z'(m)$.
2. Solve the equation $Z'(m) = 0$.
   As a practical matter, it is usually required that m be a whole number. If m does not come out to be a whole number in Exercise 2, then m^+ and m^-, the two whole numbers closest to m, must be chosen. Calculate both $Z(m^+)$ and $Z(m^-)$; the smaller of the two provides the optimum value of Z.
3. Suppose a company finds that its demand for trainees is 3 per month, that a training program requires 12 months, that the fixed cost of training a batch of trainees is \$15,000, that the variable cost per trainee per month is \$100, and that trainees are paid \$900 per month after training but before going to work. Use your result from Exercise 2 and find m.
4. Since m is not a whole number, find m^+ and m^-.
5. Calculate $Z(m^+)$ and $Z(m^-)$.
6. What is the optimum time interval between successive batches of trainees? How many trainees should be in a batch?

Applications at a glance . . .
World Population . . . strength of a habit . . . theft of vehicles in the United States . . .

CHAPTER 5

EXPONENTIAL AND LOGARITHMIC FUNCTIONS

(See page 258.)

Exponential functions may be the single most important type of functions used in practical applications. These functions, and the closely related logarithmic functions, are used to describe growth and decay, which are important ideas in management, social science, and biology. A function in which the variable appears in the exponent is an *exponential function.* Exponential functions tend to increase or decrease rapidly.

For many years *logarithms* were used primarily to assist in involved calculations. Current technology has made this use of logarithms obsolete, but *logarithmic functions* play an important role in many applications of mathematics. We shall see in this chapter that exponential and logarithmic functions are intimately related; in fact, they are inverses of each other.

5.1 EXPONENTIAL FUNCTIONS

How much interest will an investment earn?

What is the oxygen consumption of yearling salmon?

Later in this section, in Examples 5 and 6, we will see that the answers to these questions depend on *exponential functions.*

In earlier chapters we discussed functions involving expressions such as x^2, $(2x + 1)^3$, or $x^{3/4}$, where the variable or variable expression is the base of an exponential expression, and the exponent is a constant. In an exponential function, the variable is in the exponent and the base is a constant.

EXPONENTIAL FUNCTION

An **exponential function** with base a is defined as

$$\boldsymbol{f(x) = a^x}, \quad \text{where } a > 0 \text{ and } a \neq 1.$$

(If $a = 1$, the function is the constant function $f(x) = 1$.)

The exponential function $f(x) = 2^x$ can be graphed by making a table of values of x and y, as shown in Figure 1 (on the next page). Plotting these points and drawing a smooth curve through them gives the graph shown in the figure. The graph approaches the negative x-axis but will never actually touch it, since y cannot be 0, so the x-axis is a horizontal asymptote. The graph suggests that the domain is the set of all real numbers and the range is the set of all positive real numbers. The function f, where $f(x) = 2^x$, has a base greater than 1, and the function is increasing over its domain. This graph is typical of the graphs of exponential functions of the form $y = a^x$, where $a > 1$.

x	y
-3	$1/8$
-2	$1/4$
-1	$1/2$
0	1
1	2
2	4
3	8
4	16

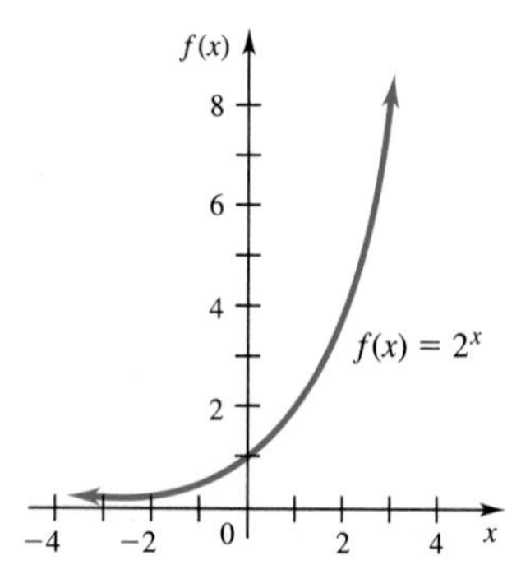

FIGURE 1

EXAMPLE 1 Graph $f(x) = 2^{-x}$.

The graph, shown in Figure 2, is the reflection about the y-axis of the graph of $f(x) = 2^x$ given in Figure 1. This graph is typical of the graphs of exponential functions of the form $y = a^x$ where $0 < a < 1$. The domain includes all real numbers and the range includes all positive numbers. Notice that this function, with $f(x) = 2^{-x} = (1/2)^x$, is decreasing over its domain. ◀

Recall from Section 1.6 that the graph of $f(-x)$ is the reflection of the graph of $f(x)$ about the y-axis.

x	y
-4	16
-3	8
-2	4
-1	2
0	1
1	$1/2$
2	$1/4$
3	$1/8$

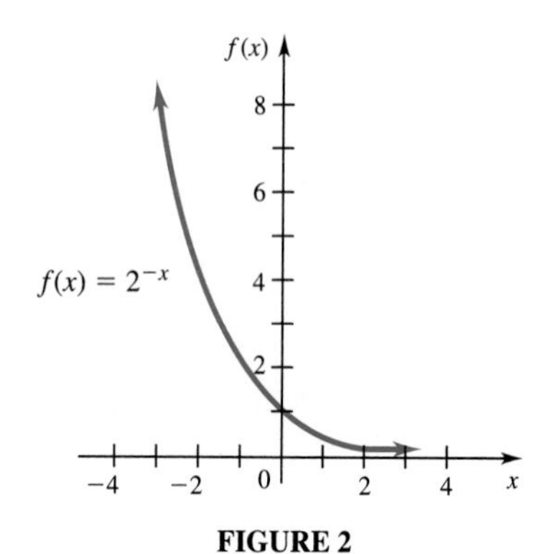

FIGURE 2

In the definition of an exponential function, notice that the base a is restricted to positive values, with negative or zero bases not allowed. For example, the function $y = (-4)^x$ could not include such numbers as $x = 1/2$ or $x = 1/4$ in the domain. The resulting graph would be at best a series of separate points having little practical use.

EXAMPLE 2 Graph $f(x) = -2^x + 3$.

The graph of $y = -2^x$ is symmetric about the x-axis to the graph of $y = 2^x$, so this is a decreasing function. The 3 indicates that the graph should be translated up 3 units, as compared to the graph of $y = -2^x$. Find some ordered pairs. Since $y = -2^x$ would have y-intercept $(0, -1)$, this function has y-intercept (0, 2), which is up 3 units. Some other ordered pairs are (1, 1), (2, −1), and (3, −5). For negative values of x, the graph approaches the line $x = 3$, which is a horizontal asymptote. The graph is shown in Figure 3. ◀

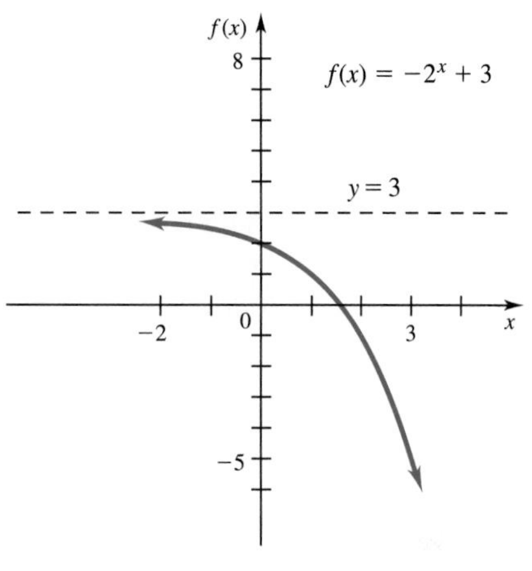

FIGURE 3

We can use the composition of functions to produce more general exponential functions. If $h(u) = ka^u$, where k is a constant and $u = g(x)$, and $f(x) = h[g(x)]$, then

$$f(x) = h[g(x)] = ka^{g(x)}.$$

In the exponential function $f(x) = 4 \cdot 7^{3x-1}$, for example, $k = 4$, $a = 7$, and $g(x) = 3x - 1$.

EXAMPLE 3 ▶ Graph $y = 2^{-x^2}$.

When $x = 1$ or $x = -1$, $y = 2^{-1} = 1/2$. As x becomes larger in magnitude, either positive or negative, y becomes smaller and approaches the horizontal asymptote $y = 0$. Using a graphing calculator or computer program, or plotting numerous points by hand, gives the graph in Figure 4. The final result looks quite different from the typical graphs of exponential functions in Figures 1–3. Graphs like this are important in probability, where the normal curve has an equation similar to the one in this example. ◀

x	y
−2	1/16
−1	1/2
0	1
1	1/2
2	1/16

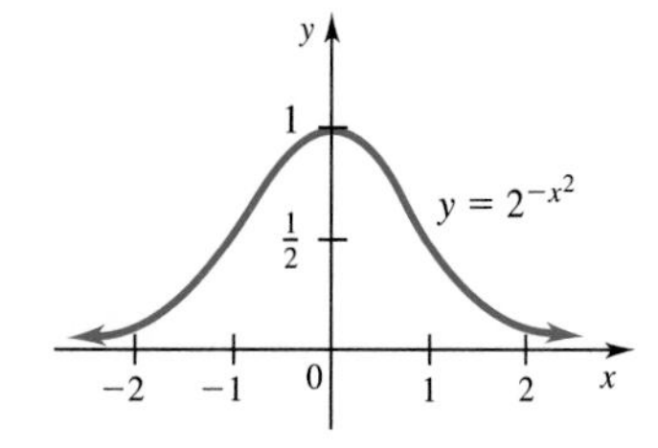

FIGURE 4

Exponential Equations In Figures 1 and 2, which are typical graphs of exponential functions, a given value of x leads to exactly one value of a^x. Because of this, an equation with a variable in the exponent, called an **exponential equation,** often can be solved using the following property.

If $a > 0$, $a \neq 1$, and $a^x = a^y$, then $x = y$. Also, if $x = y$, then $a^x = a^y$.

(Both bases must be the same.) The value $a = 1$ is excluded since $1^2 = 1^3$, for example, even though $2 \neq 3$. To solve $2^{3x} = 2^7$ using this property, work as follows.

$$\begin{aligned} 2^{3x} &= 2^7 \\ 3x &= 7 \\ x &= \frac{7}{3} \end{aligned}$$

EXAMPLE 4

(a) Solve $9^x = 27$.

First rewrite both sides of the equation so the bases are the same. Since $9 = 3^2$ and $27 = 3^3$,

$$\begin{aligned} (3^2)^x &= 3^3 \\ 3^{2x} &= 3^3 \\ 2x &= 3 \\ x &= \frac{3}{2}. \end{aligned}$$

(b) Solve $32^{2x-1} = 128^{x+3}$.

Since the bases must be the same, write 32 as 2^5 and 128 as 2^7, giving

$$\begin{aligned} 32^{2x-1} &= 128^{x+3} \\ (2^5)^{2x-1} &= (2^7)^{x+3} \\ 2^{10x-5} &= 2^{7x+21}. \end{aligned}$$

Now use the property above to get

$$\begin{aligned} 10x - 5 &= 7x + 21 \\ 3x &= 26 \\ x &= \frac{26}{3}. \end{aligned}$$

Verify this solution in the original equation. ◀

Compound Interest The calculation of compound interest is an important application of exponential functions. The cost of borrowing money or the return on an investment is called **interest.** The amount borrowed or invested is the **principal,** P. The **rate of interest** r is given as a percent per year, and t is the **time,** measured in years.

SIMPLE INTEREST

The product of the principal P, rate r, and time t gives **simple interest,** I:

$$I = Prt.$$

With **compound interest,** interest is charged (or paid) on interest as well as on principal. To find a formula for compound interest, first suppose that P dollars, the principal, is deposited at a rate of interest r per year. The interest earned during the first year is found by using the formula for simple interest:

$$\text{First-year interest} = P \cdot r \cdot 1 = Pr.$$

At the end of one year, the amount on deposit will be the sum of the original principal and the interest earned, or

$$P + Pr = P(1 + r). \tag{1}$$

If the deposit earns compound interest, the interest earned during the second year is found from the total amount on deposit at the end of the first year. Thus, the interest earned during the second year (again found by the formula for simple interest), is given by

$$[P(1 + r)](r)(1) = P(1 + r)r, \tag{2}$$

so that the total amount on deposit at the end of the second year is given by the sum of amounts from (1) and (2) above, or

$$P(1 + r) + P(1 + r)r = P(1 + r)(1 + r) = P(1 + r)^2.$$

In the same way, the total amount on deposit at the end of three years is

$$P(1 + r)^3.$$

After t years, the total amount on deposit, called the **compound amount,** is $P(1 + r)^t$.

When interest is compounded more than once a year, the compound interest formula is adjusted. For example, if interest is to be paid quarterly (four times a year), $1/4$ of the interest rate is used each time interest is calculated, so the rate becomes $r/4$, and the number of compounding periods in t years becomes $4t$. Generalizing from this idea gives the following formula.

COMPOUND AMOUNT

If P dollars is invested at a yearly rate of interest r per year, compounded m times per year for t years, the compound amount is

$$A = P\left(1 + \frac{r}{m}\right)^{tm} \text{ dollars.}$$

EXAMPLE 5 Françoise Dykler invests a bonus of \$9000 at 6% annual interest compounded semiannually for 4 years. How much interest will she earn?

Use the formula for compound interest with $P = 9000$, $r = .06$, $m = 2$, and $t = 4$.

$$A = P\left(1 + \frac{r}{m}\right)^{tm}$$

$$= 9000\left(1 + \frac{.06}{2}\right)^{4(2)} = 9000(1.03)^8$$

To find $(1.03)^8$, use a calculator. You should get 1.266770081, which gives

$$A = 9000(1.266770081) = 11400.93.$$

The investment plus the interest is \$11,400.93. The interest amounts to \$11,400.93 − \$9000 = \$2400.93. ◀

Note When using a calculator to compute the compound interest, keep each partial result in the calculator and avoid rounding off until the final answer.

The Number e Perhaps the single most useful base for an exponential function is the number e, an irrational number that occurs often in practical applications. The letter e was chosen to represent this number in honor of the Swiss mathematician Leonhard Euler (pronounced "oiler") (1707–1783). To see how the number e occurs in an application, begin with the formula for compound interest,

$$P\left(1 + \frac{r}{m}\right)^{tm}.$$

Suppose that a lucky investment produces annual interest of 100%, so that $r = 1.00 = 1$. Suppose also that you can deposit only \$1 at this rate, and for only one year. Then $P = 1$ and $t = 1$. Substituting these values into the formula for compound interest gives

$$P\left(1 + \frac{r}{m}\right)^{t(m)} = \mathbf{1}\left(1 + \frac{\mathbf{1}}{m}\right)^{1(m)} = \left(1 + \frac{1}{m}\right)^{m}.$$

As interest is compounded more and more often, m gets larger and the value of this expression will increase. For example, if $m = 1$ (interest is compounded annually),

$$\left(1 + \frac{1}{m}\right)^{m} = \left(1 + \frac{1}{1}\right)^{1} = 2^1 = 2,$$

so that your \$1 becomes \$2 in one year. Using a calculator with a y^x key gives the results shown in the following table for larger and larger values of m. These results have been rounded to five decimal places.

m	$\left(1 + \frac{1}{m}\right)^m$
1	2
2	2.25
5	2.48832
10	2.59374
25	2.66584
50	2.69159
100	2.70481
500	2.71557
1000	2.71692
10,000	2.71815
1,000,000	2.71828

The table suggests that as m increases, the value of $(1 + 1/m)^m$ gets closer and closer to some fixed number. It turns out that this is indeed the case. This fixed number is called e.

DEFINITION OF *e*

$$e = \lim_{m \to \infty} \left(1 + \frac{1}{m}\right)^m \approx 2.718281828$$

This last approximation gives the value of e to nine decimal places. Euler approximated e to 23 decimal places using the definition $\lim_{m\to\infty} (1 + 1/m)^m$. Many calculators give values of e^x, usually with a key labeled e^x. Some require two keys, either INV ln x or 2nd ln x. (We will define ln x in the next section.) In Figure 5, the functions $y = 2^x$, $y = e^x$, and $y = 3^x$ are graphed for comparison. The number e is often used as the base in an exponential equation because it provides a good model for many natural, as well as economic, phenomena. In the Applications section of this chapter, we will look at several examples of such applications.

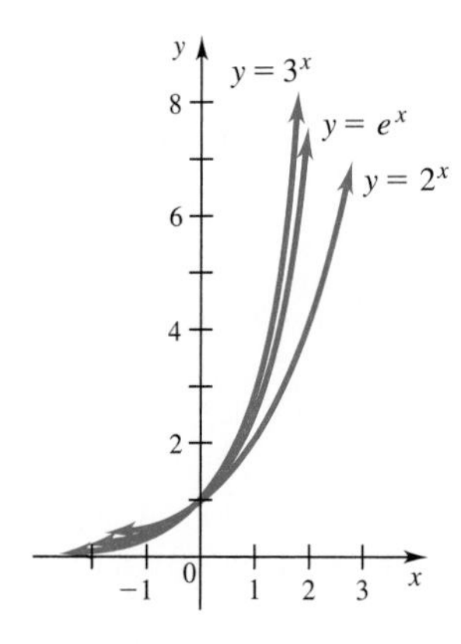

FIGURE 5

Refer to the discussion on linear regression at the end of Section 1.2. A similar process is used to fit data points to other types of functions. Many of the functions in this chapter's applications were determined in this way, including that given in Example 6.

In situations that involve growth or decay of a population, the size of the population at a given time t often is determined by an exponential function of t. The next example illustrates a typical application of this kind.

EXAMPLE 6 Biologists studying salmon have found that the oxygen consumption of yearling salmon (in appropriate units) increases exponentially with the speed of swimming according to the function

$$f(x) = 100(3)^{.6x},$$

where x is the speed in feet per second. Find each of the following.

(a) The oxygen consumption when the fish are still

When the fish are still, their speed is 0. Substitute 0 for x:

$$\begin{aligned} f(0) &= 100(3)^{(.6)(0)} = 100(3)^0 \\ &= 100 \cdot 1 = 100. \end{aligned}$$

When the fish are still, their oxygen consumption is 100 units.

(b) The oxygen consumption at a speed of 2 feet per second

Find $f(2)$ as follows.

$$f(2) = 100(3)^{(.6)(2)} = 100(3)^{1.2}$$

A calculator gives $3^{1.2} \approx 3.73719$ to five decimal places.

$$f(2) = 100(3)^{1.2} \approx 100(3.7) = 370$$

At a speed of 2 feet per second, oxygen consumption is about 370 units. ◀

Note We rounded the answer to two significant digits. Because the function is only an approximation of the real situation, further accuracy is not realistic.

5.1 Exercises

Graph each of the following.

1. $y = 3^x$
2. $y = 4^x$
3. $y = 3^{-x/2}$
4. $y = 4^{-x/3}$
5. $y = \left(\frac{1}{4}\right)^x + 1$
6. $y = \left(\frac{1}{3}\right)^x - 2$
7. $y = -3^{2x-1}$
8. $y = -3^{x^2-1}$
9. $y = e^{-x^2+1}$
10. $y = e^{x+1}$
11. $y = 10 - 5e^{-x}$
12. $y = 100 - 80e^{-x}$

Graph the following equations in the interval $[-3, 2]$ by plotting points or using a graphing calculator or a computer.

13. $y = x \cdot 2^x$
14. $y = x^2 \cdot 2^x$
15. $y = \dfrac{e^x - e^{-x}}{2}$
16. $y = \dfrac{e^x + e^{-x}}{2}$

17. Explain why the exponential equation $4^x = 6$ cannot be solved using the method described in this section.

Solve each equation.

18. $2^x = \frac{1}{8}$ **19.** $4^x = 64$ **20.** $e^x = e^2$ **21.** $e^x = \frac{1}{e^2}$

22. $4^x = 8^{x+1}$ **23.** $25^x = 125^{x-2}$ **24.** $16^{-x+1} = 8^x$ **25.** $16^{x+2} = 64^{2x-1}$

26. $(e^4)^{-2x} = e^{-x+1}$ **27.** $e^{-x} = (e^2)^{x+3}$ **28.** $2^{|x|} = 16$ **29.** $5^{-|x|} = \frac{1}{25}$

30. $2^{x^2-4x} = \left(\frac{1}{16}\right)^{x-4}$ **31.** $5^{x^2+x} = 1$ **32.** $8^{x^2} = 2^{5x+2}$ **33.** $9^{x+4} = 3^{x^2}$

34. Find the average rate of change of $f(x) = 2^x$ as x changes from 0 to x_1 for the following values of x_1.

(a) 1 **(b)** .1 **(c)** .01 **(d)** .001 **(e)** Estimate the slope of the graph of $f(x)$ at $x = 0$.

35. Repeat Exercise 34 for $f(x) = e^x$.

36. Suppose the domain of $y = 2^x$ is restricted to rational values of x (which are the only values discussed prior to this section.) Describe in words the resulting graph.

37. In our definition of exponential function, we ruled out negative values of a. The author of a textbook on mathematical economics, however, obtained a "graph" of $y = (-2)^x$ by plotting the following points and drawing a smooth curve through them.

x	-4	-3	-2	-1	0	1	2	3
y	1/16	$-1/8$	1/4	$-1/2$	1	-2	4	-8

The graph oscillates very neatly from positive to negative values of y. Comment on this approach. (This exercise shows the dangers of point plotting when drawing graphs.)

Applications

Business and Economics

38. Interest Find the interest earned on \$10,000 invested for 5 years at 6% interest compounded as follows.

(a) Annually **(b)** Semiannually (twice a year)
(c) Quarterly **(d)** Monthly

39. Interest Suppose \$26,000 is borrowed for 3 years at 12% interest. Find the interest paid over this period if the interest is compounded as follows.

(a) Annually **(b)** Semiannually **(c)** Quarterly
(d) Monthly

40. Interest Terry Wong deposits a \$9430 commission earned on a real estate sale in the bank at 6% interest compounded quarterly. How much will be on deposit in 4 years?

41. Interest Cynthia Klein-Herring borrows \$17,500 to open a new restaurant. The money is to be repaid at the end of 6 years, with 10% interest compounded semiannually. How much will she owe in 6 years?

42. Interest David Glenn needs to choose between two investments: one pays 8% compounded semiannually, and the other pays 7½% compounded monthly. If he plans to invest \$18,000 for 1½ years, which investment should he choose? How much extra interest will he earn by making the better choice?

43. Interest Find the interest rate required for an investment of \$5000 to grow to \$8000 in 4 years if interest is compounded as follows: **(a)** annually **(b)** quarterly.

44. Inflation If money loses value at the rate of 8% per year, the value of \$1 in t yr is given by

$$y = (1 - .08)^t = (.92)^t.$$

(a) Use a calculator to help complete the following table.

t	0	1	2	3	4	5	6	7	8	9	10
y	1					.66					.43

(b) Graph $y = (.92)^t$.

(c) Suppose a house costs \$165,000 today. Use the results of part (a) to estimate the cost of a similar house in 10 yr.

(d) Find the cost of a \$20 textbook in 8 yr.

45. Employee Training A person learning certain skills involving repetition tends to learn quickly at first. Then learning tapers off and approaches some upper limit. Suppose the number of symbols per minute that a person using a word processor can type is given by

$$p(t) = 250 - 120(2.8)^{-.5t},$$

where t is the number of months the operator has been in training. Find each of the following.

(a) $p(2)$ **(b)** $p(4)$ **(c)** $p(10)$ **(d)** Graph $p(t)$.

(e) What happens to the number of symbols per minute after several months of training?

(f) According to this function, what is the limit to the number of symbols typed per minute?

46. Interest On January 1, 1980, Jack deposited $1000 into Bank X to earn interest at the rate of j per annum compounded semiannually. On January 1, 1985, he transferred his account to Bank Y to earn interest at the rate of k per annum compounded quarterly. On January 1, 1988, the balance at Bank Y was $1990.76. If Jack could have earned interest at the rate of k per annum compounded quarterly from January 1, 1980, through January 1, 1988, his balance would have been $2203.76. Which of the following represents the ratio k/j? (This exercise is reprinted from an actuarial examination.)*

(a) 1.25 **(b)** 1.30 **(c)** 1.35 **(d)** 1.40

(e) 1.45

Life Sciences

47. Population Growth Since 1950, the growth in world population (in millions) closely fits the exponential function defined by

$$A(t) = 2600e^{.018t},$$

where t is the number of years since 1950.

(a) World population was about 3700 million in 1970. How closely does the function approximate this value?

(b) Use the function to approximate world population in 1990. (The actual 1990 population was about 5320 million.)

(c) Estimate world population in the year 2000.

48. Growth of Bacteria *Escherichia coli* is a strain of bacteria that occurs naturally in many different situations. Under certain conditions, the number of these bacteria present in a colony is given by

$$E(t) = 1{,}000{,}000 \cdot 2^{t/30},$$

where $E(t)$ is the number of bacteria present t min after the beginning of an experiment.

(a) How many bacteria were present initially?

(b) How many bacteria are present after 30 min?

(c) How often do the bacteria double?

(d) How quickly will the number of bacteria increase to 32,000,000?

49. Growth of Species Under certain conditions, the number of individuals of a species that is newly introduced into an area can double every year. That is, if t represents the number of years since the species was introduced into the area, and y represents the number of individuals, then

$$y = 6 \cdot 2^t.$$

Assume that the same growth rate continues.

(a) How many animals were originally introduced into the area?

(b) How long will it take for the number to grow to 24 animals?

(c) In how many years will there be 384 animals?

Social Sciences

50. City Planning City planners have determined that the population of a city is

$$P(t) = 1{,}000{,}000(2^{.2t}),$$

where t represents time measured in years. Find each of the following values. Assume the growth rate is maintained.

(a) What population will the city have in 5 years?

(b) In how many years will the population reach 8,000,000?

Physical Sciences

51. Radioactive Decay Suppose the quantity (in grams) of a radioactive substance present at time t is

$$Q(t) = 1000(5^{-.3t}),$$

where t is measured in months.

(a) How much will be present in 6 months?

(b) How long will it take to reduce the substance to 8 g?

---

*Problem 5 from "November 1989 Course 140 Examination, Mathematics of Compound Interest" of the Education and Examination Committee of The Society of Actuaries. Reprinted by permission of The Society of Actuaries.

General Interest

52. Vehicle Theft Vehicle theft in the United States has been rising exponentially since 1972. The number of stolen vehicles (in millions) is given by

$$f(x) = .88(1.03) ,$$

where $x = 0$ represents the year 1972. Find the number of vehicles stolen in the following years.

(a) 1975 **(b)** 1980 **(c)** 1985 **(d)** 1990

Computer/Graphing Calculator

Make a quick sketch of each function using the information from this section and what you learned in Chapter 1 about general graphing techniques. Then use a computer or graphing calculator to graph each function and compare with your sketch.

53. $y = 2^{1-x} + 2$ **54.** $y = 1 - 3^{2x-4}$ **55.** $y = e^{-x+1}$ **56.** $y = e^{x^2/2}$

5.2 LOGARITHMIC FUNCTIONS

With an inflation rate averaging 5% per year, how long will it take for prices to double?

The number of years it will take for prices to double under given conditions is called the **years to double.** For \$1 to double (become \$2) in t years, assuming 5% annual compounding, means that

$$A = P\left(1 + \frac{r}{m}\right)^{mt}$$

becomes

$$2 = 1\left(1 + \frac{.05}{1}\right)^{1(t)}$$

or

$$2 = (1.05)^t.$$

This equation would be easier to solve if the variable were not in the exponent. *Logarithms* are defined for just this purpose. In Example 8, we will use logarithms to answer the question posed above.

DEFINITION OF LOGARITHM

For $a > 0$ and $a \neq 1$,

$$\mathbf{y = \log_a x \quad \text{means} \quad a^y = x.}$$

(Read $y = \log_a x$ as "y is the logarithm of x to the base a.") For example, the exponential statement $2^4 = 16$ can be translated into the logarithmic statement $4 = \log_2 16$. Also, in the problem discussed above $(1.05)^t = 2$ can be rewritten with this definition as $t = \log_{1.05} 2$. Think of a logarithm as an exponent: $\log_a x$ is the exponent used with the base a to get x.

EXAMPLE 1 This example shows the same statements written in both exponential and logarithmic forms.

Exponential Form	*Logarithmic Form*
(a) $3^2 = 9$	$\log_3 9 = 2$
(b) $(1/5)^{-2} = 25$	$\log_{1/5} 25 = -2$
(c) $10^5 = 100{,}000$	$\log_{10} 100{,}000 = 5$
(d) $4^{-3} = 1/64$	$\log_4 1/64 = -3$
(e) $2^{-4} = 1/16$	$\log_2 1/16 = -4$
(f) $e^0 = 1$	$\log_e 1 = 0$

Logarithmic Functions For a given positive value of x, the definition of logarithm leads to exactly one value of y, making $y = \log_a x$ the *logarithmic function* of base a (the base a must be positive, with $a \neq 1$).

LOGARITHMIC FUNCTION If $a > 0$ and $a \neq 1$, then the logarithmic function of base a is

$$f(x) = \log_a x.$$

The graphs of the exponential function $f(x) = 2^x$ and the logarithmic function $g(x) = \log_2 x$ are shown in Figure 6. The graphs show that $f(3) = 2^3 = 8$, while $g(8) = \log_2 8 = 3$. Thus, $f(3) = 8$ and $g(8) = 3$. Also, $f(2) = 4$ and $g(4) = 2$. In fact, for any number m, if $f(m) = p$, then $g(p) = m$. Functions

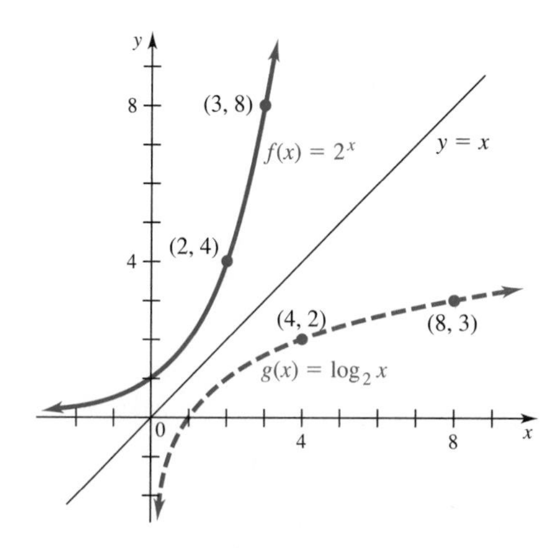

FIGURE 6

related in this way are called ***inverses*** of each other. The graphs also show that the domain of the exponential function (the set of real numbers) is the range of the logarithmic function. Also, the range of the exponential function (the set of positive real numbers) is the domain of the logarithmic function. Every logarithmic function is the inverse of some exponential function. This means that we can graph logarithmic functions by rewriting them as exponential functions using the definition of logarithm. The graphs in Figure 6 show a characteristic of inverse functions: their graphs are mirror images about the line $y = x$. A more complete discussion of inverse functions is given in most standard intermediate algebra and college algebra books.

EXAMPLE 2 Graph $y = \log_{1/2} x$.

Use the equivalent equation $(1/2)^y = x$ to find several ordered pairs. Plotting points and connecting them with a smooth curve gives the graph of $y = \log_{1/2} x$ shown in Figure 7, which also includes the graph of $y = (1/2)^x$. Again, the logarithmic and exponential graphs are mirror images with respect to the line $y = x$. ◀

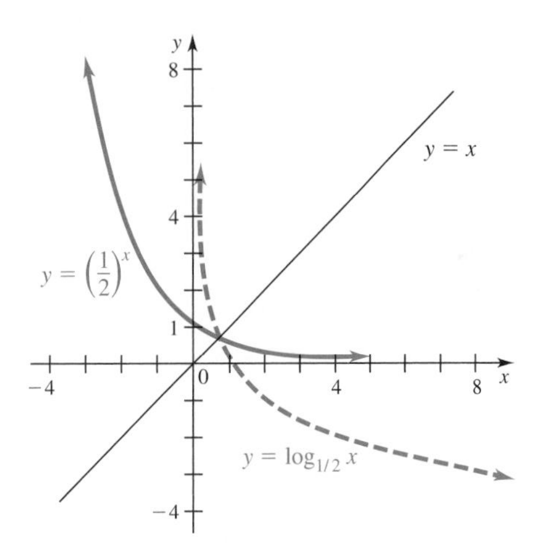

FIGURE 7

The graph of $y = \log_2 x$ shown in Figure 6 is typical of the graphs of logarithmic functions $y = \log_a x$, where $a > 1$. The graph of $y = \log_{1/2} x$, shown in Figure 7, is typical of logarithmic functions of the form $y = \log_a x$, where $0 < a < 1$. For both logarithmic graphs, the y-axis is a vertical asymptote.

Properties of Logarithms The usefulness of logarithmic functions depends in large part on the following **properties of logarithms.**

PROPERTIES OF LOGARITHMS

Let x and y be any positive real numbers and r be any real number. Let a be a positive real number, $a \neq 1$. Then

(a) $\log_a xy = \log_a x + \log_a y$ — Product rule

(b) $\log_a \frac{x}{y} = \log_a x - \log_a y$ — Quotient rule

(c) $\log_a x^r = r \log_a x$ — Power rule

(d) $\log_a a = 1$

(e) $\log_a 1 = 0$

(f) $\log_a a^r = r.$

To prove property (a), let $m = \log_a x$ and $n = \log_a y$. Then, by the definition of logarithm,

$$a^m = x \quad \text{and} \quad a^n = y.$$

Hence,

$$a^m a^n = xy.$$

By a property of exponents, $a^m a^n = a^{m+n}$, so

$$a^{m+n} = xy.$$

Now use the definition of logarithm to write

$$\log_a xy = m + n.$$

Since $m = \log_a x$ and $n = \log_a y$,

$$\log_a xy = \log_a x + \log_a y.$$

Proofs of properties (b) and (c) are left for the exercises. Properties (d) and (e) depend on the definition of a logarithm. Property (f) follows from properties (c) and (d).

EXAMPLE 3 If all the following variable expressions represent positive numbers, then for $a > 0$, $a \neq 1$, the statements in (a)–(c) are true.

(a) $\log_a x + \log_a (x - 1) = \log_a x(x - 1)$

(b) $\log_a \frac{x^2 - 4x}{x + 6} = \log_a (x^2 - 4x) - \log_a (x + 6)$

(c) $\log_a (9x^5) = \log_a 9 + \log_a (x^5) = \log_a 9 + 5 \cdot \log_a x$ ◀

Evaluating Logarithms The invention of logarithms is credited to John Napier (1550–1617), who first called logarithms "artificial numbers." Later he joined the Greek words *logos* (ratio) and *arithmos* (number) to form the word used today. The development of logarithms was motivated by a need for faster computation. Tables of logarithms and slide rule devices were developed by Napier, Henry Briggs (1561–1631), Edmund Gunter (1581–1626), and others.

Historically, one of the main applications of logarithms has been as an aid to numerical calculation. The properties of logarithms and a table of logarithms were used to simplify many numerical problems. With today's widespread use of calculators, this application of logarithms has declined. Since our number system has base 10, logarithms to base 10 were most convenient for numerical calculations, so base 10 logarithms were called **common logarithms.** Common logarithms are still useful in other applications. For simplicity,

$$\mathbf{\log_{10} x} \text{ is abbreviated } \mathbf{\log x}.$$

Most practical applications of logarithms use the number e as base. (Recall that to seven decimal places, $e = 2.7182818$.) Logarithms to base e are called **natural logarithms**, and

$$\mathbf{\log_e x} \text{ is abbreviated } \mathbf{\ln x}$$

(read ''el-en x''). A graph of $f(x) = \ln x$ is given in Figure 8.

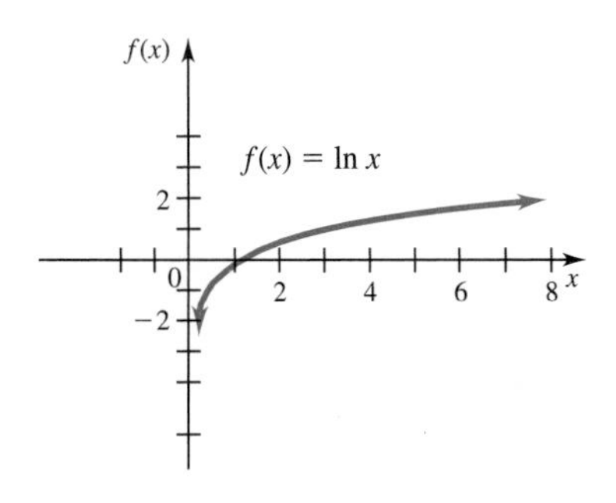

FIGURE 8

Although common logarithms may seem more ''natural'' than logarithms to base e, there are several good reasons for using natural logarithms instead. The most important reason is discussed in Section 4 of this chapter.

A calculator can be used to find both common and natural logarithms. For example, using a calculator and four decimal places, we get

$$\log 2.34 = .3692, \qquad \log 594 = 2.7738, \qquad \text{and} \qquad \log .0028 = -2.5528.$$

We also get

$$\ln 55 = 4.0073, \qquad \ln 1.9 = .6419, \qquad \text{and} \qquad \ln .4 = -.9163.$$

Notice that logarithms of numbers less than 1 are negative when the base is greater than 1. A look at the graph of $y = \log_2 x$ or $y = \ln x$ will show why.

Sometimes it is convenient to use logarithms to bases other than 10 or e. For example, some computer science applications use base 2. Also, some calculators may have only one logarithm key, either $\log x$ or $\ln x$, but not both. In such cases, the following theorem is useful for converting from one base to another.

CHANGE-OF-BASE THEOREM FOR LOGARITHMS

If x is any positive number and if a and b are positive real numbers, $a \neq 1$, $b \neq 1$, then

$$\log_a x = \frac{\log_b x}{\log_b a}.$$

To prove this result, use the definition of logarithm to write $y = \log_a x$ as $x = a^y$ or $x = a^{\log_a x}$ (for positive x and positive a, $a \neq 1$). Now take base b logarithms of both sides of this last equation. This gives

$$\log_b x = \log_b a^{\log_a x}$$

or

$$\log_b x = (\log_a x)(\log_b a), \qquad \text{Power rule}$$

from which

$$\log_a x = \frac{\log_b x}{\log_b a}.$$

If the base b is equal to e, then by the change-of-base theorem,

$$\log_a x = \frac{\log_e x}{\log_e a}.$$

Using $\ln x$ for $\log_e x$ gives the special case of the theorem using natural logarithms.

For any positive numbers a and x, $a \neq 1$,

$$\log_a x = \frac{\ln x}{\ln a}.$$

EXAMPLE 4 Use natural logarithms to find each of the following. Round to the nearest hundredth.

(a) $\log_5 27$

Let $x = 27$ and $a = 5$. Using the second form of the theorem gives

$$\log_5 27 = \frac{\ln 27}{\ln 5}.$$

Now use a calculator.

$$\log_5 27 \approx \frac{3.2958}{1.6094} \approx 2.05$$

To check, use a calculator, along with the definition of logarithm, to verify that $5^{2.05} \approx 27$.

(b) $\log_2 .1$

$$\log_2 .1 = \frac{\ln .1}{\ln 2} \approx \frac{-2.3026}{.6931} = -3.32$$ ◀

Caution As mentioned earlier, when using a calculator, do not round off intermediate results. Keep all numbers in the calculator until you have the final answer. In Example 4(a), we showed the rounded intermediate values of ln 27 and ln 5, but we used the unrounded quantities when doing the division.

Logarithmic Equations Equations involving logarithms are often solved by using the fact that exponential functions and logarithmic functions are inverses, so a logarithmic equation can be rewritten (with the definition of logarithm) as an exponential equation. In other cases, the properties of logarithms may be useful in simplifying an equation involving logarithms.

EXAMPLE 5 Solve each of the following equations for x.

(a) $\log_x \frac{8}{27} = 3$

First, use the definition of logarithm and write the expression in exponential form. To solve for x, take the cube root on both sides.

$$x^3 = \frac{8}{27}$$

$$x = \frac{2}{3}$$

(b) $\log_4 x = \frac{5}{2}$

In exponential form, the given statement becomes

$$4^{5/2} = x$$

$$(4^{1/2})^5 = x$$

$$2^5 = x$$

$$32 = x.$$

(c) $\log_2 x - \log_2 (x - 1) = 1$

By a property of logarithms,

$$\log_2 x - \log_2 (x - 1) = \log_2 \frac{x}{x - 1},$$

so the original equation becomes

$$\log_2 \frac{x}{x - 1} = 1.$$

Use the definition of logarithms to write this result in exponential form.

$$\frac{x}{x - 1} = 2^1 = 2$$

Solve this equation.

$$\frac{x}{x-1}(x-1) = 2(x-1)$$
$$x = 2(x-1)$$
$$x = 2x - 2$$
$$-x = -2$$
$$x = 2$$

It is important to check solutions when solving equations involving logarithms because $\log_a u$, where u is an expression in x, has domain $u > 0$. ◀

Exponential Equations In the previous section exponential equations like $(1/3)^x = 81$ were solved by writing each side of the equation as a power of 3. That method cannot be used to solve an equation such as $3^x = 5$, however, since 5 cannot easily be written as a power of 3. A more general method for solving these equations depends on the following property of logarithms, which is supported by the graphs of logarithmic functions (Figures 6 and 7).

For $x > 0$, $y > 0$, $b > 0$, and $b \neq 1$,

$$\text{if } x = y, \text{ then } \log_b x = \log_b y,$$

and

$$\text{if } \log_b x = \log_b y, \text{ then } x = y.$$

EXAMPLE 6

Solve $3^{2x} = 4^{x+1}$.

Taking natural logarithms (logarithms to any base could be used) on both sides gives

$$\ln 3^{2x} = \ln 4^{x+1}.$$

Now use the power rule for logarithms.

$$2x \ln 3 = (x+1)\ln 4$$
$$(2 \ln 3)x = (\ln 4)x + \ln 4$$
$$(2 \ln 3)x - (\ln 4)x = \ln 4$$
$$(2 \ln 3 - \ln 4)x = \ln 4$$
$$x = \frac{\ln 4}{2 \ln 3 - \ln 4}$$

Use a calculator to evaluate the logarithms, then divide, to get

$$x \approx \frac{1.3863}{2(1.0986) - 1.3863} \approx 1.710.$$ ◀

Just as $\log_a x$ can be written as a base e logarithm, any exponential function $y = a^x$ can be written as an exponential function with base e. For example, there exists a real number k such that

$$2 = e^k.$$

Raising both sides to the power x gives

$$2^x = e^{kx},$$

so that powers of 2 can be found by evaluating appropriate powers of e. To find the necessary number k, solve the equation $2 = e^k$ for k by first taking logarithms on both sides.

$$\begin{aligned} 2 &= e^k \\ \ln 2 &= \ln e^k \\ \ln 2 &= k \ln e \qquad \text{Power rule} \\ \ln 2 &= k \qquad \log_a a = 1 \end{aligned}$$

Thus, $k = \ln 2$. In Section 5 of this chapter, we will see why this change of base is useful. A general statement can be drawn from this example.

CHANGE-OF-BASE THEOREM FOR EXPONENTIALS

For every positive real number a,

$$\boldsymbol{a^x = e^{(\ln a)x}}.$$

Evaluate each exponential by first writing it as a power of e.

(a) $2^{3.57}$

$$2^{3.57} = e^{(\ln 2)(3.57)} \approx 11.88$$

(b) $5^{-1.38} = e^{(\ln 5)(-1.38)} \approx .1085$ ◀

The final examples in this section illustrate the use of logarithms in practical problems.

Complete the solution of the problem posed at the beginning of this section. Recall that if prices will double after t years at an inflation level of 5%, compounded annually, t is given by the equation

$$2 = (1.05)^t.$$

We can solve this equation by first taking natural logarithms on both sides.

$$\begin{aligned} \ln 2 &= \ln (1.05)^t \\ \ln 2 &= t \ln 1.05 \qquad \text{Power rule} \\ t &= \frac{\ln 2}{\ln 1.05} \approx 14.2 \end{aligned}$$

It will take about 14 years for prices to double. 

The problem solved in Example 8 can be generalized for the compound interest equation

$$A = P(1 + r)^t.$$

Solving for t as in Example 8 gives the doubling time in years as

$$t = \frac{\ln 2}{\ln (1 + r)}.$$

It can be shown that for certain values of r,

$$t = \frac{\ln 2}{\ln (1 + r)} \approx \frac{.693}{r},$$

and

$$\frac{70}{100r} \leq \frac{\ln 2}{\ln (1 + r)} \leq \frac{72}{100r}.$$

The **rule of 70** says that for $.001 \leq r \leq .05$, $70/100r$ gives a good approximation of t. The **rule of 72** says that for $.05 \leq r \leq .12$, $72/100r$ approximates t quite well.

EXAMPLE 9 Approximate the years to double at an interest rate of 6% using first the rule of 70, then the rule of 72.

By the rule of 70, money will double at 6% interest after

$$\frac{70}{100r} = \frac{70}{100(.06)} = \frac{70}{6} = 11.67 \left(\text{or } 11\frac{2}{3}\right)$$

years.

Using the rule of 72 gives

$$\frac{72}{100r} = \frac{72}{6} = 12$$

years doubling time. Since a more precise answer is given by

$$\frac{\ln 2}{\ln(1 + r)} = \frac{\ln 2}{\ln(1.06)} = \frac{.693}{.058} \approx 11.9,$$

the rule of 72 gives a better approximation than the rule of 70. This agrees with the statement that the rule of 72 works well for values of r where $.05 \leq r \leq .12$, since $r = .06$ falls into this category. ◀

EXAMPLE 10 One measure of the diversity of the species in an ecological community is given by the *index of diversity H*, where

$$H = \frac{-1}{\ln 2}[P_1 \ln P_1 + P_2 \ln P_2 + \cdots + P_n \ln P_n],$$

and $P_1, P_2, \ldots, P_n$ are the proportions of a sample belonging to each of n species found in the sample. For example, in a community with two species, where there are 90 of one species and 10 of the other, $P_1 = 90/100 = .9$ and $P_2 = 10/100 = .1$, with

$$H = \frac{-1}{\ln 2}[.9 \ln .9 + .1 \ln .1].$$

Using a calculator, we find

$$\ln 2 = .6931, \qquad \ln .9 = -.1054, \qquad \text{and} \qquad \ln .1 = -2.3026.$$

Therefore,

$$H \approx \frac{-1}{.6931}[(.9)(-.1054) + (.1)(-2.3026)]$$
$$\approx .469.$$

Verify that $H \approx .971$ if there are 60 of one species and 40 of the other, and that $H = 1$ if there are 50 of each species. As the index of diversity gets close to 1, there is more diversity, with perfect diversity equal to 1. ◀

5.2 Exercises

Write each exponential equation in logarithmic form.

1. $2^3 = 8$
2. $5^2 = 25$
3. $3^4 = 81$
4. $6^3 = 216$
5. $\left(\frac{1}{3}\right)^{-2} = 9$
6. $\left(\frac{3}{4}\right)^{-2} = \frac{16}{9}$

Write each logarithmic equation in exponential form.

7. $\log_2 128 = 7$
8. $\log_3 81 = 4$
9. $\log_{25} \frac{1}{25} = -1$
10. $\log_2 \frac{1}{8} = -3$
11. $\log 10{,}000 = 4$
12. $\log .00001 = -5$

Evaluate each logarithm.

13. $\log_5 25$
14. $\log_9 81$
15. $\log_4 64$
16. $\log_6 216$
17. $\log_2 \frac{1}{4}$
18. $\log_3 \frac{1}{27}$
19. $\log_2 \sqrt[3]{\frac{1}{4}}$
20. $\log_8 \sqrt[4]{\frac{1}{2}}$
21. $\ln e$
22. $\ln e^2$
23. $\ln e^{5/3}$
24. $\ln 1$

25. Is the logarithm to the base 3 of 4 written as $\log_4 3$ or $\log_3 4$?

26. Write a few sentences describing the relationship between e^x and $\ln x$.

Graph each of the following.

27. $y = \log_4 x$
28. $y = 1 - \log x$
29. $y = \log_{1/3}(x - 1)$
30. $y = \log_{1/2}(1 + x)$
31. $y = \ln x^2$
32. $y = \ln |x|$

Use the properties of logarithms to write each expression as a sum, difference, or product.

33. $\log_9 (7m)$

34. $\log_5 (8p)$

35. $\log_3 \frac{3p}{5k}$

36. $\log_7 \frac{11p}{13y}$

37. $\log_3 \frac{5\sqrt{2}}{\sqrt[4]{7}}$

38. $\log_2 \frac{9\sqrt[3]{5}}{\sqrt[4]{3}}$

Suppose $\log_b 2 = a$ *and* $\log_b 3 = c$. *Use the properties of logarithms to find each of the following.*

39. $\log_b 8$

40. $\log_b 24$

41. $\log_b (72b)$

42. $\log_b (4b^2)$

Use natural logarithms to evaluate each logarithm to the nearest hundredth.

43. $\log_5 20$

44. $\log_{12} 170$

45. $\log_{1.2} 5.5$

46. $\log_{2.8} .12$

47. $\log_3 1.1^{-2.4}$

48. $\log_2 4.5^{-.8}$

Solve each equation in Exercises 49–63. Round decimal answers to the nearest hundredth.

49. $\log_x 25 = -2$

50. $\log_9 27 = m$

51. $\log_8 4 = z$

52. $\log_y 8 = \frac{3}{4}$

53. $\log_r 7 = \frac{1}{2}$

54. $\log_3 (5x + 1) = 2$

55. $\log_5 (9x - 4) = 1$

56. $\log_4 x - \log_4 (x + 3) = -1$

57. $\log_9 m - \log_9 (m - 4) = -2$

58. $3^x = 5$

59. $4^x = 12$

60. $e^{k-1} = 4$

61. $e^{2y} = 12$

62. $2e^{5a+2} = 8$

63. $10e^{3z-7} = 5$

64. Prove: $\log_a \frac{x}{y} = \log_a x - \log_a y$

65. Prove: $\log_a x^r = r \log_a x$

Applications

Business and Economics

66. Inflation Assuming annual compounding, find the time it would take for the general level of prices in the economy to double at the following average inflation rates.

(a) 3% **(b)** 6% **(c)** 8%

(d) Check your answers using either the rule of 70 or the rule of 72, whichever applies.

67. Interest Laura Stowe invests $15,000 in an account paying 6% per year compounded semiannually.

(a) In how many years will the compounded amount double?

(b) In how many years will the amount triple?

(c) Check your answer to part (a) using the rule of 72.

Life Sciences

Index of Diversity *For Exercises 68 and 69, refer to Example 10.*

68. Suppose a sample of a small community shows two species with 50 individuals each. Find the index of diversity H.

69. A virgin forest in northwestern Pennsylvania has 4 species of large trees with the following proportions of each: hemlock, .521; beech, .324; birch, .081; maple, .074. Find the index of diversity H.

70. Growth of a Species The population of an animal species that is introduced into a certain area may grow rapidly at first but then grow more slowly as time goes on. A logarithmic function can provide an excellent model of such growth. Suppose that the population of foxes $F(t)$ in an area t yr after the foxes were introduced there is

$$F(t) = 50 \ln (2t + 3).$$

Find the population of foxes at times (a) through (c) below.

(a) When they are first released into the area (that is, when $t = 0$)

(b) After 3 yr

(c) After 15 yr

(d) Graph $y = F(t)$.

(e) When will the population double?

The graph for Exercise 71 is plotted on a logarithmic scale where differences between successive measurements are not always the same. Data that do not plot in a linear pattern on the usual Cartesian axes often form a linear pattern when plotted on a logarithmic scale. Notice that on the vertical scale, the distance from 1 to 2 is not the same as the distance from 2 to 3, and so on. This is characteristic of a graph drawn with logarithmic scales.*

71. Oxygen Consumption The accompanying graph gives the rate of oxygen consumption for resting guinea pigs of various sizes. This rate is proportional to body mass raised to the power .67. Estimate the oxygen consumption for a guinea pig with body mass of .3 kg. Do the same for one with body mass of .7 kg.

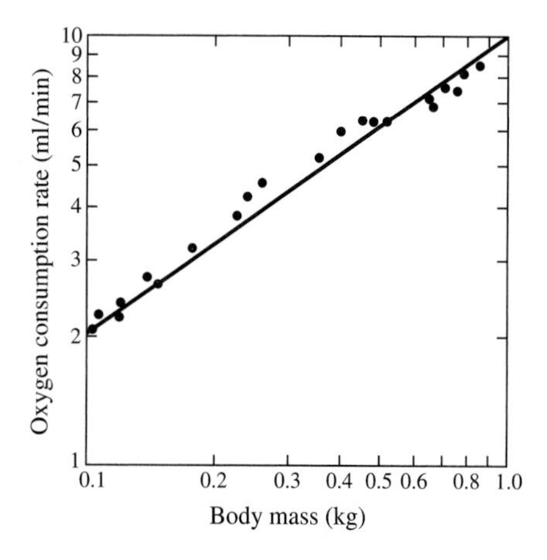

Social Sciences

72. Evolution of Languages The number of years $N(r)$ since two independently evolving languages split off from a common ancestral language is approximated by

$$N(r) = -5000 \ln r,$$

where r is the proportion of the words from the ancestral language that are common to both languages now.
Find each of the following.

(a) $N(.9)$ **(b)** $N(.5)$ **(c)** $N(.3)$

(d) How many years have elapsed since the split if 70% of the words of the ancestral language are common to both languages today?

(e) If two languages split off from a common ancestral language about 1000 years ago, find r.

Physical Sciences

For Exercises 73 and 74, recall that log x represents the common (base 10) logarithm of x.

73. Intensity of Sound The loudness of sounds is measured in a unit called a *decibel*. To do this, a very faint sound, called the *threshold sound*, is assigned an intensity I_0. If a particular sound has intensity I, then the decibel rating of this louder sound is

$$10 \log \frac{I}{I_0}.$$

Find the decibel ratings of the following sounds having intensities as given. Round answers to the nearest whole number.

(a) Whisper, $115I_0$

(b) Busy street, $9{,}500{,}000I_0$

(c) Heavy truck, 20 m away, $1{,}200{,}000{,}000I_0$

(d) Rock music, $895{,}000{,}000{,}000I_0$

(e) Jetliner at takeoff, $109{,}000{,}000{,}000{,}000I_0$

(f) A city council is considering whether outdoor amplified sound should be limited to 96 decibels or 98 decibels. What is the difference in intensity between these two decibel ratings?

74. Earthquake Intensity The intensity of an earthquake, measured on the Richter scale, is given by

$$R(I) = \log \frac{I}{I_0},$$

where I_0 is the intensity of an earthquake of a certain (small) size. Find the Richter scale ratings of earthquakes with the following intensities.

(a) $1{,}000{,}000I_0$ **(b)** $100{,}000{,}000I_0$

(c) The San Francisco earthquake of 1906 had a Richter scale rating of 8.3. Express the intensity of this earthquake as a multiple of I_0.

(d) In 1989, the San Francisco region experienced an earthquake with a Richter scale rating of 7.1. Express the intensity of this earthquake as a multiple of I_0.

(e) Compare the intensities of these two earthquakes.

75. Acidity of a Solution A common measure for the acidity of a solution is its pH. It is defined by $\text{pH} = -\log[H^+]$, where H^+ measures the concentration of hydrogen ions in the solution. The pH of pure water is 7. Solutions that are

---

*From *On Size and Life* by Thomas A. McMahon and John Tyler Bonner. Copyright © 1983 by Thomas A. McMahon and John Tyler Bonner. Reprinted by permission of W. H. Freeman and Company.

more acidic than pure water have a lower pH, while solutions that are less acidic (referred to as basic solutions) have a higher pH.

(a) Acid rain sometimes has a pH as low as 4. How much greater is the concentration of hydrogen ions in such rain than in pure water?

(b) A typical mixture of laundry soap and water for washing clothes has a pH of about 11, while black coffee has a pH of about 5. How much greater is the concentration of hydrogen ions in black coffee than in the laundry?

Computer/Graphing Calculator

Make a quick sketch of each function using the information from this section and what you learned in Chapter 1 about general graphing techniques. Then use a computer or a graphing calculator to graph each function, and compare it with your sketch.

76. $y = \log_2 (-x)$

77. $y = -\log_2 x$

78. $y = 2 + \ln(x - 1)$

79. $y = 30 \ln(2x + 1)$

5.3 APPLICATIONS: GROWTH AND DECAY; MATHEMATICS OF FINANCE

What interest rate will cause \$5000 to grow to \$7250 in 4 years if money is compounded continuously?

This is one of many situations that occur in biology, economics, and the social sciences, in which a quantity changes at a rate proportional to the amount of the quantity present. In such cases the amount present at time t is a function of t called the **exponential growth and decay function.** (The derivation of this equation is presented in a later section on differential equations.)

EXPONENTIAL GROWTH AND DECAY FUNCTION

Let y_0 be the amount or number of some quantity present at time $t = 0$. Then, under certain conditions, the amount present at any time t is given by

$$y = y_0 e^{kt},$$

where k is a constant.

If $k > 0$, then k is called the **growth constant;** if $k < 0$, then k is called the **decay constant.** A common example is the growth of bacteria in a culture. The more bacteria present, the faster the population increases.

EXAMPLE 1 Yeast in a sugar solution is growing at a rate such that 1 gram becomes 1.5 grams after 20 hours. Find the growth function, assuming exponential growth.

The values for y_0 and k in the exponential growth function $y = y_0 e^{kt}$ must be found. Since y_0 is the amount present at time $t = 0$, $y_0 = 1$. To find k, substitute $y = 1.5$, $t = 20$, and $y_0 = 1$ into the equation.

$$y = y_0 e^{kt}$$
$$1.5 = 1\ e^{k(20)}$$

Now take natural logarithms on both sides and use the power rule for logarithms and the fact that $\ln e = 1$.

$$1.5 = e^{20k}$$
$$\ln 1.5 = \ln e^{20k}$$
$$\ln 1.5 = 20k \ln e \qquad \text{Power rule}$$
$$\ln 1.5 = 20k \qquad \ln e = 1$$
$$\frac{\ln 1.5}{20} = k$$
$$k \approx .02 \text{ (to the nearest hundredth)}$$

The exponential growth function is $y = e^{.02t}$, where y is the number of grams of yeast present after t hours. ◀

The decline of a population or decay of a substance may also be described by the exponential growth function. In this case the decay constant k is negative, since an increase in time leads to a decrease in the quantity present. Radioactive substances provide a good example of exponential decay. By definition, the **half-life** of a radioactive substance is the time it takes for exactly half of the initial quantity to decay.

EXAMPLE 2 Carbon 14 is a radioactive form of carbon that is found in all living plants and animals. After a plant or animal dies, the carbon 14 disintegrates. Scientists determine the age of the remains by comparing its carbon 14 with the amount found in living plants and animals. The amount of carbon 14 present after t years is given by the exponential equation

$$A(t) = A_0 e^{kt},$$

with $k = -[(\ln 2)/5600]$.

(a) Find the half-life of carbon 14.

Let $A(t) = (1/2)A_0$ and $k = -[(\ln 2)/5600]$.

$$\frac{1}{2}A_0 = A_0 e^{-[(\ln 2)/5600]t}$$
$$\frac{1}{2} = e^{-[(\ln 2)/5600]t} \qquad \text{Divide by } A_0.$$
$$\ln \frac{1}{2} = \ln e^{-[(\ln 2)/5600]t} \qquad \text{Take logarithms of both sides.}$$
$$\ln \frac{1}{2} = -\frac{\ln 2}{5600}t \qquad \ln e^x = x$$
$$-\frac{5600}{\ln 2} \ln \frac{1}{2} = t \qquad \text{Multiply by } -\frac{5600}{\ln 2}.$$

$$-\frac{5600}{\ln 2}(\ln 1 - \ln 2) = t \qquad \text{Quotient rule}$$

$$-\frac{5600}{\ln 2}(-\ln 2) = t \qquad \ln 1 = 0$$

$$5600 = t$$

The half-life is 5600 years.

(b) Charcoal from an ancient fire pit on Java had ¼ the amount of carbon 14 found in a living sample of wood of the same size. Estimate the age of the charcoal.

Let $A(t) = \frac{1}{4}A_0$ and $k = -[(\ln 2)/5600]$.

$$\frac{1}{4}A_0 = A_0 e^{-[(\ln 2)/5600]t}$$

$$\frac{1}{4} = e^{-[(\ln 2)/5600]t}$$

$$\ln\frac{1}{4} = \ln e^{-[(\ln 2)/5600]t}$$

$$\ln\frac{1}{4} = -\frac{\ln 2}{5600}t$$

$$-\frac{5600}{\ln 2}\ln\frac{1}{4} = t$$

$$t = 11{,}200$$

The charcoal is about 11,200 years old. ◀

Recall the formula for compound amount $A = P\left(1 + \frac{r}{m}\right)^{tm}$ given in the first section of this chapter.

Continuous Compounding In economics, the formula for **continuous compounding** is a good example of an exponential growth function. As interest is compounded more and more often, the compound amount A approaches

$$\lim_{m\to\infty} P\left(1 + \frac{r}{m}\right)^{tm}.$$

For example, if $P = 1$ and $t = 1$,

$$A = \left(1 + \frac{r}{m}\right)^{m}.$$

Let $r/m = 1/s$ so that $m = sr$; then

$$\lim_{m\to\infty}\left(1 + \frac{r}{m}\right)^{m} = \lim_{s\to\infty}\left(1 + \frac{1}{s}\right)^{sr}$$

$$= \lim_{s\to\infty}\left[\left(1 + \frac{1}{s}\right)^{s}\right]^{r}$$

$$= \left[\lim_{s\to\infty}\left(1 + \frac{1}{s}\right)^{s}\right]^{r}.$$

Calculus at Work

This section shows you that calculus is not just a body of theorems, concepts, and exercises without any real human dimension; there are practicing mathematicians, scientists, and engineers using this knowledge. The following essays were written by professionals who work with mathematics daily to create products and services we commonly use.

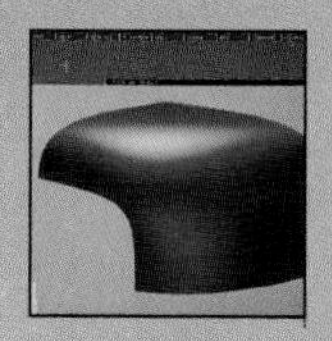

W. Weston Meyer
Mathematics Department, Research Laboratories
General Motors Corporation
"Perfecting the Skin of an Automobile"

Richard E. Klabunde, Ph.D.
Senior Cardiovascular Group Leader, Department of Pharmacology
Abbott Laboratories
Roberto Mercado, Ph.D.
Manager, Nonclinical Statistics, Pharmaceutical Products Division
Abbott Laboratories
"Drugs, Derivatives, and Heart Disease"

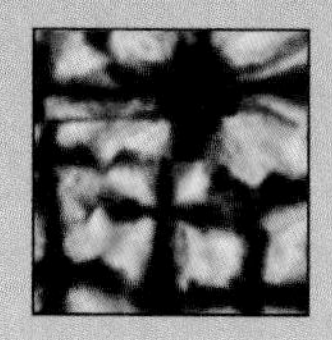

Jane Gillum (Summer Intern), Alan Opsahl, Allen Blaurock
Technology Center
Kraft General Foods
"Crystals, Integration, and . . . Margarine!"

Henry Robinson
Meteorologist, National Weather Service
National Oceanic and Atmospheric Administration
"Calculus in a Cornfield"

When you read these testaments, you may appreciate more the phenomenal versatility of calculus–how it is useful, in fact indispensable, in virtually countless aspects of our lives: from the margarine you spread on your breakfast toast, to the car you drive, to predicting the weather you experience.

The fact that calculus works as well as it does in as many ways as it does leads us to the following conclusion: learning about calculus is learning about an amazing pattern, one repeated in the most disparate and surprising places in our world.

Perfecting the Skin of an Automobile

W. Weston Meyer
Mathematics Department, Research Laboratories / General Motors Corporation, Warren, Michigan

As far as we know, it was Jacob Bernoulli (1654 – 1705) who first published a formula to measure curvature. Three centuries later, the same formula is put to uses Bernoulli could never have imagined. Designing the "skin" of an automobile — its outer panel surface — is one such use.

Bernoulli's formula, given below, provides an expression for the radius R of the circle whose curvature is identical to the curvature of the curve $y = f(x)$ at a fixed point (see Figure 1).

$$R(x) = \frac{\left[1 + f'(x)^2\right]^{3/2}}{|f''(x)|}$$

Since curvature of a circle is equal to the inverse of its radius, the curvature of f at a fixed point is the inverse of the expression for $R(x)$.

Traditionally, the outer panels of an automobile materialized from scores of longitudinal and transverse section lines painstakingly inscribed on parchment by drafting artisans, then transferred to a clay mass by means of wooden templates. The clay was then smoothed between section lines by hand to give an iconic model.

Today, the design process is largely computer-aided. The artisan sits before a graphic display console and chooses control parameters to generate section lines, plus the surface that interpolates them, as mathematical entities. Feedback information is immediate and visual, but a computer image does not always convey enough information for the designer to make an aesthetic judgment. A line or surface that seems impeccable on the computer screen may have very

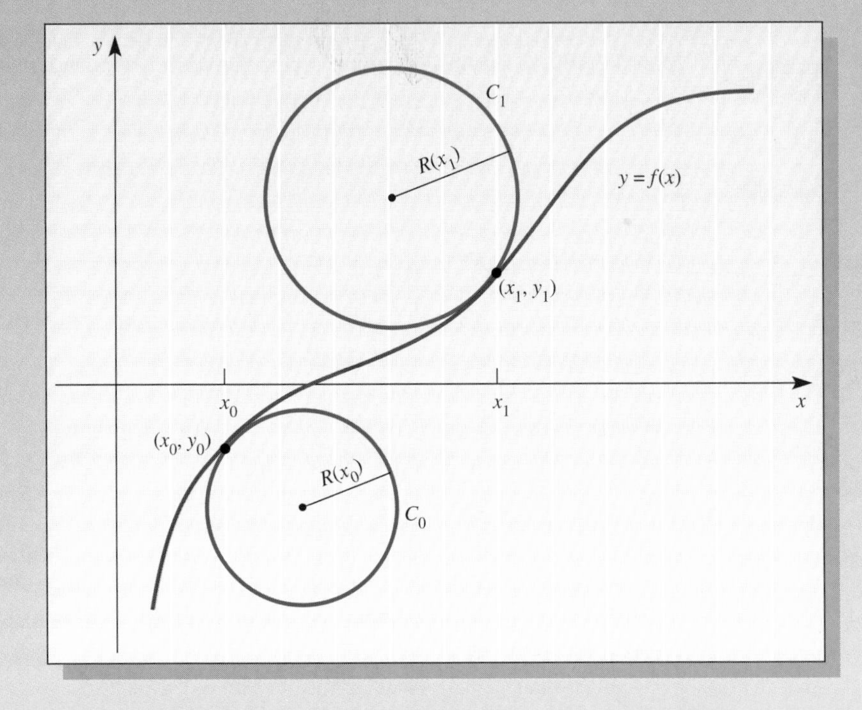

Figure 1 The curvature of circle C_0 is identical to the curvature of $y = f(x)$ at x_0; the curvature of circle C_1 is identical to the curvature of $y = f(x)$ at x_1. In each case the radius R is a function of x:

$$R(x) = \frac{\left[1 + f'(x)^2\right]^{3/2}}{|f''(x)|}.$$

minuscule undulations, flat spots, or depressions that become apparent only after full-scale manufacturing.

Curvature measurements can be especially helpful in detecting and eliminating such flaws at the design level. Figures 2(a) and 2(b) show a line "before and after" adjustment, with emanating quills to represent, by their lengths, the local curvatures according to Bernoulli. Before adjustment, the line has invisible irregularities, as evidenced by occasional spikes among the quills. After adjustment, the quill pattern is quite smooth.

Calculus at a somewhat higher level provides techniques to measure curvature of surfaces. Figure 3(a) shows a typical segment of an automobile outer panel. Superimposed on the surface in Figure 3(b) are curvature measurements, communicated by means of color coding. The presence of blue, in contrast to yellow, indicates a local flattening that needs to be eliminated. Figure 3(c) shows the result of a minor mathematical adjustment: the flattening is gone.

Figure 3 On the surface of an automobile's outer panel (a), curvature measurements coded by color indicate flattenings that need to be eliminated (b). In (c), the surface appears smooth after minor mathematical adjustments are made.

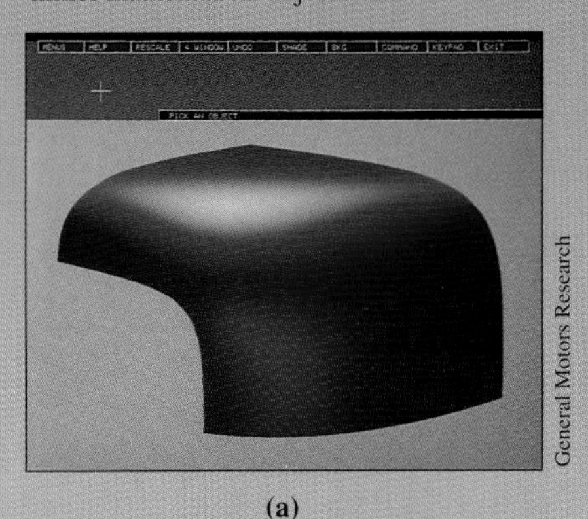

General Motors Research

(a)

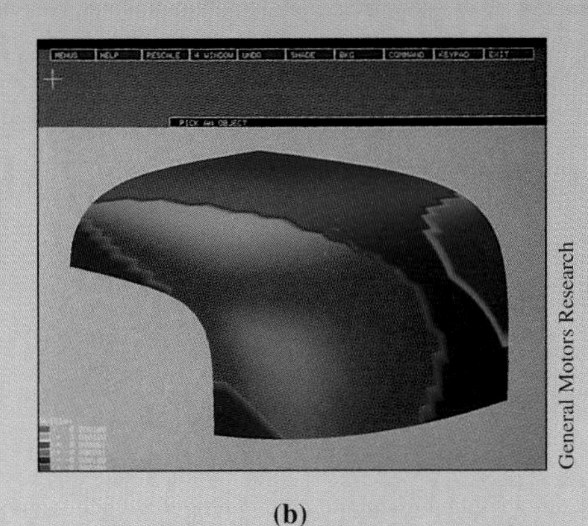

General Motors Research

(b)

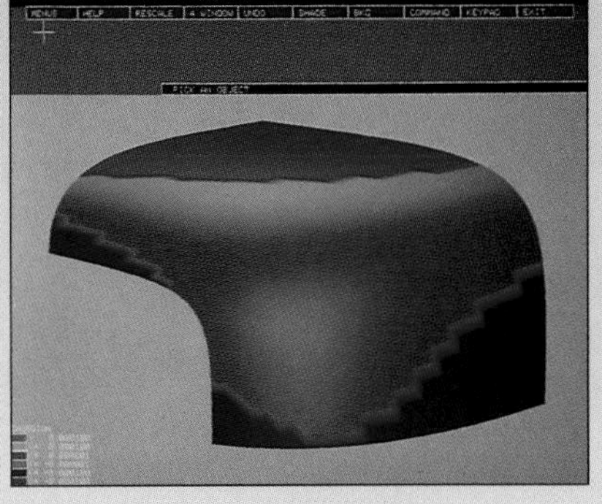

General Motors Research

(c)

Figure 2 The length of each quill emanating from the curve in (a) represents the local curvature according to Bernoulli's formula. In (b), the quills are quite regular, after adjustments have fixed the unseen undulations in the curve. (Figure from "The Porcupine Technique: Principle, Application, and Algorithm" by Klaus-Peter Beier, October 1987. Reprinted by permission of Klaus-Peter Beier.)

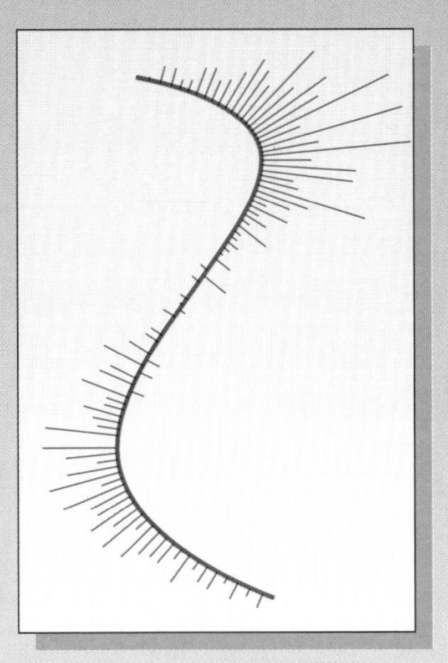

(a)

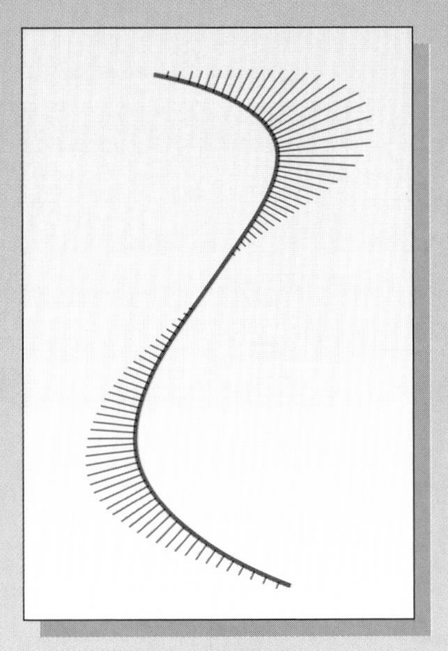

(b)

Drugs, Derivatives, and Heart Disease

Richard E. Klabunde, Ph.D.
Senior Cardiovascular Group Leader, Department of Pharmacology
Abbott Laboratories, Abbott Park, Illinois

Roberto Mercado, Ph.D.
Manager, Nonclinical Statistics, Pharmaceutical Products Division
Abbott Laboratories, Abbott Park, Illinois

(Dr. Alan Sewell served as a consultant for this article.)

A healthy heart is a flexible muscle, one that contracts vigorously to force blood out its left ventricle, then relaxes easily to repeat the process. The pressure exerted by blood leaving the heart is measured as a function of time (see Figure 1). The *rate* of change of pressure with respect to time measures the *contractility* of the heart — that is, how quickly the heart contracts and relaxes. A heart with poor contractility will demonstrate a consistently slow rate of change of pressure. Drugs are used to improve this rate.

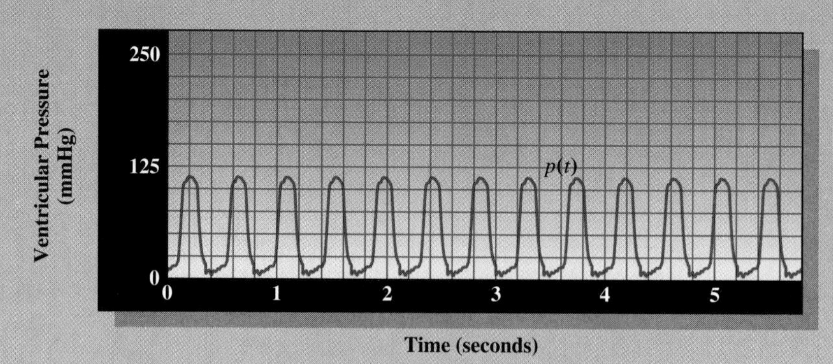

Figure 1 Left ventricular pressure is measured in mmHg as a function of time.

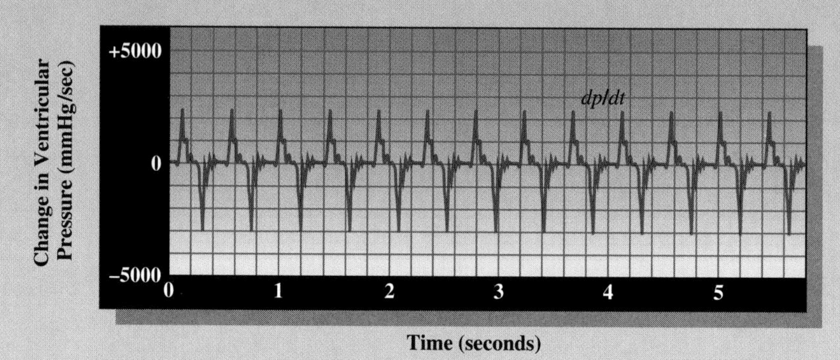

Figure 2 Change in left ventricular pressure is measured in mmHg / sec as a function of time. dp/dt is computed using an analog differentiator.

Mathematically, we can describe the pressure of blood leaving the left ventricle as $p(t)$, where p is the pressure at a given time t. The rate of change of pressure with respect to time, then, can be described as the derivative of $p(t)$, or dp/dt. Pharmacologists are interested in both the positive rate of change, measuring the contraction of the heart, and the negative rate of change, measuring the relaxation of the heart. The positive rate of change is denoted by $+dp/dt$, while the negative rate of change is denoted by $-dp/dt$. The critical measurements are the maximum positive rate of change and the maximum negative rate of change: the heart's "best efforts" in contracting and relaxing.

As is often the case when we use mathematics to model the real world, an explicit equation describing $p(t)$ does not exist. Pressure is clearly a function of time, and can be measured accurately. But we cannot express the relationship except in the form of data and graphs.

Since no equation explicitly describes $p(t)$, no equation explicitly describes dp/dt. In practice, dp/dt is determined by inserting a fluid-filled plastic tube into the ventricular cavity. The tube is hooked up to a gauge and an amplifier, and pressure is measured directly. A special type of computer, called an *analog differentiator,* receives the output signal of the amplifier and numerically approximates values for $+dp/dt$ and $-dp/dt$, particularly the maximum values (see Figure 2). With these results, a physician can judge the contractility of the heart and, when a drug is being used, the effectiveness of the drug in improving contractility.

Abbott Laboratories

Pharmacologist making amplifier adjustments to the analog differentiator.

Crystals, Integration, and . . . Margarine!

Jane Gillum, Summer Intern; Alan Opsahl; Allen Blaurock
Technology Center / Kraft General Foods, Glenview, Illinois

Crystals of gems are valuable commodities, but in the manufacturing of certain foods, it is crystals of fat that are valuable. Makers of peanut butter, for example, add a small amount of a crystallizable fat — "partially hydrogenated vegetable oil" — so that a separate layer of oil doesn't form on top.

In making margarine, the rate fat crystals form is very important, because if the crystals form too rapidly, the margarine solidifies and plugs the pipes on its way to the machine where the margarine is shaped into sticks. Of course, sticks won't form at all if fat crystals form too slowly.

Crystallization is an orderly process involving two key events: (1) the creation of a crystal "seed" and (2) the growth of the crystal as oil molecules add on to the seed. The number n of crystal seeds depends on temperature, time, and the volume of oil. The size g, in meters, of the growing crystal seed depends on tem-

Figure 1 *Beta* crystals formed by very slow cooling of trilaurin, an edible oil. The feathery growth pattern started from a *single* crystal seed. Polarized-light micrograph (x 62.5).

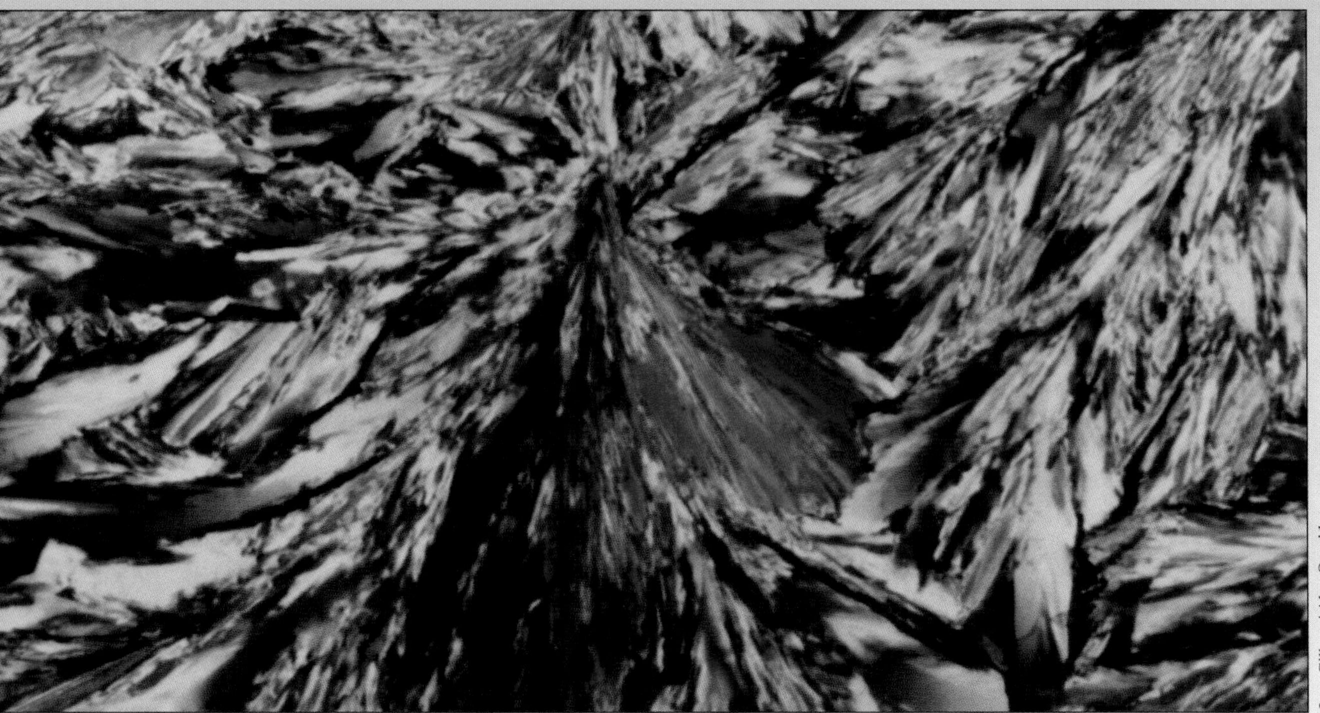

Jane Gillum / Alan Opsahl

perature and time. Given a constant temperature and a fixed volume of oil, however, both n and g are simple, constant functions of time: $n(t) = Nt$ and $g(t) = Gt$ (in meters). The rate of seed creation can thus be given as $n'(t) = N$, a constant, and the rate of size growth as $g'(t) = G$, also a constant.

The volume of the crystals is a function of the rate of seed creation, the rate of size growth, and time. It is, in fact, an integral function. Fixing the amount of oil and the temperature, the volume V, in cubic meters, of crystals can be expressed as

$$V(t) = \left(8NG^3\right)\int_0^t x^3 dx .$$

Finding the antiderivative and evaluating,

$$V(t) = 8NG^3\left[x^4/4\right]_0^t$$
$$= 2NG^3t^4 .$$

Keen-eyed readers will recognize that this expression for V cannot be valid for arbitrarily large values of t, for then the volume of the crystals would grow arbitrarily large, too. This growth does not in fact occur, however, since the amount of oil, after all, is fixed. Thus, the expression for V does not hold true for large values of t. It is valid only for the early period of crystallization, that is, when t is small.

Since the rate N at which crystal seeds appear varies with temperature, margarine makers alter temperature to create varying textures. Lower temperatures mean more crystal seeds and a finer texture; higher temperatures mean fewer crystal seeds and a coarser texture. Spreadability and "mouthfeel" are controlled, then, by adjusting temperature and, hence, texture.

Figure 2 *Beta prime* (dim circular regions) and *alpha* crystals (brighter regions) formed by more rapid cooling. Each region ("Maltese crosses") started with a single crystal seed. Polarized-light micrograph (x 125).

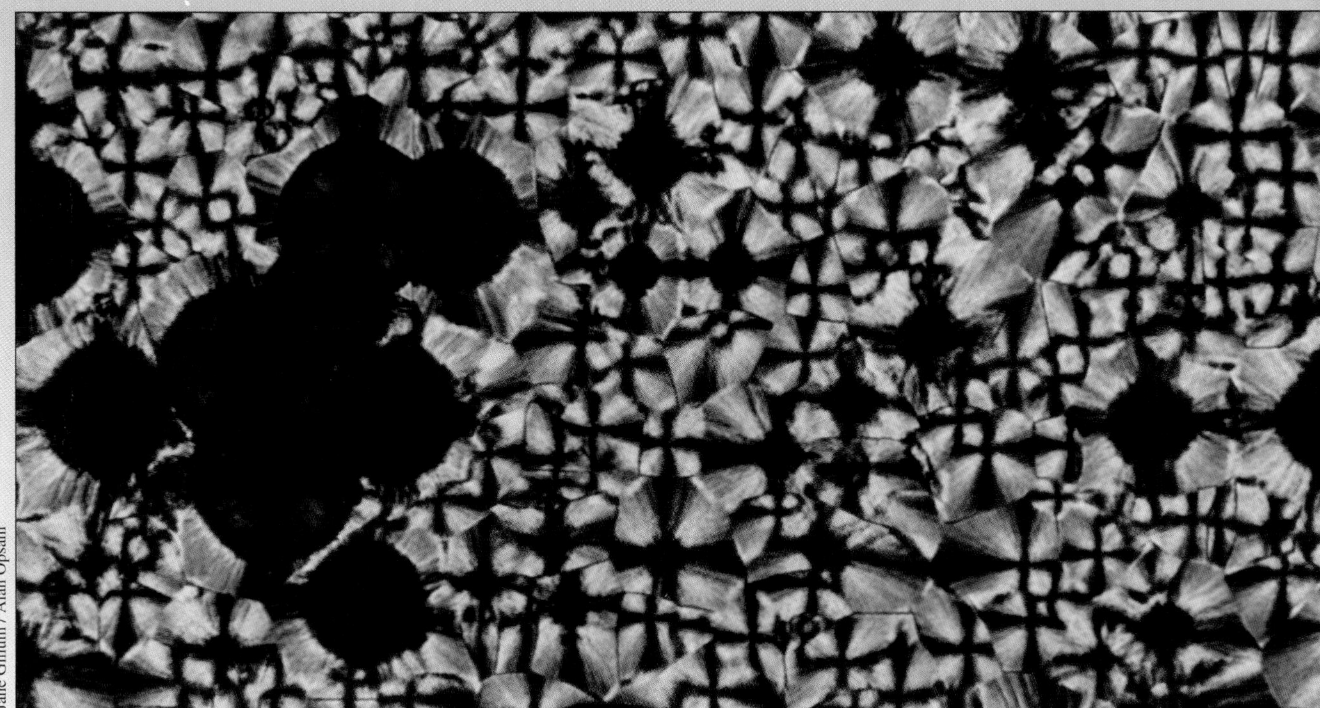

Jane Gillum / Alan Opsahl

Grant Heilman / Grant Heilman Photography

Calculus in a Cornfield

Henry Robinson

Meteorologist, National Weather Service

National Oceanic and Atmospheric Administration, Silver Springs, Maryland

Predicting the weather is important for almost everybody, but for farmers it often means the difference between success and failure.

For example, agriculture specialists have found that the amount of time a crop needs to reach peak maturity is closely related to the temperature during its growing time. Predicting the daily temperature after planting can help a farmer know the optimal harvest day and schedule a manageable workload.

The relationship between temperature and growing time is expressed in terms of *growing degree days*, determined as follows: when the average temperature on a day is one degree above a certain threshold (specific to each crop); the day amounts to one growing degree day, when the average temperature on a day is two degrees above threshold, the day amounts to two growing degree days; and so on. For peas, the threshold temperature has been found to be 39° F. Thus, if a day's average temperature is 72° F, the day amounts to 33 growing degree days for peas.

The total number of growing degree days D between the time t_p of planting and the time t_h of optimal harvest is thus a sum, expressible as an integral:

$$D = \int_{t_p}^{t_h} [T(t) - r]\,dt,$$

where $T(t)$ is the temperature at any time t and r is the threshold temperature.

Predicting t_h, the time of optimal harvest, for a crop involves knowing the threshold temperature and total number of growing degree days for that crop, and finding an explicit expression for $T(t)$, the temperature at any time t.

For virtually all cash crops, experience and experimentation has allowed agricultural specialists to determine the threshold temperature and the total number of growing degree days. This information has not gone unnoticed by investors who deal in crop futures on the various worldwide exchanges. The missing piece of information is an explicit expression for $T(t)$.

One formulation is the Cosine Day of the Year Model for Temperature. Patterns of temperature change for individual locales tend to resemble a cosine curve, as the red line in the figure shows. Modifying the cosine function by appropriate factors results in an explicit expression of temperature as a function of time (the blue line). In general the model used is

$$T(t) = a \cdot \cos(t - b) + c,$$

where T is temperature, t is the day of the year (1 to 365), and a, b, c are constants.

Using this model allows the integral expression for the total number of growing degree days D to be evaluated, and hence the time of optimal harvest t_h to be approximated.

On the national and international scene, this information helps to predict potential shortages or oversupplies of crops. Farmers can use a simpler technique to determine the optimal harvest time, since they can reasonably consider the change in time dt to be a single day. By taking each day's temperature and subtracting the threshold, the farmer keeps a running total of growing degree days. When the total gets close to D, the farmer predicts the growing degree days on the basis of the weather forecast, thereby identifying the day for harvesting.

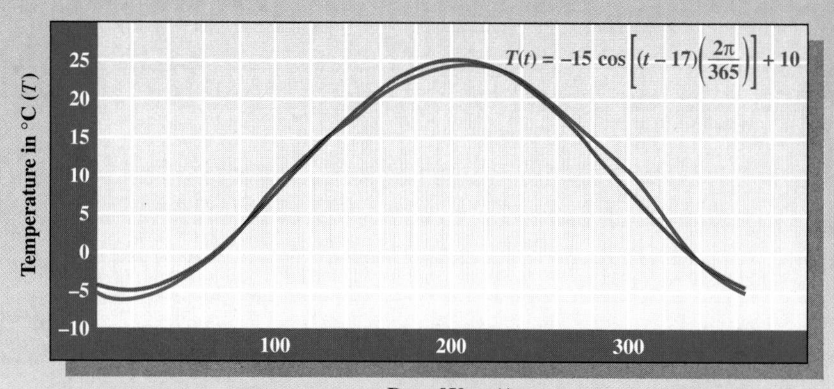

The Cosine Day of the Year Model for Temperature approximates the climatology of temperature for Des Moines, Iowa, with little error.

In Section 1 of this chapter we saw that

$$\lim_{m\to\infty}\left(1+\frac{1}{m}\right)^m = e.$$

Using s as the variable instead of m gives

$$\lim_{s\to\infty}\left(1+\frac{1}{s}\right)^s = e.$$

Thus,

$$\left[\lim_{s\to\infty}\left(1+\frac{1}{s}\right)^s\right]^r = e^r,$$

and so

$$\lim_{m\to\infty}\left(1+\frac{r}{m}\right)^m = e^r.$$

At this time, you may want to review the rules for limits at the beginning of the chapter on the derivative.

Finally,

$$\begin{aligned}\lim_{m\to\infty} P\left(1+\frac{r}{m}\right)^{mt} &= P\lim_{m\to\infty}\left(1+\frac{r}{m}\right)^{mt} && \text{Limit rule 1}\\ &= P\left[\lim_{m\to\infty}\left(1+\frac{r}{m}\right)^m\right]^t && \text{Limit rule 6}\\ &= P(e^r)^t && \text{Substitute from the work above.}\\ &= Pe^{rt}.\end{aligned}$$

This discussion is summarized as follows.

CONTINUOUS COMPOUNDING

If a deposit of P dollars is invested at a rate of interest r compounded continuously for t years, the compound amount is

$$A = Pe^{rt} \text{ dollars.}$$

Note This is the same equation as the general exponential growth equation given at the beginning of this section. Here, $y = A$, $y_0 = P$, and $k = r$.

EXAMPLE 3 Assuming continuous compounding, if the inflation rate averaged 6% per year for 5 years, how much would a \$1 item cost at the end of the 5 years?

In the formula for continuous compounding, let $P = 1$, $t = 5$, and $r = .06$ to get

$$A = 1\, e^{5(.06)} = e^{.3} \approx 1.34986.$$

An item that cost \$1 at the beginning of the 5-year period would cost \$1.35 at the end of the period, an increase of 35%, or about 1/3.

Effective Rate We could use a calculator to see that \$1 at 8% interest (per year) compounded semiannually is $1(1.04)^2 = 1.0816$ or \$1.0816. The actual increase of \$.0816 is 8.16% rather than the 8% that would be earned with interest compounded annually. To distinguish between these two amounts, 8% (the annual interest rate) is called the **nominal** or **stated** interest rate, and 8.16% is called the **effective** interest rate. We will continue to use r to designate the stated rate and we will use r_E for the effective rate.

EFFECTIVE RATE COMPOUND INTEREST

If r is the annual stated rate of interest and m is the number of compounding periods per year, the effective rate of interest is

$$r_E = \left(1 + \frac{r}{m}\right)^m - 1.$$

Effective rate is sometimes called *annual yield.*

With continuous compounding, \$1 at 8% for 1 year becomes $(1)e^{1(.08)} = e^{.08} = 1.0833$. The increase is 8.33% rather than 8%, so a stated interest rate of 8% produces an effective rate of 8.33%.

EFFECTIVE RATE CONTINUOUS COMPOUNDING

If interest is compounded continuously at an annual stated rate of r, the effective rate of interest is

$$r_E = e^r - 1.$$

Find the effective rate corresponding to the following stated rates.

(a) 6% compounded quarterly

Using the formula, we get

$$\left(1 + \frac{.06}{4}\right)^4 - 1 = (1.015)^4 - 1 = .0614.$$

The effective rate is 6.14%.

(b) 6% compounded continuously

The formula for continuous compounding gives

$$e^{.06} - 1 = .0618,$$

so the effective rate is 6.18%. ◀

The formula for interest compounded m times a year, $A = P(1 + r/m)^{tm}$, has five variables: A, P, r, m, and t. If the values of any four are known, then the value of the fifth can be found. In particular, if A, the amount of money we wish to end up with, is given as well as r, m, and t, then P can be found. Here P is the amount that should be deposited today to produce A dollars in t years. The amount P is called the **present value** of A dollars.

EXAMPLE 5

Ed Calvin has a balloon payment of \$100,000 due in 3 years. What is the present value of that amount if the money earns interest at 12% annually?

Here P in the compound interest formula is unknown, with $A = 100{,}000$, $r = .12$, $t = 3$, and $m = 1$. Substitute the known values into the formula to get $100{,}000 = P(1.12)^3$. Solve for P using a calculator to find $(1.12)^3$.

$$P = \frac{100{,}000}{(1.12)^3} = 71{,}178.02$$

The present value of \$100,000 in 3 years at 12% per year is \$71,178.02.

The equation in Example 5 could have been solved as follows.

$$100{,}000 = P(1.12)^3$$

Multiply both sides by $(1.12)^{-3}$ to solve for P, so

$$P = 100{,}000(1.12)^{-3}.$$

This suggests a general formula for present value.

PRESENT VALUE

The present value of A dollars at a rate of interest r compounded m times per year for t years is

$$\boldsymbol{P = A\left(1 + \frac{r}{m}\right)^{-tm}}.$$

The equation $A = Pe^{rt}$ also can be solved for any of the variables A, P, r, or t, as the following example shows.

EXAMPLE 6

Find the interest rate that will cause \$5000 to grow to \$7250 in 4 years if the money is compounded continuously.

Use the formula for continuous compounding, $A = Pe^{rt}$, with $A = 7250$, $P = 5000$, and $t = 4$. Solve first for e^{rt}, then for r.

$$\begin{aligned} A &= Pe^{rt} \\ 7250 &= 5000e^{4r} \\ 1.45 &= e^{4r} && \text{Divide by 5000.} \\ \ln 1.45 &= \ln e^{4r} && \text{Take logarithms of both sides.} \\ \ln 1.45 &= 4r && \ln e^x = x \\ r &= \frac{\ln 1.45}{4} \\ r &\approx .093 \end{aligned}$$

The required interest rate is 9.3%.

Limited Growth Functions The exponential growth functions discussed so far all continued to grow without bound. More realistically, many populations grow exponentially for a while, but then the growth is slowed by some external constraint that eventually limits the growth. For example, an animal population may grow to the point where its habitat can no longer support the population and the growth rate begins to dwindle until a stable population size is reached. Models that reflect this pattern are called **limited growth functions.** The next example discusses a function of this type that occurs in industry.

EXAMPLE 7 Assembly-line operations tend to have a high turnover of employees, forcing companies to spend much time and effort in training new workers. It has been found that a worker new to a task on the line will produce items according to the function defined by

$$P(x) = 25 - 25e^{-.3x},$$

where $P(x)$ items are produced by the worker on day x.

(a) What is the limit on the number of items a worker on this assembly line can produce?

Since the exponent on e is negative, write the function as

$$P(x) = 25 - \frac{25}{e^{.3x}}.$$

As $x \to \infty$, the term $(25/e^{.3x}) \to 0$, so $P(x) \to 25$. The limit is 25 items.

(b) How many days will it take for a new worker to produce 20 items?

Let $P(x) = 20$ and solve for x.

$$\begin{aligned} P(x) &= 25 - 25e^{-.3x} \\ 20 &= 25 - 25e^{-.3x} \\ -5 &= -25e^{-.3x} \\ .2 &= e^{-.3x} \end{aligned}$$

Now take natural logarithms of both sides and use properties of logarithms.

$$\begin{aligned} \ln .2 &= \ln e^{-.3x} \\ \ln .2 &= -.3x \qquad \ln e^u = u \\ x &= \frac{\ln .2}{-.3} \approx 5.4 \end{aligned}$$

In about 5½ days on the job a new worker will be producing 20 items. A graph of the function P is shown in Figure 9. ◀

Graphs such as the one in Figure 9 are called *learning curves.* According to such a graph, a new worker tends to learn quickly at first; then learning tapers off and approaches some upper limit. This is characteristic of the learning of certain types of skills involving the repetitive performance of the same task.

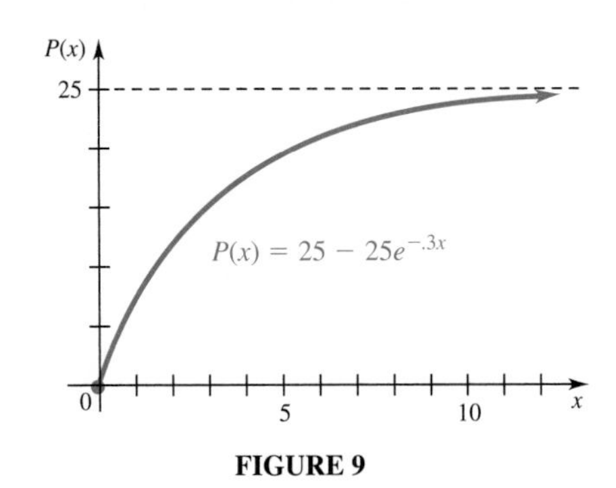

FIGURE 9

5.3 Exercises

Find the effective rate corresponding to each of the following nominal rates of interest.

1. 5% compounded monthly
2. 15% compounded quarterly
3. 10% compounded semiannually
4. 18% compounded monthly
5. 11% compounded continuously
6. 7% compounded continuously

Find the present value of each deposit.

7. \$2000 at 6% compounded semiannually for 11 yr
8. \$10,000 if interest is 12% compounded semiannually for 5 yr
9. \$10,000 at 10% compounded quarterly for 8 yr
10. \$45,678.93 if interest is 12.6% compounded monthly for 11 mo
11. \$7300 at 11% compounded continuously for 3 yr
12. \$25,000 at 9% compounded continuously for 8 yr
13. What is the difference between stated interest rate and effective interest rate?
14. In the exponential growth or decay function $y = y_0e^{kt}$, what does y_0 represent? What does k represent?
15. In the exponential growth or decay function, explain the circumstances that cause k to be positive or negative.
16. What is meant by the half-life of a quantity?
17. Show that if a radioactive substance has a half-life of T, then the corresponding constant k in the exponential decay function is given by $k = -(\ln 2)/T$.
18. Show that if a radioactive substance has a half-life of T, then the corresponding exponential decay function can be written as $y = y_0\,(1/2)^{(t/T)}$.

Applications

Business and Economics

19. **Inflation** Assuming continuous compounding, what will it cost to buy a \$10 item in 3 yr at the following inflation rates?

    (a) 3% **(b)** 4% **(c)** 5%

20. **Interest** Bert Bezzone invests a \$25,000 inheritance in a fund paying 9% per year compounded continuously. What will be the amount on deposit after each of the following time periods?

    (a) 1 yr **(b)** 5 yr **(c)** 10 yr

21. **Interest** Linda Youngman, who is self-employed, wants to invest \$60,000 in a pension plan. One investment offers 10% compounded quarterly. Another offers 9.75% compounded continuously.

    (a) Which investment will earn the most interest in 5 yr?

    (b) How much more will the better plan earn?

    (c) What is the effective rate in each case?

    (d) If Ms. Youngman chooses the plan with continuous compounding, how long will it take for her \$60,000 to grow to \$80,000?

22. **Effective Rate** Janet Tilden bought a television set with money borrowed from the bank at 9% interest compounded semiannually. What effective interest rate did she pay?

23. **Effective Rate** A firm deposits some funds in a special account at 7.2% compounded quarterly. What effective rate will they earn?

24. **Effective Rate** Virginia Nicolai deposits $7500 of lottery winnings in an account paying 6% interest compounded monthly. What effective rate does the account earn?

25. **Present Value** Matt Harr must make a balloon payment of $20,000 in 4 yr. Find the present value of the payment if it includes annual interest of 8%.

26. **Present Value** A company must pay a $307,000 settlement in 3 yr.
   - **(a)** What amount must be deposited now at 6% semiannually to have enough money for the settlement?
   - **(b)** How much interest will be earned?
   - **(c)** Suppose the company can deposit only $200,000 now. How much more will be needed in 3 yr?

27. **Present Value** A couple wants to have $20,000 in 5 yr for a down payment on a new house.
   - **(a)** How much could they deposit today, at 8% compounded quarterly, to have the required amount in 5 yr?
   - **(b)** How much interest will be earned?
   - **(c)** If they can deposit only $10,000 now, how much more will they need to complete the $20,000 after 5 yr?

28. **Sales** Experiments have shown that sales of a product, under relatively stable market conditions, but in the absence of promotional activities such as advertising, tend to decline at a constant yearly rate. This sales decline, which varies considerably from product to product, can often be expressed by an exponential function of the form

$$S(t) = S_0e^{-at},$$

where $S(t)$ is the sales at time t measured in years, S_0 is sales at time $t = 0$, and a is the sales decay constant. Suppose a certain product had sales of 50,000 at the beginning of the year and 45,000 at the end of the year.
   - **(a)** Write an equation for the sales decline.
   - **(b)** Find $S(2)$.
   - **(c)** How long will it take for sales to decrease to 40,000?
   - **(d)** Is there a limit to the sales decline? If so, what is it?

29. **Sales** Sales of a new model of compact disc player are approximated by the function $S(x) = 1000 - 800e^{-x}$, where $S(x)$ is in appropriate units and x represents the number of years the disc player has been on the market.
   - **(a)** Find the sales during year 0.
   - **(b)** In how many years will sales reach 500 units?
   - **(c)** Will sales ever reach 1000 units?
   - **(d)** Is there a limit on sales for this product? If so, what is it?

30. **Sales** Sales of a new model of word processor are approximated by

$$S(x) = 5000 - 4000e^{-x},$$

where x represents the number of years that the word processor has been on the market, and $S(x)$ represents sales in thousands.
   - **(a)** Find the sales in year 0.
   - **(b)** When will sales reach $4,500,000?
   - **(c)** Find the limit on sales.

Life Sciences

31. **Growth of Lice** A population of 100 lice is growing exponentially. After 2 mo the population has increased to 125.
   - **(a)** Write an exponential equation to express the exponential growth function y in terms of time t in months.
   - **(b)** How long will it take for the population to reach 500?

32. **Growth of Bacteria** A culture contains 25,000 bacteria, with the population increasing exponentially. The culture contains 40,000 bacteria after 10 hr.
   - **(a)** Write an exponential equation to express the growth function y in terms of time t in hours.
   - **(b)** How long will it be until there are 60,000 bacteria?

33. **Decrease in Bacteria** When a bactericide is introduced into a culture of 50,000 bacteria, the number of bacteria decreases exponentially. After 9 hr, there are only 20,000 bacteria.
   - **(a)** Write an exponential equation to express the decay function y in terms of time t in hours.
   - **(b)** In how many hours will half the number of bacteria remain?

34. **Growth of Bacteria** The growth of bacteria in food products makes it necessary to time-date some products (such as milk) so that they will be sold and consumed before the bacteria count is too high. Suppose for a certain product that the number of bacteria present is given by

$$f(t) = 500e^{.1t},$$

under certain storage conditions, where t is time in days after packing of the product and the value of $f(t)$ is in millions.

(a) If the product cannot be safely eaten after the bacteria count reaches 3,000 million, low long will this take?

(b) If $t = 0$ corresponds to January 1, what date should be placed on the product?

35. **Limited Growth** Another limited growth function is the *logistic function*

$$G(t) = \frac{mG_0}{G_0 + (m - G_0)e^{-kmt}},$$

where G_0 is the initial number present, m is the maximum possible size of the population, and k is a positive constant. Assume $G_0 = 1000$, $m = 2500$, $k = .0004$, and t is time in decades (10-year periods).

(a) Find $G(.2)$. (b) Find $G(1)$. (c) Find $G(3)$.

(d) At what time t will the population reach 2000?

36. **Population Growth** A population of fruit flies, contained in a large glass jar, is growing according to the model $y = 300 - 290e^{-.693t}$, where y is the population at time t in days.

(a) Find the initial population.

(b) Find the limit of the population.

(c) When will the population reach 150?

37. **Cancer Research** An article on cancer treatment contains the following statement: A 37% five-year survival rate for women with ovarian cancer yields an estimated annual mortality rate of 0.1989.* The authors of this article assume that the number of survivors is described by the exponential decay function given at the beginning of this chapter, where y is the number of survivors and k is the mortality rate. Verify that the given survival rate leads to the given mortality rate.

Social Sciences

38. **Skills Training** The number of words per minute that an average typist can type is given by

$$W(t) = 60 - 30e^{-.5t},$$

where t is time in months after the beginning of a typing class.

Find each of the following.

(a) $W_0 = W(0)$ (b) $W(1)$ (c) $W(4)$

(d) When will the average typist type 45 words per minute?

(e) What is the upper limit on the number of words that can be typed according to this model?

(f) Graph $W(t)$.

39. **Skills Training** Assuming that a person new to an assembly line will produce

$$P(x) = 500 - 500e^{-x}$$

items per day, where x is time measured in days, find each of the following.

(a) $P_0 = P(0)$ (b) $P(2)$ (c) $P(5)$

(d) When will a new person produce 400 items?

(e) Find the limit on the number of items produced per day.

(f) Graph $P(x)$.

40. **Spread of a Rumor** A sociologist has shown that the fraction $y(t)$ of people who have heard a rumor after t days is approximated by

$$y(t) = \frac{y_0e^{kt}}{1 - y_0(1 - e^{kt})},$$

where y_0 is the fraction of people who have heard the rumor at time $t = 0$, and k is a constant. A graph of $y(t)$ for a particular value of k is shown in the figure.

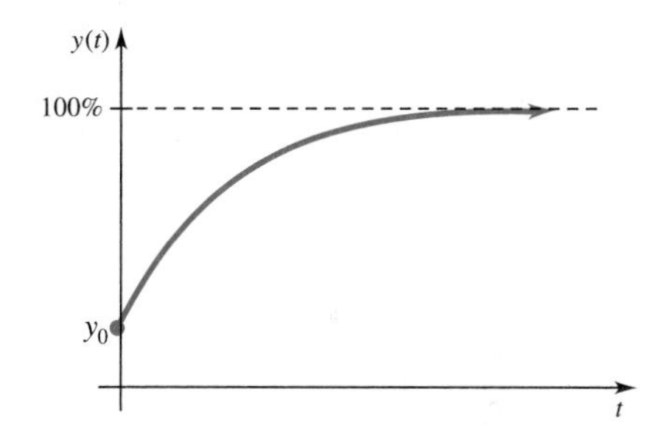

(a) If $k = .1$ and $y_0 = .05$, find $y(10)$.

(b) If $k = .2$ and $y_0 = .10$, find $y(5)$.

(c) If $k = .1$ and $y_0 = .02$, find the number of days until half the people have heard the rumor.

(d) What happens to the fraction of people who have heard the rumor as t gets larger and larger? What does this mean? (*Hint:* Look at the graph.)

Physical Science

41. **Carbon Dating** Refer to Example 2. A sample from a refuse deposit near the Strait of Magellan had 60% of the carbon 14 found in a contemporary living sample. How old was the sample?

---

*Theodore Speroff et al., "A risk-benefit analysis of elective bilateral oophorectomy: effect of changes in compliance with estrogen therapy on outcome," *American Journal of Obstetrics and Gynecology*, vol. 164 (January 1991), pp. 165–74.

Half-life *Find the half-life of the following radioactive substances. See Example 2.*

42. Plutonium 241; $A(t) = A_0e^{-.053t}$

43. Radium 226; $A(t) = A_0e^{-.00043t}$

44. Iodine 131; $A(t) = A_0e^{-.087t}$

45. Nuclear Energy Nuclear energy derived from radioactive isotopes can be used to supply power to space vehicles. The output of the radioactive power supply for a certain satellite is given by the function $y = 40e^{-.004t}$, where y is in watts and t is the time in days.

(a) How much power will be available at the end of 180 days?

(b) How long will it take for the amount of power to be half of its original strength?

(c) Will the power ever be completely gone? Explain.

46. Decay of Radioactivity A large cloud of radioactive debris from a nuclear explosion has floated over the Pacific Northwest, contaminating much of the hay supply. Consequently, farmers in the area are concerned that the cows who eat this hay will give contaminated milk. (The tolerance level for radioactive iodine in milk is 0.) The percent of the initial amount of radioactive iodine still present in the hay after t days is approximated by $P(t)$, which is given by the mathematical model

$$P(t) = 100e^{-.1t}.$$

(a) Find the percent remaining after 4 days.

(b) Find the percent remaining after 10 days.

(c) Some scientists feel that the hay is safe after the percent of radioactive iodine has declined to 10% of the original amount. Solve the equation $10 = 100e^{-.1t}$ to find the number of days before the hay may be used.

(d) Other scientists believe that the hay is not safe until the level of radioactive iodine has declined to only 1% of the original level. Find the number of days that this would take.

47. Chemical Dissolution The amount of chemical that will dissolve in a solution increases exponentially as the temperature is increased. At 0° C 10 g of the chemical dissolves, and at 10° C 11 g dissolves.

(a) Write an equation to express the amount of chemical dissolved, y, in terms of temperature, t, in degrees Celsius.

(b) At what temperature will 15 g dissolve?

Newton's Law of Cooling *Newton's law of cooling says that the rate at which a body cools is proportional to the difference in temperature between the body and an environment into which it is introduced. This leads to an equation where the temperature $f(t)$ of the body at time t after being introduced into an environment having constant temperature T_0 is*

$$f(t) = T_0 + Ce^{-kt},$$

where C and k are constants. Use this result in Exercises 48–50.

48. Find the temperature of an object when $t = 9$ if $T_0 = 18$, $C = 5$, and $k = .6$.

49. If $C = 100$, $k = .1$, and t is time in minutes, how long will it take a hot cup of coffee to cool to a temperature of 25° Celsius in a room at 20° Celsius?

50. If $C = -14.6$ and $k = .6$ and t is time in hours, how long will it take a frozen pizza to thaw to 10° Celsius in a room at 18° Celsius?

5.4 DERIVATIVES OF LOGARITHMIC FUNCTIONS

If the number of births per female in an animal population varies with the mother's age, at what age does the maximum number of births occur?

As we saw in the chapter on applications of the derivative, extrema occur when the derivative of a function does not exist or equals zero. In this section formulas will be developed for the derivatives of $y = \ln x$ and other logarithmic functions that will allow us to answer the question above.

To find the derivative of the function $y = \ln x$, assuming $x > 0$, go back to the definition of the derivative given earlier: if $y = f(x)$, then

$$y' = f'(x) = \lim_{h\to 0} \frac{f(x+h) - f(x)}{h},$$

provided this limit exists. (Remember that h represents the variable in this expression, while x is constant.)

Here, $f(x) = \ln x$, with

$$y' = f'(x) = \lim_{h\to 0} \frac{\ln(x+h) - \ln x}{h}.$$

This limit can be found using properties of logarithms. By the quotient rule for logarithms,

$$y' = \lim_{h\to 0} \frac{\ln(x+h) - \ln x}{h} = \lim_{h\to 0} \frac{\ln\left(\frac{x+h}{x}\right)}{h} = \lim_{h\to 0}\left[\frac{1}{h}\ln\left(\frac{x+h}{x}\right)\right].$$

The power rule for logarithms is used to get

$$y' = \lim_{h\to 0} \ln\left(\frac{x+h}{x}\right)^{1/h} = \lim_{h\to 0} \ln\left(1 + \frac{h}{x}\right)^{1/h}.$$

Now make a substitution: let $m = x/h$, so that $h/x = 1/m$. As $h \to 0$, with x fixed, $m \to \infty$ (since $x > 0$), so

$$\begin{aligned}\lim_{h\to 0} \ln\left(1 + \frac{h}{x}\right)^{1/h} &= \lim_{m\to\infty} \ln\left(1 + \frac{1}{m}\right)^{m/x}\\ &= \lim_{m\to\infty} \ln\left[\left(1 + \frac{1}{m}\right)^{m}\right]^{1/x}.\end{aligned}$$

By the power rule for logarithms, this becomes

$$= \lim_{m\to\infty}\left[\frac{1}{x}\cdot\ln\left(1 + \frac{1}{m}\right)^{m}\right].$$

Since x is fixed here (a constant), by the product rule for limits, this last limit becomes

$$\begin{aligned}\lim_{m\to\infty}\left[\frac{1}{x}\cdot\ln\left(1 + \frac{1}{m}\right)^{m}\right] &= \left(\lim_{m\to\infty}\frac{1}{x}\right)\lim_{m\to\infty}\left[\ln\left(1 + \frac{1}{m}\right)^{m}\right]\\ &= \frac{1}{x}\cdot\lim_{m\to\infty}\left[\ln\left(1 + \frac{1}{m}\right)^{m}\right].\end{aligned}$$

Since the logarithmic function is continuous,

$$\frac{1}{x}\cdot\lim_{m\to\infty}\left[\ln\left(1 + \frac{1}{m}\right)^{m}\right] = \frac{1}{x}\cdot\ln\left[\lim_{m\to\infty}\left(1 + \frac{1}{m}\right)^{m}\right].$$

In Section 1 of this chapter we saw that the limit on the right equals e, so that

$$\frac{1}{x}\cdot \ln\left[\lim_{m\to\infty}\left(1+\frac{1}{m}\right)^m\right] = \frac{1}{x}\ln e = \frac{1}{x}\cdot 1 = \frac{1}{x}.$$

Summarizing, if $y = \ln x$, then $y' = 1/x$.

Find the derivative of $y = \ln 6x$. Assume $x > 0$.

Use the properties of logarithms and the rules for derivatives.

$$\begin{aligned} y' &= \frac{d}{dx}(\ln 6x) \\ &= \frac{d}{dx}(\ln 6 + \ln x) \\ &= \frac{d}{dx}(\ln 6) + \frac{d}{dx}(\ln x) = 0 + \frac{1}{x} = \frac{1}{x} \end{aligned}$$ ◀

Recall the chain rule: If $y = f[g(x)]$, then $y' = f'[g(x)] \cdot g'(x)$. For review, find the following derivatives.

1. $D_x(2x^3 - 4)^5$
2. $D_x\sqrt{x^4 + 5x}$
3. $D_x(5x^2 + 3x)^{3/2}$

Answers:

1. $30x^2(2x^3 - 4)^4$
2. $\dfrac{4x^3 + 5}{2\sqrt{x^4 + 5x}}$
3. $\dfrac{3}{2}(5x^2 + 3x)^{1/2}(10x + 3)$

The chain rule can be used to find the derivative of the more general logarithmic function $y = \ln g(x)$. Let $y = f(u) = \ln u$ and $u = g(x)$, so that $f[g(x)] = \ln g(x)$. Then

$$f'[g(x)] = f'(u) = \frac{1}{u} = \frac{1}{g(x)},$$

and by the chain rule,

$$\begin{aligned} y' &= f'[g(x)] \cdot g'(x) \\ &= \frac{1}{g(x)} \cdot g'(x) = \frac{g'(x)}{g(x)}. \end{aligned}$$

EXAMPLE 2

Find the derivative of $y = \ln (x^2 + 1)$.

Here $g(x) = x^2 + 1$ and $g'(x) = 2x$. Thus,

$$y' = \frac{g'(x)}{g(x)} = \frac{2x}{x^2 + 1}.$$ ◀

If $y = \ln(-x)$, where $x < 0$, the chain rule with $g(x) = -x$ and $g'(x) = -1$ gives

$$y' = \frac{g'(x)}{g(x)} = \frac{-1}{-x} = \frac{1}{x}.$$

The derivative of $y = \ln(-x)$ is the same as the derivative of $y = \ln x$. For this reason, these two results can be combined into one rule using the absolute value of x. A similar situation holds true for $y = \ln[g(x)]$ and $y = \ln[-g(x)]$. These results are summarized on the next page.

DERIVATIVES OF ln $|x|$ AND ln $|g(x)|$

$$\frac{d}{dx} \ln |x| = \frac{1}{x}$$

$$\frac{d}{dx} \ln |g(x)| = \frac{g'(x)}{g(x)}$$

EXAMPLE 3 Find the derivative of each function.

(a) $y = \ln |5x|$

Let $g(x) = 5x$, so that $g'(x) = 5$. From the formula above,

$$y' = \frac{g'(x)}{g(x)} = \frac{5}{5x} = \frac{1}{x}.$$

Notice that the derivative of $\ln |5x|$ is the same as the derivative of $\ln |x|$. Also, in Example 1, the derivative of $\ln 6x$ was the same as that for $\ln x$. This suggests that for any constant a,

$$\frac{d}{dx} \ln |ax| = \frac{d}{dx} \ln |x|$$
$$= \frac{1}{x}.$$

Exercise 43 asks for a proof of this result.

(b) $y = \ln |3x^2 - 4x|$

$$y' = \frac{6x - 4}{3x^2 - 4x}$$

(c) $y = 3x \ln x^2$

This function is the product of the two functions $3x$ and $\ln x^2$, so use the product rule.

$$y' = (3x)\left[\frac{d}{dx} \ln x^2\right] + (\ln x^2)\left[\frac{d}{dx} 3x\right]$$
$$= 3x\left(\frac{2x}{x^2}\right) + (\ln x^2)(3)$$
$$= 6 + 3 \ln x^2$$

By the power rule for logarithms,

$$y' = 6 + \ln (x^2)^3$$
$$= 6 + \ln x^6.$$

Alternatively, write the answer as $y' = 6 + 6 \ln x$. ◀

EXAMPLE 4 Find the derivative of $y = \log |3x + 2|$.

This is a base 10 logarithm, while the derivative rule developed above applies only to natural logarithms. To find the derivative, first use the change-of-base theorem to convert the function to one involving natural logarithms.

The change-of-base theorem was given in the second section of this chapter. It allows us to change a logarithm from one base to another by the following rule.

$$\log_a x = \frac{\log_b x}{\log_b a}$$

Use the rule to change $\log_{10} x$ to $\ln x$; that is, let $a = 10$ and $b = e$ in the rule given above.

Answer: $\log_{10} x = \dfrac{\ln x}{\ln 10}$

$$y = \log_{10} |3x + 2|$$
$$= \frac{\ln |3x + 2|}{\ln 10}$$
$$= \frac{1}{\ln 10} \ln |3x + 2|$$

Now find the derivative. (Remember: $1/(\ln 10)$ is a constant.)

$$y' = \frac{1}{\ln 10} \cdot \frac{d}{dx} [\ln |3x + 2|]$$
$$= \frac{1}{\ln 10} \cdot \frac{3}{3x + 2}$$
$$y' = \frac{3}{\ln 10\,(3x + 2)}$$

The procedure in Example 4 can be used to derive the following results.

DERIVATIVES WITH OTHER BASES

For any suitable value of a,

$$D_x(\log_a |x|) = \frac{1}{\ln a} \cdot \frac{1}{x}.$$

When $y = \log_a |g(x)|$, the chain rule gives

$$D_x(\log_a |g(x)|) = \frac{1}{\ln a} \cdot \frac{g'(x)}{g(x)}.$$

EXAMPLE 5 The proportion of births per female in a certain animal population is given by $y = (\ln x)/x^2$, where x is the age of the female in years.

(a) What is the domain of the function for this application?

Since the proportion of births must be nonnegative, and $y = 0$ when $x = 1$, the domain is $x \geq 1$.

(b) Find any extrema or inflection points and sketch the graph of $y = (\ln x)/x^2$, $x \geq 1$.

To locate any extrema, begin by finding the first derivative.

$$y' = \frac{x^2(1/x) - 2x \ln x}{x^4}$$

$$= \frac{x - 2x \ln x}{x^4}$$

$$= \frac{x(1 - 2 \ln x)}{x^4}$$

$$y' = \frac{1 - 2 \ln x}{x^3}$$

The derivative function exists everywhere on the domain ($x \geq 1$), so any extrema will occur only at points where the derivative equals 0.

$$\frac{1 - 2 \ln x}{x^3} = 0$$

$$1 - 2 \ln x = 0$$

$$2 \ln x = 1$$

$$\ln x = \frac{1}{2}$$

Use the definition of logarithm to get the equivalent exponential statement

$$x = e^{1/2} \approx 1.65.$$

To locate inflection points and determine whether $x = e^{1/2}$ represents a maximum or a minimum, find the second derivative.

$$y'' = \frac{x^3(-2/x) - 3x^2(1 - 2 \ln x)}{x^6}$$

$$= \frac{-2x^2 - 3x^2 + 6x^2 \ln x}{x^6}$$

$$= \frac{-5x^2 + 6x^2 \ln x}{x^6} = \frac{x^2(-5 + 6 \ln x)}{x^6} = \frac{-5 + 6 \ln x}{x^4}$$

Use a calculator to show that the second derivative is negative for $x = e^{1/2}$ indicating a maximum where $x = e^{1/2}$. Set the second derivative equal to 0 and solve the resulting equation to identify any inflection points.

$$\frac{-5 + 6 \ln x}{x^4} = 0$$

$$-5 + 6 \ln x = 0$$

$$6 \ln x = 5$$

$$\ln x = 5/6$$

$$x = e^{5/6} \approx 2.30$$

Verify that $f''(1) < 0$ and $f''(3) > 0$, indicating an inflection point when $x = e^{5/6} \approx 2.3$. Also, verify that the graph is concave downward for $x < e^{5/6}$ and concave upward for $x > e^{5/6}$. Use this information and plot a few points as necessary to get the graph in Figure 10.

x	y
1	0
1.7	.18
2.3	.16
4	.1

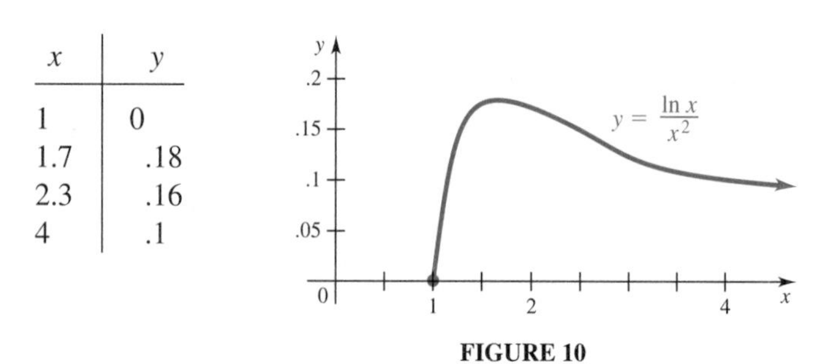

FIGURE 10

(c) At what age does the maximum number of births occur?

We saw in part (b) that the maximum occurred at $x = 1.65$, or about age 1 year, 8 months.

(d) What does the inflection point indicate?

The inflection point shows where the rate of change of the proportion of births stops decreasing and begins increasing again. However, since the function has no minimum (except at its endpoint), the proportion of births continues to decrease. ◀

5.4 Exercises

Find the derivative of each of the functions defined as follows.

1. $y = \ln (8x)$

2. $y = \ln (-4x)$

3. $y = \ln (3 - x)$

4. $y = \ln (1 + x^2)$

5. $y = \ln |2x^2 - 7x|$

6. $y = \ln |-8x^2 + 6x|$

7. $y = \ln \sqrt{x + 5}$

8. $y = \ln \sqrt{2x + 1}$

9. $y = \ln (x^4 + 5x^2)^{3/2}$

10. $y = \ln (5x^3 - 2x)^{3/2}$

11. $y = -3x \ln (x + 2)$

12. $y = (3x + 1) \ln (x - 1)$

13. $y = x^2 \ln |x|$

14. $y = x \ln |2 - x^2|$

15. $y = \dfrac{2 \ln (x + 3)}{x^2}$

16. $y = \dfrac{\ln x}{x^3}$

17. $y = \dfrac{\ln x}{4x + 7}$

18. $y = \dfrac{-2 \ln x}{3x - 1}$

19. $y = \dfrac{3x^2}{\ln x}$

20. $y = \dfrac{x^3 - 1}{2 \ln x}$

21. $y = (\ln |x + 1|)^4$

22. $y = \sqrt{\ln |x - 3|}$

23. $y = \ln |\ln x|$

24. $y = (\ln 4)(\ln |3x|)$

25. Why do we use the absolute value of x or of $g(x)$ in the derivative formulas for the natural logarithm?

26. Which of the following is the correct derivative of $\log_2 x$?

(a) $\dfrac{1}{x}$ **(b)** $\dfrac{1}{(\ln 2)x}$ **(c)** $\dfrac{\ln 2}{x}$ **(d)** $\dfrac{2}{x}$

Find the derivative of each of the following.

27. $y = \log(6x)$ **28.** $y = \log(2x - 3)$ **29.** $y = \log|1 - x|$ **30.** $y = \log|-3x|$

31. $y = \log_5 \sqrt{5x + 2}$ **32.** $y = \log_7 \sqrt{2x - 3}$ **33.** $y = \log_3 (x^2 + 2x)^{3/2}$ **34.** $y = \log_2 (2x^2 - x)^{5/2}$

Find all relative maxima and minima for the functions defined as follows. Sketch the graphs. (Hint: In Exercises 35 and 37, $\lim_{x \to 0} x \ln |x| = 0$.*)*

35. $y = x \ln x, \quad x > 0$ **36.** $y = x - \ln x, \quad x > 0$ **37.** $y = x \ln |x|$

38. $y = x - \ln |x|$ **39.** $y = \frac{\ln x}{x}, \quad x > 0$ **40.** $y = \frac{\ln x^2}{x^2}$

41. What is true about the slope of the tangent line to the graph of $f(x) = \ln x$ as $x \to \infty$? as $x \to 0$?

42. Let $f(x) = \ln x$.

(a) Compute $f'(x)$, $f''(x)$, $f'''(x)$, $f^{(4)}(x)$, and $f^{(5)}(x)$.

(b) Guess a formula for $f^{(n)}(x)$, where n is any positive integer.

43. Prove $\frac{d}{dx} \ln |ax| = \frac{d}{dx} \ln |x|$ for any constant a.

Applications

Business and Economics

44. Profit Assume that the total revenue received from the sale of x items is given by

$$R(x) = 30 \ln (2x + 1),$$

while the total cost to produce x items is $C(x) = x/2$. Find the number of items that should be manufactured so that profit, $R(x) - C(x)$, is a maximum.

45. Revenue Suppose the demand function for x units of a certain item is

$$p = 100 + \frac{50}{\ln x}, \quad x > 1,$$

where p is in dollars.

(a) Find the marginal revenue.

(b) Approximate the revenue from one more unit when 8 units are sold.

(c) How might a manager use the information from part (b)?

46. Profit If the cost function in dollars for x units of the item in Exercise 45 is $C(x) = 100x + 100$, find the following.

(a) The marginal cost **(b)** The profit function $P(x)$

(c) The profit from one more unit when 8 units are sold

(d) How might a manager use the information from part (c)?

47. Marginal Revenue Product The demand function for x units of a product is

$$p = 100 - 10 \ln x, \quad 1 < x < 20{,}000,$$

where $x = 6n$ and n is the number of employees producing the product.

(a) Find the revenue function $R(x)$.

(b) Find the marginal revenue product function. (See Example 10 in Section 6 of the chapter on the derivative.)

(c) Evaluate and interpret the marginal revenue product when 20 units are sold.

Life Sciences

48. Insect Mating Consider an experiment in which equal numbers of male and female insects of a certain species are permitted to intermingle. Assume that

$$M(t) = (.1t + 1) \ln \sqrt{t}$$

represents the number of matings observed among the insects in an hour, where t is the temperature in degrees Celsius. (*Note:* The formula is an approximation at best and holds only for specific temperature intervals.)

(a) Find the number of matings when the temperature is 15° C.

(b) Find the number of matings when the temperature is 25° C.

(c) Find the rate of change of the number of matings when the temperature is 15° C.

49. Population Growth Suppose that the population of a certain collection of rare Brazilian ants is given by

$$P(t) = (t + 100) \ln (t + 2),$$

where t represents the time in days. Find the rates of change of the population on the second day and on the eighth day.

General Interest

50. Information Content Suppose dots and dashes are transmitted over a telegraph line so that dots occur a fraction p of the time (where $0 \le p \le 1$) and dashes occur a fraction $1 - p$ of the time. The **information content** of the telegraph line is given by $I(p)$, where

$$I(p) = -p \ln p - (1 - p) \ln (1 - p).$$

(a) Show that $I'(p) = -\ln p + \ln (1 - p)$.)

(b) Let $I'(p) = 0$ and find the value of p that maximizes the information content.

(c) How might the result in part (b) be used?

Computer/Graphing Calculator

Use a computer or a graphing calculator to graph the following on the interval $(0, 9]$. *Then use the graph to find any extrema.*

51. $y = x^2 - \ln x$

52. $y = \ln x + \dfrac{1}{x}$

5.5 DERIVATIVES OF EXPONENTIAL FUNCTIONS

When does the growth rate of an insect population change from an increasing to a decreasing rate?

We will use a derivative to answer this question in Example 8 at the end of this section.

The derivatives of exponential functions can be found using the formula for the derivative of the natural logarithm function. To find the derivative of the function $y = e^x$, first take the natural logarithm of each side so that the power rule for logarithms can be used to get x out of the exponent.

$$y = e^x$$
$$\ln y = \ln e^x$$

By the power rule for logarithms and the fact that $\ln e^x = x$,

$$\ln y = x.$$

Now, using implicit differentiation, take the derivative of each side. (Since y is a function of x, we use the chain rule.)

$$\frac{1}{y} \cdot \frac{dy}{dx} = 1$$
$$\frac{dy}{dx} = y$$

Since $y = e^x$,

$$\frac{d}{dx}(e^x) = e^x \qquad \text{or} \qquad y' = e^x.$$

This result shows one of the main reasons for the widespread use of e as a base—the function $y = e^x$ is its own derivative. Furthermore, constant multiples of this function are the *only* useful functions with this property. For a more general result, let $y = e^u$ and $u = g(x)$ so that $y = e^{g(x)}$. Then

$$\frac{dy}{du} = e^u \qquad \text{and} \qquad \frac{du}{dx} = g'(x).$$

By the chain rule,

$$\begin{aligned} \frac{dy}{dx} &= \frac{dy}{du} \cdot \frac{du}{dx} \\ &= e^u \cdot g'(x) \\ &= e^{g(x)} \cdot g'(x). \end{aligned}$$

Now replace y with $e^{g(x)}$ to get

$$\frac{d}{dx}e^{g(x)} = e^{g(x)} \cdot g'(x).$$

These results are summarized below.

DERIVATIVES OF e^x AND $e^{g(x)}$

$$\frac{d}{dx}e^x = e^x \qquad \text{and} \qquad \frac{d}{dx}e^{g(x)} = g'(x)e^{g(x)}$$

Caution Notice the difference between the derivative of a variable to a constant power, such as $D_x\, x^3 = 3x^2$, and a constant to a variable power, like $D_x\, e^x = e^x$. Remember, $D_x\, e^x \neq xe^{x-1}$.

Find derivatives of the following functions.

(a) $y = e^{5x}$

Let $g(x) = 5x$, with $g'(x) = 5$. Then

$$y' = 5e^{5x}.$$

(b) $y = 3e^{-4x}$

$$y' = 3 \cdot e^{-4x}(-4) = -12e^{-4x}$$

(c) $y = 10e^{3x^2}$

$$y' = 10(e^{3x^2})(6x) = 60xe^{3x^2}$$ ◀

Let $y = e^x(\ln x)$. Find y'.

Use the product rule.

$$y' = e^x\left(\frac{1}{x}\right) + (\ln x)e^x$$

$$y' = e^x\left(\frac{1}{x} + \ln x\right)$$ ◀

EXAMPLE 3 Let $y = \dfrac{100{,}000}{1 + 100e^{-.3x}}$. Find y'.

Use the quotient rule.

$$y' = \frac{(1 + 100e^{-.3x})(0) - 100{,}000(-30e^{-.3x})}{(1 + 100e^{-.3x})^2}$$

$$= \frac{3{,}000{,}000e^{-.3x}}{(1 + 100e^{-.3x})^2}$$ ◀

Find $D_x(3^x)$.

Change 3^x to e^{kx}. Here, $k = \ln 3$ and $3^x = e^{(\ln 3)x}$.

Recall from the second section of this chapter that

$$a^x = e^{(\ln a)x}.$$

Thus, $2^x = e^{(\ln 2)x}$, $5^x = e^{(\ln 5)x}$, and so on.

$$D_x(3^x) = D_x e^{(\ln 3)x}$$

$$= (\ln 3)e^{(\ln 3)x}$$

$$= (\ln 3)3^x \qquad \text{Substitute } 3^x \text{ for } e^{(\ln 3)x}.$$ ◀

Extending this idea from Example 4 gives a general rule for the derivative of an exponential function with any appropriate base a.

$$D_x a^x = (\ln a)e^{x \ln a}$$

$$= (\ln a)e^{\ln a^x}$$

$$D_x a^x = (\ln a)a^x \qquad e^{\ln a^x} = a^x$$

Use the chain rule to get the following result.

DERIVATIVE OF $a^{g(x)}$

$$D_x[a^{g(x)}] = (\ln a)e^{(\ln a)g(x)}g'(x) = (\ln a)a^{g(x)}g'(x)$$

This formula and the formula

$$D_x \log_a |x| = \frac{1}{\ln a} \cdot \frac{1}{x}$$

are simplest when $a = e$, since $\ln e = 1$. This is a major reason for using base e exponential and logarithmic functions.

EXAMPLE 5 The amount in grams of a radioactive substance present after t years is given by

$$A(t) = 100e^{-.12t}.$$

(a) Find the rate of change of the amount present after 3 years.

The rate of change is given by the derivative dy/dt.

$$\frac{dy}{dt} = 100(e^{-.12t})(-.12) = -12e^{-.12t}$$

After 3 years ($t = 3$), the rate of change is

$$\frac{dy}{dt} = -12e^{-.12(3)} = -12e^{-.36} \approx -8.4$$

grams per year.

(b) In part (a), is the amount present increasing or decreasing when $t = 3$?

Since the rate of change is negative, the amount is decreasing.

(c) Graph $A(t)$ on [0, 20].

From part (a), $dy/dt = -12e^{-.12t}$. Since $e^{-.12t}$ is never equal to 0, there is no maximum or minimum function value. Find the second derivative to determine any points of inflection and the concavity.

$$\frac{d^2y}{dt^2} = -12(e^{-.12t})(-.12) = 1.44e^{-.12t}$$

The second derivative is always positive, so the graph of A is always concave upward. Plot a few points to get the graph shown in Figure 11. ◀

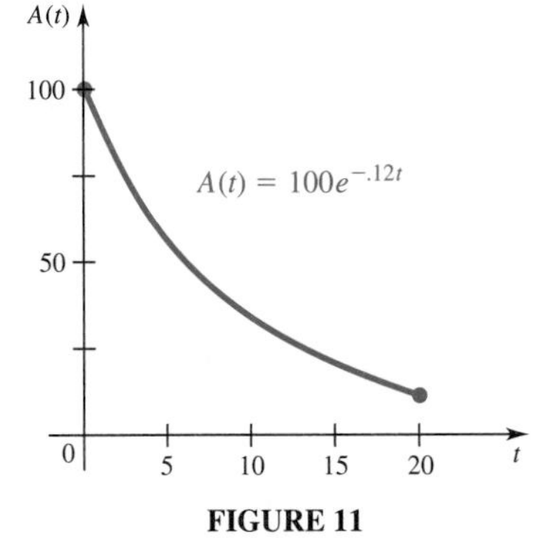

FIGURE 11

EXAMPLE 6 Use the information given by the first and second derivatives to graph $f(x) = 2^{-x^2}$.

This function was graphed in Section 1 of this chapter by point plotting. Now we are able to graph it more carefully and with greater detail.

The first derivative is

$$f'(x) = \ln 2(2^{-x^2})(-2x)$$

$$= (-2x \ln 2)(2^{-x^2}) \quad \text{or} \quad \frac{-2x \ln 2}{2^{x^2}}.$$

Set $f'(x)$ equal to 0 and solve for x.

$$(-2x \ln 2)(2^{-x^2}) = 0$$
$$-2x \ln 2 = 0$$
$$x = 0$$

Note that 2^{-x^2} can never be zero, so $x = 0$ is the only critical number. We can find the second derivative by using the product rule since $f'(x)$ is the product of two factors.

$$f''(x) = (-2x \ln 2)[\ln 2(2^{-x^2})(-2x)] + (2^{-x^2})(-2 \ln 2)$$
$$= 4x^2(\ln 2)^2(2^{-x^2}) - 2 \ln 2(2^{-x^2})$$

Factoring out the common factors gives

$$f''(x) = (2 \ln 2)(2^{-x^2})(2x^2 \ln 2 - 1).$$

Set $f''(x)$ equal to 0 and solve for x to find any inflection points.

$$2 \ln 2(2^{-x^2})(2x^2 \ln 2 - 1) = 0$$
$$2x^2 \ln 2 - 1 = 0$$
$$2x^2 \ln 2 = 1$$
$$x^2 = \frac{1}{2 \ln 2}$$
$$x = \pm\sqrt{\frac{1}{2 \ln 2}} \approx \pm .85$$

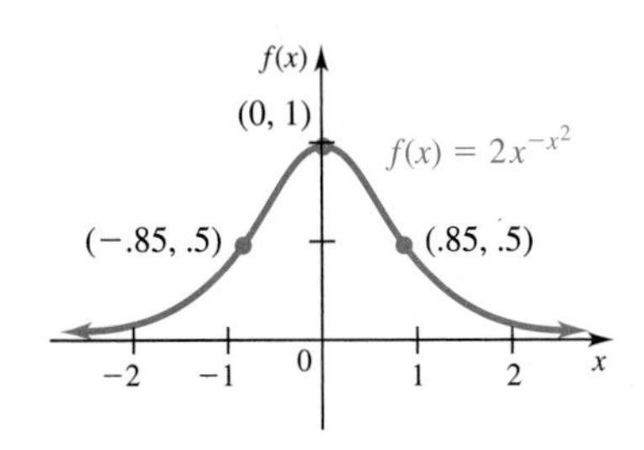

FIGURE 12

Test a value from each of the intervals

$$\left(-\infty, -\sqrt{\frac{1}{2 \ln 2}}\right), \left(-\sqrt{\frac{1}{2 \ln 2}}, \sqrt{\frac{1}{2 \ln 2}}\right), \text{ and } \left(\sqrt{\frac{1}{2 \ln 2}}, \infty\right)$$

to verify that there are inflection points at $x \approx -.85$ and $x \approx .85$. Use the second derivative test to verify that there is a maximum of 1 at $x = 0$. Plot the points where $x \approx -.85$, $x = 1$, and $x \approx .85$, and use information about concavity from testing the intervals given above to get the graph in Figure 12. ◀

Frequently a population, or the sales of a certain product, will start growing slowly, then grow more rapidly, and then gradually level off. Such growth can often be approximated by a mathematical model of the form

$$f(x) = \frac{b}{1 + ae^{kx}}$$

for appropriate constants a, b, and k.

EXAMPLE 7 Suppose that the sales of a new product can be approximated for its first few years on the market by the function

$$S(x) = \frac{100{,}000}{1 + 100e^{-.3x}},$$

where x is time in years since the introduction of the product.

(a) Find the rate of change of sales after 4 years.

The derivative of this sales function, which gives the rate of change of sales, was found in Example 3. Using that derivative,

$$S'(4) = \frac{3{,}000{,}000e^{-.3(4)}}{[1 + 100e^{-.3(4)}]^2} = \frac{3{,}000{,}000e^{-1.2}}{(1 + 100e^{-1.2})^2}.$$

By using a calculator, $e^{-1.2} \approx .301$, and

$$S'(4) \approx \frac{3{,}000{,}000(.301)}{[1 + 100(.301)]^2}$$
$$\approx \frac{903{,}000}{(1 + 30.1)^2}$$
$$= \frac{903{,}000}{967.21} \approx 934.$$

The rate of change of sales after 4 years is about 934 units per year. The positive number indicates that sales are increasing at this time.

(b) What happens to sales in the long run?

As time increases, $x \to \infty$, and

$$e^{-.3x} = \frac{1}{e^{.3x}} \to 0.$$

Thus,

$$\lim_{x \to \infty} S(x) = \frac{100{,}000}{1 + 100(0)} = 100{,}000.$$

Sales approach a horizontal asymptote of 100,000.

(c) Graph $S(x)$.

Use the first derivative to check for extrema. Set $S'(x) = 0$ and solve for x.

$$S'(x) = \frac{3{,}000{,}000e^{-.3x}}{(1 + 100e^{-.3x})^2} = 0$$
$$3{,}000{,}000e^{-.3x} = 0$$
$$\frac{3{,}000{,}000}{e^{.3x}} = 0$$

The expression on the left can never equal 0, so $S'(x) = 0$ leads to no extrema. Next, find any critical values for which $S'(x)$ does not exist; that is, values that make the denominator equal 0.

$$1 + 100e^{-.3x} = 0$$
$$100e^{-.3x} = -1$$
$$\frac{100}{e^{.3x}} = -1$$

Since the expression on the left is always positive, it cannot equal -1. There are no critical values and no extrema. Further, $e^{-.3x}$ is always positive, so $S'(x)$ is always positive. This means the graph is always increasing over the domain of the function. Use the second derivative to verify that the graph has an inflection point at approximately (15.4, 50,000) and is concave upward on $(-\infty, 15.4)$ and concave downward on $(15.4, \infty)$. Plotting a few points leads to the graph in Figure 13 (on the next page). ◀

x	y
0	990
5	4300
10	17,000
15	47,000
20	80,000
30	99,000

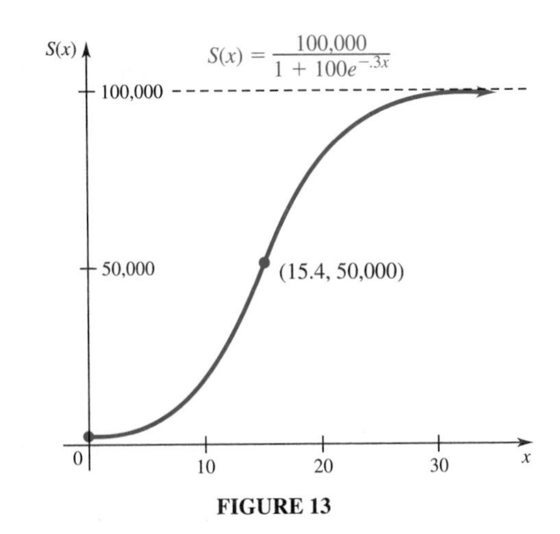

FIGURE 13

Environmental situations place effective limits on the population growth of an organism in an area. Many such limited growth situations are described by the *logistic function*

$$G(t) = \frac{mG_0}{G_0 + (m - G_0)e^{-kmt}},$$

where t represents time in appropriate units, G_0 is the initial number present, m is the maximum possible size of the population, k is a positive constant, and $G(t)$ is the population at time t. (This function is also given in Exercise 35 of Section 3 of this chapter.) Notice that the logistic function is a special case of the model discussed before Example 7.

EXAMPLE 8 The growth of a population of rare South American insects is given by the logistic function with $k = .00001$ and t in months. Assume that there are 200 insects initially and that the maximum population size is 10,000.

(a) Find the growth function $G(t)$ for these insects.

Substitute the given values for k, G_0, and m into the logistic function to get

$$G(t) = \frac{mG_0}{G_0 + (m - G_0)e^{-kmt}}$$

$$= \frac{(10{,}000)(200)}{200 + (10{,}000 - 200)e^{-(.00001)(10{,}000)t}}$$

$$= \frac{2{,}000{,}000}{200 + 9800e^{-.1t}}.$$

(b) Find the population after 5 months and after 10 months.

$$G(5) = \frac{2{,}000{,}000}{200 + 9800e^{-.5}} = 325.5$$

There are about 326 insects after 5 months.

$$G(10) = \frac{2{,}000{,}000}{200 + 9800e^{-1}} = 525.6$$

There are about 526 insects after 10 months.

(c) What happens to the population in the long run?

Notice that $G(t)$ has the same form as $S(x)$ in Example 7. Therefore, we know that

$$\lim_{t \to \infty} G(t) = \frac{2{,}000{,}000}{200 + 0} = 10{,}000,$$

so the population will approach an upper limit of 10,000 insects.

(d) Find the number of months before the rate of growth stops increasing and begins to decrease.

This will happen at the point of inflection. Verify that the point of inflection is approximately (39, 5000). The population grows quite rapidly for about 39 months (3 1/4 years) and then grows at a slower and slower pace as the population approaches 10,000.

(e) Graph $G(t)$ for $t \geq 0$.

The graph is shown in Figure 14. Compare it with the graph in Figure 13. Although these two applications are very different, the graphs are essentially the same. ◀

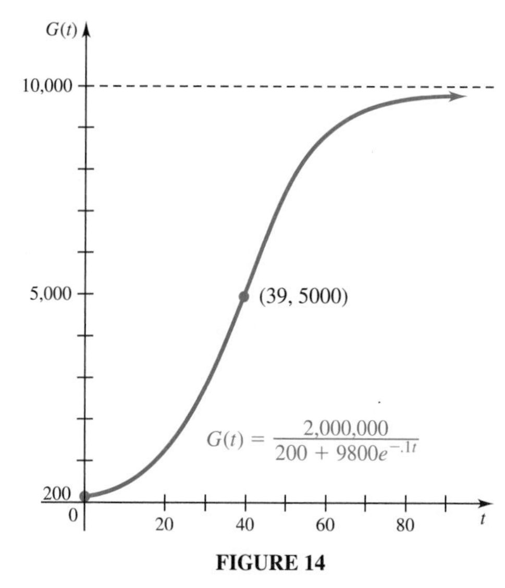

FIGURE 14

5.5 Exercises

Find derivatives of the functions defined as follows.

1. $y = e^{4x}$
2. $y = e^{-2x}$
3. $y = -8e^{2x}$
4. $y = .2e^{5x}$
5. $y = -16e^{x+1}$
6. $y = -4e^{-.1x}$
7. $y = e^{x^2}$
8. $y = e^{-x^2}$
9. $y = 3e^{2x^2}$
10. $y = -5e^{4x^3}$
11. $y = 4e^{2x^2-4}$
12. $y = -3e^{3x^2+5}$
13. $y = xe^x$
14. $y = x^2e^{-2x}$
15. $y = (x - 3)^2e^{2x}$

16. $y = (3x^2 - 4x)e^{-3x}$ **17.** $y = e^{x^2} \ln x, \quad x > 0$ **18.** $y = e^{2x-1} \ln(2x - 1), \quad x > \frac{1}{2}$

19. $y = \frac{e^x}{\ln x}, \quad x > 0$ **20.** $y = \frac{\ln x}{e^x}, \quad x > 0$ **21.** $y = \frac{x^2}{e^x}$

22. $y = \frac{e^x}{2x + 1}$ **23.** $y = \frac{e^x + e^{-x}}{x}$ **24.** $y = \frac{e^x - e^{-x}}{x}$

25. $y = \frac{5000}{1 + 10e^{.4x}}$ **26.** $y = \frac{600}{1 - 50e^{.2x}}$ **27.** $y = \frac{10{,}000}{9 + 4e^{-.2x}}$

28. $y = \frac{500}{12 + 5e^{-.5x}}$ **29.** $y = (2x + e^{-x^2})^2$ **30.** $y = (e^{2x} + \ln x)^3, \quad x > 0$

Find the derivative of each of the following.

31. $y = 8^{5x}$ **32.** $y = 2^{-x}$ **33.** $y = 3 \cdot 4^{x^2+2}$

34. $y = -10^{3x^2-4}$ **35.** $y = 2 \cdot 3^{\sqrt{x}}$ **36.** $y = 5 \cdot 7^{\sqrt{x-2}}$

37. Verify the inflection point given in Example 7.

38. Verify the inflection point given in Example 8.

Find all relative maxima, minima, and inflection points for the functions defined as follows. Sketch the graphs.

39. $y = -xe^x$ **40.** $y = xe^{-x}$ **41.** $y = x^2e^{-x}$

42. $y = (x - 1)e^{-x}$ **43.** $y = e^x + e^{-x}$ **44.** $y = -x^2e^x$

45. For $f(x) = e^x$, find $f''(x)$ and $f'''(x)$. What is the nth derivative of f with respect to x?

46. Describe the slope of the tangent line to the graph of $f(x) = e^x$ for each of the following.

(a) $x \to -\infty$ **(b)** $x \to 0$

47. Prove: $\frac{d}{dx} e^{ax} = ae^{ax}$ for any constant a.

Applications

Business and Economics

48. Sales The sales of a new personal computer (in thousands) are given by

$$S(t) = 100 - 90e^{-.3t},$$

where t represents time in years.

Find the rate of change of sales at each of the following times.

(a) After 1 yr **(b)** After 5 yr

(c) What is happening to the rate of change of sales as time goes on?

(d) Does the rate of change of sales ever equal zero?

49. Product Durability Suppose $P(x) = e^{-.02x}$ represents the proportion of shoes manufactured by a given company that are still wearable after x days of use.

Find the proportions of shoes wearable after the following periods of time.

(a) 1 day **(b)** 10 days **(c)** 100 days

(d) Calculate and interpret $P'(100)$.

(e) As the number of days increases, what happens to the proportion of shoes that are still wearable? Is this reasonable?

50. Product Durability Using data in a car magazine, we constructed the mathematical model

$$y = 100e^{-.03045t}$$

for the percent of cars of a certain type still on the road after t years.

Find the percent of cars on the road after the following numbers of years.

(a) 0 **(b)** 2 **(c)** 4 **(d)** 6

Find the rate of change of the percent of cars still on the road after the following numbers of years.

(e) 0 **(f)** 2

(g) Interpret your answers to (e) and (f).

51. Elasticity Refer to the discussion of elasticity in the previous chapter. For the demand function $p = 400e^{-.2x}$, find each of the following.

(a) The elasticity E

(b) The number of items (x) for which total revenue is maximum

Life Sciences

52. Pollution Concentration The concentration of pollutants (in grams per liter) in the east fork of the Big Weasel River is approximated by

$$P(x) = .04e^{-4x},$$

where x is the number of miles downstream from a paper mill that the measurement is taken.

Find each of the following values.

(a) The concentration of pollutants .5 mi downstream

(b) The concentration of pollutants 1 mi downstream

(c) The concentration of pollutants 2 mi downstream

Find the rate of change of concentration with respect to distance for each of the following distances.

(d) .5 mi **(e)** 1 mi **(f)** 2 mi

53. Radioactive Decay Assume that the amount (in grams) of a radioactive substance present after t years is given by

$$A(t) = 500e^{-.25t}.$$

Find the rate of change of the quantity present after each of the following years: **(a)** 4 **(b)** 6 **(c)** 10.

(d) What is happening to the rate of change of the amount present as the number of years increases?

(e) Will the substance ever be gone completely?

Social Sciences

54. Habit Strength According to work by the psychologist C. L. Hull, the strength of a habit is a function of the number of times the habit is repeated. If N is the number of repetitions and $H(N)$ is the strength of the habit, then

$$H(N) = 1000(1 - e^{-kN}),$$

where k is a constant. Find $H'(N)$ if $k = .1$ and the number of times the habit is repeated is as follows.

(a) 10 **(b)** 100 **(c)** 1000

(d) Show that $H'(N)$ is always positive. What does this mean?

Computer/Graphing Calculator

Graph the following over the given intervals. Then find any relative extrema (not endpoints) and inflection points.

55. $y = e^x \ln x$, $(0, 5]$

56. $y = \dfrac{5 \ln 10x}{e^{.5x}}$, $(0, 10]$

EXPONENTIAL AND LOGARITHMIC FUNCTIONS SUMMARY

EXPONENTIAL FUNCTIONS

Exponential Growth and Decay Function If y_0 is the number present at time $t = 0$,

$$y = y_0 e^{kt}.$$

Math of Finance Formulas If I is the interest, P is the principal or present value, r is the annual interest rate, t is time in years and m is the number of compounding periods per year:

Simple interest $I = Prt$

	Compounded m times per year	*Compounded continuously*
Compound amount	$A = P\left(1 + \frac{r}{m}\right)^{tm}$	$A = Pe^{rt}$
Effective rate	$r_E = \left(1 + \frac{r}{m}\right)^{m} - 1$	$r_E = e^r - 1$
Present value	$P = A\left(1 + \frac{r}{m}\right)^{-tm}$	$P = Ae^{-rt}$

Change of Base Theorem for Exponentials $a^x = e^{(\ln a)x}$

Derivatives $D_x(e^x) = e^x$ $\qquad D_x(a^x) = (\ln a)a^x$

$D_x(e^{g(x)}) = g'(x)e^{g(x)}$ $\qquad D_x(a^{g(x)}) = (\ln a)a^{g(x)}g'(x)$

LOGARITHMIC FUNCTIONS

Properties of Logarithms Let x and y be any positive real numbers and r be any real number. Let a be a positive real number, $a \neq 1$. Then

(a) $\log_a xy = \log_a x + \log_a y$ Product rule

(b) $\log_a \frac{x}{y} = \log_a x - \log_a y$ Quotient rule

(c) $\log_a x^r = r \log_a x$ Power rule

(d) $\log_a a = 1$

(e) $\log_a 1 = 0$

(f) $\log_a a^r = r$.

Change of Base Theorem for Logarithms $\log_a x = \dfrac{\log_b x}{\log_b a}$

Derivatives $D_x(\ln |x|) = \dfrac{1}{x}$ $\qquad D_x(\log_a |x|) = \dfrac{1}{\ln a} \cdot \dfrac{1}{x}$

$D_x(\ln |g(x)|) = \dfrac{g'(x)}{g(x)}$ $\qquad D_x(\log_a |g(x)|) = \dfrac{1}{\ln a} \cdot \dfrac{g'(x)}{g(x)}$

Chapter Summary Key Terms

5.1 **exponential function**
exponential equation
simple interest
compound interest
compound amount
e

5.2 **years to double**
logarithm
logarithmic function
properties of logarithms
common logarithms
natural logarithms
change-of-base theorem
logarithmic equation

5.3 **exponential growth and decay function**
growth constant
decay constant
half-life
continuous compounding
effective rate
nominal (stated) rate
present value
limited growth function
learning curve

Chapter 5 Review Exercises

1. Describe in words what a logarithm is.

2. Why is e a convenient base for exponential and logarithmic functions?

Solve each equation.

3. $2^{3x} = \frac{1}{8}$

4. $\left(\frac{9}{16}\right)^x = \frac{3}{4}$

5. $9^{2y-1} = 27^y$

6. $\frac{1}{2} = \left(\frac{b}{4}\right)^{1/4}$

Graph the functions defined as follows.

7. $y = 5^x$

8. $y = 5^{-x} + 1$

9. $y = \left(\frac{1}{5}\right)^{2x-3}$

10. $y = \left(\frac{1}{2}\right)^{x-1}$

11. $y = \log_2 (x - 1)$ **12.** $y = 1 + \log_3 x$ **13.** $y = -\log_3 x$ **14.** $y = 2 - \ln x^2$

Write the following equations using logarithms.

15. $2^6 = 64$ **16.** $3^{1/2} = \sqrt{3}$ **17.** $e^{.09} = 1.09417$ **18.** $10^{1.07918} = 12$

Write the following equations using exponents.

19. $\log_2 32 = 5$ **20.** $\log_{10} 100 = 2$

21. $\ln 82.9 = 4.41763$ **22.** $\log 15.46 = 1.18921$

Evaluate each expression.

23. $\log_3 81$ **24.** $\log_{1/3} 81$ **25.** $\log_{32} 16$

26. $\log_{25} 5$ **27.** $\log_{100} 1000$ **28.** $\log_{1/2} 4$

Simplify each expression using the properties of logarithms.

29. $\log_5 3k + \log_5 7k^3$ **30.** $\log_3 2y^3 - \log_3 8y^2$

31. $2 \log_2 x - 3 \log_2 m$ **32.** $5 \log_4 r - 3 \log_4 r^2$

Solve each equation. Round each answer to the nearest thousandth.

33. $8^p = 19$ **34.** $3^{z-2} = 11$ **35.** $2^{1-m} = 7$ **36.** $15^{-k} = 9$

37. $e^{-5-2x} = 5$ **38.** $e^{3x-1} = 12$ **39.** $\left(1 + \frac{m}{3}\right)^5 = 10$ **40.** $\left(1 + \frac{2p}{5}\right)^2 = 3$

Find the derivative of each function defined as follows.

41. $y = -6e^{2x}$ **42.** $y = 8e^{.5x}$ **43.** $y = e^{-2x^3}$ **44.** $y = -4e^{x^2}$

45. $y = 5xe^{2x}$ **46.** $y = -7x^2e^{-3x}$ **47.** $y = \ln (2 + x^2)$ **48.** $y = \ln (5x + 3)$

49. $y = \frac{\ln |3x|}{x - 3}$ **50.** $y = \frac{\ln |2x - 1|}{x + 3}$ **51.** $y = \frac{x e^x}{\ln (x^2 - 1)}$ **52.** $y = \frac{(x^2 + 1)e^{2x}}{\ln x}$

53. $y = (x^2 + e^x)^2$ **54.** $y = (e^{2x+1} - 2)^4$

Find all relative maxima, minima, and inflection points for the functions defined as follows. Graph each function.

55. $y = xe^x$ **56.** $y = 3xe^{-x}$ **57.** $y = \frac{e^x}{x - 1}$ **58.** $y = \frac{\ln |5x|}{2x}$

Find the amount of interest earned by each of the following deposits.

59. \$6902 at 12% compounded semiannually for 8 yr

60. \$2781.36 at 8% compounded quarterly for 6 yr

Find the compound amount if \$12,104 is invested at 8% compounded continuously for each of the following periods.

61. 2 yr **62.** 4 yr **63.** 7 yr

Find the compound amounts for the following deposits if interest is compounded continuously.

64. \$1500 at 10% for 9 yr **65.** \$12,000 at 5% for 8 yr

Find the effective rate to the nearest hundredth for each nominal interest rate.

66. 7% compounded quarterly **67.** 9% compounded monthly **68.** 11% compounded semiannually

69. 9% compounded continuously **70.** 11% compounded continuously

Find the present value of each of the following amounts.

71. \$2000 at 6% interest compounded annually for 5 yr **72.** \$10,000 at 8% interest compounded semiannually for 6 yr

73. \$43,200 at 8% interest compounded quarterly for 4 yr **74.** \$9760 at 12% interest compounded monthly for 3 yr

Applications

Business and Economics

75. Employee Training A company finds that its new workers produce

$$P(x) = 100 - 100e^{-.8x}$$

items per day after x days on the job.
Find each of the following.

(a) What is the limit of $P(x)$ as $x \to \infty$?

(b) How many items per day should an experienced worker produce?

(c) How long will it be before a new employee will produce 50 items per day?

76. Interest To help pay for college expenses, Jane Baile borrowed \$10,000 at 10% interest compounded semiannually for 8 yr. How much will she owe at the end of the 8-yr period?

77. Interest How long will it take for \$1 to triple at an annual inflation rate of 8% compounded continuously?

78. Interest Find the interest rate needed for \$6000 to grow to \$8000 in 3 yr with continuous compounding.

79. Present Value Edison Diest wants to open a camera shop. How much must he deposit now at 6% interest compounded monthly to have \$25,000 at the end of 3 yr?

Life Sciences

80. Population Growth A population of 15,000 small deer in a specific region has grown exponentially to 17,000 in 4 yr.

(a) Write an exponential equation to express the population growth y in terms of time t in years.

(b) At this rate, how long will it take for the population to reach 45,000?

81. Intensity of Light The intensity of light (in appropriate units) passing through water decreases exponentially with the depth it penetrates beneath the surface according to the function

$$I(x) = 10e^{-.3x},$$

where x is the depth in meters. A certain water plant requires light of an intensity of 1 unit. What is the greatest depth of water in which it will grow?

82. Drug Concentration The concentration of a certain drug in the bloodstream at time t (in minutes) is given by

$$c(t) = e^{-t} - e^{-2t}.$$

Find the maximum concentration and the time when it occurs.

83. Glucose Concentration When glucose is infused into a person's bloodstream at a constant rate of c grams per minute, the glucose is converted and removed from the bloodstream at a rate proportional to the amount present. The amount of glucose in the bloodstream at time t (in minutes) is given by

$$g(t) = \frac{c}{a} + \left(g_0 - \frac{c}{a}\right)e^{-at},$$

where a is a positive constant. Assume $g_0 = .08$, $c = .1$, and $a = 1.3$.

(a) At what time is the amount of glucose a maximum? What is the maximum amount of glucose in the bloodstream?

(b) When is the amount of glucose in the bloodstream .1 gram?

(c) What happens to the amount of glucose in the bloodstream after a very long time?

Physical Sciences

84. Oil Production The production of an oil well has decreased exponentially from 128,000 barrels per year five years ago to 100,000 barrels per year at present.

(a) Letting $t = 0$ represent the present time, write an exponential equation for production y in terms of time t in years.

(b) Find the time it will take for production to fall to 70,000 barrels per year.

85. Dating Rocks Geologists sometimes measure the age of rocks by using "atomic clocks." By measuring the amounts of potassium 40 and argon 40 in a rock, the age t of the specimen (in years) is found with the formula

$$t = (1.26 \times 10^9)\frac{\ln[1 + 8.33(A/K)]}{\ln 2},$$

where A and K, respectively, are the numbers of atoms of argon 40 and potassium 40 in the specimen.

(a) How old is a rock in which $A = 0$ and $K > 0$?

(b) The ratio A/K for a sample of granite from New Hampshire is .212. How old is the sample?

(c) Let $A/K = r$ and find dt/dr.

(d) What happens to t as r gets larger? smaller?

General Interest

86. Consumer Price Index The U.S. consumer price index (CPI, or cost of living index) has risen exponentially over the years. In 1960 the index was 34; in 1990 it was 113.*

(a) Write an exponential equation for the CPI in terms of time in years.

(b) Find the year in which costs will be 50% higher than in the base year (1983), when the CPI was 100.

(c) Find the year the CPI was 100 by using your equation from part (a). If you do not get 1983, explain why.

87. Living Standards One measure of living standards in the United States is given by $L(t) = 9 + 2e^{.15t}$, where t is the number of years since 1982. Find $L(t)$ for the following years.

(a) 1982 (b) 1986 (c) 1992

(d) Graph $L(t)$.

(e) What can be said about the growth of living standards in the United States according to this equation?

88. Nuclear Warheads The number of nuclear warheads in the USSR arsenal from 1965 to 1980 is approximated by $A(t) = 830e^{.15t}$, where t is the number of years since 1965.* Find the number of nuclear warheads in the following years.

(a) 1975 (b) 1980

(c) In 1985, the number of warheads was actually 10,012. Find the corresponding number using the given function. Is the discrepancy significant? If so, what might have caused it?

(d) What does the derivative of $A(t)$ tell you about the rate of change in the number of nuclear warheads?

Connections

89. Give the following properties of the exponential function $f(x) = a^x$; $a > 0$, $a \neq 1$.

(a) Domain (b) Range

(c) y-intercept (d) Discontinuities

(e) Asymptote(s) (f) Increasing if a is ____

(g) Decreasing if a is ____

90. Give the following properties of the logarithmic function $f(x) = \log_a x$; $a > 0$, $a \neq 1$.

(a) Domain (b) Range

(c) x-intercept (d) Discontinuities

(e) Asymptote(s) (f) Increasing if a is ____

(g) Decreasing if a is ____

91. Compare your answers for Exercises 89 and 90. What similarities do you notice? What differences?

The Universal Almanac, edited by John W. Wright (Kansas City and New York: Andrews and McMeel, 1990), p. 289.

Extended Application / Individual Retirement Accounts†

Money deposited in an individual retirement account (IRA) produces earnings that are not taxed until they are withdrawn from the account. Thus, an IRA effectively defers the tax on earnings, but there is a corresponding loss of liquidity since there are penalties for withdrawals from an IRA before the age of 59½ years. Suppose that you have P dollars to invest for n years in either an IRA or a regular account, both of which pay an interest rate i compounded annually. Earnings from the regular account are immediately taxable, in contrast to the IRA earnings, which are tax sheltered until withdrawal. To compare these two investment options mathematically, assume for simplicity that the same interest rate, i, and the same income tax rate, t, apply throughout the n years of the investment.

By the compound interest formula, the principal P grows to $P(1 + i)^n$ after n years in the IRA. If the entire amount is then withdrawn (and the investor is past 59½ years of age), the tax due on the *earnings* is $[P(1 + i)^n - P]t$. Thus, the amount left after taxes is

$$\begin{aligned} A &= P(1 + i)^n - [P(1 + i)^n - P]t \\ &= P[(1 + i)^n(1 - t) + t]. \end{aligned}$$

If $P = 1$, then

$$M = (1 + i)^n(1 - t) + t$$

is called the **IRA multiplier,** because one dollar will grow to M dollars (after taxes) after n years in the IRA.

†The discussion and examples about IRAs are courtesy of Thomas B. Muenzenberger, Kansas State University, Manhattan, Kansas.

Example 1

Suppose that you deposit \$2000 in an IRA that pays 10% interest compounded annually. Suppose further that the money is left in the account for 30 years, then withdrawn, and the earnings are taxed at 37%. Find the tax due and the IRA multiplier.

Here the principal grows to $2000(1 + .1)^{30} = 34{,}899$ dollars (to the nearest dollar), and the tax due on the *earnings* is

$$(\$34{,}899 - \$2000)(.37) = \$12{,}173.$$

Thus, the amount left after taxes is $A = 34{,}899 - 12{,}173 = 22{,}726$ dollars. The IRA Multiplier is $M = 11.4$. ◀

Each year in the regular account, t percent of the i percent earned is withdrawn to pay the tax on the earnings, leaving $(1 - t)i$ percent. Thus, by the compound interest formula, the principal P grows to

$$A = P[1 + (1 - t)i]^n$$

after n years in the regular account. If $P = 1$, then

$$m = [1 + (1 - t)i]^n$$

is the multiplier, because one dollar will grow to m dollars (after taxes) after n years in the regular account.

Example 2

Suppose that you deposit \$2000 in a regular account that pays 10% compounded annually. Suppose further that the income tax rate is 37% and that the money is left in the account for 30 years and then withdrawn. Find the multiplier, m.

Here, the principal grows to

$$A = 2{,}000[1 + (1 - .37)(.10)]^{30} = 12{,}503 \text{ dollars.}$$

The multiplier is 6.3. ◀

From the previous two examples, $M/m = 11.4/6.3 = 1.8$, indicating that the IRA yields almost twice as much as the regular account over a thirty-year period. The **multiplier function**

$$\frac{M}{m} = \frac{(1 + i)^n(1 - t) + t}{[1 + (1 - t)i]^n}$$

can be used in the same way to compare the two investment options for other values of n, i, and t. Note that the multiplier function is a function in three variables.* The following table contains values of M/m for $n = 30$ and various values of i and t. It is apparent from the table that for $n = 30$ and a fixed t, the multiplier function M/m is an increasing function of i (see Figure 15 for $t = .40$). In fact, for a fixed t and for *any* $n > 1$, the multiplier function is an increasing function of i. Thus, the advantage of the IRA over the regular account widens as the interest rate i increases. This effect is particularly dramatic for high income tax rates (see the table).

t \ i	.05	.10	.15	.20	.25	.30
.10	1.07	1.19	1.34	1.49	1.65	1.81
.20	1.13	1.41	1.77	2.21	2.72	3.30
.30	1.18	1.64	2.33	3.27	4.48	6.02
.40	1.23	1.89	3.02	4.77	7.33	11.0
.50	1.27	2.13	3.83	6.83	11.8	19.8
.60	1.29	2.34	4.72	9.50	18.6	35.0
.70	1.28	2.45	5.49	12.5	27.8	59.3
.80	1.23	2.37	5.79	14.9	37.6	91.4
.90	1.15	1.96	4.81	13.6	38.9	108
1	1	1	1	1	1	1

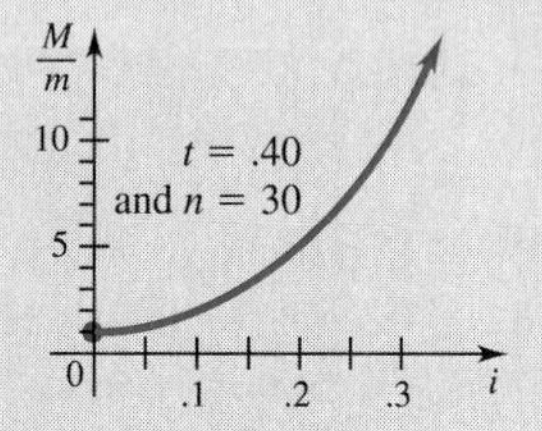

Exercises

1. **(a)** Suppose that you deposit \$2000 in an IRA that pays 10% compounded annually. Suppose further that the money is left in the account for 20 years, then withdrawn, and the earnings are taxed at 37%. Find the amount left after taxes, then find the IRA multiplier M.
   (b) Suppose instead that the money was deposited in a regular account for the same period of time at the same rates. Find the amount left after taxes; then find the multiplier m.
   (c) Find M/m. What can you conclude about investment options (a) and (b)?
2. Redo Exercise 1 for $n = 10$.
3. Draw a conclusion from Example 1 and the results of Exercises 1(a) and 2(a). (*Hint:* It is called the *time value of money.*)
4. Draw a conclusion from Example 2 and Exercises 1(b) and 2(b).
5. Construct tables for $n = 10$ and $n = 20$ and various values of i and t similar to the table for $n = 30$ given after Example 2. Draw conclusions from your data.

*Functions of more than one independent variable are discussed in the chapter entitled "Multivariable Calculus."

Applications at a glance . . .

Cold Weather in New York City . . . equitable distribution of income profit from a new technology . . . cost of maintaining a machine . . .

CHAPTER 6

INTEGRATION

(See page 360.)

In previous chapters, we studied the mathematical process of finding the derivative of a function, and we considered various applications of derivatives. The material that was covered belongs to the branch of calculus called *differential calculus*. In this chapter and the next we will study another branch of calculus, called *integral calculus*. Like the derivative of a function, the indefinite integral of a function is a special limit with many diverse applications. Geometrically, the derivative is related to the slope of the tangent line to a curve, while the definite integral is related to the area under a curve. As this chapter will show, differential and integral calculus are connected by the Fundamental Theorem of Calculus.

6.1 ANTIDERIVATIVES

If an object is thrown from the top of the Sears Tower in Chicago, how fast is it going when it hits the ground?

Using *antiderivatives*, we can answer this question.

Functions used in applications in previous chapters have provided information about a *total amount* of a quantity, such as cost, revenue, profit, temperature, gallons of oil, or distance. Working with these functions provided information about the rate of change of these quantities and allowed us to answer important questions about the extrema of the functions. It is not always possible to find ready-made functions that provide information about the total amount of a quantity, but it is often possible to collect enough data to come up with a function that gives the *rate of change* of a quantity. Since we know that derivatives give the rate of change when the total amount is known, is it possible to reverse the process and use a known rate of change to get a function that gives the total amount of a quantity? The answer is yes; this reverse process, called *antidifferentiation*, is the topic of this section. The *antiderivative* of a function is defined as follows.

ANTIDERIVATIVE

If $F'(x) = f(x)$, then $F(x)$ is an **antiderivative** of $f(x)$.

EXAMPLE 1

(a) If $F(x) = 10x$, then $F'(x) = 10$, so $F(x) = 10x$ is an antiderivative of $f(x) = 10$.

(b) For $F(x) = x^2$, $F'(x) = 2x$, making $F(x) = x^2$ an antiderivative of $f(x) = 2x$. ◀

EXAMPLE 2

Find an antiderivative of $f(x) = 5x^4$.

To find a function $F(x)$ whose derivative is $5x^4$, work backwards. Recall that the derivative of x^n is nx^{n-1}. If

$$nx^{n-1} \text{ is } 5x^4,$$

then $n - 1 = 4$ and $n = 5$, so x^5 is an antiderivative of $5x^4$. ◀

EXAMPLE **3** Suppose a population is growing at a rate given by $f(x) = e^x$, where x is time in years from some initial date. Find a function giving the population at time x.

Let the population function be $F(x)$. Then

$$f(x) = F'(x) = e^x.$$

The derivative of the function defined by $F(x) = e^x$ is $F'(x) = e^x$, so a population function with the given growth rate is $F(x) = e^x$. ◀

The function from Example 1(b), defined by $F(x) = x^2$, is not the only function whose derivative is $f(x) = 2x$; for example, both $G(x) = x^2 + 2$ and $H(x) = x^2 - 7$ have $f(x) = 2x$ as a derivative. In fact, for any real number C, the function $F(x) = x^2 + C$ has $f(x) = 2x$ as its derivative. This means that there is a *family* or *class* of functions having $2x$ as an antiderivative. As the next theorem states, if two functions $F(x)$ and $G(x)$ are antiderivatives of $f(x)$, then $F(x)$ and $G(x)$ can differ only by a constant.

If $F(x)$ and $G(x)$ are both antiderivatives of a function $f(x)$, then there is a constant C such that

$$\mathbf{F(x) - G(x) = C.}$$

(Two antiderivatives of a function can differ only by a constant.)

For example,

$$F(x) = x^2 + 2, \qquad G(x) = x^2, \qquad \text{and} \qquad H(x) = x^2 - 4$$

are all antiderivatives of $f(x) = 2x$, and any two of them differ only by a constant. The derivative of a function gives the slope of the tangent line at any x-value. The fact that these three functions have the same derivative, $f(x) = 2x$, means that their slopes at any particular value of x are the same, as shown in Figure 1. Thus, each graph can be obtained from another by a vertical shift of $|C|$ units, where C is any constant. We will represent this family of antiderivatives of $f(x)$ by $F(x) + C$.

The family of all antiderivatives of the function f is indicated by

$$\int f(x)\, dx.$$

The symbol $\int$ is the **integral sign,** $f(x)$ is the **integrand,** and $\int f(x)\, dx$ is called an **indefinite integral,** the most general antiderivative of f.

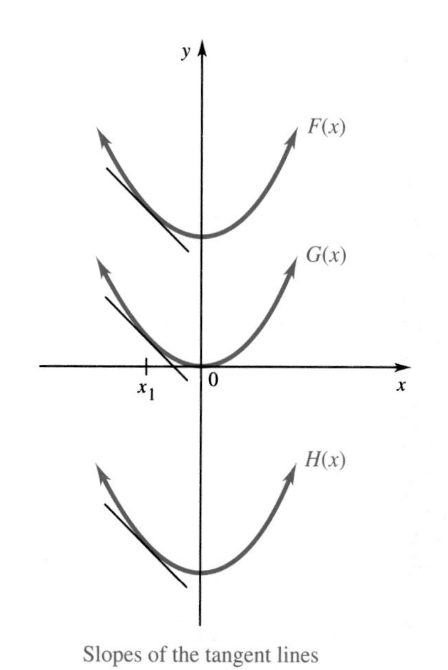

Slopes of the tangent lines at $x = x_1$ are the same.

FIGURE 1

INDEFINITE INTEGRAL

If $F'(x) = f(x)$, then

$$\int f(x)\,dx = F(x) + C,$$

for any real number C.

For example, using this notation,

$$\int (2x)\,dx = x^2 + C.$$

The dx in the indefinite integral indicates that $\int f(x)\,dx$ is the "integral of $f(x)$ *with respect to* x" just as the symbol dy/dx denotes the "derivative of y with respect to x." For example, in the indefinite integral $\int 2ax\,dx$, dx indicates that a is to be treated as a constant and x as the variable, so that

$$\int 2ax\,dx = \int a(2x)dx = ax^2 + C.$$

On the other hand,

$$\int 2ax\,da = a^2x + C = xa^2 + C.$$

A more complete interpretation of dx will be discussed later.

The symbol $\int f(x)dx$ was created by G.W. Leibniz (1646–1716) in the latter part of the seventeenth century. The $\int$ is an elongated S from *summa*, the Latin word for *sum*. The word *integral* as a term in the calculus was coined by Jakob Bernoulli (1654–1705), a Swiss mathematician who corresponded frequently with Leibniz. The relationship between sums and integrals will be clarified in the next two sections.

The method of working backwards, as above, to find an antiderivative is not very satisfactory. Some rules for finding antiderivatives are needed. Since the process of finding an indefinite integral is the inverse of the process of finding a derivative, each formula for derivatives leads to a rule for indefinite integrals.

Recall that $\frac{d}{dx}x^n = nx^{n-1}$.

As mentioned in Example 2, the derivative of x^n is found by multiplying x by n and reducing the exponent on x by 1. To find an indefinite integral—that is, to undo what was done—*increase* the exponent by 1 and *divide* by the new exponent, $n + 1$.

POWER RULE

For any real number $n \neq -1$,

$$\int x^n\,dx = \frac{1}{n+1}x^{n+1} + C.$$

This rule can be verified by differentiating the expression on the right above:

$$\frac{d}{dx}\left(\frac{1}{n+1}x^{n+1} + C\right) = \frac{n+1}{n+1}x^{(n+1)-1} + 0 = x^n.$$

(If $n = -1$, the expression in the denominator is 0, and the above rule cannot be used. Finding an antiderivative for this case is discussed later.)

EXAMPLE 4 Find each indefinite integral.

(a) $\int t^3\,dt$

Use the power rule with $n = 3$.

$$\int t^3\,dt = \frac{1}{3+1}t^{3+1} + C = \frac{1}{4}t^4 + C$$

(b) $\int \frac{1}{t^2}dt$

First, write $1/t^2$ as t^{-2}. Then

$$\int \frac{1}{t^2}dt = \int t^{-2}\,dt = \frac{t^{-1}}{-1} + C = -\frac{1}{t} + C.$$

(c) $\int \sqrt{u}\,du$

Since $\sqrt{u} = u^{1/2}$,

$$\int \sqrt{u}\,du = \int u^{1/2}\,du = \frac{1}{1/2+1}u^{3/2} + C = \frac{2}{3}u^{3/2} + C.$$

To check this, differentiate $(2/3)u^{3/2} + C$; the derivative is $u^{1/2}$, the original function.

(d) $\int dx$

Write dx as $1 \cdot dx$ and use the fact that $x^0 = 1$ for any nonzero number x to get

$$\int dx = \int 1\,dx = \int x^0\,dx = \frac{1}{1}x^1 + C = x + C.$$ ◀

Recall that $\frac{d}{dx}[f(x) \pm g(x)] = [f'(x) \pm g'(x)]$ and $\frac{d}{dx}[kf(x)] = kf'(x)$.

As shown earlier, the derivative of the product of a constant and a function is the product of the constant and the derivative of the function. A similar rule applies to indefinite integrals. Also, since derivatives of sums or differences are found term by term, indefinite integrals also can be found term by term.

CONSTANT MULTIPLE RULE AND SUM OR DIFFERENCE RULE

If all indicated integrals exist,

$$\int k \cdot f(x)dx = k\int f(x)dx \quad \text{for any real number } k,$$

and

$$\int [f(x) \pm g(x)]dx = \int f(x)dx \pm \int g(x)dx.$$

EXAMPLE 5 Find each of the following.

(a) $\int 2v^3\,dv$

By the constant multiple rule and the power rule,

$$\int 2v^3\,dv = 2\int v^3\,dv = 2\left(\frac{1}{4}v^4\right) + C = \frac{1}{2}v^4 + C.$$

(b) $\int \frac{12}{z^5}\,dz$

Use negative exponents.

$$\int \frac{12}{z^5}\,dz = \int 12z^{-5}\,dz = 12\int z^{-5}\,dz = 12\left(\frac{z^{-4}}{-4}\right) + C$$

$$= -3z^{-4} + C = \frac{-3}{z^4} + C$$

(c) $\int (3z^2 - 4z + 5)\,dz$

Using the rules in this section,

$$\int (3z^2 - 4z + 5)\,dz = 3\int z^2\,dz - 4\int z\,dz + 5\int dz$$

$$= 3\left(\frac{1}{3}z^3\right) - 4\left(\frac{1}{2}z^2\right) + 5z + C$$

$$= z^3 - 2z^2 + 5z + C.$$

Only one constant C is needed in the answer; the three constants from integrating term by term are combined. In Example 5(a), C represents any real number, so it is not necessary to multiply it by 2 in the next-to-last step. ◀

EXAMPLE 6 Find each of the following.

(a) $\int \frac{x^2 + 1}{\sqrt{x}}\,dx$

First rewrite the integrand as follows.

$$\int \frac{x^2 + 1}{\sqrt{x}}\,dx = \int \left(\frac{x^2}{\sqrt{x}} + \frac{1}{\sqrt{x}}\right)dx$$

$$= \int \left(\frac{x^2}{x^{1/2}} + \frac{1}{x^{1/2}}\right)dx$$

$$= \int (x^{3/2} + x^{-1/2})\,dx$$

Now find the antiderivative.

$$\int (x^{3/2} + x^{-1/2})\,dx = \frac{x^{5/2}}{5/2} + \frac{x^{1/2}}{1/2} + C$$
$$= \frac{2}{5}x^{5/2} + 2x^{1/2} + C$$

(b) $\int (x^2 - 1)^2\,dx$

Square the binomial first, and then find the antiderivative.

$$\int (x^2 - 1)^2\,dx = \int (x^4 - 2x^2 + 1)\,dx$$
$$= \frac{x^5}{5} - \frac{2x^3}{3} + x + C$$ ◀

To check integration, take the derivative of the result. For instance, in Example 5(c) check that $z^3 - 2z^2 + 5z + C$ is the required indefinite integral by taking the derivative

$$\frac{d}{dz}(z^3 - 2z^2 + 5z + C) = 3z^2 - 4z + 5,$$

which agrees with the original information.

It was shown earlier that the derivative of $f(x) = e^x$ is $f'(x) = e^x$. Also, the derivative of $f(x) = e^{kx}$ is $f'(x) = k \cdot e^{kx}$. These results lead to the following formulas for indefinite integrals of exponential functions.

INDEFINITE INTEGRALS OF EXPONENTIAL FUNCTIONS

$$\int e^x\,dx = e^x + C$$
$$\int e^{kx}\,dx = \frac{1}{k} \cdot e^{kx} + C, \qquad k \neq 0$$

EXAMPLE 7

(a) $\int 9e^t\,dt = 9\int e^t\,dt = 9e^t + C$

(b) $\int e^{9t}\,dt = \frac{1}{9}e^{9t} + C$

(c) $\int 3e^{(5/4)u}\,du = 3\left(\frac{1}{5/4}e^{(5/4)u}\right) + C$

$$= 3\left(\frac{4}{5}\right)e^{(5/4)u} + C$$
$$= \frac{12}{5}e^{(5/4)u} + C$$ ◀

The restriction $n \neq -1$ was necessary in the formula for $\int x^n\,dx$ since $n = -1$ made the denominator of $1/(n + 1)$ equal to 0. To find $\int x^n\,dx$ when $n = -1$, that is, to find $\int x^{-1}\,dx$, recall the differentiation formula for the logarithmic function: the derivative of $f(x) = \ln|x|$, where $x \neq 0$, is $f'(x) = 1/x = x^{-1}$. This formula for the derivative of $f(x) = \ln|x|$ gives a formula for $\int x^{-1}\,dx$.

INDEFINITE INTEGRAL OF x^{-1}

$$\int x^{-1}\,dx = \int \frac{1}{x}\,dx = \ln|x| + C$$

(a) $\displaystyle\int \frac{4}{x}\,dx = 4\int \frac{1}{x}\,dx = 4\ln|x| + C$

(b) $\displaystyle\int \left(-\frac{5}{x} + e^{-2x}\right)dx = -5\ln|x| - \frac{1}{2}e^{-2x} + C$ ◀

In all the examples above, the antiderivative family of functions was found. In many applications, however, the given information allows us to determine the value of the integration constant C. The next examples illustrate this idea.

Suppose a publishing company has found that the marginal cost at a level of production of x thousand books is given by

$$C'(x) = \frac{50}{\sqrt{x}}$$

and that the fixed cost (the costs before the first book can be produced) is \$25,000. Find the cost function $C(x)$.

Write $50/\sqrt{x}$ as $50/x^{1/2}$ or $50x^{-1/2}$, and then use the indefinite integral rules to integrate the function.

$$\int \frac{50}{\sqrt{x}}\,dx = \int 50x^{-1/2}\,dx = 50(2x^{1/2}) + k = 100x^{1/2} + k$$

(Here k is used instead of C to avoid confusion with the cost function $C(x)$.) To find the value of k, use the fact that $C(0)$ is 25,000.

$$\begin{aligned} C(x) &= 100x^{1/2} + k \\ 25{,}000 &= 100 \cdot 0 + k \\ k &= 25{,}000 \end{aligned}$$

With this result, the cost function is $C(x) = 100x^{1/2} + 25{,}000$. ◀

Suppose the marginal revenue from a product is given by $50 - 3x - x^2$. Find the demand function for the product.

The marginal revenue is the derivative of the revenue function

$$\frac{dR}{dx} = 50 - 3x - x^2,$$

so

$$R = \int (50 - 3x - x^2)\,dx$$
$$= 50x - \frac{3x^2}{2} - \frac{x^3}{3} + C.$$

If $x = 0$, then $R = 0$ (no items sold means no revenue), and

$$0 = 50(0) - \frac{3(0)^2}{2} - \frac{(0)^3}{3} + C$$
$$0 = C.$$

Thus,

$$R = 50x - \frac{3x^2}{2} - \frac{x^3}{3}$$

gives the revenue function. Now, recall that $R = xp$, where p is the demand function. Then

$$50x - \frac{3}{2}x^2 - \frac{1}{3}x^3 = xp$$
$$50 - \frac{3}{2}x - \frac{1}{3}x^2 = p,$$

which gives the demand function. 

In the next example integrals are used to find the position of a particle when the acceleration of the particle is given.

EXAMPLE 11 Recall that if the function $s(t)$ gives the position of a particle at time t, then its velocity $v(t)$ and its acceleration $a(t)$ are given by

$$v(t) = s'(t) \quad \text{and} \quad a(t) = v'(t) = s''(t).$$

(a) Suppose the velocity of an object is $v(t) = 6t^2 - 8t$ and that the object is at -5 when time is 0. Find $s(t)$.
Since $v(t) = s'(t)$, the function $s(t)$ is an antiderivative of $v(t)$:

$$s(t) = \int v(t)\,dt = \int (6t^2 - 8t)\,dt$$
$$= 2t^3 - 4t^2 + C$$

for some constant C. Find C from the given information that $s = -5$ when $t = 0$.

$$s(t) = 2t^3 - 4t^2 + C$$
$$-5 = 2(0)^3 - 4(0)^2 + C$$
$$-5 = C$$
$$s(t) = 2t^3 - 4t^2 - 5$$

(b) Many experiments have shown that when an object is dropped, its acceleration (ignoring air resistance) is constant. This constant has been found to be approximately 32 feet per second every second; that is,

$$a(t) = -32.$$

The negative sign is used because the object is falling. Suppose an object is thrown down from the top of the 1100-foot-tall Sears Tower in Chicago. If the initial velocity of the object is -20 feet per second, find $s(t)$.

First find $v(t)$ by integrating $a(t)$:

$$v(t) = \int (-32)\,dt = -32t + k.$$

When $t = 0$, $v(t) = -20$:

$$-20 = -32(0) + k$$
$$-20 = k$$

and

$$v(t) = -32t - 20.$$

Now integrate $v(t)$ to find $s(t)$.

$$s(t) = \int (-32t - 20)\,dt = -16t^2 - 20t + C$$

Since $s(t) = 1100$ when $t = 0$, we can substitute these values into the equation for $s(t)$ to get $C = 1100$ and

$$s(t) = -16t^2 - 20t + 1100$$

as the distance of the object from the ground after t seconds.

? **(c)** Use the equations derived in (b) to find out how fast the object was falling when it hit the ground and how long it took to strike the ground.

When the object strikes the ground, $s = 0$, so

$$0 = -16t^2 - 20t + 1100.$$

To solve this equation for t, factor out the common factor of -4 and then use the quadratic formula.

$$0 = -4(4t^2 + 5t - 275)$$
$$t = \frac{-5 \pm \sqrt{25 + 4400}}{8} \approx \frac{-5 \pm 66.5}{8}$$

Only the positive value of t is meaningful here: $t \approx 7.69$. It takes the object about 7.69 seconds to strike the ground. From the velocity equation, with $t = 7.69$, we find

$$v(t) = -32t - 20$$
$$v(7.69) = -32(7.69) - 20 \approx -266,$$

so the object was falling (as indicated by the negative sign) at about 266 feet per second when it hit the ground. ◀

EXAMPLE 12 Find a function f whose graph has slope $f'(x) = 6x^2 + 4$ and goes through the point (1, 1).

Since $f'(x) = 6x^2 + 4$,

$$f(x) = \int (6x^2 + 4)\,dx = 2x^3 + 4x + C.$$

The graph of f goes through (1, 1), so C can be found by substituting 1 for x and 1 for $f(x)$.

$$1 = 2(1)^3 + 4(1) + C$$
$$1 = 6 + C$$
$$C = -5$$

Finally, $f(x) = 2x^3 + 4x - 5$. ◀

6.1 Exercises

Find each of the following.

1. $\int 4x\,dx$
2. $\int 8x\,dx$
3. $\int 5t^2\,dt$
4. $\int 6x^3\,dx$
5. $\int 6\,dk$
6. $\int 2\,dy$
7. $\int (2z + 3)\,dz$
8. $\int (3x - 5)\,dx$
9. $\int (x^2 + 6x)\,dx$
10. $\int (t^2 - 2t)\,dt$
11. $\int (t^2 - 4t + 5)\,dt$
12. $\int (5x^2 - 6x + 3)\,dx$
13. $\int (4z^3 + 3z^2 + 2z - 6)\,dz$
14. $\int (12y^3 + 6y^2 - 8y + 5)\,dy$
15. $\int 5\sqrt{z}\,dz$
16. $\int t^{1/4}\,dt$
17. $\int (u^{1/2} + u^{3/2})\,du$
18. $\int (4\sqrt{v} - 3v^{3/2})\,dv$
19. $\int (15x\sqrt{x} + 2\sqrt{x})\,dx$
20. $\int (x^{1/2} - x^{-1/2})\,dx$
21. $\int (10u^{3/2} - 14u^{5/2})\,du$
22. $\int (56t^{5/2} + 18t^{7/2})\,dt$
23. $\int \left(\frac{1}{z^2}\right)dz$
24. $\int \left(\frac{4}{x^3}\right)dx$
25. $\int \left(\frac{1}{y^3} - \frac{1}{\sqrt{y}}\right)dy$
26. $\int \left(\sqrt{u} + \frac{1}{u^2}\right)du$
27. $\int (-9t^{-2} - 2t^{-1})\,dt$
28. $\int (8x^{-3} + 4x^{-1})\,dx$
29. $\int e^{2t}\,dt$
30. $\int e^{-3y}\,dy$
31. $\int 3e^{-.2x}\,dx$
32. $\int -4e^{.2v}\,dv$
33. $\int \left(\frac{3}{x} + 4e^{-.5x}\right)dx$
34. $\int \left(\frac{9}{x} - 3e^{-.4x}\right)dx$
35. $\int \frac{1 + 2t^3}{t}\,dt$
36. $\int \frac{2y^{1/2} - 3y^2}{y}\,dy$
37. $\int (e^{2u} + 4u)\,du$
38. $\int (v^2 - e^{3v})\,dv$
39. $\int (x + 1)^2\,dx$
40. $\int (2y - 1)^2\,dy$
41. $\int \frac{\sqrt{x} + 1}{\sqrt[3]{x}}\,dx$
42. $\int \frac{1 - 2\sqrt[3]{z}}{\sqrt[3]{z}}\,dz$

43. Find an equation of the curve whose tangent line has a slope of

$$f'(x) = x^{2/3},$$

given that the point $(1, 3/5)$ is on the curve.

44. The slope of the tangent line to a curve is given by

$$f'(x) = 6x^2 - 4x + 3.$$

If the point $(0, 1)$ is on the curve, find an equation of the curve.

Applications

Business and Economics

Cost *Find the cost function for each of the following marginal cost functions.*

45. $C'(x) = 4x - 5$, fixed cost is $8

46. $C'(x) = 2x + 3x^2$, fixed cost is $15

47. $C'(x) = .2x^2 + 5x$, fixed cost is $10

48. $C'(x) = .8x^2 - x$, fixed cost is $5

49. $C'(x) = .03e^{.01x}$, fixed cost is $8

50. $C'(x) = x^{1/2}$, 16 units cost $45

51. $C'(x) = x^{2/3} + 2$, 8 units cost $58

52. $C'(x) = x^2 - 2x + 3$, 3 units cost $15

53. $C'(x) = x + 1/x^2$, 2 units cost $5.50

54. $C'(x) = 1/x + 2x$, 7 units cost $58.40

55. $C'(x) = 5x - 1/x$, 10 units cost $94.20

56. $C'(x) = 1.2^x (\ln 1.2)$, 2 units cost $9.44 (*Hint:* Recall that $a^x = e^{x \ln a}$.)

57. **Profit** The marginal profit of a small fast-food stand is given by

$$P'(x) = 2x + 20,$$

where x is the sales volume in thousands of hamburgers. The "profit" is -50 dollars when no hamburgers are sold. Find the profit function.

58. **Profit** Suppose the marginal profit from the sale of x hundred items is

$$P'(x) = 4 - 6x + 3x^2,$$

and the profit on 0 items is $-\$40$. Find the profit function.

Life Sciences

59. **Biochemical Excretion** If the rate of excretion of a biochemical compound is given by

$$f'(t) = .01e^{-.01t},$$

the total amount excreted by time t (in minutes) is $f(t)$.

(a) Find an expression for $f(t)$.

(b) If 0 units are excreted at time $t = 0$, how many units are excreted in 10 minutes?

60. **Concentration of a Solute** According to Fick's law, the diffusion of a solute across a cell membrane is given by

$$c'(t) = \frac{kA}{V}[C - c(t)], \qquad (1)$$

where A is the area of the cell membrane, V is the volume of the cell, $c(t)$ is the concentration inside the cell at time t, C is the concentration outside the cell, and k is a constant. If c_0 represents the concentration of the solute inside the cell when $t = 0$, then it can be shown that

$$c(t) = (c_0 - C)e^{-kAt/V} + M. \qquad (2)$$

(a) Use the last result to find $c'(t)$.

(b) Substitute back into equation (1) to show that (2) is indeed the correct antiderivative of (1).

Physical Sciences

Exercises 61–65 refer to Example 11 in this section.

61. **Velocity** For a particular object, $a(t) = t^2 + 1$ and $v(0) = 6$. Find $v(t)$.

62. **Distance** Suppose $v(t) = 6t^2 - 2/t^2$ and $s(1) = 8$. Find $s(t)$.

63. **Time** An object is dropped from a small plane flying at 6400 feet. Assume that $a(t) = -32$ feet per second per second and $v(0)$ is 0, and find $s(t)$. How long will it take the object to hit the ground?

64. **Distance** Suppose $a(t) = 18t + 8$, $v(1) = 15$, and $s(1) = 19$. Find $s(t)$.

65. **Distance** Suppose $a(t) = \frac{15}{2}\sqrt{t} + 3e^{-t}$, $v(0) = -3$, and $s(0) = 4$. Find $s(t)$.

6.2 SUBSTITUTION

If a formula for the marginal revenue is known, how can a formula for the total revenue be found?

Using the method of substitution, this question will be answered in an exercise in this section.

We saw how to integrate a few simple functions in the previous section. More complicated functions can sometimes be integrated by *substitution*. The substitution technique depends on the idea of a differential, discussed in an earlier chapter. If $u = f(x)$, the *differential* of u, written du, is defined as

$$du = f'(x)\,dx.$$

For example, if $u = 6x^4$, then $du = 24x^3\,dx$.

Recall the chain rule for derivatives as used in the following example:

The chain rule, discussed in detail in the chapter on differentiation, states that $\frac{d}{dx}[f(g(x))] = f'(g(x)) \cdot g'(x)$.

$$\frac{d}{dx}(x^2 - 1)^5 = 5(x^2 - 1)^4(2x) = 10x(x^2 - 1)^4.$$

As in this example, the result of using the chain rule is often a product of two functions. Because of this, functions formed by the product of two functions can sometimes be integrated by using the chain rule in reverse. In the example above, working backwards from the derivative gives

$$\int 10x(x^2 - 1)^4\,dx = (x^2 - 1)^5 + C.$$

To find an antiderivative involving products, it often helps to make a substitution: let $u = x^2 - 1$, so that $du = 2x\,dx$. Now substitute u for $x^2 - 1$ and du for $2x\,dx$ in the indefinite integral above.

$$\begin{aligned}\int 10x(x^2 - 1)^4\,dx &= \int 5 \cdot 2x(x^2 - 1)^4\,dx \\ &= 5\int (x^2 - 1)^4\,(2x\,dx) \\ &= 5\int u^4\,du\end{aligned}$$

This last integral can now be found by the power rule.

$$5\int u^4\,du = 5 \cdot \frac{1}{5}u^5 + C = u^5 + C$$

Finally, substitute $x^2 - 1$ for u.

$$\int 10x(x^2 - 1)^4\,dx = (x^2 - 1)^5 + C$$

This method of integration is called **integration by substitution.** As shown above, it is simply the chain rule for derivatives in reverse. The results can always be verified by differentiation.

This discussion leads to the following integration formula, which is sometimes called the general power rule for integrals.

GENERAL POWER RULE For $u = f(x)$ and $du = f'(x)\,dx$,

$$\int u^n du = \frac{u^{n+1}}{n+1} + C.$$

EXAMPLE 1

Find $\int 6x(3x^2 + 4)^4\,dx$.

A certain amount of trial and error may be needed to decide on the expression to set equal to u. The integrand must be written as two factors, one of which is the derivative of the other. In this example, if $u = 3x^2 + 4$, then $du = 6x\,dx$. Now substitute.

$$\int 6x(3x^2 + 4)^4\,dx = \int (3x^2 + 4)^4(6x\,dx) = \int u^4\,du$$

Find this last indefinite integral.

$$\int u^4\,du = \frac{u^5}{5} + C$$

Now replace u with $3x^2 + 4$.

$$\int 6x(3x^2 + 4)^4\,dx = \frac{u^5}{5} + C = \frac{(3x^2 + 4)^5}{5} + C$$

To verify this result, find the derivative.

$$\frac{d}{dx}\left[\frac{(3x^2 + 4)^5}{5} + C\right] = \frac{5}{5}(3x^2 + 4)^4(6x) + 0 = (3x^2 + 4)^4(6x)$$

The derivative is the original function, as required. ◀

EXAMPLE 2

Find $\int x^2\sqrt{x^3 + 1}\,dx$.

An expression raised to a power is usually a good choice for u, so because of the square root or $1/2$ power, let $u = x^3 + 1$; then $du = 3x^2\,dx$. The integrand does not contain the constant 3, which is needed for du. To take care of this, multiply by $3/3$, placing 3 inside the integral sign and $1/3$ outside.

$$\int x^2\sqrt{x^3 + 1}\,dx = \frac{1}{3}\int 3x^2\sqrt{x^3 + 1}\,dx = \frac{1}{3}\int \sqrt{x^3 + 1}\,(3x^2\,dx)$$

Now substitute u for $x^3 + 1$ and du for $3x^2\,dx$, and then integrate.

$$\frac{1}{3}\int \sqrt{x^3 + 1}\, 3x^2\, dx = \frac{1}{3}\int \sqrt{u}\, du = \frac{1}{3}\int u^{1/2}\, du$$
$$= \frac{1}{3} \cdot \frac{u^{3/2}}{3/2} + C = \frac{2}{9}u^{3/2} + C$$

Since $u = x^3 + 1$,

$$\int x^2\sqrt{x^3 + 1}\, dx = \frac{2}{9}(x^3 + 1)^{3/2} + C. \blacktriangleleft$$

The substitution method given in the examples above *will not always work.* For example, you might try to find

$$\int x^3\sqrt{x^3 + 1}\, dx$$

by substituting $u = x^3 + 1$, so that $du = 3x^2\, dx$. However, there is no *constant* that can be inserted inside the integral sign to give $3x^2$. This integral, and a great many others, cannot be evaluated by substitution.

With practice, choosing u will become easy if you keep two principles in mind. First, u should equal some expression in the integral that, when replaced with u, tends to make the integral simpler. Second and most important, u must be an expression whose derivative is also present in the integral. The substitution should include as much of the integral as possible, so long as its derivative is still present. In Example 1, we could have chosen $u = 3x^2$, but $u = 3x^2 + 4$ is better, because it has the same derivative as $3x^2$ and captures more of the original integral. If we carry this reasoning further, we might try $u = (3x^2 + 4)^4$, but this is a poor choice, for $du = 4(3x^2 + 4)^3(6x)\,dx$, an expression not present in the original integral.

The exact form of the derivative of u need not be present. Any constant multiplier is easily taken care of, as in Example 2 and the next example.

Find $\displaystyle\int \frac{x + 3}{(x^2 + 6x)^2}\, dx.$

Let $u = x^2 + 6x$, so that $du = (2x + 6)\,dx = 2(x + 3)\,dx$. The integral is missing the 2, so multiply by 2/2, putting 2 inside the integral sign and 1/2 outside.

$$\int \frac{x + 3}{(x^2 + 6x)^2}\, dx = \frac{1}{2}\int \frac{2(x + 3)}{(x^2 + 6x)^2}\, dx$$
$$= \frac{1}{2}\int \frac{du}{u^2} = \frac{1}{2}\int u^{-2}\, du$$
$$= \frac{1}{2} \cdot \frac{u^{-1}}{-1} + C = \frac{-1}{2u} + C$$

Substituting $x^2 + 6x$ for u gives

$$\int \frac{x + 3}{(x^2 + 6x)^2}\,dx = \frac{-1}{2(x^2 + 6x)} + C. \blacktriangleleft$$

Recall the formula for $\frac{d}{dx}(e^u)$, where $u = f(x)$:

$$\frac{d}{dx}(e^u) = e^u \frac{d}{dx}u.$$

For example, if $u = x^2$ then $\frac{d}{dx}u = \frac{d}{dx}(x^2) = 2x$, and

$$\frac{d}{dx}(e^{x^2}) = e^{x^2} \cdot 2x.$$

Working backwards, if $u = x^2$, then $du = 2x\,dx$, so

$$\int e^{x^2} \cdot 2x\,dx = \int e^u\,du = e^u + C = e^{x^2} + C.$$

The work above suggests the following rule for the indefinite integral of e^u, where $u = f(x)$.

INDEFINITE INTEGRAL OF e^u

If $u = f(x)$, then $du = f'(x)dx$ and

$$\int e^u\,du = e^u + C.$$

EXAMPLE 4

Find $\int x^2 \cdot e^{x^3}\,dx$.

Let $u = x^3$, the exponent on e. Then $du = 3x^2\,dx$. Multiplying by $3/3$ gives

$$\int x^2 \cdot e^{x^3}\,dx = \frac{1}{3}\int e^{x^3}(3x^2\,dx)$$

$$= \frac{1}{3}\int e^u du = \frac{1}{3}e^u + C = \frac{1}{3}e^{x^3} + C. \blacktriangleleft$$

Recall that the antiderivative of $f(x) = 1/x$ is $\ln|x|$. The next example uses the fact that $\int x^{-1}\,dx = \ln|x| + C$, together with the method of substitution.

Find $\int \frac{(2x - 3)dx}{x^2 - 3x}$.

Let $u = x^2 - 3x$, so that $du = (2x - 3)dx$. Then

$$\int \frac{(2x - 3)dx}{x^2 - 3x} = \int \frac{du}{u} = \ln|u| + C = \ln|x^2 - 3x| + C. \blacktriangleleft$$

Generalizing from the results of Example 5 suggests the following rule for finding the indefinite integral of u^{-1}, where $u = f(x)$.

INDEFINITE INTEGRAL OF u^{-1}

If $u = f(x)$, then $du = f'(x)dx$ and

$$\int u^{-1}\,du = \int \frac{du}{u} = \ln|u| + C.$$

The next example shows a more complicated integral evaluated by the method of substitution.

EXAMPLE 6 Find $\int x\sqrt{1 - x}\,dx$.

Let $u = 1 - x$. Then $x = 1 - u$ and $dx = -du$. Now substitute:

$$\begin{aligned}\int x\sqrt{1-x}\,dx &= \int (1-u)\sqrt{u}(-du) = \int (u-1)u^{1/2}\,du\\ &= \int (u^{3/2} - u^{1/2})\,du = \frac{2}{5}u^{5/2} - \frac{2}{3}u^{3/2} + C\\ &= \frac{2}{5}(1-x)^{5/2} - \frac{2}{3}(1-x)^{3/2} + C.\end{aligned}$$

The substitution method is useful if the integral can be written in one of the following forms, where $u(x)$ is some function of x.

SUBSTITUTION METHOD

Form of the Integral	*Form of the Antiderivative*
1. $\int [u(x)]^n \cdot u'(x)\,dx,\ n \neq -1$	$\dfrac{[u(x)]^{n+1}}{n+1} + C$
2. $\int e^{u(x)} \cdot u'(x)\,dx$	$e^{u(x)} + C$
3. $\int \dfrac{u'(x)\,dx}{u(x)}$	$\ln\lvert u(x)\rvert + C$

EXAMPLE 7 The research department for a hardware chain has determined that at one store the marginal price of x boxes per week of a particular type of nails is

$$p'(x) = \frac{-4000}{(2x+15)^3}.$$

Find the demand equation if the weekly demand of this type of nails is 10 boxes when the price of a box of nails is \$4.

To find the demand function, first integrate $p'(x)$ as follows.

$$\begin{aligned}p(x) &= \int p'(x)\,dx\\ &= \int \frac{-4000}{(2x+15)^3}\,dx\end{aligned}$$

Let $u = 2x + 15$. Then $du = 2\,dx$, and

$$\begin{aligned} p(x) &= -2000 \int (2x + 15)^{-3}\, 2\,dx \\ &= -2000 \int u^{-3}\,du \\ &= (-2000)\frac{u^{-2}}{-2} + C \\ &= \frac{1000}{u^2} + C \end{aligned}$$

$$p(x) = \frac{1000}{(2x + 15)^2} + C. \tag{1}$$

Find the value of C by using the given information that $p = 4$ when $x = 10$.

$$\begin{aligned} 4 &= \frac{1000}{(2 \cdot 10 + 15)^2} + C \\ 4 &= \frac{1000}{35^2} + C \\ 4 &= .82 + C \\ 3.18 &= C \end{aligned}$$

Replacing C with 3.18 in equation (1) gives the demand function,

$$p(x) = \frac{1000}{(2x + 15)^2} + 3.18. \blacktriangleleft$$

With a little practice, you will find you can skip the substitution step for integrals such as that shown in Example 7, in which the derivative of u is a constant. Recall from the chain rule that when you differentiate a function, such as $p'(x) = -4000/(2x + 15)^3$ in the previous example, you multiply by 2, the derivative of $(2x + 15)$. So when taking the antiderivative, simply divide by 2:

$$\begin{aligned} \int -4000(2x + 15)^{-3}\,dx &= \frac{-4000}{2} \cdot \frac{(2x + 15)^{-2}}{-2} + C \\ &= \frac{1000}{(2x + 15)^2} + C. \end{aligned}$$

Caution This procedure is valid because of the constant multiple rule presented in the previous section, which says that constant multiples can be brought into or out of integrals, just as they can with derivatives. This procedure is *not* valid with any expression other than a constant.

EXAMPLE 8 To determine the top 100 popular songs of each year since 1956, Jim Quirin and Barry Cohen developed a function that represents the rate of change on the charts of *Billboard* magazine required for a song to earn a "star" on the *Billboard* "Hot 100" survey.* They developed the function

$$f(x) = \frac{A}{B + x},$$

where $f(x)$ represents the rate of change in position on the charts, x is the position on the "Hot 100" survey, and A and B are appropriate constants. The function

$$F(x) = \int f(x)\,dx$$

is defined as the "Popularity Index." Find $F(x)$.

Integrating $f(x)$ gives

$$\begin{aligned} F(x) &= \int f(x)\,dx \\ &= \int \frac{A}{B + x}\,dx \\ &= A\int \frac{1}{B + x}\,dx. \end{aligned}$$

Let $u = B + x$, so that $du = dx$. Then

$$\begin{aligned} F(x) &= A\int \frac{1}{u}\,du = A \ln u + C \\ &= A \ln (B + x) + C. \end{aligned}$$

(The absolute value bars are not necessary, since $B + x$ is always positive here.) ◀

6.2 Exercises

Use substitution to find the following indefinite integrals.

1. $\int 4(2x + 3)^4\,dx$
2. $\int (-4t + 1)^3\,dt$
3. $\int \frac{2\,dm}{(2m + 1)^3}$
4. $\int \frac{3\,du}{\sqrt{3u - 5}}$
5. $\int \frac{2x + 2}{(x^2 + 2x - 4)^4}\,dx$
6. $\int \frac{6x^2\,dx}{(2x^3 + 7)^{3/2}}$
7. $\int z\sqrt{z^2 - 5}\,dz$
8. $\int r\sqrt{r^2 + 2}\,dr$
9. $\int (-4e^{2p})\,dp$
10. $\int 5e^{-.3g}\,dg$
11. $\int 3x^2e^{2x^3}\,dx$
12. $\int re^{-r^2}\,dr$
13. $\int (1 - t)e^{2t - t^2}\,dt$
14. $\int (x^2 - 1)e^{x^3 - 3x}\,dx$
15. $\int \frac{e^{1/z}}{z^2}\,dz$
16. $\int \frac{e^{\sqrt{y}}}{2\sqrt{y}}\,dy$

*Formula for the "Popularity Index" from Jim Quirin and Barry Cohen, *Chartmasters' Rock 100*, 4th ed. Copyright 1987 by Chartmasters. Reprinted by permission.

17. $\int \frac{-8}{1 + 3x}\, dx$

18. $\int \frac{9}{2 + 5t}\, dt$

19. $\int \frac{dt}{2t + 1}$

20. $\int \frac{dw}{5w - 2}$

21. $\int \frac{v\, dv}{(3v^2 + 2)^4}$

22. $\int \frac{x\, dx}{(2x^2 - 5)^3}$

23. $\int \frac{x - 1}{(2x^2 - 4x)^2}\, dx$

24. $\int \frac{2x + 1}{(x^2 + x)^3}\, dx$

25. $\int \left(\frac{1}{r} + r\right)\left(1 - \frac{1}{r^2}\right) dr$

26. $\int \left(\frac{2}{A} - A\right)\left(\frac{-2}{A^2} - 1\right) dA$

27. $\int \frac{x^2 + 1}{(x^3 + 3x)^{2/3}}\, dx$

28. $\int \frac{B^3 - 1}{(2B^4 - 8B)^{3/2}}\, dB$

29. $\int p(p + 1)^5\, dp$

30. $\int x^3(1 + x^2)^{1/4}\, dx$

31. $\int t\sqrt{5t - 1}\, dt$

32. $\int 4r\sqrt{8 - r}\, dr$

33. $\int \frac{u}{\sqrt{u - 1}}\, du$

34. $\int \frac{2x}{(x + 5)^6}\, dx$

35. $\int (\sqrt{x^2 + 12x})(x + 6)\, dx$

36. $\int (\sqrt{x^2 - 6x})(x - 3)\, dx$

37. $\int \frac{t}{t^2 + 2}\, dt$

38. $\int \frac{-4x}{x^2 + 3}\, dx$

39. $\int ze^{2z^2}\, dz$

40. $\int x^2 e^{-x^3}\, dx$

41. $\int \frac{(1 + \ln x)^2}{x}\, dx$

42. $\int \frac{\sqrt{2 + \ln x}}{x}\, dx$

43. $\int x^{3/2}\sqrt{x^{5/2} + 4}\, dx$

44. $\int \frac{1}{x(\ln x)}\, dx$

Applications

Business and Economics

45. **Revenue** Suppose the marginal revenue in dollars from the sale of x jet planes is

$$R'(x) = 2x(x^2 + 50)^2.$$

(a) Find the total revenue function if the revenue from 3 planes is $206,379.

(b) How many planes must be sold for a revenue of at least $450,000?

46. **Maintenance** The rate of expenditure for maintenance of a particular machine is given by

$$M'(x) = \sqrt{x^2 + 12x}\,(2x + 12),$$

where x is time measured in years. Maintenance costs for the fourth year are $612.

(a) Find the total maintenance function.

(b) How many years must pass before the annual maintenance costs reach $2000?

47. **Profit** The rate of growth of the profit (in millions of dollars) from a new technology is approximated by

$$P'(x) = xe^{-x^2},$$

where x represents time measured in years. The total profit in the third year that the new technology is in operation is $10,000.

(a) Find the total profit function.

(b) What happens to the total amount of profit in the long run?

48. **Cost** A company has found that the marginal cost of a new production line (in thousands) is

$$C'(x) = -100(x + 10)^{-2},$$

where x is the number of years the line is in use.

(a) Find the total cost function for the production line. The fixed cost is 10 thousand dollars.

(b) The company will add the new line if the total cost is reduced to $2000 within 5 years. Should they add the new line?

Life Sciences

49. **Drug Sensitivity** If x milligrams of a certain drug are administered to a person, the rate of change in the person's temperature (in degrees Celsius) with respect to the dosage (the person's *sensitivity* to the drug) is given by

$$D'(x) = \frac{2}{x + 9}.$$

One milligram raises the person's temperature 2.5° C.

(a) Find the function giving the total change in body temperature.

(b) How much of the drug is necessary to change the person's temperature 3° C?

General Interest

50. **Border Crossing** Canadians who live near the United States border have been shopping in the United States at an increasing rate since 1986. They cross the border to

shop because large savings are possible on groceries, gasoline, cigarettes, and children's clothes. The rate of change in the number of Canadian cross-border shoppers (in millions) is given by

$$S'(t) = 4.4e^{.16t},$$

where t is the number of years since 1986.

(a) Find $S(t)$ if there were 27.3 million cross-border shoppers in 1986 (year 0).

(b) In how many years will the number of Canadian cross-border shoppers double?

6.3 AREA AND THE DEFINITE INTEGRAL

If the rate of annual maintenance charges is known, how can the total amount paid for maintenance over a 10-year period be estimated?

This section introduces a method for answering such questions.

To calculate the areas of geometric figures such as rectangles, squares, triangles, and circles, we use specific formulas. In this section we consider the problem of finding the area of a figure or region that is bounded by curves, such as the shaded region in Figure 2.

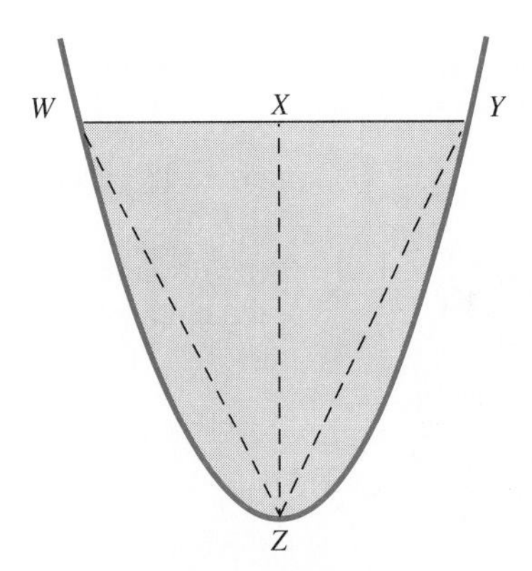

Area of parabolic segment

$= \frac{4}{3}$(area of triangle WYZ)

FIGURE 2

The brilliant Greek mathematician Archimedes (about 287 B.C.–212 B.C.) is considered one of the greatest mathematicians of all time. His development of a rigorous method known as "exhaustion" to derive results was a forerunner of the ideas of integral calculus. Archimedes used a method that would later be verified by the more rigorous theory of integration. His method involved viewing a geometric figure as a sum of other figures. For example, he thought of a plane surface area as a figure consisting of infinitely many parallel line segments. Among the results established by Archimedes' method was the fact that the area of a segment of a parabola (shown in color in Figure 2) is equal to 4/3 the area of a triangle with the same base and the same height.

Under certain conditions the area of a region can be thought of as a sum of parts. Figure 3 shows the region bounded by the y-axis, the x-axis, and the graph of $f(x) = \sqrt{4 - x^2}$. A very rough approximation of the area of this region can be found by using two rectangles as in Figure 4. The height of the rectangle on the left is $f(0) = 2$ and the height of the rectangle on the right is $f(1) = \sqrt{3}$. The width of each rectangle is 1, making the total area of the two rectangles

$$1 \cdot f(0) + 1 \cdot f(1) = 2 + \sqrt{3} \approx 3.7321 \text{ square units.}$$

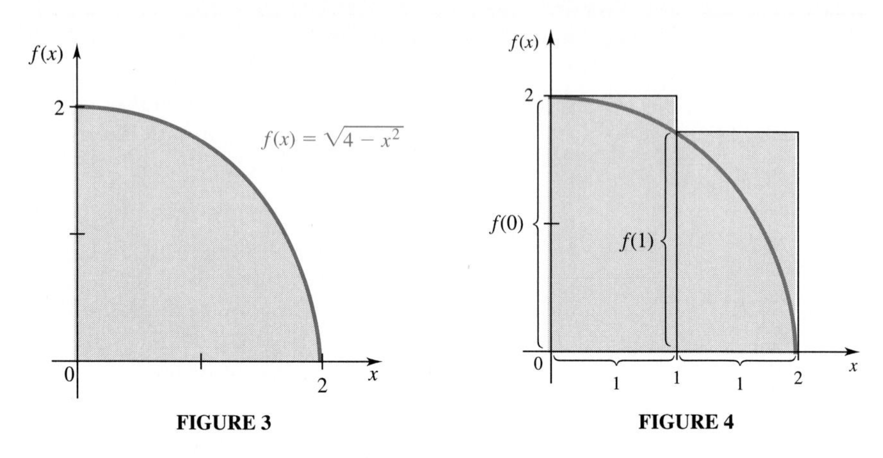

FIGURE 3

FIGURE 4

As shown in Figure 4, this approximation is greater than the actual area. To improve the accuracy of the approximation, we could divide the interval from $x = 0$ to $x = 2$ into four equal parts, each of width $1/2$, as shown in Figure 5. As before, the height of each rectangle is given by the value of f at the left side of the rectangle, and its area is the width, $1/2$, multiplied by the height. The total area of the four rectangles is

$$\begin{aligned}
&\frac{1}{2} \cdot f(0) + \frac{1}{2} \cdot f\left(\frac{1}{2}\right) + \frac{1}{2} \cdot f(1) + \frac{1}{2} \cdot f\left(1\frac{1}{2}\right) \\
&= \frac{1}{2}(2) + \frac{1}{2}\left(\frac{\sqrt{15}}{2}\right) + \frac{1}{2}(\sqrt{3}) + \frac{1}{2}\left(\frac{\sqrt{7}}{2}\right) \\
&= 1 + \frac{\sqrt{15}}{4} + \frac{\sqrt{3}}{2} + \frac{\sqrt{7}}{4} \approx 3.4957 \text{ square units.}
\end{aligned}$$

This approximation looks better, but it is still greater than the actual area desired. To improve the approximation, divide the interval from $x = 0$ to $x = 2$ into 8 parts with equal widths of $1/4$ (see Figure 6). The total area of all these rectangles is

$$\begin{aligned}
&\frac{1}{4} \cdot f(0) + \frac{1}{4} \cdot f\left(\frac{1}{4}\right) + \frac{1}{4} \cdot f\left(\frac{1}{2}\right) + \frac{1}{4} \cdot f\left(\frac{3}{4}\right) + \frac{1}{4} \cdot f(1) + \frac{1}{4} \cdot f\left(\frac{5}{4}\right) \\
&\quad + \frac{1}{4} \cdot f\left(\frac{3}{2}\right) + \frac{1}{4} \cdot f\left(\frac{7}{4}\right) \\
&\approx 3.3398 \text{ square units.}
\end{aligned}$$

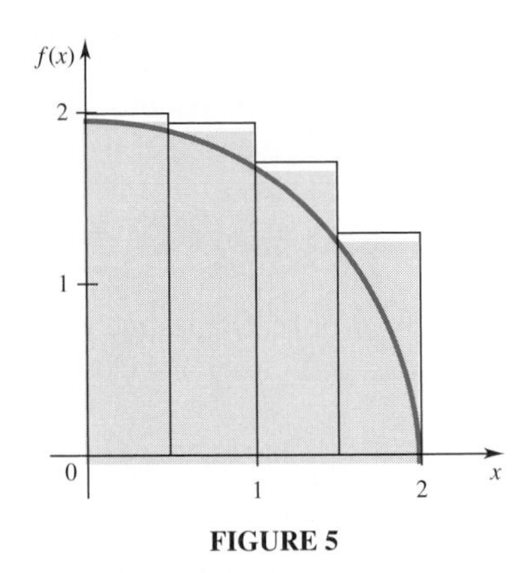

FIGURE 5

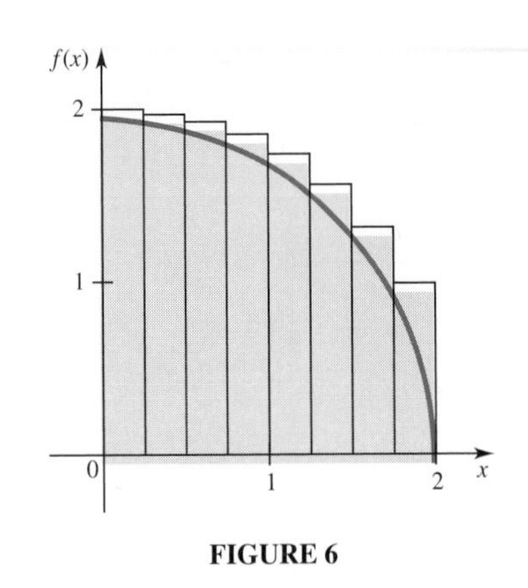

FIGURE 6

This process of approximating the area under a curve by using more and more rectangles to get a better and better approximation can be generalized. To do this, divide the interval from $x = 0$ to $x = 2$ into n equal parts. Each of these n intervals has width

$$\frac{2 - 0}{n} = \frac{2}{n},$$

so each rectangle has width $2/n$ and height determined by the function-value at the left side of the rectangle. Using a computer to find approximations to the area

for several values of n gives the results in the following table.

n	Area
2	3.7321
4	3.4957
8	3.3398
10	3.3045
20	3.2285
50	3.1783
100	3.1512
500	3.1455

The areas in the last column in the table are approximations of the area under the curve, above the x-axis, and between the lines $x = 0$ and $x = 2$. As n becomes larger and larger, the approximation is better and better, getting closer to the actual area. In this example, the exact area can be found by a formula from plane geometry. Write the given function as

$$y = \sqrt{4 - x^2},$$

then square both sides to get

$$y^2 = 4 - x^2$$
$$x^2 + y^2 = 4,$$

the equation of a circle centered at the origin with radius 2. The region in Figure 3 is the quarter of this circle that lies in the first quadrant. The actual area of this region is one-quarter of the area of the entire circle, or

$$\frac{1}{4}\pi(2)^2 = \pi \approx 3.1416.$$

As the number of rectangles increases without bound, the sum of the areas of these rectangles gets closer and closer to the actual area of the region, π. This can be written as

$$\lim_{n\to\infty} (\text{sum of areas of } n \text{ rectangles}) = \pi.$$

(The value of π was originally found by a process similar to this.)*

This approach could be used to find the area of any region bounded by the x-axis, the lines $x = a$ and $x = b$, and the graph of a function $y = f(x)$. At this point, we need some new notation and terminology to write sums concisely. We will indicate addition (or summation) by using the Greek letter sigma, Σ, as shown in the next example.

EXAMPLE 1 Find the following sums.

(a) $\sum_{i=1}^{5} i$

Replace i in turn with the integers 1 through 5 and add the resulting terms.

$$\sum_{i=1}^{5} i = 1 + 2 + 3 + 4 + 5 = 15$$

(b) $\sum_{i=1}^{4} a_i = a_1 + a_2 + a_3 + a_4$

(c) $\sum_{i=1}^{3} (6x_i - 2)$ if $x_1 = 2, x_2 = 4, x_3 = 6$

Let $i = 1, 2,$ and 3, respectively, to get

$$\sum_{i=1}^{3} (6x_i - 2) = (6x_1 - 2) + (6x_2 - 2) + (6x_3 - 2).$$

*The number π is the ratio of the circumference of a circle to its diameter. It is an example of an *irrational number*, and as such it cannot be expressed as a terminating or repeating decimal. Many approximations have been used for π over the years. A passage in the Bible (I Kings 7:23) indicates a value of 3. The Egyptians used the value 3.16, and Archimedes showed that its value must be between 22/7 and 223/71. A Hindu writer, Brahmagupta, used $\sqrt{10}$ as its value in the seventh century. The search for the digits of π has continued into modern times. Two mathematician brothers named Chudnovsky, both at Columbia University, recently computed the value to over 2 billion places.

Now substitute the given values for x_1, x_2, and x_3.

$$\sum_{i=1}^{3} (6x_i - 2) = (6 \cdot 2 - 2) + (6 \cdot 4 - 2) + (6 \cdot 6 - 2)$$
$$= 10 + 22 + 34$$
$$= 66$$

(d) $\sum_{i=1}^{4} f(x_i)\Delta x$ if $f(x) = x^2$, $x_1 = 0$, $x_2 = 2$, $x_3 = 4$, $x_4 = 6$, and $\Delta x = 2$

$$\sum_{i=1}^{4} f(x_i)\Delta x = f(x_1)\Delta x + f(x_2)\Delta x + f(x_3)\Delta x + f(x_4)\Delta x$$
$$= x_1{}^2\Delta x + x_2{}^2\Delta x + x_3{}^2\Delta x + x_4{}^2\Delta x$$
$$= 0^2(2) + 2^2(2) + 4^2(2) + 6^2(2)$$
$$= 0 + 8 + 32 + 72 = 112$$

Now we can generalize to get a method of finding the area bounded by the curve $y = f(x)$, the x-axis, and the vertical lines $x = a$ and $x = b$, as shown in Figure 7. To approximate this area, we could divide the region under the curve first into 10 rectangles (Figure 7(a)) and then into 20 rectangles (Figure 7(b)). The sum of the areas of the rectangles give approximations to the area under the curve.

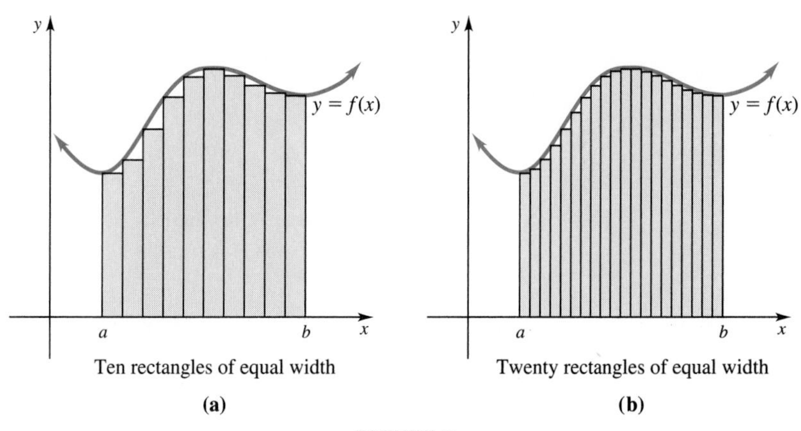

FIGURE 7

To get a number that can be defined as the *exact* area, begin by dividing the interval from a to b into n pieces of equal width, using each of these n pieces as the base of a rectangle (see Figure 8). The left endpoints of the n intervals are labeled $x_1, x_2, x_3, \ldots, x_{n+1}$, where $a = x_1$ and $b = x_{n+1}$. In the graph of Figure 8, the symbol Δx is used to represent the width of each of the intervals. The darker rectangle is an arbitrary rectangle called the ith rectangle. Its area is the product of its length and width. Since the width of the ith rectangle is Δx and the length of the ith rectangle is given by the height $f(x_i)$,

$$\text{Area of the } i\text{th rectangle} = f(x_i) \cdot \Delta x.$$

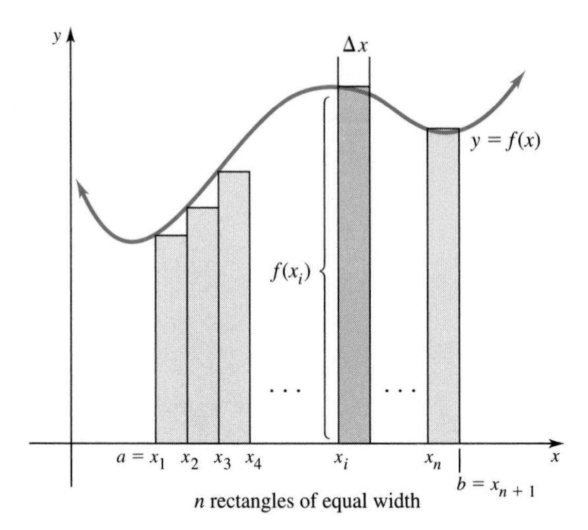

FIGURE 8

The total area under the curve is approximated by the sum of the areas of all n of the rectangles. With sigma notation, the approximation to the total area becomes

$$\text{Area of all } n \text{ rectangles} = \sum_{i=1}^{n} f(x_i) \cdot \Delta x.$$

The exact area is defined to be the limit of this sum (if the limit exists) as the number of rectangles increases without bound:

$$\text{Exact area} = \lim_{n\to\infty} \sum_{i=1}^{n} f(x_i)\Delta x.$$

This limit is called the *definite integral* of $f(x)$ from a to b and is written as follows.

THE DEFINITE INTEGRAL

If f is defined on the interval $[a, b]$, the **definite integral** of f from a to b is given by

$$\int_a^b f(x)dx = \lim_{n\to\infty} \sum_{i=1}^{n} f(x_i)\Delta x,$$

provided the limit exists, where $\Delta x = (b - a)/n$ and x_i is *any* value of x in the ith interval.

As indicated in this definition, although the left endpoint of the ith interval has been used to find the height of the ith rectangle, any number in the ith interval can be used. (A more general definition is possible in which the rectangles do not necessarily all have the same width.) The b above the integral sign is called the **upper limit** of integration, and the a is the **lower limit** of integration. This use of the word "limit" has nothing to do with the limit of the sum; it refers to the limits, or boundaries, on x.

The sum in the definition of the definite integral is an example of a Riemann sum, named for the German mathematician Georg Riemann (1826–1866) who was responsible for the early development of integrability of functions. The concepts of "Riemann sum" and "Riemann integral" are still studied in rigorous calculus textbooks.

In the example at the beginning of this section, the area bounded by the x-axis, the curve $y = \sqrt{4 - x^2}$, the lines $x = 0$ and $x = 2$ could be written as the definite integral

$$\int_0^2 \sqrt{4 - x^2}\, dx = \pi.$$

Notice that unlike the indefinite integral, which is a set of *functions*, the definite integral represents a *number*. The next section will show how antiderivatives are used in finding the definite integral and thus the area under a curve.

Keep in mind that finding the definite integral of a function can be thought of as a mathematical process that gives the sum of an infinite number of individual parts (within certain limits). The definite integral represents area only if the function involved is *nonnegative* ($f(x) \geq 0$) at every x-value in the interval $[a, b]$. There are many other interpretations of the definite integral, and all of them involve this idea of approximation by appropriate sums.

EXAMPLE 2 Approximate $\int_0^4 2x\, dx$, the area of the region under the graph of $f(x) = 2x$, above the x-axis, and between $x = 0$ and $x = 4$, by using four rectangles of equal width whose heights are the values of the function at the midpoint of each rectangle.

We want to find the area of the shaded region in Figure 9. The heights of the four rectangles given by $f(x_i)$ for $i = 1, 2, 3$, and 4 are as follows.

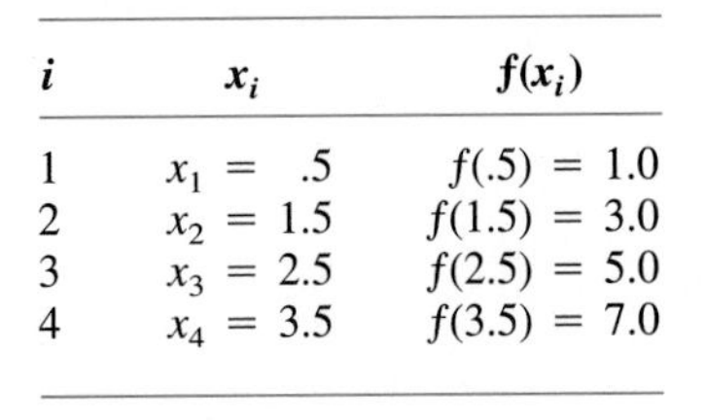

i	x_i	$f(x_i)$
1	$x_1 = .5$	$f(.5) = 1.0$
2	$x_2 = 1.5$	$f(1.5) = 3.0$
3	$x_3 = 2.5$	$f(2.5) = 5.0$
4	$x_4 = 3.5$	$f(3.5) = 7.0$

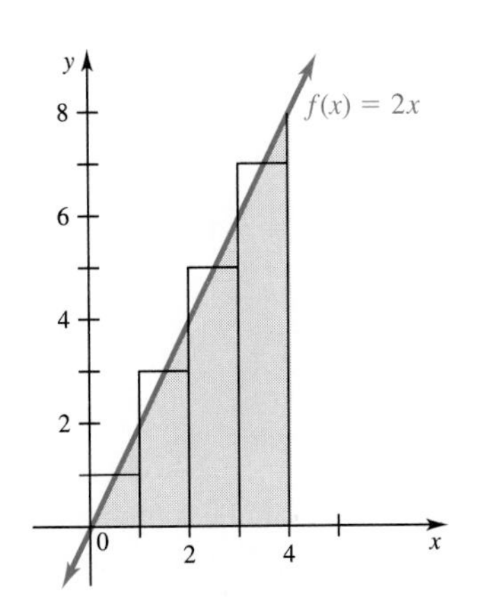

FIGURE 9

The width of each rectangle is $\Delta x = (4 - 0)/4 = 1$. The sum of the areas of the four rectangles is

$$\begin{aligned}\sum_{i=1}^{4} f(x_i)\Delta x &= f(x_1)\Delta x + f(x_2)\Delta x + f(x_3)\Delta x + f(x_4)\Delta x \\ &= f(.5)\Delta x + f(1.5)\Delta x + f(2.5)\Delta x + f(3.5)\Delta x \\ &= (1)(1) + (3)(1) + 5(1) + 7(1) \\ &= 16.\end{aligned}$$

Using the formula for the area of a triangle, $A = (1/2)bh$, with b, the length of the base, equal to 4 and h, the height, equal to 8, gives

$$A = \frac{1}{2}bh = \frac{1}{2}(4)(8) = 16,$$

the exact value of the area. The approximation equals the exact area in this case because our use of the midpoints of each subinterval distributed the error evenly above and below the graph. ◀

Total Change Suppose the function $f(x) = x^2 + 20$ gives the marginal cost of some item at a particular x-value. Then $f(2) = 24$ gives the rate of change of cost at $x = 2$. That is, a unit change in x (at this point) will produce a change of 24 units in the cost function. Also, $f(3) = 29$ means that each unit of change in x (when $x = 3$) will produce a change of 29 units in the cost function.

To find the *total* change in the cost function as x changes from 2 to 3, we could divide the interval from 2 to 3 into n equal parts, using each part as the base of a rectangle as we did above. The area of each rectangle would approximate the change in cost at the x-value that is the left endpoint of the base of the rectangle. Then the sum of the areas of these rectangles would approximate the net total change in cost from $x = 2$ to $x = 3$. The limit of this sum as $n \to \infty$ would give the exact total change.

This result produces another application of the definite integral: the area of the region under the graph of the marginal cost function $f(x)$ that is above the x-axis and between $x = a$ and $x = b$ gives the *net total change in the cost* as x goes from a to b.

TOTAL CHANGE IN $F(x)$

If $f(x)$ gives the rate of change of $F(x)$ for x in $[a, b]$, then the **total change** in $F(x)$ as x goes from a to b is given by

$$\lim_{n\to\infty} \sum_{i=1}^{n} f(x_i)\Delta x = \int_a^b f(x)\,dx.$$

In other words, the total change in a quantity can be found from the function that gives the rate of change of the quantity, using the same methods used to approximate the area under a curve.

EXAMPLE 3 Figure 10 shows the rate of change of the annual maintenance charges for a certain machine. To approximate the total maintenance charges over the 10-year life of the machine, use approximating rectangles, dividing the interval from 0 to 10 into ten equal subdivisions. Each rectangle has width 1; using the left endpoint of each rectangle to determine the height of the rectangle, the approximation becomes

$$1 \cdot 0 + 1 \cdot 500 + 1 \cdot 750 + 1 \cdot 1800 + 1 \cdot 1800 + 1 \cdot 3000 + 1 \cdot 3000 + 1 \cdot 3400 + 1 \cdot 4200 + 1 \cdot 5200 = 23{,}650.$$

About \$23,550 will be spent on maintenance over the 10-year life of the machine.

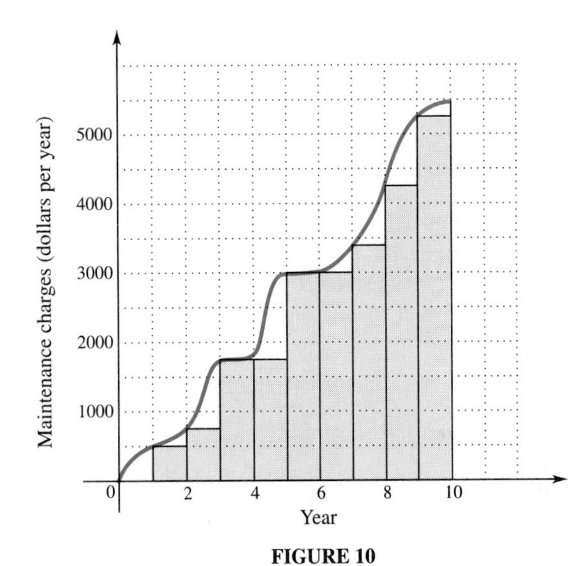

FIGURE 10

Before discussing further applications of the definite integral, we need a more efficient method for evaluating it. This method will be developed in the next section.

6.3 Exercises

Evaluate each sum.

1. $\sum_{i=1}^{3} 3i$

2. $\sum_{i=1}^{6} (-5i)$

3. $\sum_{i=1}^{5} (2i + 7)$

4. $\sum_{i=1}^{10} (5i - 8)$

5. Let $x_1 = -5$, $x_2 = 8$, $x_3 = 7$, and $x_4 = 10$. Find $\sum_{i=1}^{4} x_i$.

6. Let $x_1 = 10$, $x_2 = 15$, $x_3 = -8$, $x_4 = -12$, and $x_5 = 0$. Find $\sum_{i=1}^{5} x_i$.

7. Let $f(x) = x - 3$ and $x_1 = 4, x_2 = 6, x_3 = 7$. Find $\sum_{i=1}^{3} f(x_i)$.

8. Let $f(x) = x^2 + 1$ and $x_1 = -2, x_2 = 0, x_3 = 2, x_4 = 4$. Find $\sum_{i=1}^{4} f(x_i)$.

9. Let $f(x) = 2x + 1, x_1 = 0, x_2 = 2, x_3 = 4, x_4 = 6$, and $\Delta x = 2$.
   (a) Find $\sum_{i=1}^{4} f(x_i)\Delta x$.
   (b) The sum in part (a) approximates a definite integral using rectangles. The height of each rectangle is given by the value of the function at the left endpoint. Write the definite integral that the sum approximates.

10. Let $f(x) = 1/x, x_1 = 1/2, x_2 = 1, x_3 = 3/2, x_4 = 2$, and $\Delta x = 1/2$.
    (a) Find $\sum_{i=1}^{4} f(x_i)\Delta x$.
    (b) The sum in part (a) approximates a definite integral using rectangles. The height of each rectangle is given by the value of the function at the left endpoint. Write the definite integral that the sum approximates.

In Exercises 11–22, first approximate the area under the given curve and above the x-axis by using two rectangles. Let the height of the rectangle be given by the value of the function at the left side of the rectangle. Then repeat the process and approximate the area using four rectangles.

11. $f(x) = 3x + 2$ from $x = 1$ to $x = 5$
12. $f(x) = -2x + 1$ from $x = -4$ to $x = 0$
13. $f(x) = x + 5$ from $x = 2$ to $x = 4$
14. $f(x) = 3 + x$ from $x = 1$ to $x = 3$
15. $f(x) = x^2$ from $x = 1$ to $x = 5$
16. $f(x) = x^2$ from $x = 0$ to $x = 4$
17. $f(x) = x^2 + 2$ from $x = -2$ to $x = 2$
18. $f(x) = -x^2 + 4$ from $x = -2$ to $x = 2$
19. $f(x) = e^x - 1$ from $x = 0$ to $x = 4$
20. $f(x) = e^x + 1$ from $x = -2$ to $x = 2$
21. $f(x) = \frac{1}{x}$ from $x = 1$ to $x = 5$
22. $f(x) = \frac{2}{x}$ from $x = 1$ to $x = 9$

23. Consider the region below $f(x) = x/2$, above the x-axis, between $x = 0$ and $x = 4$. Let x_i be the left endpoint of the ith subinterval.
    (a) Approximate the area of the region using four rectangles.
    (b) Approximate the area of the region using eight rectangles.
    (c) Find $\int_0^4 f(x)dx$ by using the formula for the area of a triangle.

24. Find $\int_0^5 (5 - x)dx$ by using the formula for the area of a triangle.

Find the exact value of each of the following integrals using formulas from geometry.

25. $\int_0^3 2x\, dx$
26. $\int_1^5 \frac{7}{2}\, dx$
27. $\int_{-3}^{3} \sqrt{9 - x^2}\, dx$
28. $\int_{-4}^{0} \sqrt{16 - x^2}\, dx$
29. $\int_1^3 (5 - x)\, dx$
30. $\int_2^5 (1 + 2x)\, dx$

Applications

In Exercises 31–35, estimate the area under each curve by summing the area of rectangles. Let the function value at the left side of each rectangle give the height of the rectangle.

Business and Economics

31. **Oil Consumption** The graph on the next page shows U.S. oil production and consumption rates in millions of barrels per day. The rates for the years beyond 1991 are projected using the policy in place in 1990 (labeled "Current policy base" in the graph) and using a new policy proposed by the Bush administration in 1991 (labeled "With strategy" in the graph).*
    (a) Estimate the amount of oil produced between 1990 and 2010 using the policy in place in 1990. Use rectangles with widths of 2 yr.
    (b) Estimate the amount of oil produced between 1990 and 2010 using the proposed policy. Use rectangles with widths of 2 yr.

*Graph, "Uncle Sam's Energy Strategy" from *National Energy Strategy*, February 1991. United States Department of Energy, Washington D.C.

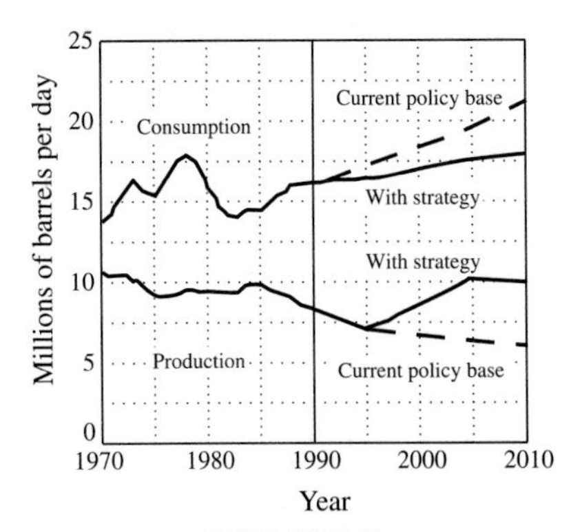

EXERCISE 31

32. **Electricity Consumption** The graph below shows the rate of use of electrical energy (in kilowatt hours) in a certain city on a very hot day. Estimate the total usage of electricity on that day. Let the width of each rectangle be 2 hr.

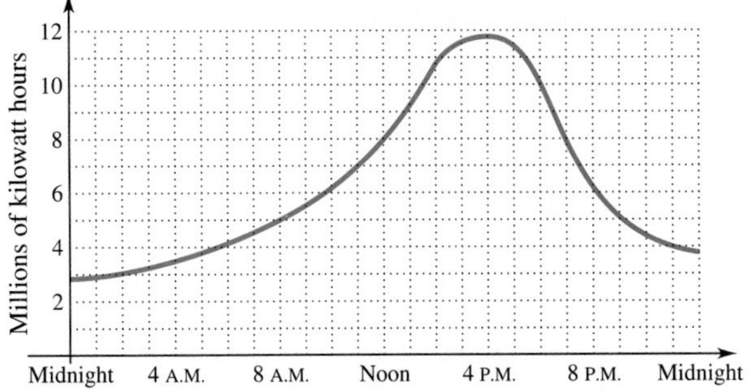

33. **Wages** The graph shows the average manufacturing hourly wage in the United States and Canada for 1987–1991.* All figures are in U.S. dollars. Assume that the average employee works 2000 hours per year.

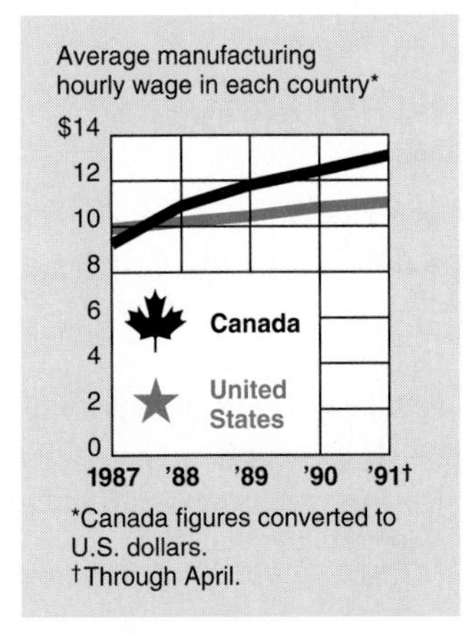

(a) Estimate the total amount earned by an average U.S. worker during the four-year period from the beginning of 1987 to the beginning of 1991. Use rectangles with widths of 1 yr.

(b) Estimate the total amount earned (in U.S. dollars) by an average Canadian worker during the four-year period from the beginning of 1987 to the beginning of 1991. Use rectangles with widths of 1 yr.

Life Sciences

34. **Alcohol Concentration** The graph shows the approximate concentration of alcohol in a person's bloodstream t hr after drinking 2 oz of alcohol. Estimate the total amount of alcohol in the bloodstream by estimating the area under the curve. Use rectangles with widths of 1 hr.

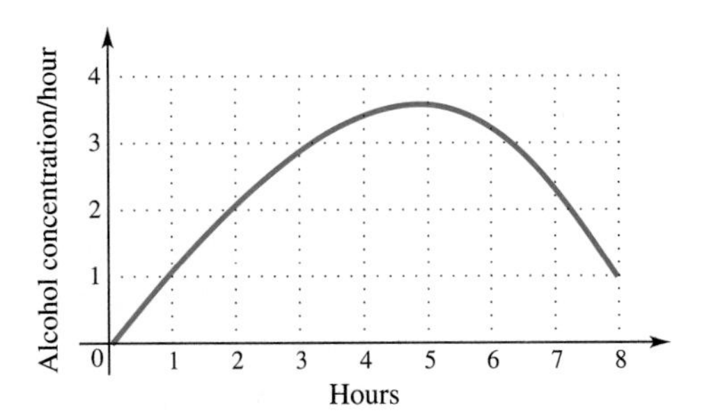

35. **Oxygen Inhalation** The graph shows the rate of inhalation of oxygen (in liters per minute) by a person riding a bicycle very rapidly for 10 min. Estimate the total volume of oxygen inhaled in the first 20 min after the beginning of the ride. Use rectangles with widths of 1 min.

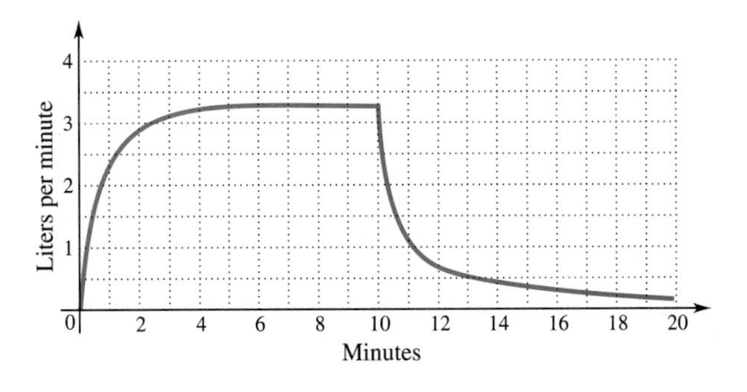

*Graph, "Comparing Wages" from *The New York Times*, 1991. Copyright © 1991 by The New York Times Company. Reprinted by permission.

Physical Sciences

Distance *The next two graphs are from* Road & Track *magazine.* The curve shows the velocity at t seconds after the car accelerates from a dead stop. To find the total distance traveled by the car in reaching* 100 *mph, we must estimate the definite integral*

$$\int_0^T v(t)\,dt,$$

where T represents the number of seconds it takes for the car to reach 100 *mph.*

Use the graphs to estimate this distance by adding the areas of rectangles with widths of 3 *sec. The last rectangle will have a width of* 1 *sec. To adjust your answer to miles per hour, divide by* 3600 *(the number of seconds in an hour). You then have the number of miles that the car traveled in reaching* 100 *mph. Finally, multiply by* 5280 *feet per mile to convert the answer to feet.*

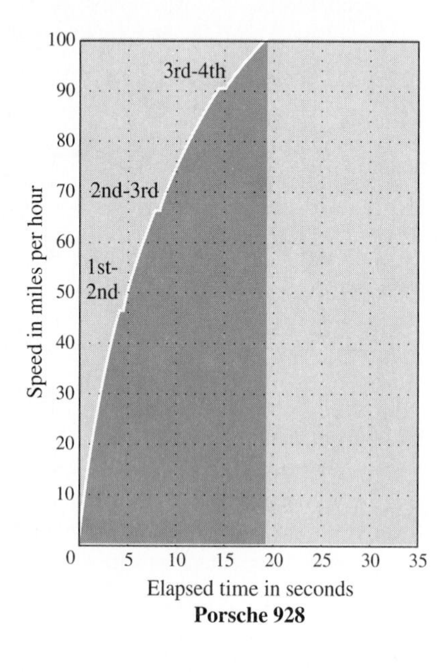

Porsche 928

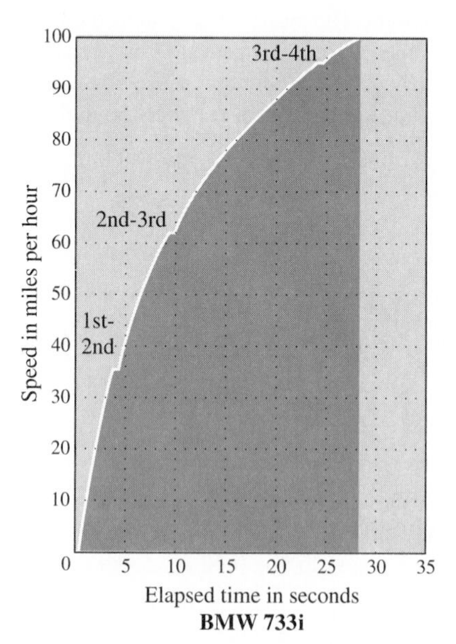

BMW 733i

36. Estimate the distance traveled by the Porsche 928, using the graph on the left.

37. Estimate the distance traveled by the BMW 733i, using the graph on the right.

Heat Gain *The graphs on the next page*[†] *show the typical heat gain, in BTUs per hour per square foot, for a window facing east and one facing south, with plain glass and with a black ShadeScreen. Estimate the total heat gain per square foot by summing the areas of the rectangles. Use rectangles with widths of* 2 *hr, and let the function value at the midpoint of the rectangle give the height of the rectangle.*

38. **(a)** Estimate the total heat gain per square foot for a plain glass window facing east.

(b) Estimate the total heat gain per square foot for a window facing east with a ShadeScreen.

39. **(a)** Estimate the total heat gain per square foot for a plain glass window facing south.

(b) Estimate the total heat gain per square foot for a window facing south with a ShadeScreen.

---

*From *Road & Track*, April and May, 1978. Reprinted with permission of *Road & Track*.
[†]Graphs courtesy of Phifer Wire Products. Reprinted by permission of Phifer Wire Products.

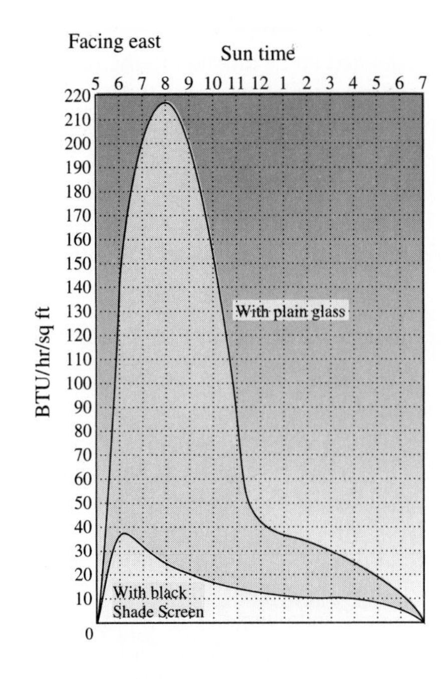

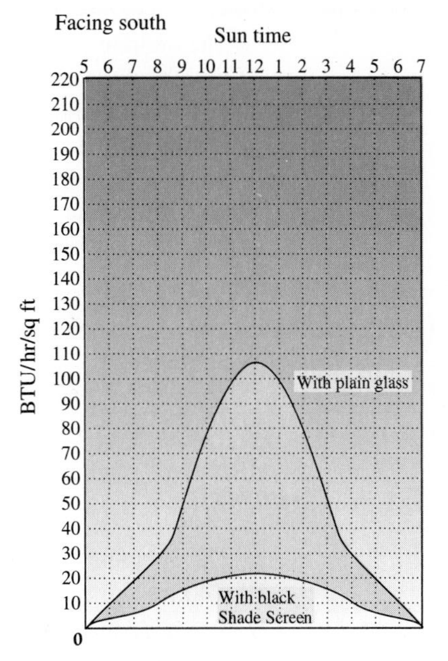

Computer/Graphing Calculator

In Exercises 40–43, use the method given in the text to approximate the area between the x-axis and the graph of each function on the given interval.

40. $f(x) = x \ln x; \quad [1, 5]$

41. $f(x) = x^2 e^{-x}; \quad [-1, 3]$

42. $f(x) = \dfrac{\ln x}{x}; \quad [1, 5]$

43. $f(x) = \dfrac{e^x - e^{-x}}{2}; \quad [0, 4]$

6.4 THE FUNDAMENTAL THEOREM OF CALCULUS

If we know how the rate of consumption of natural gas varies over time, how can we compute the total amount of natural gas used?

This section introduces a powerful theorem for answering such questions.

The work from the last two sections can now be put together. We have seen that, if $f(x) > 0$,

$$\int_a^b f(x)\, dx$$

gives the area between the graph of $f(x)$ and the x-axis, from $x = a$ to $x = b$. We can find this definite integral by using the antiderivatives discussed earlier.

The definite integral was defined and evaluated in the previous section using the limit of a sum. In that section, we also saw that if $f(x)$ gives the rate of change of $F(x)$, the definite integral $\int_a^b f(x)\,dx$ gives the total change of $F(x)$ as x changes from a to b. If $f(x)$ gives the rate of change of $F(x)$, then $F(x)$ is an antiderivative of $f(x)$. Writing the total change in $F(x)$ from $x = a$ to $x = b$ as $F(b) - F(a)$ shows the connection between antiderivatives and definite integrals. This relationship is called the **Fundamental Theorem of Calculus.**

FUNDAMENTAL THEOREM OF CALCULUS

Let f be continuous on the interval $[a, b]$, and let F be *any* antiderivative of f. Then

$$\int_a^b \mathbf{f(x)\,dx = F(b) - F(a) = F(x)}\Big|_a^b.$$

The symbol $F(x)|_a^b$ is used to represent $F(b) - F(a)$. It is important to note that the Fundamental Theorem does not require $f(x) > 0$. The condition $f(x) > 0$ is necessary only when using the Fundamental Theorem to find area. Also, note that the Fundamental Theorem does not *define* the definite integral; it just provides a method for evaluating it.

First find $\int 4t^3\,dt$ and then find $\int_1^2 4t^3\,dt$.

By the rules given earlier,

$$\int 4t^3\,dt = t^4 + C.$$

By the Fundamental Theorem, the value of $\int_1^2 4t^3\,dt$ is found by evaluating $t^4|_1^2$, with no constant C required.

$$\int_1^2 4t^3\,dt = t^4\Big|_1^2 = 2^4 - 1^4 = 15 \quad ◀$$

Example 1 illustrates the difference between the definite integral and the indefinite integral. A definite integral is a real number; an indefinite integral is a family of functions in which all the functions are antiderivatives of a function f.

To see why the Fundamental Theorem of Calculus is true for $f(x) > 0$, look at Figure 11. Define the function $A(x)$ as the area between the x-axis and the graph of $y = f(x)$ from a to x. We first show that A is an antiderivative of f; that is, $A' = f$.

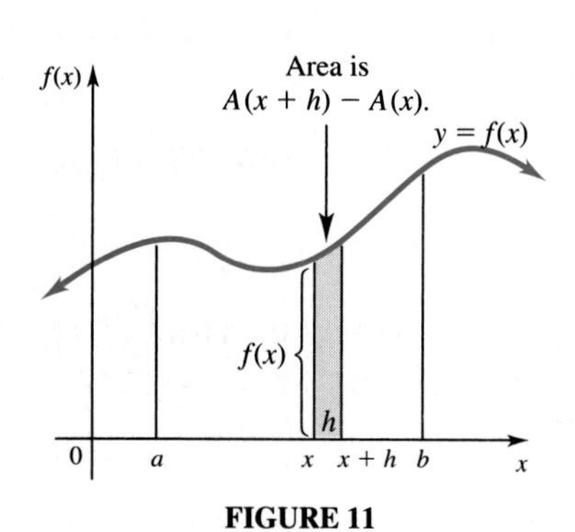

FIGURE 11

To do this, let h be a small positive number. Then $A(x + h) - A(x)$ is the shaded area in Figure 11. This area can be approximated with a rectangle having width h and height $f(x)$. The area of the rectangle is $h \cdot f(x)$, and

$$A(x + h) - A(x) \approx h \cdot f(x).$$

Dividing both sides by h gives

$$\frac{A(x + h) - A(x)}{h} \approx f(x).$$

This approximation improves as h gets smaller and smaller. Take the limit on the left as h approaches 0.

$$\lim_{h \to 0} \frac{A(x + h) - A(x)}{h} = f(x)$$

This limit is simply $A'(x)$, so

$$A'(x) = f(x).$$

This result means that A is an antiderivative of f, as we set out to show.

Since $A(x)$ is the area under the curve $y = f(x)$ from a to x, $A(a) = 0$. The expression $A(b)$ is the area from a to b, the desired area. Since A is an antiderivative of f,

$$\begin{aligned} \int_a^b f(x)\,dx &= A(x)\Big|_a^b \\ &= A(b) - A(a) \\ &= A(b) - 0 = A(b). \end{aligned}$$

This suggests the proof of the Fundamental Theorem: for $f(x) > 0$, the area under the curve $y = f(x)$ from a to b is given by $\int_a^b f(x)\,dx$.

The Fundamental Theorem of Calculus certainly deserves its name, which sets it apart as the most important theorem of calculus. It is the key connection between differential calculus and integral calculus, which originally were developed separately without knowledge of this connection between them.

The variable used in the integrand does not matter; each of the following definite integrals represents the number $F(b) - F(a)$.

$$\int_a^b f(x)\,dx = \int_a^b f(t)\,dt = \int_a^b f(u)\,du$$

Caution Because the definite integral is a number, there is no constant C added, as there is for the indefinite integral. Even if C were added to the antiderivative F, it would be eliminated in the final answer:

$$\begin{aligned} \int_a^b f(x)\,dx &= (F(x) + C)\Big|_a^b \\ &= (F(b) + C) - (F(a) + C) \\ &= F(b) - F(a). \end{aligned}$$

Key properties of definite integrals are listed on the next page. Some of them are just restatements of properties from Section 1.

PROPERTIES OF DEFINITE INTEGRALS

If all indicated definite integrals exist,

1. $\int_a^a f(x)\,dx = 0;$

2. $\int_a^b k \cdot f(x)\,dx = k \cdot \int_a^b f(x)\,dx$ for any real constant k (constant multiple of a function);

3. $\int_a^b [f(x) \pm g(x)]\,dx = \int_a^b f(x)\,dx \pm \int_a^b g(x)\,dx$ (sum or difference of functions); and

4. $\int_a^b f(x)\,dx = \int_a^c f(x)\,dx + \int_c^b f(x)\,dx$ for any real number c.

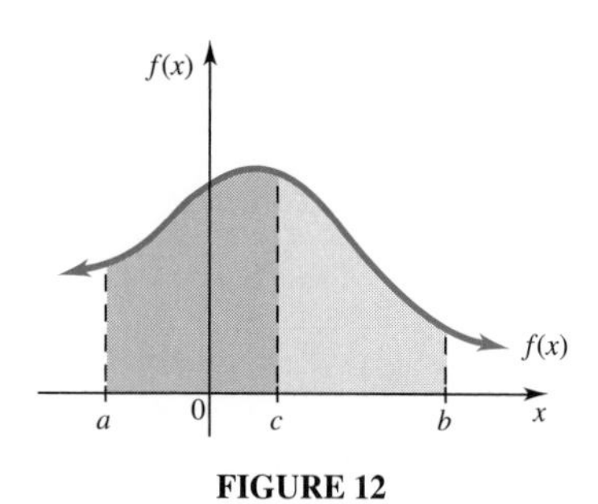

FIGURE 12

For $f(x) > 0$, since the distance from a to a is 0, the first property says that the "area" under the graph of f bounded by $x = a$ and $x = a$ is 0. Also, since $\int_a^c f(x)\,dx$ represents the darker region in Figure 12, and $\int_c^b f(x)\,dx$ represents the lighter region,

$$\int_a^b f(x)\,dx = \int_a^c f(x)\,dx + \int_c^b f(x)\,dx,$$

as stated in the fourth property. While the figure shows $a < c < b$, the property is true for any value of c where both $f(x)$ and $F(x)$ are defined.

An algebraic proof is given here for the third property; proofs of the other properties are left for the exercises. If $F(x)$ and $G(x)$ are antiderivatives of $f(x)$ and $g(x)$ respectively,

$$\begin{aligned}\int_a^b [f(x) + g(x)]\,dx &= [F(x) + G(x)]\Big|_a^b \\ &= [F(b) + G(b)] - [F(a) + G(a)] \\ &= [F(b) - F(a)] + [G(b) - G(a)] \\ &= \int_a^b f(x)\,dx + \int_a^b g(x)\,dx.\end{aligned}$$

EXAMPLE 2 Find $\int_2^5 (6x^2 - 3x + 5)\,dx$.

Use the properties above, and the Fundamental Theorem, along with properties from Section 1.

$$\begin{aligned}\int_2^5 (6x^2 - 3x + 5)\,dx &= 6\int_2^5 x^2\,dx - 3\int_2^5 x\,dx + 5\int_2^5 dx \\ &= 2x^3\Big|_2^5 - \frac{3}{2}x^2\Big|_2^5 + 5x\Big|_2^5\end{aligned}$$

$$= 2(5^3 - 2^3) - \frac{3}{2}(5^2 - 2^2) + 5(5 - 2)$$

$$= 2(125 - 8) - \frac{3}{2}(25 - 4) + 5(3)$$

$$= 234 - \frac{63}{2} + 15 = \frac{435}{2}$$

$$\int_1^2 \frac{dy}{y} = \ln|y|\Big|_1^2 = \ln|2| - \ln|1|$$

$$= \ln 2 - \ln 1 \approx .6931 - 0 = .6931$$

EXAMPLE 4

Evaluate $\int_0^5 x\sqrt{25 - x^2}\, dx$.

Use substitution. Let $u = 25 - x^2$, so that $du = -2x\, dx$. With a definite integral, the limits should be changed, too. The new limits on u are found as follows.

If $x = 5$, then $u = 25 - 5^2 = 0$.
If $x = 0$, then $u = 25 - 0^2 = 25$.

Then

$$\int_0^5 x\sqrt{25 - x^2}\, dx = -\frac{1}{2}\int_0^5 \sqrt{25 - x^2}(-2x\, dx)$$

$$= -\frac{1}{2}\int_{25}^0 \sqrt{u}\, du$$ Substitute and change limits.

$$= -\frac{1}{2}\int_{25}^0 u^{1/2}\, du$$

$$= -\frac{1}{2} \cdot \frac{u^{3/2}}{3/2}\Big|_{25}^0$$

$$= -\frac{1}{2} \cdot \frac{2}{3}[0^{3/2} - 25^{3/2}]$$

$$= -\frac{1}{3}(-125)$$

$$= \frac{125}{3}.$$

Caution When substitution is used on a definite integral, it is best to not revert from u back to the original variable after antidifferentiation, as we did for the indefinite integral. Notice in Example 4 that after changing from the old limits on x to the new limits on u, we never returned to x or its limits.

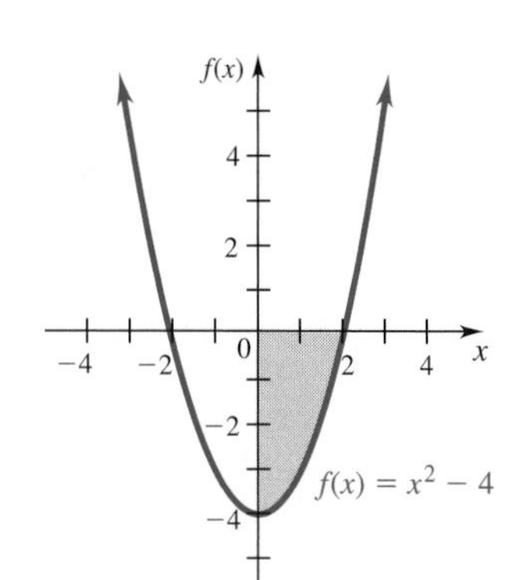

FIGURE 13

Area In the previous section we saw that, if $f(x) > 0$ in $[a, b]$, the definite integral $\int_a^b f(x)\,dx$ gives the area below the graph of the function $y = f(x)$, above the x-axis, and between the lines $x = a$ and $x = b$.

To see how to work around the requirement that $f(x) > 0$, look at the graph of $f(x) = x^2 - 4$ in Figure 13. The area bounded by the graph of f, the x-axis, and the vertical lines $x = 0$ and $x = 2$ lies below the x-axis. Using the Fundamental Theorem to find this area gives

$$\int_0^2 (x^2 - 4)\,dx = \left(\frac{x^3}{3} - 4x\right)\Bigg|_0^2$$

$$= \left(\frac{8}{3} - 8\right) - (0 - 0) = -\frac{16}{3}.$$

The result is a negative number because $f(x)$ is negative for values of x in the interval $[0, 2]$. Since Δx is always positive, if $f(x) < 0$ the product $f(x) \cdot \Delta x$ is negative, so $\int_0^2 f(x)\,dx$ is negative. Since area is nonnegative, the required area is given by $|-16/3|$ or $16/3$. Using a definite integral, the area could be written as

$$\left|\int_0^2 (x^2 - 4)\,dx\right| = \left|-\frac{16}{3}\right| = \frac{16}{3}.$$

EXAMPLE 5 Find the area of the region between the x-axis and the graph of $f(x) = x^2 - 3x$ from $x = 1$ to $x = 3$.

The region is shown in Figure 14. Since the region lies below the x-axis, the area is given by

$$\left|\int_1^3 (x^2 - 3x)\,dx\right|.$$

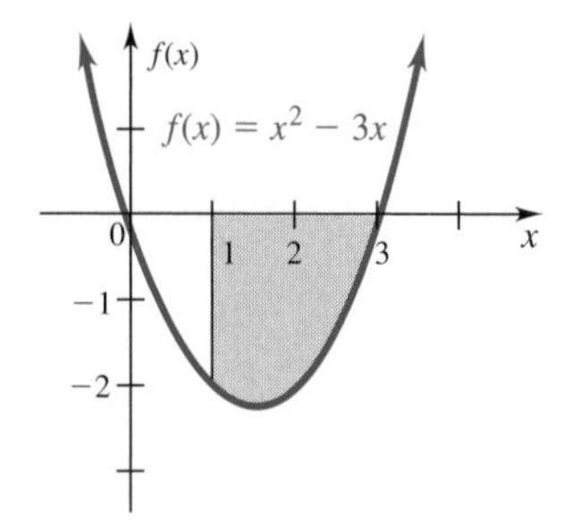

FIGURE 14

By the Fundamental Theorem,

$$\int_1^3 (x^2 - 3x)\,dx = \left(\frac{x^3}{3} - \frac{3x^2}{2}\right)\Bigg|_1^3 = \left(\frac{27}{3} - \frac{27}{2}\right) - \left(\frac{1}{3} - \frac{3}{2}\right) = -\frac{10}{3}.$$

The required area is $|-10/3| = 10/3$. ◀

EXAMPLE 6 Find the area between the x-axis and the graph of $f(x) = x^2 - 4$ from $x = 0$ to $x = 4$.

Figure 15 shows the required region. Part of the region is below the x-axis. The definite integral over that interval will have a negative value. To find the area, integrate the negative and positive portions separately and take the absolute value of the first result before combining the two results to get the total area. Start by finding the point where the graph crosses the x-axis. This is done by solving the equation

$$x^2 - 4 = 0.$$

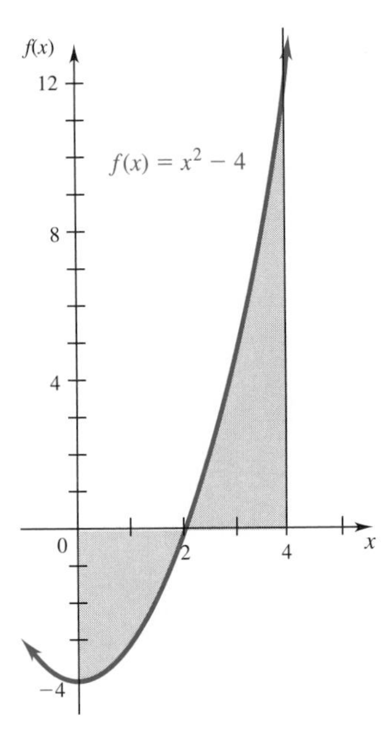

FIGURE 15

The solutions of this equation are 2 and -2. The only solution in the interval $[0, 4]$ is 2. The total area of the region in Figure 15 is

$$\left|\int_0^2 (x^2 - 4)\,dx\right| + \int_2^4 (x^2 - 4)\,dx = \left|\left(\frac{1}{3}x^3 - 4x\right)\Bigg|_0^2\right| + \left(\frac{1}{3}x^3 - 4x\right)\Bigg|_2^4$$

$$= \left|\frac{8}{3} - 8\right| + \left(\frac{64}{3} - 16\right) - \left(\frac{8}{3} - 8\right)$$

$$= 16.$$

Incorrectly using one integral over the entire interval to find the area in Example 6 would have given

$$\int_0^4 (x^2 - 4)\,dx = \left(\frac{x^3}{3} - 4x\right)\Bigg|_0^4 = \left(\frac{64}{3} - 16\right) - 0 = \frac{16}{3},$$

which is not the correct area. This definite integral represents no area, but is just a real number.

If $f(x)$ in Example 6 represents the annual rate of profit of a company, then $16/3$ represents the total profit for the company over a four-year period. The integral between 0 and 2 is $-16/3$; the negative sign indicates a loss for the first two years. The integral between 2 and 4 is $32/3$, indicating a profit. The overall profit is $32/3 - 16/3 = 16/3$, while the total shaded area is $32/3 + |-16/3| = 16$.

FINDING AREA

In summary, to find the area bounded by $f(x)$, $x = a$, $x = b$, and the x-axis, use the following steps.

1. Sketch a graph.
2. Find any x-intercepts of $f(x)$ in $[a, b]$. These divide the total region into subregions.
3. The definite integral will be *positive* for subregions above the x-axis and *negative* for subregions below the x-axis. Use separate integrals to find the areas of the subregions.
4. The total area is the sum of the areas of all of the subregions.

In the last section, we saw that the area under a rate of change function $f'(x)$ from $x = a$ to $x = b$ gives the total value of $f(x)$ on $[a, b]$. Now we can use the definite integral to solve these problems.

EXAMPLE 7 The yearly rate of consumption of natural gas (in trillions of cubic feet) for a certain city is

$$C'(t) = t + e^{.01t},$$

where t is time in years and $t = 0$ corresponds to 1990. At this consumption rate, what is the total amount the city will use in the 10-year period of the 1990s?

To find the consumption over the 10-year period from 1990 through 1999, use the definite integral.

$$\int_0^{10} (t + e^{.01t})\,dt = \left(\frac{t^2}{2} + \frac{e^{.01t}}{.01}\right)\Bigg|_0^{10}$$
$$= (50 + 100e^{.1}) - (0 + 100)$$
$$\approx -50 + 100(1.10517) \approx 60.5$$

Therefore, a total of 60.5 trillion cubic feet of natural gas will be used during the 1990s if the consumption rate remains the same. ◀

6.4 Exercises

Evaluate each definite integral.

1. $\int_{-2}^{4} (-1)\,dp$

2. $\int_{-4}^{1} 6x\,dx$

3. $\int_{-1}^{2} (3t - 1)\,dt$

4. $\int_{-2}^{2} (4z + 3)\,dz$

5. $\int_{0}^{2} (5x^2 - 4x + 2)\,dx$

6. $\int_{-2}^{3} (-x^2 - 3x + 5)\,dx$

7. $\int_{0}^{2} 3\sqrt{4u + 1}\,du$

8. $\int_{3}^{9} \sqrt{2r - 2}\,dr$

9. $\int_{0}^{1} 2(t^{1/2} - t)\,dt$

10. $\int_{0}^{4} -(3x^{3/2} + x^{1/2})\,dx$

11. $\int_{1}^{4} (5y\sqrt{y} + 3\sqrt{y})\,dy$

12. $\int_{4}^{9} (4\sqrt{r} - 3r\sqrt{r})\,dr$

13. $\int_{4}^{6} \frac{2}{(x - 3)^2}\,dx$

14. $\int_{1}^{4} \frac{-3}{(2p + 1)^2}\,dp$

15. $\int_{1}^{5} (5n^{-2} + n^{-3})\,dn$

16. $\int_{2}^{3} (3x^{-2} - x^{-4})\,dx$

17. $\int_{2}^{3} \left(2e^{-.1A} + \frac{3}{A}\right) dA$

18. $\int_{1}^{2} \left(\frac{-1}{B} + 3e^{.2B}\right) dB$

19. $\int_{1}^{2} \left(e^{5u} - \frac{1}{u^2}\right) du$

20. $\int_{.5}^{1} (p^3 - e^{4p})\,dp$

21. $\int_{-1}^{0} y(2y^2 - 3)^5\,dy$

22. $\int_{0}^{3} m^2(4m^3 + 2)^3\,dm$

23. $\int_{1}^{64} \frac{\sqrt{z} - 2}{\sqrt[3]{z}}\,dz$

24. $\int_{1}^{8} \frac{3 - y^{1/3}}{y^{2/3}}\,dy$

25. $\int_{1}^{2} \frac{\ln x}{x}\,dx$

26. $\int_{1}^{3} \frac{\sqrt{\ln x}}{x}\,dx$

27. $\int_{0}^{8} x^{1/3}\sqrt{x^{4/3} + 9}\,dx$

28. $\int_{1}^{2} \frac{3}{x(1 + \ln x)}\,dx$

29. $\int_{0}^{1} \frac{e^t}{(3 + e^t)^2}\,dt$

30. $\int_{0}^{1} \frac{e^{2z}}{\sqrt{1 + e^{2z}}}\,dz$

31. $\int_{1}^{49} \frac{(1 + \sqrt{x})^{4/3}}{\sqrt{x}}\,dx$

32. $\int_{1}^{8} \frac{(1 + x^{1/3})^6}{x^{2/3}}\,dx$

In each of the following, use the definite integral to find the area between the x-axis and f(x) over the indicated interval. Check first to see if the graph crosses the x-axis in the given interval.

33. $f(x) = 2x + 3$; $[8, 10]$

34. $f(x) = 4x - 7$; $[5, 10]$

35. $f(x) = 2 - 2x^2$; $[0, 5]$

36. $f(x) = 9 - x^2$; $[0, 6]$

37. $f(x) = x^2 + 4x - 5; \quad [-6, 3]$

38. $f(x) = x^2 - 6x + 5; \quad [-1, 4]$

39. $f(x) = x^3; \quad [-1, 3]$

40. $f(x) = x^3 - 2x; \quad [-2, 4]$

41. $f(x) = e^x - 1; \quad [-1, 2]$

42. $f(x) = 1 - e^{-x}; \quad [-1, 2]$

43. $f(x) = \frac{1}{x}; \quad [1, e]$

44. $f(x) = \frac{1}{x}; \quad [e, e^2]$

Find the area of each shaded region.

45.

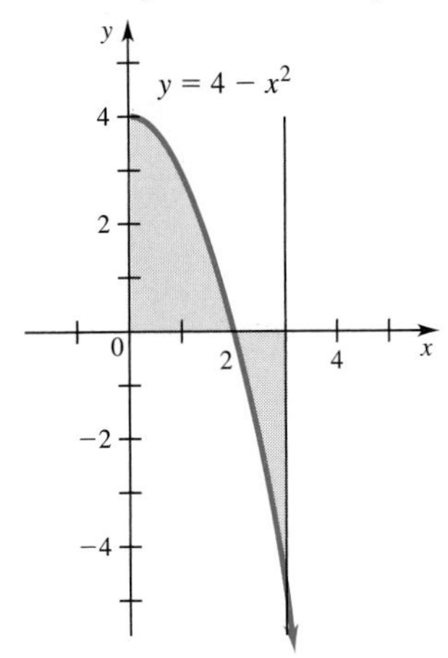

46.

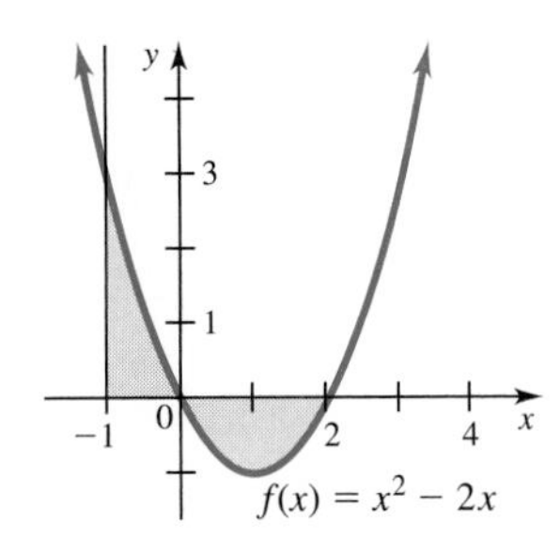

47.

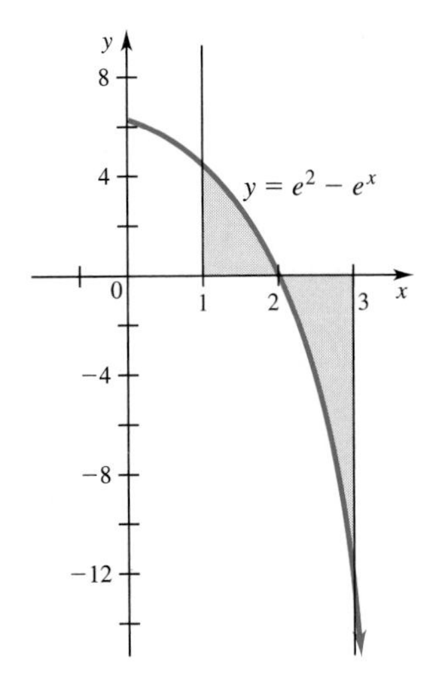

48.

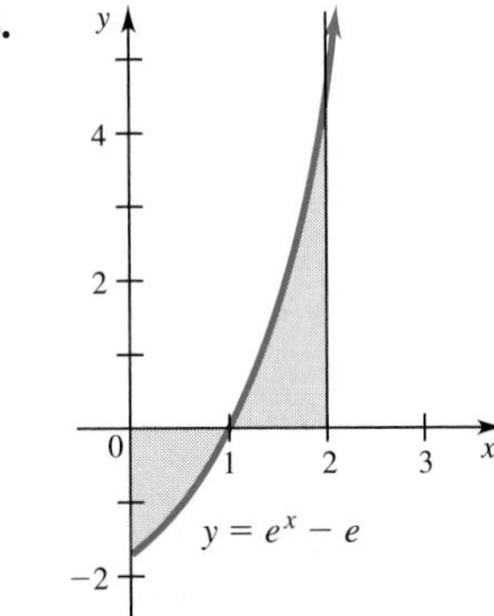

Show that each of the following is true.

49. $\int_a^a f(x)dx = 0$

50. $\int_a^b kf(x)dx = k\int_a^b f(x)dx$

51. $\int_a^b f(x)dx = \int_a^c f(x)dx + \int_c^b f(x)dx$

52. Use Exercise 51 to find $\int_{-1}^{4} f(x)dx$, given

$$f(x) = \begin{cases} 2x + 3 & \text{if } x \leq 0 \\ -\frac{x}{4} - 3 & \text{if } x > 0. \end{cases}$$

Applications

Business and Economics

53. Expenditures De Win Enterprises has found that its expenditure rate per day (in hundreds of dollars) on a certain type of job is given by

$$E'(x) = 4x + 2,$$

where x is the number of days since the start of the job.

(a) Find the total expenditure if the job takes 10 days.

(b) How much will be spent on the job from the tenth to the twentieth day?

(c) If the company wants to spend no more than \$5000 on the job, in how many days must they complete it?

54. Income De Win Enterprises (see Exercise 53) also knows that the rate of income per day (in hundreds of dollars) for the same job is

$$I'(x) = 100 - x,$$

where x is the number of days since the job was started.

(a) Find the total income for the first 10 days.

(b) Find the income from the tenth to the twentieth day.

(c) How many days must the job last for the total income to be at least \$5000?

55. Profit Karla Harby Communications, a small company of science writers, found that its rate of profits (in thousands of dollars) after t years of operation is given by

$$P'(t) = (3t + 3)(t^2 + 2t + 2)^{1/3}.$$

(a) Find the total profits in the first three years.

(b) Find the profit in the fourth year of operation.

(c) What is happening to the annual profits over the long run?

56. Worker Efficiency A worker new to a job will improve his efficiency with time so that it takes him fewer hours to produce an item with each day on the job, up to a certain point. Suppose the rate of change of the number of hours it takes a worker in a certain factory to produce the xth item is given by

$$H'(x) = 20 - 2x.$$

(a) What is the total number of hours required to produce the first 5 items?

(b) What is the total number of hours required to produce the first 10 items?

Life Sciences

57. Pollution Pollution from a factory is entering a lake. The rate of concentration of the pollutant at time t is given by

$$P'(t) = 140t^{5/2},$$

where t is the number of years since the factory started introducing pollutants into the lake. Ecologists estimate that the lake can accept a total level of pollution of 4850 units before all the fish life in the lake ends. Can the factory operate for 4 yr without killing all the fish in the lake?

58. Spread of an Oil Leak An oil tanker is leaking oil at the rate given in barrels per hour by

$$L'(t) = \frac{80 \ln (t + 1)}{t + 1},$$

where t is the time in hours after the tanker hits a hidden rock (when $t = 0$).

(a) Find the total number of barrels that the ship will leak on the first day.

(b) Find the total number of barrels that the ship will leak on the second day.

(c) What is happening over the long run to the amount of oil leaked per day?

59. Tree Growth After long study, tree scientists conclude that a eucalyptus tree will grow at the rate of $.2 + 4t^{-4}$ feet per year, where t is time in years.

(a) Find the number of feet that the tree will grow in the second year.

(b) Find the number of feet the tree will grow in the third year.

60. Growth of a Substance The rate at which a substance grows is given by

$$R'(x) = 200e^{.2x},$$

where x is the time in days. What is the total accumulated growth after 2.5 days?

61. Drug Reaction For a certain drug, the rate of reaction in appropriate units is given by

$$R'(t) = \frac{5}{t} + \frac{2}{t^2},$$

where t is time measured in hours after the drug is administered. Find the total reaction to the drug over the following time periods.

(a) From $t = 1$ to $t = 12$ **(b)** From $t = 12$ to $t = 24$

62. Blood Flow In the exercises for an earlier chapter, the speed s of the blood in a blood vessel was given as

$$s = k(R^2 - r^2),$$

where R is the (constant) radius of the blood vessel, r is the distance of the flowing blood from the center of the blood vessel, and k is a constant. Total blood flow (in millimeters per minute) is given by

$$Q(R) = \int_0^R 2\pi sr\, dr.$$

(a) Find the general formula for Q in terms of R by evaluating the definite integral given above.

(b) Evaluate $Q(.4)$.

Physical Sciences

63. Oil Consumption Suppose that the rate of consumption of a natural resource is $c(t)$, where

$$c'(t) = ke^{rt}.$$

Here t is time in years, r is a constant, and k is the consumption in the year when $t = 0$. In 1992, an oil company sold 1.2 billion barrels of oil. Assume that $r = .04$.

(a) Write $c'(t)$ for the oil company, letting $t = 0$ represent 1992.

(b) Set up a definite integral for the amount of oil that the company will sell in the next ten years.

(c) Evaluate the definite integral of part (b).

(d) The company has about 20 billion barrels of oil in reserve. To find the number of years that this amount will last, solve the equation

$$\int_0^T 1.2e^{.04t}\, dt = 20.$$

(e) Rework part (d), assuming that $r = .02$.

64. Mine Production A mine begins producing at time $t = 0$. After t yr, the mine is producing at the rate of

$$P'(t) = 10t - \frac{15}{\sqrt{t}}$$

tons per year. Write an expression for the total output of the mine from year 1 to year T.

65. Consumption of a Natural Resource The rate of consumption of one natural resource in one country is

$$C'(t) = 72e^{.014t},$$

where $t = 0$ corresponds to 1992. How much of the resource will be used, altogether, from 1992 to year T?

66. Oil Consumption The rate of consumption of oil (in billions of barrels) by the company in Exercise 63 was given as

$$1.2e^{.04t},$$

where $t = 0$ corresponds to 1992. Find the total amount of oil used by the company from 1992 to year T. At this rate, how much will be used in 5 years?

6.5 THE AREA BETWEEN TWO CURVES

If an executive knows how the savings from a new manufacturing process decline over time and how the costs of that process will increase, how can she compute when the savings will cease and what the total savings will be?

This section shows a method for answering such questions.

Many important applications of integrals require finding the area between two graphs. The method used in previous sections to find the area between the graph of a function and the x-axis from $x = a$ to $x = b$ can be generalized to find such an area. For example, the area between the graphs of $f(x)$ and $g(x)$ from $x = a$ to $x = b$ in Figure 16(a) (on the next page) is the same as the difference of the area from a to b between $f(x)$ and the x-axis, shown in Figure 16(b), and the area from a to b between $g(x)$ and the x-axis (see Figure 16(c)). That is, the area between the graphs is given by

$$\int_a^b f(x)\, dx - \int_a^b g(x)\, dx,$$

which can be written as

$$\int_a^b [f(x) - g(x)]\, dx.$$

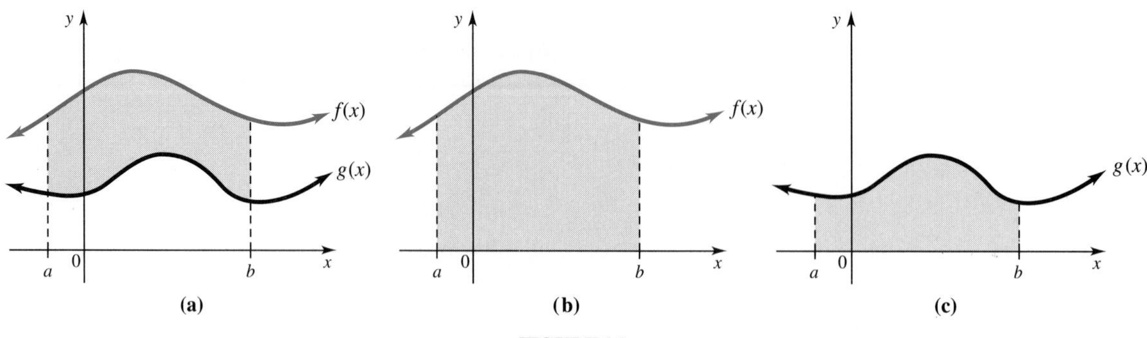

FIGURE 16

AREA BETWEEN TWO CURVES

If f and g are continuous functions and $f(x) \geq g(x)$ on $[a, b]$, then the area between the curves $f(x)$ and $g(x)$ from $x = a$ to $x = b$ is given by

$$\int_a^b [f(x) - g(x)]\, dx.$$

EXAMPLE 1 Find the area bounded by $f(x) = -x^2 + 1$, $g(x) = 2x + 4$, $x = -1$, and $x = 2$.

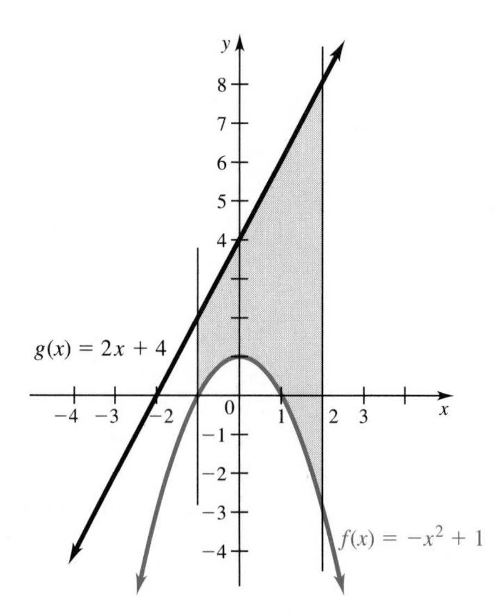

FIGURE 17

A sketch of the four equations is shown in Figure 17. In general, it is not necessary to spend time drawing a detailed sketch, but only to know whether the two functions intersect, and which function is greater between the intersections. To find out, set the two functions equal.

$$-x^2 + 1 = 2x + 4$$
$$0 = x^2 + 2x + 3$$

Verify by the quadratic formula that this equation has no roots. Since the graph of f is a parabola opening downward that does not cross the graph of g (a line), the parabola must be entirely under the line, as shown in Figure 17. Therefore $g(x) \geq f(x)$ for x in the interval $[-1, 2]$, and the area is given by

$$\begin{aligned}\int_{-1}^{2} [g(x) - f(x)]\, dx &= \int_{-1}^{2} (2x + 4) - (-x^2 + 1)\, dx \\ &= \int_{-1}^{2} 2x + 4 + x^2 - 1\, dx \\ &= \int_{-1}^{2} (x^2 + 2x + 3)\, dx \\ &= \frac{x^3}{3} + x^2 + 3x \Big|_{-1}^{2} \\ &= \left(\frac{8}{3} + 4 + 6\right) - \left(\frac{-1}{3} + 1 - 3\right) \\ &= \frac{8}{3} + 10 + \frac{1}{3} + 2 \\ &= 15.\end{aligned}$$

EXAMPLE 2 Find the area between the curves $y = x^{1/2}$ and $y = x^3$.

Let $f(x) = x^{1/2}$ and $g(x) = x^3$. As before, set the two equal to find where they intersect.

$$x^{1/2} = x^3$$
$$0 = x^3 - x^{1/2}$$
$$0 = x^{1/2}(x^{5/2} - 1)$$

The only solutions are $x = 0$ and $x = 1$. Verify that the graph of f is concave downward, while the graph of g is concave upward, so the graph of f must be greater between 0 and 1. (This may also be verified by taking a point between 0 and 1, such as .5, and verifying that $.5^{1/2} > .5^3$.) The graph is shown in Figure 18.

The area between the two curves is given by

$$\int_a^b [f(x) - g(x)]\, dx = \int_0^1 (x^{1/2} - x^3)\, dx.$$

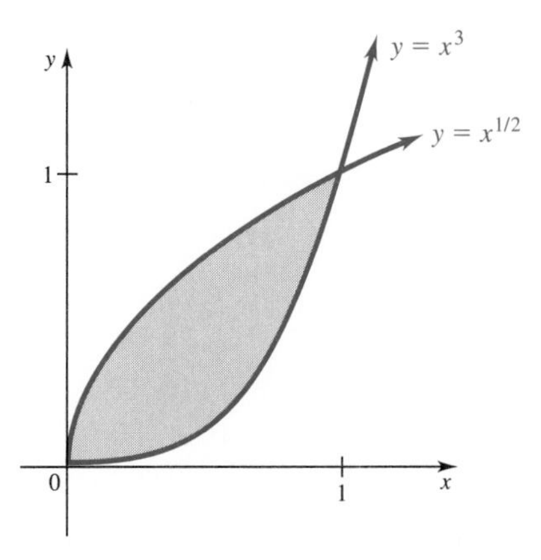

FIGURE 18

Using the Fundamental Theorem,

$$\int_0^1 (x^{1/2} - x^3)\, dx = \left(\frac{x^{3/2}}{3/2} - \frac{x^4}{4}\right)\Bigg|_0^1$$

$$= \left(\frac{2}{3}x^{3/2} - \frac{x^4}{4}\right)\Bigg|_0^1$$

$$= \frac{2}{3}(1) - \frac{1}{4}$$

$$= \frac{5}{12}.$$ ◀

The difference between two integrals can be used to find the area between the graphs of two functions even if one graph lies below the x-axis. In fact, if $f(x) \geq g(x)$ for all values of x in the interval $[a, b]$, then the area between the two graphs is always given by

$$\int_a^b [f(x) - g(x)]\, dx.$$

To see this, look at the graphs in Figure 19(a), where $f(x) \geq g(x)$ for x in $[a, b]$. Suppose a constant C is added to both functions, with C large enough so that both graphs lie above the x-axis, as in Figure 19(b). The region between the graphs is not changed. By the work above, this area is given by $\int_a^b [f(x) - g(x)]\, dx$ regardless of where the graphs of $f(x)$ and $g(x)$ are located. So long as $f(x) \geq g(x)$ on $[a, b]$, then the area between the graphs from $x = a$ to $x = b$ will equal $\int_a^b [f(x) - g(x)]\, dx$.

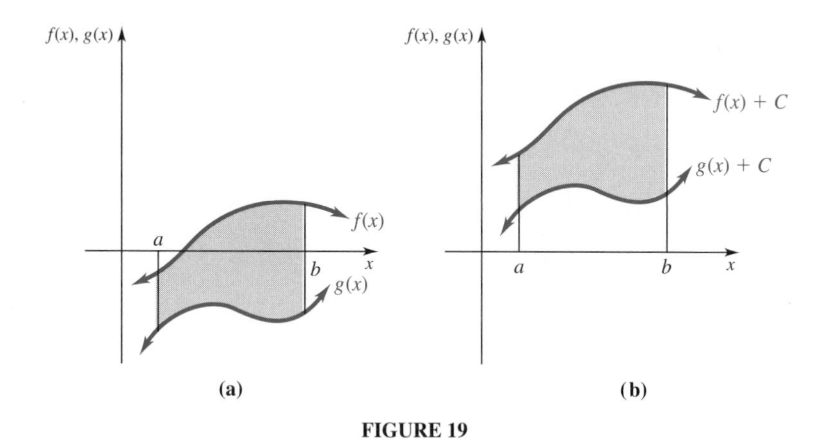

FIGURE 19

EXAMPLE 3 Find the area of the region enclosed by $y = x^2 - 2x$ and $y = x$ on $[0, 4]$.

Verify that the two graphs cross at $x = 0$ and $x = 3$. Because the first graph is a parabola opening upward, the parabola must be below the line between 0 and

3 and above the line between 3 and 4. See Figure 20. (The greater function could also be identified by checking a point between 0 and 3, such as 1, and a point between 3 and 4, such as 3.5. For each of these values of x, we could calculate the corresponding value of y for the two functions and see which is greater.) Because the graphs cross at $x = 3$, the area is found by taking the sum of two integrals as follows.

$$\begin{aligned}
\text{Area} &= \int_0^3 [x - (x^2 - 2x)]\,dx + \int_3^4 [(x^2 - 2x) - x]\,dx \\
&= \int_0^3 (-x^2 + 3x)\,dx + \int_3^4 (x^2 - 3x)\,dx \\
&= \left(\frac{-x^3}{3} + \frac{3x^2}{2}\right)\bigg|_0^3 + \left(\frac{x^3}{3} - \frac{3x^2}{2}\right)\bigg|_3^4 \\
&= \left(-9 + \frac{27}{2} - 0\right) + \left(\frac{64}{3} - 24 - 9 + \frac{27}{2}\right) \\
&= \frac{19}{3}
\end{aligned}$$

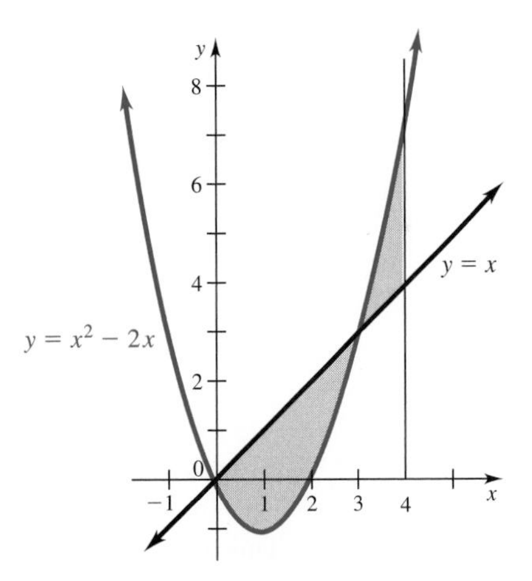

FIGURE 20

In the remainder of this section we will consider some typical applications that require finding the area between two curves.

EXAMPLE 4 A company is considering a new manufacturing process in one of its plants. The new process provides substantial initial savings, with the savings declining with time x (in years) according to the rate-of-savings function

$$S(x) = 100 - x^2,$$

where $S(x)$ is in thousands of dollars. At the same time, the cost of operating the new process increases with time x (in years), according to the rate-of-cost function (in thousands of dollars)

$$C(x) = x^2 + \frac{14}{3}x.$$

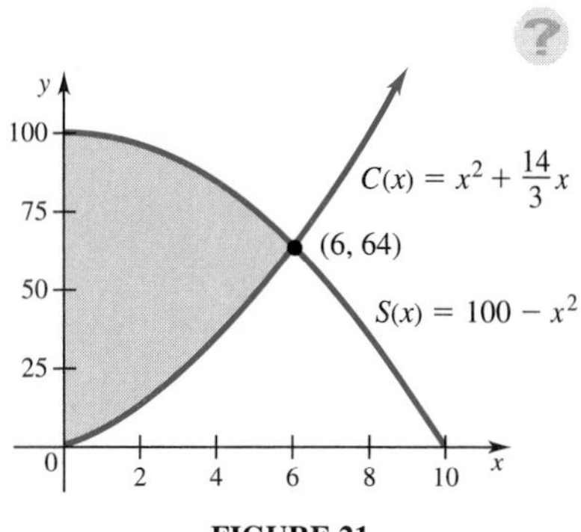

FIGURE 21

(a) For how many years will the company realize savings?

Figure 21 shows the graphs of the rate-of-savings and rate-of-cost functions. The rate-of-cost (marginal cost) is increasing, while the rate-of-savings (marginal savings) is decreasing. The company should use this new process until the difference between these quantities is zero; that is, until the time at which these graphs intersect. The graphs intersect when

$$C(x) = S(x),$$

or

$$100 - x^2 = x^2 + \frac{14}{3}x.$$

Solve this equation as follows.

$$0 = 2x^2 + \frac{14}{3}x - 100$$

$$0 = 3x^2 + 7x - 150 \qquad \text{Multiply by 3/2.}$$

$$= (x - 6)(3x + 25) \qquad \text{Factor.}$$

Set each factor equal to 0 separately to get

$$x = 6 \qquad \text{or} \qquad x = -25/3.$$

Only 6 is a meaningful solution here. The company should use the new process for 6 years.

(b) What will be the net total savings during this period?

Since the total savings over the 6-year period is given by the area under the rate-of-savings curve and the total additional cost by the area under the rate-of-cost curve, the net total savings over the 6-year period is given by the area between the rate of cost and the rate of savings curves and the lines $x = 0$ and $x = 6$. This area can be evaluated with a definite integral as follows.

$$\text{Total savings} = \int_0^6 \left[(100 - x^2) - \left(x^2 + \frac{14}{3}x\right)\right] dx$$

$$= \int_0^6 \left(100 - \frac{14}{3}x - 2x^2\right) dx$$

$$= 100x - \frac{7}{3}x^2 - \frac{2}{3}x^3 \Bigg|_0^6$$

$$= 100(6) - \frac{7}{3}(36) - \frac{2}{3}(216) = 372$$

The company will save a total of \$372,000 over the 6-year period. ◀

The answer to a problem will not always be an integer. Suppose in solving the quadratic equation in Example 4 we found the solutions to be $x = 6.7$ and $x = -7.3$. It may not be realistic to use a new process for 6.7 years; it may be necessary to choose between 6 years and 7 years. Since the mathematical model produces a result that is not in the domain of the function in this case, it is necessary to find the total savings after 6 years and after 7 years and then select the best result.

Consumers' Surplus The market determines the price at which a product is sold. As indicated earlier, the point of intersection of the demand curve and the supply curve for a product gives the equilibrium price. At the equilibrium price, consumers will purchase the same amount of the product that the manufacturers want to sell. Some consumers, however, would be willing to spend more for an item than the equilibrium price. The total of the differences between the equilibrium price of the item and the higher prices all those individuals would be willing to pay is thought of as savings realized by those individuals and is called the **consumers' surplus.**

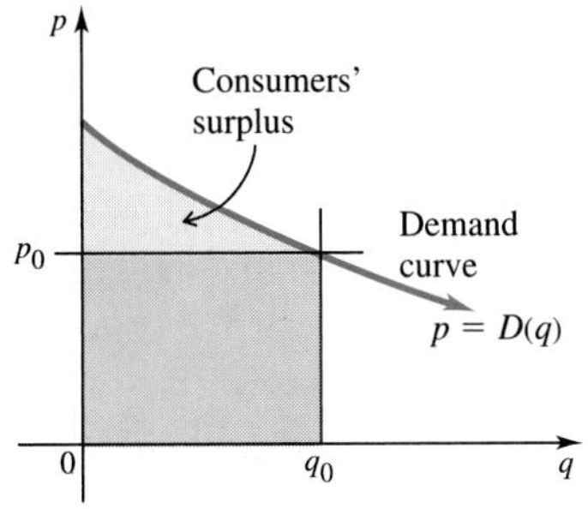

In Figure 22, the area under the demand curve is the total amount consumers are willing to spend for q_0 items. The heavily shaded area under the line $y = p_0$ shows the total amount consumers actually will spend at the equilibrium price of p_0. The lightly shaded area represents the consumers' surplus. As the figure suggests, the consumers' surplus is given by an area between the two curves $p = D(q)$ and $p = p_0$, so its value can be found with a definite integral as follows.

FIGURE 22

CONSUMERS' SURPLUS

If $D(q)$ is a demand function with equilibrium price p_0 and equilibrium demand q_0, then

$$\textbf{Consumers' surplus} = \int_0^{q_0} [D(q) - p_0]\, dq.$$

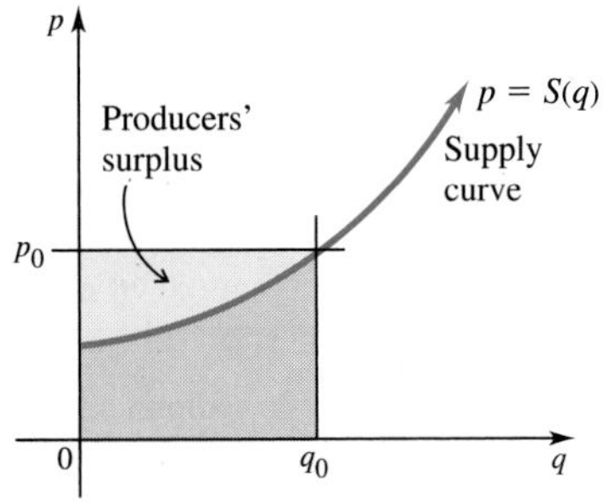

Similarly, if some manufacturers would be willing to supply a product at a price *lower* than the equilibrium price p_0, the total of the differences between the equilibrium price and the lower prices at which the manufacturers would sell the product is considered added income for the manufacturers and is called the **producers' surplus.** Figure 23 shows the (heavily shaded) total area under the supply curve from $q = 0$ to $q = q_0$, which is the minimum total amount the manufacturers are willing to realize from the sale of q_0 items. The total area under the line $p = p_0$ is the amount actually realized. The difference between these two areas, the producers' surplus, is also given by a definite integral.

FIGURE 23

PRODUCERS' SURPLUS

If $S(q)$ is a supply function with equilibrium price p_0 and equilibrium supply q_0, then

$$\textbf{Producers' surplus} = \int_0^{q_0} [p_0 - S(q)]\, dq.$$

(b) What will be the net savings over this period of time?

25. **Profit** De Win Enterprises had an expenditure rate of $E(x) = e^{.1x}$ dollars per day and an income rate of $I(x) = 98.8 - e^{.1x}$ dollars per day on a particular job, where x was the number of days from the start of the job. The company's profit on that job will equal total income less total expenditures. Profit will be maximized if the job ends at the optimum time, which is the point where the two curves meet. Find the following.

(a) The optimum number of days for the job to last.

(b) The total income for the optimum number of days.

(c) The total expenditues for the optimum number of days.

(d) The maximum profit for the job.

26. **Net Savings** A factory at Harold Levinson Industries has installed a new process that will produce an increased rate of revenue (in thousands of dollars) of

$$R(t) = 104 - .4e^{t/2},$$

where t is time measured in years. The new process produces additional costs (in thousands of dollars) at the rate of

$$C(t) = .3e^{t/2}.$$

(a) When will it no longer be profitable to use this new process?

(b) Find the total net savings.

27. **Producers' Surplus** Find the producers' surplus if the supply function for pork bellies is given by

$$S(q) = q^{5/2} + 2q^{3/2} + 50.$$

Assume supply and demand are in equilibrium at $q = 16$.

28. **Producers' Surplus** Suppose the supply function for concrete is given by

$$S(q) = 100 + 3q^{3/2} + q^{5/2},$$

and that supply and demand are in equilibrium at $q = 9$. Find the producers' surplus.

29. **Consumers' Surplus** Find the consumers' surplus if the demand function for grass seed is given by

$$D(q) = \frac{100}{(3q + 1)^2},$$

assuming supply and demand are in equilibrium at $q = 3$.

30. **Consumers' Surplus** Find the consumers' surplus if the demand function for extra virgin olive oil is given by

$$D(q) = \frac{16{,}000}{(2q + 8)^3},$$

and if supply and demand are in equilibrium at $q = 6$.

31. **Consumers' and Producers' Surplus** Suppose the supply function of a certain item is given by

$$S(q) = q^2 + 10q,$$

and the demand function is given by

$$D(q) = 900 - 20q - q^2.$$

(a) Graph the supply and demand curves.

(b) Find the point at which supply and demand are in equilibrium.

(c) Find the consumers' surplus.

(d) Find the producers' surplus.

32. **Consumers' and Producers' Surplus** Repeat the four steps in Exercise 31 for the supply function

$$S(q) = q^2 + \frac{11}{4}q$$

and the demand function

$$D(q) = 150 - q^2.$$

Social Sciences

33. **Distribution of Income** Suppose that all the people in a country are ranked according to their incomes, starting at the bottom. Let x represent the fraction of the community making the lowest income ($0 \le x \le 1$); $x = .4$, therefore, represents the lower 40% of all income producers. Let $I(x)$ represent the proportion of the total income earned by the lowest x of all people. Thus, $I(.4)$ represents the fraction of total income earned by the lowest 40% of the population. Suppose

$$I(x) = .9x^2 + .1x.$$

Find and interpret the following.

(a) $I(.1)$ (b) $I(.4)$ (c) $I(.6)$ (d) $I(.9)$

If income were distributed uniformly, we would have $I(x) = x$. The area under this line of complete equality is $1/2$. As $I(x)$ dips farther below $y = x$, there is less equality of income distribution. This inequality can be quantified by the ratio of the area between $I(x)$ and $y = x$ to $1/2$. This ratio is called the *coefficient of inequality* and equals $2\int_0^1 (x - I(x))\,dx$.

(e) Graph $I(x) = x$ and $I(x) = .9x^2 + .1x$ for $0 \le x \le 1$ on the same axes.

(f) Find the area between the curves.

Physical Sciences

34. **Metal Plate** A worker sketches the curves $y = \sqrt{x}$ and $y = x/2$ on a sheet of metal and cuts out the region between the curves to form a metal plate. Find the area of the plate.

Computer/Graphing Calculator

Approximate the area between the graphs of each pair of functions on the given interval.

35. $y = \ln x$ and $y = xe^x$; $[1, 4]$

36. $y = \ln x$ and $y = 4 - x^2$; $[2, 4]$

37. $y = \sqrt{9 - x^2}$ and $y = \sqrt{x + 1}$; $[-1, 3]$

38. $y = \sqrt{4 - 4x^2}$ and $y = \sqrt{\dfrac{9 - x^2}{3}}$; $[-1, 1]$

INTEGRATION SUMMARY

ANTIDIFFERENTIATION FORMULAS

Power Rule $\displaystyle\int x^n\,dx = \frac{x^{n+1}}{n+1} + C, \quad n \neq -1$

Constant Multiple Rule $\displaystyle\int k \cdot f(x)\,dx = k\int f(x)\,dx$ for any real number k

Sum or Difference Rule $\displaystyle\int [f(x) \pm g(x)]\,dx = \int f(x)\,dx \pm \int g(x)\,dx$

Integration of Exponential Functions $\displaystyle\int e^{kx}\,dx = \frac{1}{k} \cdot e^{kx} + C, \quad k \neq 0$

Integration of x^{-1} $\displaystyle\int x^{-1}\,dx = \ln|x| + C$

Substitution Formulas

1. $\displaystyle\int [u(x)]^n \cdot u'(x)\,dx = \frac{[u(x)]^{n+1}}{n+1} + C, \quad n \neq -1$

2. $\displaystyle\int e^{u(x)} \cdot u'(x)\,dx = e^{u(x)} + C$

3. $\displaystyle\int \frac{u'(x)}{u(x)}\,dx = \ln|u(x)| + C$

FORMULAS FOR DEFINITE INTEGRALS

Definition of the Definite Integral $\displaystyle\int_a^b f(x)\,dx = \lim_{n\to\infty} \sum_{i=1}^{n} f(x_i)\Delta x$, where $\Delta x = (b - a)/n$ and x_i is any value of x in the ith interval. If $f(x)$ gives the rate of change of $F(x)$ for x in $[a, b]$, then this represents the total change in $F(x)$ as x goes from a to b.

Properties of Definite Integrals

1. $\int_a^a f(x)\,dx = 0$
2. $\int_a^b k \cdot f(x)\,dx = k\int_a^b f(x)\,dx$
3. $\int_a^b [f(x) \pm g(x)]\,dx = \int_a^b f(x)\,dx \pm \int_a^b g(x)\,dx$
4. $\int_a^b f(x)\,dx = \int_a^c f(x)\,dx + \int_c^b f(x)\,dx$

Fundamental Theorem of Calculus $\int_a^b f(x)\,dx = F(x)\Big|_a^b = F(b) - F(a),$ where f is continuous on $[a, b]$ and F is any antiderivative of f.

Area Between Two Curves $\int_a^b [f(x) - g(x)]\,dx$, where f and g are continuous functions and $f(x) \geq g(x)$ on $[a, b]$.

Consumers' Surplus $\int_0^{q_0} [D(q) - p_0]\,dq$, where D is the demand function and p_0 and q_0 are the equilibrium price and demand.

Producers' Surplus $\int_0^{q_0} [p_0 - S(q)]\,dq$, where S is the supply function and p_0 and q_0 are the equilibrium price and supply.

Chapter Summary Key Terms

6.1 **antiderivative**
integral sign
integrand
indefinite integral

6.2 **integration by substitution**

6.3 **definite integral**
limits of integration
total change

6.4 **Fundamental Theorem of Calculus**
area between a curve and the *x*-axis

6.5 **area between two curves**
consumers' surplus
producers' surplus

Chapter 6 Review Exercises

1. Explain the differences between an indefinite integral and a definite integral.
2. Explain under what circumstances substitution is useful in integration.
3. Explain why the limits of integration are changed when u is substituted for an expression in x in a definite integral.

4. Explain why rectangles are used to approximate the area under a curve.

Find each indefinite integral.

5. $\int 6\, dx$
6. $\int (-4)dx$
7. $\int (2x + 3)dx$
8. $\int (5x - 1)dx$
9. $\int (x^2 - 3x + 2)dx$
10. $\int (6 - x^2)dx$
11. $\int 3\sqrt{x}\, dx$
12. $\int \frac{\sqrt{x}}{2}\, dx$
13. $\int (x^{1/2} + 3x^{-2/3})dx$
14. $\int (2x^{4/3} + x^{-1/2})dx$
15. $\int \frac{-4}{x^3}\, dx$
16. $\int \frac{5}{x^4}\, dx$
17. $\int -3e^{2x}\, dx$
18. $\int 5e^{-x}\, dx$
19. $\int \frac{2}{x - 1}\, dx$
20. $\int \frac{-4}{x + 2}\, dx$
21. $\int xe^{3x^2}\, dx$
22. $\int 2xe^{x^2}\, dx$
23. $\int \frac{3x}{x^2 - 1}\, dx$
24. $\int \frac{-x}{2 - x^2}\, dx$
25. $\int \frac{x^2\, dx}{(x^3 + 5)^4}$
26. $\int (x^2 - 5x)^4(2x - 5)dx$
27. $\int \frac{4x - 5}{2x^2 - 5x}\, dx$
28. $\int \frac{12(2x + 9)}{x^2 + 9x + 1}\, dx$
29. $\int \frac{x^3}{e^{3x^4}}\, dx$
30. $\int e^{3x^2+4}\, x\, dx$
31. $\int -2e^{-5x}\, dx$
32. $\int e^{-4x}\, dx$

33. Evaluate $\sum_{i=1}^{4} (i^2 - i)$.

34. Let $f(x) = 3x + 1, x_1 = -1, x_2 = 0, x_3 = 1, x_4 = 2$, and $x_5 = 3$. Find $\sum_{i=1}^{5} f(x_i)$.

35. Approximate the area under the graph of $f(x) = 2x + 3$ and above the x-axis from $x = 0$ to $x = 4$ using four rectangles. Let the height of each rectangle be the function value on the left side.

36. Find $\int_0^4 (2x + 3)\, dx$ by using the formula for the area of a trapezoid: $A = \frac{1}{2}(B + b)h$, where B and b are the lengths of the parallel sides and h is the distance between them. Compare with Exercise 35.

37. What does the Fundamental Theorem of Calculus state?

Find each definite integral.

38. $\int_1^2 (3x^2 + 5)dx$
39. $\int_1^6 (2x^2 + x)dx$
40. $\int_1^5 (3x^{-2} + x^{-3})dx$
41. $\int_2^3 (5x^{-2} + x^{-4})dx$
42. $\int_1^3 2x^{-1}\, dx$
43. $\int_1^6 8x^{-1}\, dx$
44. $\int_0^4 2e^x\, dx$
45. $\int_1^6 \frac{5}{2}e^{4x}\, dx$
46. $\int_{\sqrt{5}}^5 2x\sqrt{x^2 - 3}\, dx$
47. $\int_0^1 x\sqrt{5x^2 + 4}\, dx$

Find the area between the x-axis and $f(x)$ over each of the given intervals.

48. $f(x) = x\sqrt{x - 1};\ [1, 10]$
49. $f(x) = x(x + 2)^6;\ [-2, 0]$
50. $f(x) = e^x;\ [0, 2]$
51. $f(x) = 1 + e^{-x};\ [0, 4]$

Find the area of the region enclosed by each group of curves.

52. $f(x) = 5 - x^2,\ g(x) = x^2 - 3$
53. $f(x) = x^2 - 4x,\ g(x) = x - 6$
54. $f(x) = x^2 - 4x,\ g(x) = x + 1,\ x = 2,\ x = 4$
55. $f(x) = 5 - x^2,\ g(x) = x^2 - 3,\ x = 0,\ x = 4$
56. Explain what the consumers' surplus and the producers' surplus are.

Applications

Business and Economics

Cost *Find the cost function for each of the marginal cost functions in Exercises 57–60.*

57. $C'(x) = 10 - 2x$; fixed cost is \$4.

58. $C'(x) = 2x + 3x^2$; 2 units cost \$12.

59. $C'(x) = 3\sqrt{2x - 1}$; 13 units cost \$270.

60. $C'(x) = \dfrac{1}{x + 1}$; fixed cost is \$18.

61. **Investment** The curve shown gives the rate that an investment accumulates income (in dollars per year). Use rectangles of width 2 units and height determined by the function value at the midpoint to find the total income accumulated over 10 yr.

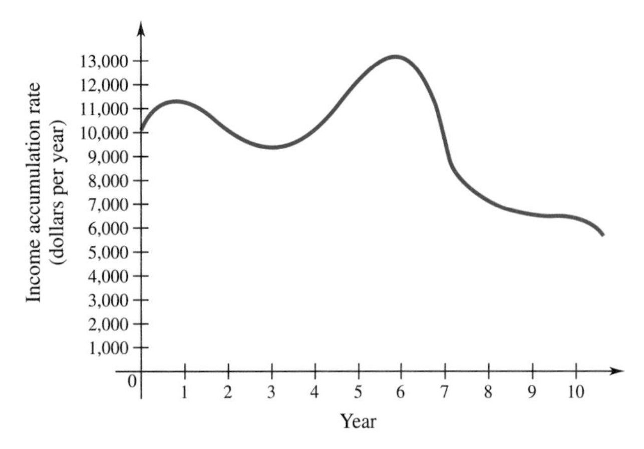

62. **Utilization of Reserves** A manufacturer of electronic equipment requires a certain rare metal. He has a reserve supply of 4,000,000 units that he will not be able to replace. If the rate at which the metal is used is given by
$$f(t) = 100{,}000e^{.03t},$$
where t is the time in years, how long will it be before he uses up the supply? (*Hint:* Find an expression for the total amount used in t years and set it equal to the known reserve supply.)

63. **Sales** The rate of change of sales of a new brand of tomato soup (in thousands) is given by
$$S'(x) = \sqrt{x} + 2,$$
where x is the time in months that the new product has been on the market. Find the total sales after 9 months.

64. **Producers' and Consumers' Surplus** Suppose that the supply function of some commodity is
$$S(q) = q^2 + 5q + 100$$
and the demand function for the commodity is
$$D(q) = 350 - q^2.$$
(a) Find the producers' surplus.
(b) Find the consumers' surplus.

65. **Net Savings** A company has installed new machinery that will produce a savings rate (in thousands of dollars) of
$$S'(x) = 225 - x^2,$$
where x is the number of years the machinery is to be used. The rate of additional costs (in thousands of dollars) to the company due to the new machinery is expected to be
$$C'(x) = x^2 + 25x + 150.$$
For how many years should the company use the new machinery? Find the net savings (in thousands of dollars) over this period.

Life Sciences

66. **Population Growth** The rate of change of the population of a rare species of Australian spider is given by
$$y' = 100 - \sqrt{2.4t + 1},$$
where y is the number of spiders present at time x, measured in months. Find the total number of spiders in the first 10 mo.

67. **Infection Rate** The rate of infection of a disease (in people per month) is given by the function
$$I'(t) = \frac{100t}{t^2 + 1},$$
where t is the time in months since the disease broke out. Find the total number of infected people over the first four months of the disease.

Physical Sciences

68. **Linear Motion** A particle is moving along a straight line with velocity $v(t) = t^2 - 2t$. Its distance from the starting point after 3 sec is 8 cm. Find $s(t)$, the distance of the particle from the starting point after t sec.

Connections

69. The following chart shows 1991 weather statistics for New York City, as well as the normal high and low tempera-

tures.* The amount of cold weather in a year is measured in heating degree-days, where 1 degree-day is added to the total for each degree that a day's average falls below 65° F. For example, if the average temperature on November 15 is 50° F, 15 degree-days are added to the year's total. Estimate the total number of heating degree-days in an average New York City year, using rectangles of width 1 mo, with the height determined by the average temperature at the middle of the rectangle. Assume that the normal average is halfway between the normal high and low.

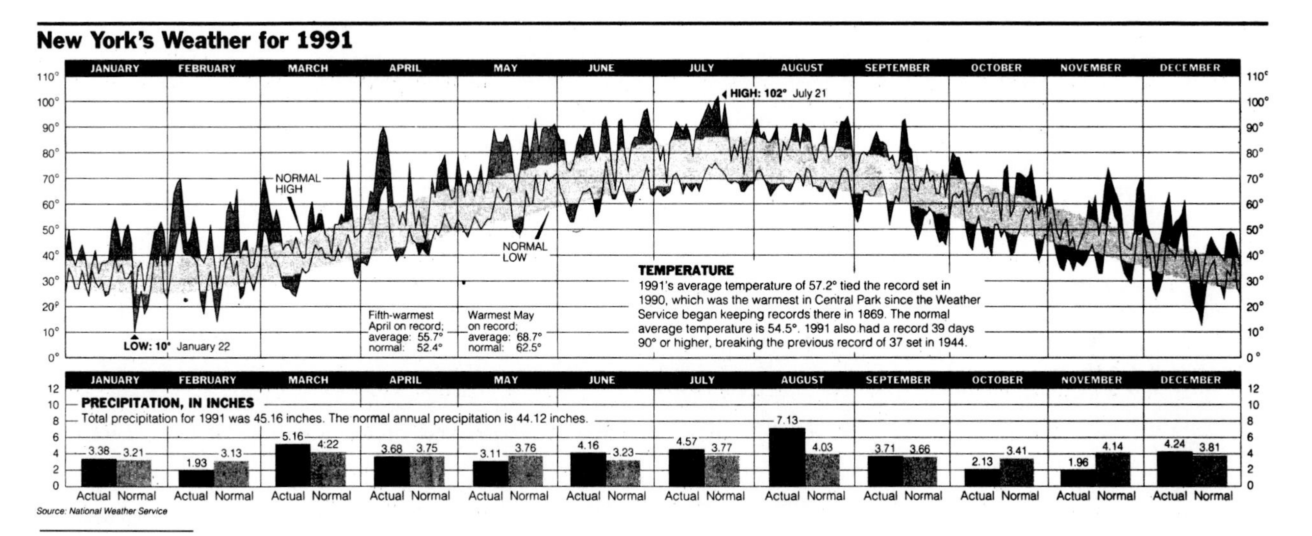

*Graph, "New York's Weather for 1991" from *The New York Times*, January 1992. Copyright © 1992 by The New York Times Company. Reprinted by permission.

Extended Application / Estimating Depletion Dates for Minerals

It is becoming more and more obvious that the earth contains only a finite quantity of minerals. The "easy and cheap" sources of minerals are being used up, forcing an ever more expensive search for new sources. For example, oil from the North Slope of Alaska would never have been used in the United States during the 1930s because a great deal of Texas and California oil was readily available.

We said in an earlier chapter that population tends to follow an exponential growth curve. Mineral usage also follows such a curve. Thus, if q represents the rate of consumption of a certain mineral at time t, while q_0 represents consumption when $t = 0$, then

$$q = q_0 e^{kt},$$

where k is the growth constant. For example, the world consumption of petroleum in a recent year was about 19,600 million barrels, with the value of k about 6%. If we let $t = 0$ correspond to this base year, then $q_0 = 19{,}600$, $k = .06$, and

$$q = 19{,}600e^{.06t}$$

is the rate of consumption at time t, assuming that all present trends continue.

Based on estimates of the National Academy of Science, 2,000,000 million barrels of oil are now in provable reserves or are likely to be discovered in the future. At the present rate of consumption, in how many years will these reserves be depleted? We can use the integral calculus of this chapter to find out.

To begin, we need to know the total quantity of petroleum that would be used between time $t = 0$ and some future time $t = t_1$. Figure 25 on the next page shows a typical graph of the function $q = q_0 e^{kt}$.

Following the work we did in Section 3, divide the time interval from $t = 0$ to $t = t_1$ into n subintervals. Let the ith subinterval have width Δt_i. Let the rate of consumption for the ith subinterval be approximated by q_i^*. Thus, the approximate total consumption for the subinterval is given by

$$q_i^* \cdot \Delta t_i,$$

and the total consumption over the interval from time $t = 0$ to $t = t_1$ is approximated by

$$\sum_{i=1}^{n} q_i^* \cdot \Delta t_i.$$

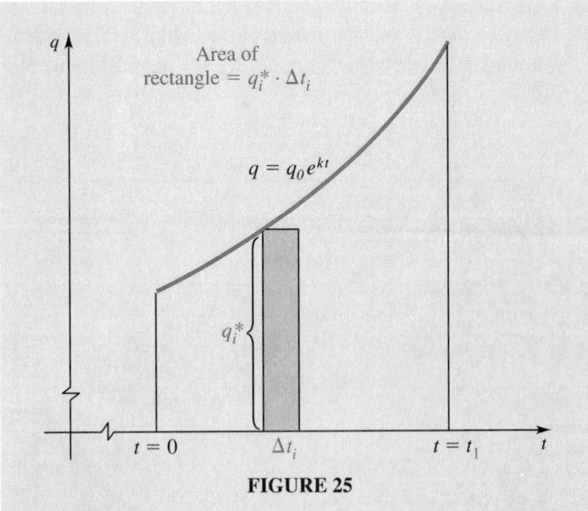

FIGURE 25

The limit of this sum as each of the Δt_i's approaches 0 gives the total consumption from time $t = 0$ to $t = t_1$. That is,

$$\text{Total consumption} = \lim_{\Delta t_i \to 0} \sum q_i^* \cdot \Delta t_i.$$

We have seen, however, that this limit is the definite integral of the function $q = q_0 e^{kt}$ from $t = 0$ to $t = t_1$, or

$$\text{Total consumption} = \int_0^{t_1} q_0 e^{kt}\, dt.$$

We can now evaluate this definite integral.

$$\begin{aligned}
\int_0^{t_1} q_0 e^{kt}\, dt &= q_0 \int_0^{t_1} e^{kt}\, dt = q_0\left(\frac{1}{k}e^{kt}\right)\Bigg|_0^{t_1} \\
&= \frac{q_0}{k}e^{kt}\Bigg|_0^{t_1} = \frac{q_0}{k}e^{kt_1} - \frac{q_0}{k}e^0 \\
&= \frac{q_0}{k}e^{kt_1} - \frac{q_0}{k}(1) \\
&= \frac{q_0}{k}(e^{kt_1} - 1) \qquad \textbf{(1)}
\end{aligned}$$

Now let us return to the numbers we gave for petroleum. We said that $q_0 = 19{,}600$ million barrels, where q_0 represents consumption in the base year. We have $k = .06$, with total petroleum reserves estimated at 2,000,000 million barrels. Thus, using equation (1) we have

$$2{,}000{,}000 = \frac{19{,}600}{.06}(e^{.06t_1} - 1).$$

Multiply both sides of the equation by .06:

$$120{,}000 = 19{,}600(e^{.06t_1} - 1).$$

Divide both sides of the equation by 19,600.

$$6.1 = e^{.06t_1} - 1$$

Add 1 to both sides.

$$7.1 = e^{.06t_1}$$

Take natural logarithms of both sides:

$$\begin{aligned}
\ln 7.1 &= \ln e^{.06t_1} \\
&= .06t_1 \ln e \\
&= .06t_1 \qquad (\text{since } \ln e = 1).
\end{aligned}$$

Finally,

$$t_1 = \frac{\ln 7.1}{.06} \approx 33.$$

By this result, petroleum reserves will last the world for 33 years.

The results of mathematical analyses such as this must be used with great caution. By the analysis above, the world would use all the petroleum that it wants in the thirty-second year after the base year, but there would be none at all in thirty-four years. This is not at all realistic. As petroleum reserves decline, the price will increase, causing demand to decline and supplies to increase.

Exercises

1. Find the number of years that the estimated petroleum reserves would last if used at the same rate as in the base year.

2. How long would the estimated petroleum reserves last if the growth constant was only 2% instead of 6%?

Estimate the length of time until depletion for each of the following minerals.

3. Bauxite (the ore from which aluminum is obtained): estimated reserves in base year 15,000,000 thousand tons; rate of consumption in base year 63,000 thousand tons; growth constant 6%

4. Bituminous coal; estimated world reserves 2,000,000 million tons; rate of consumption in base year 2200 million tons; growth constant 4%

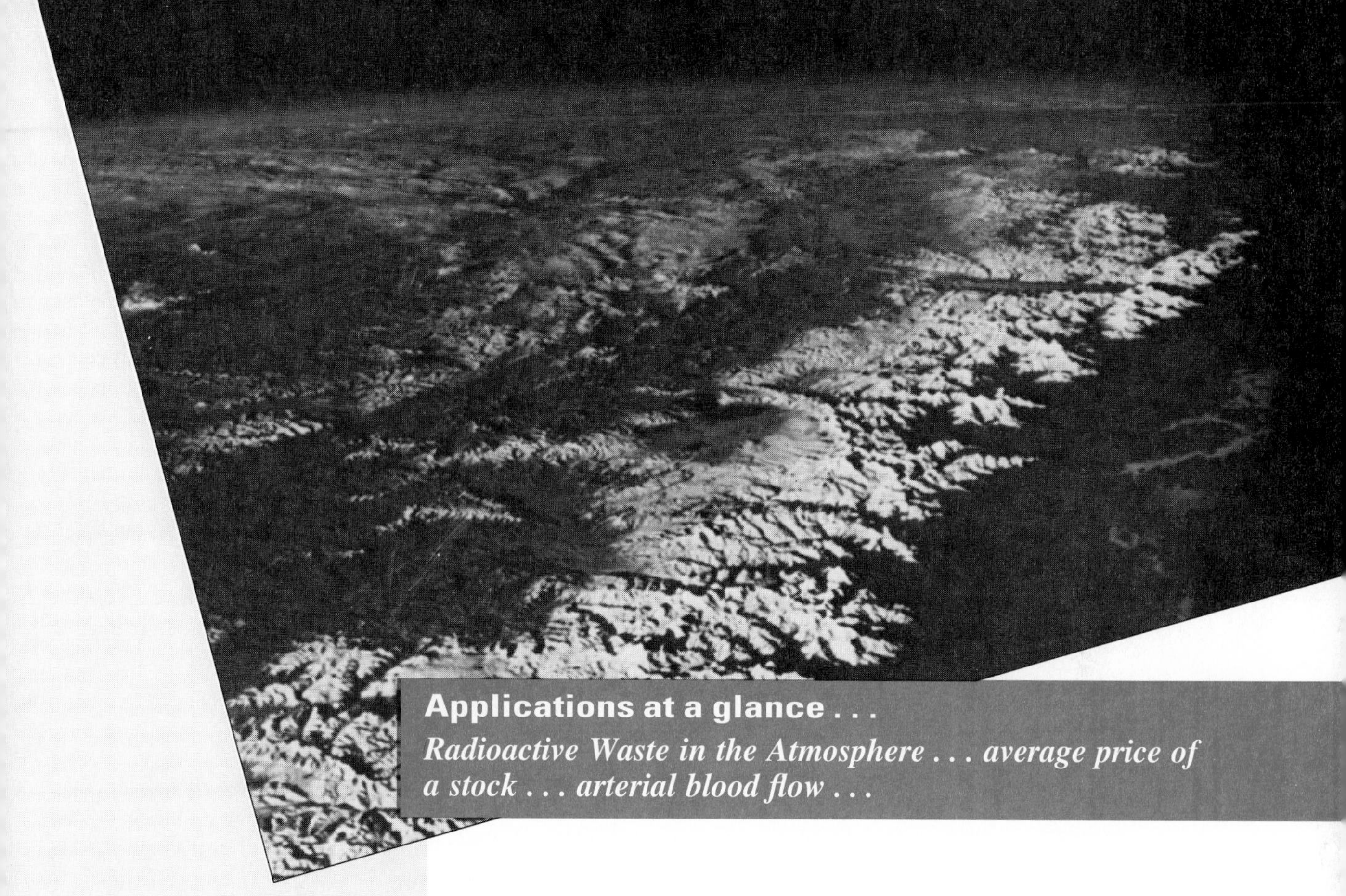

Applications at a glance . . .

Radioactive Waste in the Atmosphere . . . average price of a stock . . . arterial blood flow . . .

CHAPTER 7

FURTHER TECHNIQUES AND APPLICATIONS OF INTEGRATION

(See page 402.)

In the previous chapter we discussed indefinite and definite integrals and presented rules for finding the antiderivatives of several types of functions. We saw how these techniques could be used in various applications. In this chapter we develop additional methods of integrating functions, including numerical methods of integration. Numerical methods often are used in applications involving experimental data or with functions that cannot be integrated by other methods. In this chapter we also show how to evaluate an integral that has one or both limits at infinity. These new techniques provide additional applications of integration, such as finding volumes of solids of revolution, the average value of a function, and continuous money flow.

7.1 INTEGRATION BY PARTS; TABLES OF INTEGRALS

The technique of *integration by parts* often makes it possible to reduce a complicated integral to a simpler integral. If u and v are both differentiable functions, then uv is also differentiable and, by the product rule for derivatives,

$$\frac{d(uv)}{dx} = u\frac{dv}{dx} + v\frac{du}{dx}.$$

This expression can be rewritten, using differentials, as

$$d(uv) = u\,dv + v\,du.$$

Integrating both sides of this last equation gives

$$\int d(uv) = \int u\,dv + \int v\,du,$$

or

$$uv = \int u\,dv + \int v\,du.$$

Rearranging terms gives the following formula.

INTEGRATION BY PARTS

If u and v are differentiable functions, then

$$\int \boldsymbol{u\,dv} = \boldsymbol{uv} - \int \boldsymbol{v\,du}.$$

The process of finding integrals by this formula is called **integration by parts.** The method is shown in the following examples.

EXAMPLE 1

Find $\int xe^{5x}\,dx$.

While this integral cannot be found by using any method studied so far, it can be found with integration by parts. First write the expression $xe^{5x}\,dx$ as a product of two functions u and dv in such a way that $\int dv$ can be found. One way to do this is to choose the two functions x and e^{5x}. Both x and e^{5x} can be integrated, but $\int x\,dx$, which is $x^2/2$, is more complicated than x itself, while the derivative of x is 1, which is simpler than x. Since e^{5x} remains the same (except for the coefficient) whether it is integrated or differentiated, it is best here to choose

$$dv = e^{5x}\,dx \qquad \text{and} \qquad u = x.$$

Then

$$du = dx,$$

and v is found by integrating dv:

$$v = \int dv = \int e^{5x}\,dx = \frac{1}{5}e^{5x} + C.$$

For simplicity, ignore the constant C and add it at the end of the process of integration by parts. Now substitute into the formula for integration by parts and complete the integration.

$$\int u\,dv = uv - \int v\,du$$

$$\int \underbrace{x}_{u}\,\underbrace{e^{5x}\,dx}_{dv} = \underbrace{x}_{u}\underbrace{\left(\frac{1}{5}e^{5x}\right)}_{v} - \int \underbrace{\frac{1}{5}e^{5x}}_{v}\,\underbrace{dx}_{du}$$

$$= \frac{1}{5}xe^{5x} - \frac{1}{25}e^{5x} + C$$

The constant C was added in the last step. As before, check the answer by taking its derivative. ◀

Find $\int \ln x\,dx$ for $x > 0$.
No rule has been given for integrating $\ln x$, so choose

$$dv = dx \qquad \text{and} \qquad u = \ln x.$$

Then

$$v = x \qquad \text{and} \qquad du = \frac{1}{x}dx,$$

and

$$\int u \cdot dv = v \cdot u - \int v \cdot du$$

$$\int \overbrace{\ln x}\,\overbrace{dx} = \overbrace{x}\,\overbrace{\ln x} - \int \overbrace{x} \cdot \overbrace{\frac{1}{x}\,dx}$$

$$= x \ln x - \int dx$$

$$= x \ln x - x + C. \quad ◀$$

The preceding examples illustrate the following general principles for identifying integrals that can be found with integration by parts.

CONDITIONS FOR INTEGRATION BY PARTS

Integration by parts can be used only if the integrand satisfies the following conditions.

1. The integrand can be written as the product of two factors, u and dv.
2. It is possible to integrate dv to get v and to differentiate u to get du.
3. The integral $\int v\, du$ can be found.

With the functions discussed so far in this book, choosing u and dv is relatively simple. Before trying integration by parts, first see if the integration can be performed using substitution. If substitution does not work, see if $\ln x$ is in the integral. If it is, set $u = \ln x$ and dv equal to the rest of the integral. If $\ln x$ is not present, see if x^k is present, where k is any positive integer. If it is, set $u = x^k$ and dv equal to the rest of the integral.

A technique called **column integration,** or tabular integration, simplifies the process of integration by parts.* We begin by creating two columns. The first column, labeled D, contains u, the part to be differentiated in the original integral. The second column, labeled I, contains the rest of the integral: that is, the part to be integrated, but without the dx. To create the remainder of the first column, write the derivative of the function in the first row underneath it in the second row. Now write the derivative of the function in the second row underneath it in the third row. Proceed in this manner down the first column, taking derivatives until you get a 0. Form the second column in a similar manner, except take an antiderivative at each row, until the second column has the same number of rows as the first.

In the section on integration by substitution, we pointed out that when the chain rule is used to find the derivative of the function e^{kx}, we multiply by k, so when finding the antiderivative of e^{kx}, we divide by k. Thus $\int e^{5x}\, dx = e^{5x}/5 + C$. Keeping this technique in mind makes it simple to fill in the integration column when doing column integration.

To illustrate this process, consider Example 1, $\int xe^{5x}\, dx$. Here $u = x$, so e^{5x} is left for the second column. Taking derivatives down the first column and antiderivatives down the second column results in the following table.

D	***I***
x	e^{5x}
1	$e^{5x}/5$
0	$e^{5x}/25$

Next, draw a diagonal line from each term (except the last) in the left column to the term in the row below it in the right column. Label the first such line with "+", the next with "−", and continue alternating the signs as shown.

D		***I***
x	$+$	e^{5x}
1	$-$	$e^{5x}/5$
0		$e^{5x}/25$

*This technique appeared in the 1988 movie *Stand and Deliver.*

Then multiply the terms on opposite ends of each diagonal line. Finally, sum up the products just formed, adding the "+" terms and subtracting the "−" terms.

$$\int xe^{5x}\,dx = x(e^{5x}/5) - 1(e^{5x}/25) + C$$
$$= \frac{1}{5}xe^{5x} - \frac{1}{25}e^{5x} + C$$

Column integration works a little differently when applied to Example 2. Choose $\ln x$ as the part to differentiate, but no matter how many times $\ln x$ is differentiated, the result is not 0. In this case, stop as soon as the natural logarithm is gone. The part to be integrated must be 1. (Think of $\ln x$ as $\ln x \cdot 1$.)

D	***I***
$\ln x$	1
$1/x$	x

Draw diagonal lines with alternating + and − as before. On the last line, because the left column does not contain a 0, draw a horizontal line:

D		***I***
$\ln x$	+	1
$1/x$	−	x

The presence of a horizontal line indicates that there is an integral sign on that product:

$$\int \ln x\,dx = (\ln x)x - \int \frac{1}{x} \cdot x\,dx$$
$$= x \ln x - \int dx$$
$$= x \ln x - x + C.$$

Note that when setting up the columns, a horizontal line is drawn only when a 0 does not eventually appear in the left column.

Column integration is particularly useful when the part to be differentiated (the part labeled u) must be differentiated more than twice to get 0. Integration by parts would have to be used repeatedly, but column integration does it all at once, as in the following example.

EXAMPLE 3 Find $\int 2x^2e^{-3x}\,dx$.

Choose $2x^2$ as the part to be differentiated, and put e^{-3x} in the integration column.

D		I
$2x^2$	+	e^{-3x}
$4x$	−	$-e^{-3x}/3$
4	+	$e^{-3x}/9$
0		$-e^{-3x}/27$

Multiplying and adding as before yields

$$\int 2x^2e^{-3x}\,dx = 2x^2(-e^{-3x}/3) - 4x(e^{-3x}/9) + 4(-e^{-3x}/27) + C$$
$$= -\frac{2}{3}x^2e^{-3x} - \frac{4}{9}xe^{-3x} - \frac{4}{27}e^{-3x} + C.$$

EXAMPLE 4 Find $\int_1^e \frac{\ln x}{x^2}\,dx$.

First we will find the indefinite integral using integration by parts. (You may wish to verify this using column integration.) Whenever $\ln x$ is present, it is selected as u, so let

$$dv = \frac{1}{x^2}\,dx \quad \text{and} \quad u = \ln x.$$

Recall that $\int x^n\,dx = x^{n+1}/(n+1) + C$, $n \neq -1$, so $\int 1/x^2\,dx = \int x^{-2}\,dx = x^{-1}/(-1) + C = -1/x + C$.

Then

$$v = -\frac{1}{x} \quad \text{and} \quad du = \frac{1}{x}\,dx.$$

Substitute these values into the formula for integration by parts, and integrate the second term on the right.

$$\int u\,dv = uv - \int v\,du$$
$$\int \frac{\ln x}{x^2}\,dx = \ln x\frac{-1}{x} - \int -\frac{1}{x}\cdot\frac{1}{x}\,dx$$
$$= -\frac{\ln x}{x} + \int \frac{1}{x^2}\,dx$$
$$= -\frac{\ln x}{x} - \frac{1}{x} + C$$
$$= \frac{-\ln x - 1}{x} + C.$$

Now find the definite integral.

$$\int_1^e \frac{\ln x}{x^2}\,dx = \left.\frac{-\ln x - 1}{x}\right|_1^e$$

$$= \left(\frac{-1-1}{e}\right) - \left(\frac{0-1}{1}\right)$$

$$= \frac{-2}{e} + 1. \quad \blacktriangleleft$$

Integration Using Tables of Integrals The method of integration by parts requires choosing the factor dv so that $\int dv$ can be found. If this is not possible, or if the remaining factor, which becomes u, does not have a differential du such that $\int v\,du$ can be found, the technique cannot be used. For example, to integrate

$$\int \frac{1}{4 - x^2}\,dx,$$

we might choose $dv = dx$ and $u = (4 - x^2)^{-1}$. Integration gives $v = x$ and differentiation gives $du = 2x\,dx/(4 - x^2)^2$, with

$$\int \frac{1}{4 - x^2}\,dx = \frac{x}{4 - x^2} - \int \frac{2x^2\,dx}{(4 - x^2)^2}.$$

The integral on the right is more complicated than the original integral, however. A second use of integration by parts on the new integral would make matters even worse. Since we cannot choose $dv = (4 - x^2)^{-1}\,dx$ because it cannot be integrated by the methods studied so far, integration by parts is not possible for this problem. In fact, there are many functions whose integrals cannot be found by any of the methods described in this text. Many of these can be found by more advanced methods and are listed in **tables of integrals.** Such a table is given at the back of this book. The next examples show how to use this table.

Find $\displaystyle\int \frac{1}{\sqrt{x^2 + 16}}\,dx.$

If $a = 4$, this antiderivative is the same as entry 5 in the table,

$$\int \frac{1}{\sqrt{x^2 + a^2}}\,dx = \ln\left|x + \sqrt{x^2 + a^2}\right| + C.$$

Substituting 4 for a gives

$$\int \frac{1}{\sqrt{x^2 + 16}}\,dx = \ln\left|x + \sqrt{x^2 + 16}\right| + C.$$

This result could be verified by taking the derivative of the right side of this last equation. ◀

EXAMPLE 6 Find $\int \frac{8}{16 - x^2}\, dx.$

Convert this antiderivative into the one given in entry 7 of the table by writing the 8 in front of the integral sign (permissible only with constants) and by letting $a = 4$. Doing this gives

$$8\int \frac{1}{16 - x^2}\, dx = 8\left[\frac{1}{2 \cdot 4} \ln \left|\frac{4 + x}{4 - x}\right|\right] + C$$

$$= \ln \left|\frac{4 + x}{4 - x}\right| + C.$$

Find $\int \sqrt{9x^2 + 1}\, dx.$

This antiderivative seems most similar to entry 15 of the table. However, entry 15 requires that the coefficient of the x^2 term be 1. That requirement can be satisfied here by factoring out the 9.

$$\int \sqrt{9x^2 + 1}\, dx = \int \sqrt{9\left(x^2 + \frac{1}{9}\right)}\, dx = \int 3\sqrt{x^2 + \frac{1}{9}}\, dx = 3\int \sqrt{x^2 + \frac{1}{9}}\, dx$$

Now use entry 15 with $a = 1/3$.

$$\int \sqrt{9x^2 + 1}\, dx = 3\left[\frac{x}{2}\sqrt{x^2 + \frac{1}{9}} + \frac{(1/3)^2}{2} \ln \left|x + \sqrt{x^2 + \frac{1}{9}}\right|\right] + C$$

$$= \frac{3x}{2}\sqrt{x^2 + \frac{1}{9}} + \frac{1}{6} \ln \left|x + \sqrt{x^2 + \frac{1}{9}}\right| + C$$

7.1 Exercises

Use integration by parts or column integration to find the integrals in Exercises 1–10.

1. $\int xe^x\, dx$

2. $\int (x + 1)e^x\, dx$

3. $\int (5x - 9)e^{-3x}\, dx$

4. $\int (6x + 3)e^{-2x}\, dx$

5. $\int_0^1 \frac{2x + 1}{e^x}\, dx$

6. $\int_0^1 \frac{1 - x}{3e^x}\, dx$

7. $\int_1^4 \ln 2x\, dx$

8. $\int_1^2 \ln 5x\, dx$

9. $\int x \ln x\, dx$

10. $\int x^2 \ln x\, dx$

11. Find the area between $y = (x - 2)e^x$ and the x-axis from $x = 2$ to $x = 4$.

12. Find the area between $y = xe^x$ and the x-axis from $x = 0$ to $x = 1$.

Exercises 13–22 are mixed—some require integration by parts or column integration, while others can be integrated by using techniques discussed earlier.

13. $\int x^2e^{2x}\, dx$

14. $\int_1^2 (1 - x^2)e^{2x}\, dx$

15. $\int_0^5 x\sqrt[3]{x^2 + 2}\, dx$

16. $\int (2x - 1) \ln (3x)\, dx$

17. $\int (8x + 7) \ln (5x)\, dx$

18. $\int xe^{x^2}\, dx$

19. $\int x^2\sqrt{x + 2}\, dx$

20. $\int_0^1 \frac{x^2\, dx}{2x^3 + 1}$

21. $\int_0^1 \frac{x^3\, dx}{\sqrt{3 + x^2}}$

22. $\int \frac{x^2\, dx}{2x^3 + 1}$

Use the table of integrals to find each indefinite integral.

23. $\int \frac{-4}{\sqrt{x^2 + 36}}\, dx$

24. $\int \frac{9}{\sqrt{x^2 + 9}}\, dx$

25. $\int \frac{6}{x^2 - 9}\, dx$

26. $\int \frac{-12}{x^2 - 16}\, dx$

27. $\int \frac{-4}{x\sqrt{9 - x^2}}\, dx$

28. $\int \frac{3}{x\sqrt{121 - x^2}}\, dx$

29. $\int \frac{-2x}{3x + 1}\, dx$

30. $\int \frac{6x}{4x - 5}\, dx$

31. $\int \frac{2}{3x(3x - 5)}\, dx$

32. $\int \frac{-4}{3x(2x + 7)}\, dx$

33. $\int \frac{4}{4x^2 - 1}\, dx$

34. $\int \frac{-6}{9x^2 - 1}\, dx$

35. $\int \frac{3}{x\sqrt{1 - 9x^2}}\, dx$

36. $\int \frac{-2}{x\sqrt{1 - 16x^2}}\, dx$

37. $\int \frac{4x}{2x + 3}\, dx$

38. $\int \frac{4x}{6 - x}\, dx$

39. $\int \frac{-x}{(5x - 1)^2}\, dx$

40. $\int \frac{-3}{x(4x + 3)^2}\, dx$

41. Use integration by parts or column integration to derive the following formula from the table of integrals.

$$\int x^n \cdot \ln |x|\, dx = x^{n+1}\left[\frac{\ln |x|}{n + 1} - \frac{1}{(n + 1)^2}\right] + C, \quad n \neq -1$$

42. Use integration by parts or column integration to derive the following formula from the table of integrals.

$$\int x^n\, e^{ax}\, dx = \frac{x^n e^{ax}}{a} - \frac{n}{a}\int x^{n-1}e^{ax}\, dx + C, \quad a \neq 0$$

43. **(a)** One way to integrate $\int x\sqrt{x + 1}\, dx$ is to use integration by parts or column integration. Do so to find the antiderivative.

(b) Another way to evaluate the integral in part (a) is by using the substitution $u = x + 1$. Do so to find the antiderivative.

(c) Compare the results from the two methods. If they do not look the same, explain how this can happen. Discuss the advantages and disadvantages of each method.

Applications

Business and Economics

44. Rate of Change of Revenue The rate of change of revenue (in dollars) from the sale of x small desk calculators is

$$R'(x) = \frac{1000}{\sqrt{x^2 + 25}}.$$

Find the total revenue from the sale of the first 12 calculators.

Life Sciences

45. Reaction to a Drug The rate of reaction to a drug is given by

$$r'(x) = 2x^2e^{-x},$$

where x is the number of hours since the drug was administered. Find the total reaction to the drug from $x = 1$ to $x = 6$.

46. Growth of a Population The rate of growth of a microbe population is given by

$$m'(x) = 30xe^{2x},$$

where x is time in days. What is the total accumulated growth after 3 days?

Social Sciences

47. Production Rate The rate (in hours per item) at which a worker in a certain job produces the xth item is

$$h'(x) = \sqrt{x^2 + 16}.$$

What is the total number of hours it will take this worker to produce the first 7 items?

7.2 NUMERICAL INTEGRATION

If the velocity of a vehicle is known only at certain points in time, how can the total distance traveled by the vehicle be estimated?

Using numerical integration, we will answer this question in Example 3 of this section.

As mentioned in the previous section, some integrals cannot be evaluated by any technique, nor are they found in any table. One solution to this problem was presented in Section 3 of the previous chapter, in which the area under a curve was approximated by summing the areas of rectangles. This method is seldom used in practice because better methods exist which give more accuracy for the same amount of work. These methods are referred to as **numerical integration** methods. We shall discuss two such methods here: the trapezoidal rule and Simpson's rule.

Trapezoidal Rule To illustrate the use of the trapezoidal rule, consider the integral

$$\int_1^5 \frac{1}{x}\,dx.$$

The shaded region in Figure 1 shows the area representing that integral, the area under the graph $f(x) = 1/x$, above the x-axis, and between the lines $x = 1$ and $x = 5$. As shown in the figure, if the area under the curve is approximated with trapezoids rather than rectangles, the approximation should be improved.

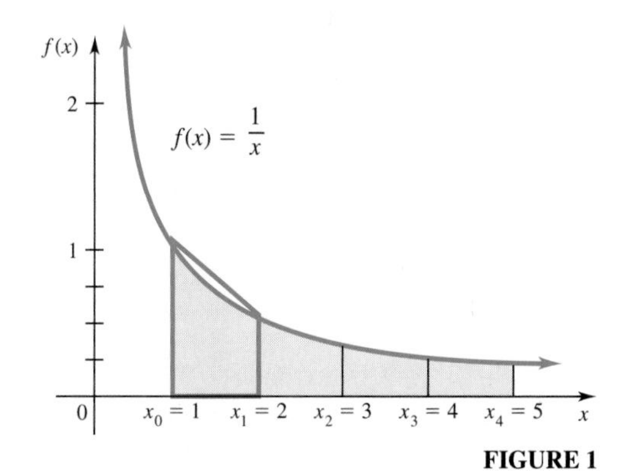

B
b
h

Left-hand trapezoid

$B = f(x_0) = f(1) = 1$

$b = f(x_1) = f(2) = \frac{1}{2}$

$h = 2 - 1 = 1$

$$\text{Area} = \frac{1}{2}(B + b)h = \frac{1}{2}\left(1 + \frac{1}{2}\right)1 = \frac{1}{2} \cdot \frac{3}{2} = \frac{3}{4}$$

FIGURE 1

Since $\int (1/x)\,dx = \ln |x| + C$,

$$\int_1^5 \frac{1}{x}\,dx = \ln |x|\Big|_1^5 = \ln 5 - \ln 1 = \ln 5 - 0 = \ln 5 \approx 1.609438.$$

As in earlier work, to approximate this area we divide the interval [1, 5] into subintervals of equal widths. To get a first approximation to ln 5 by the trapezoidal rule, find the sum of the areas of the four trapezoids shown in Figure 1. From geometry, the area of a trapezoid is half the product of the sum of the bases and the altitude. Each of the trapezoids in Figure 1 has altitude 1. (In this case, the bases of the trapezoid are vertical and the altitudes are horizontal.) Adding the areas gives

$$\ln 5 = \int_1^5 \frac{1}{x}\,dx \approx \frac{1}{2}\left(\frac{1}{1} + \frac{1}{2}\right)(1) + \frac{1}{2}\left(\frac{1}{2} + \frac{1}{3}\right)(1) + \frac{1}{2}\left(\frac{1}{3} + \frac{1}{4}\right)(1) + \frac{1}{2}\left(\frac{1}{4} + \frac{1}{5}\right)(1)$$

$$= \frac{1}{2}\left(\frac{3}{2} + \frac{5}{6} + \frac{7}{12} + \frac{9}{20}\right) \approx 1.68333.$$

To get a better approximation, divide the interval [1, 5] into more subintervals. Generally speaking, the larger the number of subintervals, the better the approximation. The results for selected values of n are shown below to 5 decimal places.

n	$\int_1^5 \frac{1}{x}\,dx = \ln 5 \approx 1.609438$
6	1.64360
8	1.62897
10	1.62204
20	1.61263
100	1.60957
1000	1.60944

When $n = 1000$, the approximation agrees with the true value to five decimal places.

Generalizing from this example, let f be a continuous function on an interval $[a, b]$. Divide the interval from a to b into n equal subintervals by the points $a = x_0, x_1, x_2, \ldots, x_n = b$, as shown in Figure 2 (on the next page). Use the subintervals to make trapezoids that approximately fill in the region under the curve. The approximate value of the definite integral $\int_a^b f(x)\,dx$ is given by the sum of the areas of the trapezoids, or

$$\int_a^b f(x)\,dx \approx \frac{1}{2}\Big[f(x_0) + f(x_1)\Big]\left(\frac{b-a}{n}\right) + \frac{1}{2}\Big[f(x_1) + f(x_2)\Big]\left(\frac{b-a}{n}\right) + \cdots + \frac{1}{2}\Big[f(x_{n-1}) + f(x_n)\Big]\left(\frac{b-a}{n}\right)$$

$$= \left(\frac{b-a}{n}\right)\left[\frac{1}{2}f(x_0) + \frac{1}{2}f(x_1) + \frac{1}{2}f(x_1) + \frac{1}{2}f(x_2) + \frac{1}{2}f(x_2) + \cdots + \frac{1}{2}f(x_{n-1}) + \frac{1}{2}f(x_n)\right]$$

$$= \left(\frac{b-a}{n}\right)\left[\frac{1}{2}f(x_0) + f(x_1) + f(x_2) + \cdots + f(x_{n-1}) + \frac{1}{2}f(x_n)\right].$$

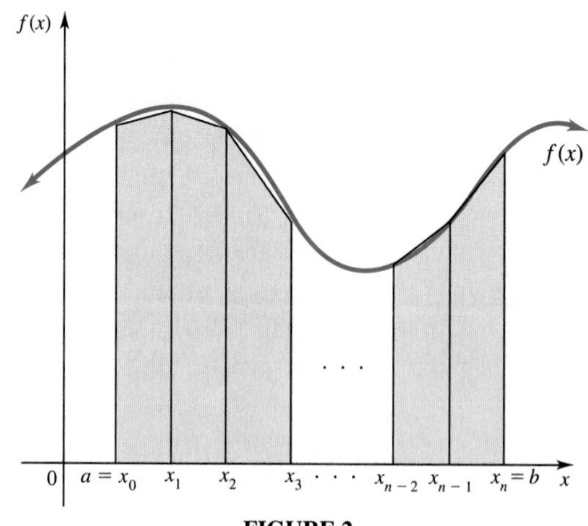

FIGURE 2

This result gives the following rule.

TRAPEZOIDAL RULE

Let f be a continuous function on $[a, b]$ and let $[a, b]$ be divided into n equal subintervals by the points $a = x_0, x_1, x_2, \ldots, x_n = b$. Then, by the **trapezoidal rule,**

$$\int_a^b f(x)\,dx \approx \left(\frac{b-a}{n}\right)\left[\frac{1}{2}f(x_0) + f(x_1) + \cdots + f(x_{n-1}) + \frac{1}{2}f(x_n)\right].$$

EXAMPLE 1 Use the trapezoidal rule with $n = 4$ to approximate

$$\int_0^2 \sqrt{x^2 + 1}\,dx.$$

Here $a = 0$, $b = 2$, and $n = 4$, with $(b - a)/n = (2 - 0)/4 = 1/2$ as the altitude of each trapezoid. Then $x_0 = 0$, $x_1 = 1/2$, $x_2 = 1$, $x_3 = 3/2$, and $x_4 = 2$. Now find the corresponding function values. The work can be organized into a table, as follows.

i	x_i	$f(x_i)$
0	0	$\sqrt{0^2 + 1} = 1$
1	1/2	$\sqrt{(1/2)^2 + 1} \approx 1.11803$
2	1	$\sqrt{1^2 + 1} \approx 1.41421$
3	3/2	$\sqrt{(3/2)^2 + 1} \approx 1.80278$
4	2	$\sqrt{2^2 + 1} \approx 2.23607$

Substitution into the trapezoidal rule gives

$$\int_0^2 \sqrt{x^2 + 1}\, dx$$

$$\approx \frac{2 - 0}{4}\left[\frac{1}{2}(1) + 1.11803 + 1.41421 + 1.80278 + \frac{1}{2}(2.23607)\right]$$

$$\approx 2.97653.$$

Using entry 15 in the table of integrals, the exact value of the integral can be found.

$$\int_0^2 \sqrt{x^2 + 1}\, dx = \frac{x}{2}\sqrt{x^2 + 1} + \frac{1}{2}\ln\left|x + \sqrt{x^2 + 1}\right|\Bigg|_0^2$$

$$= \sqrt{5} + \frac{1}{2}\ln(2 + \sqrt{5}) - (0 + 0) \approx 2.95789$$

The approximation 2.97653 found above using the trapezoidal rule with $n = 4$ differs from the true value by .01864. As mentioned above, this error would be reduced if larger values were used for n. For example, if $n = 8$, the trapezoidal rule gives an answer of 2.96254, which differs from the true value by .00465. Techniques for estimating such errors are considered in more advanced courses. ◀

Simpson's Rule Another numerical method, *Simpson's rule*, approximates consecutive portions of the curve with portions of parabolas rather than the line segments of the trapezoidal rule. Simpson's rule usually gives a better approximation than the trapezoidal rule for the same number of subintervals. As shown in Figure 3, one parabola is fitted through points A, B, and C, another through C, D, and E, and so on. Then the sum of the areas under these parabolas will approximate the area under the graph of the function. Because of the way the parabolas overlap, it is necessary to have an even number of intervals, and therefore an odd number of points, to apply Simpson's rule.

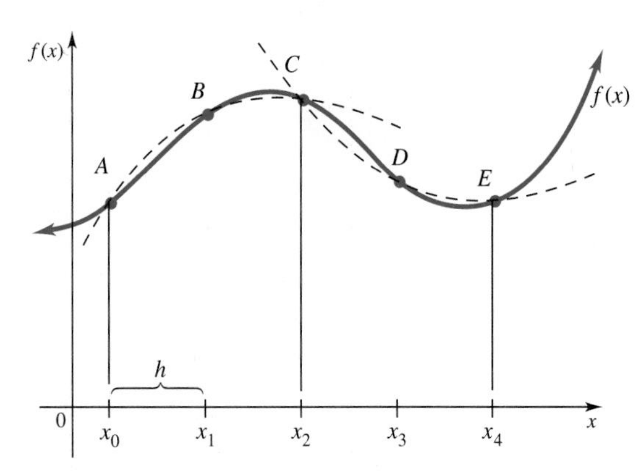

FIGURE 3

If h, the length of each subinterval, is $(b - a)/n$, the area under the parabola through points A, B, and C can be found by a definite integral. The details are omitted; the result is

$$\frac{h}{3}\left[f(x_0) + 4f(x_1) + f(x_2)\right].$$

Similarly, the area under the parabola through points C, D, and E is

$$\frac{h}{3}\left[f(x_2) + 4f(x_3) + f(x_4)\right].$$

When these expressions are added, the last term of one expression equals the first term of the next. For example, the sum of the two areas given above is

$$\frac{h}{3}\left[f(x_0) + 4f(x_1) + 2f(x_2) + 4f(x_3) + f(x_4)\right].$$

This illustrates the origin of the pattern of the terms in the following rule.

SIMPSON'S RULE

Let f be a continuous function on $[a, b]$ and let $[a, b]$ be divided into an even number n of equal subintervals by the points $a = x_0, x_1, x_2, \ldots, x_n = b$. Then by **Simpson's rule,**

$$\int_a^b f(x)\,dx \approx \frac{b-a}{3n}[f(x_0) + 4f(x_1) + 2f(x_2) + 4f(x_3) + \cdots$$
$$+ 2f(x_{n-2}) + 4f(x_{n-1}) + f(x_n)].$$

Thomas Simpson (1710–1761), a British mathematician, wrote texts on many branches of mathematics. Some of these texts went through as many as ten editions. His name became attached to this numerical method of approximating definite integrals even though the method preceded his work.

Caution In Simpson's rule, n (the number of subintervals) must be even.

Use Simpson's rule with $n = 4$ to approximate

$$\int_0^2 \sqrt{x^2 + 1}\,dx,$$

which was approximated by the trapezoidal rule in Example 1.

As in Example 1, $a = 0$, $b = 2$, and $n = 4$, and the endpoints of the four intervals are $x_0 = 0$, $x_1 = 1/2$, $x_2 = 1$, $x_3 = 3/2$, and $x_4 = 2$. The table of values is also the same.

i	x_i	$f(x_i)$
0	0	1
1	1/2	1.11803
2	1	1.41421
3	3/2	1.80278
4	2	2.23607

Since $(b - a)/(3n) = 2/12 = 1/6$, substituting into Simpson's rule gives

$$\int_0^2 \sqrt{x^2 + 1}\, dx \approx \frac{1}{6}[1 + 4(1.11803) + 2(1.41421) + 4(1.80278) + 2.23607] \approx 2.95796.$$

This differs from the true value by .00007, which is less than the trapezoidal rule with $n = 8$. If $n = 8$ for Simpson's rule, the approximation is 2.95788, which differs from the true value by only .00001. ◀

Numerical methods make it possible to approximate

$$\int_a^b f(x)\, dx$$

even when $f(x)$ is not known. The next example shows how this is done.

EXAMPLE 3 As mentioned earlier, the velocity $v(t)$ gives the rate of change of distance $s(t)$ with respect to time t. Suppose a vehicle travels an unknown distance. The passengers keep track of the velocity at 10-minute intervals (every 1/6 of an hour) with the following results.

Time in Hours, t	1/6	2/6	3/6	4/6	5/6	1	7/6
Velocity in Miles per Hour, $v(t)$	45	55	52	60	64	58	47

What is the total distance traveled in the 60-minute period from $t = 1/6$ to $t = 7/6$?

The distance traveled in t hours is $s(t)$, with $s'(t) = v(t)$. The total distance traveled between $t = 1/6$ and $t = 7/6$ is given by

$$\int_{1/6}^{7/6} v(t)\, dt.$$

Even though this integral cannot be evaluated since we do not have an expression for $v(t)$, either the trapezoid rule or Simpson's rule can be used to approximate its value and give the total distance traveled. In either case, let $n = 6$, $a = t_0 = 1/6$, and $b = t_6 = 7/6$. By the trapezoidal rule,

$$\int_{1/6}^{7/6} v(t)\, dt \approx \frac{7/6 - 1/6}{6}\left[\frac{1}{2}(45) + 55 + 52 + 60 + 64 + 58 + \frac{1}{2}(47)\right]$$

$$\approx 55.83.$$

By Simpson's rule,

$$\int_{1/6}^{7/6} v(t)\, dt \approx \frac{7/6 - 1/6}{3(6)}[45 + 4(55) + 2(52) + 4(60) + 2(64) + 4(58) + 47]$$

$$= \frac{1}{18}(45 + 220 + 104 + 240 + 128 + 232 + 47) \approx 56.44.$$

The distance traveled in the 1-hour period was about 56 miles.

As mentioned above, Simpson's rule generally gives a better approximation than the trapezoidal rule. As n increases, the two approximations get closer and closer. For the same accuracy, however, a smaller value of n generally can be used with Simpson's rule so that less computation is necessary. ◀

The branch of mathematics that studies methods of approximating definite integrals (as well as many other topics) is called *numerical analysis.* Some calculators give such approximations by using Simpson's rule, as shown in the following excerpt from the manual for model TI-60, a calculator produced by Texas Instruments.*

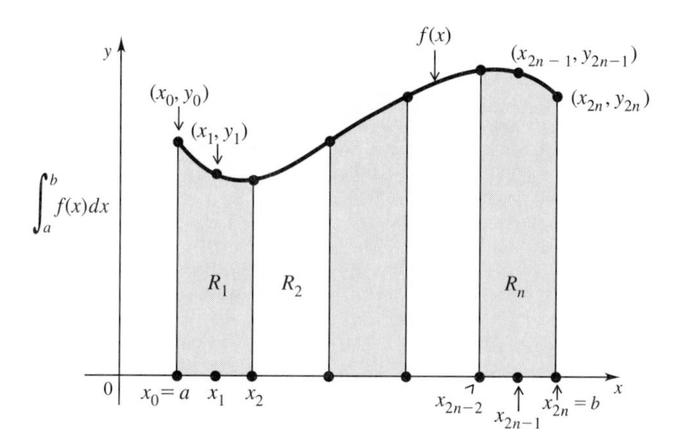

Numerical integration is useful even with functions whose antiderivatives can be computed if the antidifferentiation is complicated and a computer or calculator programmed with Simpson's rule is handy. For example, it would be easier to evaluate $\int_0^2 (x^{12} - 3x^7 + 4x + 5)e^{-.5x}\, dx$ by using Simpson's rule than by using the antiderivative. On the other hand, using the antiderivative makes it easier to see the effect of changing the upper limit, say, from 2 to 3, or changing the power of e from $-.5x$ to $-1.5x$. Numerical integration only gives information about the particular integral being evaluated.

*"Simpson's Rule" from *Texas Instruments TI-60 Guidebook*. Copyright © 1986 by Texas Instruments Incorporated. Reprinted by permission.

7.2 Exercises

In Exercises 1–10, use $n = 4$ to approximate the value of each of the given integrals by the following methods: **(a)** *the trapezoidal rule, and* **(b)** *Simpson's rule.* **(c)** *Find the exact value by integration.*

1. $\int_0^2 x^2\,dx$ **2.** $\int_0^2 (2x + 1)\,dx$ **3.** $\int_{-1}^3 \frac{1}{4 - x}\,dx$ **4.** $\int_1^5 \frac{1}{x + 1}\,dx$ **5.** $\int_{-2}^2 (2x^2 + 1)\,dx$

6. $\int_0^3 (2x^2 + 1)\,dx$ **7.** $\int_1^5 \frac{1}{x^2}\,dx$ **8.** $\int_2^4 \frac{1}{x^3}\,dx$ **9.** $\int_0^4 \sqrt{x^2 + 1}\,dx$ **10.** $\int_1^4 x\sqrt{2x - 1}\,dx$

11. Find the area under the semicircle $y = \sqrt{4 - x^2}$ and above the x-axis by using $n = 8$ with the following methods:

(a) the trapezoidal rule; **(b)** Simpson's rule.

(c) Compare the results with the area found by the formula for the area of a circle. Which of the two approximation techniques was more accurate?

12. Find the area between the x-axis and the ellipse $4x^2 + 9y^2 = 36$ by using $n = 12$ with the following methods:

(a) the trapezoidal rule; **(b)** Simpson's rule.

(*Hint:* Solve the equation for y and find the area of the semiellipse.)

(c) Compare the results with the actual area, $6\pi \approx 18.8496$ (which can be found by methods not considered in this text). Which approximation technique was more accurate?

Applications

Business and Economics

13. Total Sales A sales manager presented the following results at a sales meeting.

Year, x	1	2	3	4	5	6	7
Rate of Sales, $f(x)$	.4	.6	.9	1.1	1.3	1.4	1.6

Find the total sales over the given period as follows.

(a) Plot these points. Connect the points with line segments.

(b) Use the trapezoidal rule to find the area bounded by the broken line of part (a), the x-axis, the line $x = 1$, and the line $x = 7$.

(c) Find the same area using Simpson's rule.

14. Total Sales A company's marginal costs (in hundreds of dollars) were as follows over a certain period.

Year, x	1	2	3	4	5	6	7
Marginal Cost, $f(x)$	9.0	9.2	9.5	9.4	9.8	10.1	10.5

Repeat steps (a)–(c) of Exercise 13 for these data to find the total sales over the given period.

Life Sciences

15. Drug Reaction Rate The reaction rate to a new drug is given by

$$y = e^{-t^2} + \frac{1}{t},$$

where t is time measured in hours after the drug is administered. Find the total reaction to the drug from $t = 1$ to $t = 9$ by letting $n = 8$ and using the following methods: **(a)** the trapezoidal rule, and **(b)** Simpson's rule.

16. Growth Rate The growth rate of a certain tree (in feet) is given by

$$y = \frac{2}{t} + e^{-t^2/2},$$

where t is time in years. Find the total growth from $t = 1$ to $t = 7$ by using $n = 12$ with the following methods: **(a)** the trapezoidal rule, **(b)** Simpson's rule.

Blood Level Curves *In the study of bioavailability in pharmacy, a drug is given to a patient. The level of concentration of the drug is then measured periodically, producing blood level curves such as the ones shown in the figure. The areas under the curves give the total amount of the drug available to the patient.* Use the trapezoidal rule with $n = 10$ to find the following areas.*

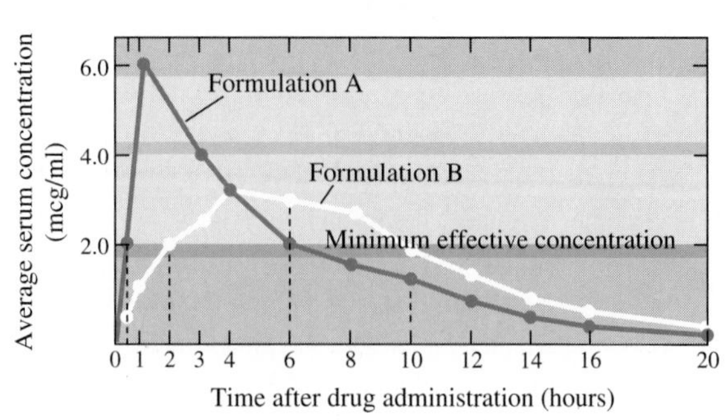

17. Find the total area under the curve for Formulation A. What does this area represent?

18. Find the total area under the curve for Formulation B. What does this area represent?

19. Find the area between the curve for Formulation A and the minimum effective concentration line. What does your answer represent?

20. Find the area between the curve for Formulation B and the minimum effective concentration line. What does this area represent?

Social Sciences

21. Educational Psychology The results from a research study in psychology were as follows.

Number of Hours of Study, x	1	2	3	4	5	6	7
Rate of Extra Points Earned on a Test, $f(x)$	4	7	11	9	15	16	23

Repeat steps (a)–(c) of Exercise 13 for these data.

Physical Sciences

22. Chemical Formation The table below shows the results from a chemical experiment.

Concentration of Chemical A, x	1	2	3	4	5	6	7
Rate of Formation of Chemical B, $f(x)$	12	16	18	21	24	27	32

Repeat steps (a)–(c) of Exercise 13 for these data.

Computer/Graphing Calculator

Use the trapezoidal rule and then use Simpson's rule to approximate each of the following integrals. Use $n = 100$.

23. $\int_4^8 \ln (x^2 - 10)\, dx$

24. $\int_{-2}^2 e^{-x^2}\, dx$

25. $\int_{-2}^2 \sqrt{9 - 2x^2}\, dx$

26. $\int_{-1}^1 \sqrt{16 + 5x^2}\, dx$

27. $\int_1^5 (2x^2 + 3x - 1)^{2/5}\, dx$

28. $\int_1^4 (x^3 + 4)^{5/4}\, dx$

---

*These graphs are from D. J. Chodos and A. R. DeSantos, *Basics of Bioavailability.* Copyright 1978 by the Upjohn Company.

Use either the trapezoidal rule or Simpson's rule with $n = 100$ for Exercises 29–31.

29. Total Revenue An electronics company analyst has determined that the rate per month at which revenue comes in from the calculator division is given by

$$R(x) = 105e^{.01x} + 32,$$

where x is the number of months the division has been in operation. Find the total revenue between the twelfth and thirty-sixth months.

30. Blood Pressure Blood pressure in an artery changes rapidly over a very short time for a healthy young adult, from a high of about 120 to a low of about 80. Suppose the blood pressure function over an interval of 1.5 sec is given by

$$f(x) = .2x^5 - .68x^4 + .8x^3 - .39x^2 + .055x + 100,$$

where x is the time in seconds after a peak reading. The area under the curve for one cycle is important in some blood pressure studies. Find the area under $f(x)$ from .1 sec to 1.1 sec.

31. Probability The most important function in probability and statistics is the density function for the standard normal distribution, which is the familiar bell-shaped curve. The function is

$$f(x) = \frac{1}{\sqrt{2\pi}}e^{-x^2/2}.$$

(a) The area under this curve between $x = -1$ and $x = 1$ represents the probability that a normal random variable is within 1 standard deviation of the mean. Find this probability.

(b) Find the area under this curve between $x = -2$ and $x = 2$, which represents the probability that a normal random variable is within 2 standard deviations of the mean.

(c) Find the probability that a normal random variable is within 3 standard deviations of the mean.

Error Analysis *The difference between the true value of an integral and the value given by the trapezoidal rule or Simpson's rule is known as the error. In numerical analysis, the error is studied to determine how large n must be for the error to be smaller than some specified amount. For both rules, the error is inversely proportional to a power of n, the number of subdivisions. In other words, the error is roughly k/n^p, where k is a constant that depends upon the function and the interval, and p is a power that depends only upon the method used. With a little experimentation, you can find out what the power p is for the trapezoidal rule and for Simpson's rule.*

32. (a) Find the exact value of $\int_0^1 x^4\,dx$.

(b) Approximate the integral in part (a) using the trapezoidal rule with $n = 4$, 8, 16, and 32. For each of these answers, find the absolute value of the error by subtracting the trapezoidal rule answer from the exact answer found in part (a).

(c) If the error is k/n^p, then the error times n^p should be approximately a constant. Multiply the errors in part (b) times n^p for $p = 1, 2$, etc., until you find a power p yielding the same answer for all four values of n.

33. Based on the results of Exercise 32, what happens to the error in the trapezoidal rule when the number of intervals is doubled?

34. Repeat Exercise 32 using Simpson's rule.

35. Based on the results of Exercise 34, what happens to the error in Simpson's rule when the number of intervals is doubled?

7.3 TWO APPLICATIONS OF INTEGRATION: VOLUME AND AVERAGE VALUE

If we have a formula giving the price of a common stock as a function of time, how can we find the average price of the stock over a certain period of time?

In this section, we will discover how to find the average value of a function, as well as how to compute the volume of a solid.

Volume Figure 4 shows the region below the graph of some function $y = f(x)$, above the x-axis, and between $x = a$ and $x = b$. We have seen how to use integrals to find the area of such a region. Now, suppose this region is revolved about the x-axis as shown in Figure 5. The resulting figure is called a **solid of revolution.** In many cases, the volume of a solid of revolution can be found by integration.

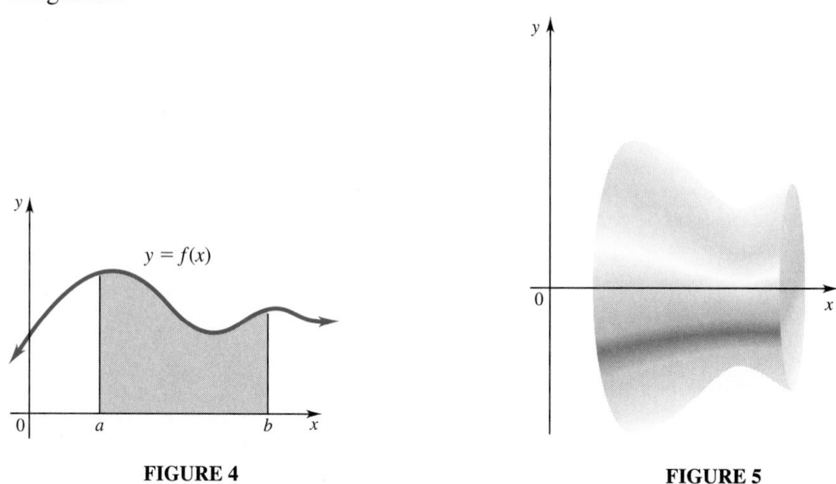

FIGURE 4

FIGURE 5

To begin, divide the interval $[a, b]$ into n subintervals of equal width Δx by the points $a = x_0, x_1, x_2, \ldots, x_i, \ldots, x_n = b$. Then think of slicing the solid into n slices of equal thickness Δx, as shown in Figure 6. If the slices are thin enough, each slice is very close to being a right circular cylinder. The formula for the volume of a right circular cylinder is $\pi r^2 h$, where r is the radius of the circular base and h is the height of the cylinder. As shown in Figure 7, the height of each slice is Δx. (The height is horizontal here, since the cylinder is on its side.) The radius of the circular base of each slice is $f(x_i)$. Thus, the volume of

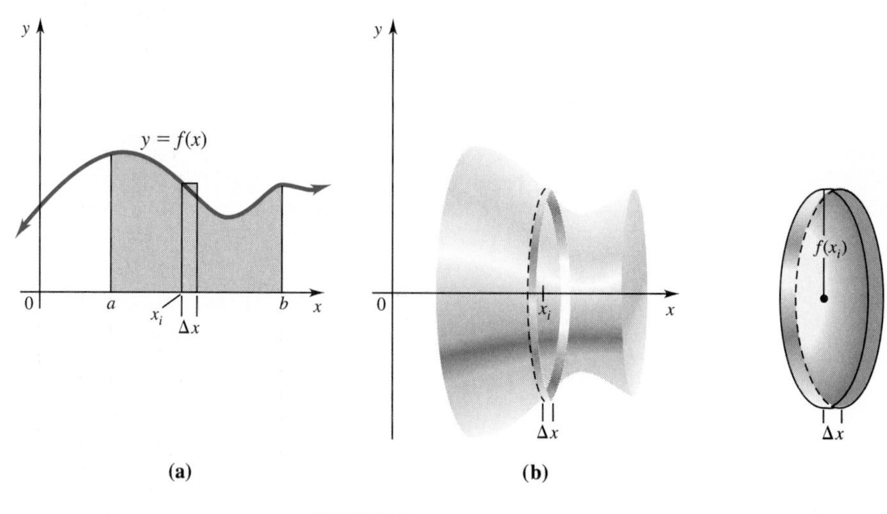

(a)

(b)

FIGURE 6

FIGURE 7

the slice is closely approximated by $\pi[f(x_i)]^2\ \Delta x$. The volume of the solid of revolution will be approximated by the sum of the volumes of the slices:

$$V \approx \sum_{i=1}^{n} \pi[f(x_i)]^2\ \Delta x.$$

By definition, the volume of the solid of revolution is the limit of this sum as the thickness of the slices approaches 0, or

$$V = \lim_{\Delta x \to 0} \sum_{i=1}^{n} \pi[f(x_i)]^2\ \Delta x.$$

This limit, like the one discussed earlier for area, is a definite integral.

VOLUME OF A SOLID OF REVOLUTION

If $f(x)$ is nonnegative and R is the region between $f(x)$ and the x-axis from $x = a$ to $x = b$, the volume of the solid formed by rotating R about the x-axis is given by

$$\mathbf{V = \lim_{\Delta x \to 0} \sum_{i=1}^{n} \pi[f(x_i)]^2\ \Delta x = \int_a^b \pi[f(x)]^2\ dx.}$$

The technique of summing disks to approximate volumes was originated by Johannes Kepler (1571–1630), a famous German astronomer who discovered three laws of planetary motion. He estimated volumes of wine casks used at his wedding by means of solids of revolution.

EXAMPLE 1

Find the volume of the solid of revolution formed by rotating about the x-axis the region bounded by $y = x + 1$, $y = 0$, $x = 1$, and $x = 4$. See Figure 8(a).

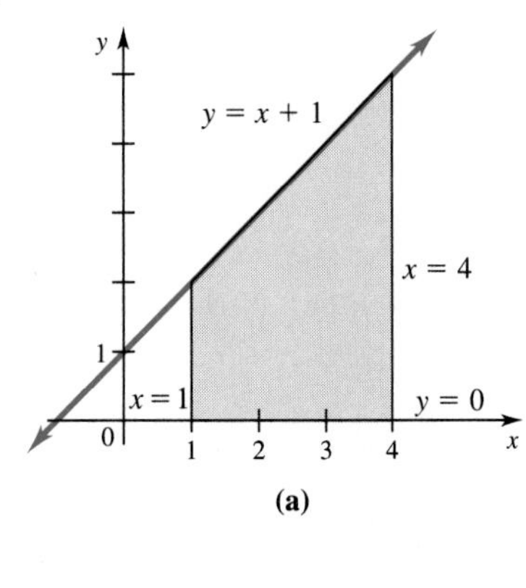

(a)

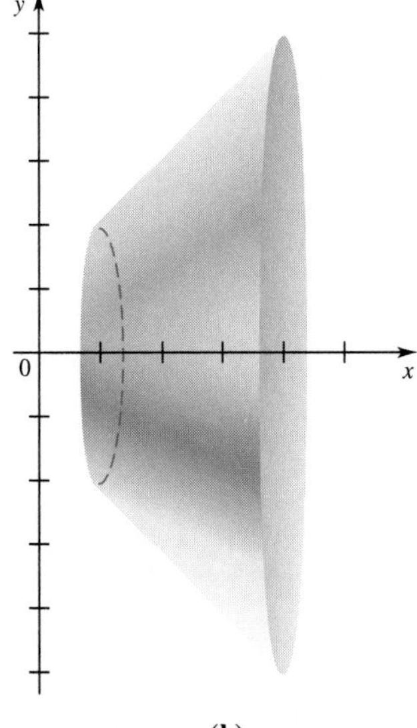

(b)

FIGURE 8

The solid is shown in Figure 8(b). Use the formula given above for the volume, with $a = 1$, $b = 4$, and $f(x) = x + 1$.

$$V = \int_1^4 \pi(x+1)^2\,dx = \pi\left(\frac{(x+1)^3}{3}\right)\Bigg|_1^4$$

$$= \frac{\pi}{3}(5^3 - 2^3) = \frac{117\pi}{3} = 39\pi$$

EXAMPLE 2 Find the volume of the solid of revolution formed by rotating about the x-axis the area bounded by $f(x) = 4 - x^2$ and the x-axis. See Figure 9(a).

The solid is shown in Figure 9(b). Find a and b from the x-intercepts. If $y = 0$, then $x = 2$ or $x = -2$, so that $a = -2$ and $b = 2$. The volume is

$$V = \int_{-2}^{2} \pi(4 - x^2)^2\,dx$$

$$= \int_{-2}^{2} \pi(16 - 8x^2 + x^4)\,dx$$

$$= \pi\left(16x - \frac{8x^3}{3} + \frac{x^5}{5}\right)\Bigg|_{-2}^{2} = \frac{512\pi}{15}.$$

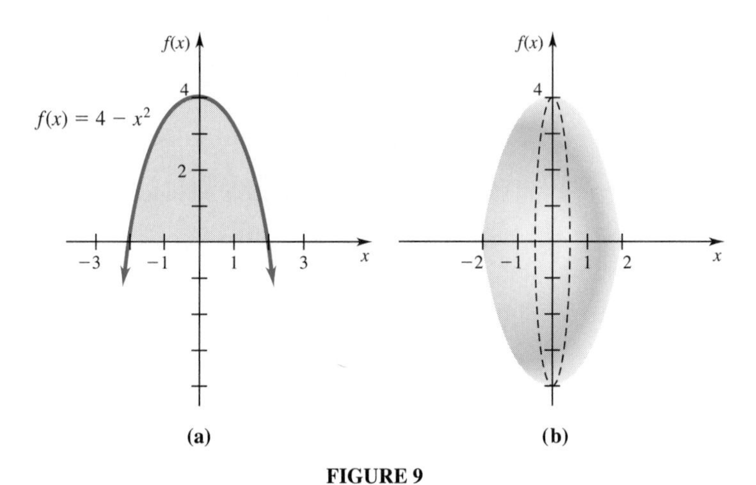

FIGURE 9

EXAMPLE 3 Find the volume of a right circular cone with height h and base radius r.

Figure 10(a) shows the required cone, while Figure 10(b) shows an area that could be rotated about the x-axis to get such a cone. See Figure 10(c). Here $y = f(x)$ is the equation (in slope-intercept form) of the line through $(0, r)$ and $(h, 0)$. The slope of this line is

$$\frac{0 - r}{h - 0} = -\frac{r}{h}.$$

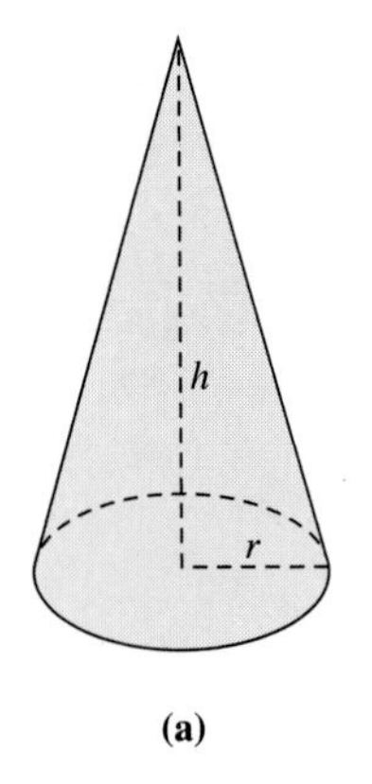

(a)

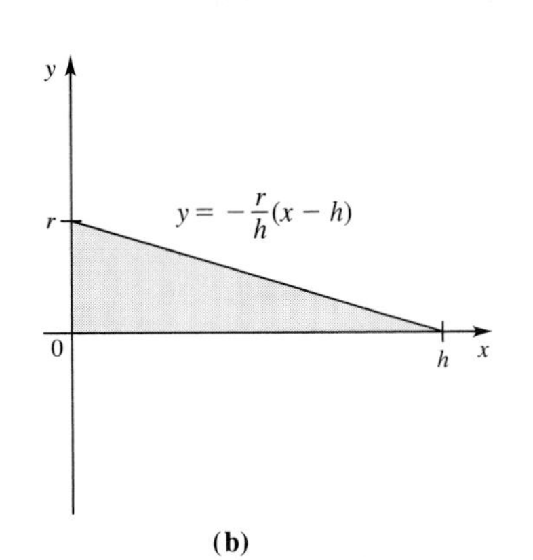

(b)

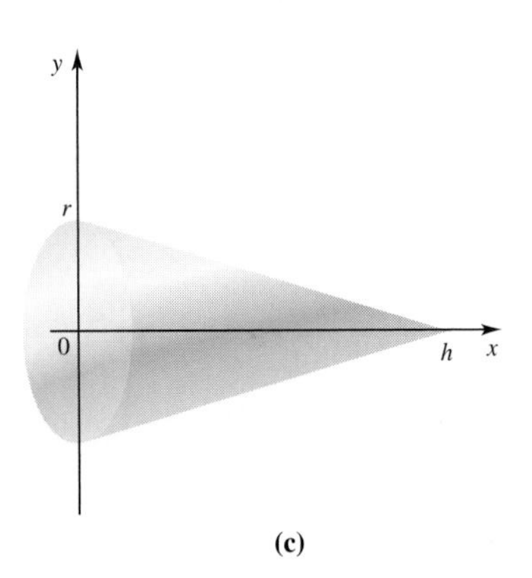

(c)

FIGURE 10

Using the point-slope formula with the point $(h, 0)$ gives

$$y - y_1 = m(x - x_1)$$
$$y = -\frac{r}{h}(x - h)$$

as the equation of the line. Then the volume is

$$V = \int_0^h \pi\left[-\frac{r}{h}(x - h)\right]^2 dx = \pi\int_0^h \frac{r^2}{h^2}(x - h)^2\, dx.$$

Since r and h are constants,

$$V = \frac{\pi r^2}{h^2}\int_0^h (x - h)^2\, dx.$$

Using substitution with $u = x - h$ and $du = dx$ gives

$$V = \frac{\pi r^2}{h^2}\left[\frac{(x - h)^3}{3}\right]\Bigg|_0^h = \frac{\pi r^2}{3h^2}[0 - (-h)^3] = \frac{\pi r^2 h}{3}.$$

This is the familiar formula for the volume of a right circular cone. ◀

Average Value of a Function The average of the n numbers $v_1, v_2, v_3, \ldots, v_i, \ldots, v_n$ is given by

$$\frac{v_1 + v_2 + v_3 + \cdots + v_n}{n} = \frac{\sum_{i=1}^{n} v_i}{n}.$$

The average value of a function f on $[a, b]$ can be defined in a similar manner; divide the interval $[a, b]$ into n subintervals, each of width Δx. Then choose an x-value, x_i, in each interval, and find $f(x_i)$. The average function value for the n subintervals and the given choices of x_i is

$$\frac{f(x_1) + f(x_2) + f(x_3) + \cdots + f(x_n)}{n} = \frac{\sum_{i=1}^{n} f(x_i)}{n}.$$

Since $(b - a)/n = \Delta x$, multiply the expression on the right side of the equation by $(b - a)/(b - a)$ and rearrange the expression to get

$$\frac{b-a}{b-a} \cdot \frac{\sum_{i=1}^{n} f(x_i)}{n} = \frac{b-a}{n} \cdot \frac{\sum_{i=1}^{n} f(x_i)}{b-a} = \Delta x \cdot \frac{\sum_{i=1}^{n} f(x_i)}{b-a} = \frac{1}{b-a} \sum_{i=1}^{n} f(x_i)\,\Delta x.$$

Now, take the limit as $n \to \infty$. If the limit exists, then

$$\lim_{n\to\infty} \frac{1}{b-a} \sum_{i=1}^{n} f(x_i)\,\Delta x = \frac{1}{b-a} \lim_{n\to\infty} \sum_{i=1}^{n} f(x_i)\,\Delta x = \frac{1}{b-a} \int_a^b f(x)\,dx.$$

The following definition summarizes this discussion.

AVERAGE VALUE OF A FUNCTION

The **average value** of a function f on the interval $[a, b]$ is

$$\frac{1}{b-a} \int_a^b f(x)\,dx,$$

provided the indicated definite integral exists.

The average value, sometimes denoted $\bar{y}$, can be thought of as the height of the rectangle with base $b - a$. See Figure 11. For $f(x) \geq 0$, this rectangle has area $\bar{y}(b - a)$, which equals the area under the graph of $f(x)$ from $x = a$ to $x = b$, so that

$$\bar{y}(b - a) = \int_a^b f(x)\,dx.$$

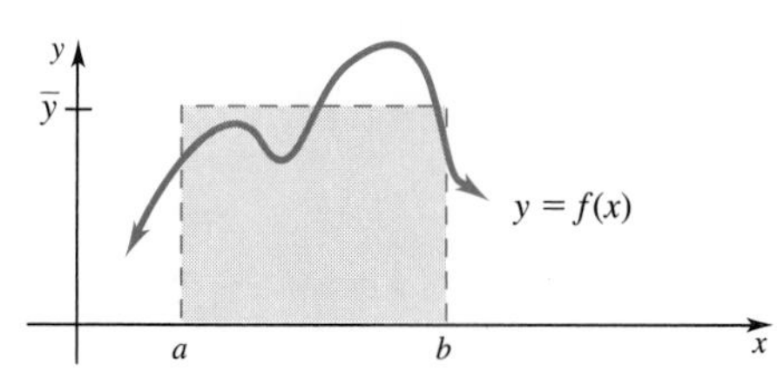

FIGURE 11

EXAMPLE 4 A stock analyst plots the price per share of a certain common stock as a function of time and finds that it can be approximated by the function

$$S(t) = 25 - 5e^{-.01t},$$

where t is the time (in years) since the stock was purchased. Find the average price of the stock over the first six years.

Use the formula for average value with $a = 0$ and $b = 6$. The average price is

$$\frac{1}{6-0}\int_0^6 (25 - 5e^{-.01t})\,dt = \frac{1}{6}\left(25t - \frac{5}{-.01}e^{-.01t}\right)\Bigg|_0^6$$
$$= \frac{1}{6}\left(25t + 500e^{-.01t}\right)\Bigg|_0^6$$
$$= \frac{1}{6}\left(150 + 500e^{-.06} - 500\right)$$
$$= 20.147,$$

or approximately \$20.15. ◀

7.3 Exercises

Find the volume of the solid of revolution formed by rotating about the x-axis each region bounded by the given curves.

1. $f(x) = x,\ y = 0,\ x = 0,\ x = 2$

2. $f(x) = 2x,\ y = 0,\ x = 0,\ x = 3$

3. $f(x) = 2x + 1,\ y = 0,\ x = 0,\ x = 4$

4. $f(x) = x - 4,\ y = 0,\ x = 4,\ x = 10$

5. $f(x) = \frac{1}{3}x + 2,\ y = 0,\ x = 1,\ x = 3$

6. $f(x) = \frac{1}{2}x + 4,\ y = 0,\ x = 0,\ x = 5$

7. $f(x) = \sqrt{x},\ y = 0,\ x = 1,\ x = 2$

8. $f(x) = \sqrt{x + 1},\ y = 0,\ x = 0,\ x = 3$

9. $f(x) = \sqrt{2x + 1},\ y = 0,\ x = 1,\ x = 4$

10. $f(x) = \sqrt{3x + 2},\ y = 0,\ x = 1,\ x = 2$

11. $f(x) = e^x,\ y = 0,\ x = 0,\ x = 2$

12. $f(x) = 2e^x,\ y = 0,\ x = -2,\ x = 1$

13. $f(x) = \frac{1}{\sqrt{x}},\ y = 0,\ x = 1,\ x = 4$

14. $f(x) = \frac{1}{\sqrt{x + 1}},\ y = 0,\ x = 0,\ x = 2$

15. $f(x) = x^2,\ y = 0,\ x = 1,\ x = 5$

16. $f(x) = \frac{x^2}{2},\ y = 0,\ x = 0,\ x = 4$

17. $f(x) = 1 - x^2,\ y = 0$

18. $f(x) = 2 - x^2,\ y = 0$

The function defined by $y = \sqrt{r^2 - x^2}$ has as its graph a semicircle of radius r with center at (0, 0) (see the figure). In Exercises 19–21, find the volume that results when each semicircle is rotated about the x-axis. (The result of Exercise 21 gives a formula for the volume of a sphere with radius r.)

19. $f(x) = \sqrt{1 - x^2}$

20. $f(x) = \sqrt{16 - x^2}$

21. $f(x) = \sqrt{r^2 - x^2}$

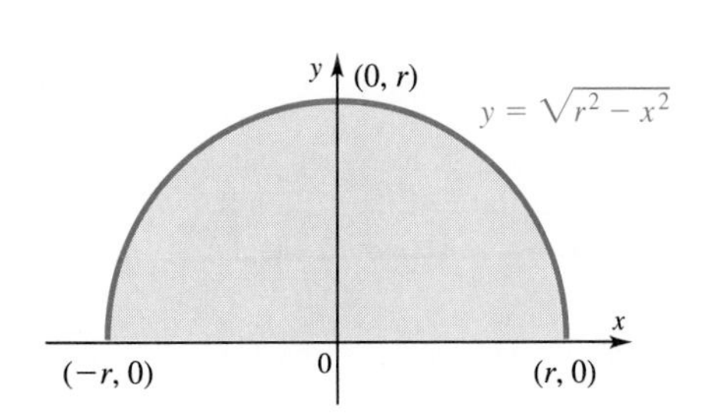

22. Find a formula for the volume of an ellipsoid. See Exercises 19–21 and the following figures.

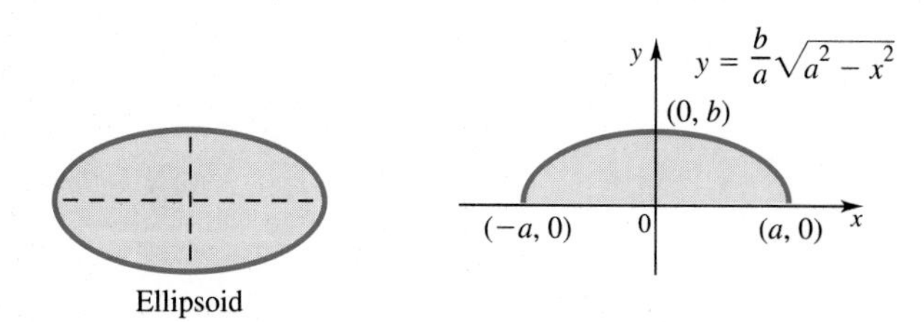

23. Use the methods of this section to find the volume of a cylinder with height h and radius r.

Find the average value of each function on the given interval.

24. $f(x) = 3 - 2x^2$; $[1, 9]$

25. $f(x) = x^2 - 2$; $[0, 5]$

26. $f(x) = (2x - 1)^{1/2}$; $[1, 13]$

27. $f(x) = \sqrt{x + 1}$; $[3, 8]$

28. $f(x) = e^{.1x}$; $[0, 10]$

29. $f(x) = e^{x/5}$; $[0, 5]$

30. $f(x) = x \ln x$; $[1, e]$

31. $f(x) = x^2e^{2x}$; $[0, 2]$

Applications

Life Sciences

32. Blood Flow The figure shows the blood flow in a small artery of the body. The flow of blood is *laminar* (in layers), with the velocity very low near the artery walls and highest in the center of the artery. In this model of blood flow, we calculate the total flow in the artery by thinking of the flow as being made up of many layers of concentric tubes sliding one on the other.

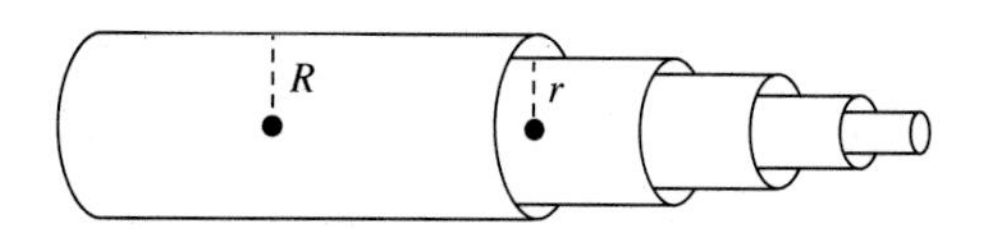

Suppose R is the radius of an artery and r is the distance from a given layer to the center. Then the velocity of blood in a given layer can be shown to equal

$$v(r) = k(R^2 - r^2),$$

where k is a numerical constant.

Since the area of a circle is $A = \pi r^2$, the change in the area of the cross section of one of the layers, corresponding to a small change in the radius, Δr, can be approximated by differentials. For $dr = \Delta r$, the differential of the area A is

$$dA = 2\pi r\, dr = 2\pi r\, \Delta r,$$

where Δr is the thickness of the layer. The total flow in the layer is defined to be the product of volume and cross-section area, or

$$F(r) = 2\pi rk(R^2 - r^2)\,\Delta r.$$

(a) Set up a definite integral to find the total flow in the artery.

(b) Evaluate this definite integral.

33. Drug Reaction The intensity of the reaction to a certain drug, in appropriate units, is given by

$$R(t) = te^{-.1t},$$

where t is time (in hours) after the drug is administered. Find the average intensity during each of the following hours.

(a) Second hour **(b)** Twelfth hour

(c) Twenty-fourth hour

Social Sciences

34. Production Rate Suppose the number of items a new worker on an assembly line produces daily after t days on the job is given by

$$I(t) = 35 \ln (t + 1).$$

Find the average number of items produced daily by this employee after the following numbers of days.

(a) 5 **(b)** 10 **(c)** 15

35. Typing Speed The function $W(t) = -6t^2 + 10t + 80$ describes a typist's speed (in words per minute) over a time interval [0, 5].

(a) Find $W(0)$.

(b) Find the maximum W value and the time t when it occurs.

(c) Find the average speed over [0, 5].

7.4 CONTINUOUS MONEY FLOW

Given the rate of change of a continuous money flow, how can we find its present value?

In an earlier chapter we looked at the concepts of present value and future value when a lump sum of money is deposited in an account and allowed to accumulate interest. In some situations, however, money flows into and out of an account almost continuously over a period of time. Examples include income in a store, bank receipts and payments, and highway tolls. Although the flow of money in such cases is not exactly continuous, it can be treated as though it were continuous, with useful results.

EXAMPLE 1 The income from a soda machine (in dollars per year) is growing exponentially. When the machine was first installed, it produced income that, if continued, would yield \$500 per year. By the end of the first year, it was producing income at a rate of \$510.10 per year. Find the total income produced by the machine during its first 3 years of operation.

Let t be the time in years since the installation of the machine. The assumption of exponential growth, coupled with the initial value of 500, implies that the rate of change of income is of the form

$$f(t) = 500e^{kt},$$

where k is some constant. To find k, use the value at the end of the first year.

$$f(1) = 500e^{k(1)} = 510.10$$

$$e^k = 1.0202 \qquad \text{Divide by 500.}$$

$$k = \ln 1.0202 \qquad \text{Take logarithms of both sides.}$$

$$\approx .02$$

We therefore have

$$f(t) = 500e^{.02t}.$$

Since the rate of change of income is given, the total income can be determined by using the definite integral.

$$\text{Total income} = \int_0^3 500e^{.02t}\,dt$$

$$= \frac{500}{.02}e^{.02t}\Big|_0^3$$

$$= 25{,}000\,e^{.02t}\Big|_0^3 = 25{,}000(e^{.06} - 1) = 1545.91$$

Thus, the soda machine will produce \$1545.91 total income in its first 3 years of operation. ◀

The money in Example 1 is not received as a one-time lump sum payment of \$1545.91. Instead, it comes in on a regular basis, perhaps daily, weekly, or monthly. In discussions of such problems it is usually assumed that the income is received continuously over a period of time, and so we speak of a "flow of money into income" and the rate of that flow.

Total Money Flow Let the continuous function $f(x)$ represent the rate of flow of money per unit of time. If x is in years and $f(x)$ is in dollars per year, the area under $f(x)$ between two points in time gives the total dollar flow over the given time interval.

The function $f(x) = 2000$, shown in Figure 12, represents a uniform rate of money flow of \$2000 per year. The graph of this money flow is a horizontal line and the total money flow over a specified time t is given by the rectangular area below the graph of $f(x)$ and above the x-axis between $x = 0$ and $x = t$. For example, the total money flow over $t = 5$ years would be $2000(5) = 10{,}000$, or \$10,000.

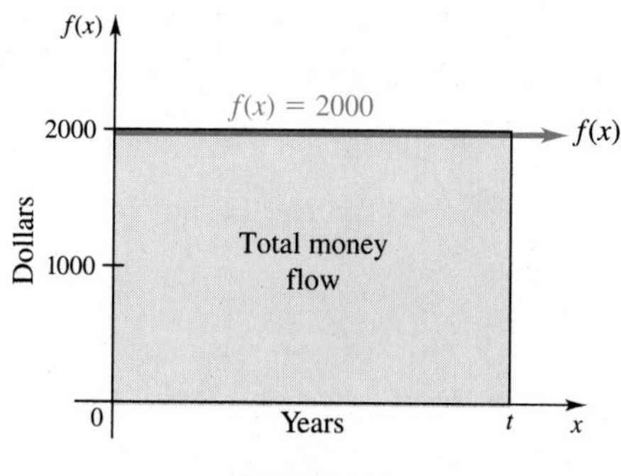

FIGURE 12

The area in the uniform rate example could be found by using an area formula from geometry. For a variable function like the function in Example 1, however, a definite integral is needed to find the total money flow over a specific time interval. For the function $f(x) = 2000e^{.08x}$, for example, the total money flow over a 5-year period would be given by

$$\int_0^5 2000e^{.08x}\,dx \approx 12{,}295.62,$$

or $12,295.62. See Figure 13.

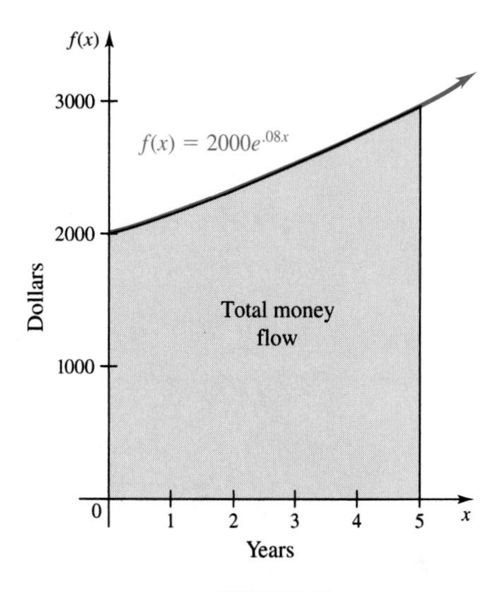

FIGURE 13

TOTAL MONEY FLOW

If $f(x)$ is the rate of money flow, then the total money flow over the time interval from $x = 0$ to $x = t$ is given by

$$\int_0^t f(x)\,dx.$$

It should be noted that this "total money flow" does not take into account the interest the money could earn after it is received. It is simply the sum of the periodic (continuous) income.

Present Value of Money Flow As mentioned earlier, an amount of money that can be deposited today to yield a given sum in the future is called the *present value* of this future sum. The future sum may be called the *future value* or *final amount*. To find the present value of a continuous money flow with interest compounded continuously, let $f(x)$ represent the rate of the continuous flow. In Figure 14, the time axis from 0 to x is divided into n subintervals, each of width Δx_i. The amount of money that flows during any interval of time is given by the area between the x-axis and the graph of $f(x)$ over the specified time interval. The area of each subinterval is approximated by the area of a rectangle with height $f(x_i)$, where x_i is the left endpoint of the ith subinterval. The area of each rectangle is $f(x_i)\,\Delta x_i$ which (approximately) gives the amount of money flow over that subinterval.

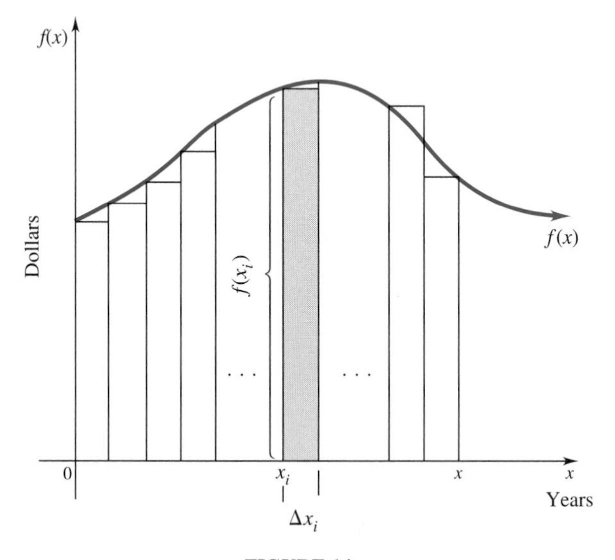

FIGURE 14

Earlier, we saw that the present value P of an amount A compounded continuously for t years at a rate of interest r is $P = Ae^{-rt}$. Letting x_i represent the time (instead of t), and replacing A with $f(x_i)\,\Delta x_i$, the present value of the money flow over the ith subinterval is approximately equal to

$$P_i = f(x_i)\,\Delta x_i e^{-rx_i}.$$

The total present value is approximately equal to the sum

$$\sum_{i=1}^{n} f(x_i)\,\Delta x_i e^{-rx_i}.$$

This approximation is improved as n increases; taking the limit of the sum as n increases without bound gives the present value

$$P = \lim_{n\to\infty} \sum_{i=1}^{n} f(x_i)\,\Delta x_i e^{-rx_i}.$$

This limit of a summation is given by the definite integral below.

PRESENT VALUE OF MONEY FLOW

If $f(x)$ is the rate of continuous money flow at an interest rate r for t years, then the present value is

$$P = \int_0^t f(x)e^{-rx}\,dx.$$

This present value of money flow is really the amount of money that would have to be deposited now in an account paying an annual interest rate r compounded continuously, so that the total sum in the account at time t would equal the continuous money flow with interest up to time t.

EXAMPLE 2 A company expects its rate of annual income during the next 3 years to be given by

$$f(x) = 75{,}000x, \quad 0 \le x \le 3.$$

What is the present value of this income over the 3-year period, assuming an annual interest rate of 8%?

Use the formula for present value given above, with $f(x) = 75{,}000x$, $t = 3$, and $r = .08$.

$$P = \int_0^3 75{,}000xe^{-.08x}\,dx = 75{,}000\int_0^3 xe^{-.08x}\,dx$$

Using integration by parts or column integration, verify that

$$\int xe^{-.08x}\,dx = -12.5xe^{-.08x} - 156.25e^{-.08x} + C.$$

Therefore,

$$\begin{aligned}75{,}000\int_0^3 xe^{-.08x}\,dx &= 75{,}000(-12.5xe^{-.08x} - 156.25e^{-.08x})\Big|_0^3\\ &= 75{,}000[-12.5(3)e^{-.08(3)} - 156.25e^{-.08(3)} - (0 - 156.25)]\\ &= 75{,}000(-29.498545 - 122.910603 + 156.25)\\ &= 288{,}064,\end{aligned}$$

or about \$288,000. Notice that the actual income over the 3-year period is given by

$$\text{Total money flow} = \int_0^3 75{,}000x\,dx = \frac{75{,}000x^2}{2}\Big|_0^3 = 337{,}500,$$

or \$337,500. This means that it would take a lump-sum deposit of \$288,064 today paying a continuously compounded interest rate of 8% over a 3-year period to equal the total cash flow of \$337,500 with interest. This approach is used as a basis for determining insurance claims involving income considerations. ◀

Accumulated Amount of Money Flow at Time t To find the amount of the money flow with interest at any time t, solve the formula $A = Pe^{rt}$ for P, yielding $P = Ae^{-rt}$. Set this equal to the value of P given in the formula for the present value of money flow.

$$Ae^{-rt} = \int_0^t f(x)e^{-rx}\,dx$$

Multiplying both sides by e^{rt} yields the following formula.

ACCUMULATED AMOUNT OF MONEY FLOW AT TIME *t*

If $f(x)$ is the rate of money flow at an interest rate r at time x, the amount of flow at time t is

$$\mathbf{A = e^{rt} \int_0^t f(x)e^{-rx}\,dx.}$$

Here, the amount of money A represents the accumulated value or final amount of the money flow *including* interest received on the money after it comes in.

It turns out that most money flows can be expressed as exponential or polynomial functions. When these are multiplied by e^{-rx}, the result is a function that can be integrated. The next example illustrates uniform flow, where $f(x)$ is a constant function. (This is a special case of the polynomial function.)

EXAMPLE 3 If money is flowing continuously at a constant rate of \$2000 per year over 5 years at 12% interest compounded continuously, find each of the following.

(a) The total amount of the flow over the 5-year period

The total amount is given by $\int_0^t f(x)\,dx$. Here $f(x) = 2000$ and $t = 5$.

$$\int_0^5 2000\,dx = 2000x\Big|_0^5 = 2000(5) = 10{,}000$$

The total money flow over the 5-year period is \$10,000.

(b) The accumulated amount, compounded continuously, at time $t = 5$

At $t = 5$ with $r = .12$, the amount is

In this example we use the following two rules for exponents.
1. $a^m \cdot a^n = a^{m+n}$
2. $a^0 = 1$

$$A = e^{rt}\int_0^t f(x)e^{-rx}\,dx = e^{(.12)5}\int_0^5 (2000)e^{-.12x}\,dx$$

$$= (e^{.6})(2000)\int_0^5 e^{-.12x}\,dx = (e^{.6})(2000)\left(\frac{1}{-.12}\right)\left(e^{-.12x}\Big|_0^5\right)$$

$$= \frac{2000e^{.6}}{-.12}(e^{-.6} - 1) = \frac{2000}{-.12}(1 - e^{.6}) \qquad (e^{.6})(e^{-.6}) = 1$$

$$= 13{,}701.98,$$

or \$13,701.98. The answer to part (a), \$10,000, was the amount of money flow over the 5-year period. The \$13,701.98 gives that amount with interest compounded continuously over the 5-year period.

(c) The present value of the amount with interest
Use $P = \int_0^t f(x)e^{-rx}\,dx$ with $f(x) = 2000$, $r = .12$, and $t = 5$.

$$P = \int_0^5 2000e^{-.12x}\,dx = 2000\left(\frac{e^{-.12x}}{-.12}\right)\Big|_0^5$$

$$= \frac{2000}{-.12}(e^{-.6} - 1) = 7519.81$$

The present value of the amount with interest in 5 years is \$7519.81, which can be checked by substituting \$13,701.98 for A in $A = Pe^{rt}$. The present value, P, could have been found by dividing the amount found in (b) by $e^{rt} = e^{.6}$. Check that this would give the same result. ◀

If the money flow is increasing or decreasing exponentially, then $f(x) = Ce^{kx}$, where C is a constant that represents the initial amount and k is the (nominal) continuous rate of change, which may be positive or negative.

EXAMPLE 4 A continuous money flow starts at \$1000 and increases exponentially at 2% per year.

(a) Find the accumulated amount at the end of 5 years at 10% interest compounded continuously.

Here $C = 1000$ and $k = .02$, so that $f(x) = 1000e^{.02x}$. Using $r = .10$ and $t = 5$,

$$A = e^{(.10)5}\int_0^5 1000e^{.02x}e^{-.10x}\,dx$$

$$= (e^{.5})(1000)\int_0^5 e^{-.08x}\,dx \qquad e^{.02x}\cdot e^{-.10x} = e^{-.08x}$$

$$= 1000e^{.5}\left(\frac{1}{-.08}e^{-.08x}\right)\Big|_0^5$$

$$= \frac{1000e^{.5}}{-.08}(e^{-.4} - 1) = \frac{1000}{-.08}(e^{.1} - e^{.5}) = 6794.38,$$

or \$6794.38.

(b) Find the present value at 5% interest compounded continuously.

Using $f(x) = 1000e^{.02x}$ with $r = .05$ and $t = 5$ in the present value expression,

$$P = \int_0^5 1000e^{.02x}e^{-.05x}\,dx$$

$$= 1000\int_0^5 e^{-.03x}\,dx = 1000\left(\frac{1}{-.03}e^{-.03x}\Big|_0^5\right)$$

$$= \frac{1000}{-.03}(e^{-.15} - 1) = 4643.07,$$

or \$4643.07. ◀

If the rate of change of the continuous money flow is given by the polynomial function $f(x) = a_n x^n + a_{n-1}x^{n-1} + \cdots + a_0$, the expressions for present value and accumulated amount can be integrated term by term using integration by parts or column integration.

The rate of change of a continuous flow of money is given by

$$f(x) = 1000x^2 + 100x.$$

Find the present value of this money flow at the end of 10 years at 10% compounded continuously.

Evaluate

$$P = \int_0^{10} (1000x^2 + 100x)e^{-.10x}\,dx.$$

Using integration by parts or column integration, verify that

$$\int (1000x^2 + 100x)e^{-.10x}\,dx = (-10{,}000x^2 - 1000x)e^{-.1x} - (200{,}000x + 10{,}000)e^{-.1x} - 2{,}000{,}000e^{-.1x} + C.$$

Thus,

$$P = (-10{,}000x^2 - 1000x)e^{-.1x} - (200{,}000x + 10{,}000)e^{-.1x} - 2{,}000{,}000e^{-.1x}\Big|_0^{10}$$

$$= (-1{,}000{,}000 - 10{,}000)e^{-1} - (2{,}000{,}000 + 10{,}000)e^{-1} - 2{,}000{,}000e^{-1} - (0 - 10{,}000 - 2{,}000{,}000)$$

$$= 163{,}245.21.$$

The present value is $163,245.21. ◀

7.4 Exercises

Each of the functions in Exercises 1–14 represents the rate of flow of money in dollars per year. Assume a 10-year period at 12% compounded continuously and find each of the following: **(a)** *the present value;* **(b)** *the accumulated amount at $t = 10$.*

1. $f(x) = 1000$

2. $f(x) = 300$

3. $f(x) = 500$

4. $f(x) = 2000$

5. $f(x) = 400e^{.03x}$

6. $f(x) = 800e^{.05x}$

7. $f(x) = 5000e^{-.01x}$

8. $f(x) = 1000e^{-.02x}$

9. $f(x) = .1x$

10. $f(x) = .5x$

11. $f(x) = .01x + 100$

12. $f(x) = .05x + 500$

13. $f(x) = 1000x - 100x^2$

14. $f(x) = 2000x - 150x^2$

Applications

Business and Economics

15. An investment is expected to yield a uniform continuous rate of change of income flow of $20,000 per year for 3 yr. Find the final amount at an interest rate of 14% compounded continuously.

16. A real estate investment is expected to produce a uniform continuous rate of change of income flow of $8000 per year for 6 yr. Find the present value at each of the following rates, compounded continuously.

(a) 12% **(b)** 10% **(c)** 15%

17. The rate of change of a continuous flow of money starts at \$5000 and decreases exponentially at 1% per year for 8 yr. Find the present value and final amount at an interest rate of 8% compounded continuously.

18. The rate of change of a continuous money flow starts at \$1000 and increases exponentially at 5% per year for 4 yr. Find the present value and final amount if interest earned is 11% compounded continuously.

19. A money market fund has a continuous flow of money that changes at a rate of $f(x) = 1500 - 60x^2$, reaching 0 in 5 yr. Find the present value of this flow if interest is 10% compounded continuously.

20. Find the amount of a continuous money flow in 3 yr if the rate of change is given by $f(x) = 1000 - x^2$ and if interest is 10% compounded continuously.

7.5 IMPROPER INTEGRALS

If we know the rate at which a pollutant is dumped into a stream, how can we compute the total amount released given that the rate of dumping is decreasing over time?

In this section we will learn how to answer such questions.

Sometimes it is useful to be able to integrate a function over an infinite period of time. For example, we might want to find the total amount of income generated by an apartment building into the indefinite future, or the total amount of pollution into a bay from a source that is continuing indefinitely. In this section we define integrals with one or more infinite limits of integration that can be used to solve such problems.

The graph in Figure 15(a) shows the area bounded by the curve $f(x) = x^{-3/2}$, the x-axis, and the vertical line $x = 1$. Think of the shaded region below the curve as extending indefinitely to the right. Does this shaded region have an area?

To see if the area of this region can be defined, introduce a vertical line at $x = b$, as shown in Figure 15(b). This vertical line gives a region with both upper and lower limits of integration. The area of this new region is given by the definite integral

$$\int_1^b x^{-3/2}\,dx.$$

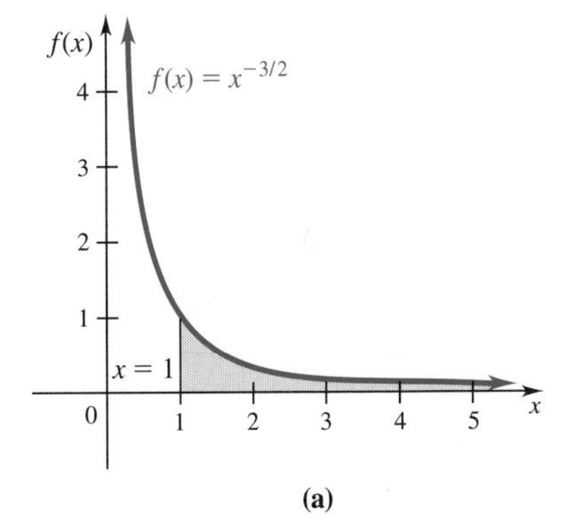

(a)

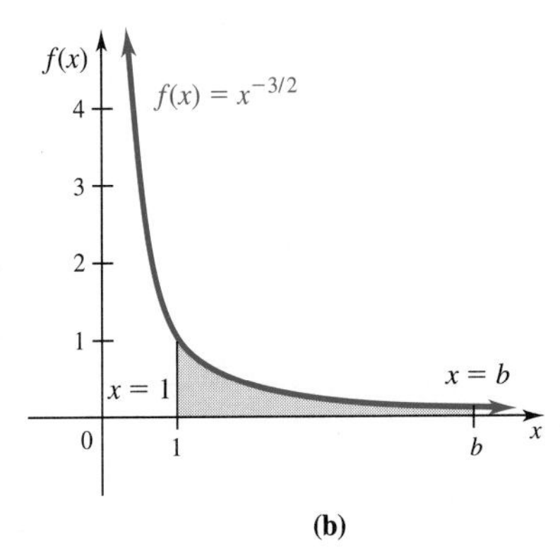

(b)

FIGURE 15

By the Fundamental Theorem of Calculus,

$$\begin{aligned}\int_1^b x^{-3/2}\,dx &= (-2x^{-1/2})\Big|_1^b \\ &= -2b^{-1/2} - (-2 \cdot 1^{-1/2}) \\ &= -2b^{-1/2} + 2 = 2 - \frac{2}{b^{1/2}}.\end{aligned}$$

In the section on limits at infinity and curve sketching, we saw that for any positive real number n,

$$\lim_{b\to\infty} \frac{1}{b^n} = \lim_{b\to-\infty} \frac{1}{b^n} = 0.$$

Suppose we now let the vertical line $x = b$ in Figure 15(b) move farther to the right. That is, suppose $b \to \infty$. The expression $-2/b^{1/2}$ would then approach 0, and

$$\lim_{b\to\infty}\left(2 - \frac{2}{b^{1/2}}\right) = 2 - 0 = 2.$$

This limit is defined to be the *area* of the region shown in Figure 15(a), so that

$$\int_1^{\infty} x^{-3/2}\,dx = 2.$$

An integral of the form

$$\int_a^{\infty} f(x)\,dx, \qquad \int_{-\infty}^{b} f(x)\,dx, \qquad \text{or} \qquad \int_{-\infty}^{\infty} f(x)\,dx$$

is called an improper integral. These **improper integrals** are defined as follows.

IMPROPER INTEGRALS

If f is continuous on the indicated interval and if the indicated limits exist, then

$$\int_a^{\infty} f(x)\,dx = \lim_{b\to\infty}\int_a^b f(x)\,dx,$$

$$\int_{-\infty}^{b} f(x)\,dx = \lim_{a\to-\infty}\int_a^b f(x)\,dx,$$

$$\int_{-\infty}^{\infty} f(x)\,dx = \int_{-\infty}^{c} f(x)\,dx + \int_c^{\infty} f(x)\,dx,$$

for real numbers a, b, and c, where c is arbitrarily chosen.

If the expressions on the right side exist, the integrals are **convergent;** otherwise, they are **divergent.**

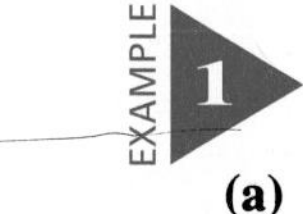

Find the following integrals.

(a) $\displaystyle\int_1^{\infty} \frac{dx}{x}$

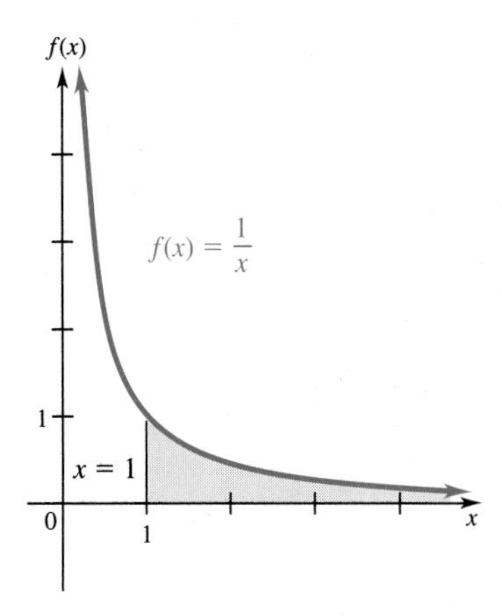

FIGURE 16

A graph of this region is shown in Figure 16. By the definition of an improper integral,

$$\int_1^{\infty} \frac{dx}{x} = \lim_{b\to\infty} \int_1^b \frac{dx}{x}.$$

Find $\int_1^b \frac{dx}{x}$ by the Fundamental Theorem of Calculus.

$$\int_1^b \frac{dx}{x} = \ln |x| \Big|_1^b = \ln |b| - \ln |1| = \ln |b| - 0 = \ln |b|$$

As $b \to \infty$, $\ln |b| \to \infty$, so

$$\lim_{b\to\infty} \ln |b| \text{ does not exist.}$$

Since the limit does not exist, $\int_1^{\infty} \frac{dx}{x}$ is divergent.

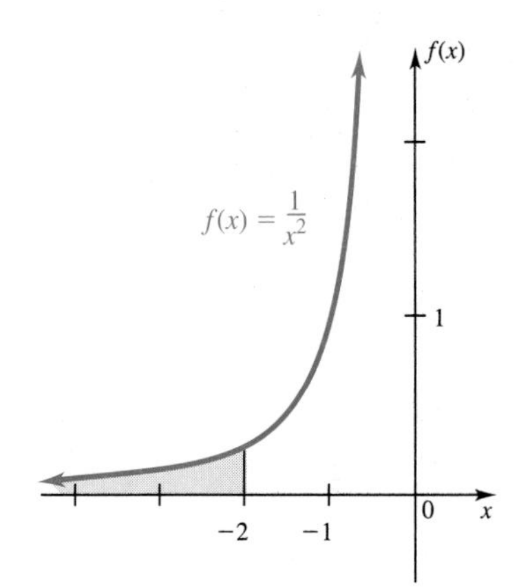

FIGURE 17

(b) $\int_{-\infty}^{-2} \frac{1}{x^2}\, dx = \lim_{a\to-\infty} \int_a^{-2} \frac{1}{x^2}\, dx = \lim_{a\to-\infty} \left(\frac{-1}{x}\right)\Big|_a^{-2}$

$$= \lim_{a\to-\infty} \left(\frac{1}{2} + \frac{1}{a}\right) = \frac{1}{2}$$

A graph of this region is shown in Figure 17. Since the limit exists, this integral converges. ◀

It may seem puzzling that the areas under the curves $f(x) = 1/x^{3/2}$ and $f(x) = 1/x^2$ are finite, while $f(x) = 1/x$ has an infinite amount of area. At first glance the graphs of these functions appear similar. The difference is that although all three functions get small as x becomes infinitely large, $f(x) = 1/x$ does not become small enough fast enough.

Find $\int_0^{\infty} 4e^{-3x}\, dx$.

By definition,

$$\int_0^{\infty} 4e^{-3x}\, dx = \lim_{b\to\infty} \int_0^b 4e^{-3x}\, dx = \lim_{b\to\infty} \left(-\frac{4}{3}e^{-3x}\right)\Big|_0^b$$

$$= \lim_{b\to\infty} \left[-\frac{4}{3}e^{-3b} - \left(-\frac{4}{3}e^0\right)\right]$$

$$= \lim_{b\to\infty} \left(\frac{-4}{3e^{3b}} + \frac{4}{3}\right) = 0 + \frac{4}{3} = \frac{4}{3}.$$

This integral converges. ◀

The following examples describe applications of improper integrals.

EXAMPLE 3 The rate at which a pollutant is being dumped into a small stream at time t is given by P_0e^{-kt}, where P_0 is the amount of pollutant initially released into the stream. Suppose $P_0 = 1000$ and $k = .06$. Find the total amount of the pollutant that will be released into the stream into the indefinite future.

Find

$$\int_0^{\infty} P_0e^{-kt}\,dt = \int_0^{\infty} 1000e^{-.06t}\,dt.$$

Work as above.

$$\int_0^{\infty} 1000e^{-.06t}\,dt = \lim_{b\to\infty} \int_0^{b} 1000e^{-.06t}\,dt$$

$$= \lim_{b\to\infty} \left(\frac{1000}{-.06}e^{-.06t}\right)\Bigg|_0^b$$

$$= \lim_{b\to\infty} \left(\frac{1000}{-.06e^{.06b}} - \frac{1000}{-.06}e^0\right) = \frac{-1000}{-.06} = 16{,}667$$

A total of 16,667 units of the pollutant eventually will be released. ◀

The *capital value* of an asset is sometimes defined as the present value of all future net earnings of the asset. If $R(t)$ gives the rate of change of the annual earnings produced by an asset at time t, as shown in Section 4, the **capital value** is

$$\int_0^{\infty} R(t)e^{-rt}\,dt,$$

where r is the annual rate of interest, compounded continuously, and the first payment is assumed to be made in one year.

EXAMPLE 4 Suppose income from a rental property is generated at the annual rate of \$4000 per year. Find the capital value of this property at an interest rate of 10% compounded continuously.

This is a continuous income stream with a rate of flow of 4000, so $R(t) = 4000$. Also, $r = .10$ or $.1$. The capital value is given by

$$\int_0^{\infty} 4000e^{-.1t}\,dt = \lim_{b\to\infty} \int_0^{b} 4000e^{-.1t}\,dt$$

$$= \lim_{b\to\infty} \left(\frac{4000}{-.1}e^{-.1t}\right)\Bigg|_0^b$$

$$= \lim_{b\to\infty} (-40{,}000e^{-.1b} + 40{,}000) = 40{,}000,$$

or \$40,000. ◀

7.5 Exercises

Determine whether the improper integrals in Exercises 1–26 converge or diverge, and find the value of each that converges.

1. $\int_2^{\infty} \frac{1}{x^2}\,dx$
2. $\int_5^{\infty} \frac{1}{x^2}\,dx$
3. $\int_1^{\infty} \frac{1}{\sqrt{x}}\,dx$
4. $\int_{16}^{\infty} \frac{-3}{\sqrt{x}}\,dx$
5. $\int_{-\infty}^{-1} \frac{2}{x^3}\,dx$
6. $\int_{-\infty}^{-4} \frac{3}{x^4}\,dx$
7. $\int_1^{\infty} \frac{1}{x^{1.0001}}\,dx$
8. $\int_1^{\infty} \frac{1}{x^{.999}}\,dx$
9. $\int_{-\infty}^{-1} x^{-2}\,dx$
10. $\int_{-\infty}^{-4} x^{-2}\,dx$
11. $\int_{-\infty}^{-1} x^{-8/3}\,dx$
12. $\int_{-\infty}^{-27} x^{-5/3}\,dx$
13. $\int_0^{\infty} 4e^{-4x}\,dx$
14. $\int_0^{\infty} 10e^{-10x}\,dx$
15. $\int_{-\infty}^{0} 4e^{x}\,dx$
16. $\int_{-\infty}^{0} 3e^{4x}\,dx$
17. $\int_{-\infty}^{-1} \ln|x|\,dx$
18. $\int_1^{\infty} \ln|x|\,dx$
19. $\int_0^{\infty} \frac{dx}{(x+1)^2}$
20. $\int_0^{\infty} \frac{dx}{(2x+1)^3}$
21. $\int_{-\infty}^{-1} \frac{2x-1}{x^2-x}\,dx$
22. $\int_1^{\infty} \frac{2x+3}{x^2+3x}\,dx$
23. $\int_2^{\infty} \frac{1}{x\ln x}\,dx$
24. $\int_2^{\infty} \frac{1}{x(\ln x)^2}\,dx$
25. $\int_0^{\infty} xe^{2x}\,dx$
26. $\int_{-\infty}^{0} xe^{3x}\,dx$

Use the table of integrals as necessary to evaluate the following integrals.

27. $\int_{-\infty}^{-1} \frac{2}{3x(2x-7)}\,dx$
28. $\int_1^{\infty} \frac{7}{2x(5x+1)}\,dx$
29. $\int_1^{\infty} \frac{4}{9x(x+1)^2}\,dx$
30. $\int_{-\infty}^{-5} \frac{5}{4x(x+2)^2}\,dx$
31. $\int_0^{\infty} \frac{1}{\sqrt{1+x^2}}\,dx$
32. $\int_9^{\infty} \frac{1}{\sqrt{x^2-4}}\,dx$

For Exercises 33–36, find the area between the graph of the given function and the x-axis over the given interval, if possible.

33. $f(x) = \frac{1}{x-1}$ for $(-\infty, 0]$
34. $f(x) = e^{-x}$ for $(-\infty, e]$
35. $f(x) = \frac{1}{(x-1)^2}$ for $(-\infty, 0]$
36. $f(x) = \frac{1}{(x-1)^3}$ for $(-\infty, 0]$.
37. Find $\int_{-\infty}^{\infty} xe^{-x^2}\,dx$.
38. Find $\int_{-\infty}^{\infty} \frac{x}{(1+x^2)^2}\,dx$.
39. Show that $\int_1^{\infty} 1/x^p\,dx$ converges if $p > 1$ and diverges if $p \le 1$.
40. Example 1(b) leads to a paradox. On the one hand, the unbounded region in that example has an area of 1/2, so theoretically it could be colored with ink. On the other hand, the boundary of that region is infinite, so it cannot be drawn with a finite amount of ink. This seems impossible, because coloring the region automatically colors the boundary. Explain why it is possible to color the region.

Applications

Business and Economics

Capital Value *Find the capital values of the properties in Exercises 41–44.*

41. A castle for which annual rent of $60,000 will be paid in perpetuity; the interest rate is 8% compounded continuously

42. A fort on a strategic peninsula in the North Sea; the annual rent is $500,000, paid in perpetuity; the interest rate is 6% compounded continuously

43. Find the capital value of an asset that generates $6000 yearly income if the interest rate is as follows.

(a) 8% compounded continuously

(b) 10% compounded continuously

44. An investment produces a perpetual stream of income with a flow rate of

$$K(t) = 1000\,e^{.02t}.$$

Find the capital value at an interest rate of 7% compounded continuously.

45. The Drucker family wants to establish an ongoing scholarship award at a college. Each year in June $3000 will be awarded, starting one year from now. What amount must the Druckers provide the college, assuming funds will be invested at 10% compounded continuously?

Life Sciences

46. Drug Reaction The rate of reaction to a drug is given by

$$r'(x) = 2x^2e^{-x},$$

where x is the number of hours since the drug was administered. Find the total reaction to the drug over all the time since it was administered, assuming this is an infinite time interval. (*Hint:* $\lim_{x\to\infty} x^k e^{-x} = 0$ for all real numbers k.)

Physical Sciences

Radioactive Waste *Radioactive waste is entering the atmosphere over an area at a decreasing rate. Use the improper integral*

$$\int_0^\infty Pe^{-kt}\,dt$$

with $P = 50$ to find the total amount of the waste that will enter the atmosphere for each of the following values of k.

47. $k = .06$

48. $k = .04$

Chapter Summary — Key Terms

7.1 **integration by parts**
column integration
tables of integrals

7.2 **numerical integration**
trapezoidal rule
Simpson's rule

7.3 **solid of revolution**
average value of a function

7.4 **total money flow**
present value of money flow

7.5 **improper integral**
convergent integral
divergent integral

Chapter 7 Review Exercises

1. Describe the type of integral for which integration by parts is useful.

2. Describe the type of integral for which tables of integrals are useful.

3. Describe the type of integral for which numerical integration is useful.

4. What is an improper integral? Explain why improper integrals must be treated in a special way.

Find each of the following integrals, using techniques from this or the previous chapter.

5. $\int x(8 - x)^{3/2}\,dx$

6. $\int \frac{3x}{\sqrt{x - 2}}\,dx$

7. $\int xe^x\,dx$

8. $\int (x + 2)e^{-3x}\,dx$

9. $\int \ln|2x + 3|\,dx$

10. $\int (x - 1)\ln|x|\,dx$

11. $\int \frac{x}{9 - 4x^2}\,dx$

12. $\int \frac{x}{\sqrt{25 + 9x^2}}\,dx$

13. $\int \frac{1}{9 - 4x^2}\,dx$

14. $\int \frac{1}{\sqrt{25 + 9x^2}}\,dx$

15. $\int_1^2 \frac{1}{x\sqrt{9 - x^2}}\,dx$

16. $\int_0^3 \sqrt{16 + x^2}\,dx$

17. $\int_1^e x^3 \ln x\,dx$

18. $\int_0^1 x^2 e^{x/2}\,dx$

19. Find the area between $y = (3 + x^2)e^{2x}$ and the x-axis from $x = 0$ to $x = 1$.

20. Find the area between $y = x^3(x^2 - 1)^{1/3}$ and the x-axis from $x = 1$ to $x = 3$.

In Exercises 21–23, use the trapezoidal rule with $n = 4$ to approximate the value of each answer. Use the table of integrals to find the exact value, and compare the two answers.

21. $\int_2^6 \frac{dx}{x^2 - 1}$

22. $\int_2^{10} \frac{x\,dx}{x - 1}$

23. $\int_1^5 \ln x\,dx$

In Exercises 24–26, use Simpson's rule with $n = 4$ to approximate the value of each integral. Compare your answers with the answers to Exercises 21–23.

24. $\int_2^6 \frac{dx}{x^2 - 1}$

25. $\int_2^{10} \frac{x\,dx}{x - 1}$

26. $\int_1^5 \ln x\,dx$

27. Find the area under the semicircle $y = \sqrt{1 - x^2}$ and above the x-axis by the trapezoidal rule, using $n = 6$.

28. Repeat Exercise 27 using Simpson's rule.

Find the volume of the solid of revolution formed by rotating each of the following bounded regions about the x-axis.

29. $f(x) = 2x - 1,\ y = 0,\ x = 3$

30. $f(x) = \sqrt{x - 2},\ y = 0,\ x = 11$

31. $f(x) = e^{-x},\ y = 0,\ x = -2,\ x = 1$

32. $f(x) = \frac{1}{\sqrt{x - 1}},\ y = 0,\ x = 2,\ x = 4$

33. $f(x) = 4 - x^2,\ y = 0,\ x = -1,\ x = 1$

34. $f(x) = \frac{x^2}{4},\ y = 0,\ x = 4$

35. A frustum is what remains of a cone when the top is cut off by a plane parallel to the base. Suppose a right circular frustum (that is, one formed from a right circular cone) has a base with radius r, a top with radius $r/2$, and a height h. (See the figure.) Find the volume of this frustum by rotating about the x-axis the region below the line segment from $(0, r)$ to $(h, r/2)$.

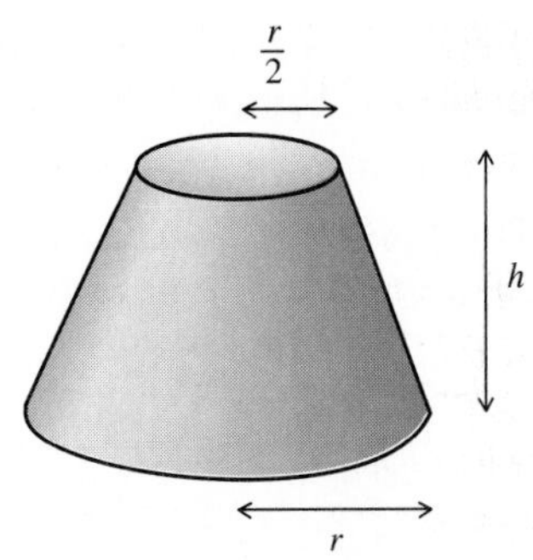

EXERCISE 35

36. How is the average value of a function found?

37. Find the average value of $f(x) = \sqrt{x + 1}$ over the interval $[0, 8]$.

38. Find the average value of $f(x) = x^2(x^3 + 1)^5$ over the interval $[0, 1]$.

Find the value of each integral that converges.

39. $\int_1^{\infty} x^{-1}\,dx$

40. $\int_{-\infty}^{-2} x^{-2}\,dx$

41. $\int_0^{\infty} \frac{dx}{(5x + 2)^2}$

42. $\int_1^{\infty} 6e^{-x}\,dx$

43. $\int_{-\infty}^{0} \frac{x}{x^2 + 3}\,dx$

44. $\int_{10}^{\infty} \ln(2x)\,dx$

Find the area between the graph of each function and the x-axis over the given interval, if possible.

45. $f(x) = \frac{3}{(x - 2)^2}$ for $(-\infty, 1]$

46. $f(x) = 3e^{-x}$ for $[0, \infty)$

47. How is the present value of money flow found? the accumulated amount of money flow?

Applications

Business and Economics

48. Total Revenue The rate of change of revenue from the sale of x toaster ovens is

$$R'(x) = x(x - 50)^{1/2}.$$

Find the total revenue from the sale of the 50th to the 75th ovens.

49. Total Sales Use the values of $f(x)$ given in the table to find total sales (in millions of dollars) for the given period by the trapezoidal rule, using $n = 6$.

Year, x	1	2	3	4	5	6	7
Rate of Growth of Sales, $f(x)$	.7	1.2	1.5	1.9	2.2	2.4	2.0

50. Total Sales Repeat Exercise 49, using Simpson's rule.

Present Value of Money Flow *Each of the functions in Exercises 51–54 represents the rate of flow of money (in dollars per year) over the given time period, compounded continuously at the given annual interest rate. Find the present value in each case.*

51. $f(x) = 5000$, 8 yr, 9%

52. $f(x) = 25{,}000$, 12 yr, 10%

53. $f(x) = 100e^{.02x}$, 5 yr, 11%

54. $f(x) = 30x$, 18 mo, 5%

Amount of Money at Time *t* *Assume that each of the following functions gives the rate of flow of money in dollars per year over the given period, with continuous compounding at the given rate. Find the accumulated amount at the end of the time period.*

55. $f(x) = 2000$, 5 mo, 1% per month

56. $f(x) = 500e^{-.03x}$, 8 yr, 10% per year

57. $f(x) = 20x$, 6 yr, 12% per year

58. $f(x) = 1000 + 200x$, 10 yr, 9% per year

59. Money Flow An investment scheme is expected to produce a continuous flow of money, starting at \$1000 and increasing exponentially at 5% a year for 7 yr. Find the present value at an interest rate of 11% compounded continuously.

60. Money Flow The proceeds from the sale of a building will yield a uniform continuous flow of \$10,000 a year for 10 yr. Find the final amount at an interest rate of 10.5% compounded continuously.

61. Capital Value Find the capital value of an office building for which annual rent of \$50,000 will be paid in perpetuity, if the interest rate is 9%.

Life Sciences

62. Drug Reaction The reaction rate to a new drug x hours after the drug is administered is

$$r'(x) = .5xe^{-x}.$$

Find the total reaction over the first 5 hours.

63. Oil Leak Pollution An oil leak from an uncapped well is polluting a bay at a rate of $f(x) = 100e^{-.05x}$ gallons per year. Use an improper integral to find the total amount of oil that will enter the bay, assuming the well is never capped.

Physical Sciences

64. Average Temperatures Suppose the temperature in a river at a point x meters downstream from a factory that is discharging hot water into the river is given by

$$T(x) = 400 - .25x^2.$$

Find the average temperature over each of the following intervals.

(a) [0, 10] **(b)** [10, 40] **(c)** [0, 40]

Connections

65. Probability In an earlier exercise, we mentioned the importance in probability and statistics of the density function for the standard normal distribution (the bell-shaped curve), which is given by

$$f(x) = \frac{1}{\sqrt{2\pi}}e^{-x^2/2}.$$

The integral of this function from $-\infty$ to ∞ has a particular significance. Parts (a)–(e) of this exercise, to be performed on a computer or a programmable calculator, are to help you guess the value of this integral. (Its exact value can only be determined by more advanced methods.)

(a) Use Simpson's rule with $n = 20$ to find the integral over the interval $[-1, 1]$.

(b) Use Simpson's rule with $n = 40$ to find the integral over the interval $[-2, 2]$.

(c) Use Simpson's rule with $n = 60$ to find the integral over the interval $[-3, 3]$.

(d) Use Simpson's rule with $n = 80$ to find the integral over the interval $[-4, 4]$.

(e) Based on your answers to parts (a)–(d), what do you think the value of the integral is over the interval $(-\infty, \infty)$?

Extended Application / How Much Does a Warranty Cost?*

This application uses some of the ideas of probability. The probability of an event is a number p, where $0 \leq p \leq 1$, such that p is equal to the number of ways that the event can happen divided by the total number of possible outcomes. For example, the probability of drawing a red card from a deck of 52 cards (of which 26 are red) is given by

$$P(\text{red card}) = \frac{26}{52} = \frac{1}{2}.$$

In the same way, the probability of drawing a black queen from a deck of 52 cards is

$$P(\text{black queen}) = \frac{2}{52} = \frac{1}{26}.$$

In this application we find the cost of a warranty program to a manufacturer. This cost depends on the quality of the products made, as we might expect. We use the following variables.

c = constant product price, per unit, including cost of warranty (We assume the price charged per unit is constant, since this price is likely to be fixed by competition.)

m = expected lifetime of the product

w = length of the warranty period

N = size of a production lot, such as a year's production

r = warranty cost per unit

$C(t)$ = pro rata customer rebate at time t

$P(t)$ = probability of product failure at any time t

$F(t)$ = number of failures occurring at time t

We assume that the warranty is of the pro rata customer rebate type, in which the customer is paid for the proportion of the warranty left. Hence, if the product has a warranty period of w, and fails at time t, then the product worked for the fraction t/w of the warranty period. The customer is reimbursed for the unused portion of the warranty, or the fraction

$$1 - \frac{t}{w}.$$

If we assume the product cost c originally, and if we use $C(t)$ to represent the customer rebate at time t, we have

$$C(t) = c\left(1 - \frac{t}{w}\right).$$

For many different types of products, it has been shown by experience that

$$P(t) = 1 - e^{-t/m}$$

provides a good estimate of the probability of product failure at time t. The total number of failures at time t is given by the product $P(t)$ and N, the total number of items per batch. If we use $F(t)$ to represent this total, we have

$$F(t) = N \cdot P(t) = N(1 - e^{-t/m}).$$

*Reprinted by permission of Warren W. Menke, "Determination of Warranty Reserves," *Management Sciences*, Vol. 15, No. 10, June 1969.

The total number of failures in some "tiny time interval" of width dt can be shown to be the derivative of $F(t)$,

$$F'(t) = \left(\frac{N}{m}\right)e^{-t/m},$$

while the cost for the failures in this "tiny time interval" is

$$C(t) \cdot F'(t) = c\left(1 - \frac{t}{w}\right)\left(\frac{N}{m}\right)e^{-t/m}.$$

The total cost for all failures during the warranty period is thus given by the definite integral

$$\int_0^w c\left(1 - \frac{t}{w}\right)\left(\frac{N}{m}\right)e^{-t/m}\,dt.$$

Using integration by parts, this definite integral can be shown to equal

$$Nc\left[-e^{-t/m} + \frac{t}{w} \cdot e^{-t/m} + \frac{m}{w}(e^{-t/m})\right]\Bigg|_0^w \quad \text{or}$$

$$Nc\left(1 - \frac{m}{w} + \frac{m}{w}e^{-w/m}\right).$$

This last quantity is the total warranty cost for all the units manufactured. Since there are N units per batch, the warranty cost per item is

$$r = \frac{1}{N}\left[Nc\left(1 - \frac{m}{w} + \frac{m}{w}e^{-w/m}\right)\right] = c\left(1 - \frac{m}{w} + \frac{m}{w}e^{-w/m}\right).$$

For example, suppose a product that costs \$100 has an expected life of 24 months, with a 12-month warranty. Then we have $c = \$100$, $m = 24$, $w = 12$, with r, the warranty cost per unit, given by

$$r = 100(1 - 2 + 2e^{-.5}) = 100[-1 + 2(.60653)]$$
$$= 100(.2131) = 21.31.$$

Exercises

1. Verify the evaluation of the definite integral using integration by parts.

Find r for each of the following.

2. $c = \$50$, $m = 48$ mo, $w = 24$ mo

3. $c = \$1000$, $m = 60$ mo, $w = 24$ mo

4. $c = \$1200$, $m = 30$ mo, $w = 30$ mo

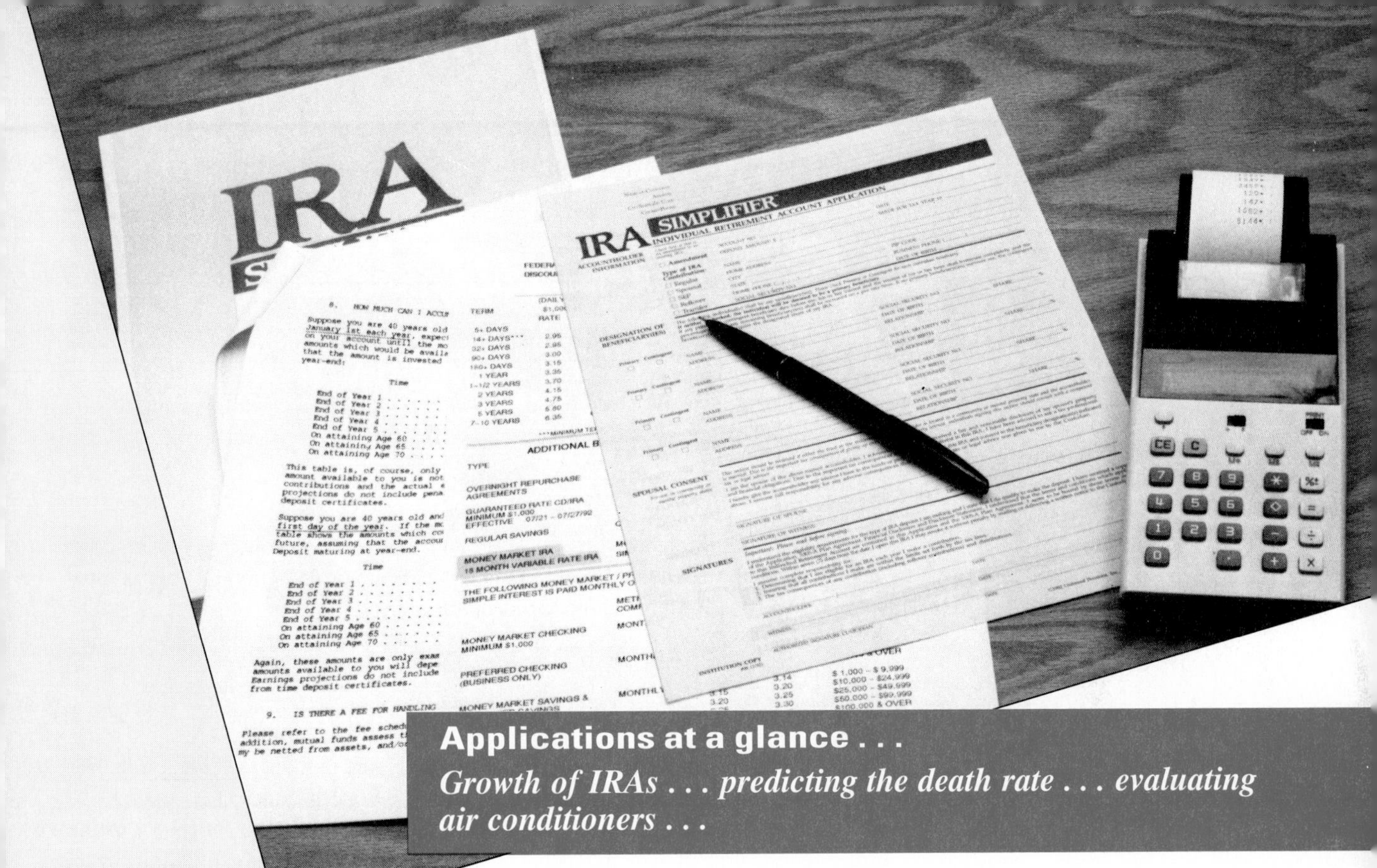

Applications at a glance . . .

Growth of IRAs . . . predicting the death rate . . . evaluating air conditioners . . .

CHAPTER 8

MULTIVARIABLE CALCULUS

(See page 416.)

Many of the ideas developed for functions of one variable also apply to functions of more than one variable. In particular, the fundamental idea of derivative generalizes in a very natural way to functions of more than one variable.

8.1 FUNCTIONS OF SEVERAL VARIABLES

How are the amounts of labor and capital needed to produce a certain number of items related?

We will be able to answer this question later in this section using a production function, which depends on the amounts of both labor and capital. That is, production is a function of two independent variables.

If a company produces x items at a cost of \$10 per item, for instance, then the total cost $C(x)$ of producing the items is given by

$$C(x) = 10x.$$

The cost is a function of one independent variable, the number of items produced. If the company produces two products, with x of one product at a cost of \$10 each, and y of another product at a cost of \$15 each, then the total cost to the firm is a function of *two* independent variables, x and y. By generalizing $f(x)$ notation, the total cost can be written as $C(x, y)$, where

$$C(x, y) = 10x + 15y.$$

When $x = 5$ and $y = 12$ the total cost is written $C(5, 12)$, with

$$C(5, 12) = 10 \cdot 5 + 15 \cdot 12 = 230.$$

A general definition follows.

FUNCTION OF TWO VARIABLES

$z = f(x, y)$ is a **function of two variables** if a unique value of z is obtained from each ordered pair of real numbers (x, y). The variables x and y are **independent variables,** and z is the **dependent variable.** The set of all ordered pairs of real numbers (x, y) such that $f(x, y)$ exists is the **domain** of f; the set of all values of $f(x, y)$ is the **range.**

Let $f(x, y) = 4x^2 + 2xy + 3/y$ and find each of the following.

(a) $f(-1, 3)$.

Replace x with -1 and y with 3.

$$f(-1, 3) = 4(-1)^2 + 2(-1)(3) + \frac{3}{3} = 4 - 6 + 1 = -1$$

(b) $f(2, 0)$

Because of the quotient $3/y$, it is not possible to replace y with 0, so $f(2, 0)$ is undefined. By inspection, we see that the domain of the function is (x, y) such that $y \neq 0$. ◀

EXAMPLE 2 Let x represent the number of milliliters (ml) of carbon dioxide released by the lungs in one minute. Let y be the change in the carbon dioxide content of the blood as it leaves the lungs (y is measured in ml of carbon dioxide per 100 ml of blood). The total output of blood from the heart in one minute (measured in ml) is given by C, where C is a function of x and y such that

$$C = C(x, y) = \frac{100x}{y}.$$

Find $C(320, 6)$.

Replace x with 320 and y with 6 to get

$$C(320, 6) = \frac{100(320)}{6}$$

$$\approx 5333 \text{ ml of blood per minute.}$$ ◀

Graph the following lines. Refer to Section 1.2 if you need review.

1. $2x + 3y = 6$
2. $x = 4$
3. $y = 2$

Answers:

1.

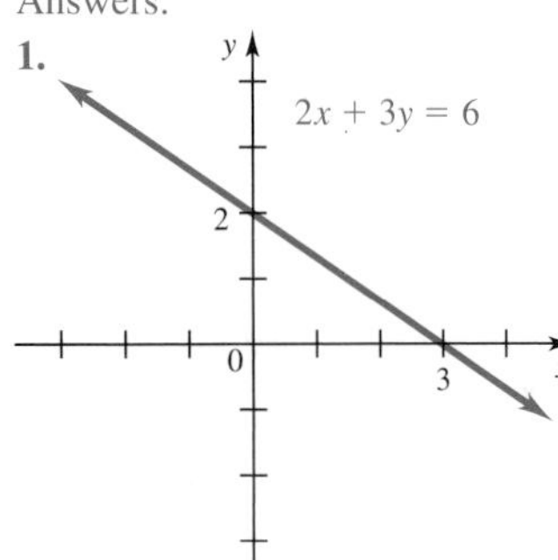

2.

y
x = 4
0
4
x

3.

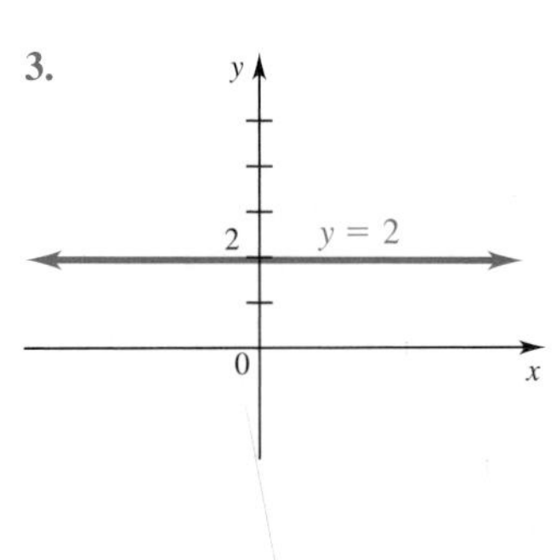

The definition given before Example 1 was for a function of two independent variables, but similar definitions could be given for functions of three, four, or more independent variables.

EXAMPLE 3 Let $f(x, y, z) = 4xz - 3x^2y + 2z^2$. Find $f(2, -3, 1)$.

Replace x with 2, y with -3, and z with 1.

$$f(2, -3, 1) = 4(2)(1) - 3(2)^2(-3) + 2(1)^2 = 8 + 36 + 2 = 46$$ ◀

Graphing Functions of Two Independent Variables Functions of one independent variable are graphed by using an x-axis and a y-axis to locate points in a plane. The plane determined by the x- and y-axes is called the *xy-plane.* A third axis is needed to graph functions of two independent variables—the z-axis, which goes through the origin in the xy-plane and is perpendicular to both the x-axis and the y-axis.

Figure 1 (on the next page) shows one possible way to draw the three axes. In Figure 1, the yz-plane is in the plane of the page, with the x-axis perpendicular to the plane of the page.

Just as we graphed ordered pairs earlier we can now graph **ordered triples** of the form (x, y, z). For example, to locate the point corresponding to the ordered triple $(2, -4, 3)$, start at the origin and go 2 units along the positive x-axis. Then go 4 units in a negative direction (to the left) parallel to the y-axis. Finally, go up 3 units parallel to the z-axis. The point representing $(2, -4, 3)$ is shown in Figure 1, together with several other points. The region of three-dimensional space where all coordinates are positive is called the **first octant.**

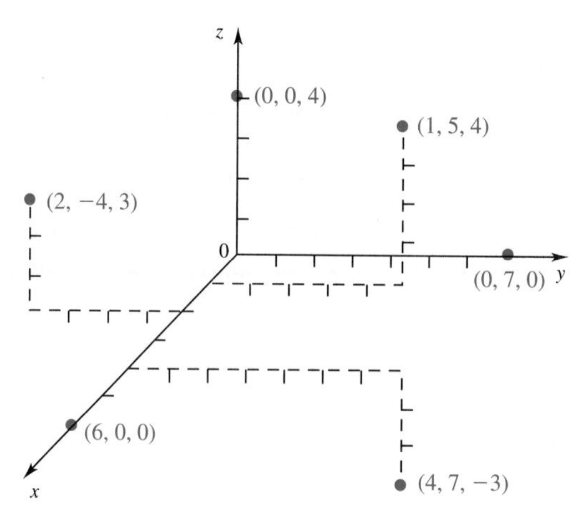

FIGURE 1

In Chapter 1 we saw that the graph of $ax + by = c$ (where a and b are not both 0) is a straight line. This result generalizes to three dimensions.

PLANE

The graph of

$$ax + by + cz = d$$

is a **plane** if a, b, and c are not all 0.

EXAMPLE 4

Graph $2x + y + z = 6$.

By the result above the graph of this equation is a plane. Earlier, we graphed straight lines by finding x- and y-intercepts. A similar idea helps in graphing a plane. To find the x-intercept, which is the point where the graph crosses the x-axis, let $y = 0$ and $z = 0$.

$$2x + 0 + 0 = 6$$
$$x = 3$$

The point (3, 0, 0) is on the graph. Letting $x = 0$ and $z = 0$ gives the point (0, 6, 0), while $x = 0$ and $y = 0$ lead to (0, 0, 6). The plane through these three points includes the triangular surface shown in Figure 2. This surface is the first-octant part of the plane that is the graph of $2x + y + z = 6$. ◀

EXAMPLE 5

Graph $x + z = 6$.

To find the x-intercept, let $z = 0$, giving (6, 0, 0). If $x = 0$, we get the point (0, 0, 6). Because there is no y in the equation $x + z = 6$, there can be no y-intercept. A plane that has no y-intercept is parallel to the y-axis. The first-octant portion of the graph of $x + z = 6$ is shown in Figure 3. ◀

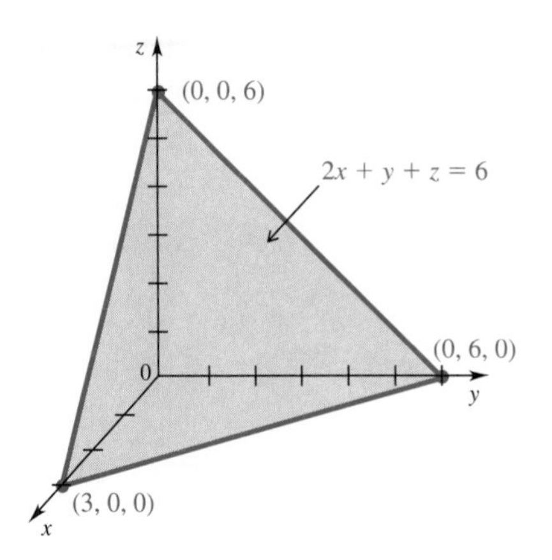

FIGURE 2

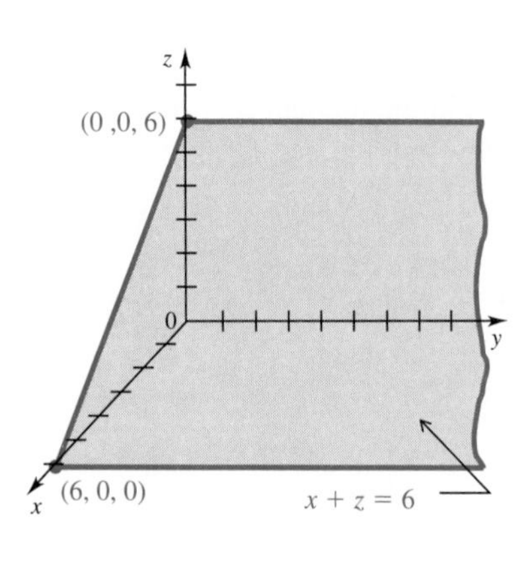

FIGURE 3

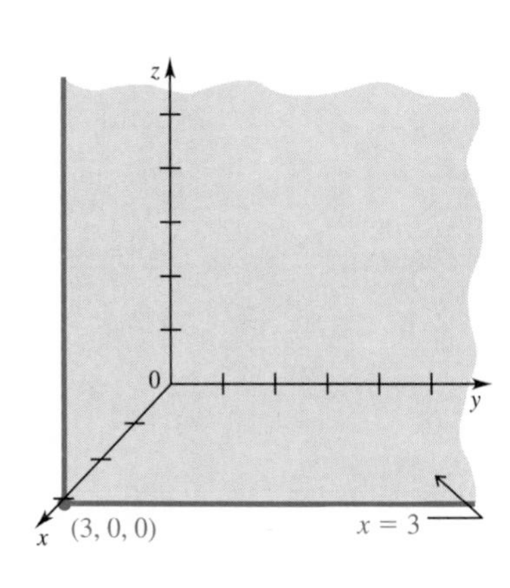

FIGURE 4

EXAMPLE 6 Graph each of the following functions in two variables.

(a) $x = 3$

This graph, which goes through (3, 0, 0), can have no y-intercept and no z-intercept. It is therefore a plane parallel to the y-axis and the z-axis and, therefore, to the yz-plane. The first-octant portion of the graph is shown in Figure 4.

(b) $y = 4$

This graph goes through (0, 4, 0) and is parallel to the xz-plane. The first-octant portion of the graph is shown in Figure 5.

(c) $z = 1$

The graph is a plane parallel to the xy-plane, passing through (0, 0, 1). Its first-octant portion is shown in Figure 6. ◀

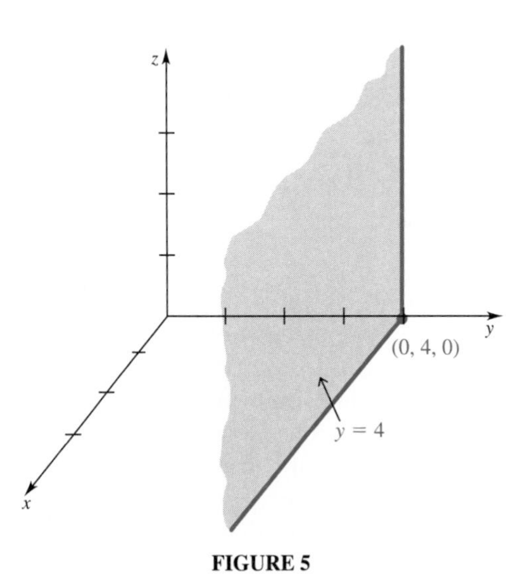

FIGURE 5

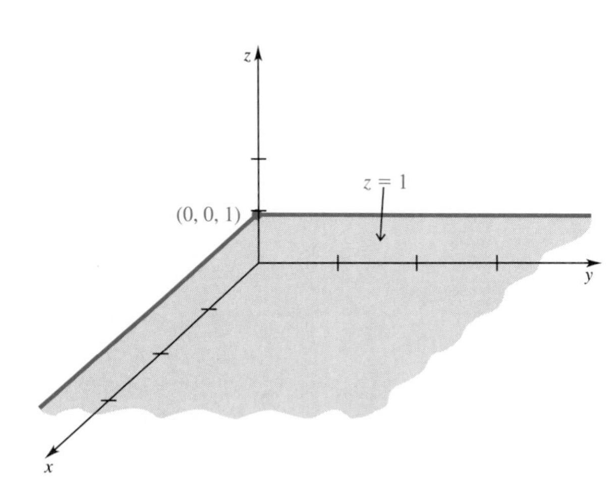

FIGURE 6

The graph of a function of one variable, $y = f(x)$, is a curve in the plane. If x_0 is in the domain of f, the point $(x_0, f(x_0))$ on the graph lies directly above or below the number x_0 on the x-axis, as shown in Figure 7.

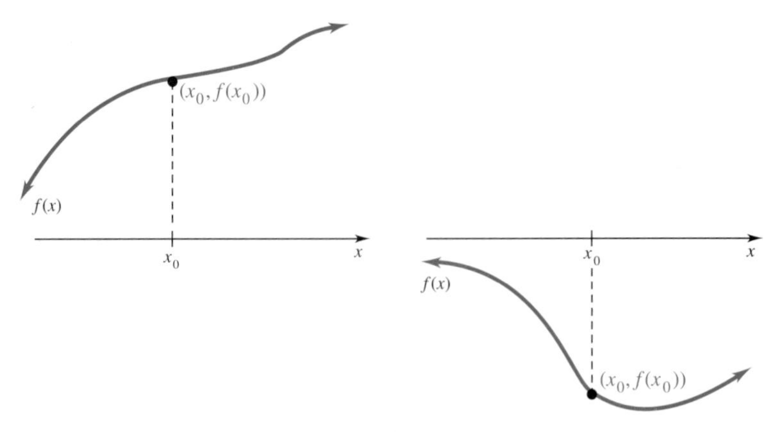

FIGURE 7

The graph of a function of two variables, $z = f(x, y)$, is a **surface** in three-dimensional space. If (x_0, y_0) is in the domain of f, the point $(x_0, y_0, f(x_0, y_0))$ lies directly above or below the point (x_0, y_0) in the xy-plane, as shown in Figure 8.

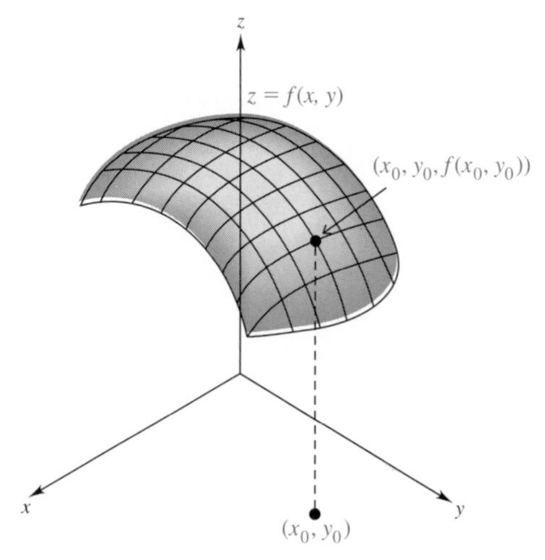

FIGURE 8

Although computer software is available for drawing the graphs of functions of two independent variables, you can often get a good picture of the graph without it by finding various **traces**—the curves that result when a surface is cut by a plane. The ***xy*-trace** is the intersection of the surface with the xy-plane. The ***yz*-trace** and ***xz*-trace** are defined similarly. You can also determine the intersection of the surface with planes parallel to the xy-plane. Such planes are of the form $z = k$, where k is a constant, and the curves that result when they cut the surface are called **level curves.**

EXAMPLE 7 Graph $z = x^2 + y^2$.

The yz-plane is the plane in which every point has a first coordinate of 0, so its equation is $x = 0$. When $x = 0$, the equation becomes $z = y^2$, which is the equation of a parabola in the yz-plane, as shown in Figure 9(a). Similarly, to find the intersection of the surface with the xz-plane (whose equation is $y = 0$), let $y = 0$ in the equation. It then becomes $z = x^2$, which is the equation of a parabola in the xz-plane, as shown in Figure 9(a). The xy-trace (the intersection of the surface with the plane $z = 0$) is the single point (0, 0, 0) because $x^2 + y^2$ is never negative, and is equal to 0 only when $x = 0$ and $y = 0$.

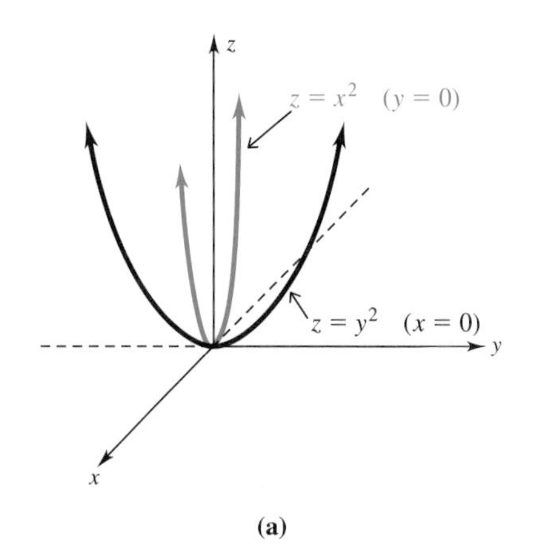

(a)

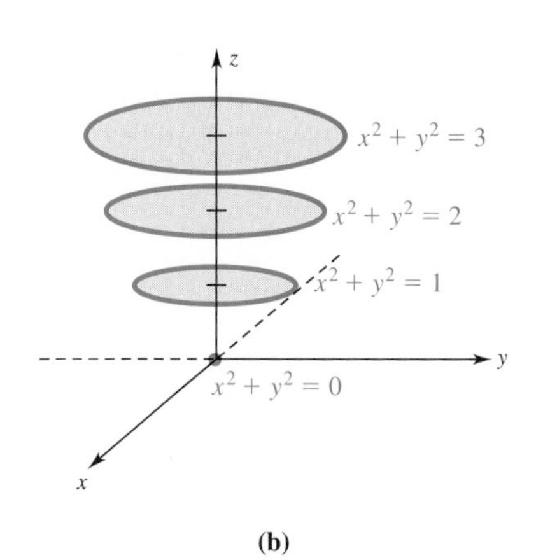

(b)

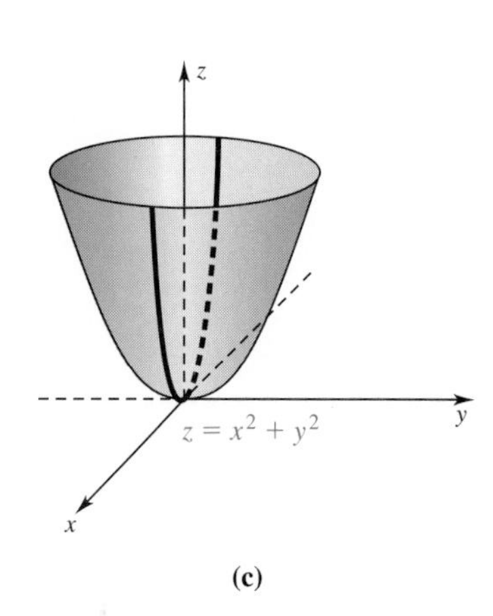

(c)

FIGURE 9

Next, we find the level curves by intersecting the surface with the planes $z = 1$, $z = 2$, $z = 3$, etc. (all of which are parallel to the xy-plane). In each case, the result is a circle:

$$x^2 + y^2 = 1, \quad x^2 + y^2 = 2, \quad x^2 + y^2 = 3,$$

and so on, as shown in Figure 9(b). Drawing the traces and level curves on the same set of axes suggests that the graph of $z = x^2 + y^2$ is the bowl-shaped figure, called a **paraboloid,** that is shown in Figure 9(c). ◀

One application of level curves in economics occurs with production functions. A **production function** $z = f(x, y)$ is a function that gives the quantity z of an item produced as a function of x and y, where x is the amount of labor and y is the amount of capital (in appropriate units) needed to produce z units. If the production function has the special form $z = P(x, y) = Ax^a y^{1-a}$, where A is a constant and $0 < a < 1$, the function is called a **Cobb-Douglas production function.**

EXAMPLE 8 Find the level curve at a production of 100 items for the Cobb-Douglas production function $z = x^{2/3}y^{1/3}$.

Let $z = 100$ to get

$$100 = x^{2/3}y^{1/3}$$

$$\frac{100}{x^{2/3}} = y^{1/3}.$$

Now cube both sides to express y as a function of x.

$$y = \frac{100^3}{x^2} = \frac{1{,}000{,}000}{x^2}$$

The level curve of height 100 found in Example 8 is shown graphed in three dimensions in Figure 10(a) and on the familiar xy-plane in Figure 10(b). The points of the graph correspond to those values of x and y that lead to production of 100 items.

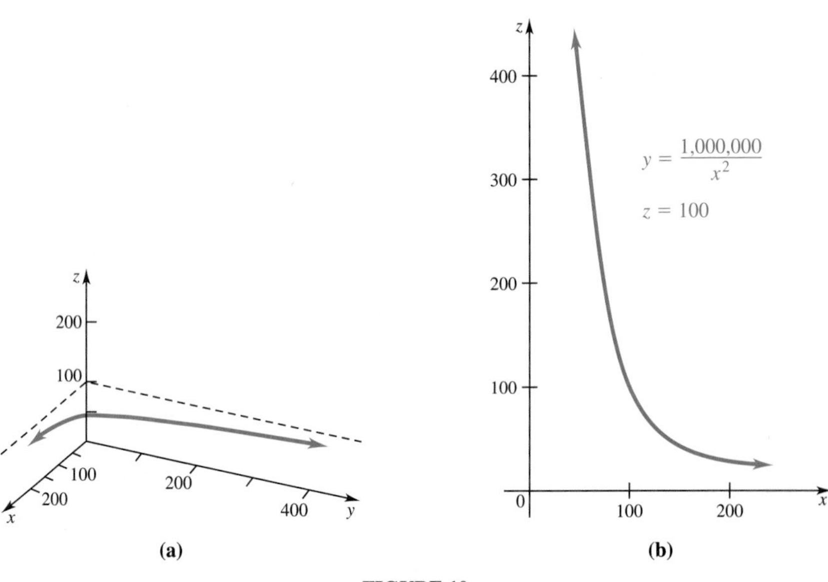

FIGURE 10

The curve in Figure 10 is called an *isoquant*, for *iso* (equal) and *quant* (amount). In Example 8, the "amounts" all "equal" 100.

Because of the difficulty of drawing the graphs of more complicated functions, we now merely list some common equations and their graphs. These graphs were drawn by computer, a very useful method of depicting three-dimensional surfaces.

Paraboloid, $z = x^2 + y^2$

xy-trace: point
yz-trace: parabola
xz-trace: parabola

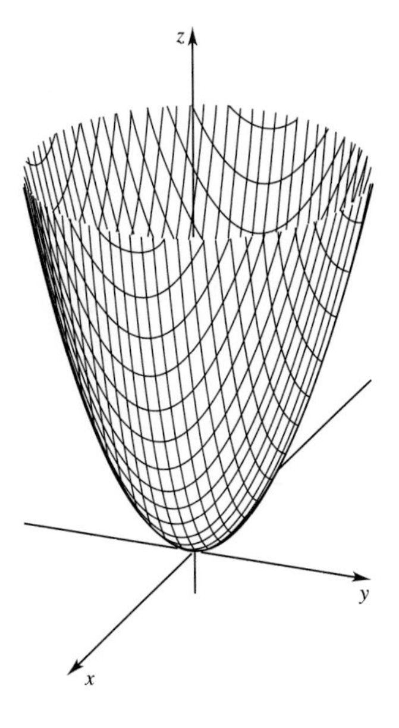

Ellipsoid, $\frac{x^2}{a^2} + \frac{y^2}{b^2} + \frac{z^2}{c^2} = 1$

xy-trace: ellipse
yz-trace: ellipse
xz-trace: ellipse

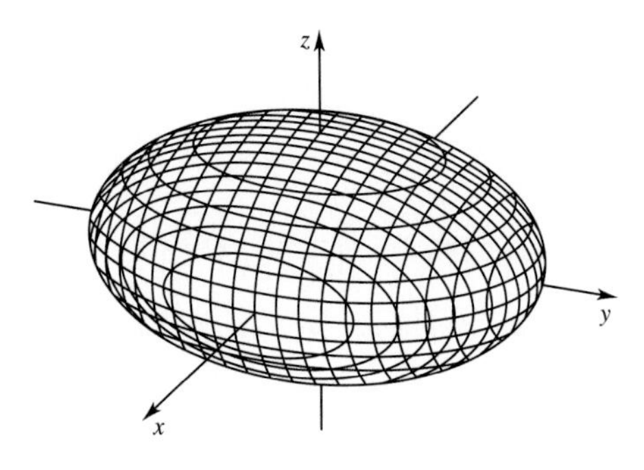

Hyperbolic Paraboloid, $x^2 - y^2 = z$
(sometimes called a **saddle**)

xy-trace: two intersecting lines
yz-trace: parabola
xz-trace: parabola

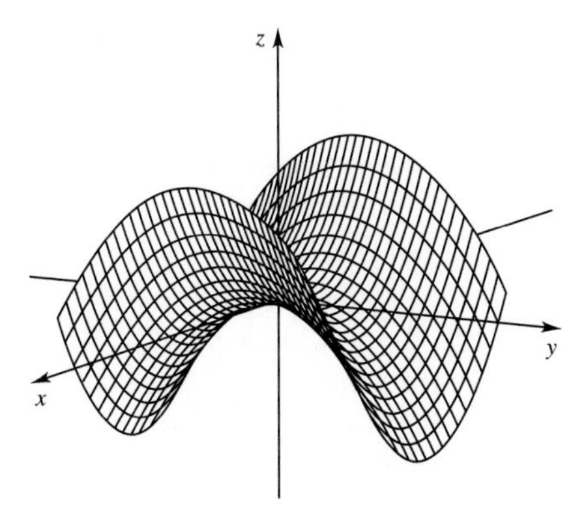

Hyperboloid of Two Sheets,
$-x^2 - y^2 + z^2 = 1$

xy-trace: none
yz-trace: hyperbola
xz-trace: hyperbola

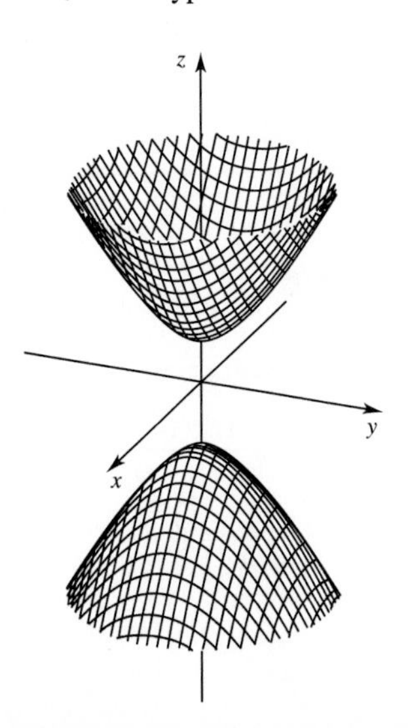

8.1 Exercises

1. Let $f(x, y) = 4x + 5y + 3$. Find the following.
   (a) $f(2, -1)$ (b) $f(-4, 1)$ (c) $f(-2, -3)$ (d) $f(0, 8)$
2. Let $g(x, y) = -x^2 - 4xy + y^3$. Find the following.
   (a) $g(-2, 4)$ (b) $g(-1, -2)$ (c) $g(-2, 3)$ (d) $g(5, 1)$
3. Let $h(x, y) = \sqrt{x^2 + 2y^2}$. Find the following.
   (a) $h(5, 3)$ (b) $h(2, 4)$ (c) $h(-1, -3)$ (d) $h(-3, -1)$
4. Let $f(x, y) = \dfrac{\sqrt{9x + 5y}}{\log x}$. Find the following.
   (a) $f(10, 2)$ (b) $f(100, 1)$ (c) $f(1000, 0)$ (d) $f\left(\frac{1}{10}, 5\right)$

Graph the first-octant portion of each plane.

5. $x + y + z = 6$
6. $x + y + z = 12$
7. $2x + 3y + 4z = 12$
8. $4x + 2y + 3z = 24$
9. $x + y = 4$
10. $y + z = 5$
11. $x = 2$
12. $z = 3$

Graph the level curves in the first octant at heights of $z = 0$, $z = 2$, and $z = 4$ for the following functions.

13. $3x + 2y + z = 18$
14. $x + 3y + 2z = 8$
15. $y^2 - x = -z$
16. $2y - \dfrac{x^2}{3} = z$

Applications

Business and Economics

17. **Production** Production of a precision camera is given by
$$P(x, y) = 100\left(\frac{3}{5}x^{-2/5} + \frac{2}{5}y^{-2/5}\right)^{-5},$$
where x is the amount of labor in work hours and y is the amount of capital. Find the following.
   (a) What is the production when 32 work hours and 1 unit of capital are provided?
   (b) Find the production when 1 work hour and 32 units of capital are provided.
   (c) If 32 work hours and 243 units of capital are used, what is the production output?

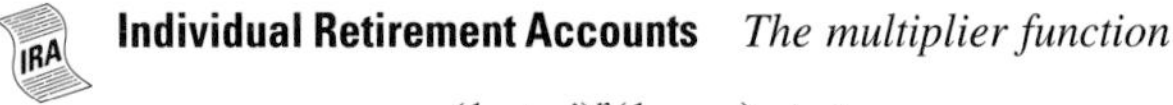

Individual Retirement Accounts *The multiplier function*
$$M = \frac{(1 + i)^n(1 - t) + t}{[1 + (1 - t)i]^n}$$
compares the growth of an Individual Retirement Account (IRA) with the growth of the same deposit in a regular savings account. The function M depends on the three variables n, i, and t, where n represents the number of years an amount is left at interest, i represents the interest rate in both types of accounts and t represents the income tax rate. Values of $M > 1$ indicate that the IRA grows faster than the savings account. Let $M = f(n, i, t)$ and find the following.

18. Find the multiplier when funds are left for 25 years at 5% interest and the income tax rate is 28%. Which account grows faster?
19. What is the multiplier when money is invested for 25 years at 6% interest and the income tax rate is 33%? Which account grows faster?

Production *Find the level curve at a production of 500 for each of the production functions in Exercises 20–21. Graph each function on the xy-plane.*

20. The production function z for the United States was once estimated as $z = x^{.7}y^{.3}$, where x stands for the amount of labor and y stands for the amount of capital.

21. If x represents the amount of labor and y the amount of capital, a production function for Canada is approximately $z = x^{.4}y^{.6}$.

22. Production For the function in Exercise 20, what is the effect on z of doubling x? Of doubling y? Of doubling both?

23. Cost If labor (x) costs \$200 per unit, materials (y) cost \$100 per unit, and capital (z) costs \$50 per unit, write a function for total cost.

Life Sciences

24. Oxygen Consumption The oxygen consumption of a well-insulated mammal that is not sweating is approximated by

$$m = \frac{2.5(T - F)}{w^{.67}},$$

where T is the internal body temperature of the animal (in °C), F is the temperature of the outside of the animal's fur (in °C), and w is the animal's weight in kilograms.* Find m for the following data.

(a) Internal body temperature = 38° C; outside temperature = 6° C; weight = 32 kg

(b) Internal body temperature = 40° C; outside temperature = 20° C; weight = 43 kg

25. Body Surface Area The surface area of a human (in square meters) is approximated by

$$A = .202W^{.425}H^{.725},$$

where W is the weight of the person in kilograms and H is the height in meters.* Find A for the following data.

(a) Weight, 72 kg; height, 1.78 m

(b) Weight, 65 kg; height, 1.40 m

(c) Weight, 70 kg; height, 1.60 m

(d) Using your weight and height, find your own surface area.

General Interest

26. Postage Rates Extra postage is charged for parcels sent by U.S. mail that are more than 84 inches in length and girth combined. (Girth is the distance around the parcel perpendicular to its length. See the figure.) Express the combined length and girth as a function of L, W, and H.

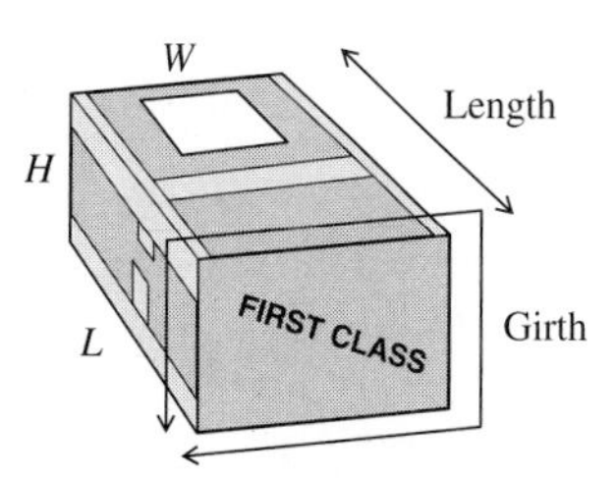

27. Required Material Refer to the figure for Exercise 26. Assume L, W, and H are in feet. Write a function in terms of L, W, and H that gives the total material required to build the box.

28. Elliptical Templates The holes cut in a roof for vent pipes require elliptical templates. A formula for determining the length of the major axis of the ellipse is $L = f(H, D) = \sqrt{H^2 + D^2}$, where D is the (outside) diameter of the pipe and H is the "rise" of the roof per D units of "run"; that is, the slope of the roof is H/D. (See the figure.) The width of the ellipse (minor axis) equals D. Find the length and width of the ellipse required to produce a hole for a vent pipe with a diameter of 3.75 inches in roofs with the following slopes.

(a) 3/4 **(b)** 2/5

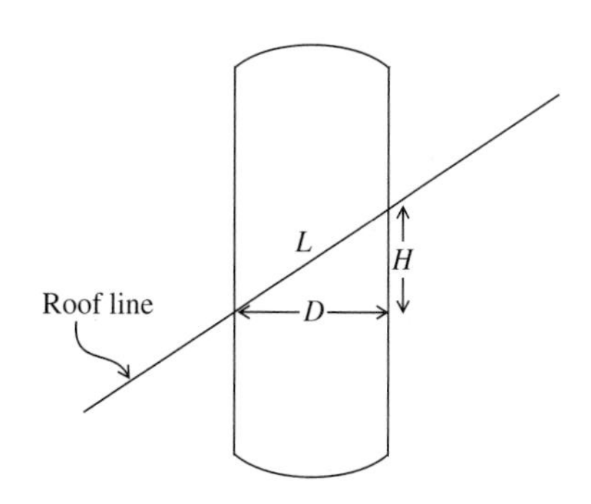

*From Duane J. Clow and N. Scott Urquhart, *Mathematics in Biology*. Copyright © 1974 by W. W. Norton & Company, Inc. Used by permission.

Match each equation in Exercises 29–34 with its graph in (a)–(f) below.

29. $z = x^2 + y^2$

30. $z^2 - y^2 - x^2 = 1$

31. $x^2 - y^2 = z$

32. $z = y^2 - x^2$

33. $\frac{x^2}{16} + \frac{y^2}{25} + \frac{z^2}{4} = 1$

34. $z = 5(x^2 + y^2)^{-1/2}$

(a)

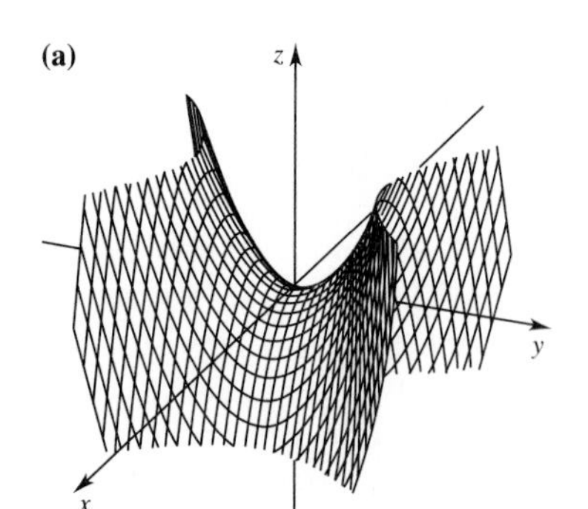

(b)

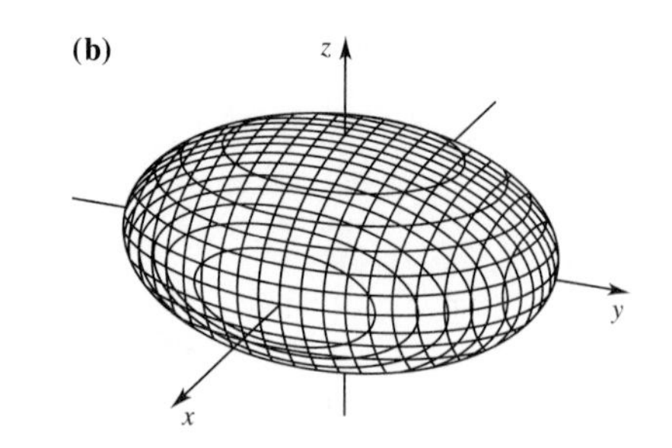

(c)

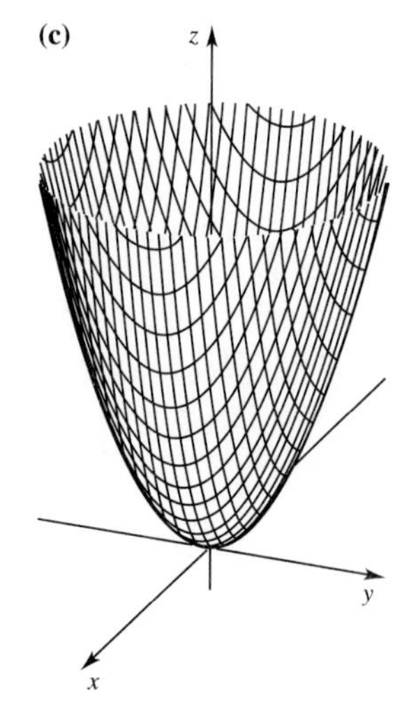

(d)

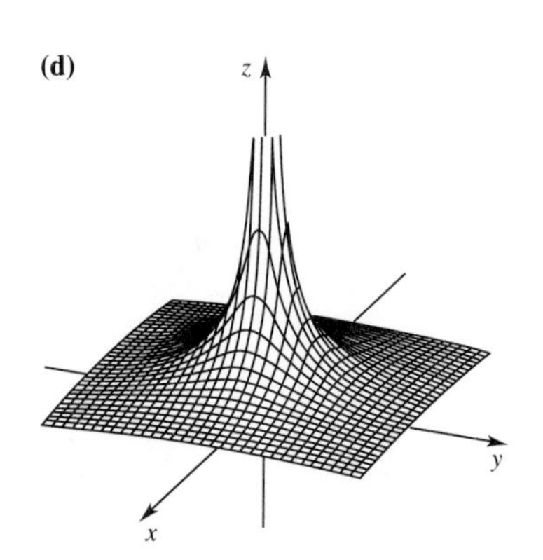

(e)

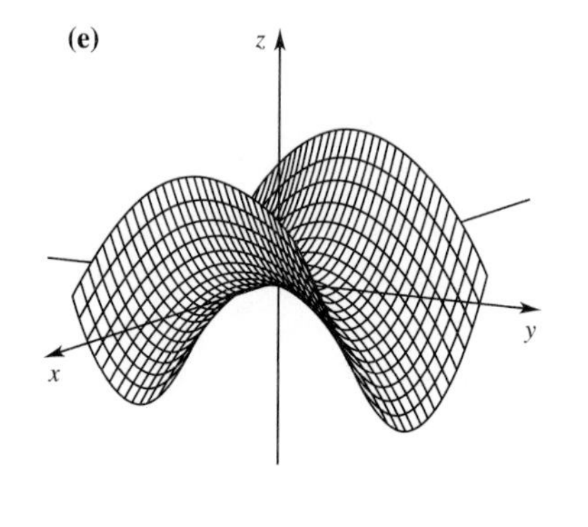

(f)

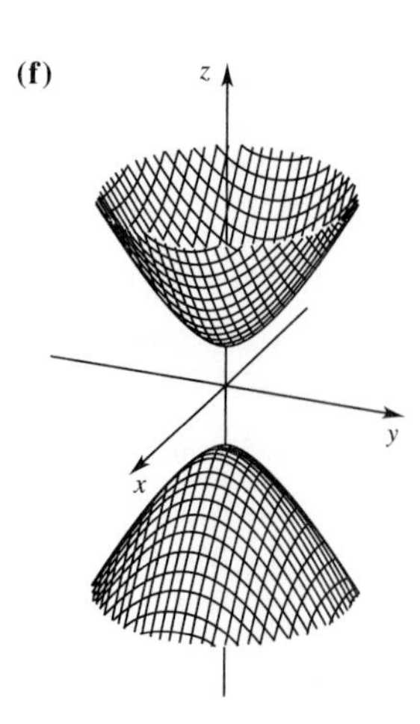

35. Let $f(x, y) = 9x^2 - 3y^2$, and find each of the following.

(a) $\frac{f(x + h, y) - f(x, y)}{h}$

(b) $\frac{f(x, y + h) - f(x, y)}{h}$

36. Let $f(x, y) = 7x^3 + 8y^2$, and find each of the following.

(a) $\frac{f(x + h, y) - f(x, y)}{h}$

(b) $\frac{f(x, y + h) - f(x, y)}{h}$

8.2 PARTIAL DERIVATIVES

What is the change in productivity if labor is increased by one work-hour? What if capital is increased by one unit?

You may want to review the chapter on the derivative for methods used to find some of the derivatives in this section.

Earlier, we found that the derivative dy/dx gives the rate of change of y with respect to x. In this section, we show how derivatives are found and interpreted for multivariable functions, and we will use that information to answer the questions posed above.

A small firm makes only two products, radios and audiocassette recorders. The profits of the firm are given by

$$P(x, y) = 40x^2 - 10xy + 5y^2 - 80,$$

where x is the number of units of radios sold and y is the number of units of recorders sold. How will a change in x or y affect P?

Suppose that sales of radios have been steady at 10 units; only the sales of recorders vary. The management would like to find the marginal profit with respect to y, the number of recorders sold. Recall that marginal profit is given by the derivative of the profit function. Here, x is fixed at 10. Using this information, we begin by finding a new function, $f(y) = P(10, y)$. Let $x = 10$ to get

$$\begin{aligned} f(y) = P(10, y) &= 40(10)^2 - 10(10)y + 5y^2 - 80 \\ &= 3920 - 100y + 5y^2. \end{aligned}$$

The function $f(y)$ shows the profit from the sale of y recorders, assuming that x is fixed at 10 units. Find the derivative df/dy to get the marginal profit with respect to y.

$$\frac{df}{dy} = -100 + 10y$$

In this example, the derivative of the function $f(y)$ was taken with respect to y only; we assumed that x was fixed. To generalize, let $z = f(x, y)$. An intuitive definition of the *partial derivatives* of f with respect to x and y follows.

PARTIAL DERIVATIVES (INFORMAL DEFINITION)

The **partial derivative of f with respect to x** is the derivative of f obtained by treating x as a variable and y as a constant.

The **partial derivative of f with respect to y** is the derivative of f obtained by treating y as a variable and x as a constant.

The symbols $f_x(x, y)$ (no prime is used), $\partial z/\partial x$, and $\partial f/\partial x$ are used to represent the partial derivative of $z = f(x, y)$ with respect to x, with similar symbols used for the partial derivative with respect to y. The symbol $f_x(x, y)$ is often abbreviated as just f_x, with $f_y(x, y)$ abbreviated as f_y.

Generalizing from the definition of the derivative given earlier, partial derivatives of a function $z = f(x, y)$ are formally defined as follows.

PARTIAL DERIVATIVES (FORMAL DEFINITION)

Let $z = f(x, y)$ be a function of two independent variables. Let all indicated limits exist. Then the partial derivative of f with respect to x is

$$\boldsymbol{f_x = \frac{\partial f}{\partial x} = \lim_{h \to 0} \frac{f(x + h, y) - f(x, y)}{h}},$$

and the partial derivative of f with respect to y is

$$\boldsymbol{f_y = \frac{\partial f}{\partial y} = \lim_{h \to 0} \frac{f(x, y + h) - f(x, y)}{h}}.$$

If the indicated limits do not exist, then the partial derivatives do not exist.

Similar definitions could be given for functions of more than two independent variables.

Let $f(x, y) = 4x^2 - 9xy + 6y^3$. Find f_x and f_y.

To find f_x, treat y as a constant and x as a variable. The derivative of the first term, $4x^2$, is $8x$. In the second term, $-9xy$, the constant coefficient of x is $-9y$, so the derivative with x as the variable is $-9y$. The derivative of $6y^3$ is zero, since we are treating y as a constant. Thus,

$$f_x = 8x - 9y.$$

Now, to find f_y, treat y as a variable and x as a constant. Since x is a constant, the derivative of $4x^2$ is zero. In the second term, the coefficient of y is $-9x$ and the derivative of $-9xy$ is $-9x$. The derivative of the third term is $18y^2$. Thus,

$$f_y = -9x + 18y^2. \blacktriangleleft$$

The next example shows how the chain rule can be used to find partial derivatives.

EXAMPLE 2

Let $f(x, y) = \ln |x^2 + y|$. Find f_x and f_y.

Recall the formula for the derivative of a natural logarithm function. If $g(x) = \ln |x|$, then $g'(x) = 1/x$. Using this formula and the chain rule,

$$f_x = \frac{1}{x^2 + y} \cdot D_x(x^2 + y) = \frac{1}{x^2 + y} \cdot 2x = \frac{2x}{x^2 + y},$$

and

$$f_y = \frac{1}{x^2 + y} \cdot D_y(x^2 + y) = \frac{1}{x^2 + y} \cdot 1 = \frac{1}{x^2 + y}. \blacktriangleleft$$

The notation

$$f_x(a, b) \qquad \text{or} \qquad \frac{\partial f}{\partial x}(a, b)$$

represents the value of a partial derivative when $x = a$ and $y = b$, as shown in the next example.

EXAMPLE 3 Let $f(x, y) = 2x^2 + 3xy^3 + 2y + 5$. Find the following.

(a) $f_x(-1, 2)$

First, find f_x by holding y constant.

$$f_x = 4x + 3y^3$$

Now let $x = -1$ and $y = 2$.

$$f_x(-1, 2) = 4(-1) + 3(2)^3 = -4 + 24 = 20$$

(b) $\dfrac{\partial f}{\partial y}(-4, -3)$

Since $\partial f/\partial y = 9xy^2 + 2$,

$$\frac{\partial f}{\partial y}(-4, -3) = 9(-4)(-3)^2 + 2 = 9(-36) + 2 = -322.$$

The derivative of a function of one variable can be interpreted as the tangent line to the graph at that point. With some modification, the same is true of partial derivatives of functions of two variables. At a point on the graph of a function of two variables, $z = f(x, y)$, there may be many tangent lines, all of which lie in the same tangent plane, as shown in Figure 11.

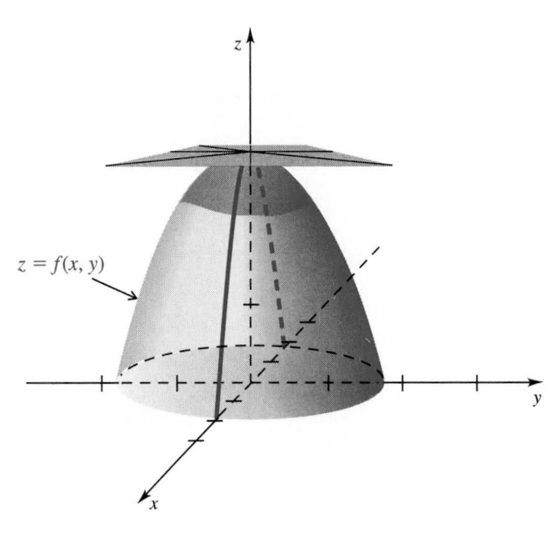

FIGURE 11

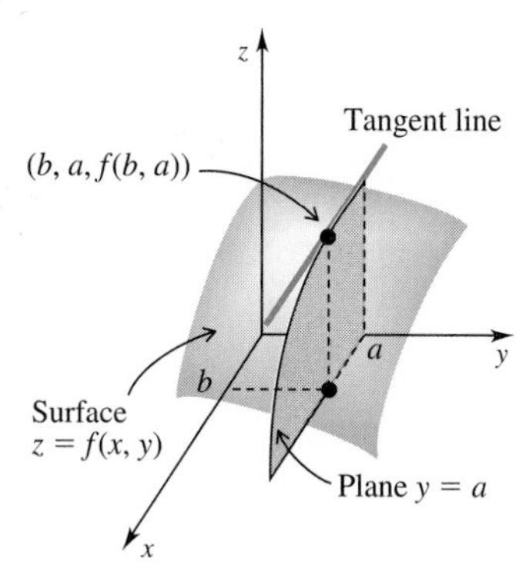

FIGURE 12

In any particular direction, however, there will be only one tangent line. We use partial derivatives to find the slope of the tangent lines in the x- and y-directions as follows.

Figure 12 shows a surface $z = f(x, y)$ and a plane that is parallel to the xz-plane. The equation of the plane is $y = a$. (This corresponds to holding y fixed.) Since $y = a$ for points on the plane, any point on the curve that represents the intersection of the plane and the surface must have the form $(x, a, f(x, a))$. Thus, this curve can be described as $z = f(x, a)$. Since a is constant, $z = f(x, a)$ is a function of one variable. When the derivative of $z = f(x, a)$ is evaluated at $x = b$, it gives the slope of the line tangent to this curve at the point $(b, a, f(b, a))$, as shown in Figure 12. Thus, the partial derivative of f with respect to x, $f_x(b, a)$, gives the rate of change of the surface $z = f(x, y)$ in the x-direction at the point $(b, a, f(b, a))$. In the same way, the partial derivative with respect to y will give the slope of the line tangent to the surface in the y-direction at the point $(b, a, f(b, a))$.

Rate of Change The derivative of $y = f(x)$ gives the rate of change of y with respect to x. In the same way, if $z = f(x, y)$, then f_x gives the rate of change of z with respect to x, if y is held constant.

EXAMPLE 4 Suppose that the temperature of the water at the point on a river where a nuclear power plant discharges its hot waste water is approximated by

$$T(x, y) = 2x + 5y + xy - 40,$$

where x represents the temperature of the river water (in degrees Celsius) before it reaches the power plant and y is the number of megawatts (in hundreds) of electricity being produced by the plant.

(a) Find and interpret $T_x(9, 5)$.

First, find the partial derivative T_x.

$$T_x = 2 + y$$

This partial derivative gives the rate of change of T with respect to x. Replacing x with 9 and y with 5 gives

$$T_x(9, 5) = 2 + 5 = 7.$$

Just as marginal cost is the approximate cost of one more item, this result, 7, is the approximate change in temperature of the output water if input water temperature changes by 1 degree, from $x = 9$ to $x = 9 + 1 = 10$, while y remains constant at 5 (500 megawatts of electricity produced).

(b) Find and interpret $T_y(9, 5)$.

The partial derivative T_y is

$$T_y = 5 + x.$$

This partial derivative gives the rate of change of T with respect to y as

$$T_y(9, 5) = 5 + 9 = 14.$$

This result, 14, is the approximate change in temperature resulting from a 1-unit increase in production of electricity from $y = 5$ to $y = 5 + 1 = 6$ (from 500 to 600 megawatts), while the input water temperature x remains constant at 9° C. ◀

As mentioned in the previous section, if $P(x, y)$ gives the output P produced by x units of labor and y units of capital, $P(x, y)$ is a production function. The partial derivatives of this production function have practical implications. For example, $\partial P/\partial x$ gives the marginal productivity of labor. This represents the rate at which the output is changing with respect to a one-unit change in labor for a fixed capital investment. That is, if the capital investment is held constant and labor is increased by 1 work hour, $\partial P/\partial x$ will yield the approximate change in the production level. Likewise, $\partial P/\partial y$ gives the marginal productivity of capital, which represents the rate at which the output is changing with respect to a one-unit change in capital for a fixed labor value. So if the labor force is held constant and the capital investment is increased by 1 unit, $\partial P/\partial y$ will approximate the corresponding change in the production level.

EXAMPLE 5 A company that manufactures computers has determined that its production function is given by

$$P(x, y) = 500x + 800y + 3x^2y - x^3 - \frac{y^4}{4},$$

where x is the size of the labor force (measured in work hours per week) and y is the amount of capital (measured in units of \$1000) invested. Find the marginal productivity of labor and capital when $x = 50$ and $y = 20$, and interpret the results.

The marginal productivity of labor is found by taking the derivative of P with respect to x.

$$\frac{\partial P}{\partial x} = 500 + 6xy - 3x^2$$

$$\frac{\partial P}{\partial x}(50, 20) = 500 + 6(50)(20) - 3(50)^2 = -1000$$

Thus, if the capital investment is held constant at \$20,000 and labor is increased from 50 to 51 work hours per week, production will decrease by about 1000 units. In the same way, the marginal productivity of capital is $\partial P/\partial y$.

$$\frac{\partial P}{\partial y} = 800 + 3x^2 - y^3$$

$$\frac{\partial P}{\partial y}(50, 20) = 800 + 3(50)^2 - (20)^3 = 300$$

If work hours are held constant at 50 hours per week and the capital investment is increased from \$20,000 to \$21,000, production will increase by about 300 units. ◀

Second-Order Partial Derivatives The second derivative of a function of one variable is very useful in determining relative maxima and minima. **Second-order partial derivatives** (partial derivatives of a partial derivative) are used in a similar way for functions of two or more variables. The situation is somewhat more complicated, however, with more independent variables. For example, $f(x, y) = 4x + x^2y + 2y$ has two first-order partial derivatives,

$$f_x = 4 + 2xy \quad \text{and} \quad f_y = x^2 + 2.$$

Since each of these has two partial derivatives, one with respect to y and one with respect to x, there are *four* second-order partial derivatives of function f. The notations for these four second-order partial derivatives are given below.

SECOND-ORDER PARTIAL DERIVATIVES

For a function $z = f(x, y)$, if the indicated partial derivative exists, then

$$\frac{\partial}{\partial x}\left(\frac{\partial z}{\partial x}\right) = \frac{\partial^2 z}{\partial x^2} = f_{xx} \qquad \frac{\partial}{\partial y}\left(\frac{\partial z}{\partial y}\right) = \frac{\partial^2 z}{\partial y^2} = f_{yy}$$

$$\frac{\partial}{\partial y}\left(\frac{\partial z}{\partial x}\right) = \frac{\partial^2 z}{\partial y \partial x} = f_{xy} \qquad \frac{\partial}{\partial x}\left(\frac{\partial z}{\partial y}\right) = \frac{\partial^2 z}{\partial x \partial y} = f_{yx}.$$

As seen above, f_{xx} is used as an abbreviation for $f_{xx}(x, y)$, with f_{yy}, f_{xy}, and f_{yx} used in a similar way. The symbol f_{xx} is read "the partial derivative of f_x with respect to x." Also, the symbol $\partial^2 z/\partial y^2$ is read "the partial derivative of $\partial z/\partial y$ with respect to y."

Note For most functions found in applications and for all of the functions in this book, the second-order partial derivatives f_{xy} and f_{yx} are equal. Therefore, it is not necessary to be particular about the order in which these derivatives are found.

Find all second-order partial derivatives for

$$f(x, y) = -4x^3 - 3x^2y^3 + 2y^2.$$

First find f_x and f_y.

$$f_x = -12x^2 - 6xy^3 \quad \text{and} \quad f_y = -9x^2y^2 + 4y$$

To find f_{xx}, take the partial derivative of f_x with respect to x.

$$f_{xx} = -24x - 6y^3$$

Take the partial derivative of f_y, with respect to y; this gives f_{yy}.

$$f_{yy} = -18x^2y + 4$$

Find f_{xy} by starting with f_x, then taking the partial derivative of f_x with respect to y.

$$f_{xy} = -18xy^2$$

Finally, find f_{yx} by starting with f_y; take its partial derivative with respect to x.

$$f_{yx} = -18xy^2$$

EXAMPLE 7

Let $f(x, y) = 2e^x - 8x^3y^2$. Find all second-order partial derivatives.

Here $f_x = 2e^x - 24x^2y^2$ and $f_y = -16x^3y$. [Recall: if $g(x) = e^x$, then $g'(x) = e^x$.] Now find the second-order partial derivatives.

$$f_{xx} = 2e^x - 48xy^2 \qquad f_{xy} = -48x^2y$$
$$f_{yy} = -16x^3 \qquad f_{yx} = -48x^2y$$

Partial derivatives of functions with more than two independent variables are found in a similar manner. For instance, to find f_z for $f(x, y, z)$, hold x and y constant and differentiate with respect to z.

Let $f(x, y, z) = 2x^2yz^2 + 3xy^2 - 4yz$. Find f_x, f_y, f_{xz}, and f_{yz}.

$$f_x = 4xyz^2 + 3y^2$$
$$f_y = 2x^2z^2 + 6xy - 4z$$

To find f_{xz}, differentiate f_x with respect to z.

$$f_{xz} = 8xyz$$

Differentiate f_y to get f_{yz}.

$$f_{yz} = 4x^2z - 4$$

8.2 Exercises

1. Let $z = f(x, y) = 12x^2 - 8xy + 3y^2$. Find each of the following.

   (a) $\frac{\partial z}{\partial x}$ (b) $\frac{\partial z}{\partial y}$ (c) $\frac{\partial f}{\partial x}(2, 3)$ (d) $f_y(1, -2)$

2. Let $z = g(x, y) = 5x + 9x^2y + y^2$. Find each of the following.

   (a) $\frac{\partial g}{\partial x}$ (b) $\frac{\partial g}{\partial y}$ (c) $\frac{\partial z}{\partial y}(-3, 0)$ (d) $g_x(2, 1)$

In Exercises 3–16, find f_x and f_y. Then find $f_x(2, -1)$ and $f_y(-4, 3)$. Leave the answers in terms of e in Exercises 7–10 and 15–16.

3. $f(x, y) = -2xy + 6y^3 + 2$
4. $f(x, y) = 4x^2y - 9y^2$
5. $f(x, y) = 3x^3y^2$
6. $f(x, y) = -2x^2y^4$
7. $f(x, y) = e^{x+y}$
8. $f(x, y) = 3e^{2x+y}$
9. $f(x, y) = -5e^{3x-4y}$
10. $f(x, y) = 8e^{7x-y}$
11. $f(x, y) = \frac{x^2 + y^3}{x^3 - y^2}$
12. $f(x, y) = \frac{3x^2y^3}{x^2 + y^2}$
13. $f(x, y) = \ln|1 + 3x^2y^3|$
14. $f(x, y) = \ln|2x^5 - xy^4|$
15. $f(x, y) = xe^{x^2y}$
16. $f(x, y) = y^2e^{x+3y}$

Find all second-order partial derivatives for the following.

17. $f(x, y) = 6x^3y - 9y^2 + 2x$
18. $g(x, y) = 5xy^4 + 8x^3 - 3y$
19. $R(x, y) = 4x^2 - 5xy^3 + 12y^2x^2$
20. $h(x, y) = 30y + 5x^2y + 12xy^2$

21. $r(x, y) = \dfrac{4x}{x + y}$ **22.** $k(x, y) = \dfrac{-5y}{x + 2y}$ **23.** $z = 4xe^y$ **24.** $z = -3ye^x$

25. $r = \ln|x + y|$ **26.** $k = \ln|5x - 7y|$ **27.** $z = x \ln|xy|$ **28.** $z = (y + 1)\ln|x^3y|$

For the functions defined as follows, find values of x and y such that both $f_x(x, y) = 0$ and $f_y(x, y) = 0$.

29. $f(x, y) = 6x^2 + 6y^2 + 6xy + 36x - 5$ **30.** $f(x, y) = 50 + 4x - 5y + x^2 + y^2 + xy$

31. $f(x, y) = 9xy - x^3 - y^3 - 6$ **32.** $f(x, y) = 2200 + 27x^3 + 72xy + 8y^2$

Find f_x, f_y, f_z, and f_{yz} for the following.

33. $f(x, y, z) = x^2 + yz + z^4$ **34.** $f(x, y, z) = 3x^5 - x^2 + y^5$

35. $f(x, y, z) = \dfrac{6x - 5y}{4z + 5}$ **36.** $f(x, y, z) = \dfrac{2x^2 + xy}{yz - 2}$

37. $f(x, y, z) = \ln|x^2 - 5xz^2 + y^4|$ **38.** $f(x, y, z) = \ln|8xy + 5yz - x^3|$

Applications

Business and Economics

39. Manufacturing Cost Suppose that the manufacturing cost of a precision electronic calculator is approximated by

$$M(x, y) = 40x^2 + 30y^2 - 10xy + 30,$$

where x is the cost of electronic chips and y is the cost of labor. Find the following.

(a) $M_y(4, 2)$ **(b)** $M_x(3, 6)$ **(c)** $(\partial M/\partial x)(2, 5)$

(d) $(\partial M/\partial y)(6, 7)$

40. Revenue The revenue from the sale of x units of a tranquilizer and y units of an antibiotic is given by

$$R(x, y) = 5x^2 + 9y^2 - 4xy.$$

Suppose 9 units of tranquilizer and 5 units of antibiotic are sold.

(a) What is the approximate effect on revenue if 10 units of tranquilizer and 5 units of antibiotic are sold?

(b) What is the approximate effect on revenue if the amount of antibiotic sold is increased to 6 units, while tranquilizer sales remain constant?

41. Sales A car dealership estimates that the total weekly sales of its most popular model is a function of the car's list price, p, and the interest rate in percent, i, offered by the manufacturer. The approximate weekly sales are given by

$$f(p, i) = 132p - 2pi - .01p^2.$$

(a) Find the weekly sales if the average list price is \$9400 and the manufacturer is offering an 8% interest rate.

(b) Find and interpret f_p and f_i.

(c) What would be the effect on weekly sales if the price is \$9400 and interest rates rise from 8% to 9%?

42. Marginal Productivity Suppose the production function of a company is given by

$$P(x, y) = 100\sqrt{x^2 + y^2},$$

where x represents units of labor and y represents units of capital. Find the following when 4 units of labor and 3 units of capital are used.

(a) The marginal productivity of labor

(b) The marginal productivity of capital

43. Marginal Productivity A manufacturer estimates that production (in hundreds of units) is a function of the amounts x and y of labor and capital used, as follows.

$$f(x, y) = \left(\frac{1}{3}x^{-1/3} + \frac{2}{3}y^{-1/3}\right)^{-3}$$

(a) Find the number of units produced when 27 units of labor and 64 units of capital are utilized.

(b) Find and interpret $f_x(27, 64)$ and $f_y(27, 64)$.

(c) What would be the approximate effect on production of increasing labor by 1 unit?

44. Marginal Productivity The production function z for the United States was once estimated as

$$z = x^{.7}y^{.3},$$

where x stands for the amount of labor and y the amount of capital. Find the marginal productivity of labor and of capital.

45. Marginal Productivity A similar production function for Canada is

$$z = x^{.4}y^{.6},$$

with x, y, and z as in Exercise 44. Find the marginal productivity of labor and of capital.

46. Marginal Productivity A manufacturer of automobile batteries estimates that his total production (in thousands of units) is given by

$$f(x, y) = 3x^{1/3}y^{2/3},$$

where x is the number of units of labor and y is the number of units of capital utilized.

(a) Find and interpret $f_x(64, 125)$ and $f_y(64, 125)$ if the current level of production uses 64 units of labor and 125 units of capital.

(b) What would be the approximate effect on production of increasing labor to 65 units while holding capital at the current level?

(c) Suppose that sales have been good and management wants to increase either capital or labor by 1 unit. Which option would result in a larger increase in production?

Life Sciences

47. Grasshopper Matings The total number of matings per day between individuals of a certain species of grasshoppers is approximated by

$$M(x, y) = 2xy + 10xy^2 + 30y^2 + 20,$$

where x represents the temperature in °C and y represents the number of days since the last rainfall. Find the following.

(a) The approximate change in matings when the temperature increases from 20° C to 21° C and the number of days since rain is constant at 4.

(b) The approximate change in matings when the temperature is constant at 24° C and the number of days since rain is changed from 10 to 11.

(c) Would an increase in temperature or an increase in days since rain cause more of an increase in matings?

48. Oxygen Consumption The oxygen consumption of a well-insulated mammal that is not sweating is approximated by

$$m = m(T, F, w) = \frac{2.5(T - F)}{w^{.67}} = 2.5(T - F)w^{-.67},$$

where T is the internal body temperature of the animal (in °C), F is the temperature of the outside of the animal's fur (in °C), and w is the animal's weight (in kilograms). Find the approximate change in oxygen consumption under the following conditions.

(a) The internal temperature increases from 38° C to 39° C, while the outside temperature remains at 12° C and the weight remains at 30 kg.

(b) The internal temperature is constant at 36° C, the outside temperature increases from 14° C to 15° C, and the weight remains at 25 kg.

49. Body Surface Area The surface area of a human (in square meters) is approximated by

$$A(W, H) = .202W^{.425}H^{.725},$$

where W is the weight of the person in kilograms and H is the height in meters. Find the approximate change in surface area under the following conditions.

(a) The weight changes from 72 kg to 73 kg, while the height remains 1.8 m.

(b) The weight remains stable at 70 kg, while the height changes from 1.6 m to 1.7 m.

50. Blood Flow In one method of computing the quantity of blood pumped through the lungs in one minute, a researcher first finds each of the following (in milliliters).

b = quantity of oxygen used by body in one minute

a = quantity of oxygen per liter of blood that has just gone through the lungs

v = quantity of oxygen per liter of blood that is about to enter the lungs

In one minute,

Amount of oxygen used = Amount of oxygen per liter × Liters of blood pumped.

If C is the number of liters of blood pumped through the lungs in one minute, then

$$b = (a - v) \cdot C \quad \text{or} \quad C = \frac{b}{a - v}.$$

(a) Find the number of liters of blood pumped through the lungs in one minute if $a = 160$, $b = 200$, and $v = 125$.

(b) Find the approximate change in C when a changes from 160 to 161, $b = 200$, and $v = 125$.

(c) Find the approximate change in C when $a = 160$, b changes from 200 to 201, and $v = 125$.

(d) Find the approximate change in C when $a = 160$, $b = 200$, and v changes from 125 to 126.

(e) A change of 1 unit in which quantity of oxygen produces the greatest change in the liters of blood pumped?

51. Health A weight-loss counselor has prepared a program of diet and exercise for a client. If the client sticks to the program, the weight loss that can be expected (in pounds per week) is given by

$$\text{Weight loss} = f(n, c) = \frac{1}{8}n^2 - \frac{1}{5}c + \frac{1937}{8},$$

where c is the average daily calorie intake for the week and n is the number of 40-min aerobic workouts per week.

(a) How many pounds can the client expect to lose by eating an average of 1200 cal per day and participating in four 40-min workouts in a week?

(b) Find and interpret $\partial f/\partial n$.

(c) The client currently averages 1100 cal per day and does three 40-minute workouts each week. What would be the approximate impact on weekly weight loss of adding a fourth workout per week?

52. **Drug Reaction** The reaction to x units of a drug t hr after it was administered is given by

$$R(x, t) = x^2(a - x)t^2e^{-t},$$

for $0 \leq x \leq a$ (where a is a constant). Find the following.

(a) $\dfrac{\partial R}{\partial x}$ (b) $\dfrac{\partial R}{\partial t}$ (c) $\dfrac{\partial^2 R}{\partial x^2}$ (d) $\dfrac{\partial^2 R}{\partial x \partial t}$

(e) Interpret your answers to parts (a) and (b).

Social Sciences

53. **Education** A developmental mathematics instructor at a large university has determined that a student's probability of success in the university's pass/fail remedial algebra course is a function of s, n, and a, where s is the student's score on the departmental placement exam, n is the number of semesters of mathematics passed in high school, and a is the student's mathematics SAT score. She estimates that p, the probability of passing the course (in percent), will be

$$p = f(s, n, a) = .003a + .1(sn)^{1/2}$$

for $200 \leq a \leq 800$, $0 \leq s \leq 10$, and $0 \leq n \leq 8$. Assuming that the above model has some merit, find the following.

(a) If a student scores 8 on the placement exam, has taken 6 semesters of high-school math, and has an SAT score of 450, what is the probability of passing the course?

(b) Find p for a student with 3 semesters of high school mathematics, a placement score of 3, and an SAT score of 320.

(c) Find and interpret $f_n(3, 3, 320)$ and $f_a(3, 3, 320)$.

Physical Sciences

54. **Gravitational Attraction** The gravitational attraction F on a body a distance r from the center of the earth, where r is greater than the radius of the earth, is a function of its mass m and the distance r as follows:

$$F = \frac{mgR^2}{r^2},$$

where R is the radius of the earth and g is the force of gravity—about 32 feet per second per second (ft/sec^2).

(a) Find and interpret F_m and F_r.

(b) Show that $F_m > 0$ and $F_r < 0$. Why is this reasonable?

8.3 MAXIMA AND MINIMA

What amounts of sugar and flavoring produce the minimum cost per batch of a soft drink? What is the minimum cost?

It may be helpful to review the section on relative extrema in the chapter on applications of the derivative at this point. The concepts presented there are basic to what will be done in this section.

One of the most important applications of calculus is in finding maxima and minima for functions. Earlier, we studied this idea extensively for functions of a single independent variable; now we will see that extrema can be found for functions of two variables. In particular, an extension of the second derivative test can be defined and used to identify maxima or minima. We begin with the definitions of relative maxima and minima.

RELATIVE MAXIMA AND MINIMA

Let (a, b) be the center of a circular region contained in the xy-plane. Then, for a function $z = f(x, y)$ defined for every (x, y) in the region, $f(a, b)$ is a **relative maximum** if

$$f(a, b) \geq f(x, y)$$

for all points (x, y) in the circular region, and $f(a, b)$ is a **relative minimum** if

$$f(a, b) \leq f(x, y)$$

for all points (x, y) in the circular region.

As before, the word *extremum* is used for either a relative maximum or a relative minimum. Examples of a relative maximum and a relative minimum are given in Figures 13 and 14.

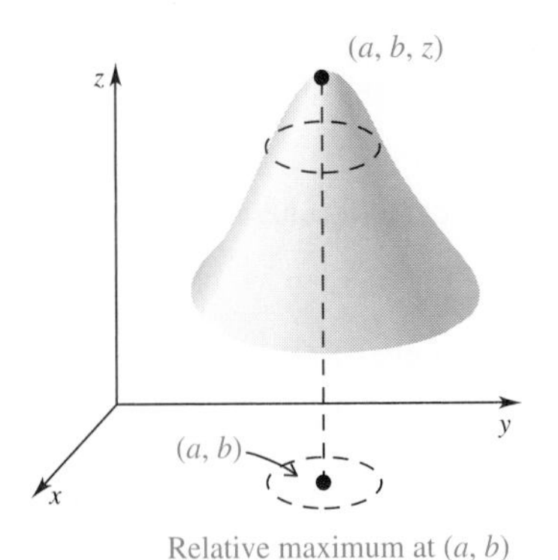

Relative maximum at (a, b)

FIGURE 13

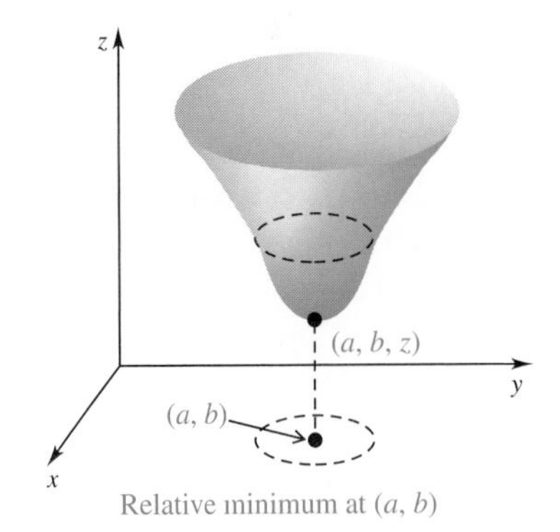

Relative minimum at (a, b)

FIGURE 14

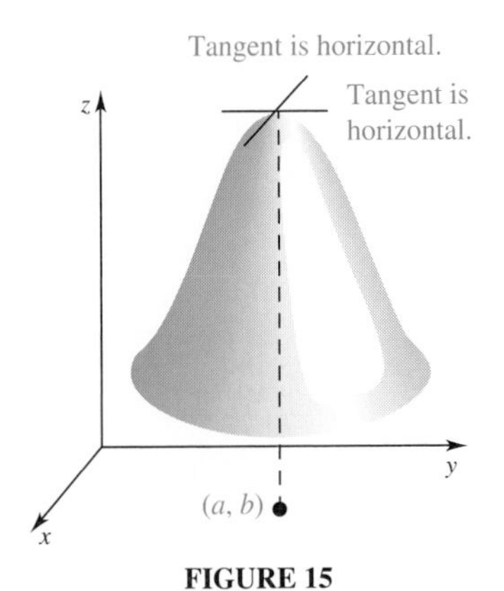

FIGURE 15

Note When functions of a single variable were discussed, a distinction was made between relative extrema and absolute extrema. The methods for finding absolute extrema are quite involved for functions of two variables, so we will discuss only relative extrema here. In most practical applications the relative extrema coincide with the absolute extrema. Also, in this brief discussion of extrema for multivariable functions, we omit cases where an extremum occurs on a boundary of the domain.

As suggested by Figure 15, at a relative maximum the tangent line parallel to the x-axis has a slope of 0, as does the tangent line parallel to the y-axis. (Notice the similarity to functions of one variable.) That is, if the function $z = f(x, y)$ has a relative extremum at (a, b), then $f_x(a, b) = 0$ and $f_y(a, b) = 0$, as stated in the next theorem.

LOCATION OF EXTREMA

Let a function $z = f(x, y)$ have a relative maximum or relative minimum at the point (a, b). Let $f_x(a, b)$ and $f_y(a, b)$ both exist. Then

$$f_x(a, b) = 0 \quad \text{and} \quad f_y(a, b) = 0.$$

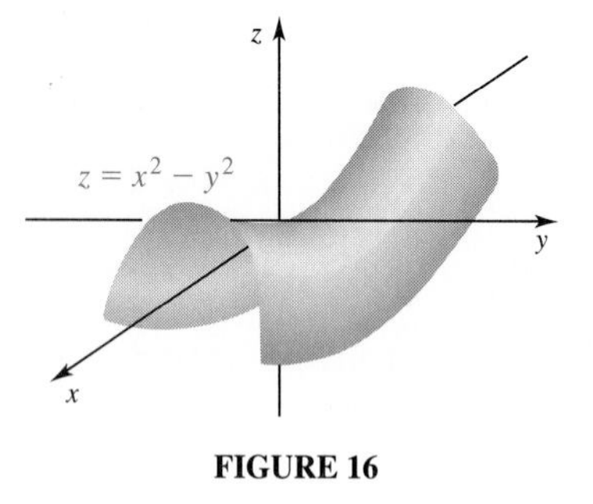

FIGURE 16

Just as with functions of one variable, the fact that the slopes of the tangent lines are 0 is no guarantee that a relative extremum has been located. For example, Figure 16 shows the graph of $z = f(x, y) = x^2 - y^2$. Both $f_x(0, 0) = 0$ and $f_y(0, 0) = 0$, and yet $(0, 0)$ leads to neither a relative maximum nor a relative minimum for the function. The point $(0, 0, 0)$ on the graph of this function is called a **saddle point;** it is a minimum when approached from one direction but a maximum when approached from another direction. A saddle point is neither a maximum nor a minimum.

The theorem on location of extrema suggests a useful strategy for finding extrema. First, locate all points (a, b) where $f_x(a, b) = 0$ and $f_y(a, b) = 0$. Then test each of these points separately, using the test given after the next example. For a function $f(x, y)$, the points (a, b) such that $f_x(a, b) = 0$ and $f_y(a, b) = 0$ are called **critical points.**

Find all critical points for

$$f(x, y) = 6x^2 + 6y^2 + 6xy + 36x - 5.$$

Find all points (a, b) such that $f_x(a, b) = 0$ and $f_y(a, b) = 0$. Here

$$f_x = 12x + 6y + 36 \quad \text{and} \quad f_y = 12y + 6x.$$

Set each of these two partial derivatives equal to 0.

$$12x + 6y + 36 = 0 \quad \text{and} \quad 12y + 6x = 0$$

These two equations make up a system of linear equations. We can use the substitution method to solve this system. First, rewrite $12y + 6x = 0$ as follows:

$$\begin{aligned} 12y + 6x &= 0 \\ 6x &= -12y \\ x &= -2y. \end{aligned}$$

Now substitute $-2y$ for x in the other equation.

$$\begin{aligned} 12x + 6y + 36 &= 0 \\ 12(-2y) + 6y + 36 &= 0 \\ -24y + 6y + 36 &= 0 \\ -18y + 36 &= 0 \\ -18y &= -36 \\ y &= 2 \end{aligned}$$

From the equation $x = -2y$, $x = -2(2) = -4$. The solution of the system of equations is $(-4, 2)$. Since this is the only solution of the system, $(-4, 2)$ is the only critical point for the given function. By the theorem above, if the function has a relative extremum, it will occur at $(-4, 2)$. ◀

The results of the next theorem can be used to decide whether $(-4, 2)$ in Example 1 leads to a relative maximum, a relative minimum, or neither.

TEST FOR RELATIVE EXTREMA

For a function $z = f(x, y)$, let f_{xx}, f_{yy}, and f_{xy} all exist. Let (a, b) be a point for which

$$f_x(a, b) = 0 \quad \text{and} \quad f_y(a, b) = 0.$$

Define the number D by

$$D = f_{xx}(a, b) \cdot f_{yy}(a, b) - [f_{xy}(a, b)]^2.$$

Then

(a) $f(a, b)$ is a relative maximum if $D > 0$ and $f_{xx}(a, b) < 0$;
(b) $f(a, b)$ is a relative minimum if $D > 0$ and $f_{xx}(a, b) > 0$;
(c) $f(a, b)$ is a saddle point (neither a maximum nor a minimum) if $D < 0$;
(d) if $D = 0$, the test gives no information.

This test is comparable to the second derivative test for extrema of functions of one independent variable. The chart below summarizes the conclusions of the theorem.

	$f_{xx}(a, b) < 0$	$f_{xx}(a, b) > 0$
$D > 0$	Relative maximum	Relative minimum
$D = 0$	No information	
$D < 0$	Saddle point	

The previous example showed that the only critical point for the function

$$f(x, y) = 6x^2 + 6y^2 + 6xy + 36x - 5$$

is $(-4, 2)$. Does $(-4, 2)$ lead to a relative maximum, a relative minimum, or neither?

Find out by using the test above. From Example 1,

$$f_x(-4, 2) = 0 \quad \text{and} \quad f_y(-4, 2) = 0.$$

Now find the various second partial derivatives used in finding D. From $f_x = 12x + 6y + 36$ and $f_y = 12y + 6x$,

$$f_{xx} = 12, \quad f_{yy} = 12, \quad \text{and} \quad f_{xy} = 6.$$

(If these second-order partial derivatives had not all been constants, they would have had to be evaluated at the point $(-4, 2)$.) Now

$$D = f_{xx}(-4, 2) \cdot f_{yy}(-4, 2) - [f_{xy}(-4, 2)]^2 = 12 \cdot 12 - 6^2 = 108.$$

Since $D > 0$ and $f_{xx}(-4, 2) = 12 > 0$, part (b) of the theorem applies, showing that $f(x, y) = 6x^2 + 6y^2 + 6xy + 36x - 5$ has a relative minimum at $(-4, 2)$. This relative minimum is $f(-4, 2) = -77$. ◀

Find all points where the function

$$f(x, y) = 9xy - x^3 - y^3 - 6$$

has any relative maxima or relative minima.

First find any critical points. Here

$$f_x = 9y - 3x^2 \qquad \text{and} \qquad f_y = 9x - 3y^2.$$

Set each of these partial derivatives equal to 0.

$$\begin{aligned} f_x &= 0 \\ 9y - 3x^2 &= 0 \\ 9y &= 3x^2 \\ 3y &= x^2 \end{aligned} \qquad\qquad \begin{aligned} f_y &= 0 \\ 9x - 3y^2 &= 0 \\ 9x &= 3y^2 \\ 3x &= y^2 \end{aligned}$$

In the first equation $(3y = x^2)$, notice that since $x^2 \geq 0$, $y \geq 0$. Also, in the second equation $(3x = y^2)$, $y^2 \geq 0$, so $x \geq 0$.

The substitution method can be used again to solve the system of equations

$$\begin{aligned} 3y &= x^2 \\ 3x &= y^2. \end{aligned}$$

The first equation, $3y = x^2$, can be rewritten as $y = x^2/3$. Substitute this into the second equation to get

$$3x = y^2 = \left(\frac{x^2}{3}\right)^2 = \frac{x^4}{9}.$$

Solve this equation as follows.

$$\begin{aligned} 27x &= x^4 \\ x^4 - 27x &= 0 \\ x(x^3 - 27) &= 0 && \text{Factor.} \\ x = 0 \quad \text{or} \quad x^3 - 27 &= 0 && \text{Set each factor equal to 0.} \\ x^3 &= 27 \\ x &= 3 && \text{Take the cube root on each side.} \end{aligned}$$

Use these values of x, along with the equation $3x = y^2$, to find y.

If $x = 0$,	If $x = 3$,
$3x = y^2$	$3x = y^2$
$3(0) = y^2$	$3(3) = y^2$
$0 = y^2$	$9 = y^2$
$0 = y$.	$3 = y$ or $-3 = y$.

The points (0, 0), (3, 3), and (3, -3) appear to be critical points; however, (3, -3) does not satisfy $y \geq 0$. The only possible relative extrema for $f(x, y) = 9xy - x^3 - y^3 - 6$ occur at the critical points (0, 0) or (3, 3). To identify any extrema, use the test. Here

$$f_{xx} = -6x, \qquad f_{yy} = -6y, \qquad \text{and} \qquad f_{xy} = 9.$$

Test each of the possible critical points.

For (0, 0):	For (3, 3):
$f_{xx}(0, 0) = -6(0) = 0$	$f_{xx}(3, 3) = -6(3) = -18$
$f_{yy}(0, 0) = -6(0) = 0$	$f_{yy}(3, 3) = -6(3) = -18$
$f_{xy}(0, 0) = 9$	$f_{xy}(3, 3) = 9$
$D = 0 \cdot 0 - 9^2 = -81.$	$D = -18(-18) - 9^2 = 243.$
Since $D < 0$, there is a saddle point at (0, 0).	Here $D > 0$ and $f_{xx}(3, 3) = -18 < 0$; there is a relative maximum at (3, 3). ◀

EXAMPLE 4 A company is developing a new soft drink. The cost in dollars to produce a batch of the drink is approximated by

$$C(x, y) = 2200 + 27x^3 - 72xy + 8y^2,$$

where x is the number of kilograms of sugar per batch and y is the number of grams of flavoring per batch.

(a) Find the amounts of sugar and flavoring that result in minimum cost per batch.
Start with the following partial derivatives.

$$C_x = 81x^2 - 72y \qquad \text{and} \qquad C_y = -72x + 16y$$

Set each of these equal to 0 and solve for y.

$81x^2 - 72y = 0$	$-72x + 16y = 0$
$-72y = -81x^2$	$16y = 72x$
$y = \dfrac{9}{8}x^2$	$y = \dfrac{9}{2}x$

From the equation on the left, $y \geq 0$. Since $(9/8)x^2$ and $(9/2)x$ both equal y, they are equal to each other. Set them equal, and solve the resulting equation for x.

$$\frac{9}{8}x^2 = \frac{9}{2}x$$

$$9x^2 = 36x$$

$$9x^2 - 36x = 0 \quad \text{Get 0 on one side.}$$

$$9x(x - 4) = 0 \quad \text{Factor.}$$

$$9x = 0 \quad \text{or} \quad x - 4 = 0 \quad \text{Set each factor equal to 0.}$$

The equation $9x = 0$ leads to $x = 0$, which is not a useful answer for the problem. Substitute $x = 4$ into $y = (9/2)x$ to find y.

$$y = \frac{9}{2}x = \frac{9}{2}(4) = 18$$

Now check to see whether the critical point (4, 18) leads to a relative minimum. For (4, 18),

$$C_{xx} = 162x = 162(4) = 648, \qquad C_{yy} = 16, \qquad \text{and} \qquad C_{xy} = -72.$$

Also,

$$D = (648)(16) - (-72)^2 = 5184.$$

Since $D > 0$ and $C_{xx}(4, 18) > 0$, the cost at (4, 18) is a minimum.

(b) What is the minimum cost?

To find the minimum cost, go back to the cost function and evaluate $C(4, 18)$.

$$C(x, y) = 2200 + 27x^3 - 72xy + 8y^2$$

$$C(4, 18) = 2200 + 27(4)^3 - 72(4)(18) + 8(18)^2 = 1336$$

The minimum cost for a batch of soft drink is \$1336.00. ◀

8.3 Exercises

Find all points where the functions defined as follows have any relative extrema. Identify any saddle points.

1. $f(x, y) = xy + x - y$

2. $f(x, y) = 4xy + 8x - 9y$

3. $f(x, y) = x^2 - 2xy + 2y^2 + x - 5$

4. $f(x, y) = x^2 + xy + y^2 - 6x - 3$

5. $f(x, y) = x^2 - xy + y^2 + 2x + 2y + 6$

6. $f(x, y) = x^2 + xy + y^2 + 3x - 3y$

7. $f(x, y) = x^2 + 3xy + 3y^2 - 6x + 3y$

8. $f(x, y) = 5xy - 7x^2 - y^2 + 3x - 6y - 4$

9. $f(x, y) = 4xy - 10x^2 - 4y^2 + 8x + 8y + 9$

10. $f(x, y) = x^2 + xy + 3x + 2y - 6$

11. $f(x, y) = x^2 + xy - 2x - 2y + 2$

12. $f(x, y) = x^2 + xy + y^2 - 3x - 5$

13. $f(x, y) = 2x^3 + 3y^2 - 12xy + 4$

14. $f(x, y) = 5x^3 + 2y^2 - 60xy - 3$

15. $f(x, y) = x^2 + 4y^3 - 6xy - 1$

16. $f(x, y) = 3x^2 + 7y^3 - 42xy + 5$

17. $f(x, y) = e^{xy}$

18. $f(x, y) = x^2 + e^y$

19. Describe the procedure for finding critical points of a function in two independent variables.

20. How are second-order partial derivatives used in finding extrema?

Figures (a)–(f) show the graphs of the functions defined in Exercises 21–26. Find all relative extrema for each function, and then match the equation to its graph.

21. $z = -3xy + x^3 - y^3 + \frac{1}{8}$

22. $z = \frac{3}{2}y - \frac{1}{2}y^3 - x^2y + \frac{1}{16}$

23. $z = y^4 - 2y^2 + x^2 - \frac{17}{16}$

24. $z = -2x^3 - 3y^4 + 6xy^2 + \frac{1}{16}$

25. $z = -x^4 + y^4 + 2x^2 - 2y^2 + \frac{1}{16}$

26. $z = -y^4 + 4xy - 2x^2 + \frac{1}{16}$

(a)

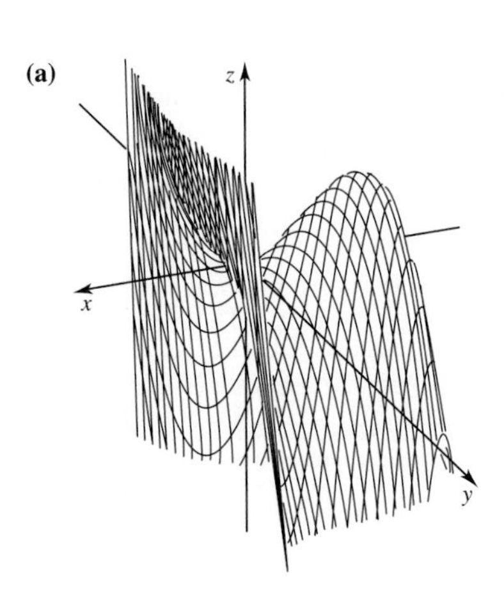

(b)

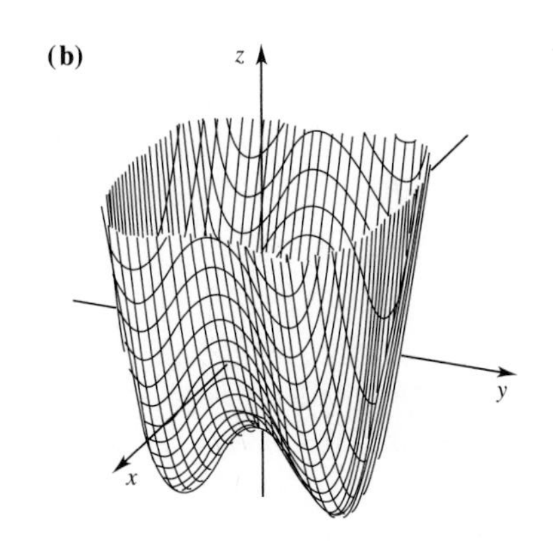

(c)

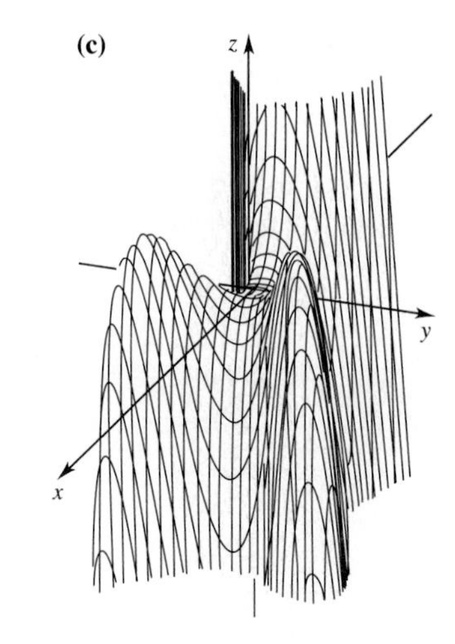

(d)

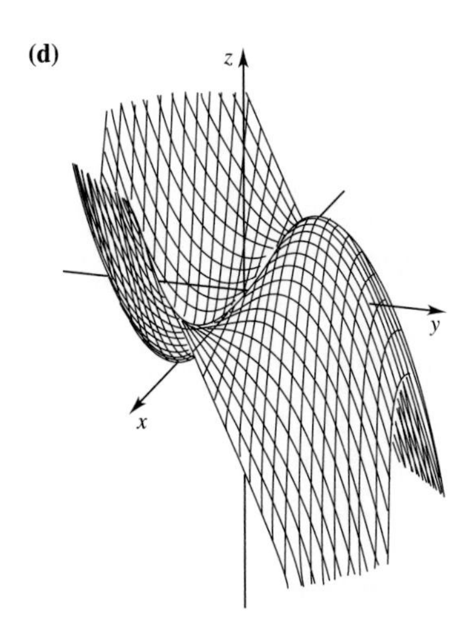

(e)

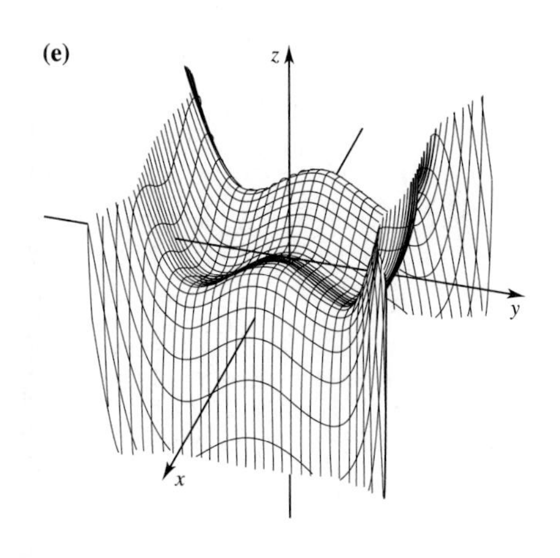

(f)

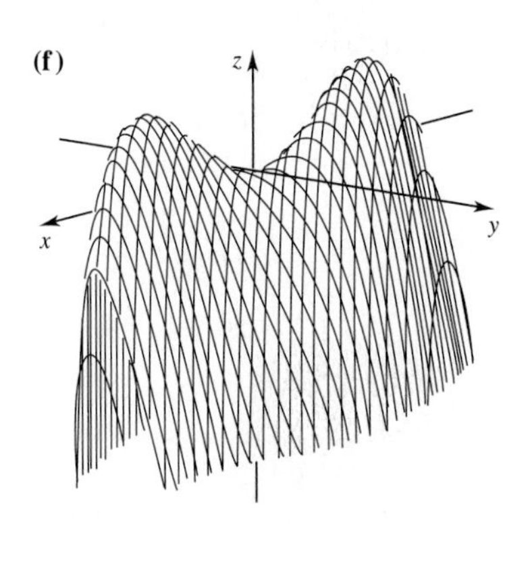

27. Show that $f(x, y) = 1 - x^4 - y^4$ has a relative maximum, even though D in the theorem is 0.

28. Show that $D = 0$ for $f(x, y) = x^3 + (x - y)^2$ and that the function has no relative extrema.

Applications

Business and Economics

29. Profit Suppose that the profit of a certain firm (in hundreds of dollars) is approximated by

$$P(x, y) = 1000 + 24x - x^2 + 80y - y^2,$$

where x is the cost of a unit of labor and y is the cost of a unit of goods. Find values of x and y that maximize profit. Find the maximum profit.

30. Labor Costs Suppose the labor cost (in dollars) for manufacturing a precision camera can be approximated by

$$L(x, y) = \frac{3}{2}x^2 + y^2 - 2x - 2y - 2xy + 68,$$

where x is the number of hours required by a skilled craftsperson and y is the number of hours required by a semiskilled person. Find values of x and y that minimize the labor cost. Find the minimum labor cost.

31. Cost The total cost (in dollars) to produce x units of electrical tape and y units of packing tape is given by

$$C(x, y) = 2x^2 + 3y^2 - 2xy + 2x - 126y + 3800.$$

Find the number of units of each kind of tape that should be produced so that the total cost is a minimum. Find the minimum total cost.

32. Revenue The total revenue (in hundreds of dollars) from the sale of x spas and y solar heaters is approximated by

$$R(x, y) = 12 + 74x + 85y - 3x^2 - 5y^2 - 5xy.$$

Find the number of each that should be sold to produce maximum revenue. Find the maximum revenue.

8.4 LAGRANGE MULTIPLIERS; CONSTRAINED OPTIMIZATION

What dimensions for a new building will maximize the floor space at a fixed cost?

In the previous section, we showed how partial derivatives are used to solve extrema problems. Another approach that works well when there is an additional restriction, called a **constraint,** uses an additional variable, called the **Lagrange multiplier**. For example, in the opening question, suppose a builder wants to maximize the floor space in a new building while keeping the costs fixed at \$500,000. The building will be 40 feet high, with a rectangular floor plan and three stories. The costs, which depend on the dimensions of the rectangular floor plan, are given by

$$\text{Costs} = xy + 20y + 20x + 474{,}000,$$

where x is the width and y the length of the rectangle. Thus, the builder wishes to maximize the area $A(x, y) = xy$ and satisfy the condition

$$xy + 20y + 20x + 474{,}000 = 500{,}000.$$

In addition to maximizing area, then, the builder must keep costs at (or below) \$500,000.

A typical problem of this type might require the smallest possible value of the function $z = x^2 + y^2$, subject to the constraint $x + y = 4$. To see how to find this minimum value, we might first graph both the surface $z = x^2 + y^2$ and the plane $x + y = 4$, as in Figure 17. The required minimum value is found on the curve formed by the intersection of the two graphs.

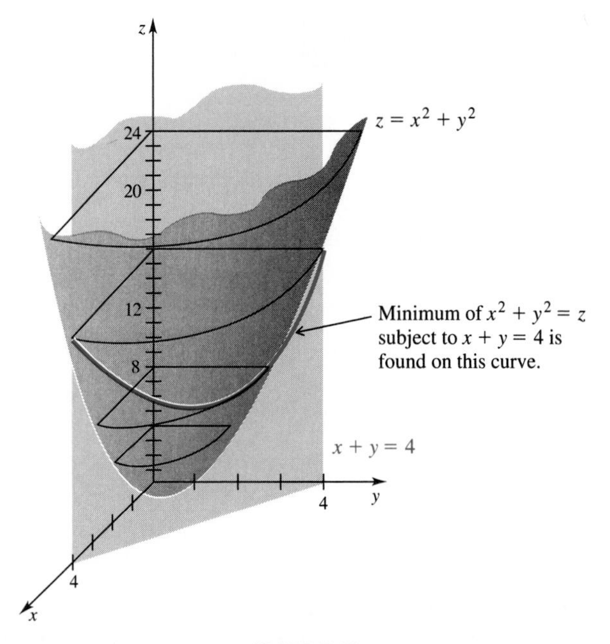

FIGURE 17

Problems with constraints are often solved by the method of Lagrange multipliers, named for the French mathematician Joseph Louis Lagrange (1736–1813). The proof for the method is complicated and is not given here. The method of Lagrange multipliers is used for problems of the form:

$$\text{Find the relative extrema for } z = f(x, y),$$
$$\text{subject to } g(x, y) = 0.$$

We state the method only for functions of two independent variables, but it is valid for any number of variables.

LAGRANGE MULTIPLIERS

All relative extrema of the function $z = f(x, y)$, subject to a constraint $g(x, y) = 0$, will be found among those points (x, y) for which there exists a value of λ such that

$$\boldsymbol{F_x(x, y, \lambda) = 0, \qquad F_y(x, y, \lambda) = 0, \qquad F_\lambda(x, y, \lambda) = 0,}$$

where $$\boldsymbol{F(x, y, \lambda) = f(x, y) + \lambda \cdot g(x, y),}$$

and all indicated partial derivatives exist.

In the theorem, the function $F(x, y, \lambda) = f(x, y) + \lambda \cdot g(x, y)$ is called the Lagrange function; λ, the Greek letter *lambda*, is the *Lagrange multiplier*.

EXAMPLE 1

Find the minimum value of

$$f(x, y) = 5x^2 + 6y^2 - xy,$$

subject to the constraint $x + 2y = 24$.

Go through the following steps.

Step 1 Rewrite the constraint in the form $g(x, y) = 0$.

In this example, the constraint $x + 2y = 24$ becomes

$$x + 2y - 24 = 0,$$

with

$$g(x, y) = x + 2y - 24.$$

Step 2 Form the Lagrange function $F(x, y, \lambda)$, the sum of the function $f(x, y)$ and the product of λ and $g(x, y)$.

Here,

$$\begin{aligned} F(x, y, \lambda) &= f(x, y) + \lambda \cdot g(x, y) \\ &= 5x^2 + 6y^2 - xy + \lambda(x + 2y - 24) \\ &= 5x^2 + 6y^2 - xy + \lambda x + 2\lambda y - 24\lambda. \end{aligned}$$

Step 3 Find F_x, F_y, and F_λ.

$$\begin{aligned} F_x &= 10x - y + \lambda \\ F_y &= 12y - x + 2\lambda \\ F_\lambda &= x + 2y - 24 \end{aligned}$$

Step 4 Form the system of equations $F_x = 0$, $F_y = 0$, and $F_\lambda = 0$.

$$10x - y + \lambda = 0 \qquad \textbf{(1)}$$

$$12y - x + 2\lambda = 0 \qquad \textbf{(2)}$$

$$x + 2y - 24 = 0 \qquad \textbf{(3)}$$

Step 5 Solve the system of equations from Step 4 for x, y, and λ.

Solve each of the first two equations for λ, then set the two results equal and simplify, as follows.

$$10x - y + \lambda = 0 \quad \text{becomes} \quad \lambda = -10x + y$$

$$12y - x + 2\lambda = 0 \quad \text{becomes} \quad \lambda = \frac{x - 12y}{2}$$

$$-10x + y = \frac{x - 12y}{2} \qquad \text{Set the expressions for } \lambda \text{ equal.}$$

$$-20x + 2y = x - 12y$$

$$-21x = -14y$$

$$x = \frac{2y}{3}$$

Now substitute $(2y)/3$ for x in equation (3).

$$x + 2y - 24 = 0$$

$$\frac{2y}{3} + 2y - 24 = 0 \qquad \text{Let } x = \frac{2y}{3}.$$

$$2y + 6y - 72 = 0$$

$$8y = 72$$

$$y = 9$$

Since $x = (2y)/3$ and $y = 9$, $x = 6$. It is not necessary to find the value of λ.

Thus, the minimum value for $f(x, y) = 5x^2 + 6y^2 - xy$, subject to the constraint $x + 2y = 24$, is at the point (6, 9). The minimum value is $f(6, 9) = 612$. The second derivative test for relative extrema from the previous section can be used to verify that 612 is indeed a minimum: since $f_{xx} = 10$, $f_{yy} = 12$, and $f_{xy} = -1$, $D = 10 \cdot 12 - (-1)^2 > 0$, so part (b) applies and indicates a minimum. ◀

Note In Example 1, we solved the system of equations by solving each equation with λ in it for λ. We then set these expressions for λ equal and solved for one of the original variables. This is a good general approach to use in solving these systems of equations, since we are not interested in the value of λ.

Before looking at applications of Lagrange multipliers, let us summarize the steps involved in solving a problem by this method.

USING LAGRANGE MULTIPLIERS

1. Write the constraint in the form $g(x, y) = 0$.
2. Form the Lagrange function

$$F(x, y, \lambda) = f(x, y) + \lambda \cdot g(x, y).$$

3. Find F_x, F_y, and F_λ.
4. Form the system of equations

$$F_x = 0, \qquad F_y = 0, \qquad F_\lambda = 0.$$

5. Solve the system in Step 4; the relative extrema for f are among the solutions of the system.

EXAMPLE 2 Complete the solution of the problem given in the introduction to this section. Maximize the area, $A(x, y) = xy$, subject to the cost constraint

$$xy + 20y + 20x + 474{,}000 = 500{,}000.$$

? Go through the five steps presented above.

Step 1 $g(x, y) = xy + 20y + 20x - 26{,}000 = 0$

Step 2 $F(x, y, \lambda) = xy + \lambda(xy + 20y + 20x - 26{,}000)$

Step 3 $F_x = y + \lambda y + 20\lambda$

$F_y = x + \lambda x + 20\lambda$

$F_\lambda = xy + 20y + 20x - 26{,}000$

Step 4 $y + \lambda y + 20\lambda = 0$ **(6)**

$x + \lambda x + 20\lambda = 0$ **(7)**

$xy + 20y + 20x - 26{,}000 = 0$ **(8)**

Step 5 Solving equations (6) and (7) for λ gives

$$\lambda = \frac{-y}{y + 20} \quad \text{and} \quad \lambda = \frac{-x}{x + 20}$$

$$\frac{-y}{y + 20} = \frac{-x}{x + 20}$$

$$y(x + 20) = x(y + 20)$$

$$xy + 20y = xy + 20x$$

$$x = y.$$

Now substitute y for x in equation (8) to get

$$y^2 + 20y + 20y - 26{,}000 = 0$$

$$y^2 + 40y - 26{,}000 = 0.$$

Use the quadratic formula to find $y \approx 142.5$. Since $x = y$, $x \approx 142.5$.

The maximum area of $(142.5)^2 \approx 20{,}306$ square feet will be achieved if the floor plan is a square with a side of 142.5 feet. ◀

As mentioned earlier, the method of Lagrange multipliers works for more than two independent variables. The next example shows how to find extrema for a function of three independent variables.

EXAMPLE 3 Find the dimensions of the closed rectangular box of maximum volume that can be produced from 6 square feet of material.

Let x, y, and z represent the dimensions of the box, as shown in Figure 18. The volume of the box is given by

$$f(x, y, z) = xyz.$$

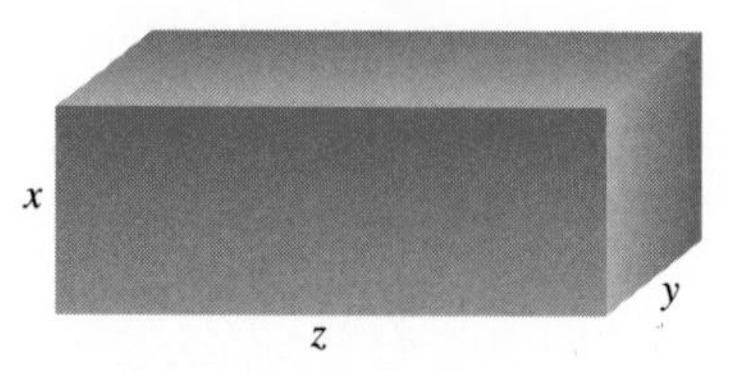

FIGURE 18

As shown in Figure 18, the total amount of material required for the two ends of the box is $2xy$, the total needed for the sides is $2xz$, and the total needed for the top and bottom is $2yz$. Since 6 square feet of material is available,

$$2xy + 2xz + 2yz = 6 \quad \text{or} \quad xy + xz + yz = 3.$$

In summary, $f(x, y, z) = xyz$ is to be maximized subject to the constraint $xy + xz + yz = 3$. Go through the steps that were given above.

Step 1 $g(x, y, z) = xy + xz + yz - 3 = 0$

Step 2 $F(x, y, z, \lambda) = xyz + \lambda(xy + xz + yz - 3)$

Step 3 $F_x = yz + \lambda y + \lambda z$

$F_y = xz + \lambda x + \lambda z$

$F_z = xy + \lambda x + \lambda y$

$F_\lambda = xy + xz + yz - 3$

Step 4 $yz + \lambda y + \lambda z = 0$

$xz + \lambda x + \lambda z = 0$

$xy + \lambda x + \lambda y = 0$

$xy + xz + yz - 3 = 0$

Step 5 Solve each of the first three equations for λ. You should get

$$\lambda = \frac{-yz}{y + z}, \quad \lambda = \frac{-xz}{x + z}, \quad \text{and} \quad \lambda = \frac{-xy}{x + y}.$$

Set these expressions for λ equal, and simplify as follows.

$$\frac{-yz}{y + z} = \frac{-xz}{x + z} \quad \text{and} \quad \frac{-xz}{x + z} = \frac{-xy}{x + y}$$

$$\frac{y}{y + z} = \frac{x}{x + z} \qquad \frac{z}{x + z} = \frac{y}{x + y}$$

$$xy + yz = xy + xz \qquad zx + zy = yx + yz$$

$$yz = xz \qquad zx = yx$$

$$y = x \qquad z = y$$

(Setting the first and third expressions equal gives no additional information.) Thus $x = y = z$. From the fourth equation in Step 4, with $x = y$ and $z = y$,

$$xy + xz + yz - 3 = 0$$
$$y^2 + y^2 + y^2 - 3 = 0$$
$$3y^2 = 3$$
$$y^2 = 1$$
$$y = \pm 1.$$

The negative solution is not applicable, so the solution of the system of equations is $x = 1$, $y = 1$, $z = 1$. In other words, the box is a cube that measures 1 foot on a side. ◀

8.4 Exercises

Find the relative maxima or minima in Exercises 1–10.

1. Maximum of $f(x, y) = 2xy$, subject to $x + y = 12$
2. Maximum of $f(x, y) = 4xy + 2$, subject to $x + y = 24$
3. Maximum of $f(x, y) = x^2y$, subject to $2x + y = 4$
4. Maximum of $f(x, y) = 4xy^2$, subject to $3x - 2y = 5$
5. Minimum of $f(x, y) = x^2 + 2y^2 - xy$, subject to $x + y = 8$
6. Minimum of $f(x, y) = 3x^2 + 4y^2 - xy - 2$, subject to $2x + y = 21$
7. Maximum of $f(x, y) = x^2 - 10y^2$, subject to $x - y = 18$
8. Maximum of $f(x, y) = 12xy - x^2 - 3y^2$, subject to $x + y = 16$
9. Maximum of $f(x, y, z) = xyz^2$, subject to $x + y + z = 6$
10. Maximum of $f(x, y, z) = xy + 2xz + 2yz$, subject to $xyz = 32$
11. Find positive numbers x and y such that $x + y = 18$ and xy^2 is maximized.
12. Find positive numbers x and y such that $x + y = 36$ and x^2y is maximized.
13. Find three positive numbers whose sum is 90 and whose product is a maximum.
14. Find three positive numbers whose sum is 240 and whose product is a maximum.
15. Explain the difference between the two methods we used in Sections 3 and 4 to solve extrema problems.
16. Why is it unnecessary to find the value of λ when using the method explained in this section?

Applications

Business and Economics

17. **Maximum Area for Fixed Expenditure** Because of terrain difficulties, two sides of a fence can be built for $6 per foot, while the other two sides cost $4 per foot. (See the sketch.) Find the field of maximum area that can be enclosed for $1200.

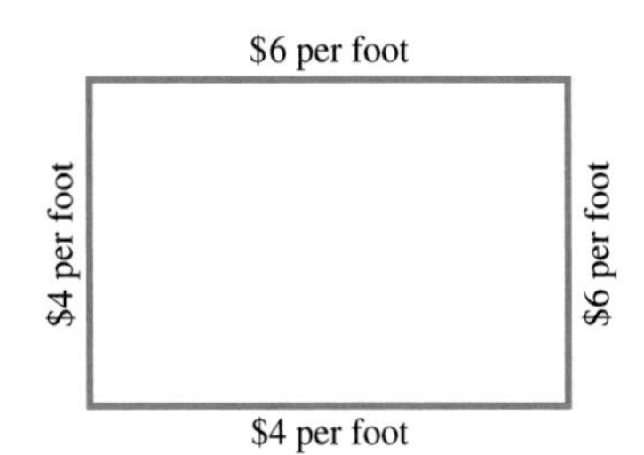

18. **Maximum Area for Fixed Expenditure** To enclose a yard, a fence is built against a large building, so that fencing material is used only on three sides. Material for the ends cost $8 per foot; material for the side opposite the building costs $6 per foot. Find the dimensions of the yard of maximum area that can be enclosed for $1200.
19. **Cost** The total cost to produce x large needlepoint kits and y small ones is given by
$$C(x, y) = 2x^2 + 6y^2 + 4xy + 10.$$
If a total of ten kits must be made, how should production be allocated so that total cost is minimized?
20. **Profit** The profit from the sale of x units of radiators for automobiles and y units of radiators for generators is given by
$$P(x, y) = -x^2 - y^2 + 4x + 8y.$$
Find values of x and y that lead to a maximum profit if the firm must produce a total of 6 units of radiators.
21. **Production** A manufacturing firm estimates that its total production of automobile batteries in thousands of units is
$$f(x, y) = 3x^{1/3}y^{2/3},$$
where x is the number of units of labor and y is the number of units of capital utilized. Labor costs are $80 per unit, and capital costs are $150 per unit. How many units each of labor and capital will maximize production, if the firm can spend $40,000 for these costs?
22. **Production** For another product, the manufacturing firm in Exercise 21 estimates that production is a function of labor x and capital y as follows:
$$f(x, y) = 12x^{3/4}y^{1/4}.$$
If $25,200 is available for labor and capital, and if the firm's costs are $100 and $180 per unit, respectively, how many units of labor and capital will give maximum production?

General Interest

23. **Area** A farmer has 200 m of fencing. Find the dimensions of the rectangular field of maximum area that can be enclosed by this amount of fencing.
24. **Area** Find the area of the largest rectangular field that can be enclosed with 600 m of fencing. Assume that no fencing is needed along one side of the field.
25. **Surface Area** A cylindrical can is to be made that will hold 250π cubic inches of candy. Find the dimensions of the can with minimum surface area.
26. **Surface Area** An ordinary 12-oz beer or soda pop can holds about 25 cubic inches. Use a calculator and find the dimensions of a can with minimum surface area. Measure a can and see how close its dimensions are to the results you found.
27. **Volume** A rectangular box with no top is to be built from 500 square meters of material. Find the dimensions of such a box that will enclose the maximum volume.
28. **Surface Area** A 1-lb soda cracker box has a volume of 185 cubic inches. The end of the box is square. Find the dimensions of such a box that has minimum surface area.
29. **Cost** A rectangular closed box is to be built at minimum cost to hold 27 cubic meters. Since the cost will depend on the surface area, find the dimensions that will minimize the surface area of the box.
30. **Cost** Find the dimensions that will minimize the surface area (and hence the cost) of a rectangular fish aquarium, open on top, with a volume of 32 cubic feet.

8.5 THE LEAST SQUARES LINE: A MINIMIZATION APPLICATION

How can we express the relationship between the size of a room and the energy needed to cool the room?

Throughout this book we have studied functions that relate various quantities. In almost every case, the function was given. An important question is, How are these functions determined? This section shows how one kind of function (linear) can be developed. We will return to the introductory question in the exercises for this section.

In trying to predict the future sales of a product or the total number of matings between two species of insects, it is common to gather as much data as possible and then draw a graph showing the data. This graph, called a **scatter diagram,** can then be inspected to see if a reasonably simple mathematical curve will fit fairly well through all the given points.

As an example, suppose a firm gathers data showing the relationship between the price y of an item (in dollars) and the number x of units sold, with results as follows.

Units Sold, x	10	15	20
Price, y	80	68	62

A graph of these data is shown in Figure 19.

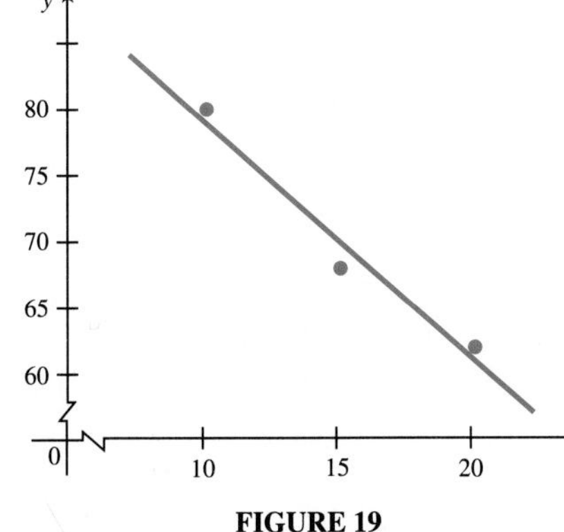

FIGURE 19

The graph suggests that a straight line fits reasonably well through these data points. If all the data points were to lie on the straight line, the point-slope form of the equation of a line could be used to find the equation of the line through the points.

In practice, the points on a scatter diagram almost never fit a straight line exactly. Usually we must decide on the "best" straight line through the points. One way to define the "best" line is as follows. Figure 20 shows the same scattered points from Figure 19. This time, however, vertical line segments indicate the distances of these points from a possible line. The "best" straight line is often defined as the one for which the sum of the squares of the distances, $d_1^2 + d_2^2 + d_3^2$ here, is minimized.

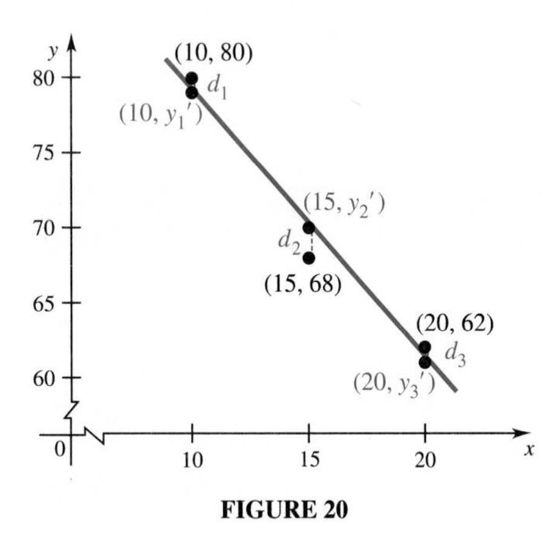

FIGURE 20

To find the equation of the "best" line, let $y' = mx + b$ be the equation of the line in Figure 20. (The purpose of using y' here is to distinguish the calculated y-values from the given y-values. This use of y' has nothing to do with the derivative.) Then for the point with distance d_1, $x = 10$ and $y = 80$, so that $y' = mx + b$ becomes $y_1' = 10m + b$, and

$$d_1 = 80 - y_1' = 80 - (10m + b).$$

We can find d_2 and d_3 in the same way. Thus,

$$d_1^2 = [80 - (10m + b)]^2,$$
$$d_2^2 = [68 - (15m + b)]^2,$$
$$d_3^2 = [62 - (20m + b)]^2.$$

Values of m and b must be found that will minimize $d_1^2 + d_2^2 + d_3^2$. If $d_1^2 + d_2^2 + d_3^2 = S$, then

$$S = [80 - (10m + b)]^2 + [68 - (15m + b)]^2 + [62 - (20m + b)]^2.$$

To minimize S using the method shown earlier in this chapter, the partial derivatives $\partial S/\partial m$ and $\partial S/\partial b$ are needed.

$$\frac{\partial S}{\partial m} = -20[80 - (10m + b)] - 30[68 - (15m + b)] - 40[62 - (20m + b)]$$

$$\frac{\partial S}{\partial b} = -2[80 - (10m + b)] - 2[68 - (15m + b)] - 2[62 - (20m + b)]$$

Simplify each of these, obtaining

$$\frac{\partial S}{\partial m} = -6120 + 1450m + 90b$$

and

$$\frac{\partial S}{\partial b} = -420 + 90m + 6b.$$

Set $\partial S/\partial m = 0$ and $\partial S/\partial b = 0$ and simplify.

$$-6120 + 1450m + 90b = 0 \quad \text{or} \quad 145m + 9b = 612$$

$$-420 + 90m + 6b = 0 \quad \text{or} \quad 15m + b = 70$$

Solving the system on the right leads to the solution $m = -1.8$, $b = 97$. The equation that best fits through the points in Figure 19 is $y' = mx + b$, or

$$\mathbf{y' = -1.8x + 97.}$$

The second derivative test can be used to show that this line *minimizes* the sum of the squares of the distances. Thus, $y' = -1.8x + 97$ is the "best" straight line that will fit through the data points. This line is called the **least squares line,** or sometimes the **regression line.**

Once a least squares line is obtained from a set of data, the equation can be used to predict a value of one variable, given a value of the other. In the example above, management might want an estimate of the number of units that would be sold at a price of \$75. To get this estimate, replace y' with 75.

$$\begin{aligned} y' &= -1.8x + 97 \\ 75 &= -1.8x + 97 \\ -22 &= -1.8x \\ 12.2 &\approx x \end{aligned}$$

About 12 units would be sold at a price of \$75.

The example above used three data points on the scatter diagram. A practical problem, however, might well involve a very large number of points. The method used above can be extended to find the least squares line for any finite number of data points. To simplify the notation for this more complicated case, *summation notation,* or *sigma notation*, is used as explained in the chapter on integration.

$$\sum_{i=1}^{n} x_i = x_1 + x_2 + x_3 + \cdots + x_n$$

Let $y' = mx + b$ be the least squares line for the set of known data points $(x_1, y_1), (x_2, y_2), \ldots, (x_n, y_n)$. See Figure 21. As above, the sum of squares,

$$d_1^2 + d_2^2 + \cdots + d_n^2,$$

is to be minimized. The square of the distance of the first point (x_1, y_1) from the line $y' = mx + b$ is given by

$$d_1^2 = [y_1 - (mx_1 + b)]^2.$$

The square of the distance of the point (x_i, y_i) from the line is

$$d_i^2 = [y_i - (mx_i + b)]^2.$$

Thus, the sum

$$\sum_{i=1}^{n} [y_i - (mx_i + b)]^2$$

is to be minimized.

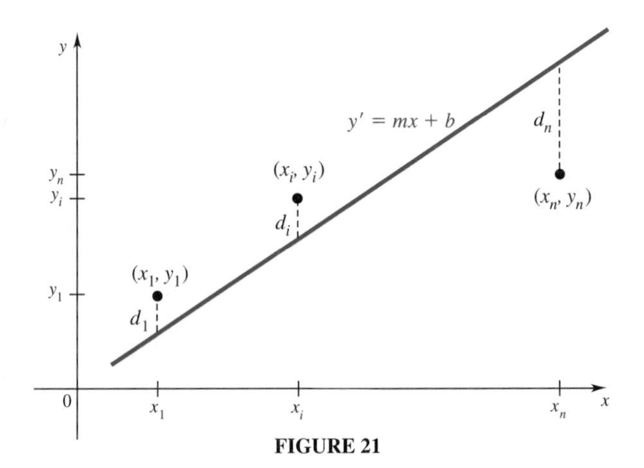

FIGURE 21

Since the x_i and y_i values represent known data points, the unknowns in this sum are the numbers m and b. To emphasize this fact, write the sum as a function of m and b:

$$\begin{aligned} f(m, b) &= \sum_{i=1}^{n} [y_i - (mx_i + b)]^2 \\ &= \sum_{i=1}^{n} [y_i - mx_i - b]^2 \\ &= (y_1 - mx_1 - b)^2 + (y_2 - mx_2 - b)^2 \\ &\quad + \cdots + (y_n - mx_n - b)^2. \end{aligned}$$

To find the minimum value of this function, find the partial derivatives with respect to m and to b and set each equal to 0. (Recall that all the x's and y's are constants here.)

$$\frac{\partial f}{\partial m} = -2x_1(y_1 - mx_1 - b) - 2x_2(y_2 - mx_2 - b) - \cdots$$
$$-2x_n(y_n - mx_n - b) = 0$$

$$\frac{\partial f}{\partial b} = -2(y_1 - mx_1 - b) - 2(y_2 - mx_2 - b) - \cdots$$
$$-2(y_n - mx_n - b) = 0$$

Using some algebra and rearranging terms, these two equations become

$$(x_1^2 + x_2^2 + \cdots + x_n^2)m + (x_1 + x_2 + \cdots + x_n)b$$
$$= x_1y_1 + x_2y_2 + \cdots + x_ny_n,$$

and $$(x_1 + x_2 + \cdots + x_n)m + nb = y_1 + y_2 + \cdots + y_n.$$

These last two equations can be rewritten using abbreviated sigma notation as follows. (Remember that n terms are being added.)

$$\left(\sum x^2\right)m + \left(\sum x\right)b = \sum xy \qquad \textbf{(1)}$$

$$\left(\sum x\right)m + nb = \sum y \qquad \textbf{(2)}$$

To solve this system of equations, multiply the first equation on both sides by $-n$ and the second on both sides by Σx.

$$-n\left(\sum x^2\right)m - n\left(\sum x\right)b = -n\left(\sum xy\right)$$
$$\left(\sum x\right)\left(\sum x\right)m + \left(\sum x\right)nb = \left(\sum x\right)\left(\sum y\right)$$

Add the two new equations to eliminate the term with b.

$$\left(\sum x\right)\left(\sum x\right)m - n\left(\sum x^2\right)m = \left(\sum x\right)\left(\sum y\right) - n\left(\sum xy\right)$$

Write the product $(\Sigma x)(\Sigma x)$ as $(\Sigma x)^2$. Using this notation and solving the last equation for m gives

$$m = \frac{\left(\sum x\right)\left(\sum y\right) - n\left(\sum xy\right)}{\left(\sum x\right)^2 - n\left(\sum x^2\right)}. \qquad \textbf{(3)}$$

The easiest way to find b is to solve equation (2) for b:

$$\left(\sum x\right)m + nb = \sum y \tag{2}$$

$$nb = \sum y - \left(\sum x\right)m$$

$$b = \frac{\sum y - m\left(\sum x\right)}{n}.$$

A summary of the formulas for m and b in the least squares equation follows.

LEAST SQUARES EQUATION

The **least squares equation** for the n points $(x_1, y_1), (x_2, y_2), \ldots, (x_n, y_n)$ is given by

$$\mathbf{y' = mx + b,}$$

where

$$m = \frac{(\Sigma x)(\Sigma y) - n(\Sigma xy)}{(\Sigma x)^2 - n(\Sigma x^2)} \quad \text{and} \quad b = \frac{\Sigma y - m(\Sigma x)}{n}.$$

As the formulas suggest, find m first. Next, use the value of m and the second formula to find b.

Find the least squares line for the data in the following chart.

x	1	2	3	5	6
y	−2	2	5	12	14

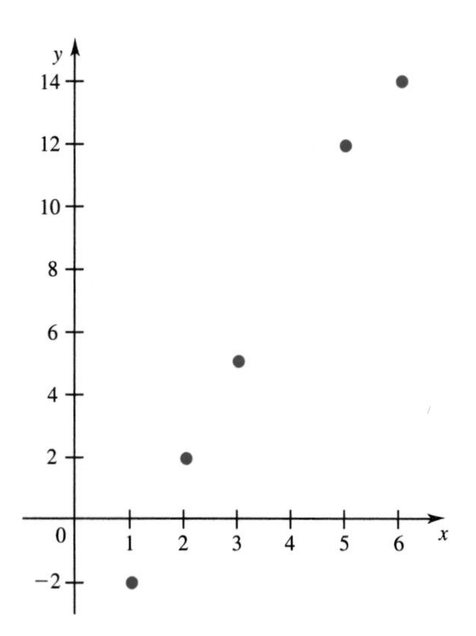

FIGURE 22

Start by drawing the scatter diagram of Figure 22. Since the points in the scatter diagram lie approximately in a straight line, it is appropriate to find the least squares line for the data.

The formulas for m and b require Σx, Σy, Σx^2, and Σxy. Organize the work as follows.

	x	y	x^2	xy
	1	−2	1	−2
	2	2	4	4
	3	5	9	15
	5	12	25	60
	6	14	36	84
Totals	17	31	75	161

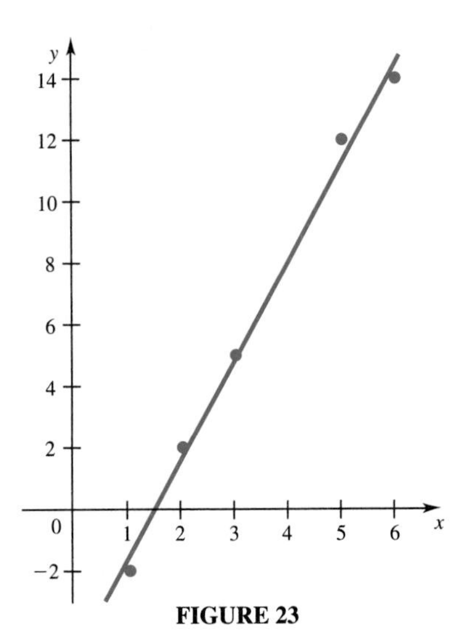

FIGURE 23

The chart shows that $\Sigma x = 17$, $\Sigma y = 31$, $\Sigma x^2 = 75$, $\Sigma xy = 161$, and $n = 5$. Using the equation for m given above,

$$m = \frac{17(31) - 5(161)}{17^2 - 5(75)}$$

$$= \frac{527 - 805}{289 - 375} = \frac{-278}{-86} \approx 3.2.$$

Now find b:

$$b = \frac{\Sigma y - m(\Sigma x)}{n}$$

$$= \frac{31 - 3.2(17)}{5} \approx -4.7.$$

The least squares equation for the given data is $y' = mx + b$, or

$$y' = 3.2x - 4.7.$$

This line is graphed in Figure 23. ◀

The equation $y' = 3.2x - 4.7$ can be used to estimate values of y for given values of x. For example, if $x = 4$,

$$y' = 3.2x - 4.7 = 3.2(4) - 4.7 = 12.8 - 4.7 = 8.1.$$

Also, if $x = 7$, then $y' = 3.2(7) - 4.7 = 17.7$.

Sir Francis Galton (1822–1911), a British scientist, is credited with setting forth the ideas of regression and correlation in his statistical work *Natural Inheritance*. His interest in statistics developed as a result of his research on weather prediction.

8.5 Exercises

1. Suppose a linear correlation is found between two quantities. Does this mean that increasing one of the quantities causes the other to increase? If not, what does it mean?
2. Given a set of points, the least squares line formed by letting x be the independent variable will not necessarily be the same as the least squares line formed by letting y be the independent variable. Give an example to show why this is true.

In Exercises 3 and 4, draw a scatter diagram for the given set of data points and then find the least squares equation.

3.

x	4	5	8	12	14
y	3	7	17	28	35

Estimate x when y is 12; 32.

4.

x	3	4	5	6	8
y	8	12	16	18	28

Estimate x when y is 17; 26.

Applications

Business and Economics

5. **Sales and Profit** A fast-food chain wishes to find the relationship between annual store sales (in thousands of dollars) and percent of pretax profit in order to estimate increases in profit due to increased sales volume. The data shown below were obtained from a sample of stores across the country.

Annual Sales, x	Pretax Profit, y
250	9.3
300	10.8
375	14.8
425	14.0
450	14.2
475	15.3
500	15.9
575	19.1
600	19.2
650	21.0

(a) Plot the ten pairs of values on a scatter diagram.

(b) Find the equation of the least squares line and graph it on the scatter diagram of part (a).

(c) Using the equation of part (b), predict the percent of pretax profit for annual sales of \$700,000; of \$750,000.

6. **Insurance Sales** Records show that the annual sales for the Sweet Palms Life Insurance Company in five-year periods for the last 20 years were as follows.

Year	Year in Coded Form, x	Sales in Millions, y
1970	1	1.0
1975	2	1.3
1980	3	1.7
1985	4	1.9
1990	5	2.1

(a) Plot the five points on a scatter diagram.

(b) Find the least squares equation and graph it on the scatter diagram from part (a).

(c) Predict the company's sales for 1995.

7. **Repair Costs and Production** The following data were used to determine whether there is a relationship between repair costs and barrels of beer produced by a brewery. The data (in thousands) are given for a 10-month period.

Month	Barrels of Beer, x	Repairs, y
Jan.	369	299
Feb.	379	280
Mar.	482	393
April	493	388
May	496	385
June	567	423
July	521	374
Aug.	482	357
Sept.	391	106
Oct.	417	332

(a) Find the equation of the least squares line.

(b) If 500,000 barrels of beer are produced, what will the equation from part (a) give as the predicted repair costs?

(c) Is there a relationship between repair costs and barrels of beer produced? If so, what kind of relationship exists?

8. **Sales Analysis** Sales (in thousands of dollars) of a certain company are shown here.

Year, x	0	1	2	3	4	5
Sales, y	48	59	66	75	80	90

(a) Find the equation of the least squares line.

(b) Predict sales in year 7.

It is sometimes possible to get a better prediction for a variable by considering its relationship with more than one other variable. For example, one should be able to predict college GPAs more precisely if both high school GPAs and scores on the ACT are considered. To do this, we alter the equation used to find a least squares line by adding a term for the new variable as follows. If y represents college GPAs, x_1 high school GPAs, and x_2 ACT scores, then y', the predicted GPA, is given by

$$y' = ax_1 + bx_2 + c.$$

This equation represents a ***least squares plane.*** *The equations for the constants a, b, and c are more complicated than those given in the text for m and b, so calculating a least squares equation for three variables is more likely to require the aid of a computer.*

9. **Revenue and Price** Alcoa* used a least squares plane with two independent variables, x_1 and x_2, to predict the effect on revenue of the price of aluminum forged truck wheels, as follows.

x_1 = the average price per wheel

x_2 = large truck production (in thousands)

y = sales of aluminum forged truck wheels (in thousands)

Using data for the past eleven years, the company found the equation of the least squares plane to be

$$y' = 49.2755 - 1.1924x_1 + .1631x_2.$$

The following figures were then forecast for truck production.

1987	1988	1989	1990	1991	1992
160.0	165.0	170.0	175.0	180.0	185.0

Three possible price levels per wheel were considered: $42, $45, and $48.

(a) Use the least squares plane equation given above to find the estimated sales of wheels (y) for 1989 at each of the three price levels.

(b) Repeat part (a) for 1992.

(c) For which price level, on the basis of the 1989 and 1992 figures, are total estimated sales greatest?

(By comparing total estimated sales for the years 1987 through 1992 at each of the three price levels, the company found that the selling price of $42 per wheel would generate the greatest sales volume over the six-year period.)

Life Sciences

10. **Crickets Chirping** Biologists have observed a linear relationship between the temperature and the frequency with which a cricket chirps. The following data were measured for the striped ground cricket.†

Temperature in °F (x)	Chirps per second (y)
88.6	20.0
71.6	16.0
93.3	19.8
84.3	18.4
80.6	17.1
75.2	15.5
69.7	14.7
82.0	17.1
69.4	15.4
83.3	16.2
79.6	15.0
82.6	17.2
80.6	16.0
83.5	17.0
76.3	14.4

(a) Find the least squares line for these data.

(b) Use the results of part (a) to determine how many chirps per second you would expect to hear from the striped ground cricket if the temperature is 73° F.

(c) Use the results of part (a) to determine what the temperature is when the striped ground cricket is chirping 18 times per second.

11. **Bacterial Growth** Sometimes the scatter diagram of the data does not have a linear pattern. This is particularly true in some biological and chemical applications. In these applications, however, often the scatter diagram of the *logarithms* of the data has a linear pattern. A least squares line can then be used to predict the logarithm of any desired value, from which the value itself can be found. Suppose that a certain kind of bacterium grows in numbers as shown in Table A. The actual number of bacteria present at each time period is replaced with the common logarithm of that number in Table B.

Table A

Time (in hours)	Number of Bacteria
0	1000
1	1649
2	2718
3	4482
4	7389
5	12182

Table B

Time, x	log, y
0	3.0000
1	3.2172
2	3.4343
3	3.6515
4	3.8686
5	4.0857

*This example supplied by John H. Van Denender, Public Relations Department, Aluminum Company of America.

†From George W. Pierce, *The Songs of Insects* (Harvard University Press, 1948), p. 20.

We can now find a least squares line that will predict y, given x.

(a) Plot the original pairs of numbers. The pattern should be nonlinear.

(b) Plot the log values against the time values. The pattern should be almost linear.

(c) Find the equation of the least squares line. (First round off the log values to the nearest hundredth.)

(d) Predict the log value for a time of 7 hr. Find the number whose logarithm is your answer. This will be the predicted number of bacteria.

12. **Height and Weight** A sample of 10 adult men gave the following data on their heights (in inches) and weights (in pounds).

Height, x	62	62	63	65	66	67	68	68	70	72
Weight, y	120	140	130	150	142	130	135	175	149	168

(a) Find the equation of the least squares line.

Using the results above, predict the weight of a man whose height is as follows.

(b) 60 inches (c) 70 inches

Social Sciences

13. **Educational Expenditures** A report issued by the U.S. Department of Education in 1991 gave the expenditure per pupil and the average mathematics proficiency in grade 8 for 37 states and the District of Columbia. Letting x equal the expenditure per pupil (ranging from \$2838 in Idaho to \$7850 in Washington, D.C.) and y equal the average mathematics proficiency score (ranging from 231 in Washington, D. C. to 281 in North Dakota), the data can be summarized as follows. (From an article in *Newsday*, June 7, 1991.)

$$n = 38 \qquad \Sigma\, x = 175{,}878 \qquad \Sigma\, y = 9989$$

$$\Sigma\, xy = 46{,}209{,}266 \qquad \Sigma\, x^2 = 872{,}066{,}218$$

(a) Find the least squares equation for these data.

(b) Assuming a linear relationship, estimate the average mathematics proficiency scores for Idaho and Washington, D.C.

(c) What does the negative sign in the equation indicate?

14. **Death Rate** The death rate in the U.S. has gradually decreased between 1910 and 1988, as shown in the data at the top of the next column.

Year, x	Death Rate, y
1910	14.7
1920	13.0
1930	11.3
1940	10.8
1950	9.6
1960	9.5
1970	9.5
1980	8.8
1988	8.8

(a) Assume this is a linear decrease and find the least squares equation for the data. Use 10 for 1910, 20 for 1920, and so on.

(b) Using the equation from part (a), find the death rates for 1920, 1950, and 1980.

(c) Compare the results from part (b) with the given data. Is the linear equation a reasonable fit? Why?

(d) Use the equation from part (a) to predict the death rate in 1994.

15. **SAT Scores** At Hofstra University, all students take the SAT before entrance, and most students take a mathematics placement test (designed by the Mathematical Association of America) before registration. In Fall 1991, one professor collected the following data for 19 students in his finite mathematics class.

Math SAT	Placement Test	Math SAT	Placement Test	Math SAT	Placement Test
540	20	580	8	440	10
510	16	680	15	520	11
490	10	560	8	620	11
560	8	560	13	680	8
470	12	500	14	550	8
600	11	470	10	620	7
540	10				

(a) Find an equation for the least squares line.

(b) Use your answer from part (a) to predict the mathematics placement test score for a student with a math SAT score of 420.

(c) Use your answer from part (a) to predict the mathematics placement test score for a student with a math SAT score of 620.

Physical Sciences

16. Temperature In an experiment to determine the linear relationship between temperatures on the Celsius scale (y) and on the Fahrenheit scale (x), a student got the following results.

$$n = 5 \qquad \Sigma x = 376 \qquad \Sigma y = 120$$

$$\Sigma xy = 28{,}050 \qquad \Sigma x^2 = 62{,}522$$

(a) Find an equation for the least squares line.

(b) Find the reading on the Celsius scale that corresponds to a reading of 120° Fahrenheit, using the equation from part (a).

General Interest

17. Air Conditioning While shopping for a new air conditioner, Adam Bryer consulted the following table to determine the right model to buy. The table shows the number of BTUs needed to cool rooms having specified areas. Find a least squares equation for these data.

Area of Room in Square Feet, x	BTUs of Air Conditioner Needed, y
150	5000
175	5500
215	6000
250	6500
280	7000
310	7500
350	8000
370	8500
420	9000
450	9500

(a) To check the fit of the data to the line, use the equation to find the number of BTUs required to cool rooms of 150 ft^2, 280 ft^2, 420 ft^2. How well do the actual data agree with the predicted values?

(b) Suppose your room measures 230 ft^2. Use the equation to decide how many BTUs are required to cool the air. If air conditioners are available only with the BTU choices in the table, how many BTUs do you need?

(c) Why do you think the table gives ft^2 instead of cubic feet, which would give the volume of the room?

8.6 TOTAL DIFFERENTIALS AND APPROXIMATIONS

How will a change in the production of fertilizer affect profit?

In the chapter on applications of the derivative, we introduced the differential. Recall that the differential of a function defined by $y = f(x)$ is

$$dy = f'(x) \cdot dx,$$

where dx, the differential of x, is any real number (usually small). We saw that the differential dy is often a good approximation of Δy, where $\Delta y = f(x + \Delta x) - f(x)$ and $\Delta x = dx$.

In the second section of this chapter we used partial derivatives to find the marginal productivity of labor and of capital for a production function. The marginal productivity gives the rate of change of production for a one-unit change in labor or for a one-unit change in capital. To find the change in productivity for any (usually small) change in labor and/or in capital, we need the concept of a *differential.* Since the differential dy equals $f'(x)\,dx$, it is a function of *two* variables, x and dx. We can extend this idea by defining a *total differential.*

TOTAL DIFFERENTIAL FOR TWO VARIABLES

Let $z = f(x, y)$ be a function of x and y. Let dx and dy be real numbers. Then the **total differential** of f is

$$\boldsymbol{df = f_x(x, y) \cdot dx + f_y(x, y) \cdot dy.}$$

(Sometimes df is written dz.)

By this definition, df is a function of *four* variables, x, dx, y, and dy.

EXAMPLE 1 Find df for each function.

(a) $f(x, y) = 9x^3 - 8x^2y + 4y^3$

First find f_x and f_y.

$$f_x = 27x^2 - 16xy \qquad \text{and} \qquad f_y = -8x^2 + 12y^2$$

By the definition,

$$df = (27x^2 - 16xy)\,dx + (-8x^2 + 12y^2)\,dy.$$

(b) $z = f(x, y) = \ln (x^3 + y^2)$

Since

$$f_x = \frac{3x^2}{x^3 + y^2} \qquad \text{and} \qquad f_y = \frac{2y}{x^3 + y^2},$$

the total differential is

$$dz = df = \left(\frac{3x^2}{x^3 + y^2}\right) dx + \left(\frac{2y}{x^3 + y^2}\right) dy.$$

EXAMPLE 2 Let $f(x, y) = 9x^3 - 8x^2y + 4y^3$. Find df when $x = 1$, $y = 3$, $dx = .01$, and $dy = -.02$.

Example 1(a) gave the total differential

$$df = (27x^2 - 16xy)\,dx + (-8x^2 + 12y^2)\,dy.$$

Replace x with 1, y with 3, dx with .01, and dy with $-.02$.

$$\begin{aligned} df &= [27(1)^2 - 16(1)(3)](.01) + [-8(1)^2 + 12(3)^2](-.02) \\ &= (27 - 48)(.01) + (-8 + 108)(-.02) \\ &= (-21)(.01) + (100)(-.02) \\ &= -2.21 \end{aligned}$$

This result indicates that an increase of .01 in x and a decrease of .02 in y, when $x = 1$ and $y = 3$, will produce a *decrease* of 2.21 in $f(x, y)$.

The idea of a total differential can be extended to include functions of three independent variables.

TOTAL DIFFERENTIAL FOR THREE VARIABLES

If $w = f(x, y, z)$, then the total differential dw is

$$dw = f_x(x, y, z)\,dx + f_y(x, y, z)\,dy + f_z(x, y, z)\,dz,$$

provided all indicated partial derivatives exist.

As this definition shows, dw is a function of six variables, x, y, z, dx, dy, and dz. Similar definitions could be given for functions of more than three independent variables.

Approximations Recall that with a function of one variable, $y = f(x)$, the differential dy can be used to approximate the change in y, Δy, corresponding to a change in x, Δx. A similar approximation can be made for a function of two variables.

APPROXIMATIONS

For small values of Δx and Δy,

$$dz \approx \Delta z,$$

where $\Delta z = f(x + \Delta x, y + \Delta y) - f(x, y)$.

EXAMPLE 3 Let $f(x, y) = 6x^2 + xy + y^3$. Find Δz and dz when $x = 2$, $y = -1$, $\Delta x = -.03$, and $\Delta y = .02$.

Here $x + \Delta x = 2 + (-.03) = 1.97$ and $y + \Delta y = -1 + .02 = -.98$. Find Δz with the definition above.

$$\begin{aligned}\Delta z &= f(x + \Delta x, y + \Delta y) - f(x, y)\\ &= f(1.97, -.98) - f(2, -1)\\ &= [6(1.97)^2 + (1.97)(-.98) + (-.98)^3] - [6(2)^2 + 2(-1) + (-1)^3]\\ &= [23.2854 - 1.9306 - .941192] - [24 - 2 - 1]\\ &= 20.413608 - 21\\ &= -.586392\end{aligned}$$

To find dz, the total differential, first find

$$f_x = 12x + y \qquad \text{and} \qquad f_y = x + 3y^2.$$

Then

$$dz = (12x + y)\,dx + (x + 3y^2)\,dy.$$

Since $dx = \Delta x$ and $dy = \Delta y$, substitution gives

$$\begin{aligned}dz &= [12 \cdot 2 + (-1)](-.03) + [2 + 3 \cdot (-1)^2](.02)\\ &= (23)(-.03) + (5)(.02)\\ &= -.59.\end{aligned}$$

In this example the values of Δz and dz are very close. (Compare the amount of work needed to find Δz with that needed for dz.) ◀

For small values of dx and dy, the values of Δz and dz are approximately equal. Since $\Delta z = f(x + dx, y + dy) - f(x, y)$,

$$f(x + dx, y + dy) = f(x, y) + \Delta z$$

or

$$f(x + dx, y + dy) \approx f(x, y) + dz.$$

Replacing dz with the expression for the total differential gives the following result.

APPROXIMATIONS BY DIFFERENTIALS

For a function f having all indicated partial derivatives, and for small values of dx and dy,

$$\boldsymbol{f(x + dx, y + dy) \approx f(x, y) + dz,}$$

or

$$\boldsymbol{f(x + dx, y + dy) \approx f(x, y) + f_x(x, y) \cdot dx + f_y(x, y) \cdot dy.}$$

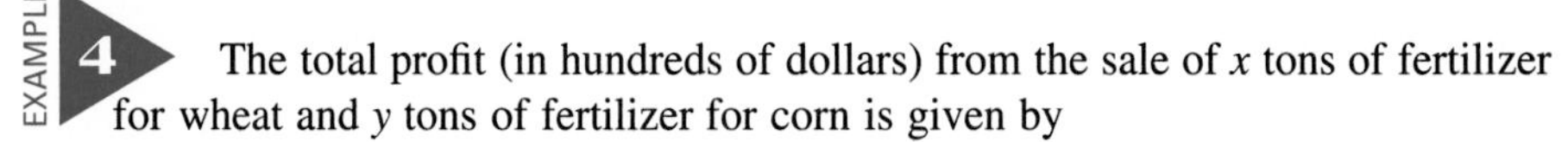

EXAMPLE 4 The total profit (in hundreds of dollars) from the sale of x tons of fertilizer for wheat and y tons of fertilizer for corn is given by

$$P(x, y) = 6x^{1/2} + 9y^{2/3} + \frac{5000}{xy}.$$

Suppose a firm is producing 64 tons of fertilizer for wheat and 125 tons of fertilizer for corn, and it is considering changing production to 63 tons and 127 tons, respectively. Before making the change, the firm wants to estimate how this would affect total profit. Use the total differential to estimate the change in profit.

The approximate change in profit, dP, is

$$dP = P_x(x, y) \cdot dx + P_y(x, y) \cdot dy.$$

Finding the necessary partial derivatives gives

$$dP = \left(3x^{-1/2} - \frac{5000}{x^2y}\right) dx + \left(6y^{-1/3} - \frac{5000}{xy^2}\right) dy.$$

Substitute 64 for x, 125 for y, -1 for dx, and 2 for dy.

$$\begin{aligned} dP &= \left(3 \cdot 64^{-1/2} - \frac{5000}{64^2 \cdot 125}\right)(-1) + \left(6 \cdot 125^{-1/3} - \frac{5000}{64 \cdot 125^2}\right)(2) \\ &= \left(3 \cdot \frac{1}{8} - \frac{5000}{512{,}000}\right)(-1) + \left(6 \cdot \frac{1}{5} - \frac{5000}{1{,}000{,}000}\right)(2) \\ &= (.375 - .0098)(-1) + (1.2 - .005)(2) \\ &\approx 2.0248 \end{aligned}$$

A change in production from 64 tons of fertilizer for wheat and 125 tons for corn to 63 and 127 tons, respectively, will increase profits by about 2.02 hundred dollars, or \$202.

To estimate the total profit from the production of 63 tons and 127 tons, use the approximation formula given above.

$$P(63, 127) \approx P(64, 125) + dP$$
$$= \left(6 \cdot 64^{1/2} + 9 \cdot 125^{2/3} + \frac{5000}{64 \cdot 125}\right) + 2.0248$$
$$= \left(6 \cdot 8 + 9 \cdot 25 + \frac{5000}{8000}\right) + 2.0248 = 275.65$$

hundred dollars, or a total profit of \$27,565. ◀

EXAMPLE 5 A short length of blood vessel is in the shape of a right circular cylinder (see Figure 24). The length of the vessel is measured as 42 millimeters, and the radius is measured as 2.5 millimeters. Suppose the maximum error in the measurement of the length is .9 millimeters, with an error of no more than .2 millimeters in the measurement of the radius. Find the maximum possible error in calculating the volume of the blood vessel.

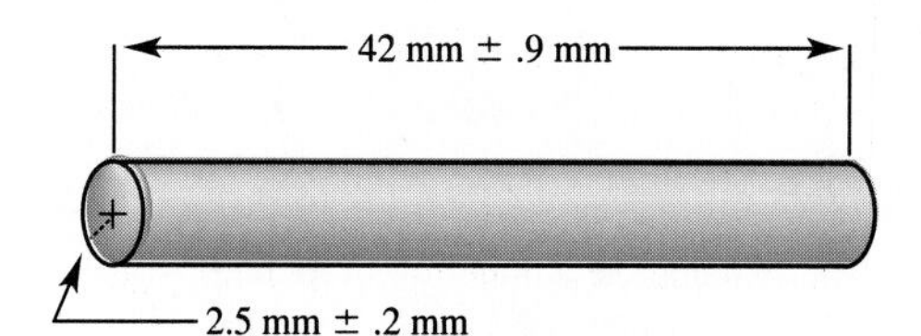

FIGURE 24

The volume of a right circular cylinder is given by $V = \pi r^2 h$. To approximate the error in the volume, find the total differential, dV.

$$dV = (2\pi rh) \cdot dr + (\pi r^2) \cdot dh$$

Here, $r = 2.5$, $h = 42$, $dr = .2$, and $dh = .9$. Substitution gives

$$dV = [(2\pi)(2.5)(42)(.2)] + [\pi(2.5)^2](.9) \approx 149.6.$$

The maximum possible error in calculating the volume is 149.6 cubic millimeters. ◀

8.6 Exercises

Find dz or dw, as appropriate, for each of the following.

1. $z = 9x^4 - 5y^3$

2. $z = x^2 + 7y^4$

3. $z = \dfrac{x + y}{x - y}$

4. $z = \dfrac{x + y^2}{y - 2}$

5. $z = 2\sqrt{xy} - \sqrt{x + y}$

6. $z = \sqrt{x^2 + y^2} + \sqrt{xy}$

7. $z = (3x + 2)\sqrt{1 - 2y}$

8. $z = (5x^2 + 6)\sqrt{4 + 3y}$

9. $z = \ln(x^2 + 2y^4)$

10. $z = \ln\left(\dfrac{8 + x}{8 - y}\right)$

11. $z = xy^2e^{x+y}$

12. $z = (x + y)e^{-x^2}$

13. $z = x^2 - y \ln x$

14. $z = x^2 + 3y - x \ln y$

15. $w = x^4yz^3$

16. $w = 6x^3y^2z^5$

Evaluate dz using the given information.

17. $z = x^2 + 3xy + y^2$; $x = 4, y = -2, dx = .02, dy = -.03$

18. $z = 8x^3 + 2x^2y - y$; $x = 1, y = 3, dx = .01, dy = .02$

19. $z = \frac{x - 4y}{x + 2y}$; $x = 0, y = 5, dx = -.03, dy = .05$

20. $z = \frac{y^2 + 3x}{y^2 - x}$; $x = 4, y = -4, dx = .01, dy = .03$

21. $z = \ln(x^2 + y^2)$; $x = 2, y = 3, dx = .02, dy = -.03$

22. $z = \ln\left(\frac{x + y}{x - y}\right)$; $x = 4, y = -2, dx = .03, dy = .02$

Evaluate dw using the given information.

23. $w = \frac{5x^2 + y^2}{z + 1}$; $x = -2, y = 1, z = 1, dx = .02, dy = -.03, dz = .02$

24. $w = x \ln(yz) - y \ln \frac{x}{z}$; $x = 2, y = 1, z = 4, dx = .03, dy = .02, dz = -.01$

Applications

Business and Economics

25. Manufacturing Approximate the amount of aluminum needed for a beverage can of radius 2.5 cm and height 14 cm. Assume the walls of the can are .08 cm thick.

26. Manufacturing Approximate the amount of material needed to make a water tumbler of diameter 3 cm and height 9 cm. Assume the walls of the tumbler are .2 cm thick.

27. Volume of a Coating An industrial coating .2 inches thick is applied to all sides of a box of dimensions 10 inches by 9 inches by 14 inches. Estimate the volume of the coating used.

28. Manufacturing Cost The manufacturing cost of a precision electronic calculator is approximated by

$$M(x, y) = 40x^2 + 30y^2 - 10xy + 30,$$

where x is the cost of the chips and y is the cost of labor. Right now, the company spends \$4 on chips and \$7 on labor. Use differentials to approximate the change in cost if the company spends \$5 on chips and \$6.50 on labor.

29. Production The production function for one country is

$$z = x^{.65}y^{.35},$$

where x stands for units of labor and y for units of capital. At present, 50 units of labor and 29 units of capital are available. Use differentials to estimate the change in production if the number of units of labor is increased to 52 and capital is decreased to 27 units.

30. Production The production function for another country is

$$z = x^{.8}y^{.2},$$

where x stands for units of labor and y for units of capital. At present, 20 units of labor and 18 units of capital are being provided. Use differentials to estimate the change in production if an additional unit of labor is provided and if capital is decreased to 16 units.

Life Sciences

31. Bone Preservative Volume A piece of bone in the shape of a right circular cylinder is 7 cm long and has a radius of 1.4 cm. It is coated with a layer of preservative .09 cm thick. Estimate the volume of preservative used.

32. Blood Vessel Volume A portion of a blood vessel is measured as having length 7.9 cm and radius .8 cm. If each measurement could be off by as much as .15 cm, estimate the maximum possible error in calculating the volume of the vessel.

33. Blood Volume In Exercise 50 of Section 2 in this chapter, we found that the number of liters of blood pumped through the lungs in one minute is given by

$$C = \frac{b}{a - v}.$$

Suppose $a = 160$, $b = 200$, $v = 125$. Estimate the change in C if a becomes 145, b becomes 190, and v changes to 130.

34. **Oxygen Consumption** In Exercise 48 of Section 2 of this chapter, we found that the oxygen consumption of a mammal is

$$m = \frac{2.5(T - F)}{w^{.67}}.$$

Suppose T is 38° C, F is 12° C, and w is 30 kg. Approximate the change in m if T changes to 36° C, F changes to 13° C, and w becomes 31 kg.

General Interest

35. **Estimating Area** The height of a triangle is measured as 42.6 cm, with the base measured as 23.4 cm. The measurement of the height can be off by as much as 1.2 cm, and that of the base by no more than .9 cm. Estimate the maximum possible error in calculating the area of the triangle.

36. **Estimating Volume** The height of a cone is measured as 8.4 cm and the radius as 2.9 cm. Each measurement could be off by as much as .1 cm. Estimate the maximum possible error in calculating the volume of the cone.

8.7 DOUBLE INTEGRALS

How can we find the volume of a bottle with curved sides?

You may wish to review the key ideas of indefinite and definite integrals from the chapter on integration before continuing with this section. See the review problems at the end of that chapter.

In an earlier chapter, we saw how integrals of functions with one variable may be used to find area. In this section, this idea is extended and used to find volume. We found partial derivatives of functions of two or more variables at the beginning of this chapter by holding constant all variables except one. A similar process is used in this section to find antiderivatives of functions of two or more variables. For example, in

$$\int (5x^3y^4 - 6x^2y + 2)\,dy$$

the notation dy indicates integration with respect to y, so we treat y as the variable and x as a constant. Using the rules for antiderivatives gives

$$\int (5x^3y^4 - 6x^2y + 2)\,dy = x^3y^5 - 3x^2y^2 + 2y + C(x).$$

The constant C used earlier must be replaced with $C(x)$ to show that the ''constant of integration'' here can be any function involving only the variable x. Just as before, check this work by taking the derivative (actually the partial derivative) of the answer:

$$\frac{\partial}{\partial y}\left[x^3y^5 - 3x^2y^2 + 2y + C(x)\right] = 5x^3y^4 - 6x^2y + 2 + 0,$$

which shows that the antiderivative is correct.

EXAMPLE 1 Find each indefinite integral.

(a) $\int x(x^2 + y)\,dx$

Multiply x and $x^2 + y$. Then (because of the dx) integrate each term with x as the variable and y as a constant.

$$\int x(x^2 + y)\,dx = \int (x^3 + xy)\,dx$$
$$= \frac{x^4}{4} + \frac{x^2}{2} \cdot y + f(y) = \frac{1}{4}x^4 + \frac{1}{2}x^2y + f(y)$$

(b) $\int x(x^2 + y)\,dy$

Since y is the variable and x is held constant,

$$\int x(x^2 + y)\,dy = \int (x^3 + xy)\,dy = x^3y + \frac{1}{2}xy^2 + g(x).$$

The analogy to integration of functions of one variable can be continued for evaluating definite integrals. We do this by holding one variable constant and using the Fundamental Theorem of Calculus with the other variable.

EXAMPLE 2 Evaluate each definite integral.

(a) $\int_3^5 (6xy^2 + 12x^2y + 4y)\,dx$

First, find an antiderivative:

$$\int (6xy^2 + 12x^2y + 4y)\,dx = 3x^2y^2 + 4x^3y + 4xy + h(y).$$

Now replace each x with 5, and then with 3, and subtract the results.

$$\left[3x^2y^2 + 4x^3y + 4xy + h(y)\right]\Big|_3^5 = [3 \cdot 5^2 \cdot y^2 + 4 \cdot 5^3 \cdot y + 4 \cdot 5 \cdot y + h(y)]$$
$$- [3 \cdot 3^2 \cdot y^2 + 4 \cdot 3^3 \cdot y + 4 \cdot 3 \cdot y + h(y)]$$
$$= 75y^2 + 500y + 20y + h(y)$$
$$- [27y^2 + 108y + 12y + h(y)]$$
$$= 48y^2 + 400y$$

The "function of integration," $h(y)$, drops out, just as the constant of integration does with definite integrals of functions of one variable.

(b) $\int_1^2 (6xy^2 + 12x^2y + 4y)\, dy$

Integrate with respect to y; then substitute 2 and 1 for y and subtract.

$$\int_1^2 (6xy^2 + 12x^2y + 4y)\, dy = (2xy^3 + 6x^2y^2 + 2y^2)\Big|_1^2$$
$$= (2x \cdot 2^3 + 6x^2 \cdot 2^2 + 2 \cdot 2^2) - (2x \cdot 1^3 + 6x^2 \cdot 1^2 + 2 \cdot 1^2)$$
$$= 16x + 24x^2 + 8 - (2x + 6x^2 + 2)$$
$$= 14x + 18x^2 + 6$$

As Example 2 suggests, an integral of the form

$$\int_a^b f(x, y)\, dy$$

produces a result that is a function of x, while

$$\int_a^b f(x, y)\, dx$$

produces a function of y. These resulting functions of one variable can themselves be integrated, as in the next example.

Evaluate each integral.

(a) $\int_1^2 \left[\int_3^5 (6xy^2 + 12x^2y + 4y)\, dx\right] dy$

In Example 2(a), we found the quantity in brackets to be $48y^2 + 400y$. Thus,

$$\int_1^2 \left[\int_3^5 (6xy^2 + 12x^2y + 4y)\, dx\right] dy = \int_1^2 (48y^2 + 400y)\, dy$$
$$= (16y^3 + 200y^2)\Big|_1^2$$
$$= 16 \cdot 2^3 + 200 \cdot 2^2 - (16 \cdot 1^3 + 200 \cdot 1^2)$$
$$= 128 + 800 - (16 + 200)$$
$$= 712.$$

(b) $\int_3^5 \left[\int_1^2 (6xy^2 + 12x^2y + 4y)\, dy\right] dx.$

(This is the same integrand, with the same limits of integration as in part (a), but the order of integration is reversed.)

Use the result from Example 2(b).

$$\int_3^5 \left[\int_1^2 (6xy^2 + 12x^2y + 4y)\,dy\right] dx = \int_3^5 (14x + 18x^2 + 6)\,dx$$
$$= (7x^2 + 6x^3 + 6x)\Big|_3^5$$
$$= 7 \cdot 5^2 + 6 \cdot 5^3 + 6 \cdot 5 - (7 \cdot 3^2 + 6 \cdot 3^3 + 6 \cdot 3)$$
$$= 175 + 750 + 30 - (63 + 162 + 18) = 712$$ ◀

The answers in the two parts of Example 3 are equal. It can be proved that for a large class of functions, including most functions that occur in applications, the following equation holds true.

$$\int_a^b \left[\int_c^d f(x, y)\,dx\right] dy = \int_c^d \left[\int_a^b f(x, y)\,dy\right] dx$$

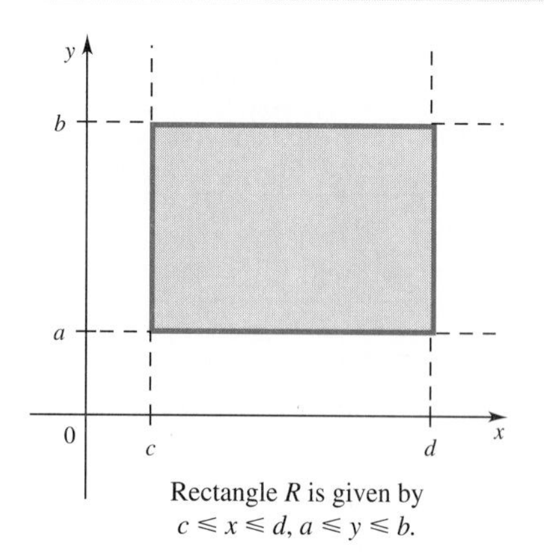

Rectangle R is given by $c \leq x \leq d, a \leq y \leq b$.

FIGURE 25

Because these two integrals are equal, the brackets are not needed, and either of the integrals is given by

$$\int_a^b \int_c^d f(x, y)\,dx\,dy.$$

This integral is called an **iterated integral** since it is evaluated by integrating twice, first using one variable and then using the other. (The order in which dx and dy are written tells the order of integration, with the innermost differential used first.)

The fact that the iterated integrals above are equal makes it possible to define a *double integral*. First, the set of points (x, y), with $c \leq x \leq d$ and $a \leq y \leq b$, defines a rectangular region R in the plane, as shown in Figure 25. Then, the *double integral over R* is defined as follows.

DOUBLE INTEGRAL

The **double integral** of $f(x, y)$ over a rectangular region R is written

$$\iint_R f(x, y)\,dx\,dy \quad \text{or} \quad \iint_R f(x, y)\,dy\,dx,$$

and equals either

$$\int_a^b \int_c^d f(x, y)\,dx\,dy \quad \text{or} \quad \int_c^d \int_a^b f(x, y)\,dy\,dx.$$

Extending earlier definitions, $f(x, y)$ is the **integrand** and R is the **region of integration.**

EXAMPLE 4 Find $\iint_R \sqrt{x} \cdot \sqrt{y-2}\, dx\, dy$ over the rectangular region R defined by $0 \leq x \leq 4$, $3 \leq y \leq 11$.

Integrate first with respect to x; then integrate the result with respect to y.

$$\begin{aligned}
\iint_R \sqrt{x} \cdot \sqrt{y-2}\, dx\, dy &= \int_3^{11} \left[\int_0^4 \sqrt{x} \cdot \sqrt{y-2}\, dx \right] dy \\
&= \int_3^{11} \left(\frac{2}{3} x^{3/2} \sqrt{y-2} \right) \Bigg|_0^4 dy \\
&= \int_3^{11} \left[\frac{2}{3}(4^{3/2})\sqrt{y-2} - \frac{2}{3}(0^{3/2})\sqrt{y-2} \right] dy \\
&= \int_3^{11} \left(\frac{16}{3}\sqrt{y-2} - 0 \right) dy = \int_3^{11} \left(\frac{16}{3}\sqrt{y-2} \right) dy \\
&= \frac{32}{9}(y-2)^{3/2} \Bigg|_3^{11} = \frac{32}{9}(9)^{3/2} - \frac{32}{9}(1)^{3/2} \\
&= 96 - \frac{32}{9} = \frac{832}{9}
\end{aligned}$$

As a check, integrate with respect to y first. The answer should be the same.

Volume As shown earlier, the definite integral $\int_a^b f(x)\, dx$ can be used to find the area under a curve. In a similar manner, double integrals are used to find the *volume under a surface.* Figure 26 shows that portion of a surface $f(x, y)$ directly over a rectangle R in the xy-plane. Just as areas were approximated by a large

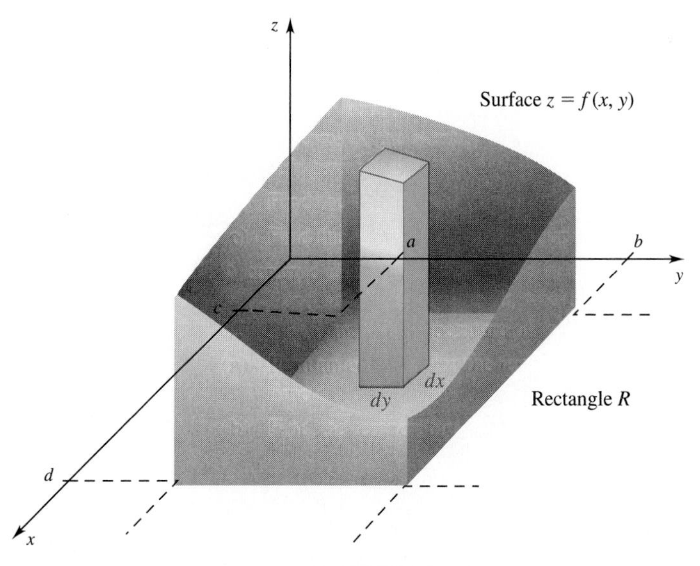

FIGURE 26

number of small rectangles, volume could be approximated by adding the volumes of a large number of properly drawn small boxes. The height of a typical box would be $f(x, y)$ with the length and width given by dx and dy. The formula for the volume of a box would then suggest the following result.

VOLUME

Let $z = f(x, y)$ be a function that is never negative on the rectangular region R defined by $c \leq x \leq d$, $a \leq y \leq b$. The volume of the solid under the graph of f and over the region R is

$$\iint_R f(x, y)\, dx\, dy.$$

Find the volume under the surface $z = x^2 + y^2$ shown in Figure 27.

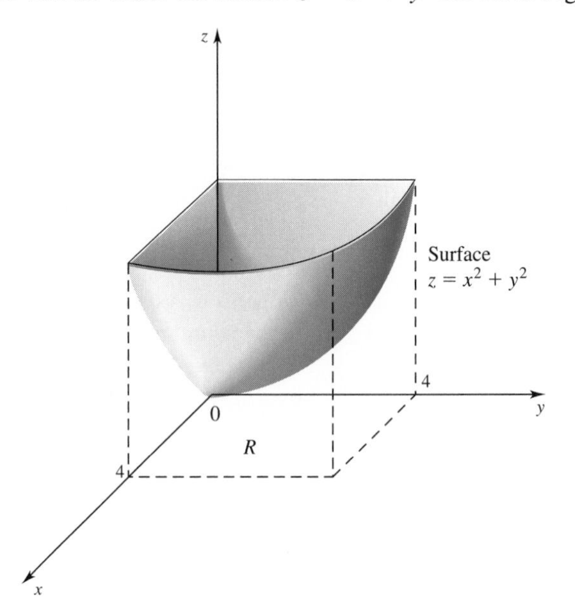

FIGURE 27

By the results given above, the volume is

$$\iint_R f(x, y)\, dx\, dy,$$

where $f(x, y) = x^2 + y^2$ and R is the region $0 \leq x \leq 4$, $0 \leq y \leq 4$. By definition,

$$\iint_R f(x, y)\, dx\, dy = \int_0^4 \left[\int_0^4 (x^2 + y^2)\, dx \right] dy$$

$$= \int_0^4 \left(\frac{1}{3}x^3 + xy^2 \right) \Big|_0^4 \, dy$$

$$= \int_0^4 \left(\frac{64}{3} + 4y^2\right) dy = \left(\frac{64}{3}y + \frac{4}{3}y^3\right)\Bigg|_0^4$$

$$= \frac{64}{3} \cdot 4 + \frac{4}{3} \cdot 4^3 - 0 = \frac{512}{3}.$$

EXAMPLE 6 A product design consultant for a cosmetics company has been asked to design a bottle for the company's newest perfume. The thickness of the glass is to vary so that the outside of the bottle has straight sides and the inside has curved sides, as shown in Figure 28. Before presenting the design to management, the consultant needs to make a reasonably accurate estimate of the amount each bottle will hold. If the base of the bottle is to be 4 centimeters by 3 centimeters, and if a cross section of its interior is to be a parabola of the form $z = -y^2 + 4y$, what is its internal volume?

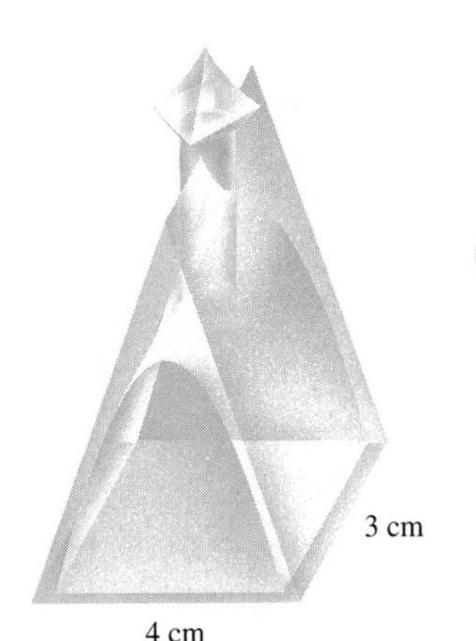

FIGURE 28

The interior of the bottle can be graphed in three-dimensional space, as shown in Figure 29, where $z = 0$ corresponds to the base of the bottle. Its volume is simply the volume above the region R in the xy plane and below the graph of $f(x, y) = -y^2 + 4y$. This volume is given by the double integral

$$\int_0^3 \int_0^4 (-y^2 + 4y)\, dy\, dx = \int_0^3 \left(\frac{-y^3}{3} + \frac{4y^2}{2}\right)\Bigg|_0^4 dx$$

$$= \int_0^3 \left(\frac{-64}{3} + 32 - 0\right) dx$$

$$= \frac{32}{3}x \Bigg|_0^3$$

$$= 32 - 0 = 32.$$

The bottle holds 32 cubic centimeters.

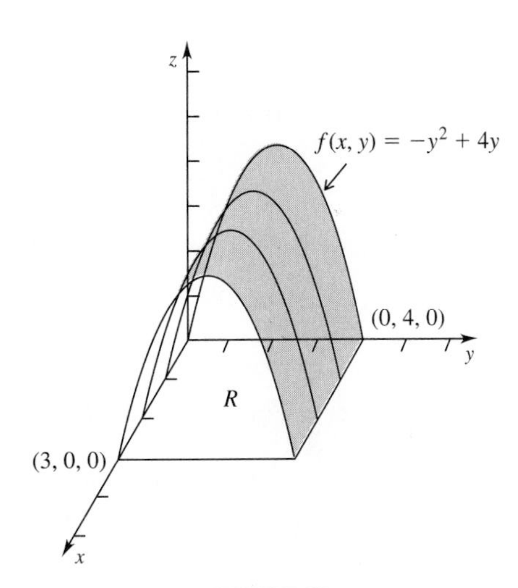

FIGURE 29

Double Integrals over Other Regions In the work in this section, we found double integrals over rectangular regions, with constant limits of integration. Now this work can be extended to include *variable* limits of integration. (Notice in the following examples that the variable limits always go on the *inner* integral sign.)

The use of variable limits of integration permits evaluation of double integrals over the types of regions shown in Figure 30. Double integrals over more complicated regions are discussed in more advanced books. Integration over regions such as those of Figure 30 is done with the results of the following theorem.

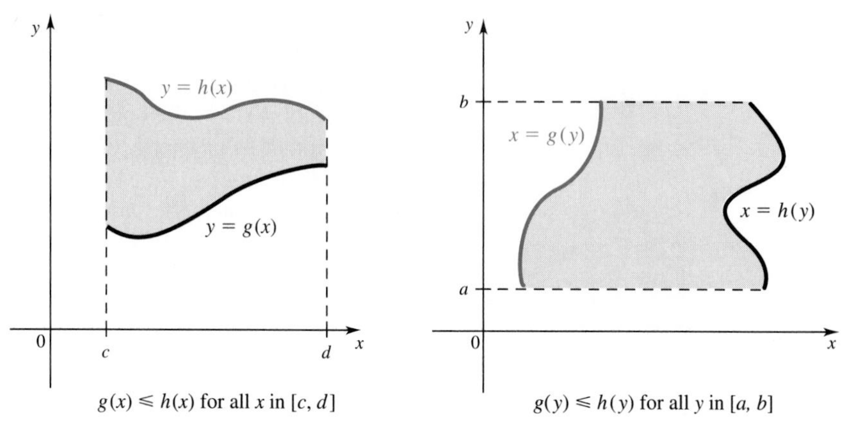

FIGURE 30

DOUBLE INTEGRALS OVER VARIABLE REGIONS

Let $z = f(x, y)$ be a function of two variables.
If R is the region (in Figure 30(a)) defined by $c \leq x \leq d$ and $g(x) \leq y \leq h(x)$, then

$$\iint_R f(x, y)\, dy\, dx = \int_c^d \left[\int_{g(x)}^{h(x)} f(x, y)\, dy \right] dx.$$

If R is the region (in Figure 30(b)) defined by $g(y) \leq x \leq h(y)$ and $a \leq y \leq b$, then

$$\iint_R f(x, y)\, dx\, dy = \int_a^b \left[\int_{g(y)}^{h(y)} f(x, y)\, dx \right] dy.$$

Evaluate $\displaystyle\int_1^2 \int_y^{y^2} xy\, dx\, dy.$

The region of integration is shown in Figure 31. Integrate first with respect to x, then with respect to y.

$$\int_1^2 \int_y^{y^2} xy\, dx\, dy = \int_1^2 \left[\int_y^{y^2} xy\, dx \right] dy = \int_1^2 \left(\frac{1}{2}x^2 y \right) \Bigg|_y^{y^2} dy$$

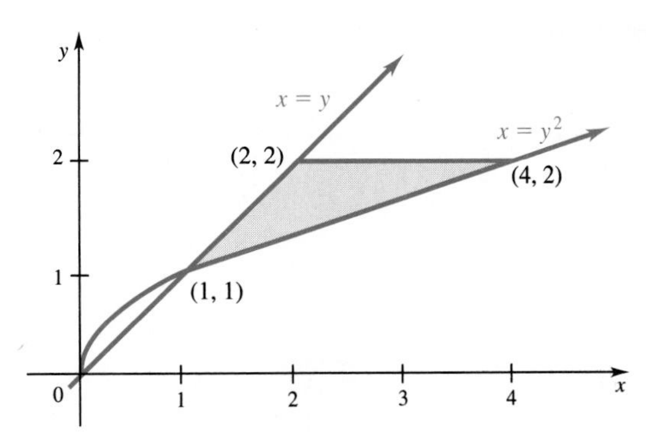

FIGURE 31

Replace x first with y^2 and then with y, and subtract.

$$\int_1^2 \int_y^{y^2} xy\,dx\,dy = \int_1^2 \left[\frac{1}{2}(y^2)^2 y - \frac{1}{2}(y)^2 y\right] dy$$

$$= \int_1^2 \left(\frac{1}{2}y^5 - \frac{1}{2}y^3\right) dy = \left(\frac{1}{12}y^6 - \frac{1}{8}y^4\right)\Bigg|_1^2$$

$$= \left(\frac{1}{12} \cdot 2^6 - \frac{1}{8} \cdot 2^4\right) - \left(\frac{1}{12} \cdot 1^6 - \frac{1}{8} \cdot 1^4\right)$$

$$= \frac{64}{12} - \frac{16}{8} - \frac{1}{12} + \frac{1}{8} = \frac{27}{8}$$ ◀

EXAMPLE 8 Let R be the shaded region in Figure 32, and evaluate

$$\iint_R (x + 2y)\,dy\,dx.$$

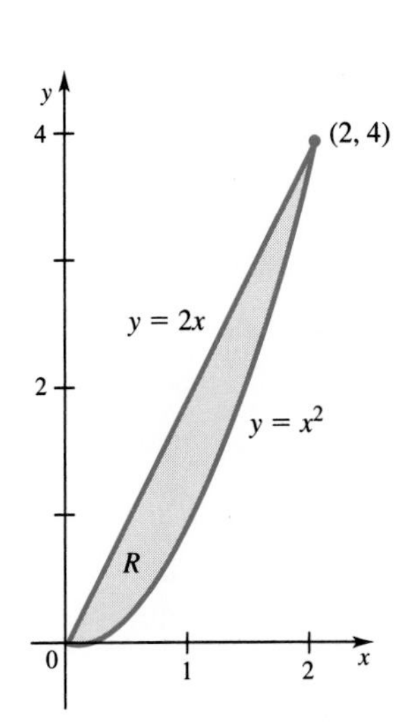

FIGURE 32

Region R is bounded by $h(x) = 2x$ and $g(x) = x^2$, with $0 \leq x \leq 2$. By the first result in the theorem above,

$$\iint_R (x + 2y)\,dy\,dx = \int_0^2 \int_{x^2}^{2x} (x + 2y)\,dy\,dx$$

$$= \int_0^2 (xy + y^2)\Bigg|_{x^2}^{2x} dx$$

$$= \int_0^2 (x(2x) + (2x)^2 - [x \cdot x^2 + (x^2)^2])\,dx$$

$$= \int_0^2 [2x^2 + 4x^2 - (x^3 + x^4)]\,dx$$

$$= \int_0^2 (6x^2 - x^3 - x^4)\,dx$$

$$= \left(2x^3 - \frac{1}{4}x^4 - \frac{1}{5}x^5\right)\Bigg|_0^2$$

$$= 2 \cdot 2^3 - \frac{1}{4} \cdot 2^4 - \frac{1}{5} \cdot 2^5 - 0$$

$$= 16 - 4 - \frac{32}{5} = \frac{28}{5}.$$

Note In Example 8, the same result would be found if we evaluated the double integral

$$\iint_R (x + 2y)\, dx\, dy.$$

In that case, we would need to define the equations of the boundaries in terms of y rather than x, so R would be defined by $y/2 \le x \le \sqrt{y}$, $0 \le y \le 4$.

Interchanging Limits of Integration Sometimes it is easier to integrate first with respect to x, and then y, while with other integrals the reverse process is easier. The limits of integration can be reversed whenever the region R is like the region in Figure 32. The next example shows how this process works.

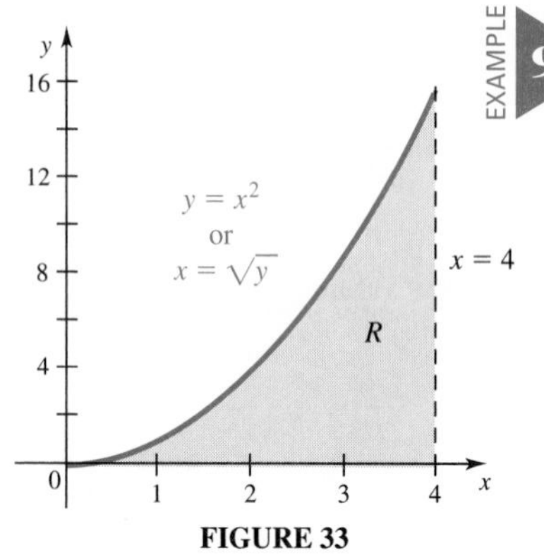

FIGURE 33

EXAMPLE 9 Interchange the limits of integration in

$$\int_0^{16}\int_{\sqrt{y}}^{4} f(x, y)\, dx\, dy.$$

For this integral, region R is given by $\sqrt{y} \le x \le 4$, $0 \le y \le 16$. A graph of R is shown in Figure 33.

The same region R can be written in an alternate way. As Figure 33 shows, one boundary of R is $x = \sqrt{y}$. Solving for y gives $y = x^2$. Also, Figure 33 shows that $0 \le x \le 4$. Since R can be written as $0 \le y \le x^2$, $0 \le x \le 4$, the double integral above can be written

$$\int_0^4\int_0^{x^2} f(x, y)\, dy\, dx.$$

8.7 Exercises

Evaluate the following integrals.

1. $\int_0^3 (x^3y + y)\, dx$
2. $\int_1^4 (xy^2 - x)\, dy$
3. $\int_4^8 \sqrt{6x + y}\, dx$
4. $\int_3^7 \sqrt{x + 5y}\, dy$
5. $\int_4^5 x\sqrt{x^2 + 3y}\, dy$
6. $\int_3^6 x\sqrt{x^2 + 3y}\, dx$
7. $\int_4^9 \frac{3 + 5y}{\sqrt{x}}\, dx$
8. $\int_2^7 \frac{3 + 5y}{\sqrt{x}}\, dy$
9. $\int_{-1}^1 e^{x+4y}\, dy$
10. $\int_2^6 e^{x+4y}\, dx$
11. $\int_0^5 xe^{x^2+9y}\, dx$
12. $\int_1^6 xe^{x^2+9y}\, dy$

Evaluate the following iterated integrals. (Many of these use results from Exercises 1–12.)

13. $\int_1^2 \left[\int_0^3 (x^3y + y)\,dx\right] dy$

14. $\int_0^3 \left[\int_1^4 (xy^2 - x)\,dy\right] dx$

15. $\int_0^1 \left[\int_3^6 x\sqrt{x^2 + 3y}\,dx\right] dy$

16. $\int_0^3 \left[\int_4^5 x\sqrt{x^2 + 3y}\,dy\right] dx$

17. $\int_1^2 \left[\int_4^9 \frac{3 + 5y}{\sqrt{x}}\,dx\right] dy$

18. $\int_{16}^{25} \left[\int_2^7 \frac{3 + 5y}{\sqrt{x}}\,dy\right] dx$

19. $\int_1^2 \int_1^2 \frac{dx\,dy}{xy}$

20. $\int_1^4 \int_2^5 \frac{dy\,dx}{x}$

21. $\int_2^4 \int_3^5 \left(\frac{x}{y} + \frac{y}{3}\right) dx\,dy$

22. $\int_3^4 \int_1^2 \left(\frac{6x}{5} + \frac{y}{x}\right) dx\,dy$

Find each double integral over the rectangular region R with the given boundaries.

23. $\iint_R (x + 3y^2)\,dx\,dy;\quad 0 \le x \le 2,\ 1 \le y \le 5$

24. $\iint_R (4x^3 + y^2)\,dx\,dy;\quad 1 \le x \le 4,\ 0 \le y \le 2$

25. $\iint_R \sqrt{x + y}\,dy\,dx;\quad 1 \le x \le 3,\ 0 \le y \le 1$

26. $\iint_R x^2\sqrt{x^3 + 2y}\,dx\,dy;\quad 0 \le x \le 2,\ 0 \le y \le 3$

27. $\iint_R \frac{2}{(x + y)^2}\,dy\,dx;\quad 2 \le x \le 3,\ 1 \le y \le 5$

28. $\iint_R \frac{y}{\sqrt{6x + 5y^2}}\,dx\,dy;\quad 0 \le x \le 3,\ 1 \le y \le 2$

29. $\iint_R ye^{(x+y^2)}\,dx\,dy;\quad 2 \le x \le 3,\ 0 \le y \le 2$

30. $\iint_R x^2e^{(x^3+2y)}\,dx\,dy;\quad 1 \le x \le 2,\ 1 \le y \le 3$

Find the volume under the given surface $z = f(x, y)$ and above the rectangle with the given boundaries.

31. $z = 6x + 2y + 5;\quad -1 \le x \le 1,\ 0 \le y \le 3$

32. $z = 9x + 5y + 12;\quad 0 \le x \le 3,\ -2 \le y \le 1$

33. $z = x^2;\quad 0 \le x \le 1,\ 0 \le y \le 4$

34. $z = \sqrt{y};\quad 0 \le x \le 4,\ 0 \le y \le 9$

35. $z = x\sqrt{x^2 + y};\quad 0 \le x \le 1,\ 0 \le y \le 1$

36. $z = yx\sqrt{x^2 + y^2};\quad 0 \le x \le 4,\ 0 \le y \le 1$

37. $z = \frac{xy}{(x^2 + y^2)^2};\quad 1 \le x \le 2,\ 1 \le y \le 4$

38. $z = e^{x+y};\quad 0 \le x \le 1,\ 0 \le y \le 1$

While it is true that a double integral can be evaluated by using either dx or dy first, sometimes one choice over the other makes the work easier. Evaluate the double integrals in Exercises 39 and 40 in the easiest way possible.

39. $\iint_R xe^{xy}\,dx\,dy;\quad 0 \le x \le 2,\ 0 \le y \le 1$

40. $\iint_R x^2e^{2x^3+6y}\,dx\,dy;\quad 0 \le x \le 1,\ 0 \le y \le 1$

Evaluate each double integral.

41. $\int_2^4 \int_2^{x^2} (x^2 + y^2)\,dy\,dx$

42. $\int_0^5 \int_0^{2y} (x^2 + y)\,dx\,dy$

43. $\int_0^4 \int_0^x \sqrt{xy}\,dy\,dx$

44. $\int_1^4 \int_0^x \sqrt{x + y}\,dy\,dx$

45. $\int_1^2 \int_y^{3y} \frac{1}{x}\,dx\,dy$

46. $\int_1^4 \int_x^{x^2} \frac{1}{y}\,dy\,dx$

47. $\int_0^4 \int_1^{e^x} \frac{x}{y}\,dy\,dx$

48. $\int_0^1 \int_{2x}^{4x} e^{x+y}\,dy\,dx$

Use the region R with the indicated boundaries to evaluate each of the following double integrals.

49. $\iint_R (4x + 7y)\,dy\,dx;\quad 1 \le x \le 3,\ 0 \le y \le x + 1$

50. $\iint_R (3x + 9y)\,dy\,dx;\quad 2 \le x \le 4,\ 2 \le y \le 3x$

51. $\iint_R (4 - 4x^2)\,dy\,dx;\quad 0 \le x \le 1,\ 0 \le y \le 2 - 2x$

52. $\iint_R \frac{dy\,dx}{x};\quad 1 \le x \le 2,\ 0 \le y \le x - 1$

53. $\iint_R e^{x/y^2}\,dx\,dy;\quad 1 \le y \le 2,\ 0 \le x \le y^2$

54. $\iint_R (x^2 - y)\,dy\,dx;\quad -1 \le x \le 1,\ -x^2 \le y \le x^2$

55. $\iint_R x^3y\,dx\,dy$; R bounded by $y = x^2$, $y = 2x$

56. $\iint_R x^2y^2\,dx\,dy$; R bounded by $y = x$, $y = 2x$, $x = 1$

57. $\iint_R \frac{dy\,dx}{y}$; R bounded by $y = x$, $y = \frac{1}{x}$, $x = 2$

The idea of the average value of a function, discussed earlier for functions of the form $y = f(x)$, can be extended to functions of more than one independent variable. For a function $z = f(x, y)$, the ***average value*** *of f over a region R is defined as*

$$\frac{1}{A}\iint_R f(x, y)\,dx\,dy,$$

where A is the area of the region R. Find the average value for each of the following functions over the regions R having the given boundaries.

58. $f(x, y) = 5xy + 2y$; $1 \le x \le 4, 1 \le y \le 2$

59. $f(x, y) = x^2 + y^2$; $0 \le x \le 2, 0 \le y \le 3$

60. $f(x, y) = e^{-5y+3x}$; $0 \le x \le 2, 0 \le y \le 2$

61. $f(x, y) = e^{2x+y}$; $1 \le x \le 2, 2 \le y \le 3$

Applications

Business and Economics

62. Packaging The manufacturer of a fruit juice drink has decided to try innovative packaging in order to revitalize sagging sales. The fruit juice drink is to be packaged in containers in the shape of tetrahedra in which three edges are perpendicular, as shown in the figure. Two of the perpendicular edges will be 3 inches long, and the third edge will be 6 inches long. Find the volume of the container. (*Hint:* The equation of the plane shown in the figure is $z = f(x, y) = 6 - 2x - 2y$.)

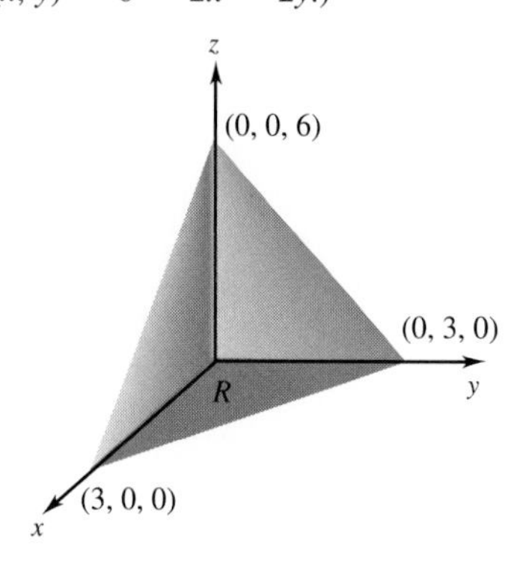

63. Average Cost A company's total cost for operating its two warehouses is

$$C(x, y) = \frac{1}{9}x^2 + 2x + y^2 + 5y + 100$$

dollars, where x represents the number of units stored at the first warehouse and y represents the number of units stored at the second. Find the average cost to store a unit if the first warehouse has between 48 and 75 units, and the second has between 20 and 60 units.

64. Average Production A production function is given by

$$P(x, y) = 500x^{.2}y^{.8},$$

where x is the number of units of labor and y is the number of units of capital. Find the average production level if x varies from 10 to 50 and y from 20 to 40.

65. Average Profit The profit (in dollars) from selling x units of one product and y units of a second product is

$$P = -(x - 100)^2 - (y - 50)^2 + 2000.$$

The weekly sales for the first product vary from 100 units to 150 units, and the weekly sales for the second product vary from 40 units to 80 units. Estimate average weekly profit for these two products.

66. Average Revenue A company sells two products. The demand functions of the products are given by

$$q_1 = 300 - 2p_1 \quad \text{and} \quad q_2 = 500 - 1.2p_2,$$

where q_1 units of the first product are demanded at price p_1 and q_2 units of the second product are demanded at price p_2. The total revenue will be given by

$$R = q_1p_1 + q_2p_2.$$

Find the average revenue if the price p_1 varies from \$25 to \$50 and the price p_2 varies from \$50 to \$75.

Physical Sciences

67. Effects of Nuclear Weapons *The Effects of Nuclear Weapons*, prepared by the U.S. Department of Defense, contains these remarks on the computation of the radioactive dose after a 1-kiloton atomic explosion.*

"If all the residues from a 1-kiloton fission yield were deposited on a smooth surface in varying concentrations typical of an early fallout pattern, instead of uniformly, the product of the dose rate at 1 hour and the area would be replaced by the 'area integral' of the 1-hour dose rate defined by

$$\text{Area integral} = \int_A R_1 dA,$$

where R_1 is the 1-hour dose rate over an element of area dA, and A square miles is the total area covered by the residues."

Explain why an integral is involved. (*Note:* In this case R_1 denotes the function and A the region. In the diverse applications of integrals, many notations are employed.)

Chapter Summary — Key Terms

8.1 function of two variables
ordered triple
first octant
plane
surface
trace
level curves
paraboloid
production function
ellipsoid
hyperbolic paraboloid
hyperboloid of two sheets

8.2 partial derivative
second-order partial derivative

8.3 saddle point
critical point

8.4 constraint
Lagrange multiplier

8.5 scatter diagram
least squares line
regression line

8.6 total differential

8.7 iterated integral
double integral
region of integration

Chapter 8 Review Exercises

Find $f(-1, 2)$ and $f(6, -3)$ for each of the following.

1. $f(x, y) = -4x^2 + 6xy - 3$

2. $f(x, y) = 3x^2y^2 - 5x + 2y$

3. $f(x, y) = \dfrac{x - 3y}{x + 4y}$

4. $f(x, y) = \dfrac{\sqrt{x^2 + y^2}}{x - y}$

Graph the first-octant portion of each plane.

5. $x + y + z = 4$

6. $x + y + 4z = 8$

7. $5x + 2y = 10$

8. $3x + 5z = 15$

9. $x = 3$

10. $y = 2$

11. Let $z = f(x, y) = -5x^2 + 7xy - y^2$. Find each of the following.

(a) $\dfrac{\partial z}{\partial x}$ **(b)** $\left(\dfrac{\partial z}{\partial y}\right)(-1, 4)$ **(c)** $f_{xy}(2, -1)$

12. Let $z = f(x, y) = \dfrac{x + y^2}{x - y^2}$. Find each of the following.

(a) $\dfrac{\partial z}{\partial y}$ **(b)** $\left(\dfrac{\partial z}{\partial x}\right)(0, 2)$ **(c)** $f_{xx}(-1, 0)$

---

*From Sherman K. Stein and Anthony Barcellos, *Calculus and Analytic Geometry*, 5th ed. Copyright © 1992, p. 801, by McGraw-Hill, Inc. Reprinted by permission.

Find f_x and f_y.

13. $f(x, y) = 9x^3y^2 - 5x$ **14.** $f(x, y) = 6x^5y - 8xy^9$ **15.** $f(x, y) = \sqrt{4x^2 + y^2}$

16. $f(x, y) = \dfrac{2x + 5y^2}{3x^2 + y^2}$ **17.** $f(x, y) = x^2 \cdot e^{2y}$ **18.** $f(x, y) = (y - 2)^2 \cdot e^{(x+2y)}$

19. $f(x, y) = \ln|2x^2 + y^2|$ **20.** $f(x, y) = \ln|2 - x^2y^3|$

Find f_{xx} and f_{xy}.

21. $f(x, y) = 4x^3y^2 - 8xy$ **22.** $f(x, y) = -6xy^4 + x^2y$ **23.** $f(x, y) = \dfrac{2x}{x - 2y}$

24. $f(x, y) = \dfrac{3x + y}{x - 1}$ **25.** $f(x, y) = x^2e^y$ **26.** $f(x, y) = ye^{x^2}$

27. $f(x, y) = \ln|2 - x^2y|$ **28.** $f(x, y) = \ln|1 + 3xy^2|$

Find all points where the functions defined below have any relative extrema. Find any saddle points.

29. $z = x^2 + 2y^2 - 4y$ **30.** $z = x^2 + y^2 + 9x - 8y + 1$

31. $f(x, y) = x^2 + 5xy - 10x + 3y^2 - 12y$ **32.** $z = x^3 - 8y^2 + 6xy + 4$

33. $z = \frac{1}{2}x^2 + \frac{1}{2}y^2 + 2xy - 5x - 7y + 10$ **34.** $f(x, y) = 3x^2 + 2xy + 2y^2 - 3x + 2y - 9$

35. $z = x^3 + y^2 + 2xy - 4x - 3y - 2$ **36.** $f(x, y) = 7x^2 + y^2 - 3x + 6y - 5xy$

Use Lagrange multipliers to find the extrema of the functions defined in Exercises 37 and 38.

37. $f(x, y) = x^2y; \quad x + y = 4$ **38.** $f(x, y) = x^2 + y^2; \quad x = y + 2$

39. Find positive numbers x and y, whose sum is 80, such that x^2y is maximized.

40. Find positive numbers x and y, whose sum is 50, such that xy^2 is maximized.

41. A closed box with square ends must have a volume of 125 cubic inches. Find the dimensions of such a box that has minimum surface area.

42. Find the maximum rectangular area that can be enclosed with 400 ft of fencing, if no fencing is needed along one side.

Find dz or dw, as appropriate, for each of the following.

43. $z = 7x^3y - 4y^3$ **44.** $z = 3x^2 + \sqrt{x + y}$ **45.** $z = x^2ye^{x-y}$

46. $z = \ln|x + 4y| + y^2 \ln x$ **47.** $w = x^5 + y^4 - z^3$ **48.** $w = \dfrac{3 + 5xy}{2 - z}$

Evaluate dz using the given information.

49. $z = 2x^2 - 4y^2 + 6xy; \quad x = 2, y = -3, dx = .01, dy = .05$

50. $z = \dfrac{x + 5y}{x - 2y}; \quad x = 1, y = -2, dx = -.04, dy = .02$

Evaluate each of the following.

51. $\int_0^4 (x^2y^2 + 5x)\,dx$ **52.** $\int_0^3 (x + 5y + y^2)\,dy$ **53.** $\int_2^5 \sqrt{6x + 3y}\,dx$ **54.** $\int_1^3 6y^4\sqrt{8x + 3y}\,dx$

55. $\int_4^9 \dfrac{6y - 8}{\sqrt{x}}\,dx$ **56.** $\int_3^5 e^{2x-7y}\,dx$ **57.** $\int_0^5 \dfrac{6x}{\sqrt{4x^2 + 2y^2}}\,dx$ **58.** $\int_1^3 \dfrac{y^2}{\sqrt{7x + 11y^3}}\,dy$

Evaluate each iterated integral.

59. $\int_0^2 \left[\int_0^4 (x^2y^2 + 5x)\,dx\right] dy$ **60.** $\int_0^2 \left[\int_0^3 (x + 5y + y^2)\,dy\right] dx$ **61.** $\int_3^4 \left[\int_2^5 \sqrt{6x + 3y}\,dx\right] dy$

62. $\int_1^2 \left[\int_3^5 e^{2x-7y}\,dx\right] dy$

63. $\int_2^4 \int_2^4 \frac{dx\,dy}{y}$

64. $\int_1^2 \int_1^2 \frac{dx\,dy}{x}$

Find each double integral over the region R with boundaries as indicated.

65. $\iint_R (x^2 + y^2)\,dx\,dy;\quad 0 \le x \le 2,\ 0 \le y \le 3$

66. $\iint_R \sqrt{2x + y}\,dx\,dy;\quad 1 \le x \le 3,\ 2 \le y \le 5$

67. $\iint_R \sqrt{y + x}\,dx\,dy;\quad 0 \le x \le 7,\ 1 \le y \le 9$

68. $\iint_R ye^{y^2+x}\,dx\,dy;\quad 0 \le x \le 1,\ 0 \le y \le 1$

Find the volume under the given surface $z = f(x, y)$ and above the given rectangle.

69. $z = x + 9y + 8;\quad 1 \le x \le 6, 0 \le y \le 8$

70. $z = x^2 + y^2;\quad 3 \le x \le 5, 2 \le y \le 4$

Evaluate each double integral.

71. $\int_0^1 \int_0^{2x} xy\,dy\,dx$

72. $\int_0^1 \int_0^{x^3} y\,dy\,dx$

73. $\int_0^1 \int_{x^2}^{x} x^3y\,dy\,dx$

74. $\int_0^1 \int_y^{\sqrt{y}} x\,dx\,dy$

Use the region R, with boundaries as indicated, to evaluate the given double integral.

75. $\iint_R (2x + 3y)\,dx\,dy;\quad 0 \le y \le 1,\ y \le x \le 2 - y$

76. $\iint_R (2 - x^2 - y^2)\,dy\,dx;\quad 0 \le x \le 1,\ x^2 \le y \le x$

Applications

Business and Economics

77. **Charge for Auto Painting** The charge (in dollars) for painting a sports car is given by

$$C(x, y) = 2x^2 + 4y^2 - 3xy + \sqrt{x},$$

where x is the number of hours of labor needed and y is the number of gallons of paint and sealant used. Find each of the following.

(a) The charge for 10 hr and 5 gal of paint and sealant

(b) The charge for 15 hr and 10 gal of paint and sealant

(c) The charge for 20 hr and 20 gal of paint and sealant

78. **Manufacturing Costs** The manufacturing cost (in dollars) for a medium-sized business computer is given by

$$c(x, y) = 2x + y^2 + 4xy + 25,$$

where x is the memory capacity of the computer in megabytes (Mb) and y is the number of hours of labor required. For 640Mb and 6 hr of labor, find the following.

(a) The approximate change in cost for an additional 1Mb of memory

(b) The approximate change in cost for an additional hour of labor

79. **Productivity** The production function z for one country is

$$z = x^{.6}y^{.4},$$

where x represents the amount of labor and y the amount of capital. Find the marginal productivity of each of the following.

(a) Labor **(b)** Capital

80. **Cost** The cost (in dollars) to manufacture x solar cells and y solar collectors is

$$c(x, y) = x^2 + 5y^2 + 4xy - 70x - 164y + 1800.$$

(a) Find values of x and y that produce minimum total cost.

(b) Find the minimum total cost.

81. **Earnings** **(a)** Find the least squares line for the following data, which give the earnings (in ten-thousands of dollars) for a certain company after x yr.

x	3	5	7	8
y	4	11	20	23

(b) Predict the earnings after 6 yr.

82. Cost The cost (in dollars) to produce x satellite receiving dishes and y transmitters is given by

$$C(x, y) = \ln (x^2 + y) + e^{xy/20}.$$

Production schedules now call for 15 receiving dishes and 9 transmitters. Use differentials to approximate the change in costs if 1 more dish and 1 fewer transmitter are made.

83. Profit The profit (in hundreds of dollars) from the sale of x small computers and y electronic games is given by

$$P(x, y) = \frac{x}{x + 5y} + \frac{y + x}{y}.$$

Right now, 75 small computers and 50 games are being sold. Use differentials to approximate the change in profit if 3 fewer computers and 2 more games were sold.

84. Production Materials Approximate the amount of material needed to manufacture a cone of radius 2 cm, height 8 cm, and wall thickness .21 cm.

85. Production Materials A sphere of radius 2 ft is to receive an insulating coating 1 inch thick. Approximate the volume of the coating needed.

86. Production Error The height of a sample cone from a production line is measured as 11.4 cm, while the radius is measured as 2.9 cm. Each of these measurements could be off by .2 cm. Approximate the maximum possible error in the volume of the cone.

Life Sciences

87. Blood Sugar and Cholesterol Levels The following data show the blood sugar levels, x, and cholesterol levels, y, for 8 different diabetic patients.

Patient	1	2	3	4	5	6	7	8
Blood Sugar Level, x	130	138	142	159	165	200	210	250
Cholesterol Level, y	170	160	173	181	201	192	240	290

For this data, $\Sigma x = 1394$; $\Sigma y = 1607$; $\Sigma xy = 291{,}990$; and $\Sigma x^2 = 255{,}214$.

(a) Find the equation of the least squares line.

(b) Predict the cholesterol level for a person whose blood sugar level is 190.

(c) Use the equation from part (a) to predict cholesterol levels when the blood sugar level is 130, 142, and 200. How do these results compare with the data in the table? Does there appear to be a close linear relationship here?

88. Pace of Life In an attempt to measure how the pace of city life is related to the size of the city, two researchers assessed the mean velocity of pedestrians in 15 cities by measuring the mean time to walk 50 ft.*

City	Population, x	Velocity, y
Brno, Czechoslovakia	341,948	4.81
Prague, Czechoslovakia	1,092,759	5.88
Corte, Corsica	5,491	3.31
Bastia, France	49,375	4.90
Munich, Germany	1,340,000	5.62
Psychro, Crete	365	2.67
Itea, Greece	2500	2.27
Iraklion, Greece	78,200	3.85
Athens, Greece	867,023	5.21
Safed, Israel	14,000	3.70
Dimona, Israel	23,700	3.27
Netanya, Israel	70,700	4.31
Jerusalem, Israel	304,500	4.42
New Haven, U.S.A.	138,000	4.39
Brooklyn, U.S.A.	2,602,000	5.05

(a) Plot the original pairs of numbers. The pattern should be nonlinear.

(b) Plot y against $\log x$, using a calculator to compute $\log x$. Is the data more linear than in part (a)?

(c) Find the equation of the least squares line for y' using $\log x$ (instead of x) and y.

89. Blood Vessel Volume A length of blood vessel is measured as 2.7 cm, with the radius measured as .7 cm. If each of these measurements could be off by .1 cm, estimate the maximum possible error in the volume of the vessel.

General Interest

90. Area The bottom of a planter is to be made in the shape of an isosceles triangle, with the two equal sides 3 ft long and the third side 2 ft long. The area of an isosceles triangle with two equal sides of length a and third side of length b is

$$f(a, b) = \frac{1}{4}b\sqrt{4a^2 - b^2}.$$

(a) Find the area of the bottom of the planter.

(b) The manufacturer is considering changing the shape so that the third side is 2.5 ft long. What would be the approximate effect on the area?

*From Marc H. Bornstein and Helen G. Bornstein, "The Pace of Life," *Nature* 259 (February 19, 1976), pp. 557–59.

Connections

91. Profit The total profit from 1 acre of a certain crop depends on the amount spent on fertilizer, x, and on hybrid seed, y, according to the model

$$P(x, y) = .01(-x^2 + 3xy + 160x - 5y^2 + 200y + 2600).$$

The budget for fertilizer and seed is limited to \$280.

(a) Use the budget constraint to express one variable in terms of the other. Then substitute into the profit function to get a function with one independent variable. Use the method shown in the chapter on applications of the derivative to find the amounts spent on fertilizer and seed that will maximize profit. What is the maximum profit per acre? (*Hint:* Throughout this problem you may ignore the coefficient of .01 until you need to find the maximum profit.)

(b) Find the amounts spent on fertilizer and seed that will maximize profit using the first method shown in this chapter. (*Hint:* You will not need to use the budget constraint.)

(c) Use the Lagrange multiplier method to solve the original problem.

(d) Look for the relationships among these methods.

Extended Application / Lagrange Multipliers for a Predator[†]

A predator is an animal that feeds upon another animal. Foxes, coyotes, wolves, weasels, lions, and tigers are well-known predators, but many other animals also fall into this category. In this case, we set up a mathematical model for predation, and then use Lagrange multipliers to minimize the difference between the desired and the actual levels of food consumption. There are several research studies which show that animals *do* control their activities so as to maximize or minimize variables—lobsters orient their bodies by minimizing the discharge rate from certain organs, for example.

In this case, we assume that the predator has a diet consisting of only two foods, food 1 and food 2. We also assume that the predator will hunt only in two locations, location 1 and location 2. To make this mathematical model meaningful, we would have to gather data on the predators and the prey that we wished to study. For example, suppose that we have gathered the following data:

$u_{11} = .4 =$ rate of feeding on food 1 in location 1

$u_{12} = .1 =$ rate of feeding on food 1 in location 2

$u_{21} = .3 =$ rate of feeding on food 2 in location 1

$u_{22} = .3 =$ rate of feeding on food 2 in location 2.

Let $x_1 =$ proportion of time spent feeding in location 1

$x_2 =$ proportion of time spent feeding in location 2

$x_3 =$ proportion of time spent on nonfeeding activities.

Using these variables, the total quantity of food 1 consumed is given by Y_1, where

$$Y_1 = u_{11}x_1 + u_{12}x_2 = .4x_1 + .1x_2. \quad (1)$$

The total quantity of food 2 consumed is given by Y_2, where

$$Y_2 = u_{21}x_1 + u_{22}x_2 = .3x_1 + .3x_2. \quad (2)$$

The total amount of food consumed is thus given by

$$Y_1 + Y_2 = .4x_1 + .1x_2 + .3x_1 + .3x_2 = .7x_1 + .4x_2. \quad (3)$$

Again, by gathering experimental data, suppose that

$z_t = .4 =$ desired level of total food consumption

$z_1 = .15 =$ desired level of consumption of food 1

$z_2 = .25 =$ desired level of consumption of food 2.

The predator wishes to find values of x_1 and x_2 such that the difference between the desired level of consumption (the z's) and the actual level of consumption (the Y's) is minimized. However, much experience has shown that the difference between desired food consumption and actual consumption is not perceived by the animal as linear, but rather perhaps as a square. That is, a 5% shortfall in the actual consumption of food, as

[†]Based on Gerald G. Martens, "An Optimization Equation for Predation," *Ecology*, Winter 1973, pp. 92–101.

compared to the desired consumption, may be perceived by the predator as a 20–25% shortfall. (This phenomenon is known as *Weber's Law*—many psychology books discuss it.) Thus, to have our model approximate reality, we must find values of x_1 and x_2 that will minimize

$$G = [z_t - (Y_1 + Y_2)]^2 + w_1[z_1 - Y_1]^2 + w_2[z_2 - Y_2]^2 + w_3(1 - x_3)^2. \quad \textbf{(4)}$$

The term $w_3(1 - x_3)^2$ represents the fact that the predator does not wish to spend the total available time in searching for food—recall that x_3 represents the proportion of available time that is spent on nonfeeding activities. The variables w_1, w_2, and w_3 represent the relative importance, or weights, assigned by the animal to food 1, food 2, and nonfood activities, respectively. Again, it is necessary to gather experimental data: reasonable values of w_1, w_2, and w_3 are as follows:

$$w_1 = .5 \qquad w_2 = .4 \qquad w_3 = .03.$$

If we substitute .4 for z_t, .15 for z_1, .25 for z_2, .5 for w_1, .4 for w_2, and .03 for w_3, the results of equation (1) for Y_1, (2) for Y_2, and (3) for $Y_1 + Y_2$ into equation (4), we have

$$G = [.4 - .7x_1 - .4x_2]^2 + .5[.15 - .4x_1 - .1x_2]^2 + .4[.25 - .3x_1 - .3x_2]^2 + .03(1 - x_3)^2.$$

We want to minimize G, subject to the constraint $x_1 + x_2 + x_3 = 1$, or $x_1 + x_2 + x_3 - 1 = 0$. This can be done with Lagrange multipliers. First, form the function F:

$$F = [.4 - .7x_1 - .4x_2]^2 + .5[.15 - .4x_1 - .1x_2]^2 + .4[.25 - .3x_1 - .3x_2]^2 + .03(1 - x_3)^2 + \lambda(x_1 + x_2 + x_3 - 1).$$

Now we must find the partial derivatives of F with respect to x_1, x_2, x_3, and λ. Doing this we have

$$\frac{\partial F}{\partial x_1} = 2(-.7)(.4 - .7x_1 - .4x_2) + 2(.5)(-.4)(.15 - .4x_1 - .1x_2) + 2(.4)(-.3)(.25 - .3x_1 - .3x_2) + \lambda$$

$$\frac{\partial F}{\partial x_2} = 2(-.4)(.4 - .7x_1 - .4x_2) + 2(.5)(-.1)(.15 - .4x_1 - .1x_2) + 2(.4)(-.3)(.25 - .3x_1 - .3x_2) + \lambda$$

$$\frac{\partial F}{\partial x_3} = -2(.03)(1 - x_3) + \lambda$$

$$\frac{\partial F}{\partial \lambda} = x_1 + x_2 + x_3 - 1.$$

Both $\partial F/\partial x_1$ and $\partial F/\partial x_2$ can be simplified, using some rather tedious algebra. Doing this, and placing each partial derivative equal to 0, we get the following system of equations:

$$\frac{\partial F}{\partial x_1} = -.68 + 1.212x_1 + .672x_2 + \lambda = 0$$

$$\frac{\partial F}{\partial x_2} = -.395 + .672x_1 + .402x_2 + \lambda = 0$$

$$\frac{\partial F}{\partial x_3} = -.06 + .06x_3 + \lambda = 0$$

$$\frac{\partial F}{\partial \lambda} = x_1 + x_2 + x_3 - 1 = 0.$$

Although we shall not go through the details of the solution here, it can be found from the system above that the values of the x's that minimize G are given by

$$x_1 = .48 \qquad x_2 = .09 \qquad x_3 = .43.$$

(These values have been rounded to the nearest hundredth.) Thus, the predator should spend about .48 of the available time searching in location 1, and about .09 of the time searching in location 2. This will leave about .43 of the time free for nonfeeding activities.

Exercises

1. Verify the simplification of $\partial F/\partial x_1$ and $\partial F/\partial x_2$.
2. Verify the solution given in the text by solving the system of equations.
3. Suppose $w_1 = .4$ and $w_2 = .5$ and all other values remain the same. What proportion of time should the predator spend on feeding in each of locations 1 and 2? What proportion of time will be spent on nonfeeding activities?

Applications at a glance . . .
Terminal Velocity in Free Fall . . . growth of algae on phosphates . . . rate that a rumor spreads . . .

CHAPTER 9

DIFFERENTIAL EQUATIONS

Suppose that an economist wants to develop an equation that will forecast interest rates. By studying data on previous changes in interest rates, she hopes to find a relationship between the level of interest rates and their rate of change. A function giving the rate of change of interest rates would be the derivative of the function describing the level of interest rates. A **differential equation** is an equation that involves an unknown function, where $y = f(x)$, and a finite number of its derivatives. Solving the differential equation for y would give the unknown function to be used for forecasting interest rates.

Differential equations have been important in the study of physical science and engineering since the eighteenth century. Among the pioneers in the field of differential equations was the French mathematician Alexis Claude Clairaut (1713–1765). A particular type of equation studied in elementary courses on differential equations bears his name.

More recently, differential equations have become useful in social sciences, life sciences, and economics for solving problems about population growth, ecological balance, and interest rates. In this chapter, we will introduce some methods for solving differential equations and give examples of their applications.

9.1 SOLUTIONS OF ELEMENTARY AND SEPARABLE DIFFERENTIAL EQUATIONS

How can we predict the future population of a flock of mountain goats?

Using differential equations, we will learn to answer such questions.

Usually a solution of an equation is a *number*. A solution of a differential equation, however, is a *function*. For example, the solutions of a differential equation such as

$$\frac{dy}{dx} = 3x^2 - 2x \qquad \textbf{(1)}$$

consist of all expressions for y that satisfy the equation. Since the left side of the equation is the derivative of y with respect to x, we can solve the equation for y by finding an antiderivative on each side. On the left, the antiderivative is $y + C_1$. On the right side,

$$\int (3x^2 - 2x)\, dx = x^3 - x^2 + C_2.$$

The solutions of equation (1) are given by

$$y + C_1 = x^3 - x^2 + C_2$$

or

$$y = x^3 - x^2 + C_2 - C_1.$$

Replacing the constant $C_2 - C_1$ with the single constant C gives

$$y = x^3 - x^2 + C. \tag{2}$$

(From now on we will add just one constant, with the understanding that it represents the difference between the two constants obtained in the two integrations.)

Each different value of C in equation (2) leads to a different solution of equation (1), showing that a differential equation can have an infinite number of solutions. Equation (2) is the **general solution** of the differential equation (1). Some of the solutions of equation (1) are graphed in Figure 1.

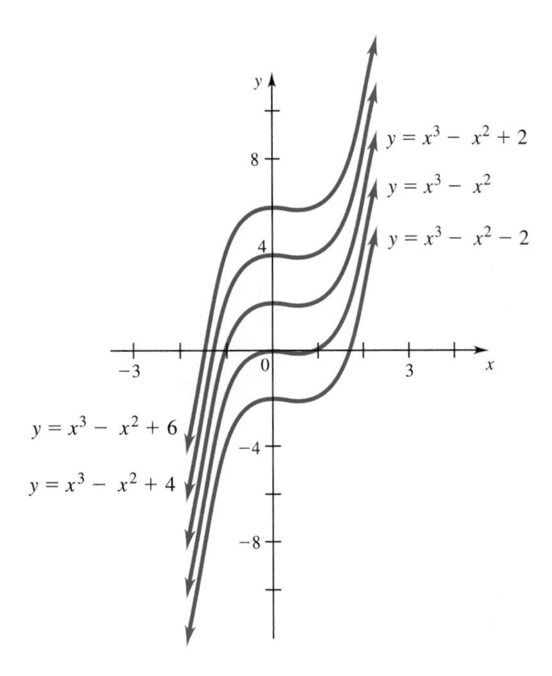

FIGURE 1

The simplest kind of differential equation has the form

$$\frac{dy}{dx} = f(x).$$

Since equation (1) has this form, the solution of equation (1) suggests the following generalization.

GENERAL SOLUTION OF $\frac{dy}{dx} = f(x)$

The general solution of the differential equation $dy/dx = f(x)$ is

$$y = \int f(x)\, dx.$$

EXAMPLE 1 The population P of a flock of birds is growing exponentially so that

$$\frac{dP}{dx} = 20e^{.05x},$$

where x is time in years. Find P in terms of x if there were 20 birds in the flock initially.

Solve the differential equation:

$$P = \int 20e^{.05x}\,dx = \frac{20}{.05}e^{.05x} + C = 400e^{.05x} + C.$$

Since P is 20 when x is 0,

$$20 = 400e^0 + C$$
$$-380 = C,$$

and

$$P = 400e^{.05x} - 380.$$ ◀

In Example 1, the given information was used to produce a solution with a specific value of C. Such a solution is called a **particular solution** of the given differential equation. The given information, $P = 20$ when $t = 0$, is called an **initial condition** because $t = 0$.

Sometimes a differential equation must be rewritten in the form

$$\frac{dy}{dx} = f(x)$$

before it can be solved.

Find the particular solution of

$$\frac{dy}{dx} - 2x = 5,$$

given that $y = 2$ when $x = -1$.

Add $2x$ to both sides of the equation to get

$$\frac{dy}{dx} = 2x + 5.$$

The general solution is

$$y = \frac{2x^2}{2} + 5x + C = x^2 + 5x + C.$$

Substituting 2 for y and -1 for x gives

$$2 = (-1)^2 + 5(-1) + C$$
$$C = 6.$$

The particular solution is $y = x^2 + 5x + 6$. ◀

So far in this section, we have used a method that is essentially the same as that used in the section on antiderivatives, when we first started the topic of integration. But not all differential equations can be solved so easily. For example, if interest on an investment is compounded continuously, then the investment grows at a rate proportional to the amount of money present.* If A is the amount in an account at time t, then for some constant k, the differential equation

$$\frac{dA}{dt} = kA$$

gives the rate of growth of A with respect to t. This differential equation is different from those discussed previously, which had the form

$$\frac{dy}{dx} = f(x).$$

Caution Since the right-hand side of the differential equation for compound interest is a function of A, rather than a function of t, it would be completely invalid to simply integrate both sides as we did before. The previous method only works when the side opposite the derivative is simply a function of the independent variable.

The differential equation for compound interest is an example of a more general differential equation we will now learn to solve; namely, those that can be written in the form

$$\frac{dy}{dx} = \frac{f(x)}{g(y)}.$$

Using differentials and multiplying on both sides by $g(y)\,dx$ gives

$$g(y)\,dy = f(x)\,dx.$$

In this form all terms involving y (including dy) are on one side of the equation, and all terms involving x (and dx) are on the other side. A differential equation in this form is said to be **separable,** since the variables x and y can be separated. A separable differential equation may be solved by integrating each side.

EXAMPLE 3

Find the general solution of $y\dfrac{dy}{dx} = x^2$.

Begin by separating the variables to get

$$y\,dy = x^2\,dx.$$

The general solution is found by taking antiderivatives on each side.

*See Section 3 in the chapter on exponential and logarithmic functions.

$$\int y\,dy = \int x^2\,dx$$

$$\frac{y^2}{2} = \frac{x^3}{3} + C$$

$$y^2 = \frac{2}{3}x^3 + 2C = \frac{2}{3}x^3 + K$$

The constant K was substituted for $2C$ in the last step. The solution is left in implicit form, not solved explicitly for y. ◀

Find the general solution of $\frac{dy}{dx} = ky$, where k is a constant.

Separating variables leads to

$$\frac{1}{y}\,dy = k\,dx.$$

To solve this equation, take antiderivatives on each side.

$$\int \frac{1}{y}\,dy = \int k\,dx$$

$$\ln|y| = kx + C$$

Use the definition of logarithm to write the equation in exponential form as

$$|y| = e^{kx+C}.$$

By properties of exponents,

$$|y| = e^{kx}e^{C}.$$

Finally, use the definition of absolute value to get

$$y = e^{kx}e^{C} \quad \text{or} \quad y = -e^{kx}e^{C}.$$

Since e^C and $-e^C$ are constants, replace them with the constant M, which may have any nonzero real-number value, to get the single equation

$$y = Me^{kx}.$$

This equation, $y = Me^{kx}$, defines the exponential growth or decay function that was discussed previously (in the chapter on exponential and logarithmic functions). ◀

Recall that equations of the form $y = Me^{kx}$ arise in situations where the rate of change of a quantity is proportional to the amount present at time x; that is, where

$$\frac{dy}{dx} = ky.$$

The constant k is called the **growth rate constant,** while M represents the amount present at time $x = 0$. (A positive value of k indicates growth, while a negative value of k indicates decay.)

In the chapter on exponential and logarithmic functions, we saw that the amount of money in an account with interest compounded continuously is given by

$$A = Pe^{rt},$$

where P is the initial amount in the account, r is the annual interest rate, and t is the time in years. Observe that this is the same as the equation for the amount of money in an account derived here, where P and r have been replaced with A_0 and k, respectively.

Applying the results of Example 4 to the equation discussed earlier,

$$\frac{dA}{dt} = kA,$$

shows that the amount in the account at time t is

$$A = A_0 e^{kt},$$

where A_0 is the amount originally invested.

As a model of population growth, the equation $y = Me^{kx}$ is not realistic over the long run for most populations. As shown by graphs of functions of the form $y = Me^{kx}$, with both M and k positive, growth would be unbounded. Additional factors, such as space restrictions or a limited amount of food, tend to inhibit growth of populations as time goes on. In an alternative model that assumes a maximum population of size N, the rate of growth of a population is proportional to how close the population is to that maximum. These assumptions lead to the differential equation

$$\frac{dy}{dx} = k(N - y),$$

the limited growth function mentioned in an earlier chapter. Graphs of limited growth functions look like the graph in Figure 2, where y_0 is the initial population.

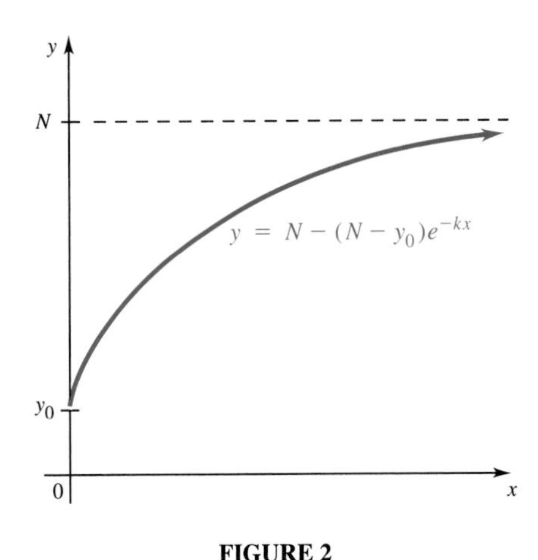

FIGURE 2

EXAMPLE 5 A certain area can support no more than 4000 mountain goats. There are 1000 goats in the area at present, with a growth constant of .20.

(a) Write a differential equation for the rate of growth of this population.

Let $N = 4000$ and $k = .20$. The rate of growth of the population is given by

$$\frac{dy}{dx} = .20(4000 - y).$$

To solve for y, first separate the variables.

$$\frac{dy}{4000 - y} = .2\,dx$$

$$\int \frac{dy}{4000 - y} = \int .2\,dx$$

$$-\ln(4000 - y) = .2x + C$$

$$\ln(4000 - y) = -.2x - C$$

$$4000 - y = e^{-.2x - C} = (e^{-.2x})(e^{-C})$$

The absolute value bars are not needed for $\ln(4000 - y)$ since y must be less than 4000 for this population, so that $4000 - y$ is always nonnegative. Let $e^{-C} = B$. Then

$$4000 - y = Be^{-.2x}$$

$$y = 4000 - Be^{-.2x}.$$

Find B by using the fact that $y = 1000$ when $x = 0$.

$$1000 = 4000 - B$$

$$B = 3000$$

Notice that the value of B is the difference between the maximum population and the initial population. Substituting 3000 for B in the equation for y gives

$$y = 4000 - 3000e^{-.2x}.$$

(b) What will the goat population be in 5 years?

In 5 years, the population will be

$$y = 4000 - 3000e^{-(.2)(5)} = 4000 - 3000e^{-1}$$
$$= 4000 - 1103.6 = 2896.4,$$

or about 2900 goats. ◀

Logistic Growth Let y be the size of a certain population at time x. In the standard model for unlimited growth,

$$\frac{dy}{dx} = ky, \tag{3}$$

the rate of growth is proportional to the current population size. The constant k, the growth rate constant, is the difference between the birth and death rates of the population. The unlimited growth model predicts that the population's growth rate is a constant, k.

Growth usually is not unlimited, however, and the population's growth rate is usually not constant because the population is limited by environmental factors to a maximum size N, called the **carrying capacity** of the environment for the species. In the limited growth model given above,

$$\frac{dy}{dx} = k(N - y),$$

the rate of growth is proportional to the remaining room for growth, $N - y$. In the **logistic growth model**

$$\frac{dy}{dx} = \frac{k}{N}(N - y)y, \tag{4}$$

the rate of growth is proportional to both the current population size y and the remaining room for growth $N - y$. Equation (4) is called the **logistic equation.** The logistic growth model predicts that the population's growth rate is a function of the population size y:

$$\frac{dy}{dx} = ky - \frac{k}{N}y^2.$$

As $y \to 0$, y^2 is insignificant when compared to y, and so

$$\frac{dy}{dx} = ky - \frac{k}{N}y^2 \approx ky.$$

That is, the growth of the population is essentially unlimited when y is small. On the other hand, when $y \to N$,

$$\frac{dy}{dx} = ky - \frac{k}{N}y^2 = \left(k - \frac{k}{N}y\right)y = \left(\frac{kN}{N} - \frac{k}{N}y\right)y = \frac{k}{N}(N - y)y \to 0.$$

That is, population growth levels off as y nears the maximum population size N. Thus, the logistic equation

$$\frac{dy}{dx} = ky - \frac{k}{N}y^2$$

is the unlimited growth equation (3) with a damping term ky^2/N subtracted to account for limiting environmental factors when y nears N. Let y_0 denote the initial population size. Under the assumption $0 < y < N$, the general solution of equation (4) is

$$y = \frac{N}{1 + be^{-kx}}, \tag{5}$$

where $b = (N - y_0)/y_0$ (see Exercise 31). This solution, called a **logistic curve,** is shown in Figure 3 on the next page.

As expected, the logistic curve begins exponentially and subsequently levels off. Another important feature is the point of inflection $((\ln b)/k, N/2)$, at which the growth rate dy/dx is a maximum (see Exercise 33).

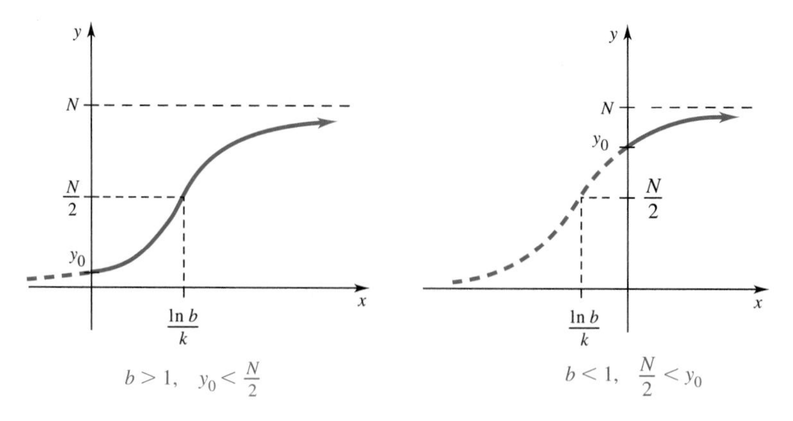

FIGURE 3

Logistic equations arise frequently in the study of populations. In about 1840 the Belgian sociologist P. F. Verhurst fitted a logistic curve to United States census figures and made predictions about the population that were subsequently proven to be quite accurate. American biologist Raymond Pearl (circa 1920) found that the growth of a population of fruit flies in a limited space could be modeled by the logistic equation

$$\frac{dy}{dx} = .2y - \frac{.2}{1035}y^2.$$

Logistic growth is an example of how a model is modified over time as new insights occur. The model for population growth changed from the early exponential curve $y = Me^{kx}$ to the logistic curve

$$y = \frac{N}{1 + be^{-kx}}.$$

Many other quantities besides population grow logistically. That is, their initial rate of growth is slow, but as time progresses, their rate of growth increases to a maximum value and subsequently begins to decline and to approach zero.

EXAMPLE 6 Rapid technological advancements in the last 20 years have made many products obsolete practically overnight. J. C. Fisher and R. H. Pry* successfully described the phenomenon of a technically superior new product replacing another product by the logistic equation

$$\frac{dz}{dx} = k(1 - z)z, \tag{6}$$

*From J. C. Fisher and R. H. Pry, "A Simple Substitution Model of Technological Change," *Technological Forecasting and Social Change,* vol. 3, 1971–1972. Copyright © 1972 by Elsevier Science Publishing Co., Inc. Reprinted by permission of the publisher.

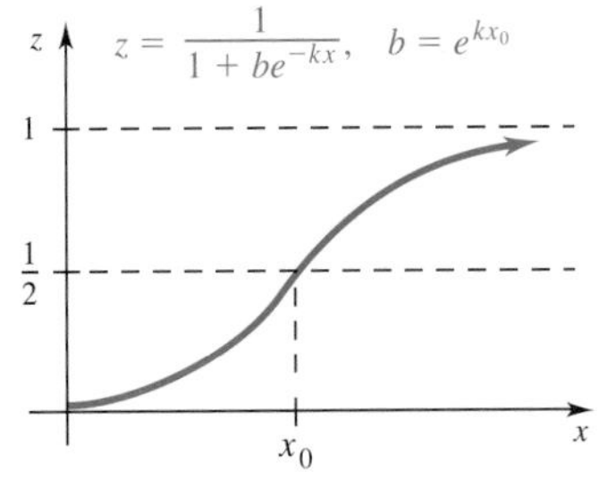

FIGURE 4

where z is the market share of the new product and $1 - z$ is the market share of the other product. The new product will initially have little or no market share; that is, $z_0 \approx 0$. Thus, the constant b in equation (5) will have to be determined in a different way. Let x_0 be the time at which $z = 1/2$. Under the assumption $0 < z < 1$, the general solution of equation (6) is

$$z = \frac{1}{1 + be^{-kx}},$$

where $b = e^{kx_0}$ (see Exercise 32). This solution is shown in Figure 4.

The market share of the new product will be growing most rapidly when the new product has captured exactly half the market, and the market share of the older product will be shrinking most rapidly at the same time. Notice that the logistic equation (4) can be transformed into the simpler logistic equation (6) by the change of variable $z = y/N$. ◀

9.1 Exercises

Find general solutions for the following differential equations.

1. $\frac{dy}{dx} = -2x + 3x^2$
2. $\frac{dy}{dx} = 3e^{-2x}$
3. $3x^3 - 2\frac{dy}{dx} = 0$
4. $3x^2 - 3\frac{dy}{dx} = 2$
5. $y\frac{dy}{dx} = x$
6. $y\frac{dy}{dx} = x^2 - 1$
7. $\frac{dy}{dx} = 2xy$
8. $\frac{dy}{dx} = x^2y$
9. $\frac{dy}{dx} = 3x^2y - 2xy$
10. $(y^2 - y)\frac{dy}{dx} = x$
11. $\frac{dy}{dx} = \frac{y}{x}, x > 0$
12. $\frac{dy}{dx} = \frac{y}{x^2}$
13. $\frac{dy}{dx} = y - 5$
14. $\frac{dy}{dx} = 3 - y$
15. $\frac{dy}{dx} = y^2e^x$
16. $\frac{dy}{dx} = \frac{e^x}{e^y}$

Find particular solutions for the following equations.

17. $\frac{dy}{dx} + 2x = 3x^2$; $y = 2$ when $x = 0$
18. $\frac{dy}{dx} = 4x^3 - 3x^2 + x$; $y = 0$ when $x = 1$
19. $2\frac{dy}{dx} = 4xe^{-x}$; $y = 42$ when $x = 0$
20. $x\frac{dy}{dx} = x^2e^{3x}$; $y = \frac{8}{9}$ when $x = 0$
21. $\frac{dy}{dx} = \frac{x^2}{y}$; $y = 3$ when $x = 0$
22. $x^2\frac{dy}{dx} = y$; $y = -1$ when $x = 1$
23. $(2x + 3)y = \frac{dy}{dx}$; $y = 1$ when $x = 0$
24. $x\frac{dy}{dx} - y\sqrt{x} = 0$; $y = 1$ when $x = 0$
25. $\frac{dy}{dx} = \frac{y^2}{x}$; $y = 5$ when $x = e$
26. $\frac{dy}{dx} = x^{1/2}y^2$; $y = 12$ when $x = 4$
27. $\frac{dy}{dx} = \frac{2x + 1}{y - 3}$; $y = 4$ when $x = 0$
28. $\frac{dy}{dx} = \frac{x^2 + 5}{2y - 1}$; $y = 11$ when $x = 0$
29. $\frac{dy}{dx} = (y - 1)^2e^x$; $y = 2$ when $x = 0$
30. $\frac{dy}{dx} = (x + 2)^2e^y$; $y = 0$ when $x = 1$

31. (a) Solve the logistic equation (4) in this section by observing that

$$\frac{1}{y} + \frac{1}{N - y} = \frac{N}{(N - y)y}.$$

(b) Assume $0 < y < N$. Verify that $b = (N - y_0)/y_0$ in equation (5), where y_0 is the initial population size.

(c) Assume $0 < N < y$ for all y. Verify that $b = (y_0 - N)/y_0$.

32. Suppose that $0 < z < 1$ for all z. Solve the logistic equation (6) as in Exercise 31. Verify that $b = e^{kx_0}$, where x_0 is the time at which $z = 1/2$.

Connections

33. Suppose that $0 < y_0 < N$. Let $b = (N - y_0)/y_0$, and let $y(x) = \dfrac{N}{1 + be^{-kx}}$ for all x. Show the following.

(a) $0 < y(x) < N$ for all x.

(b) The lines $y = 0$ and $y = N$ are horizontal asymptotes of the graph.

(c) $y(x)$ is an increasing function.

(d) $((\ln b)/k, N/2)$ is a point of inflection of the graph.

(e) dy/dx is a maximum at $x_0 = (\ln b)/k$.

34. Suppose that $0 < N < y_0$. Let $b = (y_0 - N)/y_0$ and let

$$y(x) = \frac{N}{1 - be^{-kx}} \quad \text{for all } x \neq \frac{\ln b}{k}.$$

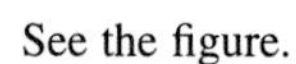

See the figure.

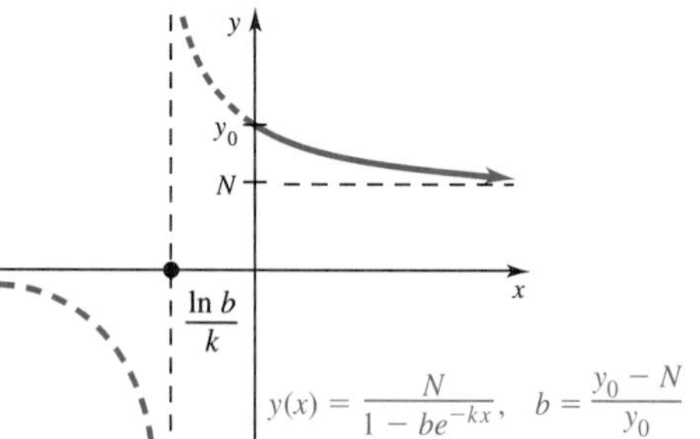

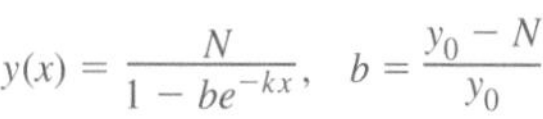

Show the following.

(a) $0 < b < 1$

(b) The lines $y = 0$ and $y = N$ are horizontal asymptotes of the graph.

(c) The line $x = (\ln b)/k$ is a vertical asymptote of the graph.

(d) $y(x)$ is decreasing on $((\ln b)/k, \infty)$ and on $(-\infty, (\ln b)/k)$.

(e) $y(x)$ is concave upward on $((\ln b)/k, \infty)$ and concave downward on $(-\infty, (\ln b)/k)$.

Applications

Business and Economics

35. Profit The marginal profit of a certain company is given by

$$\frac{dy}{dx} = \frac{100}{32 - 4x},$$

where x represents the amount of money (in thousands of dollars) that the company spends on advertising.

Find the profit for each of the following advertising expenditures if the profit is \$1000 when nothing is spent on advertising.

(a) \$3000 **(b)** \$5000

(c) Can advertising expenditures ever reach \$8000 according to this model? Why?

36. Sales Decline Sales (in thousands) of a certain product are declining at a rate proportional to the amount of sales, with a decay constant of 25% per year.

(a) Write a differential equation to express the rate of sales decline.

(b) Find a general solution to the equation in part (a).

(c) How much time will pass before sales become 30% of their original value?

37. Inflation If inflation grows continuously at a rate of 6% per year, how long will it take for \$1 to lose half its value?

38. **Gross National Product** Suppose that the gross national product (GNP) of a particular country increases exponentially, with a growth constant of 2% per year. Ten years ago, the GNP was 10^5 dollars. What will be the GNP in 5 years?

39. **Bankruptcy** In a certain area, 1500 small business firms are threatened by bankruptcy. Assume the rate of change in the number of bankruptcies is proportional to the number of small firms that are not yet bankrupt. If the growth constant is 6%, and if 100 firms are bankrupt initially, how many will be bankrupt in 2 years?

40. **New Car Sales** The sales of a new model of car are expected to follow the logistic curve

$$y = \frac{100{,}000}{1 + 100e^{-x}},$$

where x is the number of months since the introduction of the model. Find the following.

(a) The initial sales level

(b) The maximum sales level

(c) The month in which the maximum sales level will be achieved

(d) The rate at which sales are growing 6 months after the introduction of the model

Elasticity of Demand *Elasticity of demand was discussed in the chapter on applications of the derivative, where it was defined as*

$$E = -\frac{p}{q} \cdot \frac{dq}{dp},$$

*for demand q and price p. Find the general demand equation $q = f(p)$ for each of the following elasticity functions. (*Hint: *Set each elasticity function equal to $-\frac{p}{q} \cdot \frac{dq}{dp}$, then solve for q. Write the constant of integration as* ln *C.)*

41. $E = \dfrac{4p^2}{q^2}$

42. $E = 2$

Life Sciences

43. **Tracer Dye** The amount of a tracer dye injected into the bloodstream decreases exponentially, with a decay constant of 3% per minute. If 6 cc are present initially, how many cubic centimeters are present after 10 min? (Here k will be negative.)

44. **Bacterial Population** The rate at which the number of bacteria (in thousands) in a culture is changing after the introduction of a bactericide is given by

$$\frac{dy}{dx} = 50 - y,$$

where y is the number of bacteria (in thousands) present at time x. Find the number of bacteria present at each of the following times if there were 1000 thousand bacteria present at time $x = 0$.

(a) $x = 2$ **(b)** $x = 5$ **(c)** $x = 10$ **(d)** $x = 15$

45. **Fish Population** An isolated fish population is limited to 5000 by the amount of food available. If there are now 150 fish and the population is growing with a growth constant of 1% a year, find the expected population at the end of 5 yr.

46. **Mite Population** A population of mites (in hundreds) increases exponentially, with a growth constant of 5% per week. At the beginning of an observation period there were 3000 mites. How many are present 4 weeks later?

Social Sciences

47. **Spread of a Rumor** Suppose the rate at which a rumor spreads—that is, the number of people who have heard the rumor over a period of time—increases with the number of people who have heard it. If y is the number of people who have heard the rumor, then

$$\frac{dy}{dt} = ky,$$

where t is the time in days.

(a) If y is 1 when $t = 0$, and y is 5 when $t = 2$, find k.

Using the value of k from part (a), find y for each of the following times.

(b) $t = 3$ **(c)** $t = 5$ **(d)** $t = 10$

48. **Worker Productivity** A company has found that the rate at which a person new to the assembly line produces items is

$$\frac{dy}{dx} = 7.5e^{-.3y},$$

where x is the number of days the person has worked on the line. How many items can a new worker be expected to produce on the eighth day if he produces none when $x = 0$?

Physical Sciences

49. **Radioactive Decay** The amount of a radioactive substance decreases exponentially, with a decay constant of 5% per month.

(a) Write a differential equation to express the rate of change.

(b) Find a general solution to the differential equation from part (a).

(c) If there are 90 g at the start of the decay process, find a particular solution for the differential equation from part (a).

(d) Find the amount left after 10 mo.

50. **Snowplow** One morning snow began to fall at a heavy and constant rate. A snowplow started out at 8:00 A.M. At 9:00 A.M. it had traveled 2 mi. By 10:00 A.M. it had traveled 3 mi. Assuming that the snowplow removes a constant volume of snow per hour, determine the time at which it started snowing. (*Hint:* Let t denote the time since the snow started to fall, and let T be the time when the snowplow started out. Let x, the distance the snowplow has traveled, and h, the height of the snow, be functions of t. The assumption that a constant volume of snow per hour is removed implies that the speed of the snowplow times the height of the snow is a constant. Set up and solve differential equations involving dx/dt and dh/dt.)*

9.2 LINEAR FIRST-ORDER DIFFERENTIAL EQUATIONS

What happens over time to the glucose level in a patient's bloodstream?

The solution to a linear differential equation gives us an answer.

Recall that $f^{(n)}(x)$ represents the nth derivative of $f(x)$, and that $f^{(n)}(x)$ is called an *nth-order* derivative. By this definition, the derivative $f'(x)$ is first-order, $f''(x)$ is second-order, and so on. The **order of a differential equation** is that of the highest-order derivative in the equation. In this section only first-order differential equations are discussed.

A **linear first-order differential equation** is an equation of the form

$$y' + P(x)y = Q(x).$$

Many useful models produce such equations. In this section we develop a general method for solving first-order linear differential equations. For simplicity, we shall use the notation y' for the derivative in this section.

The first step in solving the equation

$$xy' + 6y + 2x^4 = 0, \tag{1}$$

for example, is to divide both sides of the equation by x and rearrange the terms to get the linear differential equation

$$y' + \frac{6}{x}y = -2x^3.$$

This equation does not have separable variables and cannot be solved by the methods discussed so far. (Verify this.) Instead, multiply both sides of the equation by x^6 (the reason will be explained shortly) to get

$$x^6y' + 6x^5y = -2x^9. \tag{2}$$

On the left, $6x^5$, the coefficient of y, is the derivative of x^6, the coefficient of y'. Recall the product rule for derivatives:

$$D_x(uv) = uv' + u'v.$$

*This problem first appeared in the *American Mathematical Monthly*, vol. 44, December 1937.

If $u = x^6$ and $v = y$, the product rule gives

$$D_x(x^6y) = x^6y' + 6x^5y,$$

which is the left side of equation (2). Substituting $D_x(x^6y)$ for the left side of equation (2) gives

$$D_x(x^6y) = -2x^9.$$

Assuming $y = f(x)$, as usual, both sides of this equation can be integrated with respect to x and the result solved for y to get

$$x^6y = \int -2x^9\,dx = -2\left(\frac{x^{10}}{10}\right) + C = -\frac{x^{10}}{5} + C$$

$$y = -\frac{x^4}{5} + \frac{C}{x^6}. \tag{3}$$

Equation (3) is the general solution of equation (2) and therefore of equation (1).

This procedure has given us a solution, but what motivated our choice of the multiplier x^6? To see where x^6 came from, let $I(x)$ represent the multiplier, and multiply both sides of the general equation

$$y' + P(x)y = Q(x)$$

by $I(x)$:

$$I(x)y' + I(x)P(x)y = I(x)Q(x). \tag{4}$$

The method illustrated above will work only if the left side of the equation is the derivative of the product function $I(x) \cdot y$, which is

$$I(x)y' + I'(x)y. \tag{5}$$

Comparing the coefficients of y in equations (4) and (5) shows that $I(x)$ must satisfy

$$I'(x) = I(x)P(x),$$

or

$$\frac{I'(x)}{I(x)} = P(x).$$

Integrating both sides of this last equation gives

$$\ln|I(x)| = \int P(x)\,dx + C$$

$$|I(x)| = e^{\int P(x)dx + C}$$

or

$$I(x) = \pm e^C e^{\int P(x)dx}.$$

Only one value of $I(x)$ is needed, so let $C = 0$, so that $e^C = 1$, and use the positive result, giving

$$I(x) = e^{\int P(x)dx}.$$

In summary, choosing $I(x)$ as $e^{\int P(x)dx}$ and multiplying both sides of a linear first-order differential equation by $I(x)$ puts the equation in a form that can be solved by integration.

INTEGRATING FACTOR

$I(x) = e^{\int P(x)dx}$ is called an **integrating factor** for the differential equation $y' + P(x)y = Q(x)$.

For equation (1) above, written as the linear differential equation

$$y' + \frac{6}{x}y = -2x^3,$$

$P(x) = 6/x$, and the integrating factor is

$$I(x) = e^{\int(6/x)dx} = e^{6\ln|x|} = e^{\ln|x|^6} = e^{\ln x^6} = x^6.$$

This last step used the fact that, for all a, $e^{\ln a} = a$.

In summary, we solve a linear first-order differential equation with the following steps.

SOLVING A LINEAR FIRST-ORDER DIFFERENTIAL EQUATION

1. Put the equation in the linear form $y' + P(x)y = Q(x)$.
2. Find the integrating factor $I(x) = e^{\int P(x)dx}$.
3. Multiply each term of the equation from Step 1 by $I(x)$.
4. Replace the sum of terms on the left with $D_x[I(x)y]$.
5. Integrate both sides of the equation and solve for y if possible.

Give the general solution of $y' + 2xy = x$.

Step 1 This equation is already in the required form.

Step 2 The integrating factor is

$$I(x) = e^{\int 2x dx} = e^{x^2}.$$

Step 3 Multiplying each term by e^{x^2} gives

$$e^{x^2}y' + 2xe^{x^2}y = xe^{x^2}.$$

Step 4 The sum of terms on the left can now be replaced with $D_x(e^{x^2}y)$, to get

$$D_x(e^{x^2}y) = xe^{x^2}.$$

Step 5 Integrating on both sides leads to the general solution

$$e^{x^2}y = \int xe^{x^2}\,dx = \frac{1}{2}e^{x^2} + C$$

$$y = \frac{1}{2} + Ce^{-x^2}.$$ ◀

EXAMPLE 2

Solve $2y' - 6y - e^x = 0$ if $y = 5$ when $x = 0$.
Write the equation in the required form:

$$y' - 3y = \frac{1}{2}e^x.$$

The integrating factor is

$$I(x) = e^{\int(-3)dx} = e^{-3x}.$$

Multiplying each term by $I(x)$ gives

$$e^{-3x}y' - 3e^{-3x}y = \frac{1}{2}e^x e^{-3x}$$

$$e^{-3x}y' - 3e^{-3x}y = \frac{1}{2}e^{-2x}.$$

The left side can now be replaced by $D_x(e^{-3x}y)$ to get

$$D_x(e^{-3x}y) = \frac{1}{2}e^{-2x}.$$

Integrating on both sides gives

$$e^{-3x}y = \int \frac{1}{2}e^{-2x}\,dx = \frac{1}{2}\left(-\frac{1}{2}\right)e^{-2x} + C.$$

Now, multiply both sides by e^{3x} to get

$$y = -\frac{1}{4}e^x + Ce^{3x},$$

the general solution. Find the particular solution by substituting 0 for x and 5 for y:

$$5 = -\frac{1}{4}e^0 + Ce^0 = -\frac{1}{4} + C$$

or

$$\frac{21}{4} = C,$$

which leads to the particular solution

$$y = -\frac{1}{4}e^x + \frac{21}{4}e^{3x}.$$ ◀

EXAMPLE 3

Suppose glucose is infused into a patient's bloodstream at a constant rate of a grams per minute. At the same time, glucose is removed from the bloodstream at a rate proportional to the amount of glucose present. Then the amount of glucose, $G(t)$, present at time t satisfies

$$\frac{dG}{dt} = a - KG$$

for some constant K. Solve this equation for G. Does the glucose concentration eventually reach a constant? That is, what happens to G as $t \to \infty$?*

The equation can be written in the form of the linear first-order differential equation

$$G' + KG = a. \qquad \textbf{(6)}$$

The integrating factor is

$$I(t) = e^{\int K\,dt} = e^{Kt}.$$

Multiply both sides of equation (6) by $I(t) = e^{Kt}$.

$$e^{Kt}G' + Ke^{Kt}G = ae^{Kt}$$

Write the left side as $D_t(e^{Kt}G)$ and solve for G by integrating on each side.

$$D_t(e^{Kt}G) = ae^{Kt}$$

$$e^{Kt}G = \int ae^{Kt}\,dt = \frac{a}{K}e^{Kt} + C$$

Multiply both sides by e^{-Kt} to get

$$G = \frac{a}{K} + Ce^{-Kt}.$$

As $t \to \infty$,

$$\lim_{t\to\infty} G = \lim_{t\to\infty}\left(\frac{a}{K} + Ce^{-Kt}\right) = \lim_{t\to\infty}\left(\frac{a}{K} + \frac{C}{e^{Kt}}\right) = \frac{a}{K}.$$

Thus, the glucose concentration stabilizes at a/K. ◀

9.2 Exercises

Find general solutions for the following differential equations.

1. $y' + 2y = 5$
2. $y' + 4y = 10$
3. $y' + xy = 3x$
4. $y' + 2xy = x$
5. $xy' - y - x = 0, \quad x > 0$
6. $xy' + 2xy - x^2 = 0$
7. $2y' - 2xy - x = 0$
8. $3y' + 6xy + x = 0$
9. $xy' + 2y = x^2 + 3x, \quad x > 0$
10. $x^2y' + xy = x^3 - 2x^2, \quad x > 0$
11. $y - x\dfrac{dy}{dx} = x^3, \quad x > 0$
12. $2xy + x^3 = x\dfrac{dy}{dx}$

Solve each differential equation, subject to the given initial condition.

13. $y' + y = 2e^x; \quad y = 100$ when $x = 0$
14. $y' + 2y = e^{3x}; \quad y = 50$ when $x = 0$
15. $\dfrac{dy}{dx} - xy - x = 0; \quad y = 10$ when $x = 1$
16. $x\dfrac{dy}{dx} - 3y + 2 = 0; \quad y = 8$ when $x = 1$

---

*This example is from *Ordinary Differential Equations with Applications* by Larry C. Andrews, Scott, Foresman and Company, 1982, p. 79.

17. $x\frac{dy}{dx} + 5y = x^2$; $y = 12$ when $x = 2$

18. $2\frac{dy}{dx} - 4xy = 5x$; $y = 10$ when $x = 1$

19. $xy' + (1 + x)y = 3$; $y = 50$ when $x = 4$

20. $y' + 2xy - e^{-x^2} = 0$; $y = 100$ when $x = 0$

Applications

Life Sciences

21. **Population Growth** The logistic equation introduced in Section 1,

$$\frac{dy}{dx} = \frac{k}{N}(N - y)y, \qquad (7)$$

can be written as

$$\frac{dy}{dx} = cy - py^2,$$

where c and p are positive constants. Although this is a nonlinear differential equation, it can be reduced to a linear equation by a suitable substitution for the variable y.

(a) Letting $y = 1/z$ and $y' = -z'/z^2$, rewrite equation (7) in terms of z and solve.

(b) Let $z(0) = 1/y_0$ in part (a) and find a particular solution for y.

(c) Find the limit of y as $x \to \infty$. This is the saturation level of the population.

22. **Glucose Level** Solve the glucose level example (Example 3) using separation of variables.

Social Sciences

Immigration and Emigration *If population is changed either by immigration or emigration, the population model discussed in Section 1 is modified to*

$$\frac{dy}{dt} = ky + f(t),$$

where y is the population at time t and $f(t)$ is some (other) function of t that describes the net effect of the emigration/immigration. Assume $k = .02$ and $y = 10{,}000$ when $t = 0$. Solve this differential equation for y, given the following functions $f(t)$.

23. $f(t) = e^t$

24. $f(t) = e^{-t}$

25. $f(t) = -t$

26. $f(t) = t$

Physical Sciences

Newton's Law of Cooling *Newton's law of cooling states that the rate of change of temperature of an object is proportional to the difference in temperature between the object and the surrounding medium. Thus, if T is the temperature of the object after t hours and T_F is the (constant) temperature of the surrounding medium, then*

$$\frac{dT}{dt} = -k(T - T_F),$$

where k is a constant. Use this equation in Exercises 27–31.

27. Use the method of this section to show that the solution of this differential equation is

$$T = ce^{-kt} + T_F,$$

where c is a constant.

28. Solve the differential equation for Newton's law of cooling using separation of variables.

29. According to the solution of the differential equation for Newton's law of cooling, what happens to the temperature of an object after it has been in a surrounding medium with constant temperature for a long period of time? How well does this agree with reality?

30. Suppose a container of soup with a temperature of 180° F is put into a refrigerator where the temperature is 40° F. When checked 1/2 hr later, the temperature of the soup is 110° F.

(a) Find an equation for the temperature of the soup in terms of the number of hours it has been in the refrigerator.

(b) Find the temperature of the soup after 1 hr.

(c) Find the temperature of the soup after 5 hr.

31. A ceramic jar is taken from a kiln at a temperature of 50° C and set to cool in a room where the temperature is 22° C. In 1 hr, the temperature of the jar is 35° C.

(a) Find an equation for the temperature of the jar at time t.

(b) Find the temperature of the jar after 2 hr.

9.3 EULER'S METHOD

Applications sometimes involve differential equations such as

$$\frac{dy}{dx} = \frac{x + y}{y}$$

that cannot be solved by the methods discussed so far, but approximate solutions to these equations often can be found by numerical methods. For many applications, these numerical methods are quite adequate. In this section we introduce Euler's method, which is only one of numerous mathematical contributions made by Leonhard Euler (1707–1783) of Switzerland. (His name is pronounced "oiler.") He also introduced the $f(x)$ notation used throughout this text. In spite of becoming blind during his later years, he was the most prolific mathematician of his era. He published in all mathematical fields. The equation

$$e^{i\theta} = \cos\,\theta + i\,\sin\,\theta,$$

which shows how concepts from three branches of mathematics are interrelated, was developed by him. The number e (in honor of Euler) is the base for the natural logarithm; the number i represents the imaginary unit $\sqrt{-1}$, and the symbols cos and sin are found in the study of trigonometry. (Trigonometric applications are given in Chapter 11 of this book).

Euler's method of solving differential equations gives approximate solutions to differential equations involving $y = f(x)$ where the initial values of x and y are known: that is, equations of the form

$$y' = g(x, y), \quad \text{with} \quad f(x_0) = y_0.$$

Geometrically, Euler's method approximates the graph of the solution $y = f(x)$ with a polygonal line whose first segment is tangent to the curve at the point (x_0, y_0), as shown in Figure 5.

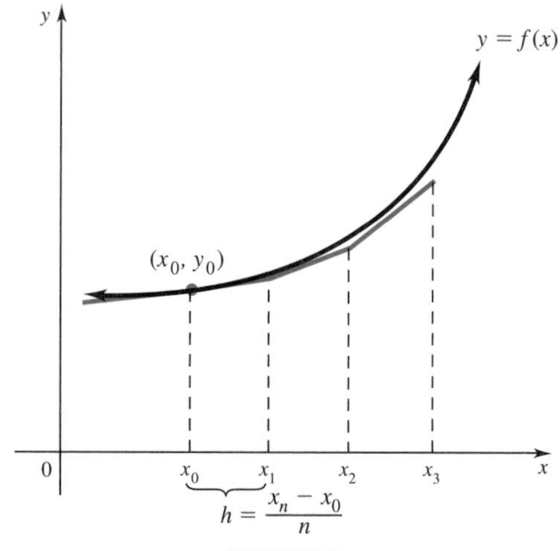

FIGURE 5

In the section on differentials, we defined Δy to be the actual change in y as x changed by an amount Δx:

$$\Delta y = f(x + \Delta x) - f(x).$$

The differential dy is an approximation to Δy. We find dy by following the tangent line from the point $(x, f(x))$, rather than by following the actual function. Then dy is found by using the formula $dy = (dy/dx)dx$ where $dx = \Delta x$. For example, let $f(x) = x^3$, $x = 1$, and $dx = \Delta x = .2$. Then $dy = f'(x)\,dx = 3x^2\,dx = 3(1^2)(.2) = .6$. The actual change in y as x changes from 1 to 1.2 is

$$\begin{aligned} f(x + \Delta x) - f(x) &= f(1.2) - f(1) \\ &= 1.2^3 - 1 \\ &= .728. \end{aligned}$$

To use Euler's method, divide the interval from x_0 to another point x_n into n subintervals of equal width (see Figure 5). The width of each subinterval is $h = (x_n - x_0)/n$.

Recall from the section on differentials that if Δx is a small change in x, then the corresponding small change in y, Δy, is approximated by

$$\Delta y \approx dy = \frac{dy}{dx} \cdot \Delta x.$$

The differential dy is the change in y along the tangent line. On the interval from x_i to x_{i+1}, dy is just $y_{i+1} - y_i$, where y_i is the approximate solution at x_i. We also have $dy/dx = g(x_i, y_i)$ and $\Delta x = h$. Putting these into the above equation yields

$$\begin{aligned} y_{i+1} - y_i &= g(x_i, y_i)h \\ y_{i+1} &= y_i + g(x_i, y_i)h. \end{aligned}$$

Because y_0 is given, we can use the equation just derived with $i = 0$ to get y_1. We can then use y_1 and the same equation with $i = 1$ to get y_2, and continue in this manner until we get y_n. A summary of Euler's method follows.

EULER'S METHOD

Let $y = f(x)$ be the solution of the differential equation

$$y' = g(x, y), \quad \text{with} \quad f(x_0) = y_0,$$

for $x_0 \le x \le x_n$. Let $x_{i+1} - x_i = h$ and

$$y_{i+1} = y_i + g(x_i, y_i)h,$$

for $0 \le i \le n$. Then

$$\boldsymbol{f(x_{i+1}) \approx y_{i+1}}.$$

As the following examples will show, the accuracy of the approximation varies for different functions. As h gets smaller, however, the approximation improves.

Use Euler's method to approximate the solution of $y' + 2xy = x$, with $f(0) = 1.5$, for $[0, 1]$. Use $h = .1$.

The general solution of this equation was found in Example 1 of the last section, so the results using Euler's method can be compared with the actual solution. Begin by writing the differential equation in the required form as

$$y' = x - 2xy, \qquad \text{so that} \qquad g(x, y) = x - 2xy.$$

Since $x_0 = 0$ and $y_0 = 1.5$, $g(x_0, y_0) = 0 - 2(0)(1.5) = 0$, and

$$y_1 = y_0 + g(x_0, y_0)h = 1.5 + 0(.1) = 1.5.$$

Now $x_1 = .1$, $y_1 = 1.5$ and $g(x_1, y_1) = .1 - 2(.1)(1.5) = -.2$. Then

$$y_2 = 1.5 + (-.2)(.1) = 1.48.$$

The ten values for x_i and y_i for $0 \le i \le 10$ are shown in Table 1, together with the actual values using the result from Example 1 in the last section. (Since the result was only a general solution, replace x with 0 and y with 1.5 to get the particular solution $y = 1/2 + e^{-x^2}$.)

Table 1

	Euler's Method	Actual Solution	Difference
x_i	y_i	$f(x_i)$	$y_i - f(x_i)$
0	1.5	1.5	0
.1	1.5	1.4900498	.0099502
.2	1.48	1.4607894	.0192106
.3	1.4408	1.4139312	.0268688
.4	1.384352	1.3521438	.0322082
.5	1.3136038	1.2788008	.034803
.6	1.232243	1.1976763	.0345667
.7	1.1443742	1.1126264	.0317102
.8	1.0541619	1.0272924	.0268695
.9	.96549595	.94485807	.02063788
1	.88170668	.86787944	.01382724

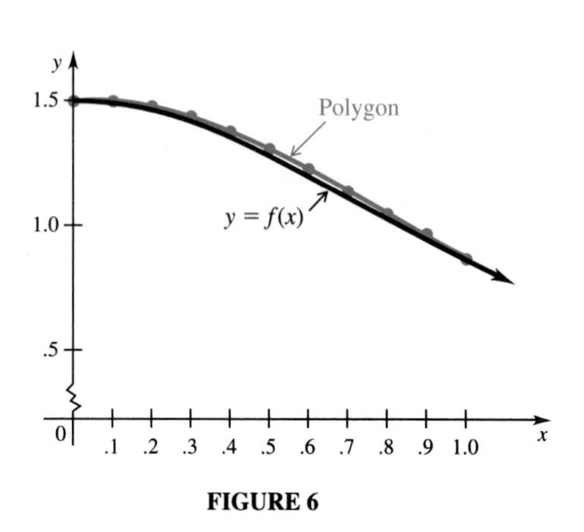

FIGURE 6

The results in Table 1 look quite good. Figure 6 shows that the polygonal line follows the actual graph of $f(x)$ closely. ◀

Euler's method produces a very good approximation for this differential equation because the slope of the solution $f(x)$ is not steep in the interval under investigation. The next example shows that such good results cannot always be expected.

EXAMPLE 2 Use Euler's method to solve $y' = 3y + (1/2)e^x$, with $y(0) = 5$, for $[0, 1]$, using 10 subintervals.

This is the differential equation of Example 2 in the last section. The general solution found there, with the initial condition given above, leads to the particular solution

$$y = -\frac{1}{4}e^x + \frac{21}{4}e^{3x}.$$

To solve by Euler's method, start with $g(x, y) = 3y + (1/2)e^x$, $x_0 = 0$, and $y_0 = 5$. For $n = 10$, $h = (1 - 0)/10 = .1$ again, and

$$y_{i+1} = y_i + g(x_i, y_i)h = y_i + \left(3y_i + \frac{1}{2}e^{x_i}\right)h.$$

For y_1, this gives

$$\begin{aligned} y_1 &= y_0 + \left(3y_0 + \frac{1}{2}e^{x_0}\right)h \\ &= 5 + \left[3(5) + \frac{1}{2}(e^0)\right](.1) \\ &= 5 + (15.5)(.1) = 6.55. \end{aligned}$$

Similarly,

$$\begin{aligned} y_2 &= 6.55 + [3(6.55) + \frac{1}{2}e^{.1}](.1) \\ &= 6.55 + (19.65 + .55258546)(.1) \\ &= 8.57025855. \end{aligned}$$

These and the remaining values for the interval [0, 1] are shown in Table 2.

Table 2 Approximate Solution Using $h = .1$

	Euler's Method	Actual Solution	Difference
x_i	y_i	$f(x_i)$	$y_i - f(x_i)$
0	5	5	0
.1	6.55	6.8104660	− .260466
.2	8.570259	9.2607730	− .690514
.3	11.20241	12.575452	−1.373042
.4	14.63062	17.057658	−2.427038
.5	19.09440	23.116687	−4.022287
.6	24.90516	31.305120	−6.399960
.7	32.46781	42.368954	−9.901144
.8	42.30884	57.315291	−15.006451
.9	55.11277	77.503691	−22.390921
1	71.76958	104.76950	−32.999920

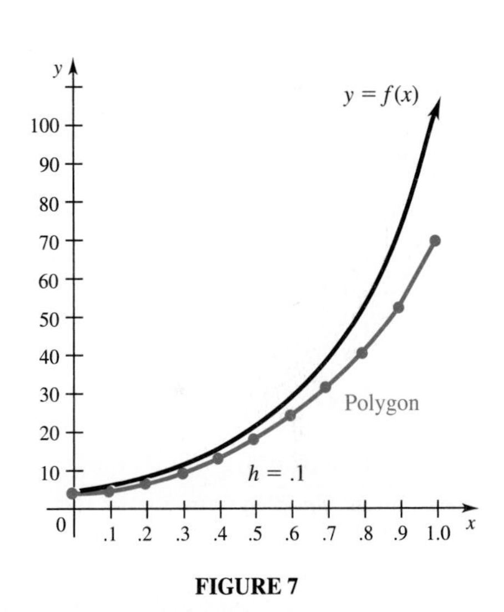

FIGURE 7

In this example the absolute value of the differences grows very rapidly as x_i gets farther from x_0. See Figure 7. These large differences come from the term e^{3x} in the solution; this term grows very quickly as x increases. ◀

As these examples show, numerical methods may produce large errors. The error often can be reduced by using more subintervals of smaller width—letting $n = 100$ or 1000, for example. Approximations for the function in Example 2 with $n = 100$ and $h = (1 - 0)/100 = .01$ are shown in Table 3 on the next page. The approximations are considerably improved.

Table 3 Approximate Solution Using $h = .01$

	Euler's Method	Actual Solution	Difference
x_i	y_i	$f(x_i)$	$y_i - f(x_i)$
0	5	5	0
.1	6.779419	6.810466	−.031047
.2	9.177102	9.260773	−.083671
.3	12.40634	12.575452	−.169112
.4	16.75386	17.057658	−.303798
.5	22.60505	23.116687	−.511637
.6	30.47795	31.305120	−.827170
.7	41.06884	42.368954	−1.30011
.8	55.31358	57.315291	−2.001711
.9	74.46998	77.503691	−3.033711
1	100.22860	104.769498	−4.540898

We could improve the accuracy of Euler's method by using a smaller h, but there are two difficulties. First, this requires more calculations, and consequently more time. Such calculations are usually done by computer, so the increased time may not matter. But this introduces a second difficulty: the increased number of calculations causes more round-off error, so there is a limit to how small we can make h and still get improvement. The preferred way to get greater accuracy is to use a more sophisticated procedure, such as the Runge-Kutta method. Such methods are beyond the scope of this book but are discussed in numerical analysis or differential equations courses.*

9.3 Exercises

Use Euler's method to approximate the indicated function value to three decimal places, using $h = .1$.

1. $y' = x^2 + y^2$; $f(0) = 1$; find $f(.5)$
2. $y' = xy + 2$; $f(0) = 0$; find $f(.5)$
3. $y' = 1 + y$; $f(0) = 2$; find $f(.6)$
4. $y' = x + y^2$; $f(0) = 0$; find $f(.6)$
5. $y' = x + \sqrt{y}$; $f(0) = 1$; find $f(.4)$
6. $y' = 1 + \frac{y}{x}$; $f(1) = 0$; find $f(1.4)$
7. $y' = 1 - e^{-x}$; $f(0) = 0$; find $f(.5)$
8. $y' = e^{-y} + x$; $f(0) = 0$; find $f(.5)$

Use Euler's method to approximate the indicated function value to three decimal places, using $h = .1$. Next, solve the differential equation and find the indicated function value to three decimal places. Compare the result with the approximation.

9. $y' = -4 + x$; $f(0) = 1$; find $f(.4)$
10. $y' = 4x + 3$; $f(1) = 0$; find $f(1.5)$
11. $y' = x^2$; $f(0) = 2$; find $f(.5)$
12. $y' = \frac{1}{x}$; $f(1) = 1$; find $f(1.4)$

*For a discussion of other methods, see Larry C. Andrews, *Introduction to Differential Equations with Boundary Value Problems*, HarperCollins Publishers Inc., 1991.

13. $y' = 2xy$; $f(1) = 1$; find $f(1.6)$

14. $y' = x^2y$; $f(0) = 1$; find $f(.6)$

15. $y' = x^2 - x$; $f(0) = 0$; find $f(.3)$

16. $y' = \frac{x}{y}$; $f(0) = 2$; find $f(.3)$

In each of the following, construct a table like the ones in the examples for $0 \le x \le 1$, *with* $h = .2$.

17. $y' = \sqrt[3]{x}$; $f(0) = 0$

18. $y' = y$; $f(0) = 1$

19. $y' = 1 - y$; $f(0) = 0$

20. $y' = x - xy$; $f(0) = .5$

Solve each differential equation and graph the function $y = f(x)$ *and the polygonal approximation on the same axes. (The approximations were found in Exercises 17–20.)*

21. $y' = \sqrt[3]{x}$; $f(0) = 0$

22. $y' = y$; $f(0) = 1$

23. $y' = 1 - y$; $f(0) = 0$

24. $y' = x - xy$; $f(0) = .5$

25. (a) Use Euler's method with $h = .2$ to approximate $f(1)$, where $f(x)$ is the solution to the differential equation

$$y' = y^2; \quad f(0) = 1.$$

(b) Solve the differential equation in part (a) using separation of variables, and discuss what happens to $f(x)$ as x approaches 1.

(c) Based on what you learned from parts (a) and (b), discuss what might go wrong when using Euler's method. (More advanced courses on differential equations discuss the question of whether a differential equation has a solution for a given interval in x.)

Applications

Solve Exercises 26–31 by using Euler's method.

Business and Economics

26. Bankruptcy Suppose 150 small business firms are threatened by bankruptcy. If y is the number bankrupt by time t, then $150 - y$ is the number not yet bankrupt by time t. The rate of change of y is proportional to both y and $150 - y$. Let 1990 correspond to $t = 0$. Assume 20 firms are bankrupt at $t = 0$.

(a) Write a differential equation using the given information. Use .002 for the constant of proportionality.

(b) Approximate the number of firms that are bankrupt in 1994, using $h = 1$.

Life Sciences

27. Growth of Algae The phosphate compounds found in many detergents are highly water soluble and are excellent fertilizers for algae. Assume that there are 3000 algae present at time $t = 0$ and conditions will support at most 100,000 algae. Assume that the rate of growth of algae, in the presence of sufficient phosphates, is proportional both to the number present (in thousands) and to the difference between 100,000 and the number present (in thousands).

(a) Write a differential equation using the given information. Use .01 for the constant of proportionality.

(b) Approximate the number present when $t = 2$, using $h = .5$.

28. Immigration An island is colonized by immigration from the mainland, where there are 100 species. Let the number of species on the island at time t (in years) equal y, where $y = f(t)$. Suppose the rate at which new species immigrate to the island is

$$\frac{dy}{dt} = .02(100 - y^{1/2}).$$

Use Euler's method with $h = .5$ to approximate y when $t = 5$ if there were 10 species initially.

29. Insect Population A population of insects, y, living in a circular colony grows at a rate

$$\frac{dy}{dt} = .05y - .1y^{1/2},$$

where t is time in weeks. If there were 60 insects initially, use Euler's method with $h = 1$ to approximate the number of insects after 6 wk.

30. **Whale Population** Under certain conditions a population may exhibit a polynomial growth rate function. A population of blue whales is growing according to the function

$$\frac{dy}{dt} = -y + .02y^2 + .003y^3.$$

Here y is the population in thousands and t is measured in years. Use Euler's method with $h = 1$ to approximate the population in 4 yr if the initial population is 15,000.

Social Sciences

31. **Spread of Rumors** A rumor spreads through a community of 500 people at the rate

$$\frac{dN}{dt} = .2(500 - N)N^{1/2},$$

where N is the number of people who have heard the rumor at time t (in hours). Use Euler's method with $h = .5$ to find the number who have heard the rumor after 3 hr, if only 2 people heard it initially.

Computer/Graphing Calculator

Use Euler's method to approximate y to three decimal places for each function, when $x = 1.5$. Use $h = .1$.

32. $y' = e^{.02x} + y;\quad f(0) = .2$

33. $y' = e^x - e^{-y};\quad f(1) = .8$

34. $y' = \dfrac{x - y}{x + y};\quad f(1) = 1.5$

35. $y' = \dfrac{x^2 + y}{y};\quad f(0) = 2.3$

9.4 APPLICATIONS OF DIFFERENTIAL EQUATIONS

How do the populations of a predator and its prey change over time?

The Austrian mathematician A. J. Lotka (1880–1949) and the Italian mathematician Vito Volterra (1860–1940) proposed the following simple model for the way in which the fluctuations of populations of a predator and its prey affect each other.* Let $x = f(t)$ denote the population of the predator and $y = g(t)$ denote the population of the prey at time t. The predator might be a type of insect and its prey a type of scale, or the predator might be a type of bird and the prey a type of insect.

Assume that if there were no predators present, the population of prey would increase at a rate py proportional to their number, but that the predators consume the prey at a rate qxy proportional to the product of the number of prey and the number of predators. The net rate of change dy/dt of y is the rate of increase of the prey minus the rate at which the prey are eaten, that is,

$$\frac{dy}{dt} = py - qxy, \tag{1}$$

with positive constants p and q.

*From A. J. Lotka, *Elements of Mathematical Biology*, Dover, 1956.

11.

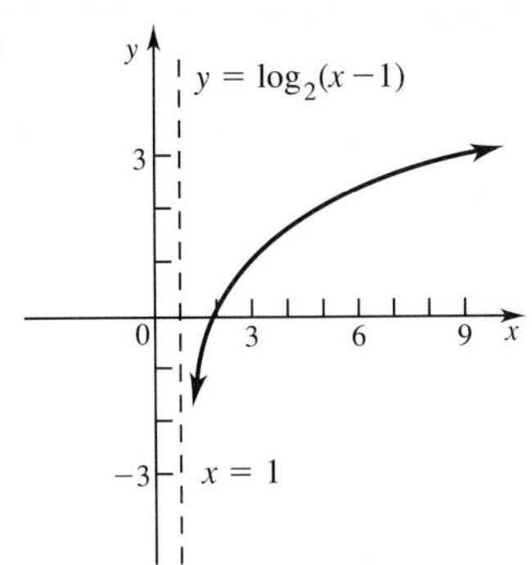

13.

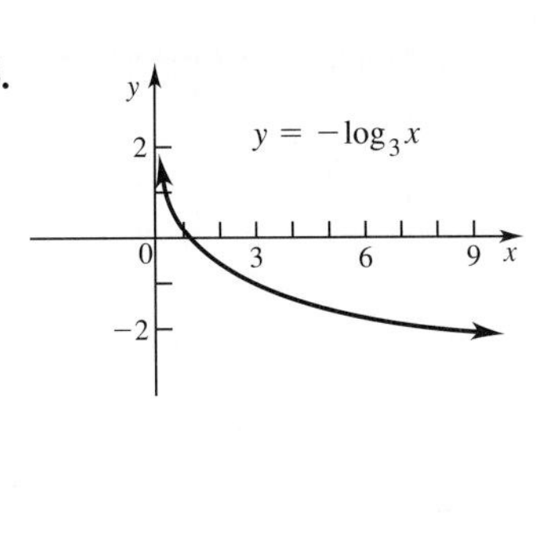

15. $\log_2 64 = 6$ **17.** $\ln 1.09417 = .09$ **19.** $2^5 = 32$ **21.** $e^{4.41763} = 82.9$ **23.** 4 **25.** $4/5$ **27.** $3/2$
29. $\log_5 (21k^4)$ **31.** $\log_2 (x^2/m^3)$ **33.** $p = 1.416$ **35.** $m = -1.807$ **37.** $x = -3.305$ **39.** $m = 1.7547$
41. $y' = -12e^{2x}$ **43.** $y' = -6x^2e^{-2x^3}$ **45.** $y' = 10xe^{2x} + 5e^{2x} = 5e^{2x}(2x + 1)$ **47.** $y' = 2x/(2 + x^2)$
49. $y' = (x - 3 - x \ln |3x|)/[x(x - 3)^2]$ **51.** $y' = [e^x(x + 1)(x^2 - 1) \ln (x^2 - 1) - 2x^2e^x]/[(x^2 - 1)(\ln (x^2 - 1))^2]$
53. $y' = 2(x^2 + e^x)(2x + e^x)$
55. relative minimum of $-e^{-1} = -.368$ at $x = -1$; inflection point at $(-2, -.27)$

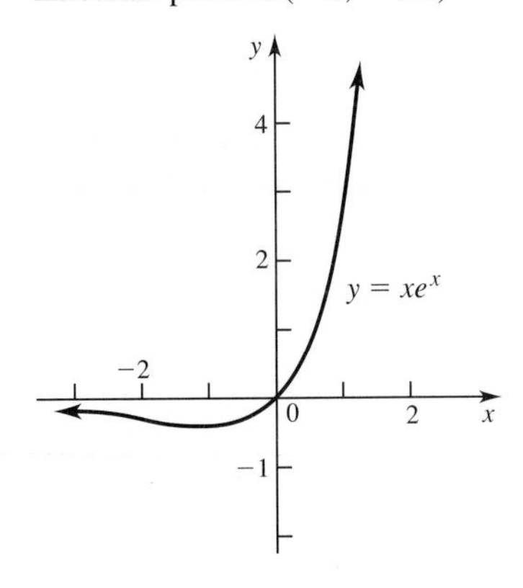

57. relative minimum of e^2 at $x = 2$; no inflection point

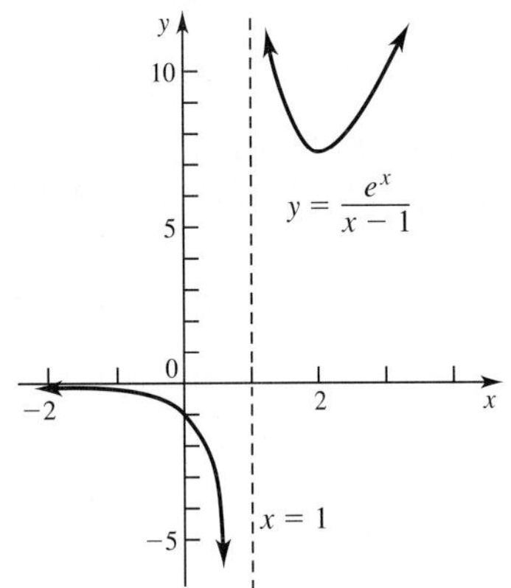

59. \$10,631.51 **61.** \$14,204.18 **63.** \$21,190.14 **65.** \$17,901.90 **67.** 9.38% **69.** 9.42% **71.** \$1494.52
73. \$31,468.86 **75.** **(a)** 100 **(b)** 100 **(c)** 1 day **77.** about 13.7 yr **79.** \$20,891.12 **81.** about 7.7 m
83. **(a)** when it is first injected; .08 g **(b)** never **(c)** It approaches $c/a = .0769$ g **85.** **(a)** 0 yr **(b)** 1.85×10^9 yr
(c) $\dfrac{dt}{dr} = \dfrac{10.4958 \times 10^9}{\ln 2\,(1 + 8.33r)}$ **(d)** As r increases, t increases, but at a slower and slower rate. As r decreases, t decreases at a faster and faster rate. **87.** **(a)** 11 **(b)** 12.6 **(c)** 18.0 **(d)**

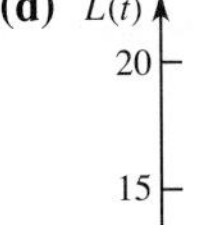
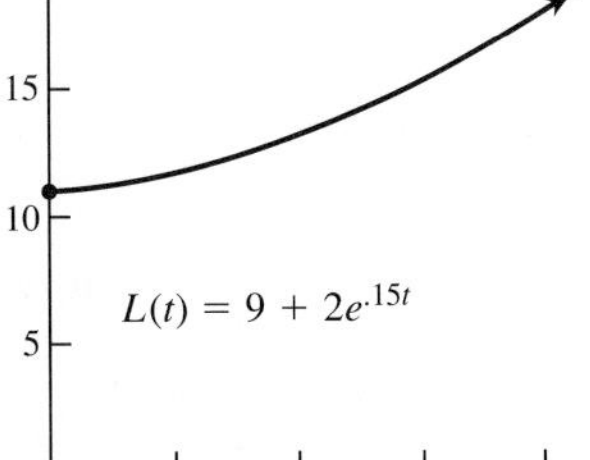

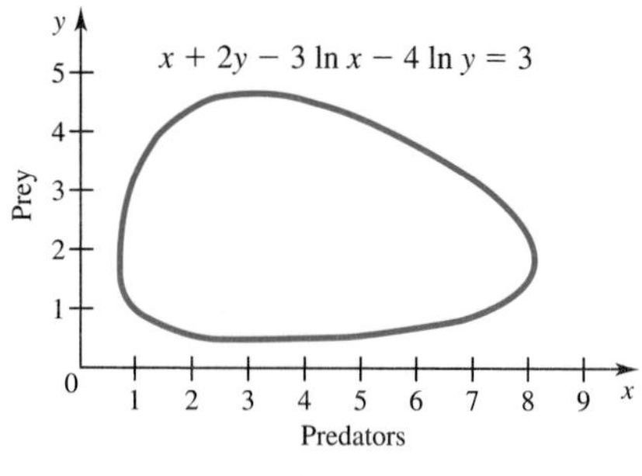

FIGURE 8

A graph of equation (4), in Figure 8, shows that the variations in the populations of the predator and prey are cyclic. The population of prey tends to rise when the population of predators is low. Then, with the resulting abundance of food, the population of predators increases. As a result, the population of prey decreases again, and this forces the population of predators to decline because of the lack of food. The pattern repeats indefinitely. ◀

Arms Race Lewis F. Richardson* proposed a pair of linear first-order differential equations as a model of an arms race between two rival countries X and Y. In order to simplify the situation, the following assumptions are made.

1. X and Y are roughly equal in size.
2. T_0, the annual trade flow from X to Y (in dollars) is a constant.
3. U_0, the annual trade flow from Y to X (in dollars) is a constant.
4. The arms race between X and Y is not affected by the policies of other countries.

Let

t = the time in years,

x = the armaments budget of X in year t (in dollars) minus T_0,

and y = the armaments budget of Y in year t (in dollars) minus U_0.

Richardson assumed that x stimulates the growth of y, but the sheer size of x would eventually inhibit its own growth. He made similar assumptions about y. These assumptions led him to the following model:

$$\frac{dx}{dt} = ky - mx + g \quad \text{and} \quad \frac{dy}{dt} = kx - my + h, \tag{5}$$

where k, m, g, and h are constants. The constants g and h account for the flow of trade between the two nations and their underlying attitudes toward each other.

*From Lewis F. Richardson, *Arms and Insecurity*, The Boxwood Press, 1960.

For example, if the two countries have long-standing grievances and ideological differences that outweigh the value of the constant trade flow, then $g > 0$, $h > 0$, and the armament budgets will grow even when $x = 0 = y$. On the other hand, if the constant trade flow is the dominant factor, then $g < 0$, $h < 0$, and the armament budgets will shrink when $x = 0 = y$. Richardson found that the constant k was roughly proportional to the size of the country, but the constant m was roughly the same for all countries.

Richardson set $z = x + y$ and added the two equations in (5), obtaining

$$
\begin{aligned}
\frac{dz}{dt} &= \frac{dx}{dt} + \frac{dy}{dt} \\
&= ky - mx + g + kx - my + h \\
&= k(x + y) - m(x + y) + g + h \\
&= kz - mz + g + h
\end{aligned}
$$

$$\frac{dz}{dt} = g + h - (m - k)z. \tag{6}$$

Solving this differential equation as in Example 3 in Section 9.2 gives the general solution of (6) as

$$z = \frac{g + h}{m - k} + Ce^{-(m-k)t}. \tag{7}$$

Richardson tested his model on the growth of armaments in the years preceding World War I for

X = the bloc consisting of Russia and France,

and Y = the bloc consisting of Germany and Austro-Hungary.

He found a remarkably good fit of equations (6) and (7) to the data.

R. Taagepera, et al.,[†] proposed the following model for the arms race (within a finite time span) between Israel and the Arab countries:

$$\frac{dx}{dt} = kx \quad \text{and} \quad \frac{dy}{dt} = my, \tag{8}$$

where k and m are constants. By Example 4 in Section 9.1 the general solution of (8) is

$$x = x_0 e^{kt} \quad \text{and} \quad y = y_0 e^{mt}. \tag{9}$$

This model asserts that x stimulates the growth of x, but not of y, and vice versa. Here we are supposing that self-stimulation is the controlling factor, whereas the Richardson model presupposes mutual stimulation. Taagepera, et al., found that the self-stimulation model (8)–(9) fit the Israeli-Arab arms race better than the Richardson model (5)–(7). This result is very surprising, for it challenges the common assumption that the driving force in any arms race is mutual stimulation.

[†]From R. Taagepera, G. M. Shiffler, R. T. Perkins, and D. L. Wagner, "Soviet-American and Israeli-Arab Arms Races and the Richardson Model," *General Systems*, vol. 20, 1975, pp. 151–58.

Mixing Problems The mixing of two solutions can lead to a first-order differential equation, as the next example shows.

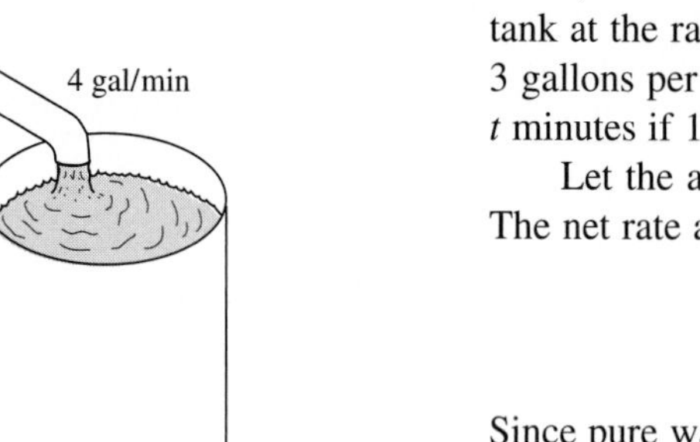

FIGURE 9

EXAMPLE 2 Suppose a tank contains 100 gallons of a solution of dissolved salt and water, which is kept uniform by stirring. If pure water is allowed to flow into the tank at the rate of 4 gallons per minute, and the mixture flows out at the rate of 3 gallons per minute (see Figure 9), how much salt will remain in the tank after t minutes if 15 pounds of salt are in the mixture initially?*

Let the amount of salt present in the tank at any specific time be $y = f(t)$. The net rate at which y changes is given by

$$\frac{dy}{dt} = (\text{Rate of salt in}) - (\text{Rate of salt out}).$$

Since pure water is coming in, the rate of salt entering the tank is zero. The rate at which salt is leaving the tank is the product of the amount of salt per gallon (in V gallons) and the number of gallons per minute leaving the tank:

$$\text{Rate of salt out} = \left(\frac{y}{V}\ \text{lb/gal}\right)(3\ \text{gal/min}).$$

The differential equation, therefore, can be written as

$$\frac{dy}{dt} = -\frac{3y}{V}; \quad y(0) = 15,$$

where $y(0)$ is the initial amount of salt in the solution. We must take into account the fact that the volume, V, of the mixture is not constant but is determined by

$$\frac{dV}{dt} = (\text{Rate of liquid in}) - (\text{Rate of liquid out}) = 4 - 3 = 1,$$

or

$$\frac{dV}{dt} = 1,$$

from which

$$V(t) = t + C_1.$$

Because the volume is known to be 100 at time $t = 0$, we have $C_1 = 100$, and

$$\frac{dy}{dt} = \frac{-3y}{t + 100}; \qquad y(0) = 15,$$

a separable equation with solution

$$\frac{dy}{y} = \frac{-3}{t + 100}\,dt$$

$$\ln y = -3 \ln (t + 100) + C.$$

---

*This example is from *Ordinary Differential Equations with Boundary Value Problems* by Larry C. Andrews, HarperCollins Publishers Inc., 1991, pp. 85–86.

where $b = (N - y_0)/y_0$ and $k = aN$. Since just one individual is infected initially, $y_0 = 1$ here. Substituting these values into equation (11) gives

$$y = \frac{N}{1 + (N - 1)e^{-aNt}} \tag{12}$$

as the general solution of equation (10).

The infection rate will be a maximum when its derivative is 0, that is, when $d^2y/dt^2 = 0$. Since

$$\frac{dy}{dt} = a(N - y)y,$$

we have

$$\begin{aligned}\frac{d^2y}{dt^2} &= a[(N - y)(1) + y(-1)]\\ &= aN - 2ay.\end{aligned}$$

Set $d^2y/dt^2 = 0$ to get

$$\begin{aligned}0 &= aN - 2ay\\ 2ay &= aN\\ y &= \frac{N}{2}.\end{aligned}$$

The maximum infection rate occurs when exactly half the total population is still uninfected and equals

$$\frac{dy}{dt} = a\left(N - \frac{N}{2}\right)\frac{N}{2} = \frac{aN^2}{4}.$$

Letting $y = N/2$ in equation (12) and solving for t shows that the maximum infection rate occurs at time

$$t_m = \frac{\ln (N - 1)}{aN}.$$

The graph of dy/dt, shown in Figure 10, is called the *epidemic curve.* It is symmetric about the line $t = t_m$. ◀

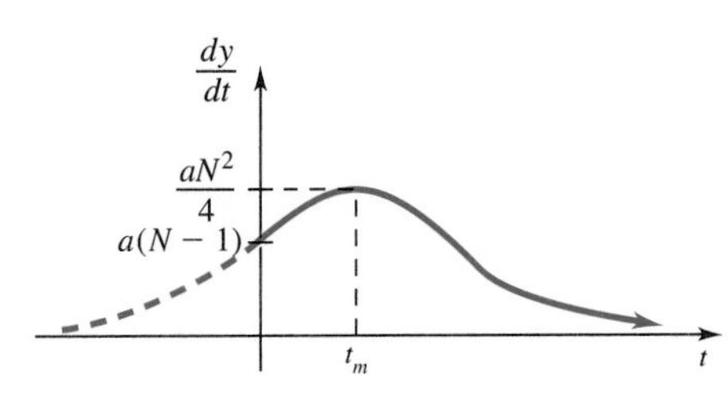

FIGURE 10

9.4 Exercises

Applications

Business and Economics

1. **Continuous Deposits** Jessie deposits $2000 in an IRA at 6% interest compounded continuously for her retirement in 10 years. She intends to make continuous deposits at the rate of $2000 a year until she retires. How much will she have accumulated at that time?
2. **Continuous Deposits** In Exercise 1, how long will it take Jessie to accumulate $20,000 in her retirement account?
3. **Continuous Deposits** To provide for a future expansion, a company plans to make continuous deposits to a savings account at the rate of $10,000 per year, with no initial deposit. The managers want to accumulate $70,000. How long will it take if the account earns 7% interest compounded continuously?
4. **Continuous Deposits** Suppose the company in Exercise 3 wants to accumulate $50,000 in 3 years. Find the approximate yearly deposit that will be required.

Life Sciences

5. **Competing Species** The system of equations

$$\frac{dy}{dt} = 3y - 2xy$$

$$\frac{dx}{dt} = -2x + 3xy$$

describes the influence of the populations of two competing species on their growth rates.

(a) Following Example 1 in the text, find an equation relating x and y, assuming $y = 2$ when $x = 1$.

(b) Find values of x and y so that both populations are constant. (*Hint:* Set both differential equations equal to 0.)

6. **Symbiotic Species** When two species coexist in a symbiotic (dependent) relationship, they either increase together or decrease together. Typical equations for the growth rates of two such species might be

$$\frac{dx}{dt} = -3x + 4xy$$

$$\frac{dy}{dt} = -3y + xy.$$

(a) Find an equation relating x and y if $x = 3$ when $y = 1$.

(b) Find values of x and y so that both populations are constant. (See Exercise 5.)

7. **Spread of an Epidemic** The native Hawaiians lived for centuries in isolation from other peoples. When foreigners finally came to the islands they brought with them diseases such as measles, whooping cough, and smallpox, which decimated the population. Suppose such an island has a native population of 5000, and a sailor from a visiting ship introduces measles, which has an infection rate of .00005.

(a) Write an equation for the number of natives who remain uninfected. Let t represent time in days.

(b) How many are uninfected after 30 days?

(c) How many are uninfected after 50 days?

(d) When will the maximum infection rate occur?

8. **Spread of an Epidemic** In Example 4, the number of infected individuals is given by equation (12).

(a) Show that the number of uninfected individuals is given by

$$N - y = \frac{N(N-1)}{N - 1 + e^{aNt}}.$$

(b) Graph the equation in (a) and equation (12) on the same axes when $N = 100$ and $a = .01$.

(c) Find the common inflection point of the two graphs.

(d) What is the significance of the common inflection point?

(e) What are the limiting values of y and $N - y$?

9. **Spread of an Epidemic** An influenza epidemic spreads at a rate proportional to the product of the number of people infected and the number not yet infected. Assume that 50 people are infected at the beginning of the epidemic in a community of 10,000 people, and 300 are infected 10 days later.

(a) Write an equation for the number of people infected, y, after t days.

(b) When will half the community be infected?

10. **Spread of an Epidemic** The Gompertz growth law,

$$\frac{dy}{dt} = kye^{-at},$$

for constants k and a, is another model used to describe the growth of an epidemic. Repeat Exercise 9, using this differential equation with $a = .02$.

11. Spread of Gonorrhea Gonorrhea is spread by sexual contact, takes 3 to 7 days to incubate, and can be treated with antibiotics. There is no evidence that a person ever develops immunity. One model proposed for the rate of change in the number of men infected by this disease is

$$\frac{dy}{dt} = -ay + b(f - y)Y,$$

where y is the fraction of men infected, f is the fraction of men who are promiscuous, Y is the fraction of women infected, and a and b are appropriate constants.*

(a) Assume $a = 1$, $b = 1$, $f = .5$, $Y = .01$, and solve for y using $y = .02$ when t is 0 as an initial condition. Round your answer to three decimal places.

(b) A comparable model for women is

$$\frac{dY}{dt} = -AY + B(F - Y)y,$$

where F is the fraction of women who are promiscuous and A and B are constants. Assume $A = 1$, $B = 1$, $y = .1$, and $F = .03$, and solve for Y, using $Y = .01$ as an initial condition.

Social Sciences

Spread of a Rumor *The equation developed in the text for the spread of an epidemic also can be used to describe diffusion of information. In a population of size N, let y be the number who have heard a particular piece of information. Then*

$$\frac{dy}{dt} = a(N - y)y$$

for a positive constant a. Use this model in Exercises 12–14.

12. Suppose a rumor starts among 3 people in a certain office building. That is, $y_0 = 3$. Suppose 500 people work in the building and 50 people have heard the rumor in 2 days. Using equation (11), write an equation for the number who have heard the rumor in t days. How many people will have heard the rumor in 5 days?

13. A rumor spreads at a rate proportional to the product of the number of people who have heard it and the number who have not heard it. Assume that 5 people in an office with 50 employees heard the rumor initially, and 15 people have heard it 3 days later.

(a) Write an equation for the number, y, of people who have heard the rumor in t days.

(b) When will 30 employees have heard the rumor?

14. A news item is heard on the late news by 5 of the 100 people in a small community. By the end of the next day 20 people have heard the news. Using equation (11), write an equation for the number of people who have heard the news in t days. How many have heard the news after 3 days?

15. Repeat Exercise 13 using the Gompertz growth law,

$$\frac{dy}{dt} = kye^{-at},$$

for constants k and a, with $a = .1$.

Physical Sciences

16. Salt Concentration A tank holds 100 gal of water that contains 20 lb of dissolved salt. A brine (salt) solution is flowing into the tank at the rate of 2 gal/min while the solution flows out of the tank at the same rate. The brine solution entering the tank has a salt concentration of 2 lb/gal.

(a) Find an expression for the amount of salt in the tank at any time.

(b) How much salt is present after 1 hr?

(c) As time increases, what happens to the salt concentration?

17. Solve Exercise 16 if the brine solution is introduced at the rate of 3 gal/min while the rate of outflow remains the same.

18. Solve Exercise 16 if the brine solution is introduced at the rate of 1 gal/min while the rate of outflow stays the same.

19. Solve Exercise 16 if pure water is added instead of brine.

20. Chemical in a Solution Five grams of a chemical is dissolved in 100 liters of alcohol. Pure alcohol is added at the rate of 2 liters/min and at the same time the solution is being drained at the rate of 1 liter/min.

(a) Find an expression for the amount of the chemical in the mixture at any time.

(b) How much of the chemical is present after 30 min?

21. Solve Exercise 20 if a 10% solution of the same mixture is added instead of pure alcohol.

22. Soap Concentration A prankster puts 4 lb of soap in a fountain that contains 200 gal of water. To clean up the mess a city crew runs clear water into the fountain at the rate of 8 gal/min allowing the excess solution to drain off at the same rate. How long will it be before the amount of soap in the mixture is reduced to 1 lb?

---

*From Edward A. Bender, *An Introduction to Mathematical Modeling*. Copyright © 1978 by John Wiley and Sons, Inc. Reprinted by permission.

Chapter Summary — Key Terms

differential equation

9.1 **general solution**
particular solution
initial condition
separable differential equation
growth rate constant
carrying capacity
logistic growth model
logistic equation
logistic curve

9.2 **linear first-order differential equation**
integrating factor

9.3 **Euler's method**

9.4 **Lotka-Volterra equations**
mixing problems
continuous deposits

Chapter 9 — Review Exercises

1. What is a differential equation? What is it used for?
2. What is the difference between a particular solution and a general solution to a differential equation?
3. How can you tell that a differential equation is separable? That it is linear?
4. Can a differential equation be both separable and linear? Explain why not, or give an example of an equation that is both.

Find general solutions for the following differential equations.

5. $\frac{dy}{dx} = 2x^3 + 6x$
6. $\frac{dy}{dx} = x^2 + 5x^4$
7. $\frac{dy}{dx} = 4e^x$
8. $\frac{dy}{dx} = \frac{1}{2x + 3}$
9. $\frac{dy}{dx} = \frac{3x + 1}{y}$
10. $\frac{dy}{dx} = \frac{e^x + x}{y - 1}$
11. $\frac{dy}{dx} = \frac{2y + 1}{x}$
12. $\frac{dy}{dx} = \frac{3 - y}{e^x}$
13. $\frac{dy}{dx} + 5y = 12$
14. $\frac{dy}{dx} + xy = 4x$
15. $3\frac{dy}{dx} + xy - x = 0$
16. $x\frac{dy}{dx} + y - e^x = 0$

Find particular solutions for the following differential equations. (Some solutions may give y implicitly.)

17. $\frac{dy}{dx} = x^2 - 5x;\quad y = 1$ when $x = 0$
18. $\frac{dy}{dx} = 4x^3 + 2;\quad y = 3$ when $x = 1$
19. $\frac{dy}{dx} = 5(e^{-x} - 1);\quad y = 17$ when $x = 0$
20. $\frac{dy}{dx} = \frac{x}{x^2 - 3};\quad y = 52$ when $x = 2$
21. $(5 - 2x)y = \frac{dy}{dx};\quad y = 2$ when $x = 0$
22. $\sqrt{x}\frac{dy}{dx} = xy;\quad y = 4$ when $x = 1$
23. $\frac{dy}{dx} = \frac{1 - 2x}{y + 3};\quad y = 16$ when $x = 0$
24. $\frac{dy}{dx} = (3x + 2)^2 e^y;\quad y = 0$ when $x = 0$
25. $y' + x^2 y = x^2;\quad y = 8$ when $x = 0$
26. $e^x y' - e^x y = x^2 - 1;\quad y = 42$ when $x = 0$
27. $x\frac{dy}{dx} - 2x^2 y + 3x^2 = 0;\quad y = 15$ when $x = 0$
28. $x^2\frac{dy}{dx} + 4xy - e^{2x^3} = 0;\quad y = e^2$ when $x = 1$

29. When is Euler's method useful?

Use Euler's method to approximate the indicated function value for $y = f(x)$ to three decimal places, using $h = .2$.

30. $y' = x + y^{-1}$; $f(0) = 1$; find $f(1)$

31. $y' = e^x + 1$; $f(0) = 1$; find $f(.6)$

32. Let $y = f(x)$ and $y' = (x/2) + 4$, with $f(0) = 0$. Use Euler's method with $h = .1$ to approximate $f(.3)$ to three decimal places. Then solve the differential equation and find $f(.3)$ to three decimal places. Also, find $y_3 - f(x_3)$.

33. Let $y = f(x)$ and $y' = 3 + \sqrt{y}$, with $f(0) = 0$. Construct a table like the one in Section 9.3, Example 2, for $[0, 1]$, with $h = .2$. Then graph the polygonal approximation of the graph of $y = f(x)$.

34. What is the logistic equation? Why is it useful?

Applications

Business and Economics

35. Marginal Sales The marginal sales (in hundreds of dollars) of a computer software company are given by

$$\frac{dy}{dx} = 5e^{.2x},$$

where x is the number of months the company has been in business. Assume that sales were 0 initially.

(a) Find the sales after 6 months.

(b) Find the sales after 12 months.

36. Continuous Withdrawals A deposit of $10,000 is made to a savings account at 5% interest compounded continuously. Assume that continuous *withdrawals* of $1000 per year are made.

(a) Write a differential equation to describe the situation.

(b) How much will be left in the account after 1 year?

37. In Exercise 36, approximately how long will it take to use up the account?

Life Sciences

38. Effect of Insecticide After use of an experimental insecticide, the rate of decline of an insect population is

$$\frac{dy}{dt} = \frac{-10}{1 + 5t},$$

where t is the number of hours after the insecticide is applied. Assume that there were 50 insects initially.

(a) How many are left after 24 hours?

(b) How long will it take for the entire population to die?

39. Growth of a Mite Population A population of mites grows at a rate proportional to the number present, y. If the growth constant is 10% and 120 mites are present at time $t = 0$ (in weeks) find the number present after 6 weeks.

40. Competing Species Find an equation relating x to y given the following equations, which describe the interaction of two competing species and their growth rates.

$$\frac{dx}{dt} = .2x - .5xy$$

$$\frac{dy}{dt} = -.3y + .4xy$$

Find the values of x and y for which both growth rates are 0.

41. Smoke Content in a Room The air in a meeting room of 15,000 cubic feet has a smoke content of 20 parts per million (ppm). An air conditioner is turned on, which brings fresh air (with no smoke) into the room at a rate of 1200 cubic feet per minute and forces the smoky air out at the same rate. How long will it take to reduce the smoke content to 5 ppm?

42. Smoke Content in a Room In Exercise 41, how long will it take to reduce the smoke content to 10 ppm if smokers in the room are adding smoke at the rate of 5 ppm per minute?

43. Spread of Influenza A small, isolated mountain community with a population of 700 is visited by an outsider who carries influenza. After 6 weeks, 300 people are uninfected.

(a) Write an equation for the number of people who remain uninfected at time t (in weeks).

(b) Find the number still uninfected after 7 weeks.

(c) When will the maximum infection rate occur?

Social Sciences

44. Production Rate The rate at which a new worker in a certain factory produces items is given by

$$\frac{dy}{dx} = .2(125 - y),$$

where y is the number of items produced by the worker per day, x is the number of days worked, and the maximum production per day is 125 items. Assume the worker produced 20 items the first day on the job ($x = 0$).

(a) Find the number of items the new worker will produce in 10 days.

(b) According to the function that is the solution of the differential equation, can the worker ever produce 125 items in a day?

45. Spread of a Rumor A rumor spreads through the offices of a company with 100 employees, starting in a meeting with 4 people. After 3 days, 15 people have heard the rumor.

(a) Write an equation for the number of people who have heard the rumor in x days.

(b) How many people have heard the rumor in 5 days?

46. Population Growth Let

$$y = \frac{N}{1 + be^{-kx}}.$$

If y is y_1, y_2, and y_3 at times x_1, x_2, and $x_3 = 2x_2 - x_1$ (that is, at three equally spaced times), then prove that

$$N = \frac{1/y_1 + 1/y_3 - 2/y_2}{1/(y_1y_3) - 1/y_2^2}.$$

Population Growth *In the following table of United States census figures, y is population in millions.*

Year	y	Year	y
1790	4	1900	76
1800	5	1910	92
1810	7	1920	106
1820	10	1930	123
1830	13	1940	132
1840	17	1950	151
1850	23	1960	179
1860	31	1970	204
1870	40	1980	227
1880	50	1990	249
1890	63		

47. Use Exercise 46 and the above data to find the following.

(a) Find N using the years 1800, 1850, and 1900.

(b) Find N using the years 1850, 1900, and 1950.

(c) Find N using the years 1870, 1920, and 1970.

(d) Explain why different values of N were obtained in (a)–(c). What does this suggest about the validity of this model and others?

48. Let $x = 0$ correspond to 1870, and let every decade correspond to an increase in x of 1.

(a) Find b and k to two decimal places using the population of the United States in 1920 or 1970 and

$$y = \frac{N}{1 + be^{-kx}}.$$

(b) Estimate the population of the United States in 1990 and compare your estimate to the actual population in 1990.

(c) Predict the populations of the United States in 2000 and 2050.

Physical Sciences

49. Newton's Law of Cooling A roast at a temperature of 40° is put in a 300° oven. After 1 hr the roast has reached a temperature of 150°. Newton's law of cooling states that

$$\frac{dT}{dt} = k(T - T_F),$$

where T is the temperature of an object, the surrounding medium has temperature T_F at time t, and k is a constant. Use Newton's law to find the temperature of the roast after 2 hr.

50. In Exercise 49, how long does it take for the roast to reach a temperature of 250°?

Connections

51. Air Resistance In the section on antiderivatives, we saw that the acceleration of gravity is a constant if air resistance is ignored. But air resistance cannot always be ignored, or parachutes would be of little use. In the presence of air resistance, the equation for acceleration also contains a term roughly proportional to the velocity squared. Since acceleration forces a falling object downward and air resistance pushes it upward, the air resistance term is opposite in sign to the acceleration of gravity. Thus,

$$a(t) = \frac{dv}{dt} = b - kv^2,$$

where b and k are positive constants. Future calculations will be simpler if we replace b and k by the squared constants B^2 and K^2, giving

$$\frac{dv}{dt} = B^2 - K^2v^2.$$

(a) Use the table of integrals and separation of variables to solve the differential equation above. Assume $v < B/K$, which is certainly true when the object starts falling (with $v = 0$). Write your solution in the form of v as a function of t.

(b) Find $\lim_{t\to\infty} v(t)$, where $v(t)$ is the solution you found in part (a). What does this tell you about a falling object in the presence of air resistance?

Extended Application / Pollution of the Great Lakes*

Industrial nations are beginning to face the problems of water pollution. Lakes present a problem, because a polluted lake contains a considerable amount of water that must somehow be cleaned. The main cleanup mechanism is the natural process of gradually replacing the water in the lake. This application deals with pollution in the Great Lakes. The basic idea is to regard the flow in the Great Lakes as a mixing problem.

We make the following assumptions.

1. Rainfall and evaporation balance each other, so the average rates of inflow and outflow are equal.
2. The average rates of inflow and outflow do not vary much seasonally.
3. When water enters the lake, perfect mixing occurs, so that the pollutants are uniformly distributed.
4. Pollutants are not removed from the lake by decay, sedimentation, or in any other way except outflow.
5. Pollutants flow freely out of the lake; they are not retained (as DDT is).

(The first two are valid assumptions; however, the last three are questionable.)

We will use the following variables in the discussion to follow.

V = volume of the lake

P_L = pollution concentration in the lake at time t

P_i = pollution concentration in the inflow to the lake at time t

r = rate of flow

t = time in years

By the assumptions stated above, the net change in total pollutants during the time interval Δt is (approximately)

$$VP_L = (P_i - P_L)(r \cdot \Delta t).$$

Dividing this equation by Δt and taking the limit as $\Delta t \to 0$, we get the differential equation

$$P_L' = \frac{(P_i - P_L)r}{V}.$$

Since we are treating V and r as constants, we replace r/V with k, so the equation can be written as the first-order linear equation

$$P_L' + kP_L = kP_i.$$

The solution is

$$P_L(t) = e^{-kt}[P_L(0) + k\int_0^t P_i(x)e^{kx}\,dx]. \qquad (1)$$

Figure 11 shows values of $1/k$ for each lake (except Huron) measured in years. *If the model is reasonable*, the numbers in the figure can be used in equation (1) to determine the effect of various pollution abatement schemes. Lake Ontario is excluded from the discussion because about 84% of its inflow comes from Erie and can be controlled only indirectly.

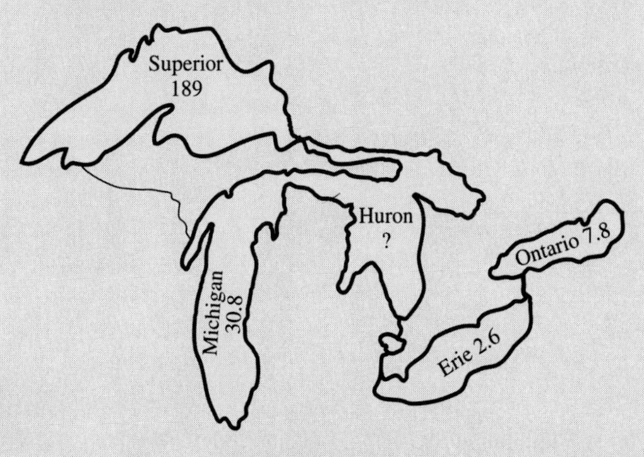

FIGURE 11

*From Edward A. Bender, *An Introduction to Mathematical Modeling*. Copyright © 1978 by John Wiley & Sons, Inc. Reprinted by permission.

The fastest possible cleanup will occur if all pollution inflow ceases. This means that $P_i = 0$. In this case, equation (1) leads to

$$t = \frac{1}{k} \ln \left(\frac{P_L(0)}{P_L(t)} \right).$$

From this we can tell the length of time necessary to reduce pollution to a given percentage of its present level. For example, from the figure, for Lake Superior $1/k = 189$. Thus, to reduce pollution to 50% of its present level, $P_L(0)$, we want

$$\frac{P_L(t)}{P_L(0)} = .5 \qquad \text{or} \qquad \frac{P_L(0)}{P_L(t)} = 2,$$

from which

$$t = 189 \ln 2 \approx 131.$$

The following figures, representing years, were found in this way.

Lake	50%	20%	10%	5%
Erie	2	4	6	8
Michigan	21	50	71	92
Superior	131	304	435	566

Fortunately, the pollution in Lake Superior is quite low at present.

As mentioned before, assumptions 3, 4, and 5 are questionable. For persistent pollutants like DDT, the estimated cleanup times may be too low. For other pollutants, how assumptions 4 and 5 affect cleanup times is unclear. However, the values of $1/k$ given in the figure probably provide rough lower bounds for the cleanup times of persistent pollutants.

Exercises

1. Calculate the number of years to reduce pollution in Lake Erie to each of the following levels.
   (a) 40% **(b)** 30%
2. Repeat Exercise 1 for Lake Michigan.
3. Repeat Exercise 1 for Lake Superior.

Applications at a glance . . .

Time a Predator Needs to Find Prey . . . insect life span . . . age of a drunk driver . . .

CHAPTER 10

PROBABILITY AND CALCULUS

(See page 542.)

In recent years, probability has become increasingly useful in fields ranging from manufacturing to medicine, as well as in all types of research. The foundations of probability were laid in the seventeenth century by Blaise Pascal (1623–1662) and Pierre de Fermat (1601–1665), who investigated *the problem of the points.* This problem dealt with the fair distribution of winnings in an interrupted game of chance between two equally matched players whose scores were known at the time of the interruption.

Probability has advanced from a study of gambling to a well-developed, deductive mathematical system. In this chapter we give a brief introduction to the use of calculus in probability.

10.1 CONTINUOUS PROBABILITY MODELS

What is the probability that there is a bird's nest within .5 kilometers of a given point?

In this section, we show how calculus is used to find the probability of certain events. Later in the section, we will answer the question posed above. Before discussing probability, however, we need to introduce some new terminology.

Suppose that a bank is studying the transaction times of its tellers. The lengths of time spent on observed transactions, rounded to the nearest minute, are shown in Table 1.

Table 1

Time	1	2	3	4	5	6	7	8	9	10	
Frequency	3	5	9	12	15	11	10	6	3	1	(Total: 75)

The table shows, for example, that 9 of the 75 transactions in the study took 3 minutes, 15 transactions took 5 minutes, and 1 transaction took 10 minutes. Because the time for any particular transaction is a random event, the number of minutes for a transaction is called a **random variable.** The frequencies can be converted to probabilities by dividing each frequency by the total number of transactions (75) to get the results shown in Table 2 on the facing page.*

*One definition of the *probability of an event* is the number of outcomes that favor the event divided by the total number of outcomes in an experiment.

Table 2

Time	1	2	3	4	5	6	7	8	9	10
Probability	.04	.07	.12	.16	.20	.15	.13	.08	.04	.01

Because each value of the random variable is associated with just one probability, Table 2 defines a function. Such a function is called a **probability distribution function** or simply a **probability function.** The special properties of a probability distribution function are given below.

PROBABILITY FUNCTION OF A RANDOM VARIABLE

If the function $f(x)$ is a probability distribution function with domain $\{x_1, x_2, \ldots, x_n\}$, then for $1 \leq i \leq n$,

$$\mathbf{0 \leq f(x_i) \leq 1,}$$

and

$$\mathbf{f(x_1) + f(x_2) + \cdots + f(x_n) = 1.}$$

The information in Table 2 can be displayed graphically with a special kind of bar graph called a **histogram.** The bars of a histogram have the same width, and their heights are determined by the probabilities of the various values of the random variable.

The probability distribution function in Table 2 is a **discrete probability function** because it has a finite domain—the integers from 1 to 10, inclusive. A discrete probability function has a finite domain or an infinite domain that can be listed. For example, if we flip a coin until we get heads, and let the random variable be the number of flips, then the domain is 1, 2, 3, 4, ... On the other hand, the distribution of heights (in inches) of college women includes infinitely many possible measurements, such as 53, 54.2, 66.5, $72.\overline{3}$, and so on, *within some real number interval.* Probability distribution functions with such domains are called *continuous probability distributions.*

CONTINUOUS PROBABILITY DISTRIBUTION

A **continuous random variable** can take on any value in some interval of real numbers. The distribution of this random variable is called a **continuous probability distribution.**

In the bank example discussed earlier in this section, it would have been possible to time the teller transactions with greater precision—to the nearest tenth of a minute or even to the nearest second (one-sixtieth of a minute), if desired. Theoretically, at least, t could take on any positive real-number value between, say, 0 and 11 minutes. The graph of the probabilities $f(t)$ of these transaction times can be thought of as the continuous curve shown in Figure 1 on the next page. As indicated in Figure 1, the curve was derived from Table 2 by connecting

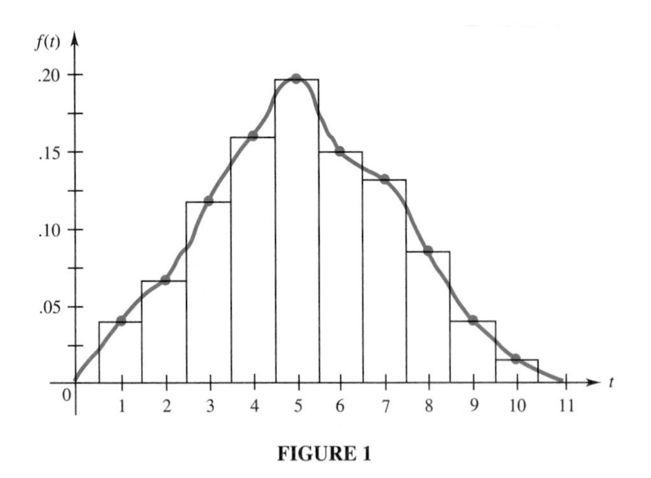

FIGURE 1

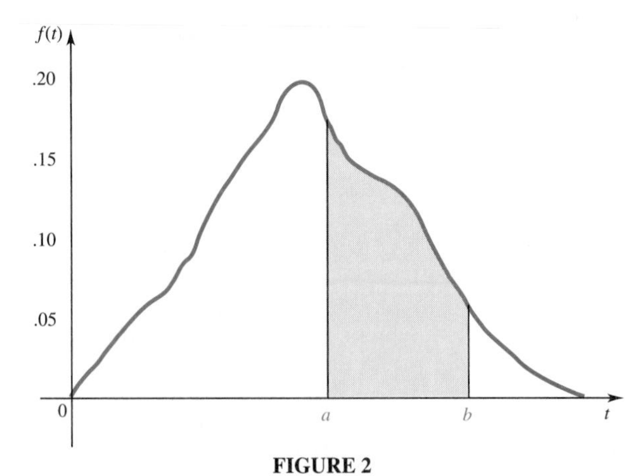

FIGURE 2

the points at the tops of the bars in the corresponding histogram and smoothing the resulting polygon into a curve.

For a discrete probability function, the area of each bar (or rectangle) gives the probability of a particular transaction time. Thus, by considering the possible transaction times t as all the real numbers between 0 and 11, the area under the curve of Figure 2 between any two values of t can be interpreted as the probability that a transaction time will be between those two numbers. For example, the shaded region in Figure 2 corresponds to the probability that t is between a and b, written $P(a \leq t \leq b)$.

The connection between area and the definite integral is discussed in the chapter on integration. For example, in that chapter we solved such problems as the following.

Find the area between the x-axis and the graph of $f(x) = x^2$ from $x = 1$ to $x = 4$.

Answer: $\int_1^4 x^2\,dx = 21$

It was shown earlier that the definite integral of a continuous function $f(x)$, where $f(x) \geq 0$, gives the area under the graph of $f(x)$ from $x = a$ to $x = b$. If a function $f(x)$ can be found to describe a continuous probability function, then the definite integral can be used to find the area under the curve from a to b that represents the probability that x will be between a and b.

If x is a continuous probability function in $[a, b]$, then

$$P(a \leq x \leq b) = \int_a^b f(x)\,dx.$$

Probability Density Functions A function $f(x)$ that describes a continuous probability distribution is called a *probability density function.* Such a function must satisfy the following conditions.

PROBABILITY DENSITY FUNCTION

The function $f(x)$ is a **probability density function** of a random variable x in the interval $[a, b]$ if

1. $\int_a^b f(x)\,dx = 1$; and

2. $f(x) \geq 0$ for all x in the interval $[a, b]$.

Intuitively, condition 1 says that the total probability for the interval must be 1; *something* must happen. Condition 2 says that the probability of a particular event can never be negative.

EXAMPLE 1 **(a)** Show that the function defined by $f(x) = (3/26)x^2$ is a probability density function for the interval [1, 3].

First, show that condition 1 holds.

$$\int_1^3 \frac{3}{26}x^2\,dx = \frac{3}{26}\left(\frac{x^3}{3}\right)\Bigg|_1^3 = \frac{3}{26}\left(9 - \frac{1}{3}\right) = 1$$

Next, show that condition 2 holds; that is, $f(x) \geq 0$ for the interval [1, 3]. Since x^2 is always positive, $f(x)$ is a probability density function.

(b) Find the probability that x will be between 1 and 2.

The desired probability is given by the area under the graph of $f(x)$ between $x = 1$ and $x = 2$, as shown in Figure 3. The area is found by using a definite integral.

$$P(1 \leq x \leq 2) = \int_1^2 \frac{3}{26}x^2\,dx = \frac{3}{26}\left(\frac{x^3}{3}\right)\Bigg|_1^2 = \frac{7}{26}$$ ◀

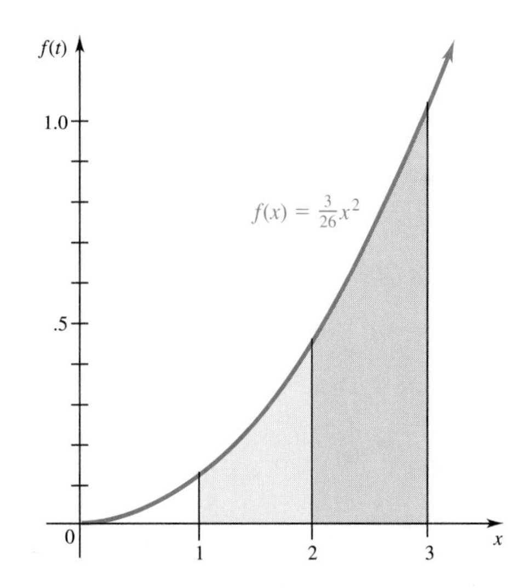

FIGURE 3

Earlier, we noted that determining a suitable function is the most difficult part of applying mathematics to actual situations. Sometimes a function appears to model an application well, but does not satisfy the requirements for a probability density function. In such cases, we may be able to change the function into a probability density function by multiplying it by a suitable constant, as shown in the next example.

EXAMPLE 2 Is the function defined by $f(x) = 3x^2$ a probability density function for the interval [0, 4]? If not, convert it to one.

First,

$$\int_0^4 3x^2\,dx = x^3\Big|_0^4 = 64.$$

Since the integral is not equal to 1, the function is not a probability density function. To convert it to one, multiply $f(x)$ by $\frac{1}{64}$. The function defined by $\frac{3}{64}x^2$ for [0, 4] will be a probability density function, since

$$\int_0^4 \frac{3}{64}x^2 = 1,$$

and $(3/64)x^2 \geq 0$ for all x in [0, 4]. ◀

An important distinction is made between a discrete probability distribution and a probability density function (which is continuous). In a discrete distribution, the probability that the random variable, x, will assume a specific value is given in the distribution for every possible value of x. In a probability density function, however, the probability that x equals a specific value, say, c, is

$$P(x = c) = \int_c^c f(x)\,dx = 0.$$

For a probability density function, only probabilities of *intervals* can be found. For example, suppose the random variable is the annual rainfall for a given region. The amount of rainfall in one year can take on any value within some continuous interval that depends on the region; however, the probability that the rainfall in a given year will be some specific amount, say 33.25 inches, is actually zero.

The above definition of a probability density function is extended to intervals such as $(-\infty, b]$, $(-\infty, b)$, $[a, \infty)$, (a, ∞), or $(-\infty, \infty)$ by using improper integrals, as follows.

PROBABILITY DENSITY FUNCTIONS ON $(-\infty, \infty)$

If $f(x)$ is a probability density function for a continuous random variable x on $(-\infty, \infty)$, then

$$\boldsymbol{P(x \leq b) = P(x < b) = \int_{-\infty}^{b} f(x)\,dx,}$$

$$\boldsymbol{P(x \geq a) = P(x > a) = \int_{a}^{\infty} f(x)\,dx,}$$

$$\boldsymbol{P(-\infty < x < \infty) = \int_{-\infty}^{\infty} f(x)\,dx = 1.}$$

The total area under the graph of a probability density function of this type must still equal 1.

Improper integrals, those with one or more infinite limits, were discussed in the chapter on further techniques and applications of integration. The type of improper integral we shall need was defined as

$$\int_a^{\infty} f(x)\,dx = \lim_{b\to\infty}\int_a^b f(x)\,dx.$$

For example,

$$\begin{aligned}\int_1^{\infty} x^{-2}\,dx &= \lim_{b\to\infty}\int_1^b x^{-2}\,dx\\ &= \lim_{b\to\infty}\left(-\frac{1}{x}\bigg|_1^b\right)\\ &= \lim_{b\to\infty}\left(-\frac{1}{b}+\frac{1}{1}\right)\\ &= 0 + 1 = 1.\end{aligned}$$

EXAMPLE 3 Suppose the random variable x is the distance (in kilometers) from a given point to the nearest bird's nest, with the probability density function of the distribution given by $f(x) = 2xe^{-x^2}$ for $x \geq 0$.

(a) Show that $f(x)$ is a probability density function.

Since $e^{-x^2} = 1/e^{x^2}$ is always positive, and $x \geq 0$,

$$f(x) = 2xe^{-x^2} \geq 0.$$

Use substitution to evaluate the definite integral $\int_0^{\infty} 2xe^{-x^2}\,dx$. Let $u = -x^2$, so that $du = -2x\,dx$, and

$$\begin{aligned}\int 2xe^{-x^2}\,dx &= -\int e^{-x^2}(-2x\,dx)\\ &= -\int e^u\,du = -e^u = -e^{-x^2}.\end{aligned}$$

Then

$$\begin{aligned}\int_0^{\infty} 2xe^{-x^2}\,dx &= \lim_{b\to\infty}\int_0^b 2xe^{-x^2}\,dx = \lim_{b\to\infty}(-e^{-x^2})\bigg|_0^b\\ &= \lim_{b\to\infty}\left(-\frac{1}{e^{b^2}} + e^0\right) = 0 + 1 = 1.\end{aligned}$$

The function defined by $f(x) = 2xe^{-x^2}$ satisfies the two conditions required of a probability density function.

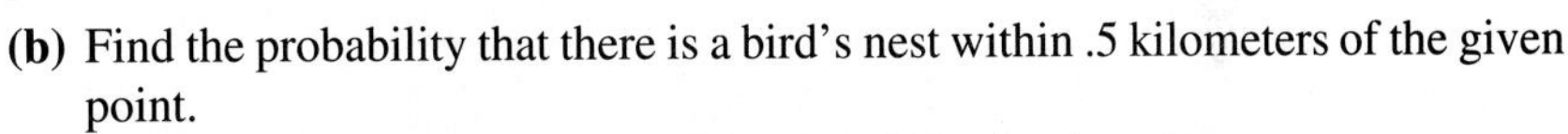

(b) Find the probability that there is a bird's nest within .5 kilometers of the given point.

Find $P(x \leq .5)$ where $x \geq 0$. This probability is given by

$$P(0 \leq x \leq .5) = \int_0^{.5} 2xe^{-x^2}\,dx.$$

Now evaluate the integral. The indefinite integral was found in part (a).

$$\begin{aligned}P(0 \leq x \leq .5) &= \int_0^{.5} 2xe^{-x^2}\,dx = (-e^{-x^2})\bigg|_0^{.5}\\ &= -e^{-(.5)^2} - (-e^0) = -e^{-.25} + 1\\ &\approx -.78 + 1 = .22\end{aligned}$$

The probability that a bird's nest will be found within .5 kilometers of the point is about .22. ◀

10.1 Exercises

Decide whether the functions defined as follows are probability density functions on the indicated intervals. If not, tell why.

1. $f(x) = \frac{1}{9}x - \frac{1}{18}$; $[2, 5]$

2. $f(x) = \frac{1}{3}x - \frac{1}{6}$; $[3, 4]$

3. $f(x) = \frac{1}{21}x^2$; $[1, 4]$

4. $f(x) = \frac{3}{98}x^2$; $[3, 5]$

5. $f(x) = 4x^3$; $[0, 3]$

6. $f(x) = \frac{x^3}{81}$; $[0, 3]$

7. $f(x) = \frac{x^2}{16}$; $[-2, 2]$

8. $f(x) = 2x^2$; $[-1, 1]$

Find a value of k that will make f a probability density function on the indicated interval.

9. $f(x) = kx^{1/2}$; $[1, 4]$

10. $f(x) = kx^{3/2}$; $[4, 9]$

11. $f(x) = kx^2$; $[0, 5]$

12. $f(x) = kx^2$; $[-1, 2]$

13. $f(x) = kx$; $[0, 3]$

14. $f(x) = kx$; $[2, 3]$

15. $f(x) = kx$; $[1, 5]$

16. $f(x) = kx^3$; $[2, 4]$

17. The total area under the graph of a probability density function always equals ________ .

18. In your own words, define a random variable.

19. What is the difference between a discrete probability function and a probability density function?

20. Why is $P(x = c) = 0$ for any number c in the domain of a probability density function?

Show that each function defined as follows is a probability density function on the given interval; then find the indicated probabilities.

21. $f(x) = \frac{1}{2}(1 + x)^{-3/2}$; $[0, \infty)$

(a) $P(0 \le x \le 2)$ **(b)** $P(1 \le x \le 3)$ **(c)** $P(x \ge 5)$

22. $f(x) = e^{-x}$; $[0, \infty)$

(a) $P(0 \le x \le 1)$ **(b)** $P(1 \le x \le 2)$ **(c)** $P(x \le 2)$

23. $f(x) = (1/2)e^{-x/2}$; $[0, \infty)$

(a) $P(0 \le x \le 1)$ **(b)** $P(1 \le x \le 3)$ **(c)** $P(x \ge 2)$

24. $f(x) = \frac{20}{(x + 20)^2}$; $[0, \infty)$

(a) $P(0 \le x \le 1)$ **(b)** $P(1 \le x \le 5)$ **(c)** $P(x \ge 5)$

Applications

Business and Economics

25. Life Span of a Computer Part The life (in months) of a certain electronic computer part has a probability density function defined by

$$f(x) = \frac{1}{2}e^{-x/2} \quad \text{for } [0, \infty).$$

Find the probability that a randomly selected component will last each of the following lengths of time.

(a) At most 12 mo

(b) Between 12 and 20 mo

26. Machine Life A machine has a useful life of 4 to 9 years, and its life (in years) is a probability density function defined by

$$f(x) = \frac{1}{11}\left(1 + \frac{3}{\sqrt{x}}\right).$$

Find the probabilities that the useful life of such a machine selected at random will be the following.

(a) Longer than 6 yr **(b)** Less than 5 yr

(c) Between 4 and 7 yr

Life Sciences

27. Petal Length The length of a petal on a certain flower varies from 1 cm to 4 cm and is a probability density function defined by

$$f(x) = \frac{1}{2\sqrt{x}}.$$

Find the probabilities that the length of a randomly selected petal will be as follows.

(a) Greater than or equal to 3 cm

(b) Less than or equal to 2 cm

(c) Between 2 cm and 3 cm

28. Clotting Time of Blood The clotting time of blood is a random variable x with values from 1 second to 20 seconds and probability density function defined by

$$f(x) = \frac{1}{(\ln 20)x}.$$

Find the following probabilities for a person selected at random.

(a) The probability that the clotting time is between 1 and 5 sec

(b) The probability that the clotting time is greater than 10 sec

Social Sciences

29. Time to Learn a Task The time x required for a person to learn a certain task is a random variable with probability density function defined by

$$f(x) = \frac{8}{7(x - 2)^2}.$$

The time required to learn the task is between 3 and 10 min. Find the probabilities that a randomly selected person will learn the task in the following lengths of time.

(a) Less than 4 min **(b)** More than 5 min

Physical Sciences

30. Annual Rainfall The annual rainfall in a remote Middle Eastern country varies from 0 to 5 inches and is a random variable with probability density function defined by

$$f(x) = \frac{5.5 - x}{15}.$$

Find the following probabilities for the annual rainfall in a randomly selected year.

(a) The probability that the annual rainfall is greater than 3 inches

(b) The probability that the annual rainfall is less than 2 inches

(c) The probability that the annual rainfall is between 1 inch and 4 inches

General Interest

31. Drunk Drivers According to an article in the journal *Traffic Safety*, the percentage of alcohol-related traffic fatalities for teenagers and young adults has dropped significantly since 1982. Based on data given in the article, the age of a randomly selected, alcohol-impaired driver in a fatal car crash is a random variable with probability density function given by

$$f(x) = \frac{105}{4x^2} \quad \text{for } [15, 35].$$

Find the following probabilities of the age of such a driver.

(a) Less than 25 **(b)** Less than 21

(c) Between 21 and 25

32. Length of a Telephone Call The length of a telephone call (in minutes), x, for a certain town is a continuous random variable with probability density function defined by

$$f(x) = 3x^{-4} \quad \text{for } [1, \infty).$$

Find the probabilities for the following situations.

(a) The call lasts between 1 and 2 min.

(b) The call lasts between 3 and 5 min.

(c) The call lasts longer than 3 min.

10.2 EXPECTED VALUE AND VARIANCE OF CONTINUOUS RANDOM VARIABLES

What is the average age of a drunk driver in a fatal car crash?

It often is useful to have a single number, a typical or "average" number, that represents a random variable. The *mean* or *expected value* for a discrete random variable is found by multiplying each value of the random variable by its corresponding probability, as follows.

EXPECTED VALUE

Suppose the random variable x can take on the n values, $x_1, x_2, x_3, \ldots, x_n$. Also, suppose the probabilities that each of these values occurs are, respectively, $p_1, p_2, p_3, \ldots, p_n$. Then the **mean**, or **expected value,** of the random variable is

$$\mu = x_1p_1 + x_2p_2 + x_3p_3 + \cdots + x_np_n = \sum_{i=1}^{n} x_ip_i.$$

This definition can be extended to continuous random variables by using definite integrals. Suppose a continuous random variable has probability function $f(x)$ on $[a, b]$. We can divide the interval from a to b into n subintervals of length Δx, where $\Delta x = (b - a)/n$. In the ith subinterval, the probability that the random variable takes the value x_i is $f(x_i)\Delta x_i$, and so

$$\mu = \sum_{i=1}^{n} x_i f(x_i)\Delta x_i.$$

As $n \to \infty$, the limit of this sum gives the expected value

$$\mu = \int_a^b xf(x)\,dx.$$

The **variance** of a probability distribution is a measure of the *spread* of the values of the distribution. For a discrete distribution, the variance is found by taking the expected value of the squares of the differences of the values of the random variable and the mean. If the random variable x takes the values $x_1, x_2, x_3, \ldots, x_n$, with respective probabilities $p_1, p_2, p_3, \ldots, p_n$ and mean μ, then the variance of x is

$$\text{Var}(x) = \sum_{i=1}^{n} (x_i - \mu)^2 p_i.$$

The **standard deviation** of x is defined as

$$\sigma = \sqrt{\text{Var}(x)}.$$

Like the mean or expected value, the variance of a continuous random variable is an integral.

$$\text{Var}(x) = \int_a^b (x - \mu)^2 f(x)\,dx$$

To find the standard deviation of a continuous probability distribution, like that of a discrete distribution, we find the square root of the variance. The formulas for the expected value, variance, and standard deviation of a continuous probability distribution are summarized on the next page.

EXPECTED VALUE, VARIANCE, AND STANDARD DEVIATION

If x is a continuous random variable with probability density function $f(x)$ on $[a, b]$, then the expected value of x is

$$E(x) = \mu = \int_a^b x f(x)\, dx.$$

The variance of x is

$$\mathrm{Var}(x) = \int_a^b (x - \mu)^2 f(x)\, dx,$$

and the standard deviation of x is

$$\sigma = \sqrt{\mathrm{Var}(x)}.$$

Geometrically, the expected value (or mean) of a probability distribution represents the "center" of the distribution. If a fulcrum were placed at μ on the x-axis, the figure would be in balance. See Figure 4.

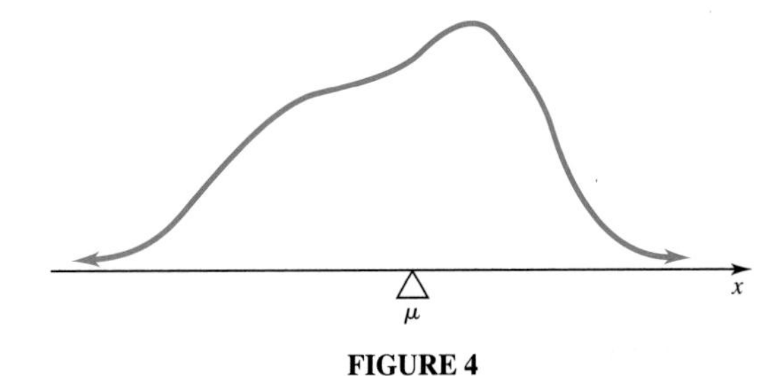

FIGURE 4

The variance or standard deviation of a probability distribution indicates how closely the values of the distribution cluster about the mean. These measures are most useful for comparing different distributions, as in Figure 5.

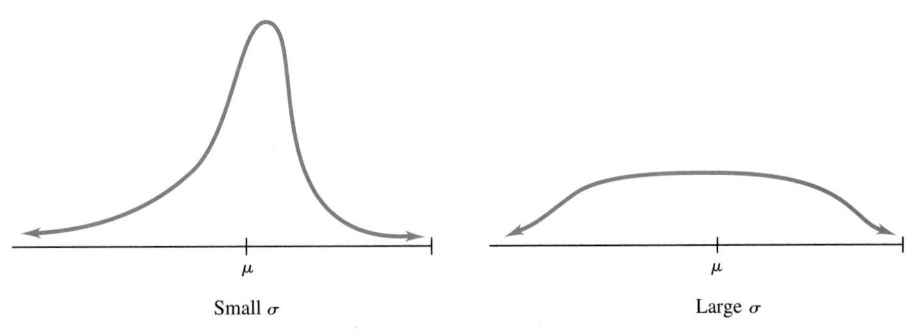

FIGURE 5

EXAMPLE 1 Find the expected value and variance of the random variable x with probability density function defined by $f(x) = (3/26)x^2$ on $[1, 3]$.

By the definition of expected value given above,

$$\begin{aligned}
\mu &= \int_1^3 xf(x)\,dx \\
&= \int_1^3 x\left(\frac{3}{26}x^2\right)dx \\
&= \frac{3}{26}\int_1^3 x^3\,dx \\
&= \frac{3}{26}\left(\frac{x^4}{4}\right)\Bigg|_1^3 = \frac{3}{104}(81 - 1) = \frac{30}{13},
\end{aligned}$$

or about 2.31.

The variance is

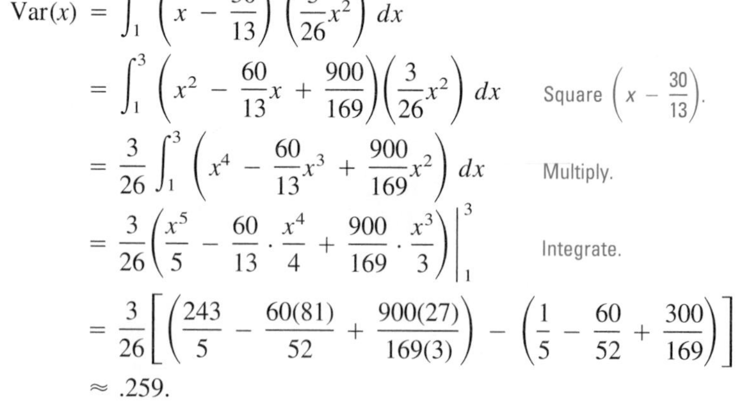

$$\begin{aligned}
\text{Var}(x) &= \int_1^3 \left(x - \frac{30}{13}\right)^2\left(\frac{3}{26}x^2\right)dx \\
&= \int_1^3 \left(x^2 - \frac{60}{13}x + \frac{900}{169}\right)\left(\frac{3}{26}x^2\right)dx && \text{Square } \left(x - \frac{30}{13}\right). \\
&= \frac{3}{26}\int_1^3 \left(x^4 - \frac{60}{13}x^3 + \frac{900}{169}x^2\right)dx && \text{Multiply.} \\
&= \frac{3}{26}\left(\frac{x^5}{5} - \frac{60}{13}\cdot\frac{x^4}{4} + \frac{900}{169}\cdot\frac{x^3}{3}\right)\Bigg|_1^3 && \text{Integrate.} \\
&= \frac{3}{26}\left[\left(\frac{243}{5} - \frac{60(81)}{52} + \frac{900(27)}{169(3)}\right) - \left(\frac{1}{5} - \frac{60}{52} + \frac{300}{169}\right)\right] \\
&\approx .259.
\end{aligned}$$

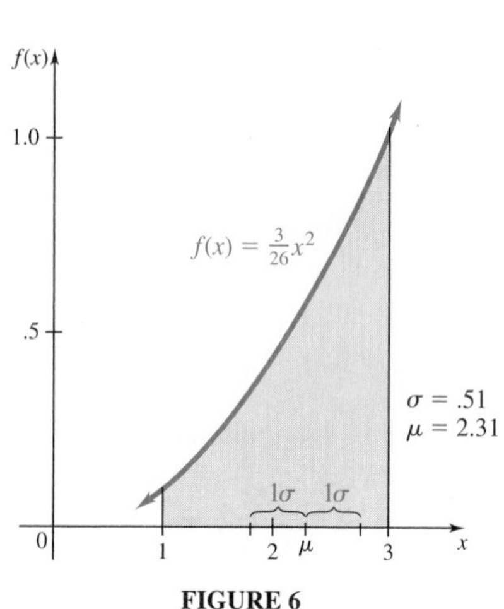

FIGURE 6

From the variance, the standard deviation is $\sigma \approx \sqrt{.259} \approx .51$. The expected value and standard deviation are shown on the graph of the probability density function in Figure 6. ◀

Calculating the variance in the last example was a messy job. An alternative version of the formula for the variance is easier to compute. This alternative formula is derived as follows.

$$\begin{aligned}
\text{Var}(x) &= \int_a^b (x - \mu)^2 f(x)\,dx \\
&= \int_a^b (x^2 - 2\mu x + \mu^2)f(x)\,dx \\
&= \int_a^b x^2 f(x)\,dx - 2\mu\int_a^b xf(x)\,dx + \mu^2\int_a^b f(x)\,dx \qquad (1)
\end{aligned}$$

By definition,

$$\int_a^b xf(x)\,dx = \mu,$$

and, since $f(x)$ is a probability density function,

$$\int_a^b f(x)\,dx = 1.$$

Substitute back into equation (1) to get the alternative formula,

$$\text{Var}(x) = \int_a^b x^2 f(x)\,dx - 2\mu^2 + \mu^2 = \int_a^b x^2 f(x)\,dx - \mu^2.$$

ALTERNATIVE FORMULA FOR VARIANCE

If x is a random variable with probability density function $f(x)$ on $[a, b]$, and if $E(x) = \mu$, then

$$\textbf{Var}(x) = \int_a^b x^2 f(x)\,dx - \mu^2.$$

Caution Notice that the term μ^2 comes *after* the differential dx, and so is *not* integrated.

EXAMPLE 2

Use the alternative formula for variance to compute the variance of the random variable x with probability density function defined by $f(x) = 3/x^4$ for $x \geq 1$.

To find the variance, first find the expected value:

$$\mu = \int_1^\infty xf(x)\,dx = \int_1^\infty x \cdot \frac{3}{x^4}\,dx = \int_1^\infty \frac{3}{x^3}\,dx$$

$$= \lim_{b\to\infty} \int_1^b \frac{3}{x^3}\,dx = \lim_{b\to\infty} \left(\frac{3}{-2x^2}\right)\Bigg|_1^b = \frac{3}{2},$$

or 1.5. Now find the variance by the alternative formula for variance:

$$\text{Var}(x) = \int_1^\infty x^2\left(\frac{3}{x^4}\right) dx - \left(\frac{3}{2}\right)^2$$

$$= \int_1^\infty \frac{3}{x^2}\,dx - \frac{9}{4}$$

$$= \lim_{b\to\infty} \int_1^b \frac{3}{x^2}\,dx - \frac{9}{4}$$

$$= \lim_{b\to\infty} \left(\frac{-3}{x}\right)\Bigg|_1^b - \frac{9}{4}$$

$$= 3 - \frac{9}{4} = \frac{3}{4}, \quad \text{or } .75.$$ ◀

EXAMPLE 3 A recent study has shown that on any given day, the proportion of airline passengers who arrive at the gate at least 1/2 hour before the scheduled flight time is a probability density function, with $f(x) = 6x - 6x^2$ for $0 \le x \le 1$.

(a) Find and interpret the expected value for this distribution.

The expected value is

$$\mu = \int_0^1 x(6x - 6x^2)\,dx = \int_0^1 (6x^2 - 6x^3)\,dx$$

$$= \left(2x^3 - \frac{3}{2}x^4\right)\Bigg|_0^1 = \frac{1}{2},$$

or .5. This result indicates that over a long period of time, about one-half of all airline passengers arrive at the gate at least 1/2 hour before the scheduled flight time.

(b) Compute the standard deviation.

First compute the variance. We use the alternative formula.

$$\text{Var}(x) = \int_0^1 x^2(6x - 6x^2)\,dx - \left(\frac{1}{2}\right)^2$$

$$= \int_0^1 (6x^3 - 6x^4)\,dx - \left(\frac{1}{2}\right)^2$$

$$= \left(\frac{3}{2}x^4 - \frac{6}{5}x^5\right)\Bigg|_0^1 - \frac{1}{4}$$

$$= \frac{3}{10} - \frac{1}{4} = \frac{1}{20} = .05$$

The standard deviation is $\sigma = \sqrt{.05} \approx .22$. ◀

10.2 Exercises

In Exercises 1–8, a probability density function of a random variable is defined. Find the expected value, the variance, and the standard deviation. Round answers to the nearest hundredth.

1. $f(x) = \frac{1}{4}$; $[3, 7]$

2. $f(x) = \frac{1}{10}$; $[0, 10]$

3. $f(x) = \frac{x}{8} - \frac{1}{4}$; $[2, 6]$

4. $f(x) = 2(1 - x)$; $[0, 1]$

5. $f(x) = 1 - \frac{1}{\sqrt{x}}$; $[1, 4]$

6. $f(x) = \frac{1}{11}\left(1 + \frac{3}{\sqrt{x}}\right)$; $[4, 9]$

7. $f(x) = 4x^{-5}$; $[1, \infty)$

8. $f(x) = 3x^{-4}$; $[1, \infty)$

9. What information does the mean (expected value) of a continuous random variable give?

10. Suppose two random variables have standard deviations of .10 and .23, respectively. What does this tell you about their distributions?

In Exercises 11–14, the probability density function of a random variable is defined.

(a) *Find the expected value to the nearest hundredth.*

(b) *Find the variance to the nearest hundredth.*

(c) *Find the standard deviation. Round to the nearest hundredth.*

(d) *Find the probability that the random variable has a value greater than the mean.*

(e) *Find the probability that the value of the random variable is within one standard deviation of the mean.*

11. $f(x) = \frac{\sqrt{x}}{18}$; $[0, 9]$

12. $f(x) = \frac{x^{-1/3}}{6}$; $[0, 8]$

13. $f(x) = \frac{1}{2}x$; $[0, 2]$

14. $f(x) = \frac{3}{2}(1 - x^2)$; $[0, 1]$

If x is a random variable with probability density function f(x) on [a, b], then the **median** *of x is the number m such that*

$$\int_a^m f(x)\,dx = \frac{1}{2}.$$

(a) *Find the median of each random variable for the probability density functions in Exercises 15–20.*

(b) *In each case, find the probability that the random variable is between the expected value (mean) and the median. The expected value for each of these functions was found in Exercises 1–8.*

15. $f(x) = \frac{1}{4}$; $[3, 7]$

16. $f(x) = \frac{1}{10}$; $[0, 10]$

17. $f(x) = \frac{x}{8} - \frac{1}{4}$; $[2, 6]$

18. $f(x) = 2(1 - x)$; $[0, 1]$

19. $f(x) = 4x^{-5}$; $[1, \infty)$

20. $f(x) = 3x^{-4}$; $[1, \infty)$

Applications

Business and Economics

21. Life of a Light Bulb The life (in hours) of a certain kind of light bulb is a random variable with probability density function defined by

$$f(x) = \frac{1}{58\sqrt{x}} \quad \text{for } [1, 900].$$

(a) What is the expected life of such a bulb?

(b) Find σ.

(c) Find the probability that one of these bulbs lasts longer than one standard deviation above the mean.

22. Machine Life The life (in years) of a certain machine is a random variable with probability density function defined by

$$f(x) = \frac{1}{11}\left(1 + \frac{3}{\sqrt{x}}\right) \quad \text{for } [4, 9].$$

(a) Find the mean life of this machine.

(b) Find the standard deviation of the distribution.

(c) Find the probability that a particular machine of this kind will last longer than the mean number of years.

23. Life of an Automobile Part The life span of a certain automobile part (in months) is a random variable with probability density function defined by

$$f(x) = \frac{1}{2}e^{-x/2} \quad \text{for } [0, \infty).$$

(a) Find the expected life of this part.

(b) Find the standard deviation of the distribution.

(c) Find the probability that one of these parts lasts less than the mean number of months.

Life Sciences

24. Blood Clotting Time The clotting time of blood (in seconds) is a random variable with probability density function defined by

$$f(x) = \frac{1}{(\ln 20)x} \quad \text{for } [1, 20].$$

(a) Find the mean clotting time.

(b) Find the standard deviation.

(c) Find the probability that a person's blood clotting time is within one standard deviation of the mean.

25. **Length of a Leaf** The length of a leaf on a tree is a random variable with probability density function defined by

$$f(x) = \frac{3}{32}(4x - x^2) \quad \text{for } [0, 4].$$

(a) What is the expected leaf length?

(b) Find σ for this distribution.

(c) Find the probability that the length of a given leaf is within one standard deviation of the expected value.

26. **Petal Length** The length (in centimeters) of a petal on a certain flower is a random variable with probability density function defined by

$$f(x) = \frac{1}{2\sqrt{x}} \quad \text{for } [1, 4].$$

(a) Find the expected petal length.

(b) Find the standard deviation.

(c) Find the probability that a petal selected at random has a length more than two standard deviations above the mean.

Physical Sciences

27. **Annual Rainfall** The annual rainfall in a remote Middle Eastern country is a random variable with probability density function defined by

$$f(x) = \frac{5.5 - x}{15} \quad \text{for } [0, 5].$$

(a) Find the mean annual rainfall.

(b) Find the standard deviation.

(c) Find the probability of a year with rainfall less than one standard deviation below the mean.

General Interest

28. **Drunk Drivers** The age of a randomly selected, alcohol-impaired driver in a fatal car crash is a random variable with probability density function given by

$$f(x) = \frac{105}{4x^2} \quad \text{for } [15, 35].$$

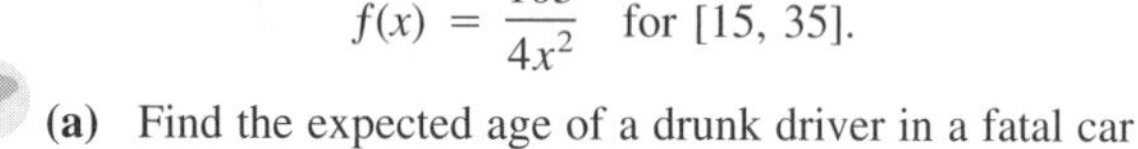

(a) Find the expected age of a drunk driver in a fatal car crash.

(b) Find the standard deviation of the distribution.

(c) Find the probability that the age of such a driver will be less than one standard deviation below the mean.

10.3 SPECIAL PROBABILITY DENSITY FUNCTIONS

What is the probability that the maximum temperature will be higher than 24 C?

What is the probability that a battery will last longer than 40 hours?

These questions can be answered if the probability density function for the maximum temperature and for the life of the battery are known. In practice, however, it is not feasible to construct a probability density function for each experiment. Instead, a researcher uses one of several probability density functions that are well known, matching the shape of the experimental distribution to one of the known distributions. In this section we discuss some of the most commonly used probability distributions.

Uniform Distribution The probability density function for the **uniform distribution** is defined by

$$f(x) = \frac{1}{b - a} \quad \text{for } [a, b],$$

where a and b are nonnegative real numbers. The graph of $f(x)$ is shown in Figure 7 on the facing page.

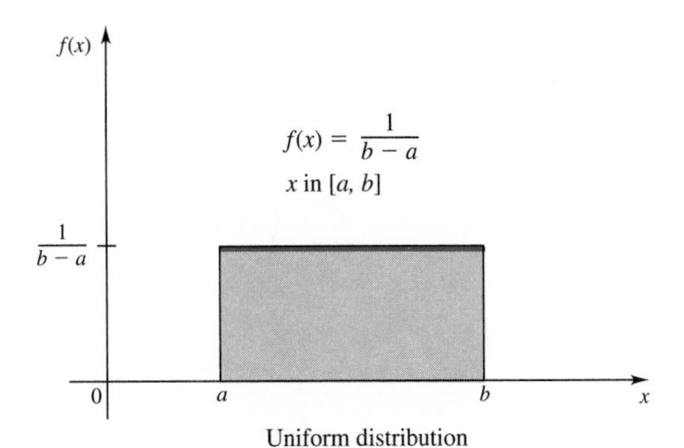

Uniform distribution

FIGURE 7

Since $b - a$ is positive, $f(x) \geq 0$, and

$$\int_a^b \frac{1}{b-a}\,dx = \frac{1}{b-a}x\Big|_a^b = \frac{1}{b-a}(b-a) = 1.$$

Therefore, the function is a probability density function.

The expected value for the uniform distribution is

$$\mu = \int_a^b \left(\frac{1}{b-a}\right)x\,dx = \left(\frac{1}{b-a}\right)\frac{x^2}{2}\Big|_a^b$$

$$= \frac{1}{2(b-a)}(b^2 - a^2) = \frac{1}{2}(b+a). \qquad b^2 - a^2 = (b-a)(b+a)$$

The variance is given by

$$\mathrm{Var}(x) = \int_a^b \left(\frac{1}{b-a}\right)x^2\,dx - \left(\frac{b+a}{2}\right)^2$$

$$= \left(\frac{1}{b-a}\right)\frac{x^3}{3}\Big|_a^b - \frac{(b+a)^2}{4}$$

$$= \frac{1}{3(b-a)}(b^3 - a^3) - \frac{1}{4}(b+a)^2$$

$$= \frac{b^2 + ab + a^2}{3} - \frac{b^2 + 2ab + a^2}{4} \qquad b^3 - a^3 = (b-a)(b^2 + ab + a^2)$$

$$= \frac{b^2 - 2ab + a^2}{12}. \qquad \text{Get a common denominator; subtract.}$$

Thus,

$$\mathrm{Var}(x) = \frac{1}{12}(b-a)^2, \qquad \text{Factor.}$$

and

$$\sigma = \frac{1}{\sqrt{12}}(b-a).$$

These properties of the uniform distribution are summarized below.

UNIFORM DISTRIBUTION If x is a random variable with probability density function

$$f(x) = \frac{1}{b - a} \quad \text{for } [a, b],$$

then

$$\mu = \frac{1}{2}(b + a) \quad \text{and} \quad \sigma = \frac{1}{\sqrt{12}}(b - a).$$

EXAMPLE 1

A couple is planning to vacation in San Francisco. They have been told that the maximum daily temperature during the time they plan to be there ranges from 15° C to 27° C. Assume that the probability of any temperature between 15° C and 27° C is equally likely for any given day during the specified time period.

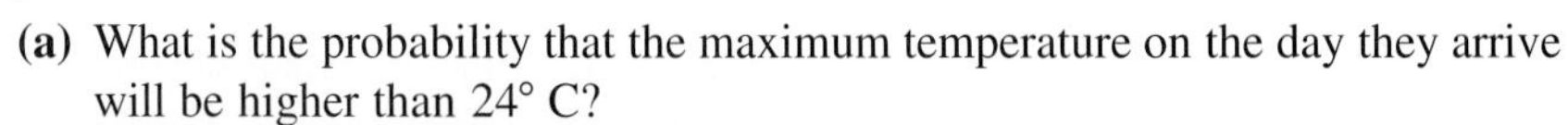

(a) What is the probability that the maximum temperature on the day they arrive will be higher than 24° C?

If the random variable t represents the maximum temperature on a given day, then the uniform probability density function for t is defined by $f(t) = 1/12$ for the interval [15, 27]. By definition,

$$P(t > 24) = \int_{24}^{27} \frac{1}{12}\, dt = \frac{1}{12}t \Big|_{24}^{27} = \frac{1}{4}.$$

(b) What average maximum temperature can they expect?

The expected maximum temperature is

$$\mu = \frac{1}{2}(27 + 15) = 21,$$

or 21° C.

(c) What is the probability that the maximum temperature on a given day will be at least one standard deviation below the mean?

First find σ:

$$\sigma = \frac{1}{\sqrt{12}}(27 - 15) = \frac{12}{\sqrt{12}} = \sqrt{12} = 2\sqrt{3} \approx 3.5.$$

One standard deviation below the mean indicates a temperature of $21 - 3.5 = 17.5$° C.

$$P(T \leq 17.5) = \int_{15}^{17.5} \frac{1}{12}\, dt = \frac{1}{12}t \Big|_{15}^{17.5} \approx .21$$

The probability is .21 that the temperature will not exceed 17.5° C.

Exponential Distribution The probability density function for the **exponential distribution** is defined by

$$f(x) = ae^{-ax} \quad \text{for } [0, \infty),$$

where a is a positive real number. The graph of $f(x)$ is shown in Figure 8.

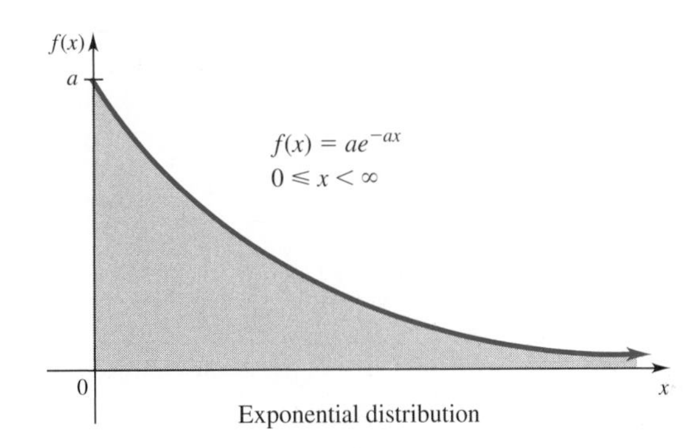

FIGURE 8

Here $f(x) \geq 0$, since e^{-ax} and a are both positive for all values of x. Also,

$$\begin{aligned}\int_0^{\infty} ae^{-ax}\,dx &= \lim_{b\to\infty}\int_0^b ae^{-ax}\,dx\\ &= \lim_{b\to\infty}(-e^{-ax})\Big|_0^b = \lim_{b\to\infty}\left(\frac{-1}{e^{ab}} + \frac{1}{e^0}\right) = 1,\end{aligned}$$

so the function is a probability density function.

The expected value and the standard deviation of the exponential distribution can be found using integration by parts. The results are given below.

EXPONENTIAL DISTRIBUTION

If x is a random variable with probability density function

$$\boldsymbol{f(x) = ae^{-ax} \quad \text{for } [0, \infty),}$$

then

$$\boldsymbol{\mu = \frac{1}{a} \quad \text{and} \quad \sigma = \frac{1}{a}.}$$

The exponential distribution is very important in reliability analysis; when manufactured items have a constant failure rate over a period of time, the exponential distribution is used to describe their probability of failure, as in the following example.

EXAMPLE 2 Suppose the useful life (in hours) of a flashlight battery is the random variable t, with probability density function given by the exponential distribution

$$f(t) = \frac{1}{20}e^{-t/20} \quad \text{for } t \geq 0.$$

(a) Find the probability that a particular battery, selected at random, has a useful life of less than 100 hours.

The probability is given by

$$P(t \leq 100) = \int_0^{100} \frac{1}{20}e^{-t/20}\,dt = \frac{1}{20}\left(-20e^{-t/20}\right)\Big|_0^{100}$$
$$= -(e^{-100/20} - e^0) = -(e^{-5} - 1)$$
$$\approx 1 - .0067 = .9933.$$

(b) Find the expected value and standard deviation of the distribution.

Use the formulas given above. Both μ and σ equal $1/a$, and since $a = 1/20$ here,

$$\mu = 20 \quad \text{and} \quad \sigma = 20.$$

This means that the average life of a battery is 20 hours, and no battery lasts less than one standard deviation below the mean.

(c) What is the probability that a battery will last longer than 40 hours?

The probability is given by

$$P(t > 40) = \int_{40}^{\infty} \frac{1}{20}e^{-t/20}\,dt = \lim_{b\to\infty} (-e^{-t/20})\Big|_{40}^{b} = \frac{1}{e^2} \approx .1353,$$

or about 14%. ◀

Normal Distribution The **normal distribution,** with its well-known bell-shaped graph, is undoubtedly the most important probability density function. It is widely used in various applications of statistics. The probability density function for the normal distribution has the following characteristics.

NORMAL DISTRIBUTION

If μ and σ are real numbers, $\sigma \geq 0$, and if x is a random variable with probability density function defined by

$$f(x) = \frac{1}{\sigma\sqrt{2\pi}}e^{-(x-\mu)^2/(2\sigma^2)} \quad \text{for } (-\infty, \infty),$$

then

$$E(x) = \mu \quad \text{and} \quad \text{Var}(x) = \sigma^2, \text{ with **standard deviation** } \sigma.$$

Advanced techniques can be used to show that

$$\int_{-\infty}^{\infty} \frac{1}{\sigma\sqrt{2\pi}} e^{-(x-\mu)^2/(2\sigma^2)}\, dx = 1.$$

Deriving the expected value and standard deviation for the normal distribution also requires techniques beyond the scope of this text.

Each normal probability distribution has associated with it a bell-shaped curve, called a **normal curve,** such as the one in Figure 9. Each normal curve is symmetric about a vertical line through the mean, μ. Vertical lines at points $+1\sigma$ and -1σ from the mean show the inflection points of the graph. A normal curve never touches the x-axis; it extends indefinitely in both directions.

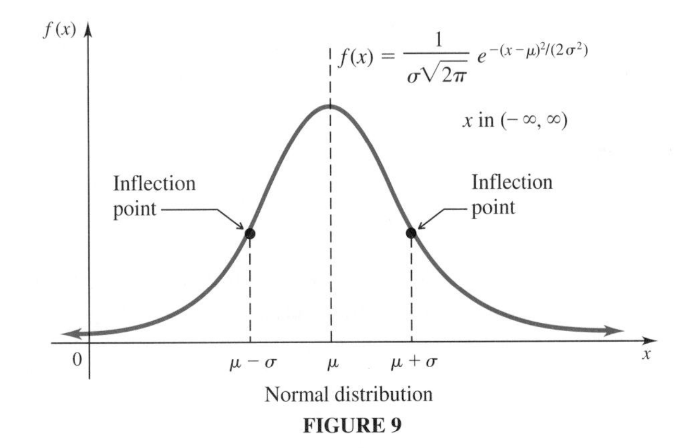

Normal distribution

FIGURE 9

The development of the normal curve is credited to the Frenchman Abraham de Moivre (1667–1754). Three of his publications dealt with probability and associated topics: *Annuities upon Lives* (which contributed to the development of actuarial studies), *Doctrine of Chances,* and *Miscellanea Analytica.*

Many different normal curves have the same mean. In such cases, a larger value of σ produces a "flatter" normal curve, while smaller values of σ produce more values near the mean, resulting in a "taller" normal curve. See Figure 10.

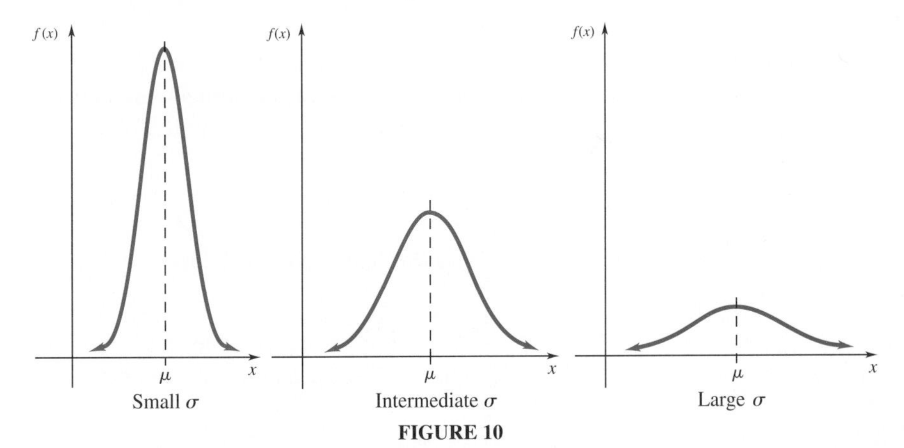

FIGURE 10

It would be far too much work to calculate values for the normal probability distribution for various values of μ and σ. Instead, values are calculated for the **standard normal distribution,** which has $\mu = 0$ and $\sigma = 1$. The graph of the standard normal distribution is shown in Figure 11.

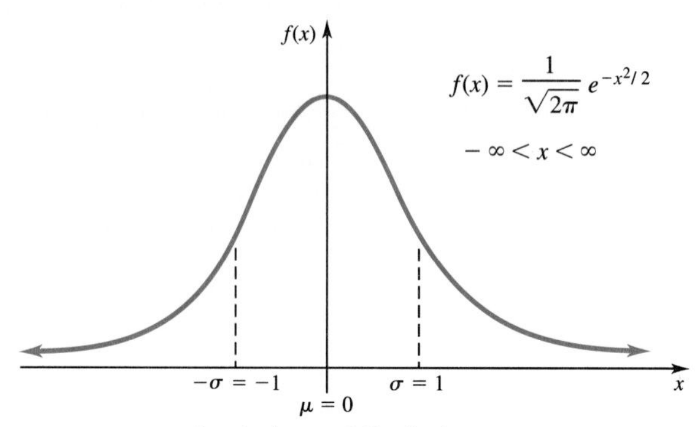

Standard normal distribution

FIGURE 11

Probabilities for the standard normal distribution come from the definite integral

$$\int_a^b \frac{1}{\sqrt{2\pi}} e^{-x^2/2}\, dx.$$

Since $f(x) = e^{-x^2/2}$ does not have an antiderivative that can be expressed in terms of functions used in this course, numerical methods are used to find values of this definite integral. A table in the appendix of this book gives areas under the standard normal curve, along with a sketch of the curve. Each value in this table is the total area under the standard normal curve to the left of the number z.

If a normal distribution does not have $\mu = 0$ and $\sigma = 1$, we use the following theorem, which is stated without proof.

z-SCORES THEOREM

Suppose a normal distribution has mean μ and standard deviation σ. The area under the associated normal curve that is to the left of the value x is exactly the same as the area to the left of

$$z = \frac{x - \mu}{\sigma}$$

for the standard normal curve.

Using this result, the table can be used for *any* normal distribution, regardless of the values of μ and σ. The number z in the theorem is called a **z-score.**

EXAMPLE 3 A normal distribution has mean 35 and standard deviation 5.9. Find the following areas under the associated normal curve.

(a) The area of the region to the left of 40

Find the appropriate z-score using $x = 40$, $\mu = 35$, and $\sigma = 5.9$. Round to the nearest hundredth.

$$z = \frac{40 - 35}{5.9} = \frac{5}{5.9} \approx .85$$

Look up .85 in the normal curve table in the Appendix. The corresponding area is .8023. Thus, the shaded area shown in Figure 12 is .8023. This area represents 80.23% of the total area under the normal curve.

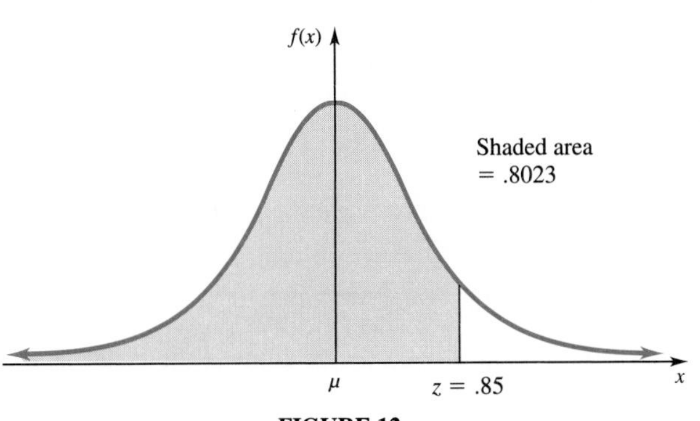

FIGURE 12

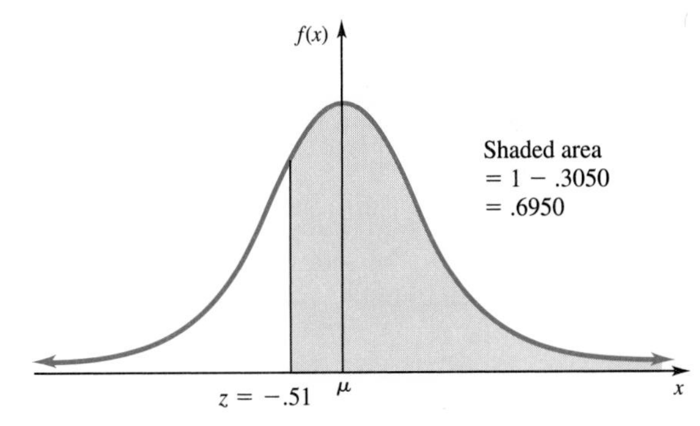

FIGURE 13

(b) The area of the region to the right of 32

$$z = \frac{32 - 35}{5.9} = \frac{-3}{5.9} \approx -.51$$

The area to the *left* of $z = -.51$ is .3050, so the area to the *right is* $1 - .3050 = .6950$. See Figure 13.

(c) The area of the region between 30 and 33

Find z-scores for both values.

$$z = \frac{30 - 35}{5.9} = \frac{-5}{5.9} \approx -.85 \quad \text{and} \quad z = \frac{33 - 35}{5.9} = \frac{-2}{5.9} \approx -.34$$

Start with the area to the left of $z = -.34$ and subtract the area to the left of $z = -.85$:

$$.3669 - .1977 = .1692.$$

The required area is shaded in Figure 14 on the next page. ◀

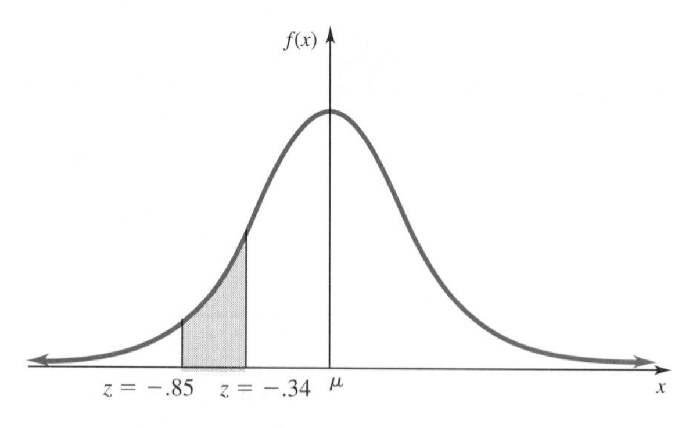

FIGURE 14

The z-scores are actually standard deviation multiples; that is, a z-score of 2.5 corresponds to a value 2.5 standard deviations above the mean. For example, looking up $z = 1.00$ and $z = -1.00$ in the table shows that

$$.8413 - .1587 = .6826,$$

so that 68.26% of the area under a normal curve is within one standard deviation of the mean. Also, using $z = 2.00$ and $z = -2.00$,

$$.9772 - .0228 = .9544,$$

meaning 95.44% of the area is within two standard deviations of the mean. These results, summarized in Figure 15, can be used to get a quick estimate of results when working with normal curves.

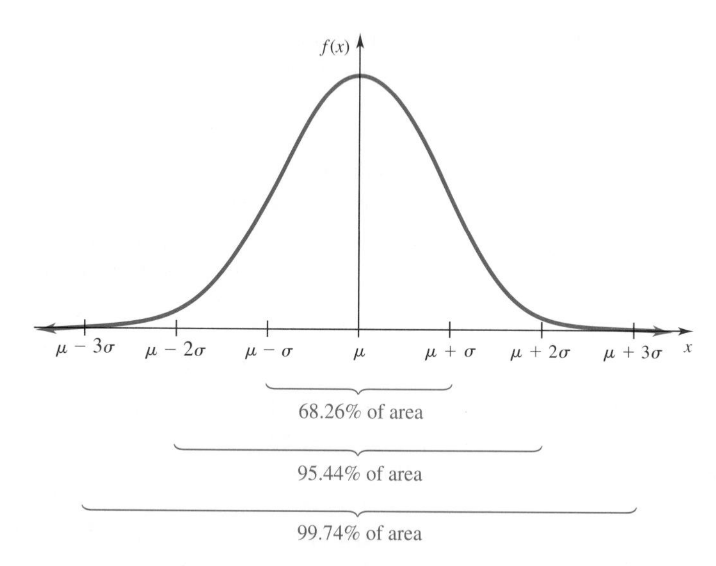

FIGURE 15

10.3 Exercises

For Exercises 1–6, find each of the following:

(a) *the mean of the distribution;*

(b) *the standard deviation of the distribution;*

(c) *the probability that the random variable is between the mean and one standard deviation above the mean.*

1. The length (in centimeters) of the leaf of a certain plant is a continuous random variable with probability density function defined by

$$f(x) = \frac{5}{4} \quad \text{for } [4, 4.8].$$

2. The price of an item (in dollars) is a continuous random variable with probability density function defined by

$$f(x) = 2 \quad \text{for } [1.25, 1.75].$$

3. The length of time (in years) until a particular radioactive particle decays is a random variable t with probability density function defined by

$$f(t) = .03e^{-.03t} \quad \text{for } [0, \infty).$$

4. The length of time (in years) that a seedling tree survives is a random variable t with probability density function defined by

$$f(t) = .05e^{-.05t} \quad \text{for } [0, \infty).$$

5. The length of time (in days) required to learn a certain task is a random variable t with probability density function defined by

$$f(t) = e^{-t} \quad \text{for } [0, \infty).$$

6. The distance (in meters) that seeds are dispersed from a certain kind of plant is a random variable x with probability density function defined by

$$f(x) = .1e^{-.1x} \quad \text{for } [0, \infty).$$

Find the percent of the area under a normal curve between the mean and the number of standard deviations above the mean given in Exercises 7 and 8.

7. 3.50

8. 1.68

Find the percent of the total area under the normal curve between the z-scores given in Exercises 9 and 10.

9. 1.28 and 2.05

10. -2.13 and $-.04$

Find a z-score satisfying the conditions given in Exercises 11–14. (Hint: *Use the table backwards.*)

11. 10% of the total area is to the left of z

12. 2% of the total area is to the left of z

13. 18% of the total area is to the right of z

14. 22% of the total area is to the right of z

15. Describe the standard normal distribution. What are its characteristics?

16. What is meant by a z-score? How is it used?

17. Describe the shape of the graph of each of the following probability distributions.

(a) Uniform **(b)** Exponential **(c)** Normal

In the exercises for Section 9.2, we defined the median of a probability distribution as an integral. The median also can be defined as the number m such that $P(x \leq m) = P(x \geq m)$.

18. Find an expression for the median of the uniform distribution.

19. Find an expression for the median of the exponential distribution.

20. Verify the expected value and standard deviation of the exponential distribution given in the text.

Applications

Business and Economics

21. Insurance Sales The amount of insurance (in thousands of dollars) sold in a day by a particular agent is uniformly distributed over the interval [10, 85].

(a) What amount of insurance does the agent sell on an average day?

(b) Find the probability that the agent sells more than $50,000 of insurance on a particular day.

22. Fast-Food Outlets The number of new fast-food outlets opening during June in a certain city is exponentially distributed, with a mean of 5.

(a) Give the probability density function for this distribution.

(b) What is the probability that the number of outlets opening is between 2 and 6?

23. **Sales Expense** A salesperson's monthly expenses (in thousands of dollars) are exponentially distributed, with an average of 4.25 (thousand dollars).
    (a) Give the probability density function for the expenses.
    (b) Find the probability that the expenses are more than $10,000.

In Exercises 24–26, assume a normal distribution.

24. **Machine Accuracy** A machine that fills quart bottles with apple juice averages 32.8 oz per bottle, with a standard deviation of 1.1 oz. What are the probabilities that the amount of juice in a bottle is as follows?
    (a) Less than 1 qt
    (b) At least 1 oz more than a quart
25. **Machine Accuracy** A machine produces screws with a mean length of 2.5 cm and a standard deviation of .2 cm. Find the probabilities that a screw produced by this machine has lengths as follows.
    (a) Greater than 2.7 cm
    (b) Within 1.2 standard deviations of the mean
26. **Customer Expenditures** Customers at a certain pharmacy spend an average of $54.40, with a standard deviation of $13.50. What are the largest and smallest amounts spent by the middle 50% of these customers?

Life Sciences

27. **Insect Life Span** The life span of a certain insect (in days) is uniformly distributed over the interval [20, 36].
    (a) What is the expected life of this insect?
    (b) Find the probability that one of these insects, randomly selected, lives longer than 30 days.
28. **Location of a Bee Swarm** A swarm of bees is released from a certain point. The proportion of the swarm located at least 2 m from the point of release after 1 hr is a random variable that is exponentially distributed, with $a = 2$ over the interval $[0, \infty)$.
    (a) Find the expected proportion under the given conditions.
    (b) Find the probability that fewer than 1/3 of the bees are located at least 2 m from the release point after 1 hr.
29. **Digestion Time** The digestion time (in hours) of a fixed amount of food is exponentially distributed, with $a = 1$.
    (a) Find the mean digestion time.
    (b) Find the probability that the digestion time is less than 30 min.
30. **Pygmy Heights** The average height of a member of a certain tribe of pygmies is 3.2 ft, with a standard deviation of .2 ft. If the heights are normally distributed, what are the largest and smallest heights of the middle 50% of this population?
31. **Finding Prey** H. R. Pulliam found that the time (in minutes) required by a predator to find a prey is a random variable that is exponentially distributed, with $\mu = 25$.*
    (a) According to this distribution, what is the longest time within which the predator will be 90% certain of finding a prey?
    (b) What is the probability that the predator will have to spend more than one hour looking for a prey?

Physical Sciences

32. **Rainfall** The rainfall (in inches) in a certain region is uniformly distributed over the interval [32, 44].
    (a) What is the expected number of inches of rainfall?
    (b) What is the probability that the rainfall will be between 38 and 40 inches?

Computer/Graphing Calculator

Use Simpson's rule, with $n = 100$, to approximate the following integrals.

33. $\int_0^{50} .5e^{-.5x}\,dx$
34. $\int_0^{50} .5xe^{-.5x}\,dx$
35. $\int_0^{50} .5x^2e^{-.5x}\,dx$
36. Use your results from Exercises 34 and 35 to verify that, for the exponential distribution, $\mu = 1/a$ and $\sigma = 1/a$.
37. The standard normal probability density function is defined by

$$f(x) = \frac{1}{\sqrt{2\pi}}e^{-x^2/2}.$$

Use Simpson's rule, with $n = 100$, and the formulas for the mean and standard deviation of a probability density function to approximate the following for the standard normal probability distribution. Use limits of -4 and 4 (instead of $-\infty$ and ∞).
    (a) The mean (b) The standard deviation

*From H. R. Pulliam, "On the Theory of Optimal Diets," *American Naturalist*, vol. 108, 1974, pp. 59–74.

Chapter Summary — Key Terms

10.1 **random variable**
probability distribution function
histogram
discrete probability function
continuous random variable
continuous probability distribution
probability density function

10.2 **mean**
expected value
variance
standard deviation

10.3 **uniform distribution**
exponential distribution
normal distribution
normal curve
standard normal distribution
***z*-score**

Chapter 10 Review Exercises

1. In a probability function, the y-values (or function values) represent __________ .

2. Define a continuous random variable.

3. Give the two conditions that a probability density function for $[a, b]$ must satisfy.

Decide whether each function defined as follows is a probability density function for the given interval.

4. $f(x) = \frac{1}{27}(2x + 4)$; $[1, 4]$

5. $f(x) = \sqrt{x}$; $[4, 9]$

6. $f(x) = .1$; $[0, 10]$

7. $f(x) = e^{-x}$; $[0, \infty)$

In Exercises 8 and 9, find a value of k that will make f(x) define a probability density function for the indicated interval.

8. $f(x) = k\sqrt{x}$; $[1, 4]$

9. $f(x) = kx^2$; $[0, 3]$

10. The probability density function of a random variable x is defined by

$$f(x) = 1 - \frac{1}{\sqrt{x - 1}} \quad \text{for } [2, 5].$$

Find the following probabilities.

(a) $P(x \geq 3)$ **(b)** $P(x \leq 4)$ **(c)** $P(3 \leq x \leq 4)$

11. The probability density function of a random variable x is defined by

$$f(x) = \frac{1}{10} \quad \text{for } [10, 20].$$

Find the following probabilities.

(a) $P(x \leq 12)$ **(b)** $P(x \geq 31/2)$ **(c)** $P(10.8 \leq x \leq 16.2)$

12. Describe what the expected value or mean of a probability distribution represents geometrically.

13. The probability density functions shown in the graphs below have the same mean. Which has the smallest standard deviation?

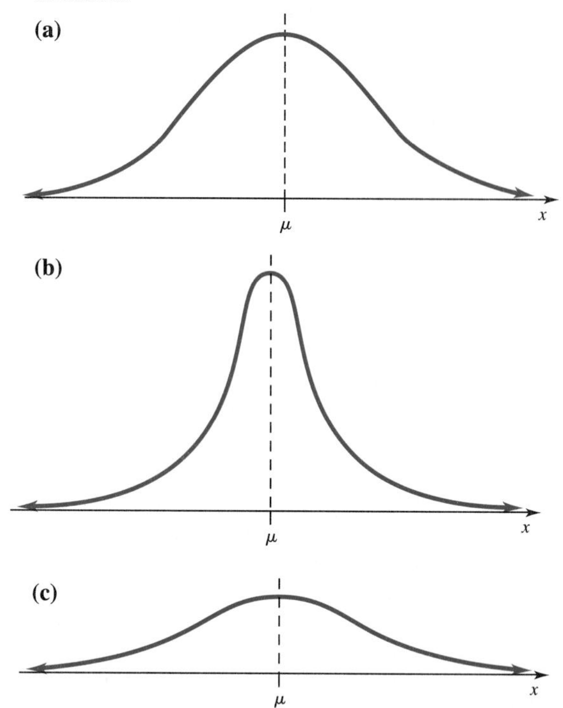

For the probability density functions defined in Exercises 14–17, find the expected value, the variance, and the standard deviation.

14. $f(x) = \frac{1}{5}$; $[4, 9]$

15. $f(x) = \frac{2}{9}(x - 2)$; $[2, 5]$

16. $f(x) = \frac{1}{7}\left(1 + \frac{2}{\sqrt{x}}\right)$; $[1, 4]$

17. $f(x) = 5x^{-6}$; $[1, \infty)$

18. The probability density function of a random variable is defined by $f(x) = 4x - 3x^2$ for $[0, 1]$.
Find each of the following for the distribution.

(a) The mean **(b)** The standard deviation

(c) The probability that the value of the random variable will be less than the mean

(d) The probability that the value of the random variable will be within one standard deviation of the mean

19. Find the median of the random variable of Exercise 18. (See Exercises 15–20 in Section 10.2.) Then find the probability that the value of the random variable will lie between the median and the mean of the distribution.

For Exercises 20–21, find **(a)** *the mean of the distribution,* **(b)** *the standard deviation of the distribution, and* **(c)** *the probability that the value of the random variable is within one standard deviation of the mean.*

20. $f(x) = \frac{5}{112}(1 - x^{-3/2})$ for $[1, 25]$

21. $f(x) = .01e^{-.01x}$ for $[0, \infty)$

In Exercises 22–27, find the percent of the area under a normal curve for each of the following.

22. The region to the right of $z = 1.53$

23. The region to the left of $z = -.49$

24. The region between $z = -1.47$ and $z = 1.03$

25. The region between $z = -.98$ and $z = -.15$

26. The region that is up to 2.5 standard deviations above the mean

27. The region that is up to 1.2 standard deviations below the mean

28. Find a z-score so that 21% of the area under the normal curve is to the left of z.

29. Find a z-score so that 52% of the area under the normal curve is to the right of z.

Applications

Business and Economics

30. Mutual Funds The price per share (in dollars) of a particular mutual fund is a random variable x with probability density function defined by

$$f(x) = \frac{3}{4}(x^2 - 16x + 65) \quad \text{for } [8, 9].$$

Find the probability that the price will be less than $8.50.

31. Machine Repairs The time (in years) until a certain machine requires repairs is a random variable t with probability density function defined by

$$f(t) = \frac{5}{112}(1 - t^{-3/2}) \quad \text{for } [1, 25].$$

Find the probability that no repairs are required in the first three years by finding the probability that a repair will be needed in years 4 through 25.

32. **Product Repairs** The number of repairs required by a new product each month is exponentially distributed, with an average of 8.
 (a) What is the probability density function for this distribution?
 (b) Find the expected number of repairs per month.
 (c) Find the standard deviation.
 (d) What is the probability that the number of repairs per month will be between 5 and 10?

33. **Retail Outlets** The number of new outlets for a clothing manufacturer is an exponential distribution with probability density function defined by

$$f(x) = \frac{1}{6}e^{-x/6} \quad \text{for } [0, \infty).$$

Find each of the following for this distribution.
 (a) The mean
 (b) The standard deviation
 (c) The probability that the number of new outlets will be greater than the mean

34. **Useful Life of an Appliance Part** The useful life of a certain appliance part (in hundreds of hours) is 46.2, with a standard deviation of 15.8. Find the probability that one such part would last for at least 6000 (60 hundred) hr. Assume a normal distribution.

Life Sciences

35. **Movement of a Released Animal** The distance (in meters) that a certain animal moves away from a release point is a random variable with probability density function defined by

$$f(x) = .01e^{-.01x} \quad \text{for } [0, \infty).$$

Find the probability that the animal will move no farther than 100 m away.

36. **Weight Gain of Rats** The weight gain (in grams) of rats fed a certain vitamin supplement is a continuous random variable with probability density function defined by

$$f(x) = \frac{8}{7}x^{-2} \quad \text{for } [1, 8].$$

 (a) Find the mean of the distribution.
 (b) Find the standard deviation of the distribution.
 (c) Find the probability that the value of the random variable is within one standard deviation of the mean.

37. **Body Temperature of a Bird** The body temperature (in degrees Celsius) of a particular species of bird is a continuous random variable with probability density function defined by

$$f(x) = \frac{6}{15{,}925}(x^2 + x) \quad \text{for } [20, 25].$$

 (a) What is the expected body temperature of this species?
 (b) Find the probability of a body temperature below the mean.

38. **Snowfall** The snowfall (in inches) in a certain area is uniformly distributed over the interval [2, 40].
 (a) What is the expected snowfall?
 (b) What is the probability of getting more than 20 inches of snow?

39. **Heart Muscle Tension** In a pilot study on tension of the heart muscle in dogs, the mean developed tension was 2.4 g, with a standard deviation of .4 g. Find the probability of a tension of less than 1.9 g. Assume a normal distribution.

40. **Average Birth Weight** The average birth weight of infants in the United States is 7.8 lb, with a standard deviation of 1.1 lb. Assuming a normal distribution, what is the probability that a newborn will weigh more than 9 lb?

General Interest

41. **State-Run Lotteries** The average state "take" on lotteries is 40%, with a standard deviation of 13%. Assuming a normal distribution, what is the probability that a state-run lottery will have a "take" of more than 50%?

Connections

The topics in this short chapter involved much of the material studied earlier in this book, including functions, domain and range, exponential functions, area and integration, improper integrals, integration by parts, and numerical integration. For each of the following special probability density functions, give
 (a) *the type of distribution;*
 (b) *the domain and range;*
 (c) *the graph;*
 (d) *the mean and standard deviation;*
 (e) $P(\mu - \sigma \le x \le \mu + \sigma)$.

42. $f(x) = .05$ for $[10, 30]$

43. $f(x) = e^{-x}$ for $[0, \infty)$

44. $f(x) = \dfrac{e^{-x^2}}{\sqrt{\pi}}$ for $(-\infty, \infty)$
(*Hint:* $\sigma = 1/\sqrt{2}$.)

Extended Application / A Crop-Planting Model*

Many firms in food processing, seed production, and similar industries face a problem every year deciding how many acres of land to plant in each of various crops. Demand for the crop is unknown, as is the actual yield per acre. In this application, we set up a mathematical model for determining the optimum number of acres to plant in a crop.

This model is designed to tell the company the number of acres of seed that it should plant. The model uses the following variables:

D = number of tons of seed demanded;

$f(D)$ = continuous probability density function for the quantity of seed demanded, D;

X = quantity of seed produced per acre of land;

Q = quantity of seed carried over in inventory from previous years;

S = selling price per ton of seed;

C_p = variable costs of production, marketing, etc., per ton of seed;

C_c = cost to carry over a ton of seed from previous years;

A = number of acres of land to be planted;

C_A = variable cost per acre of land contracted;

T = total number of tons of seed available for sale;

a = lower limit of the domain of $f(D)$; and

b = upper limit of the domain of $f(D)$.

To find the optimum number of acres to plant, it is necessary to calculate the expected value of the profit from the planting of A acres.

Based on the definition of the variables above, the total number of tons of seed that will be available for sale is given by the product of the number of acres planted, A, and the yield per acre, X, added to the carryover, Q. If T represents this total, then

$$T = AX + Q.$$

The variable here is A; we assume X and Q are known and fixed.

The expected profit can be broken down into several parts. The first portion comes from multiplying the profit per ton and the average number of tons demanded. The profit per ton is found by subtracting the variable cost per ton, C_p, from the selling price per ton, S:

$$\text{Profit per ton} = S - C_p.$$

The average number of tons demanded for our interval of concern is given by

$$\int_a^T D \cdot f(D)\, dD.$$

Thus, this portion of the expected profit is

$$(S - C_p) \cdot \int_a^T D \cdot f(D)\, dD. \qquad (1)$$

A second portion of expected profit is found by multiplying the profit per ton, $S - C_p$, the total number of tons available, T (recall that this is a variable), and the probability that T or more tons will be demanded by the marketplace.

$$(S - C_p)(T) \int_T^b f(D)\, dD \qquad (2)$$

If T is greater than D, there will be costs associated with carrying over the excess seed. The expected value of these costs is given by the product of the carrying cost per ton, C_c, and the number of tons to be carried over, or

$$-C_c \int_a^T (T - D) f(D)\, dD. \qquad (3)$$

The minus sign shows that these costs reduce profit. If $T < D$, this term would be omitted. Finally, the total cost of producing the seed is given by the product of the variable cost per acre and the number of acres:

$$-C_A \cdot A. \qquad (4)$$

The expected profit is the sum of the expressions in (1)–(4), or

$$\begin{aligned}\text{Expected profit} = {} & (S - C_p) \cdot \int_a^T D \cdot f(D)\, dD \\ & + (S - C_p)(T) \int_T^b f(D)\, dD \\ & - C_c \int_a^T (T - D) f(D)\, dD \\ & - C_A \cdot A. \qquad (5)\end{aligned}$$

*Based on work by David P. Rutten, Senior Mathematician, The Upjohn Company, Kalamazoo, Michigan. Reprinted with permission.

For a specific crop, once the values for all the variables but A (the number of acres to plant) are determined, the expected profit can be expressed as a function of A. Then the number of acres to plant for maximum expected profit can be found by taking the derivative, setting it equal to 0, and solving for A. We leave these steps for the exercises.

Exercises

Suppose that we have the following information for a particular crop.

Probability density function $= f(D) = \dfrac{1}{1000}$

for $500 \le D \le 1500$ tons

$a = 500$

$b = 1500$

Selling price $= S =$ \$10,000 per ton

Variable cost $= C_p =$ \$5000 per ton

Carrying cost $= C_C =$ \$3000 per ton

Variable cost per acre $= C_A =$ \$100

Inventory carryover $= Q = 200$ tons

Yield per acre $= X = .1$ ton

$T = AX + Q = .1A + 200$

1. Find the expected profit as a function of A.
2. Find the number of acres to plant for maximum expected profit.

Applications at a glance . . .

Revenue from Seasonal Merchandise . . . fluctuation of air pollution . . . best depth to build a wine cellar . . .

CHAPTER 11

THE TRIGONOMETRIC FUNCTIONS

(See page 566.)

Throughout this book we have discussed many different types of functions, including linear, quadratic, exponential, and logarithmic functions. In this chapter we introduce the *trigonometric functions*, which differ in a fundamental way from those previously studied: the trigonometric functions describe periodic or repetitive relationships.

An example of a periodic relationship is given by an electrocardiogram (EKG), a graph of a human heartbeat. The EKG in Figure 1 shows electrical impulses from a heart.* Each small square represents .04 second, and each large square represents .2 second. How often does this heart beat?

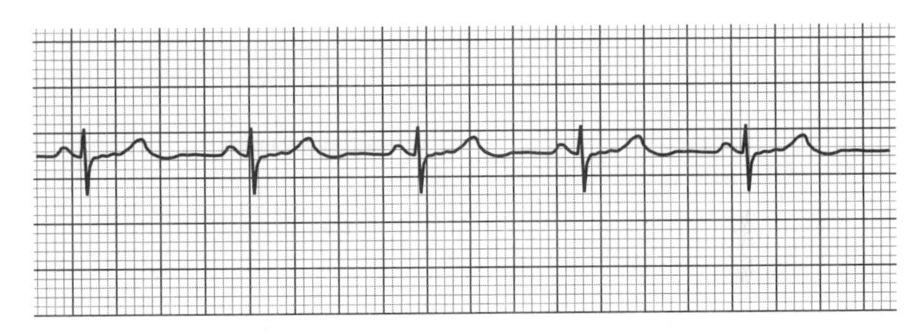

FIGURE 1

Trigonometric functions describe many natural phenomena and are important in the study of optics, heat, electronics, acoustics, and seismology. Also, many algebraic functions have integrals involving trigonometric functions.

11.1 DEFINITIONS OF THE TRIGONOMETRIC FUNCTIONS

How far from a camera should an object be to put it in focus?

In Exercise 73 in this section, we will use trigonometry to answer this question.

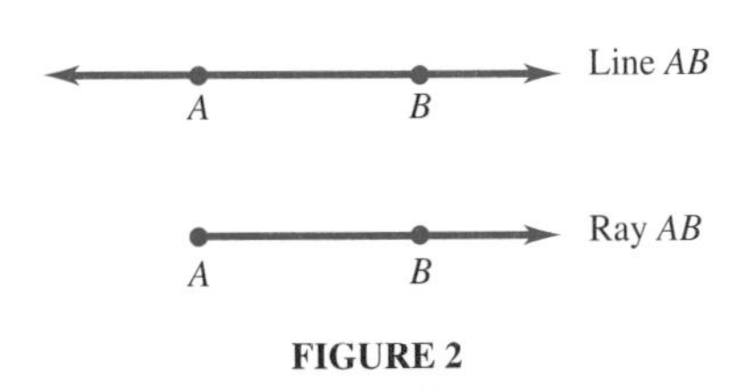

FIGURE 2

The angle is one of the basic concepts of trigonometry. The definition of an angle depends on that of a ray: a **ray** is the portion of a line that starts at a given point and continues indefinitely in one direction. For example, Figure 2 shows a line through the two points A and B. The portion of the line AB that starts at A and continues through and past B is called ray AB. Point A is the **endpoint** of the ray.

An **angle** is formed by rotating a ray about its endpoint. The initial position of the ray is called the **initial side** of the angle, and the endpoint of the ray is

*EKG courtesy of Nancy Schiller, Phoenix, Arizona.

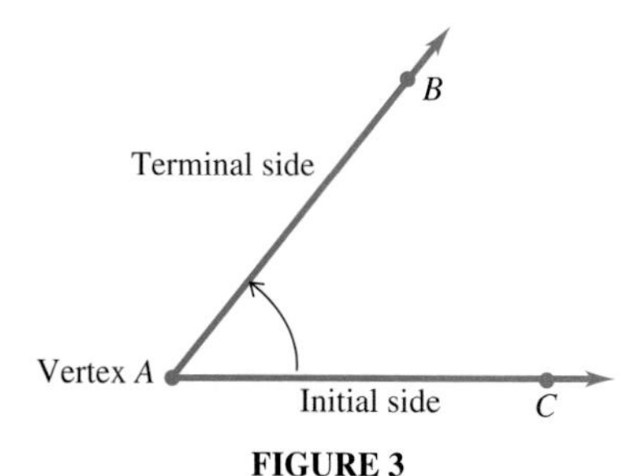

FIGURE 3

called the **vertex** of the angle. The location of the ray at the end of its rotation is called the **terminal side** of the angle. Figure 3 shows the initial and terminal sides of an angle with vertex A.

An angle can be named by its vertex. For example, the angle in Figure 3 can be called angle A. An angle also can be named by using three letters. For example, the angle in Figure 3 could be named angle BAC or angle CAB. (The vertex is always the middle letter.)

An angle is in **standard position** if its vertex is at the origin of a coordinate system and if its initial side is along the positive x-axis. The angles in Figures 4 and 5 are in standard position. An angle in standard position is said to be in the quadrant of its terminal side. For example, the angle in Figure 4(a) is in quadrant I, while the angle in Figure 4(b) is in quadrant II.

Notice that the angles in Figures 3, 4, and 5 are measured counterclockwise from the positive x-axis. This is true for any positive angle. A negative angle is measured clockwise from the positive x-axis, as we shall see in Example 5.

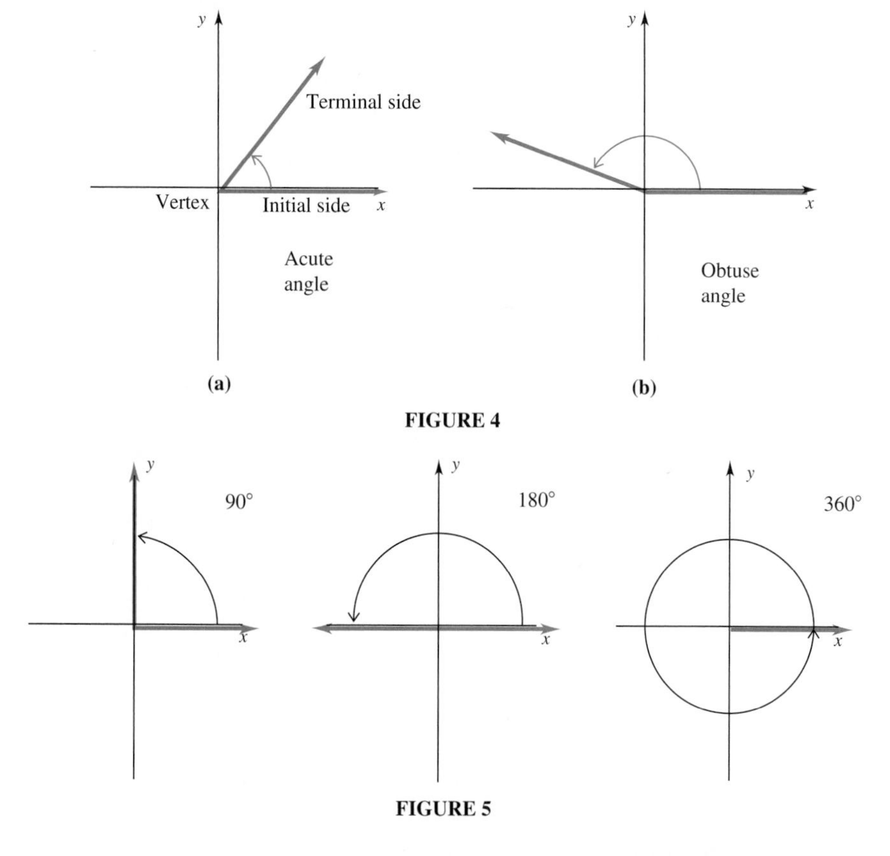

FIGURE 5

Degree Measure The sizes of angles are often indicated in *degrees*. Degree measure has remained unchanged since the Babylonians developed it 4000 years ago. In degree measure, 360 degrees represents a complete rotation of a ray. **One degree**, written 1°, is $1/360$ of a rotation. Also, 90° is $90/360$ or $1/4$ of a rotation, and 180° is $180/360$ or $1/2$ of a rotation. See Figure 5.

There are various theories about why the Babylonians chose to divide a complete revolution into 360 parts. One theory suggests that 360 was chosen because there are approximately 360 days in one calendar year. Another theory says that because the Babylonians used a sexagesimal (base sixty) system of numeration (as opposed to our familiar decimal, or base ten, system), they used 360 since it is a multiple of 60. In any event, the practice has continued for thousands of years.

An angle having a degree measure between 0° and 90° is called an **acute angle.** An angle of 90° is a **right angle.** An angle having measure more than 90° and less than 180° is an **obtuse angle,** while an angle of 180° is a **straight angle.** See Figures 4 and 5.

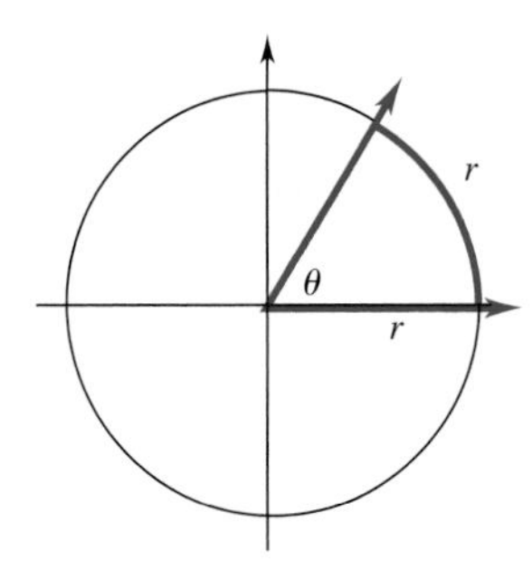

FIGURE 6

Radian Measure While degree measure is best for some applications, this system of angle measurement is not the best for calculus. To keep the formulas for derivatives as simple as possible, it is better to measure angles with *radian measure*. To see how this alternative system for measuring angles is obtained, look at angle θ (the Greek letter *theta*) in Figure 6. The angle θ is in standard position; Figure 6 also shows a circle of radius r, centered at the origin.

The vertex of θ is at the center of the circle in Figure 6. Angle θ cuts a piece of the circle called an **arc.** Because the length of this arc is equal to the radius of the circle, angle θ has a measure of 1 radian. **One radian** is the measure of an angle that has its vertex at the center of a circle and that cuts an arc on the circle equal in length to the radius of the circle. The term *radian* comes from the phrase *radial angle*. Two nineteenth-century scientists, mathematician Thomas Muir and physicist James Thomson, are credited with the development of the radian as a unit of angular measure.

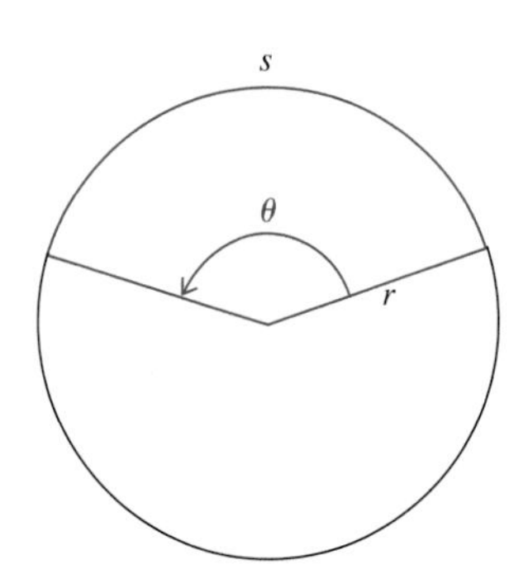

FIGURE 7

Generalizing, the radian measure of a central angle θ (see Figure 7) cutting off an arc of length s in a circle of radius r is defined as follows:

$$\text{Radian measure of } \theta = \frac{\text{Length of arc}}{\text{Radius}} = \frac{s}{r}.$$

In this definition, the units of measure of the length of the arc and the radius cancel, leaving a quotient without units. For this reason, a real number can be thought of as the radian measure of some angle.

Since the circumference of a circle is 2π times the radius of the circle, the radius could be marked off 2π times around the circle. Therefore, an angle of 360°—that is, a complete circle—cuts off an arc equal in length to 2π times the radius of the circle, or

$$360° = 2\pi \text{ radians}.$$

This result gives a basis for comparing degree and radian measure.

Since an angle of 180° is half the size of an angle of 360°, an angle of 180° would have half the radian measure of an angle of 360°, or

$$180° = \frac{1}{2}(2\pi) \text{ radians} = \pi \text{ radians}.$$

$$\mathbf{180° = \pi \text{ radians}}$$

Since π radians $= 180°$, divide both sides by π to find the degree measure of 1 radian.

1 RADIAN

$$\mathbf{1 \text{ radian} = \left(\frac{180°}{\pi}\right)}$$

This quotient is approximately 57.29578°. Since $180° = \pi$ radians, we can find the radian measure of 1 degree by dividing by 180° on both sides.

1 DEGREE

$$\mathbf{1° = \frac{\pi}{180} \text{ radians}}$$

One degree is approximately equal to .0174533 radians.

Convert degree measures to radians and radian measures to degrees.

(a) 45°

Since $1° = \pi/180$ radians,

$$45° = 45\left(\frac{\pi}{180}\right) \text{ radians} = \frac{45\pi}{180} \text{ radians} = \frac{\pi}{4} \text{ radians.}$$

The word *radian* is often omitted, so the answer could be written as just $45° = \pi/4$.

(b) $240° = 240\left(\frac{\pi}{180}\right) = \frac{4\pi}{3}$

(c) $\frac{9\pi}{4}$

Since 1 radian $= 180°/\pi$,

$$\frac{9\pi}{4} \text{ radians} = \frac{9\pi}{4}\left(\frac{180°}{\pi}\right) = 405°.$$

(d) $\frac{11\pi}{3}$ radians $= \frac{11\pi}{3}\left(\frac{180°}{\pi}\right) = 660°$ ◀

The following chart shows the equivalent radian and degree measure for several angles that we will encounter frequently.

Degrees	0°	30°	45°	60°	90°	180°	270°	360°
Radians	0	$\pi/6$	$\pi/4$	$\pi/3$	$\pi/2$	π	$3\pi/2$	2π

The Trigonometric Functions The origins of trigonometry, while somewhat obscure, are tied closely to the development of another science: astronomy. The early foundations of trigonometry consisted of tables of shadow lengths corresponding to various times of day, and lengths of chords of arbitrary circular arcs.* Based upon the work of the Greek astronomer Hipparchus, the *Almagest* of Claudius Ptolemy was a source of information for astronomers such as Copernicus and Kepler. This work consisted of tables of chords and identities that could be used for computing chord lengths. With the invention of the calculus, trigonometry became more than simply a tool for calculations. The work of Newton, Johann Bernoulli (1667–1748), and Euler paved the way for new applications by expanding the study of trigonometry into a theory of functions of real numbers with no dependence on angles or arc lengths.

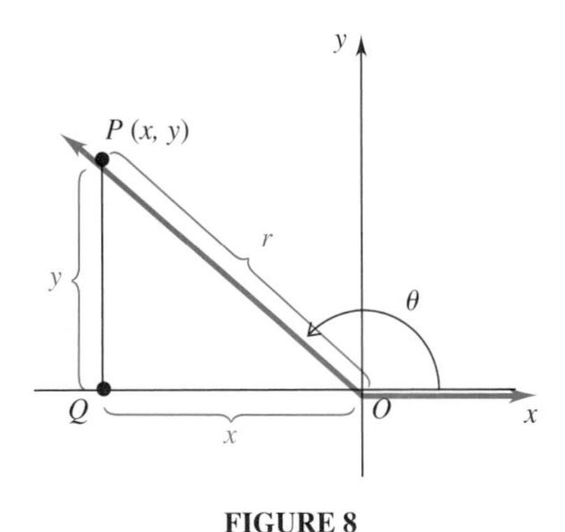

FIGURE 8

To define the six basic trigonometric functions, we start with an angle θ in standard position, as shown in Figure 8. Next, we choose an arbitrary point P having coordinates (x, y), located on the terminal side of angle θ. (The point P must not be the vertex of θ.)

Drawing a line segment perpendicular to the x-axis from P to point Q sets up a right triangle having vertices at O (the origin), P, and Q. The distance from P to O is r. Since the distance from P to O can never be negative, $r > 0$. The six **trigonometric functions** of angle θ are defined as follows.

TRIGONOMETRIC FUNCTIONS

Let (x, y) be a point other than the origin on the terminal side of an angle θ in standard position. Let r be the distance from the origin to (x, y). Then

$$\textbf{sine } \theta = \sin\theta = \frac{y}{r} \qquad \textbf{cosecant } \theta = \csc\theta = \frac{r}{y}$$

$$\textbf{cosine } \theta = \cos\theta = \frac{x}{r} \qquad \textbf{secant } \theta = \sec\theta = \frac{r}{x}$$

$$\textbf{tangent } \theta = \tan\theta = \frac{y}{x} \qquad \textbf{cotangent } \theta = \cot\theta = \frac{x}{y}.$$

*A *chord* is a line segment in a circle with endpoints on the circle itself. (The diameter of a circle is the length of a chord that passes through the center of the circle.)

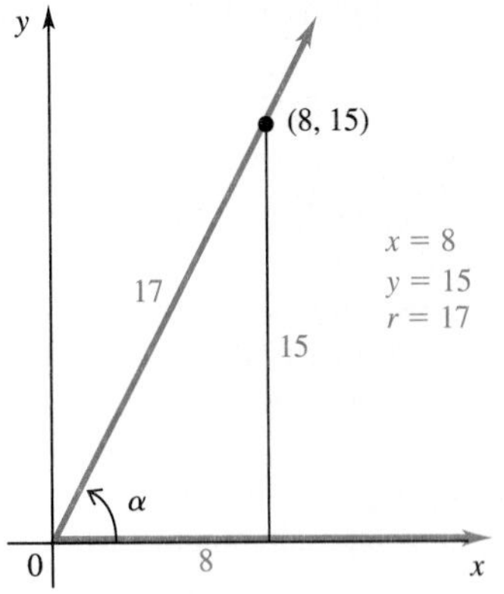

EXAMPLE 2 The terminal side of an angle α (the Greek letter alpha) goes through the point (8, 15). Find the values of the six trigonometric functions of angle α.

Figure 9 shows angle α and the triangle formed by dropping a perpendicular from the point (8, 15). To reach the point (8, 15), begin at the origin and go 8 units to the right and 15 units up, so that $x = 8$ and $y = 15$. To find the radius r, use the Pythagorean theorem*: in a triangle with a right angle, if the longest side of the triangle is r and the shorter sides are x and y, then

$$r^2 = x^2 + y^2,$$

or

$$r = \sqrt{x^2 + y^2}.$$

(Recall that $\sqrt{a}$ represents the *positive* square root of a.)

Substituting the known values $x = 8$ and $y = 15$ in the equation gives

$$r = \sqrt{8^2 + 15^2} = \sqrt{64 + 225} = \sqrt{289} = 17.$$

FIGURE 9

We have $x = 8$, $y = 15$, and $r = 17$. The values of the six trigonometric functions of angle α are found by using the definitions given above.

$$\sin \alpha = \frac{y}{r} = \frac{15}{17} \qquad \tan \alpha = \frac{y}{x} = \frac{15}{8} \qquad \sec \alpha = \frac{r}{x} = \frac{17}{8}$$

$$\cos \alpha = \frac{x}{r} = \frac{8}{17} \qquad \cot \alpha = \frac{x}{y} = \qquad \csc \alpha = \frac{r}{y} = \frac{17}{15}$$

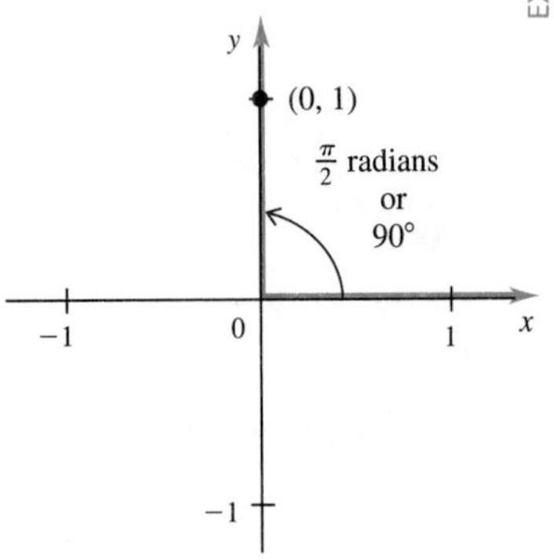

EXAMPLE 3 Find the values of the six trigonometric functions for an angle of $\pi/2$.

Select any point on the terminal side of an angle of measure $\pi/2$ radians (or 90°). See Figure 10. Selecting the point (0, 1) gives $x = 0$ and $y = 1$. Check that $r = 1$ also. Then

$$\sin \frac{\pi}{2} = \frac{1}{1} = 1 \qquad \tan \frac{\pi}{2} = \frac{1}{0} \text{ (undefined)} \qquad \sec \frac{\pi}{2} = \frac{1}{0} \text{ (undefined)}$$

$$\cos \frac{\pi}{2} = \frac{0}{1} = 0 \qquad \cot \frac{\pi}{2} = \frac{0}{1} = 0 \qquad \csc \frac{\pi}{2} = \frac{1}{1} = 1.$$

FIGURE 10

Methods similar to the procedure in Example 3 can be used to find the values of the six trigonometric functions for the angles with measures 0, π, and $3\pi/2$. These results are summarized in the table on the facing page. The table shows that the results for 2π are the same as those for 0.

*Although one of the most famous theorems in mathematics is named after the Greek mathematician Pythagoras, there is much evidence that the relationship between the sides of a right triangle was known long before his time. The Babylonian mathematical tablet identified as *Plimpton 322* has been determined to be essentially a list of *Pythagorean triples*—sets of three numbers a, b, and c that satisfy the equation $a^2 + b^2 = c^2$.

θ (in radians)	θ (in degrees)	$\sin\theta$	$\cos\theta$	$\tan\theta$	$\cot\theta$	$\sec\theta$	$\csc\theta$
0	0°	0	1	0	Undefined	1	Undefined
$\pi/2$	90°	1	0	Undefined	0	Undefined	1
π	180°	0	-1	0	Undefined	-1	Undefined
$3\pi/2$	270°	-1	0	Undefined	0	Undefined	-1
2π	360°	0	1	0	Undefined	1	Undefined

Special Angles The values of the trigonometric functions for most angles must be found by using a calculator with trigonometric keys. For a few angles called *special angles*, however, the function values can be found exactly. These values are found with the aid of two kinds of right triangles that will be described in this section.

30°–60°–90° TRIANGLE

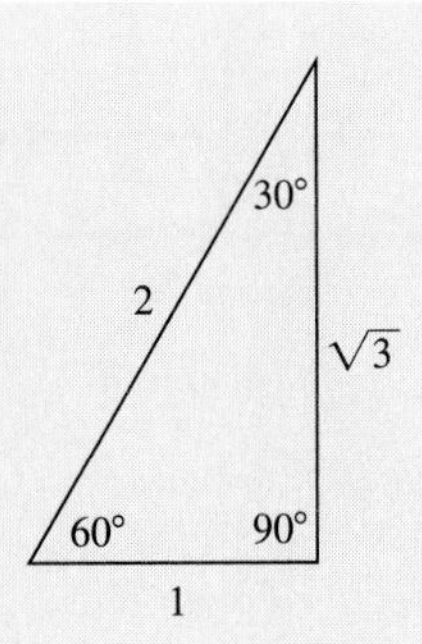

In a right triangle having angles of 30°, 60°, and 90°, the hypotenuse is always twice as long as the shortest side, and the middle side has a length that is $\sqrt{3}$ times as long as that of the shortest side. Also, the shortest side is opposite the 30° angle.

EXAMPLE 4 Find the values of the trigonometric functions for an angle of $\pi/6$ radians.

Since $\pi/6$ radians $= 30°$, find the necessary values by placing a 30° angle in standard position, as in Figure 11. Choose a point P on the terminal side of the angle so that $r = 2$. From the description of 30°–60°–90° triangles, P will have coordinates $(\sqrt{3}, 1)$, with $x = \sqrt{3}$, $y = 1$, and $r = 2$. Using the definitions of the trigonometric functions gives the following results.

$$\sin\frac{\pi}{6} = \frac{1}{2} \qquad \tan\frac{\pi}{6} = \frac{1}{\sqrt{3}} = \frac{\sqrt{3}}{3} \qquad \sec\frac{\pi}{6} = \frac{2}{\sqrt{3}} = \frac{2\sqrt{3}}{3}$$

$$\cos\frac{\pi}{6} = \frac{\sqrt{3}}{2} \qquad \cot\frac{\pi}{6} = \sqrt{3} \qquad \csc\frac{\pi}{6} = 2$$

FIGURE 11

We can find the trigonometric function values for 45° angles by using the properties of a right triangle having two sides of equal length.

45°–45°–90° TRIANGLE

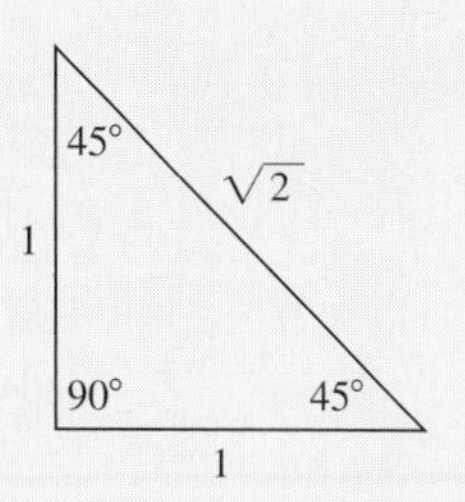

In a 45°–45° right triangle, the hypotenuse has a length that is $\sqrt{2}$ times as long as the length of either of the shorter (equal) sides.

EXAMPLE 5

Find the trigonometric function values for an angle of $-\pi/4$.

Place an angle of $-\pi/4$ radians, or $-45°$, in standard position, as in Figure 12. Choose point P on the terminal side so that $r = \sqrt{2}$. By the description of 45°–45°–90° triangles, P has coordinates $(1, -1)$, with $x = 1$, $y = -1$, and $r = \sqrt{2}$.

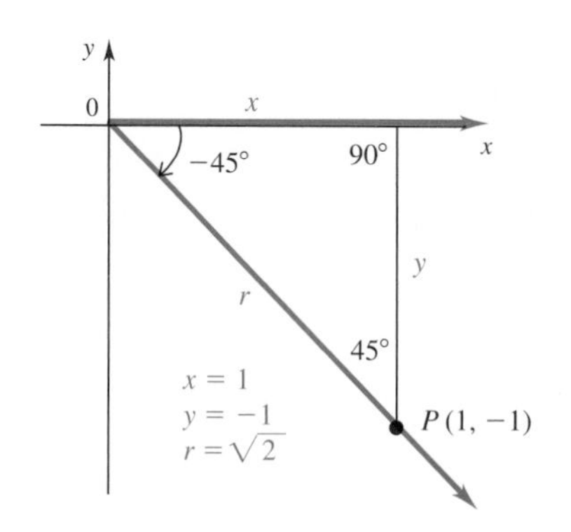

FIGURE 12

$$\sin -\frac{\pi}{4} = -\frac{1}{\sqrt{2}} = -\frac{\sqrt{2}}{2} \qquad \tan -\frac{\pi}{4} = -1 \qquad \sec -\frac{\pi}{4} = \sqrt{2}$$

$$\cos -\frac{\pi}{4} = \frac{1}{\sqrt{2}} = \frac{\sqrt{2}}{2} \qquad \cot -\frac{\pi}{4} = -1 \qquad \csc -\frac{\pi}{4} = -\sqrt{2}$$

For angles other than the special angles of 30°, 45°, 60°, and their multiples, a calculator should be used. Many calculators have keys labeled sin, cos, and tan. To get the other trigonometric functions, use the fact that $\sec x = 1/\cos x$, $\csc x = 1/\sin x$, and $\cot x = 1/\tan x$. (The x^{-1} key is also useful here.)

Caution Whenever you use a calculator to compute trigonometric functions, check whether the calculator is set on radians or degrees. If you want one and your calculator is set on the other, you will get erroneous answers. Most calculators have a way of switching back and forth; check the calculator manual for details.

EXAMPLE 6 Use a calculator to verify the following results.

(a) $\sin 10° = .1736$

(b) $\cos 48° = .6691$

(c) $\tan 82° = 7.1154$

(d) $\sin .2618 = .2588$

(e) $\cot 1.2043 = 1/\tan 1.2043 = 1/2.6053 = .3838$

(f) $\sec .7679 = 1/\cos .7679 = 1/.71937 = 1.3901$ ◀

Graphs of the Trigonometric Functions Because of the way the trigonometric functions are defined (using a circle), the same function values will be obtained for any two angles that differ by 2π radians (or 360°). For example,

$$\sin(x + 2\pi) = \sin x \quad \text{and} \quad \cos(x + 2\pi) = \cos x$$

for any value of x. Because of this property, the trigonometric functions are *periodic functions.*

PERIODIC FUNCTION

A function $y = f(x)$ is **periodic** if there exists a positive real number a such that

$$f(x) = f(x + a)$$

for all values of x in the domain of the function. The smallest possible value of a is called the **period** of the function.

When graphing, it is customary to use x (rather than θ) for the domain elements, as we did with earlier functions, and to write $y = \sin x$ instead of $y = \sin \theta$. Because sine is periodic, with period 2π, we can graph $y = \sin x$ by finding y for values of x between 0 and 2π. This portion of the graph can then be repeated as many times as necessary.

Think of a point moving counterclockwise around a circle, tracing out an arc for angle x. The value of $\sin x$ gradually increases from 0 to 1 as x increases from 0 to $\pi/2$. The values of $\sin x$ then decrease back to 0 as x goes from $\pi/2$ to π. For $\pi < x < 2\pi$, $\sin x$ is negative. A few typical values from these intervals are given in the following table, where decimals have been rounded to the nearest tenth.

x	0	$\pi/4$	$\pi/2$	$3\pi/4$	π	$5\pi/4$	$3\pi/2$	$7\pi/4$	2π
$\sin x$	0	.7	1	.7	0	−.7	−1	−.7	0

Plotting the points from the table of values and connecting them with a smooth line gives the solid portion of the graph in Figure 13 on the next page. Since $y = \sin x$ is periodic, the graph continues in both directions indefinitely, as suggested by the dashed lines.

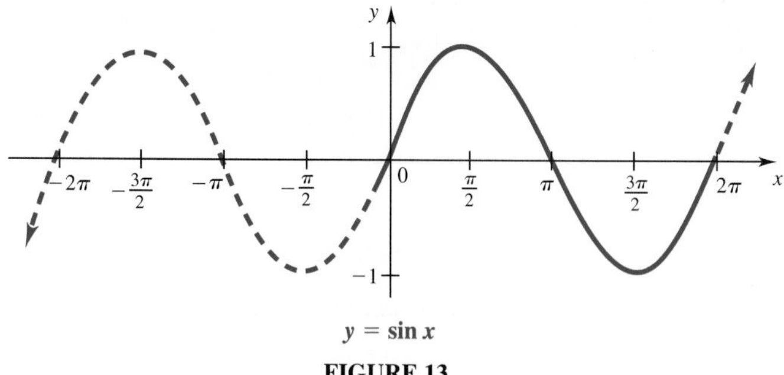

$y = \sin x$

FIGURE 13

The graph of $y = \cos x$ in Figure 14 can be found in much the same way. Again, the period is 2π. (These graphs could also be drawn using a graphing calculator or a computer.)

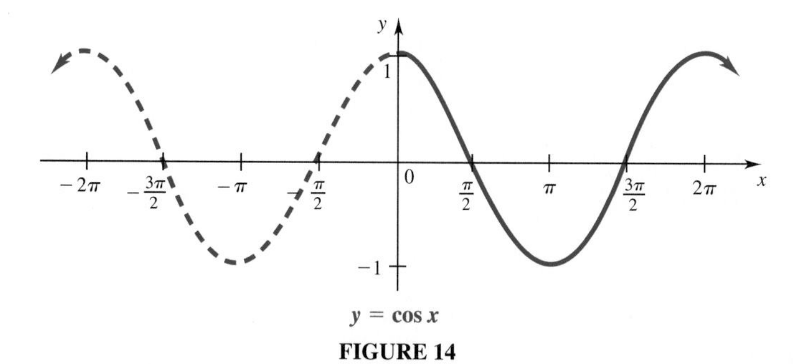

$y = \cos x$

FIGURE 14

Finally, Figure 15 shows the graph of $y = \tan x$. Since $\tan x$ is undefined (because of zero denominators) for $x = \pi/2, 3\pi/2, -\pi/2$, and so on, the graph has vertical asymptotes at these values. As the graph suggests, the tangent function is periodic, with a period of π.

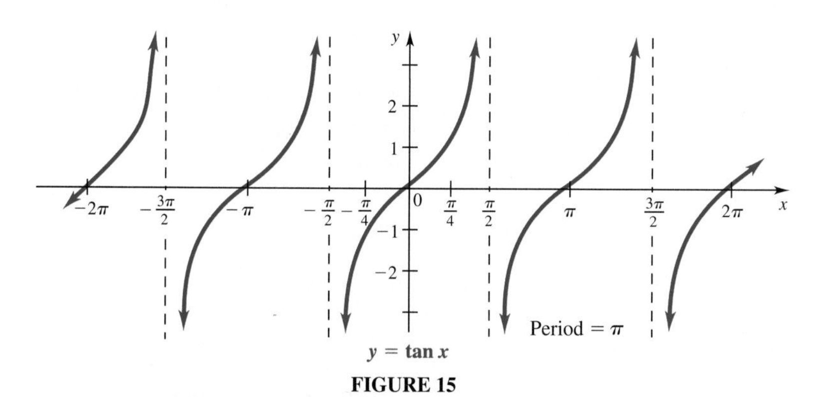

$y = \tan x$

FIGURE 15

The graphs of the other trigonometric functions are not as useful as these three, so they are not given here.

11.1 Exercises

Convert the following degree measures to radians. Leave answers as multiples of π.

1. $60°$ **2.** $90°$ **3.** $150°$ **4.** $135°$

5. $210°$ **6.** $300°$ **7.** $390°$ **8.** $480°$

Convert the following radian measures to degrees.

9. $\frac{7\pi}{4}$ **10.** $\frac{2\pi}{3}$ **11.** $\frac{11\pi}{6}$ **12.** $\frac{-\pi}{4}$

13. $\frac{8\pi}{5}$ **14.** $\frac{7\pi}{10}$ **15.** $\frac{4\pi}{15}$ **16.** 5π

Find the values of the six trigonometric functions for the angles in standard position having the points in Exercises 17–20 on their terminal sides.

17. $(-3, 4)$ **18.** $(-12, -5)$ **19.** $(6, 8)$ **20.** $(-7, 24)$

In quadrant I, x, y, and r are all positive, so that all six trigonometric functions have positive values. In quadrant II, x is negative and y is positive (r is always positive). Thus, in quadrant II, sine is positive, cosine is negative, and so on. For Exercises 21–24, complete the following table of values for the signs of the trigonometric functions.

	Quadrant of θ	$\sin\theta$	$\cos\theta$	$\tan\theta$	$\cot\theta$	$\sec\theta$	$\csc\theta$
21.	I	+					
22.	II						
23.	III						
24.	IV						

For Exercises 25–32, complete the following chart. Do not use a calculator.

	θ	$\sin\theta$	$\cos\theta$	$\tan\theta$	$\cot\theta$	$\sec\theta$	$\csc\theta$
25.	$30°$	$1/2$	$\sqrt{3}/2$			$2\sqrt{3}/3$	
26.	$45°$			1	1		
27.	$60°$		$1/2$	$\sqrt{3}$		2	
28.	$120°$	$\sqrt{3}/2$		$-\sqrt{3}$			$2\sqrt{3}/3$
29.	$135°$	$\sqrt{2}/2$	$-\sqrt{2}/2$			$-\sqrt{2}$	$\sqrt{2}$
30.	$150°$		$-\sqrt{3}/2$	$-\sqrt{3}/3$			2
31.	$210°$	$-1/2$		$\sqrt{3}/3$	$\sqrt{3}$		-2
32.	$240°$	$-\sqrt{3}/2$	$-1/2$			-2	$-2\sqrt{3}/3$

Find the following function values without using a calculator.

33. $\sin\frac{\pi}{3}$ **34.** $\cos\frac{\pi}{6}$ **35.** $\tan\frac{\pi}{4}$ **36.** $\cot\frac{\pi}{3}$

37. $\sec\frac{\pi}{6}$ **38.** $\sin\frac{\pi}{2}$ **39.** $\cos 3\pi$ **40.** $\sec\pi$

41. $\sin\frac{4\pi}{3}$ **42.** $\tan\frac{3\pi}{4}$ **43.** $\csc\frac{5\pi}{4}$ **44.** $\cos 5\pi$

45. $\tan -\frac{\pi}{3}$ **46.** $\cot -\frac{2\pi}{3}$ **47.** $\sin -\frac{7\pi}{6}$ **48.** $\cos -\frac{\pi}{6}$

Use a calculator to find the following function values.

49. sin 39° **50.** cos 58° **51.** tan 82° **52.** tan 54°

53. sin .4014 **54.** tan 1.0123 **55.** cos 1.4137 **56.** sin 1.5359

Graph each function defined as follows over a two-period interval.

57. $y = 2 \sin x$ **58.** $y = 2 \cos x$ **59.** $y = \cos 2x$ **60.** $y = \sin 2x$

61. $y = -\sin x$ **62.** $y = -\frac{1}{2} \cos x$ **63.** $y = \frac{1}{2} \tan x$ **64.** $y = -3 \tan x$

Applications

Business and Economics

65. Sales Sales of snowblowers are seasonal. Suppose the sales of snowblowers in one region of the country are approximated by

$$S(t) = 500 + 500 \cos \frac{\pi}{6} t,$$

where t is time in months, with $t = 0$ corresponding to November. Find the sales for each of the following months.

(a) November **(b)** January **(c)** February

(d) May **(e)** August

(f) Graph $y = S(t)$.

Physical Sciences

Light Rays *When a light ray travels from one medium, such as air, to another medium, such as water or glass, the speed of the light changes, and the direction that the ray is traveling changes. (This is why a fish under water is in a different position from the place at which it appears to be.) These changes are given by Snell's law,*

$$\frac{c_1}{c_2} = \frac{\sin \theta_1}{\sin \theta_2},$$

where c_1 is the speed in the first medium, c_2 is the speed in the second medium, and θ_1 and θ_2 are the angles shown in the figure.

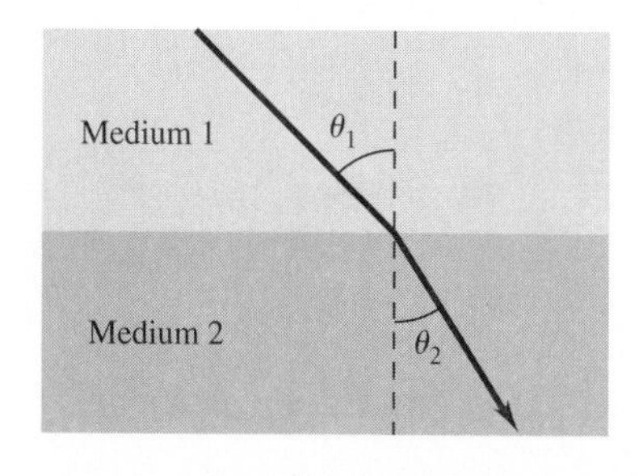
Medium 1
θ_1
Medium 2
θ_2
If this medium is less dense, light travels at a faster speed, c_1.
If this medium is more dense, light travels at a slower speed, c_2.

In Exercises 66 and 67, assume that $c_1 = 3 \times 10^8$ meters per second, and find the speed of light in the second medium.

66. $\theta_1 = 39°, \theta_2 = 28°$ **67.** $\theta_1 = 46°, \theta_2 = 31°$

Sound *Pure sounds produce single sine waves on an oscilloscope. Find the period of each sine wave in the photographs in Exercises 68 and 69. On the vertical scale each square represents .5, and on the horizontal scale each square represents 30°.*

68.

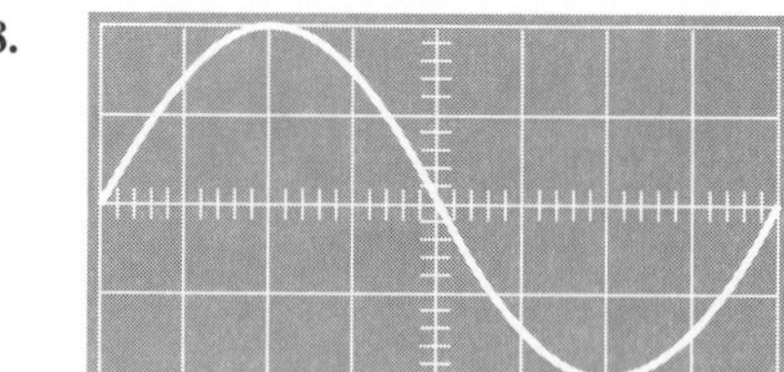

69.

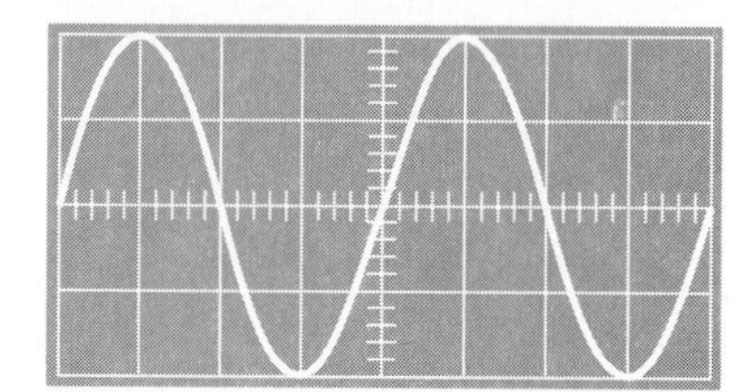

70. Air Pollution The amount of pollution in the air fluctuates with the seasons. It is lower after heavy spring rains and higher after periods of little rain. In addition to this seasonal fluctuation, the long-term trend is upward. An idealized graph of this situation is shown in the figure at the top of the facing page. Trigonometric functions can be used to describe the fluctuating part of the pollution levels. Powers of the number e can be used to show the long-term growth. In fact, the pollution level in a certain area might be given by

$$P(t) = 7(1 - \cos 2\pi t)(t + 10) + 100e^{.2t},$$

where t is time in years, with $t = 0$ representing January 1 of the base year. Thus, July 1 of the same year would be represented by $t = .5$, while October 1 of the following year would be represented by $t = 1.75$. Find the pollution levels on the following dates.

(a) January 1, base year **(b)** July 1, base year

(c) January 1, following year

(d) July 1, following year

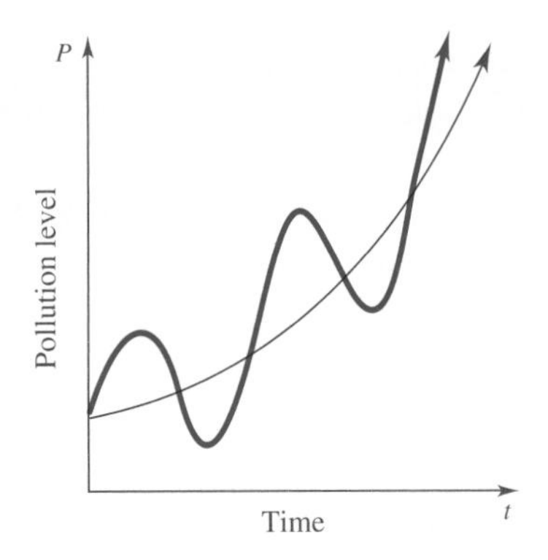

71. Temperature The maximum afternoon temperature (in degrees Fahrenheit) in a given city is approximated by

$$T(x) = 60 - 30 \cos (x/2),$$

where x represents the month, with $x = 0$ representing January, $x = 1$ representing February, and so on. Use a calculator with trigonometric function keys to find the maximum afternoon temperature for each of the following months.

(a) January **(b)** March **(c)** October

(d) June **(e)** August

72. Temperature A mathematical model for the temperature in Fairbanks is

$$T(x) = 37 \sin \left[\frac{2\pi}{365} (x - 101) \right] + 25,*$$

where $T(x)$ is the temperature (in degrees Celsius) on day x, with $x = 0$ corresponding to January 1 and $x = 365$ corresponding to December 31.

Use a calculator with trigonometric function keys to estimate the temperature on each of the following days.

(a) March 1 (Day 60) **(b)** April 1 (Day 91)

(c) Day 101 **(d)** Day 150

(e) Find maximum and minimum values of T.

*From Barbara Lando and Clifton Lando, ''Is the Graph of Temperature Variation a Sine Curve?'' *The Mathematics Teacher*, vol. 70, September 1977, pp. 534–37.

73. Cameras In the Kodak Customer Service Pamphlet AA-26, entitled *Optical Formulas and Their Applications*, the near and far limits of the depth of field (how close or how far away an object can be placed and still be in focus) are given by

$$w_1 = \frac{u^2 (\tan \theta)}{L + u (\tan \theta)} \quad \text{and} \quad w_2 = \frac{u^2 (\tan \theta)}{L - u (\tan \theta)}.$$

In these equations, θ represents the angle between the lens and the ''circle of confusion,'' which is the circular image on the film of a point that is not exactly in focus. (The pamphlet suggests letting $\theta = \frac{1}{30}°$.) L is the diameter of the lens opening, which is found by dividing the focal length by the f-stop. (This is camera jargon you need not worry about here.) For this problem, let the focal length be 50 millimeters, or .05 meters; if the lens is set at f/8, then $L = .05/8 = .00625$ meters. Finally, u is the distance to the object being photographed. Find the near and far limits of the depth of field when the object being photographed is 6 meters from the camera.

74. Measurement A surveyor standing 48 m from the base of a building measures the angle to the top of the building and finds it to be 37.4°. (See the figure.) Use trigonometry to find the height of the building.

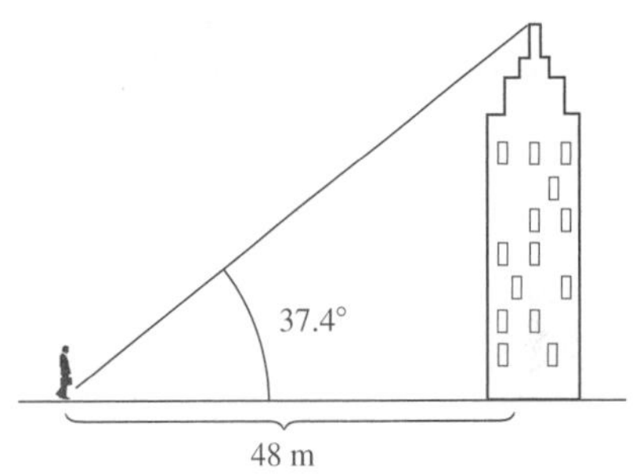

75. Measurement Elizabeth Linton stands on a cliff at the edge of a canyon. On the opposite side of the canyon is another cliff equal in height to the one she is on. (See the figure.) By dropping a rock and timing its fall, she determines that it is 80 ft to the bottom of the canyon. She also determines that the angle to the base of the opposite cliff is 24°. How far is it to the opposite side of the canyon?

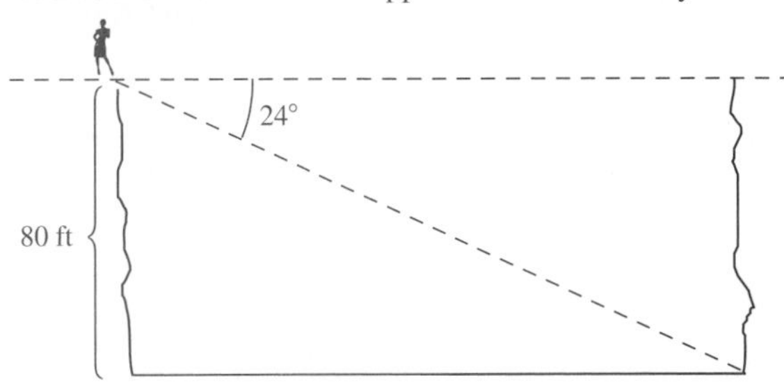

Computer/Graphing Calculator

76. (a) Air Pollution Using a computer or a graphing calculator, sketch the function for air pollution given in Exercise 70 over the interval $0 \le t \le 6$.

(b) Change the function in Exercise 70 so the amplitude of the oscillations are half of what they were in your graph from part (a). Sketch the graph on a computer or a graphing calculator to verify your answer.

11.2 DERIVATIVES OF TRIGONOMETRIC FUNCTIONS

How long must a ladder be to reach over a 9-foot-high fence and lean against a nearby building?

In Exercise 34 in this section, we will use trigonometry to answer this question. First we derive formulas for the derivatives of some of the trigonometric functions. All these derivatives can be found from the formula for the derivative of $y = \sin x$.

We will need to use the following identities, which are listed without proof, to find the derivatives of the trigonometric functions.

BASIC IDENTITIES

$$\sin^2 x + \cos^2 x = 1$$

$$\tan x = \frac{\sin x}{\cos x}$$

$$\sin(x + y) = \sin x \cos y + \cos x \sin y$$

$$\sin(x - y) = \sin x \cos y - \cos x \sin y$$

$$\cos(x + y) = \cos x \cos y - \sin x \sin y$$

$$\cos(x - y) = \cos x \cos y + \sin x \sin y$$

The derivative of $y = \sin x$ also depends on the value of

$$\lim_{x \to 0} \frac{\sin x}{x}.$$

To estimate this limit, find the quotient $(\sin x)/x$ for various values of x close to 0. (Be sure that your calculator is set for radian measure.)

x	.1	.01	.001	$-.001$	$-.01$	$-.1$
$\frac{\sin x}{x}$	.998	.99998	.99999983	.99999983	.99998	.998

x approaches 0.

Values approach 1.

These results suggest that

$$\lim_{x \to 0} \frac{\sin x}{x} = 1.$$

In Example 1, this limit is used to obtain another limit. Then the derivative of $y = \sin x$ can be found.

EXAMPLE 1 Find $\lim_{h \to 0} \frac{\cos h - 1}{h}$.

Use the limit above and some trigonometric identities.

$$\lim_{h \to 0} \frac{\cos h - 1}{h} = \lim_{h \to 0} \frac{(\cos h - 1)}{h} \cdot \frac{(\cos h + 1)}{(\cos h + 1)}$$

$$= \lim_{h \to 0} \frac{\cos^2 h - 1}{h(\cos h + 1)}$$

$$= \lim_{h \to 0} \frac{-\sin^2 h}{h(\cos h + 1)} \qquad \cos^2 h = 1 - \sin^2 h$$

$$= \lim_{h \to 0} (-\sin h)\left(\frac{\sin h}{h}\right)\left(\frac{1}{\cos h + 1}\right)$$

$$= (0)(1)\left(\frac{1}{1 + 1}\right)$$

$$\lim_{h \to 0} \frac{\cos h - 1}{h} = 0$$ ◀

Recall from the section on limits: when taking the limit of a product, if the limit of each factor exists, the limit of the product is simply the product of the limits.

We can now find the derivative of $y = \sin x$ by using the general definition for the derivative of a function f given in Chapter 2:

$$f'(x) = \lim_{h \to 0} \frac{f(x + h) - f(x)}{h},$$

provided this limit exists. By this definition, the derivative of $f(x) = \sin x$ is

$$f'(x) = \lim_{h \to 0} \frac{\sin(x + h) - \sin x}{h}$$

$$= \lim_{h \to 0} \frac{\sin x \cdot \cos h + \cos x \cdot \sin h - \sin x}{h} \qquad \text{Identity for } \sin(x + h)$$

$$= \lim_{h \to 0} \frac{(\sin x \cdot \cos h - \sin x) + \cos x \cdot \sin h}{h} \qquad \text{Rearranging terms}$$

$$= \lim_{h \to 0} \frac{\sin x(\cos h - 1) + \cos x \cdot \sin h}{h} \qquad \text{Factoring}$$

$$f'(x) = \lim_{h\to 0}\left(\sin x\,\frac{\cos h - 1}{h}\right) + \lim_{h\to 0}\left(\cos x\,\frac{\sin h}{h}\right) \quad \text{Limit rule for sums}$$
$$= (\sin x)(0) + (\cos x)(1)$$
$$= \cos x.$$

This result is summarized below.

DERIVATIVE OF sin *x*

$$D_x(\sin x) = \cos x$$

Recall that the symbol $D_x[f(x)]$ means the derivative of $f(x)$ with respect to x.

We can use the chain rule to find derivatives of other sine functions, as shown in the following examples.

EXAMPLE 2 Find the derivatives of the functions defined as follows.

(a) $y = \sin 6x$

By the chain rule,

$$y' = (\cos 6x) \cdot D_x(6x)$$
$$= (\cos 6x) \cdot 6$$
$$y' = 6 \cos 6x.$$

(b) $y = 5 \sin (9x^2 + 2)$

$$y' = [5 \cos (9x^2 + 2)] \cdot D_x(9x^2 + 2)$$
$$= [5 \cos (9x^2 + 2)]18x$$
$$y' = 90x \cos (9x^2 + 2)$$

EXAMPLE 3 Find $D_x(\sin^4 x)$.

The expression $\sin^4 x$ means $(\sin x)^4$. By the generalized power rule,

$$D_x(\sin^4 x) = 4 \cdot \sin^3 x \cdot D_x(\sin x)$$
$$= 4 \sin^3 x \cos x.$$

The derivative of $y = \cos x$ is found from trigonometric identities and from the fact that $D_x(\sin x) = \cos x$. First, use the identity for $\sin (x - y)$ to get

$$\sin\left(\frac{\pi}{2} - x\right) = \sin \frac{\pi}{2} \cdot \cos x - \cos \frac{\pi}{2} \cdot \sin x$$
$$= 1 \cdot \cos x - 0 \cdot \sin x$$
$$= \cos x.$$

In the same way, $\cos\left(\frac{\pi}{2} - x\right) = \sin x$. Therefore,

$$D_x(\cos x) = D_x\left[\sin\left(\frac{\pi}{2} - x\right)\right].$$

By the chain rule,

$$\begin{aligned} D_x\left[\sin\left(\frac{\pi}{2} - x\right)\right] &= \cos\left(\frac{\pi}{2} - x\right) \cdot D_x\left(\frac{\pi}{2} - x\right) \\ &= \cos\left(\frac{\pi}{2} - x\right) \cdot (-1) \qquad \pi/2 \text{ is constant.} \\ &= -\cos\left(\frac{\pi}{2} - x\right) \\ &= -\sin x. \end{aligned}$$

DERIVATIVE OF cos *x*

$$D_x(\cos x) = -\sin x$$

EXAMPLE 4 Find the following derivatives.

(a) $D_x[\cos(3x)] = -\sin(3x) \cdot D_x(3x) = -3\sin 3x$

(b) $D_x(\cos^4 x) = 4\cos^3 x \cdot D_x(\cos x) = 4\cos^3 x(-\sin x)$
$= -4\sin x\cos^3 x$

(c) $D_x(3x \cdot \cos x)$

Use the product rule.

$$\begin{aligned} D_x(3x \cdot \cos x) &= 3x(-\sin x) + (\cos x)(3) \\ &= -3x\sin x + 3\cos x \end{aligned}$$

As mentioned in the list of basic identities at the beginning of this section, $\tan x = (\sin x)/\cos x$. The derivative of $y = \tan x$ can be found by using the quotient rule to find the derivative of $y = (\sin x)/\cos x$.

$$\begin{aligned} D_x(\tan x) = D_x\left(\frac{\sin x}{\cos x}\right) &= \frac{\cos x \cdot D_x(\sin x) - \sin x \cdot D_x(\cos x)}{\cos^2 x} \\ &= \frac{\cos x(\cos x) - \sin x(-\sin x)}{\cos^2 x} \\ &= \frac{\cos^2 x + \sin^2 x}{\cos^2 x} \\ &= \frac{1}{\cos^2 x} = \sec^2 x \end{aligned}$$

The last step follows from the definitions of the trigonometric functions, which could be used to show that $1/\cos x = \sec x$. A similar calculation leads to the derivative of $\cot x$.

DERIVATIVES OF tan *x* AND cot *x*

$$D_x(\tan x) = \sec^2 x$$

$$D_x(\cot x) = -\csc^2 x$$

EXAMPLE 5 Find the derivatives of the following functions.

(a) $D_x(\tan 9x) = \sec^2 9x \cdot D_x(9x) = 9 \sec^2 9x$

(b) $D_x(\cot^6 x) = 6 \cot^5 x \cdot D_x(\cot x) = -6 \cot^5 x \csc^2 x$

(c) $D_x(\ln|6 \tan x|) = \dfrac{D_x(6 \tan x)}{6 \tan x} = \dfrac{6 \sec^2 x}{6 \tan x} = \dfrac{\sec^2 x}{\tan x}$ ◀

11.2 Exercises

Find the derivatives of the functions defined as follows.

1. $y = 2 \sin 6x$
2. $y = -\cos 4x$
3. $y = 12 \tan(9x + 1)$
4. $y = -3 \cos(8x^2 + 2)$
5. $y = \cos^4 x$
6. $y = -9 \sin^5 x$
7. $y = \tan^5 x$
8. $y = 2 \cot^4 x$
9. $y = -5x \cdot \sin 4x$
10. $y = 6x \cdot \cos 3x$
11. $y = \dfrac{\sin x}{x}$
12. $y = \dfrac{\tan x}{2x + 4}$
13. $y = \sin e^{5x}$
14. $y = \cos 4e^{2x}$
15. $y = e^{\sin x}$
16. $y = -8e^{\tan x}$
17. $y = \sin(\ln 4x^2)$
18. $y = \cos(\ln|2x^3|)$
19. $y = \ln|\sin x^2|$
20. $y = \ln|\tan^2 x|$
21. $y = \dfrac{2 \sin x}{3 - 2 \sin x}$
22. $y = \dfrac{4 \cos x}{2 - \cos x}$
23. $y = \sqrt{\dfrac{\sin x}{\sin 3x}}$
24. $y = \sqrt{\dfrac{\cos 4x}{\cos x}}$

25. Find the derivative of $\cot x$ by using the quotient rule and the fact that $\cot x = \cos x / \sin x$.

26. Find the derivative of $\sec x$ using the fact that $\sec x = 1/\cos x$.

27. Find the derivative of $\csc x$ using the fact that $\csc x = 1/\sin x$.

Applications

Business and Economics

28. Revenue from Seasonal Merchandise The revenue received from the sale of electric fans is seasonal, with maximum revenue in the summer. Let the revenue received from the sale of fans be approximated by

$$R(x) = 100 \cos 2\pi x,$$

where x is time in years, measured from July 1.

(a) Find $R'(x)$.

(b) Find $R'(x)$ for August 1. (*Hint:* August 1 is 1/12 of a year from July 1.)

(c) Find $R'(x)$ for January 1.

(d) Find $R'(x)$ for June 1.

Physical Sciences

29. Motion of a Particle A particle moves along a straight line. The distance of the particle from the origin at time t is given by

$$s(t) = \sin t + 2 \cos t.$$

Find the velocity at each of the following times.

(a) $t = 0$ **(b)** $t = \pi/2$ **(c)** $t = \pi$

Find the acceleration at each of the following times.

(d) $t = 0$ **(e)** $t = \pi/2$ **(f)** $t = \pi$

Life Sciences

30. Swing of a Runner's Arm A runner's arm swings rhythmically according to the equation

$$y = \frac{\pi}{8} \cos 3\pi\left(t - \frac{1}{3}\right),$$

where y denotes the angle between the actual position of the upper arm and the downward vertical position (as shown in the figure*) and where t denotes time in seconds.

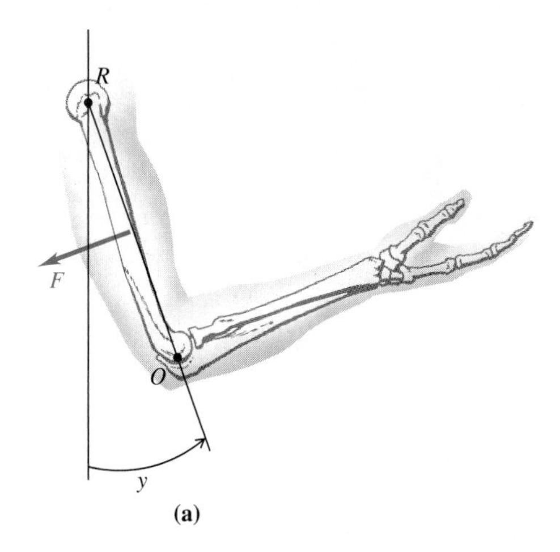

(a)

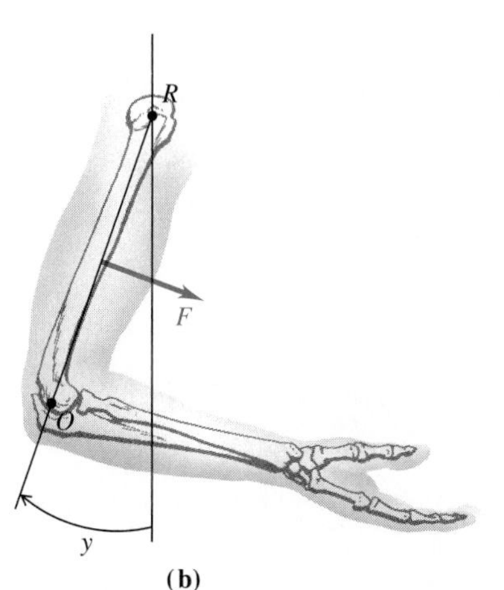

(b)

(a) Make a rough sketch of the graph of y as a function of t. (A computer or graphing calculator may be helpful.)

(b) Calculate the velocity and the acceleration of the arm.

(c) Verify that the angle y and the acceleration d^2y/dt^2 are related by the differential equation

$$\frac{d^2y}{dt^2} + 9\pi^2 y = 0.$$

(d) Apply the fact that the force exerted by the muscle as the arm swings is proportional to the acceleration of y, with a positive constant of proportionality, to find the direction of the force (counterclockwise or clockwise) at $t = 1$ sec, $t = 4/3$ sec, and $t = 5/3$ sec. What is the position of the arm at each of these times?

31. Swing of a Jogger's Arm A jogger's arm swings according to the equation

$$y = \frac{1}{5} \sin \pi(t - 1).$$

Proceed as directed in (a), (b), (c), and (d) of the preceding exercise, with the following exceptions: in (c), replace the differential equation with

$$\frac{d^2y}{dt^2} + \pi^2 y = 0,$$

and in (d), consider the times $t = 1.5$ seconds, $t = 2.5$ seconds, and $t = 3.5$ seconds.

General Interest

32. (a) Rotating Lighthouse The beacon on a lighthouse 40 m from a straight shoreline rotates twice per minute. (See the figure.) How fast is the beam moving along

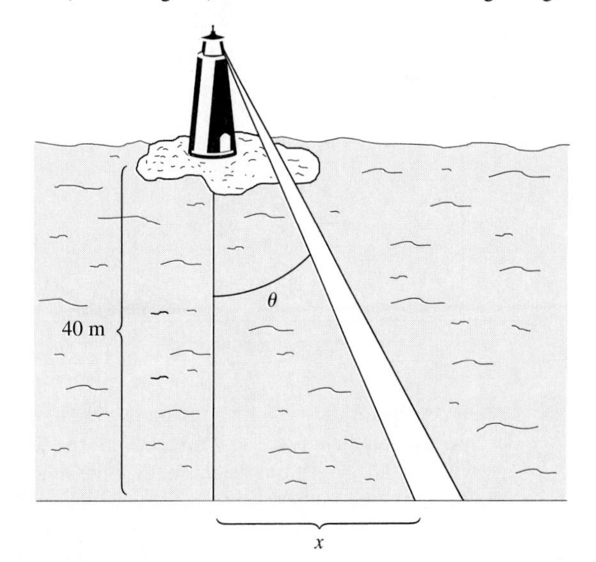

*Art for Exercises 30 and 31 from Rodolfo De Sapio, *Calculus for the Life Sciences*. Copyright © 1976, 1978 by W. H. Freeman and Company. Reprinted by permission.

the shoreline at the moment when the light beam and the shoreline are at right angles? (*Hint:* This is a related rate exercise. Find an equation relating θ, the angle between the beam of light and the line from the lighthouse to the shoreline, and x, the distance along the shoreline from the point on the shoreline closest to the lighthouse and the point where the beam hits the shoreline. You need to express $d\theta/dt$ in radians per minute.)

(b) In part (a), how fast is the beam moving along the shoreline when the beam hits the shoreline 40 m from the point on the shoreline closest to the lighthouse?

33. (a) Rotating Camera A television camera on a tripod 60 ft from a road is filming a car carrying the President of the United States. (See the figure.) The car is moving along the road at 600 ft/min. How fast is the camera rotating (in revolutions per minute) when the car is at the point on the road closest to the camera? (See the hint for Exercise 32.)

(b) How fast is the camera rotating 6 sec after the moment in part (a)?

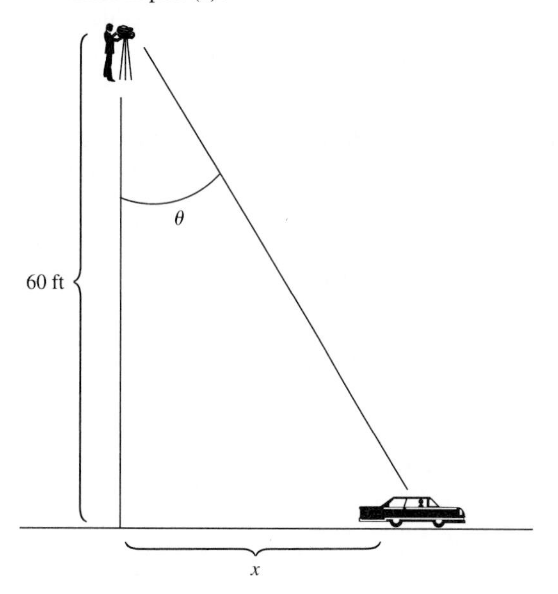

34. Ladder A thief tries to enter a building by placing a ladder over a 9-foot-high fence so it rests against the building, which is 2 ft back from the fence. (See the figure.) What length is the shortest ladder that can be used? (*Hint:* Let θ be the angle between the ladder and the ground. Express the length of the ladder in terms of θ, and then find the value of θ that minimizes the length of the ladder.)

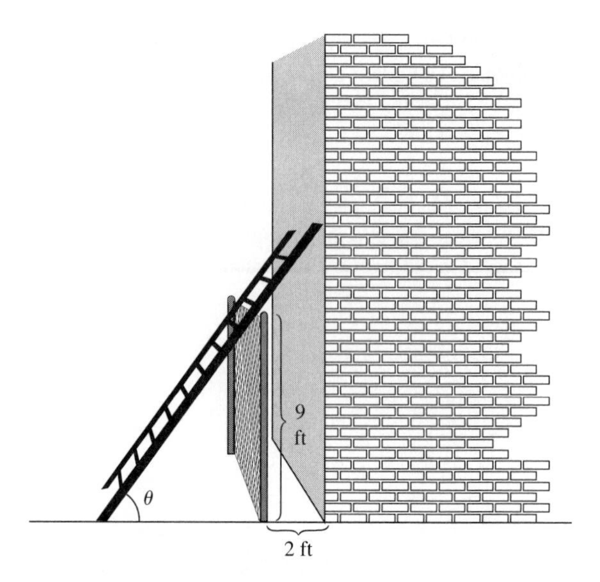

EXERCISE 34

35. Ladder A janitor in a hospital needs to carry a ladder around a corner connecting a 10-foot-wide corridor and a 5-foot-wide corridor. (See the figure.) What is the longest such ladder that can make it around the corner? (*Hint:* Find the narrowest point in the corridor by minimizing the length of the ladder as a function of θ, the angle the ladder makes with the 5-foot-wide corridor.)

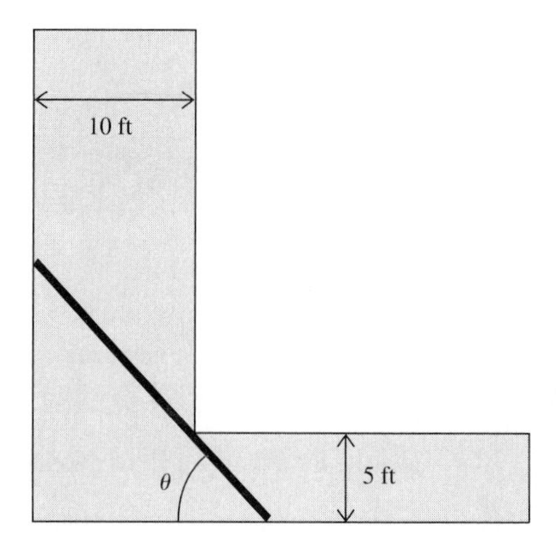

11.3 INTEGRATION

Given a sales equation, how many snowblowers are sold in a year?

In Exercise 37 in this section, we will use trigonometry to answer this question.

Any differentiation formula leads to a corresponding formula for integration. In particular, the formulas of the last section lead to the following indefinite integrals.

BASIC TRIGONOMETRIC INTEGRALS

$$\int \sin x\,dx = -\cos x + C \qquad \int \cos x\,dx = \sin x + C$$

$$\int \sec^2 x\,dx = \tan x + C \qquad \int \csc^2 x\,dx = -\cot x + C$$

EXAMPLE 1 Find each integral.

(a) $\int \sin 7x\,dx$

Use substitution. Let $u = 7x$, so that $du = 7\,dx$. Then

$$\begin{aligned}\int \sin 7x\,dx &= \frac{1}{7}\int \sin 7x\,(7\,dx)\\ &= \frac{1}{7}\int \sin u\,du\\ &= -\frac{1}{7}\cos u + C\\ &= -\frac{1}{7}\cos 7x + C.\end{aligned}$$

(b) $\int \cos \frac{2}{3}x\,dx = \frac{3}{2}\int \cos \frac{2}{3}x\left(\frac{2}{3}\,dx\right) = \frac{3}{2}\sin \frac{2}{3}x + C$

(c) $\int \sin^2 x \cos x\,dx$

Let $u = \sin x$, with $du = \cos x\,dx$. This gives

$$\int \sin^2 x \cos x\,dx = \int u^2\,du = \frac{1}{3}u^3 + C.$$

Replacing u with $\sin x$ gives

$$\int \sin^2 x \cos x\,dx = \frac{1}{3}\sin^3 x + C.$$

(d) $\int \frac{\sin x}{\sqrt{\cos x}}\,dx$

Rewrite the integrand as

$$\int (\cos x)^{-1/2} \sin x\,dx.$$

If $u = \cos x$, then $du = -\sin x\,dx$, with

$$\begin{aligned}\int (\cos x)^{-1/2} \sin x\,dx &= -\int (\cos x)^{-1/2}(-\sin x\,dx)\\ &= -\int u^{-1/2}\,du\\ &= -2u^{1/2} + C\\ &= -2\cos^{1/2} x + C.\end{aligned}$$

(e) $\int \sec^2 12x\,dx = \frac{1}{12}\int \sec^2 12x\,(12\,dx) = \frac{1}{12}\tan 12x + C$ ◀

As in Chapter 6, we can find the area under a curve by setting up an appropriate definite integral.

EXAMPLE 2 Find the shaded area in Figure 16.

The shaded area in Figure 16 is bounded by $y = \cos x$, $y = 0$, $x = -\pi/2$, and $x = \pi/2$. By the Fundamental Theorem of Calculus, this area is given by

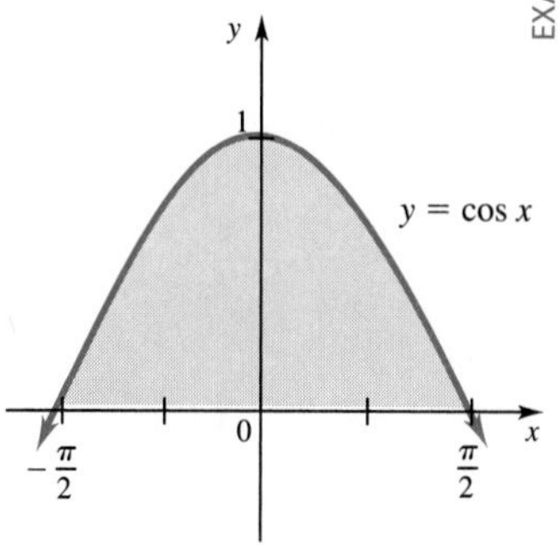

$$\begin{aligned}\int_{-\pi/2}^{\pi/2} \cos x\,dx &= \sin x\Big|_{-\pi/2}^{\pi/2}\\ &= \sin\frac{\pi}{2} - \sin\left(-\frac{\pi}{2}\right)\\ &= 1 - (-1)\\ &= 2.\end{aligned}$$

FIGURE 16

By symmetry, the same area could be found by evaluating

$$2\int_0^{\pi/2} \cos x\,dx.$$ ◀

The method of integration by parts discussed in Chapter 7 is often useful for finding certain integrals involving trigonometric functions.

EXAMPLE 3 Find $\int 2x \sin x\,dx$.

Let $u = 2x$ and $dv = \sin x\,dx$. Then $du = 2\,dx$ and $v = -\cos x$. Use the formula for integration by parts,

$$\int u\,dv = uv - \int v\,du,$$

to get

$$\begin{aligned}\int 2x \sin x\, dx &= -2x \cos x - \int (-\cos x)(2\, dx) \\ &= -2x \cos x + 2 \int \cos x\, dx \\ &= -2x \cos x + 2 \sin x + C.\end{aligned}$$

Check the result by differentiating. (This integral could also have been found by using column integration.) ◀

As mentioned earlier, $\tan x = (\sin x)/\cos x$, so that

$$\int \tan x\, dx = \int \frac{\sin x}{\cos x}\, dx.$$

To find $\int \tan x\, dx$, let $u = \cos x$, with $du = -\sin x\, dx$. Then

$$\int \tan x\, dx = \int \frac{\sin x}{\cos x}\, dx = -\int \frac{du}{u} = -\ln |u| + C.$$

Replacing u with $\cos x$ gives the formula for integrating $\tan x$. The integral for $\cot x$ is found in a similar way.

INTEGRALS OF tan *x* AND cot *x*

$$\int \tan x\, dx = -\ln |\cos x| + C$$

$$\int \cot x\, dx = \ln |\sin x| + C$$

EXAMPLE 4

(a) $\int \tan 6x\, dx = \frac{1}{6} \int \tan 6x\,(6\, dx) = -\frac{1}{6} \ln |\cos 6x| + C$

(b) $\int x \cot x^2\, dx = \frac{1}{2} \int (\cot x^2)(2x\, dx) = \frac{1}{2} \ln |\sin x^2| + C$ ◀

11.3 Exercises

Find the following integrals.

1. $\int \cos 5x\, dx$

2. $\int \sin 8x\, dx$

3. $\int (5 \cos x + 2 \sin x)\, dx$

4. $\int (7 \sin x - 8 \cos x)\, dx$

5. $\int x \sin x^2\, dx$

6. $\int 2x \cos x^2\, dx$

7. $-\int 6 \sec^2 2x\, dx$

8. $-\int 2 \csc^2 8x\, dx$

9. $\int \sin^7 x \cos x\, dx$

10. $\int \sin^6 x \cos x\, dx$

11. $\int \sqrt{\sin x}(\cos x)\, dx$

12. $\int \frac{\cos x}{\sqrt{\sin x}}\, dx$

13. $\int \frac{\sin x}{1 + \cos x}\, dx$

14. $\int \frac{\cos x}{1 - \sin x}\, dx$

15. $\int x^5 \cos x^6\, dx$

16. $\int (x + 2)^4 \sin (x + 2)^5\, dx$

17. $\int \tan \frac{1}{4}x\, dx$

18. $\int \cot \left(-\frac{3}{8}x\right)\, dx$

19. $\int x^2 \cot x^3\, dx$

20. $\int \frac{x}{4} \tan \left(\frac{x}{4}\right)^2\, dx$

21. $\int e^x \sin e^x \, dx$ **22.** $\int e^{-x} \tan e^{-x} \, dx$ **23.** $\int -6x \cos 5x \, dx$ **24.** $\int 9x \sin 2x \, dx$

25. $\int 8x \sin x \, dx$ **26.** $\int -11x \cos x \, dx$ **27.** $\int -6x^2 \cos 8x \, dx$ **28.** $\int 10x^2 \sin \frac{1}{2}x \, dx$

Evaluate the following definite integrals.

29. $\int_0^{\pi/4} \sin x \, dx$ **30.** $\int_{-\pi/2}^{0} \cos x \, dx$ **31.** $\int_0^{\pi/3} \tan x \, dx$

32. $\int_{\pi/4}^{\pi/2} \cot x \, dx$ **33.** $\int_{\pi/2}^{2\pi/3} \cos x \, dx$ **34.** $\int_{\pi/4}^{3\pi/4} \sin x \, dx$

Find each improper integral (if you have studied improper integrals). (Hint: *Use the table of integrals.*)

35. $\int_0^{\infty} e^{-x} \sin x \, dx$ **36.** $\int_0^{\infty} e^{-x} \cos x \, dx$

Applications

Business and Economics

37. Sales Sales of snowblowers are seasonal. Suppose the sales of snowblowers in one region of the country are approximated by

$$S(t) = 500 + 500 \cos\left(\frac{\pi}{6}t\right),$$

where t is time in months, with $t = 0$ corresponding to November. The figure below shows a graph of S. Use a definite integral to find total sales over a year.

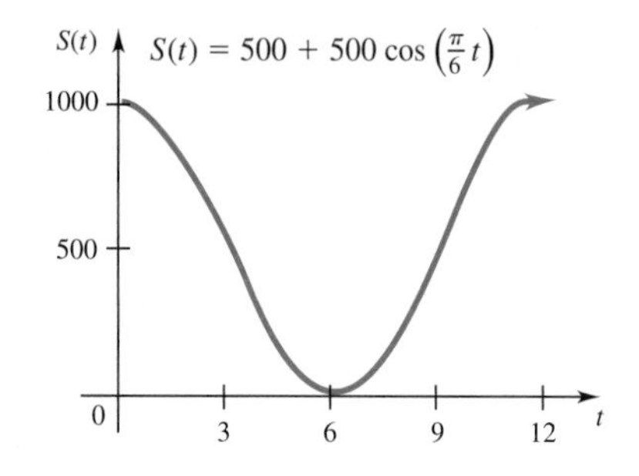

Life Sciences

38. Migratory Animals The number of migratory animals (in hundreds) counted at a certain checkpoint is given by

$$T(t) = 50 + 50 \cos\left(\frac{\pi}{6}t\right),$$

where t is time in months, with $t = 0$ corresponding to July. The figure shows a graph of T. Use a definite integral to find the number of animals passing the checkpoint in a year.

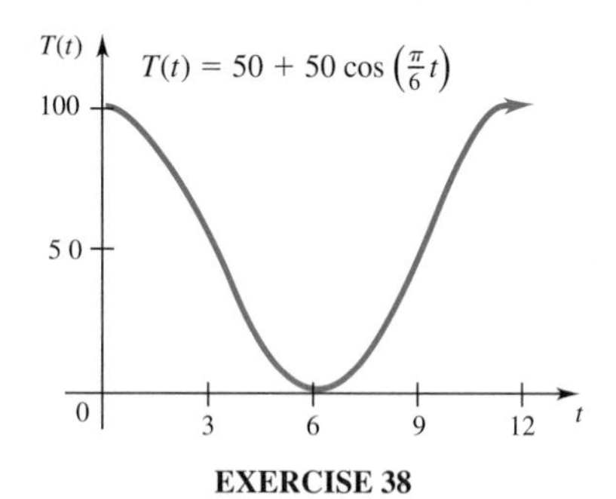

EXERCISE 38

Physical Sciences

39. Voltage The electrical voltage from a standard wall outlet is given as a function of time t by

$$V(t) = 170 \sin(120\pi t).$$

This is an example of alternating current, which is electricity that reverses direction at regular intervals. The common method for measuring the level of voltage from an alternating current is the root mean square, which is given by

$$\text{Root mean square} = \sqrt{\frac{\int_0^T V^2(t)\,dt}{T}},$$

where T is one period of the current.

(a) Verify that $T = \frac{1}{60}$ sec for $V(t)$ given above.

(b) You may have seen that the voltage from a standard wall outlet is 120 volts. Verify that this is the root mean square value for $V(t)$ given above. (*Hint:* Use the trigonometric identity $\sin^2 x = (1 - \cos 2x)/2$. This identity can be derived by letting $y = x$ in the basic identity for $\cos(x + y)$, and then eliminating $\cos^2 x$ by using the identity $\cos^2 x = 1 - \sin^2 x$.)

11.4 INVERSE TRIGONOMETRIC FUNCTIONS

How far back should you stand from a painting to get the best view?

In Exercise 52 in this section, we will use trigonometry to answer this question.

To solve the equation $y = e^x$ for x, we introduced a new function, the logarithmic function. Similarly, we introduce the *inverse sine* function in order to solve the equation $y = \sin x$ for x. The inverse sine function determines an angle from the sine of the angle. For example, since $\sin(\pi/2) = 1$, the inverse sine function must start with 1 and produce the result $\pi/2$. One difficulty is that there are many values of x such that $\sin x = 1$: $\pi/2$ is one, as are $5\pi/2$, $9\pi/2$, and $13\pi/2$. To be sure that the inverse sine produces only one answer, it is necessary to restrict the range of the inverse sine function. The range is commonly restricted to the interval $[-\pi/2, \pi/2]$. With this restriction, the inverse sine function is defined as follows.

INVERSE SINE

Let $x = \sin y$, for y in $[-\pi/2, \pi/2]$. Then

$$\mathbf{y = \sin^{-1} x.}$$

EXAMPLE 1 Find each of the following.

(a) $\sin^{-1}(1/2)$

Let $y = \sin^{-1}(1/2)$. Then, by the definition of the inverse sine function, $\sin y = 1/2$. Since $\sin(\pi/6) = 1/2$, and $\pi/6$ is in $[-\pi/2, \pi/2]$,

$$\sin^{-1}\frac{1}{2} = \frac{\pi}{6}.$$

(b) $\sin^{-1}\left(-\dfrac{\sqrt{3}}{2}\right) = -\dfrac{\pi}{3}$

(c) $\sin^{-1} 1 = \dfrac{\pi}{2}$ ◀

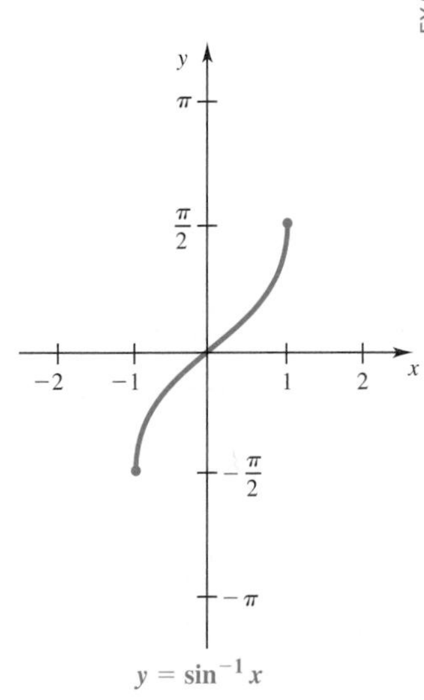

FIGURE 17

A graph of $y = \sin^{-1} x$ is shown in Figure 17. (Sometimes the notation $y = \arcsin x$ is used for $y = \sin^{-1} x$.)

For each of the other trigonometric functions, an inverse function can be defined by putting a suitable restriction on the range of the inverse function. The three most commonly used **inverse trigonometric functions** and their ranges are listed on the next page.

INVERSE TRIGONOMETRIC FUNCTIONS

$$y = \sin^{-1} x, \qquad -\frac{\pi}{2} \le y \le \frac{\pi}{2}$$

$$y = \cos^{-1} x, \qquad 0 \le y \le \pi$$

$$y = \tan^{-1} x, \qquad -\frac{\pi}{2} < y < \frac{\pi}{2}$$

Graphs of $y = \cos^{-1} x$ and $y = \tan^{-1} x$ are shown in Figures 18 and 19, respectively.

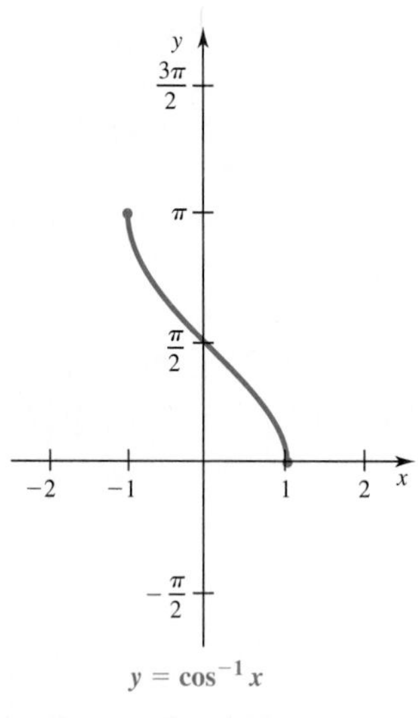

$y = \cos^{-1} x$

FIGURE 18

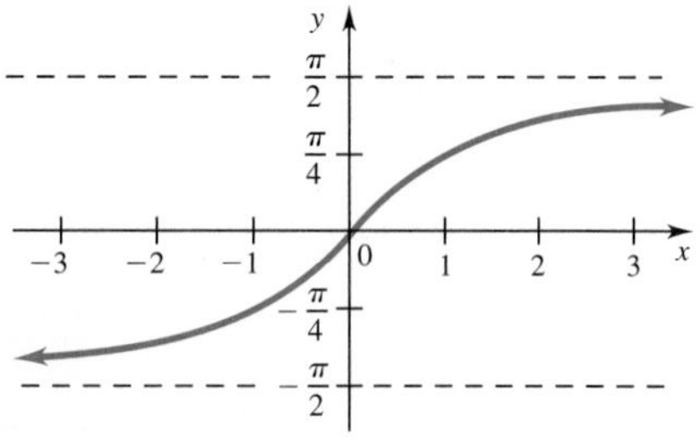

$y = \tan^{-1} x$

FIGURE 19

Caution Even though $\sin^2 x = (\sin x)^2$, $\sin^{-1} x$ is *not* the same as $(\sin x)^{-1} = 1/\sin x = \csc x$.

EXAMPLE 2 Find $\cos^{-1}(-\sqrt{2}/2)$.

The values of $\cos^{-1} x$ are in quadrants I and II, according to the definition given above. Since $-\sqrt{2}/2$ is negative, $\cos^{-1}(-\sqrt{2}/2)$ is restricted to quadrant II. Let $y = \cos^{-1}(-\sqrt{2}/2)$. Then $\cos y = -\sqrt{2}/2$. In quadrant II, $\cos(3\pi/4) = -\sqrt{2}/2$, so

$$\cos^{-1}\left(-\frac{\sqrt{2}}{2}\right) = \frac{3\pi}{4}.$$

Derivatives of Inverse Trigonometric Functions We found that the derivatives of the trigonometric functions were themselves trigonometric functions. The derivatives of the inverse trigonometric functions, however, are *algebraic functions*. To find the derivative of $y = \sin^{-1} x$, use the definition of $y = \sin^{-1} x$ to write

$$x = \sin y.$$

Use implicit differentiation to find $D_x\, y$. Taking the derivative with respect to x on both sides of $x = \sin y$ gives

$$D_x(x) = (\cos y)\, D_x\, y.$$

Since $D_x(x) = 1$,

$$1 = (\cos y)\, D_x\, y$$

or

$$D_x\, y = \frac{1}{\cos y}.$$

Since $\sin^2 y + \cos^2 y = 1$, $\cos y = \pm\sqrt{1 - \sin^2 y}$, and

$$D_x\, y = \frac{1}{\pm\sqrt{1 - \sin^2 y}}.$$

Choose the positive square root, since $\cos y > 0$ for y in $[-\pi/2, \pi/2]$. Replace $\sin y$ with x to get

$$D_x\, y = \frac{1}{\sqrt{1 - x^2}}.$$

This result is summarized below.

DERIVATIVE OF $\sin^{-1} x$

$$\boldsymbol{D_x\,(\sin^{-1} x) = \frac{1}{\sqrt{1 - x^2}}}$$

Find each derivative.

(a) $y = \sin^{-1} 9x$

Use the above formula with the chain rule to get

$$D_x\,(\sin^{-1} 9x) = \frac{1}{\sqrt{1 - (9x)^2}}\, D_x\,(9x) = \frac{9}{\sqrt{1 - 81x^2}}.$$

(b) $$D_x\,[\sin^{-1}(5x^2 + 1)] = \frac{1}{\sqrt{1 - (5x^2 + 1)^2}}\, D_x\,(5x^2 + 1)$$

$$= \frac{10x}{\sqrt{1 - (5x^2 + 1)^2}}$$ ◀

Starting with $y = \cos^{-1} x$ and $y = \tan^{-1} x$ and again using implicit differentiation gives the following derivatives.

DERIVATIVES OF $\cos^{-1} x$ AND $\tan^{-1} x$

$$D_x (\cos^{-1} x) = \frac{-1}{\sqrt{1 - x^2}}$$

$$D_x (\tan^{-1} x) = \frac{1}{1 + x^2}$$

EXAMPLE 4 Find each derivative.

(a) $y = \cos^{-1} (3x^2)$

Using the formula for the derivative of inverse cosine, together with the chain rule, gives

$$D_x [\cos^{-1} (3x^2)] = \frac{-1}{\sqrt{1 - (3x^2)^2}} (6x) = \frac{-6x}{\sqrt{1 - 9x^4}}.$$

(b) $D_x [\tan^{-1} (8x^2)] = \dfrac{1}{1 + (8x^2)^2} D_x (8x^2) = \dfrac{16x}{1 + 64x^4}$

(c) $D_x [\tan^{-1} (e^{2x})] = \dfrac{1}{1 + (e^{2x})^2} D_x (e^{2x}) = \dfrac{2e^{2x}}{1 + e^{4x}}$ ◀

Integrals The three derivatives above lead to the following integrals.

$$\int \frac{1}{\sqrt{1 - x^2}} \, dx = \sin^{-1} x + C$$

$$\int \frac{-1}{\sqrt{1 - x^2}} \, dx = \cos^{-1} x + C$$

$$\int \frac{1}{1 + x^2} \, dx = \tan^{-1} x + C$$

The second integral is seldom used, since we could just as well have written the antiderivative as $-\sin^{-1} x + C$.

EXAMPLE 5 Find each integral.

(a) $\displaystyle\int \frac{2x}{\sqrt{1 - x^4}} \, dx$

Let $u = x^2$ to get $du = 2x\,dx$, so that

$$\begin{aligned}\int \frac{2x}{\sqrt{1 - x^4}}\,dx &= \int \frac{2x\,dx}{\sqrt{1 - (x^2)^2}}\\ &= \int \frac{du}{\sqrt{1 - u^2}}\\ &= \sin^{-1} u + C\\ &= \sin^{-1}(x^2) + C.\end{aligned}$$

(b) $\displaystyle\int \frac{e^x}{1 + e^{2x}}\,dx$

Let $u = e^x$, with $du = e^x\,dx$. Then

$$\begin{aligned}\int \frac{e^x}{1 + e^{2x}}\,dx &= \int \frac{e^x\,dx}{1 + (e^x)^2}\\ &= \int \frac{du}{1 + u^2}\\ &= \tan^{-1} u + C\\ &= \tan^{-1} e^x + C.\end{aligned}$$ ◀

11.4 Exercises

Give the value of y in radians.

1. $y = \sin^{-1}(-\sqrt{3}/2)$
2. $y = \cos^{-1}(\sqrt{3}/2)$
3. $y = \tan^{-1} 1$
4. $y = \tan^{-1}(-1)$
5. $y = \sin^{-1}(-1)$
6. $y = \cos^{-1}(-1)$
7. $y = \cos^{-1}(1/2)$
8. $y = \sin^{-1}(-\sqrt{2}/2)$
9. $y = \cos^{-1}(-\sqrt{2}/2)$
10. $y = \tan^{-1}(\sqrt{3}/3)$
11. $y = \tan^{-1}(-\sqrt{3})$
12. $y = \cos^{-1}(-1/2)$

Use a calculator to give each value in degrees.

13. $\sin^{-1}(-.1392)$
14. $\cos^{-1}(-.1392)$
15. $\cos^{-1}(-.8988)$
16. $\sin^{-1} .7880$
17. $\cos^{-1} .9272$
18. $\tan^{-1} 1.7321$
19. $\tan^{-1} 1.111$
20. $\sin^{-1} .8192$
21. $\tan^{-1}(-.9004)$
22. $\tan^{-1}(-.2867)$
23. $\sin^{-1} .9272$
24. $\cos^{-1} .4384$

Find the derivative of each inverse trigonometric function.

25. $y = \sin^{-1} 12x$
26. $y = \cos^{-1} 10x$
27. $y = \tan^{-1} 3x$
28. $y = \sin^{-1}(1/x)$
29. $y = \cos^{-1}(-2/x)$
30. $y = \cos^{-1}\sqrt{x}$
31. $y = \tan^{-1}(\ln|7x|)$
32. $y = \tan^{-1}(\ln|x + 2|)$
33. $y = \ln|\tan^{-1}(x + 1)|$
34. $y = \ln|\tan^{-1}(3x - 5)|$

35. Verify the formulas for the derivatives of $\cos^{-1} x$ and $\tan^{-1} x$ by implicitly differentiating $x = \cos y$ and $x = \tan y$.

36. Since $\displaystyle\int \frac{-1}{\sqrt{1 - x^2}}\,dx = \cos^{-1} x + C$ and $\displaystyle\int \frac{-1}{\sqrt{1 - x^2}}\,dx = -\sin^{-1} x + C$, one might conclude that $\cos^{-1} x = -\sin^{-1} x$. Explain why this reasoning is incorrect.

Find each integral.

37. $\int \frac{x^3}{1 + x^8} dx$

38. $\int \frac{x^2}{\sqrt{1 - x^6}} dx$

39. $\int \frac{e^x}{\sqrt{1 - e^{2x}}} dx$

40. $\int \frac{-e^{2x}}{1 + e^{4x}} dx$

41. $\int \frac{4}{\sqrt{1 - 9x^2}} dx$

42. $\int \frac{\cos x}{1 + \sin^2 x} dx$

43. $\int \frac{1/x}{1 + (\ln x)^2} dx$

44. $\int \frac{1}{\sqrt{25 - x^2}} dx$

45. $\int_0^1 \frac{1}{1 + x^2} dx$

46. $\int_0^{.5} \frac{1}{\sqrt{1 - x^2}} dx$

47. $\int_0^2 \frac{1}{\sqrt{16 - x^2}} dx$

48. $\int_0^1 \frac{x^{1/3}}{1 + x^{8/3}} dx$

49. $\int_0^{\infty} \frac{1}{1 + x^2} dx$

Applications

General Interest

Viewing Art *While visiting a museum, Patricia Quinlan views a painting that is 3 ft high and hanging 6 ft above the ground. (See the figure.) Assume Patricia's eyes are 5 ft above the ground, and let x be the distance from the spot where Patricia is standing to the wall displaying the painting.*

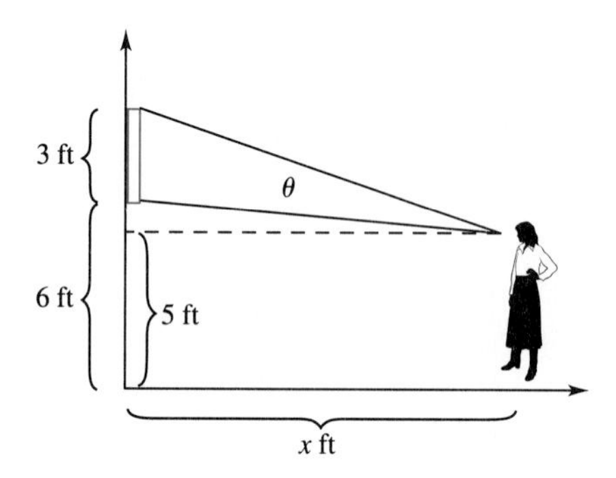

50. Show that θ, the viewing angle subtended by the painting, is given by

$$\theta = \tan^{-1}\frac{4}{x} - \tan^{-1}\frac{1}{x}.$$

(*Hint:* $\tan^{-1}\frac{4}{x}$ is an angle whose tangent is $4/x$.)

51. Find the value of θ for each of the following values of x. Round to the nearest degree.

(a) 1 **(b)** 2 **(c)** 3 **(d)** 4

52. Find how far Patricia should stand from the wall to maximize θ, which should give her the best view of the painting.

53. Later Patricia views a painting that is 4 feet high and hanging 7 feet above the ground. How far should she stand from the wall to get the best view of this painting?

11.5 APPLICATIONS

How does temperature vary beneath the earth's surface?

Using trigonometry, we will answer this question in Example 2.

While introducing the concepts of trigonometry in this chapter, we have included in the exercise sets a few applications, such as measuring distances, describing music and electricity, and determining a camera's depth of field. But trigonometry has many more applications, a few of which we will now explore.

EXAMPLE 1 Many biological populations, both plant and animal, experience seasonal growth. For example, an animal population might flourish during the spring and summer and die back in the fall. If $f(t)$ represents the population at time t, then the rate of change of the population with respect to time might be governed by the differential equation

$$\frac{df}{dt} = c \cdot f(t) \cdot \cos t. \qquad c \text{ is a positive constant.}$$

Since $\cos t$ ranges from -1 to 1, df/dt will change from negative to positive, so that $f(t)$ is both increasing and decreasing. The solution of the differential equation above is

$$f(t) = f(0)e^{c \sin t},$$

where $f(0)$ is the size of the population when $t = 0$. ◀

EXAMPLE 2 Mathematical models of ground temperature variation usually involve Fourier series or other sophisticated methods. However, Mary Kay Corbitt and C. H. Edwards, Jr., have given the elementary model

$$u(x, t) = T_0 + A_0 e^{-ax} \cos(wt - ax) \qquad \textbf{(1)}$$

for the temperature $u(x, t)$ at a given location at a given time t and a given depth x beneath the earth's surface.* In equation (1),

T_0 = the annual average surface temperature;

A_0 = the amplitude of the seasonal surface temperature variation;

w = a constant chosen to make the period one year in length;

k = the thermal conductivity of the ground;

and $a = \sqrt{w/2k}$.

Figure 20 shows $u(0, t)$, which models the temperature at the earth's surface.

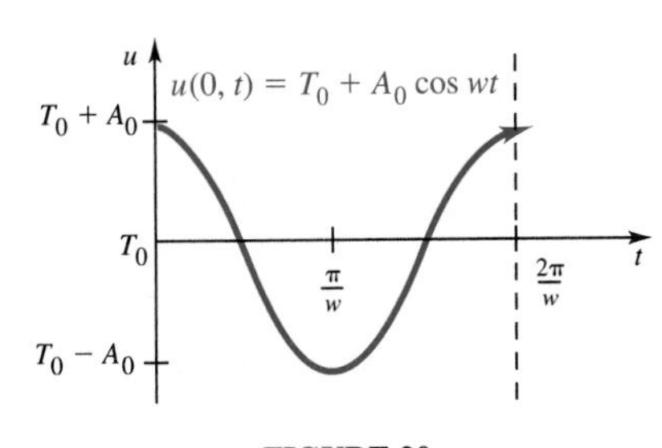

FIGURE 20

*Figure and equation from "Mathematical Modeling and Cool Buttermilk in the Summer" by Mary Kay Corbitt and C. H. Edwards, Jr. in *Applications in School Mathematics: 1979 Yearbook*, edited by Sidney Sharron and Robert E. Hays, p. 221. Reprinted by permission of National Council of Teachers of Mathematics.

In deriving equation (1), geothermal heat was ignored and the density of the earth was assumed to be uniform, so that k is constant and the annual average temperature *at any depth* is T_0. The factor e^{-ax} in equation (1) accounts for the relationship between temperature at depth x and surface temperature. The phase shift $-ax$ in equation (1) accounts for the seasons at depth x lagging behind those at the surface. ◀

EXAMPLE 3 Suppose a light source is hanging at the end of a pendulum, as shown in Figure 21. As the pendulum swings back and forth, the light oscillates on the floor. If a roll of photographic paper is moved at a constant speed under the light, the curve traced on the paper is very close to a sine curve. This curve will have an equation of the form

$$s = A \sin (Bt + C)$$

for constants A, B, and C and time t. As we saw in Section 1.6, different values of A cause the range of the graph to stretch or contract vertically, while B stretches or compresses the function horizontally, and C shifts the whole graph horizontally. In $s = A \sin (Bt + C)$, the *amplitude* of the graph is $|A|$, the *period* is $2\pi/|B|$, and the *phase shift* is the value of t that makes $Bt + C = 0$. Motion of this type, called **simple harmonic motion,** occurs in such diverse areas as sound and wave mechanics.

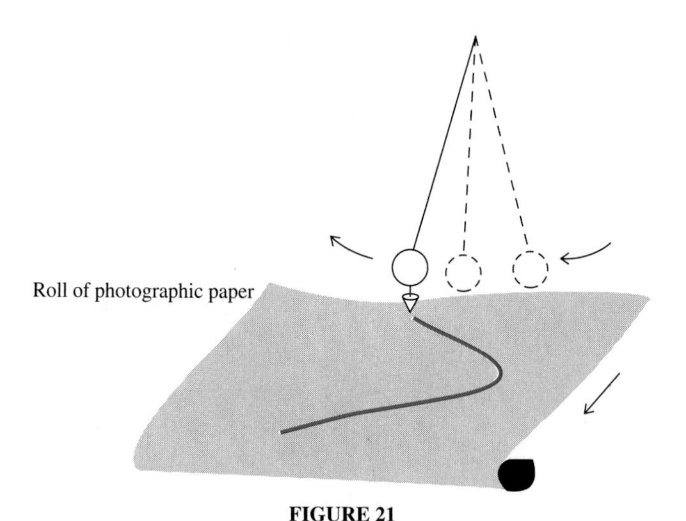

FIGURE 21

The reason a pendulum moves in this way is related to Newton's Second Law of Motion, which states that the force on an object is equal to its mass times its acceleration, or $F = ma$. Recall that acceleration is given by $a(t) = s''(t)$. Verify that $s'(t) = BA \cos (Bt + C)$ and $s''(t) = -B^2A \sin (Bt + C) = -B^2s(t)$. Thus, for the pendulum, $F = -mB^2s(t)$. Substituting $ms''(t)$ for F yields

$$ms''(t) = -mB^2s(t).$$

Canceling m from both sides leads to the second-order differential equation

$$s''(t) = -B^2 s(t).$$

This equation tells us two things about the force on the pendulum. First, because of the negative sign, the acceleration $s''(t)$ is opposite the displacement $s(t)$. Of course, if you push a pendulum to the left and upward, the force of gravity will accelerate it to the right and downward. Second, the acceleration is proportional to the displacement. In other words, the further you push the pendulum from its resting position, the more quickly it will accelerate back. It turns out that the force is not quite $-mB^2 s(t)$, but $-mB^2 \sin[s(t)]$. Recall from the first section of this chapter that $\lim_{x \to 0} (\sin x)/x = 1$, so $\sin x \approx x$ when x is small, and the resulting motion is approximately sinusoidal if the displacement of the pendulum is not too great. If the pendulum is making large swings back and forth, the motion must be described by the equation $s''(t) = -B^2 \sin[s(t)]$, which is much more difficult to analyze. ◀

11.5 Exercises

In Exercises 1–6, recall that the slope of the tangent line to a graph is given by the derivative of the function. Find the slope of the tangent line to the graph of each equation at the given point.

1. $y = \sin x; \quad x = 0$

2. $y = \sin x; \quad x = \pi/4$

3. $y = \cos x; \quad x = \pi/2$

4. $y = \cos x; \quad x = -\pi/4$

5. $y = \tan x; \quad x = 0$

6. $y = \cot x; \quad x = \pi/2$

Applications

Life Sciences

Seasonal Growth *Example 1 gave the following mathematical model for seasonal growth:*

$$f(t) = f(0)e^{c \sin t}.$$

Suppose that $f(0) = 1000$ *and* $c = 2$. *Find the function values in Exercises 7–16 using a calculator with trigonometric functions and powers of e.*

7. $f(.2)$

8. $f(.4)$

9. $f(.5)$

10. $f(.8)$

11. $f(1)$

12. $f(1.4)$

13. $f(1.8)$

14. $f(2.3)$

15. $f(3)$

16. $f(3.1)$

17. Graph $f(t)$ using the information from Exercises 7–16.

18. Find the maximum and minimum values of $f(t)$ and the values of t where they occur.

Physical Sciences

19. Simple Harmonic Motion Verify that

$$s(t) = A \cos(Bt + C)$$

also satisfies the differential equation for simple harmonic motion, $s''(t) = -B^2 s(t)$.

Ground Temperature *In Exercises 20–22, use the ground temperature model in Example 2 and assume that* $T_0 = 16°$ C *and* $A_0 = 11°$ C *at a certain location. Also assume that* $a = .00706$ *in cgs (centimeter-gram-second) units.*

20. At what minimum depth x (in centimeters) is the amplitude of $u(x, t)$ at most 1° C?

21. Suppose we wish to construct a cellar to keep wine at a temperature between 14° C and 18° C. What minimum depth (in centimeters) will accomplish this?

22. At what minimum depth x does the ground temperature model predict that it will be winter when it is summer at the surface, and vice versa? That is, when will the phase shift correspond to 1/2 year?

23. Show that the ground temperature model

$$u(x, t) = T_0 + A_0e^{-ax} \cos (wt - ax)$$

satisfies the *heat equation*

$$\frac{\partial u}{\partial t} = k \frac{\partial^2 u}{\partial x^2}.$$

Connections

Blood Vessel System *The body's system of blood vessels is made up of arteries, arterioles, capillaries, and veins. The transport of blood from the heart through all organs of the body and back to the heart should be as efficient as possible. One way this can be done is by having large enough blood vessels to avoid turbulence, with blood cells small enough to minimize viscosity.**

In Exercises 24–37, we will find the value of angle θ (see the figure) such that total resistance to the flow of blood is minimized. Assume that a main vessel of radius r_1 runs along the horizontal line from A to B. A side artery, of radius r_2, heads for a point C. Choose point B so that CB is perpendicular to AB. Let $CB = s$ and let D be the point where the axis of the branching vessel cuts the axis of the main vessel.

According to Poiseuille's law, the resistance R in the system is proportional to the length L of the vessel and inversely proportional to the fourth power of the radius r. That is,

$$R = k \cdot \frac{L}{r^4}, \tag{1}$$

where k is a constant determined by the viscosity of the blood. Let $AB = L_0$, $AD = L_1$, and $DC = L_2$.

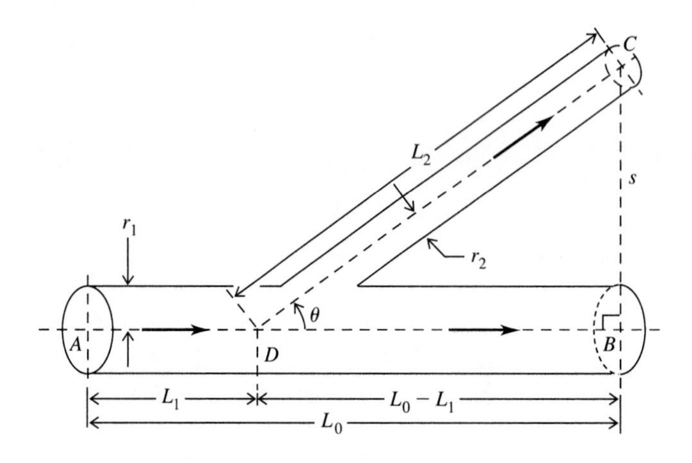

24. Use right triangle BDC and find $\sin \theta$.

25. Solve the result of Exercise 24 for L_2.

26. Find $\cot \theta$ in terms of s and $L_0 - L_1$.

27. Solve the result of Exercise 26 for L_1.

28. Write an expression similar to equation (1) for the resistance R_1 along AD.

29. Write a formula for the resistance along DC.

---

*Art from Edward Batschelet, *Introduction to Mathematics for Life Scientists.* Copyright © 1971 by Springer-Verlag New York, Inc. Reprinted by permission.

30. The total resistance R is given by the sum of the resistances along AD and DC. Use your answers to Exercises 28 and 29 to write an expression for R.
31. In your formula for R, replace L_1 with the result of Exercise 27 and L_2 with the result of Exercise 25.
32. Find R'. (Remember that k, L_1, L_0, s, r_1, and r_2 are constants.)
33. Set R' equal to 0.
34. Multiply through by $(\sin^2 \theta)/s$.
35. Solve for $\cos \theta$.
36. Suppose $r_1 = 1$ cm and $r_2 = 1/4$ cm. Find $\cos \theta$ and then find θ.
37. Find θ if $r_1 = 1.4$ cm and $r_2 = .8$ cm.

Chapter Summary — Key Terms

11.1 **ray**
angle
vertex
standard position
degree measure
acute angle
right angle
obtuse angle
straight angle
arc
radian measure
trigonometric functions
sine
cosine
tangent
cotangent
secant
cosecant
special angles
periodic functions
period

11.4 **inverse trigonometric functions**

11.5 **simple harmonic motion**

Chapter 11 — Review Exercises

1. What is the relationship between the degree measure and the radian measure of an angle?
2. Under what circumstances should radian measure be used instead of degree measure? Degree measure instead of radian measure?
3. Describe in words how each of the six trigonometric functions is defined.
4. At what angles (given as rational multiples of π) can you determine the exact values for the trigonometric functions?

Convert the following degree measures to radians. Leave answers as multiples of π.

5. $90°$
6. $120°$
7. $210°$
8. $270°$
9. $360°$
10. $420°$

Convert the following radian measures to degrees.

11. 7π
12. $\frac{3\pi}{4}$
13. $\frac{9\pi}{20}$
14. $\frac{7\pi}{15}$
15. $\frac{13\pi}{20}$
16. $\frac{11\pi}{15}$

Find each function value without using a calculator.

17. $\sin 60°$
18. $\tan 120°$
19. $\cos(-30°)$
20. $\sec 45°$
21. $\csc 120°$
22. $\cot 300°$
23. $\sin \frac{\pi}{6}$
24. $\cos \frac{2\pi}{3}$
25. $\sec \frac{5\pi}{4}$
26. $\csc \frac{7\pi}{3}$

Find each function value.

27. sin 47° **28.** cos 59° **29.** tan 81° **30.** sin (−32°)

31. sin 1.4661 **32.** cos .3142 **33.** cos .5934 **34.** tan 1.2915

35. Why are the inverse trigonometric functions useful?

Give each value in radians without using a calculator.

36. $\sin^{-1}(-1/2)$ **37.** $\tan^{-1}(-1)$ **38.** $\cos^{-1} 1$

39. $\sin^{-1}(\sqrt{3}/2)$ **40.** $\sin^{-1}(-1)$ **41.** $\tan^{-1}\sqrt{3}$

42. Because the derivative of $y = \sin x$ is $y' = \cos x$, the slope of $y = \sin x$ varies from ________ to ________ .

Graph one period of each function.

43. $y = 3 \cos x$ **44.** $y = \frac{1}{2} \tan x$ **45.** $y = -\tan x$ **46.** $y = -2 \sin x$

Find the derivative of each function.

47. $y = -4 \sin 7x$ **48.** $y = 6 \tan 3x$ **49.** $y = \tan(4x^2 + 3)$ **50.** $y = \cot(9 - x^2)$

51. $y = 3 \cos^6 x$ **52.** $y = 2 \sin^4(4x^2)$ **53.** $y = \cot(4x^5)$ **54.** $y = \cos(1 + x^2)$

55. $y = x^2 \cos x$ **56.** $y = e^{-x} \sin x$ **57.** $y = \dfrac{\sin x - 1}{\sin x + 1}$ **58.** $y = \dfrac{\cos^2 x}{1 - \cos x}$

59. $y = \dfrac{x - 2}{\sin x}$ **60.** $y = \dfrac{\tan x}{1 + x}$ **61.** $y = \ln|\cos x|$ **62.** $y = \ln|5 \sin x|$

63. $y = \sin^{-1}(-3x)$ **64.** $y = \tan^{-1}(1 + x)$ **65.** $y = \tan^{-1}(5x^2)$ **66.** $y = \sin^{-1}(2 - x)$

Find each integral.

67. $\int \sin 2x\, dx$ **68.** $\int \cos 3x\, dx$ **69.** $\int \tan 9x\, dx$

70. $\int \sec^2 5x\, dx$ **71.** $\int 5 \sec^2 x\, dx$ **72.** $\int 4 \csc^2 x\, dx$

73. $\int x \sin 3x^2\, dx$ **74.** $\int 5x \cos 2x^2\, dx$ **75.** $\int \sqrt{\cos x} \sin x\, dx$

76. $\int \sin^4 x \cos x\, dx$ **77.** $\int x \tan 11x^2\, dx$ **78.** $\int x^2 \cot 8x^3\, dx$

79. $\int (\sin x)^{5/2} \cos x\, dx$ **80.** $\int (\cos x)^{-4/3} \sin x\, dx$ **81.** $\displaystyle\int \frac{3}{\sqrt{1 - x^2}}\, dx$

82. $\displaystyle\int \frac{4}{1 + x^2}\, dx$ **83.** $\displaystyle\int \frac{-6x^4}{1 + x^{10}}\, dx$ **84.** $\displaystyle\int \frac{-12x^2}{\sqrt{1 - x^6}}\, dx$

Find each definite integral.

85. $\displaystyle\int_0^{\pi/2} \cos x\, dx$ **86.** $\displaystyle\int_{\pi/2}^{\pi} \sin x\, dx$ **87.** $\displaystyle\int_0^{2\pi} (10 + 10 \cos x)\, dx$ **88.** $\displaystyle\int_0^{2\pi} (5 + 5 \sin x)\, dx$

89. What is wrong with the following derivation?

$$\frac{d \sin^{-1} x}{dx} = (-1) \sin^{-2} x \cos x$$

90. Someone claims that since the derivative of a trigonometric function is a trigonometric function, then the derivative of an inverse trigonometric function should be an inverse trigonometric function. How would you respond to this statement?

91. The same person who gave the argument in Exercise 90 says that $\displaystyle\int \frac{1}{1 + x^2}\, dx$ cannot possibly involve trigonometry, since the original problem has nothing to do with trigonometry, circles, or angles. How would you respond to this statement?

Applications

Life Sciences

92. Blood Pressure A person's blood pressure at time t (in seconds) is given by

$$P(t) = 90 + 15 \sin 144\pi t.$$

Find the maximum and minimum values of P on the interval $[0, 1/72]$. Graph one period of $y = P(t)$.

General Interest

93. Area A 6-ft board is placed against a wall as shown in the figure, forming a triangular-shaped area beneath it. At what angle θ should the board be placed to make the triangular area as large as possible?

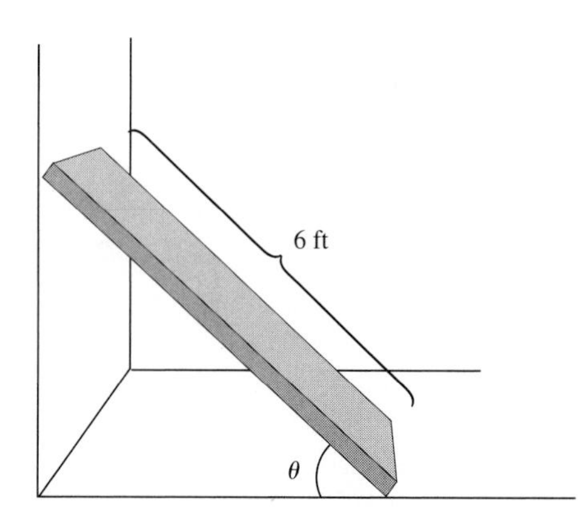

94. Movie Viewing A 10-foot-high movie screen is located 2 ft above the eyes of the viewers, all of whom are sitting at the same level. How far away from the screen should a viewer sit for the maximum viewing area? (See Section 11.4, Exercises 50–53.)

Connections

95. Mercator's World Map Before Gerardus Mercator designed his map of the world in 1569, sailors who traveled in a fixed compass direction could follow a straight line on a map only over short distances. Over long distances, such a course would be a curve on existing maps, which tried to make area on the map proportional to the actual area. Mercator's map greatly simplified navigation: even over long distances, straight lines on the map corresponded to fixed compass bearings. This was accomplished by distorting distances. On Mercator's map, the distance of an object from the equator to a parallel at latitude θ is given by

$$D(\theta) = k \int_0^{\theta} \sec x \, dx,$$

where k is a constant of proportionality. Calculus had not yet been discovered when Mercator designed his map; he approximated the distance between parallels of latitude by hand.*

(a) Verify that

$$\frac{d}{dx} \ln |\sec x + \tan x| = \sec x.$$

(b) Verify that

$$\frac{d}{dx}(-\ln |\sec x - \tan x|) = \sec x.$$

(c) Using parts (a) and (b), give two different formulas for $\int \sec x \, dx$. Explain how they can both be correct.

(d) Los Angeles has a latitude of 34°03′N. (The 03′ represents 3 minutes of latitude. Each minute of latitude is $\frac{1}{60}$ of a degree.) If Los Angeles is to be 7 inches from the equator on a Mercator map, how far from the equator should we place New York City, which has a latitude of 40°45′N?

(e) Repeat part (d) for Miami, which has a latitude of 25°46′N.

(f) If you do not live in Los Angeles, New York City, or Miami, repeat part (d) for your town or city.

---

*From V. Frederick Rickey and Philip M. Tuchinsky, ''An Application of Geography to Mathematics: History of the Integral of the Secant,'' *Mathematics Magazine*, vol. 53, May 1980, pp. 162–166.

Extended Application / The Mathematics of a Honeycomb*

The bee's cell is a regular hexagonal prism with one open end and one trihedral apex (see Figure 22).† We may construct the surface by starting with a regular hexagonal base *abcdef* with side *s*. Over the base we raise a right prism of height *h* and top *ABCDEF*. The corners *B*, *D*, and *F* are cut off by planes through the lines *AC*, *CE*, and *EA*, meeting in a point *V* on the axis *VN* of the prism, and intersecting *Bb*, *Dd*, and *Ff* in *X*, *Y*, and *Z*. The three cutoff pieces are the tetrahedra *ABCX, CDEY*, and *EFAZ*. We put these pieces on top of the remaining solid such that *X*, *Y*, and *Z* coincide with *V*. Hereby, the lines *AC*, *CE*, and *EA* act as "hinges." The faces *AXCV*, *CYEV*, *EZAV* are rhombuses; that is, quadrilaterals with equal sides. The new body is the bee's cell and has the same volume as the original prism. The hexagonal base *abcdef* is the open end.

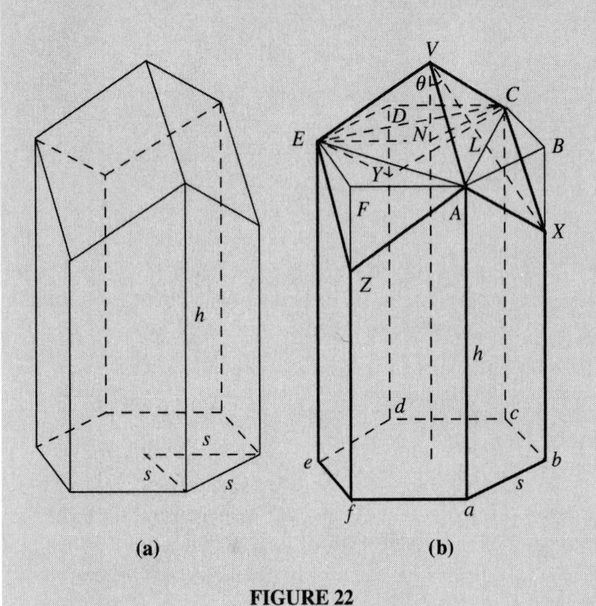

FIGURE 22

The bees form the faces by using wax. When the volume is given, it is economic to spare wax and therefore to choose the angle of inclination, $\theta = NVX$, in such a way that the *surface of the bee's cell is minimized.*

The problem can be solved mathematically as follows. Let *L* be the intersection of *CA* and *VX*. Then *L* bisects the segment *NB* and, hence, $NL = s/2$. The segment *CL* is the height of the equilateral triangle *BCN*. Therefore,

$$CL = \frac{s}{2}\sqrt{3}. \tag{1}$$

In the triangle *NLV* we have the relationship

$$VL = \frac{s}{2 \sin \theta}. \tag{2}$$

The rhombus *AXCV* has its center in *L* and consists of four congruent right triangles with legs equal to *CL* and *VL*. Therefore, from (1) and (2) we get

$$\text{Area } AXCV = 4 \cdot \frac{1}{2} \cdot \frac{s}{2}\sqrt{3} \cdot \frac{s}{2 \sin \theta} = \frac{s^2\sqrt{3}}{2 \sin \theta}.$$

The surface of the bee's cell contains three such areas.

The six lateral faces of the bee's cell, such as *abXA*, are congruent trapezoids. Since $BX = VN$, we obtain from triangle *VNL*:

$$BX = \frac{s}{2} \cot \theta.$$

Hence,

$$\text{Area } abXA = \frac{s}{2}(aA + bX) = \frac{s}{2}(h + h - BX)$$

$$= hs - \frac{s^2}{4} \cot \theta.$$

The total area made of wax amounts to

$$6hs - \frac{3}{2}s^2 \cot \theta + \frac{3s^2\sqrt{3}}{2 \sin \theta}.$$

*From Edward Batschelet, *Introduction to Mathematics for Life Scientists*. Copyright © 1971 by Springer-Verlag New York, Inc. Reprinted by permission.

†"The Bee's Cell" from d'Arcy W. Thompson, *On Growth and Form*, Cambridge University Press, as adapted in *Introduction to Mathematics for Life Scientists* by Edward Batschelet, Springer-Verlag New York, Inc. Reprinted by permission.

This area is a function of the variable angle θ, and thus we denote it by $f(\theta)$. We may rewrite $f(\theta)$ in the form

$$f(\theta) = 6hs + \frac{3}{2}s^2\left(-\cot\theta + \frac{\sqrt{3}}{\sin\theta}\right).$$

Only the expression in the parentheses contains the variable θ. Some numerical values (rounded off to two decimals) are given in the following table.

θ	$-\cot\theta + \dfrac{\sqrt{3}}{\sin\theta}$
10°	4.30
20°	2.32
30°	1.73
40°	1.50
50°	1.42
60°	1.42
70°	1.48
80°	1.58
90°	1.73

The minimum of $f(\theta)$ is reached somewhere between $\theta = 50°$ and $\theta = 60°$. To get the optimal angle, say θ_0, we differentiate $f(\theta)$.

$$f'(\theta) = \frac{3}{2}s^2\left(\frac{1}{\sin^2\theta} - \frac{\sqrt{3}\cos\theta}{\sin^2\theta}\right) \tag{3}$$

The derivative vanishes if, and only if,

$$1 = \sqrt{3}\cos\theta. \tag{4}$$

Hence, $\cos\theta_0 = 1/\sqrt{3} = .57735$, and $\theta_0 = 54.7°$. Notice that the optimal angle θ_0 is independent of the choice of s and h.

It is worth comparing the result with the actual angle chosen by the bees. It is difficult to measure this angle; however, the average of all measurements does not differ significantly from the theoretical value $\theta_0 = 54.7°$. Therefore, the bees strongly prefer the optimal angle. It is unlikely that the result is due to chance. Rather, we may suppose that selection pressure had an effect on the angle θ.

Exercises

1. Show how equation (3) above was found.

2. Show how the result given in equation (4) was obtained.

TABLES

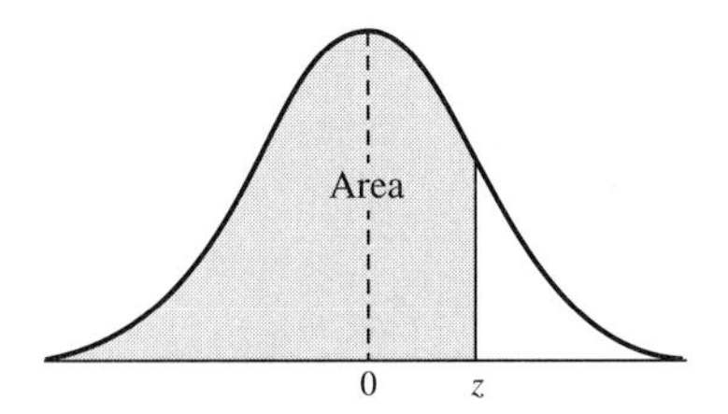

Table 1: Area Under a Normal Curve to the Left of z, Where $z = \dfrac{x - \mu}{\sigma}$

z	.00	.01	.02	.03	.04	.05	.06	.07	.08	.09
−3.4	.0003	.0003	.0003	.0003	.0003	.0003	.0003	.0003	.0003	.0002
−3.3	.0005	.0005	.0005	.0004	.0004	.0004	.0004	.0004	.0004	.0003
−3.2	.0007	.0007	.0006	.0006	.0006	.0006	.0006	.0005	.0005	.0005
−3.1	.0010	.0009	.0009	.0009	.0008	.0008	.0008	.0008	.0007	.0007
−3.0	.0013	.0013	.0013	.0012	.0012	.0011	.0011	.0011	.0010	.0010
−2.9	.0019	.0018	.0017	.0017	.0016	.0016	.0015	.0015	.0014	.0014
−2.8	.0026	.0025	.0024	.0023	.0023	.0022	.0021	.0021	.0020	.0019
−2.7	.0035	.0034	.0033	.0032	.0031	.0030	.0029	.0028	.0027	.0026
−2.6	.0047	.0045	.0044	.0043	.0041	.0040	.0039	.0038	.0037	.0036
−2.5	.0062	.0060	.0059	.0057	.0055	.0054	.0052	.0051	.0049	.0048
−2.4	.0082	.0080	.0078	.0075	.0073	.0071	.0069	.0068	.0066	.0064
−2.3	.0107	.0104	.0102	.0099	.0096	.0094	.0091	.0089	.0087	.0084
−2.2	.0139	.0136	.0132	.0129	.0125	.0122	.0119	.0116	.0113	.0110
−2.1	.0179	.0174	.0170	.0166	.0162	.0158	.0154	.0150	.0146	.0143
−2.0	.0228	.0222	.0217	.0212	.0207	.0202	.0197	.0192	.0188	.0183
−1.9	.0287	.0281	.0274	.0268	.0262	.0256	.0250	.0244	.0239	.0233
−1.8	.0359	.0352	.0344	.0336	.0329	.0322	.0314	.0307	.0301	.0294
−1.7	.0446	.0436	.0427	.0418	.0409	.0401	.0392	.0384	.0375	.0367
−1.6	.0548	.0537	.0526	.0516	.0505	.0495	.0485	.0475	.0465	.0455
−1.5	.0668	.0655	.0643	.0630	.0618	.0606	.0594	.0582	.0571	.0559
−1.4	.0808	.0793	.0778	.0764	.0749	.0735	.0722	.0708	.0694	.0681
−1.3	.0968	.0951	.0934	.0918	.0901	.0885	.0869	.0853	.0838	.0823
−1.2	.1151	.1131	.1112	.1093	.1075	.1056	.1038	.1020	.1003	.0985
−1.1	.1357	.1335	.1314	.1292	.1271	.1251	.1230	.1210	.1190	.1170
−1.0	.1587	.1562	.1539	.1515	.1492	.1469	.1446	.1423	.1401	.1379

Table 1: Area Under a Normal Curve (continued)

z	.00	.01	.02	.03	.04	.05	.06	.07	.08	.09
−.9	.1841	.1814	.1788	.1762	.1736	.1711	.1685	.1660	.1635	.1611
−.8	.2119	.2090	.2061	.2033	.2005	.1977	.1949	.1922	.1894	.1867
−.7	.2420	.2389	.2358	.2327	.2296	.2266	.2236	.2206	.2177	.2148
−.6	.2743	.2709	.2676	.2643	.2611	.2578	.2546	.2514	.2483	.2451
−.5	.3085	.3050	.3015	.2981	.2946	.2912	.2877	.2843	.2810	.2776
−.4	.3446	.3409	.3372	.3336	.3300	.3264	.3228	.3192	.3156	.3121
−.3	.3821	.3783	.3745	.3707	.3669	.3632	.3594	.3557	.3520	.3483
−.2	.4207	.4168	.4129	.4090	.4052	.4013	.3974	.3936	.3897	.3859
−.1	.4602	.4562	.4522	.4483	.4443	.4404	.4364	.4325	.4286	.4247
−.0	.5000	.4960	.4920	.4880	.4840	.4801	.4761	.4721	.4681	.4641
.0	.5000	.5040	.5080	.5120	.5160	.5199	.5239	.5279	.5319	.5359
.1	.5398	.5438	.5478	.5517	.5557	.5596	.5636	.5675	.5714	.5753
.2	.5793	.5832	.5871	.5910	.5948	.5987	.6026	.6064	.6103	.6141
.3	.6179	.6217	.6255	.6293	.6331	.6368	.6406	.6443	.6480	.6517
.4	.6554	.6591	.6628	.6664	.6700	.6736	.6772	.6808	.6844	.6879
.5	.6915	.6950	.6985	.7019	.7054	.7088	.7123	.7157	.7190	.7224
.6	.7257	.7291	.7324	.7357	.7389	.7422	.7454	.7486	.7517	.7549
.7	.7580	.7611	.7642	.7673	.7704	.7734	.7764	.7794	.7823	.7852
.8	.7881	.7910	.7939	.7967	.7995	.8023	.8051	.8078	.8106	.8133
.9	.8159	.8186	.8212	.8238	.8264	.8289	.8315	.8340	.8365	.8389
1.0	.8413	.8438	.8461	.8485	.8508	.8531	.8554	.8577	.8599	.8621
1.1	.8643	.8665	.8686	.8708	.8729	.8749	.8770	.8790	.8810	.8830
1.2	.8849	.8869	.8888	.8907	.8925	.8944	.8962	.8980	.8997	.9015
1.3	.9032	.9049	.9066	.9082	.9099	.9115	.9131	.9147	.9162	.9177
1.4	.9192	.9207	.9222	.9236	.9251	.9265	.9278	.9292	.9306	.9319
1.5	.9332	.9345	.9357	.9370	.9382	.9394	.9406	.9418	.9429	.9441
1.6	.9452	.9463	.9474	.9484	.9495	.9505	.9515	.9525	.9535	.9545
1.7	.9554	.9564	.9573	.9582	.9591	.9599	.9608	.9616	.9625	.9633
1.8	.9641	.9649	.9656	.9664	.9671	.9678	.9686	.9693	.9699	.9706
1.9	.9713	.9719	.9726	.9732	.9738	.9744	.9750	.9756	.9761	.9767
2.0	.9772	.9778	.9783	.9788	.9793	.9798	.9803	.9808	.9812	.9817
2.1	.9821	.9826	.9830	.9834	.9838	.9842	.9846	.9850	.9854	.9857
2.2	.9861	.9864	.9868	.9871	.9875	.9878	.9881	.9884	.9887	.9890
2.3	.9893	.9896	.9898	.9901	.9904	.9906	.9909	.9911	.9913	.9916
2.4	.9918	.9920	.9922	.9925	.9927	.9929	.9931	.9932	.9934	.9936
2.5	.9938	.9940	.9941	.9943	.9945	.9946	.9948	.9949	.9951	.9952
2.6	.9953	.9955	.9956	.9957	.9959	.9960	.9961	.9962	.9963	.9964
2.7	.9965	.9966	.9967	.9968	.9969	.9970	.9971	.9972	.9973	.9974
2.8	.9974	.9975	.9976	.9977	.9977	.9978	.9979	.9979	.9980	.9981
2.9	.9981	.9982	.9982	.9983	.9984	.9984	.9985	.9985	.9986	.9986
3.0	.9987	.9987	.9987	.9988	.9988	.9989	.9989	.9989	.9990	.9990
3.1	.9990	.9991	.9991	.9991	.9992	.9992	.9992	.9992	.9993	.9993
3.2	.9993	.9993	.9994	.9994	.9994	.9994	.9994	.9995	.9995	.9995
3.3	.9995	.9995	.9995	.9996	.9996	.9996	.9996	.9996	.9996	.9997
3.4	.9997	.9997	.9997	.9997	.9997	.9997	.9997	.9997	.9997	.9998

Table 2: Integrals

(C is an arbitrary constant.)

1. $\int x^n \, dx = \frac{1}{n+1}x^{n+1} + C \qquad (\text{if } n \neq -1)$

2. $\int e^{kx} \, dx = \frac{1}{k}e^{kx} + C$

3. $\int \frac{a}{x} dx = a \ln |x| + C$

4. $\int \ln |ax| \, dx = x (\ln |ax| - 1) + C$

5. $\int \frac{1}{\sqrt{x^2 + a^2}} \, dx = \ln |x + \sqrt{x^2 + a^2}| + C$

6. $\int \frac{1}{\sqrt{x^2 - a^2}} \, dx = \ln |x + \sqrt{x^2 - a^2}| + C$

7. $\int \frac{1}{a^2 - x^2} \, dx = \frac{1}{2a} \cdot \ln \left| \frac{a + x}{a - x} \right| + C \qquad (a \neq 0)$

8. $\int \frac{1}{x^2 - a^2} \, dx = \frac{1}{2a} \cdot \ln \left| \frac{x - a}{x + a} \right| + C \qquad (a \neq 0)$

9. $\int \frac{1}{x\sqrt{a^2 - x^2}} \, dx = -\frac{1}{a} \cdot \ln \left| \frac{a + \sqrt{a^2 - x^2}}{x} \right| + C \qquad (a \neq 0)$

10. $\int \frac{1}{x\sqrt{a^2 + x^2}} \, dx = -\frac{1}{a} \cdot \ln \left| \frac{a + \sqrt{a^2 + x^2}}{x} \right| + C \qquad (a \neq 0)$

11. $\int \frac{x}{ax + b} \, dx = \frac{x}{a} - \frac{b}{a^2} \cdot \ln |ax + b| + C \qquad (a \neq 0)$

12. $\int \frac{x}{(ax + b)^2} \, dx = \frac{b}{a^2(ax + b)} + \frac{1}{a^2} \cdot \ln |ax + b| + C \qquad (a \neq 0)$

13. $\int \frac{1}{x(ax + b)} \, dx = \frac{1}{b} \cdot \ln \left| \frac{x}{ax + b} \right| + C \qquad (b \neq 0)$

14. $\int \frac{1}{x(ax + b)^2} \, dx = \frac{1}{b(ax + b)} + \frac{1}{b^2} \cdot \ln \left| \frac{x}{ax + b} \right| + C \qquad (b \neq 0)$

15. $\int \sqrt{x^2 + a^2} \, dx = \frac{x}{2} \sqrt{x^2 + a^2} + \frac{a^2}{2} \cdot \ln |x + \sqrt{x^2 + a^2}| + C$

16. $\int x^n \cdot \ln |x| \, dx = x^{n+1} \left[\frac{\ln |x|}{n + 1} - \frac{1}{(n + 1)^2} \right] + C \qquad (n \neq -1)$

17. $\int x^n e^{ax} \, dx = \frac{x^n e^{ax}}{a} - \frac{n}{a} \cdot \int x^{n-1} e^{ax} \, dx + C \qquad (a \neq 0)$

Table 3: Integrals Involving Trigonometric Functions

18. $\int \sin u \, du = -\cos u + C$

19. $\int \cos u \, du = \sin u + C$

20. $\int \sec^2 u \, du = \tan u + C$

21. $\int \csc^2 u \, du = -\cot u + C$

22. $\int \sec u \tan u \, du = \sec u + C$

23. $\int \csc u \cot u \, du = -\csc u + C$

24. $\int \tan u \, du = \ln |\sec u| + C$

25. $\int \cot u \, du = \ln |\sin u| + C$

26. $\int \sec u \, du = \ln |\sec u + \tan u| + C$

27. $\int \csc u \, du = \ln |\csc u - \cot u| + C$

28. $\int \frac{du}{\sqrt{a^2 - u^2}} = \text{Arcsin}\frac{u}{a} + C \quad (a \neq 0)$

29. $\int \frac{du}{a^2 + u^2} = \frac{1}{a}\text{Arctan}\frac{u}{a} + C \quad (a \neq 0)$

30. $\int \frac{du}{u\sqrt{u^2 - a^2}} = \frac{1}{a}\text{Arcsec}\frac{u}{a} + C \quad (a \neq 0)$

31. $\int \sqrt{a^2 - u^2} \, du = \frac{u}{2}\sqrt{a^2 - u^2} + \frac{a^2}{2}\text{Arcsin}\frac{u}{a} + C \quad (a \neq 0)$

32. $\int u^2\sqrt{a^2 - u^2} \, du = \frac{u}{8}(2u^2 - a^2)\sqrt{a^2 - u^2} + \frac{a^4}{8}\text{Arcsin}\frac{u}{a} + C \quad (a \neq 0)$

33. $\int \frac{\sqrt{a^2 - u^2}}{u^2} du = -\frac{1}{u}\sqrt{a^2 - u^2} - \text{Arcsin}\frac{u}{a} + C \quad (a \neq 0)$

34. $\int \frac{u^2 du}{\sqrt{a^2 - u^2}} = -\frac{u}{2}\sqrt{a^2 - u^2} + \frac{a^2}{2}\text{Arcsin}\frac{u}{a} + C \quad (a \neq 0)$

35. $\int \frac{\sqrt{u^2 - a^2}}{u} du = \sqrt{u^2 - a^2} - a \, \text{Arccos}\frac{a}{u} + C$

Table 3: Integrals Involving Trigonometric Functions (continued)

36. $\int \sin^n u \, du = -\frac{1}{n} \sin^{n-1} u \cos u + \frac{n-1}{n} \int \sin^{n-2} u \, du \quad (n \neq 0)$

37. $\int \cos^n u \, du = \frac{1}{n} \cos^{n-1} u \sin u + \frac{n-1}{n} \int \cos^{n-2} u \, du \quad (n \neq 0)$

38. $\int \tan^n u \, du = \frac{1}{n-1} \tan^{n-1} u - \int \tan^{n-2} u \, du \quad (n \neq 1)$

39. $\int \sec^n u \, du = \frac{1}{n-1} \tan u \sec^{n-2} u + \frac{n-2}{n-1} \int \sec^{n-2} u \, du \quad (n \neq 1)$

40. $\int \sin au \sin bu \, du = \frac{\sin (a-b)u}{2(a-b)} - \frac{\sin (a+b)u}{2(a+b)} + C, \quad |a| \neq |b|$

41. $\int \cos au \, bu \, du = \frac{\sin (a-b)u}{2(a-b)} + \frac{\sin (a+b)u}{2(a+b)} + C, \quad |a| \neq |b|$

42. $\int \sin au \cos bu \, du = -\frac{\cos(a-b)u}{2(a-b)} - \frac{\cos (a+b)u}{2(a+b)} + C, \quad |a| \neq |b|$

43. $\int u \sin u \, du = \sin u - u \cos u + C$

44. $\int u^n \sin u \, du = -u^n \cos u + n \int u^{n-1} \cos u \, du$

45. $\int u \operatorname{Arccos} u \, du = \frac{2u^2 - 1}{4} \operatorname{Arccos} u - \frac{u\sqrt{1-u^2}}{4} + C$

46. $\int u \operatorname{Arcsin} u \, du = \frac{2u^2 - 1}{4} \operatorname{Arcsin} u + \frac{u\sqrt{1-u^2}}{4} + C$

47. $\int e^{au} \sin bu \, du = \frac{e^{au}}{a^2 + b^2}(a \sin bu - b \cos bu) + C$

48. $\int e^{au} \cos bu \, du = \frac{e^{au}}{a^2 + b^2}(a \cos bu + b \sin bu) + C$

ANSWERS TO SELECTED EXERCISES

To the Student

If you want further help with this course, you may wish to buy a copy of the *Student's Solution Manual* that accompanies this textbook. This manual provides detailed, step-by-step solutions to the odd-numbered exercises in the textbook. Also included are practice tests for each chapter, with answers given so you can check your work. Your college bookstore either has this manual or can order it for you.

ALGEBRA REFERENCE

Section R.1 (page xxii)

1. $-x^2 + x + 9$ **2.** $-6y^2 + 3y + 10$ **3.** $-14q^2 + 11q - 14$ **4.** $9r^2 - 4r + 19$ **5.** $-.327x^2 - 2.805x - 1.458$ **6.** $-2.97r^2 - 8.083r + 7.81$ **7.** $-18m^3 - 27m^2 + 9m$ **8.** $12k^2 - 20k + 3$ **9.** $25r^2 + 5rs - 12s^2$ **10.** $18k^2 - 7kq - q^2$ **11.** $\frac{6}{25}y^2 + \frac{11}{40}yz + \frac{1}{16}z^2$ **12.** $\frac{15}{16}r^2 - \frac{7}{12}rs - \frac{2}{9}s^2$ **13.** $144x^2 - 1$ **14.** $36m^2 - 25$ **15.** $27p^3 - 1$ **16.** $6p^3 - 11p^2 + 14p - 5$ **17.** $8m^3 + 1$ **18.** $12k^4 + 21k^3 - 5k^2 + 3k + 2$ **19.** $m^2 + mn - 2n^2 - 2km + 5kn - 3k^2$ **20.** $2r^2 - 7rs + 3s^2 + 3rt - 4st + t^2$

Section R.2 (page xxiv)

1. $8a(a^2 - 2a + 3)$ **2.** $3y(y^2 + 8y + 3)$ **3.** $5p^2(5p^2 - 4pq + 20q^2)$ **4.** $10m^2(6m^2 - 12mn + 5n^2)$ **5.** $(m + 7)(m + 2)$ **6.** $(x + 5)(x - 1)$ **7.** $(z + 4)(z + 5)$ **8.** $(b - 7)(b - 1)$ **9.** $(a - 5b)(a - b)$ **10.** $(s - 5t)(s + 7t)$ **11.** $(y - 7z)(y + 3z)$ **12.** $6(a - 10)(a + 2)$ **13.** $3m(m + 3)(m + 1)$ **14.** $(2x + 1)(x - 3)$ **15.** $(3a + 7)(a + 1)$ **16.** $(2a - 5)(a - 6)$ **17.** $(5y + 2)(3y - 1)$ **18.** $(7m + 2n)(3m + n)$ **19.** $2a^2(4a - b)(3a + 2b)$ **20.** $4z^3(8z + 3a)(z - a)$ **21.** $(x + 8)(x - 8)$ **22.** $(3m + 5)(3m - 5)$ **23.** $(11a + 10)(11a - 10)$ **24.** prime **25.** $(z + 7y)^2$ **26.** $(m - 3n)^2$ **27.** $(3p - 4)^2$ **28.** $(a - 6)(a^2 + 6a + 36)$ **29.** $(2r - 3s)(4r^2 + 6rs + 9s^2)$ **30.** $(4m + 5)(16m^2 - 20m + 25)$

Section R.3 (page xxvii)

1. $z/2$ **2.** $5p/2$ **3.** $8/9$ **4.** $3/(t - 3)$ **5.** $2(x + 2)/x$ **6.** $4(y + 2)$ **7.** $(m - 2)/(m + 3)$ **8.** $(r + 2)/(r + 4)$ **9.** $(x + 4)/(x + 1)$ **10.** $(z - 3)/(z + 2)$ **11.** $(2m + 3)/(4m + 3)$ **12.** $(2y + 1)/(y + 1)$ **13.** $3k/5$ **14.** $25p^2/9$ **15.** $6/(5p)$ **16.** 2 **17.** $2/9$ **18.** $3/10$ **19.** $2(a + 4)/(a - 3)$ **20.** $2/(r + 2)$ **21.** $(k + 2)/(k + 3)$ **22.** $(m + 6)/(m + 3)$ **23.** $(m - 3)/(2m - 3)$ **24.** $(2n - 3)/(2n + 3)$ **25.** 1 **26.** $(6 + p)/(2p)$ **27.** $(8 - y)/(4y)$ **28.** $137/(30m)$ **29.** $(3m - 2)/[m(m - 1)]$ **30.** $(r - 12)/[r(r - 2)]$ **31.** $14/[3(a - 1)]$ **32.** $23/[20(k - 2)]$ **33.** $(7x + 9)/[(x - 3)(x + 1)(x + 2)]$ **34.** $y^2/[(y + 4)(y + 3)(y + 2)]$ **35.** $k(k - 13)/[(2k - 1)(k + 2)(k - 3)]$ **36.** $m(3m - 19)/[(3m - 2)(m + 3)(m - 4)]$

Section R.4 (page xxxiii)

1. 12 **2.** $-2/7$ **3.** $-7/8$ **4.** -1 **5.** $-11/3, 7/3$ **6.** $-11/7, 19/7$ **7.** $-12, -4/3$ **8.** $-5, 3/4$ **9.** $-3, -2$ **10.** $-1, 3$ **11.** 4 **12.** $-2, 5/2$ **13.** $-1/2, 4/3$ **14.** 2, 5 **15.** $-4/3, 4/3$ **16.** $-4, 1/2$ **17.** 0, 4 **18.** $(5 + \sqrt{13})/6 \approx 1.434, (5 - \sqrt{13})/6 \approx .232$ **19.** $(1 + \sqrt{33})/4 \approx 1.686, (1 - \sqrt{33})/4 \approx -1.186$ **20.** $(-1 + \sqrt{5})/2 \approx .618, (-1 - \sqrt{5})/2 \approx -1.618$ **21.** $5 + \sqrt{5} \approx 7.236, 5 - \sqrt{5} \approx 2.764$ **22.** $(-6 + \sqrt{26})/2 \approx -.450, (-6 - \sqrt{26})/2 \approx -5.550$ **23.** 1, 5/2 **24.** no real-number solutions **25.** $(-1 + \sqrt{73})/6 \approx 1.257, (-1 - \sqrt{73})/6 \approx -1.591$ **26.** $-1, 0$ **27.** 3 **28.** 12 **29.** $-59/6$ **30.** $-11/5$ **31.** no real-number solutions **32.** $-5/2$ **33.** 2/3 **34.** 1 **35.** $(-13 - \sqrt{185})/4 \approx -6.650, (-13 + \sqrt{185})/4 \approx .150$

Section R.5 (page xxxviii)

1. $(-\infty, -1]$

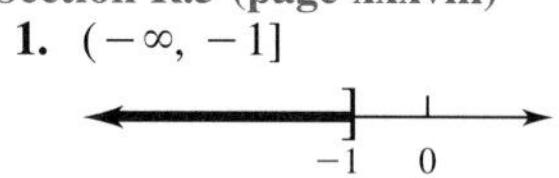

2. $(-\infty, 1)$ 0 1

3. $(-1, \infty)$ −1 0

4. $(-\infty, 1]$ 0 1

5. $(1/5, \infty)$ 1/5 0 1

6. $(1/3, \infty)$

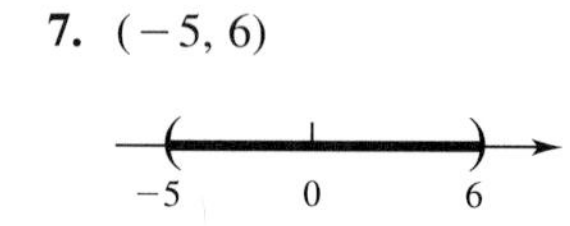

7. $(-5, 6)$ −5 0 6

8. $[7/3, 4]$ 0 7/3 4

9. $[-11/2, 7/2]$ −11/2 0 7/2

10. $[-1, 2]$ −1 0 2

11. $[-17/7, \infty)$ −17/7 0

12. $(-\infty, 50/9]$ 0 50/9

13. $(-2, 4)$ −2 0 4

14. $(-\infty, -6] \cup [1, \infty)$ −6 0 1

15. $(1, 2)$ 0 1 2

16. $(-\infty, -4) \cup (1/2, \infty)$

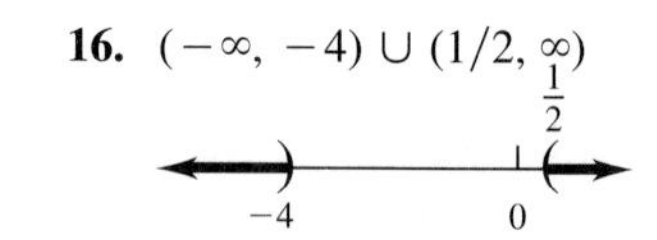

17. $[1, 6]$ 0 1 6

18. $[-3/2, 5]$ −3/2 0 5

19. $(-\infty, -1/2) \cup (1/3, \infty)$ −1/2 0 1/3

20. $[-1/2, 2/5]$

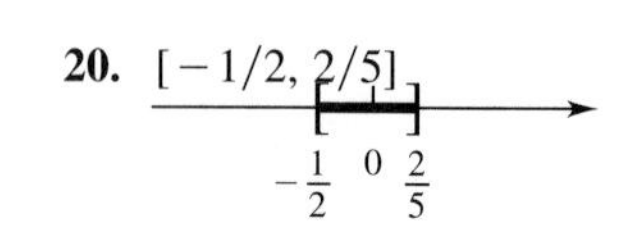

21. $[-3, 1/2]$ −3 0 1/2

22. $(-\infty, -2) \cup (5/3, \infty)$ −2 0 5/3

23. $[-5, 5]$ −5 0 5

24. $(-\infty, 0) \cup (16, \infty)$ 0 16

25. $(-5, 3]$ **26.** $(-\infty, -1] \cup (1, \infty)$ **27.** $(-\infty, -2)$ **28.** $(-2, 3/2)$ **29.** $[-8, 5)$ **30.** $(-\infty, -3/2) \cup [-13/9, \infty)$ **31.** $(-2, \infty)$ **32.** $(-\infty, -1)$ **33.** $(-\infty, -1) \cup (-1/2, 1) \cup (2, \infty)$ **34.** $(-4, -2) \cup (0, 2)$ **35.** $(1, 3/2]$ **36.** $(-\infty, -2) \cup (-2, 2) \cup [4, \infty)$

Section R.6 (page xli)

1. 1/64 **2.** 1/81 **3.** 1/216 **4.** 1 **5.** 1 **6.** 3/4 **7.** $-1/16$ **8.** 1/16 **9.** $-1/9$ **10.** 1/9 **11.** 25/64 **12.** 216/343 **13.** 8 **14.** 125 **15.** 49/4 **16.** 27/64 **17.** $1/7^4$ **18.** $1/3^6$ **19.** $1/2^3$ **20.** $1/6^2$ **21.** 4^3 **22.** 8^5 **23.** $1/10^8$ **24.** 5 **25.** x^2 **26.** y^3 **27.** 2^3k^3 **28.** $1/(3z^7)$ **29.** $x^2/(2y)$ **30.** $m^3/5^4$ **31.** a^3b^6 **32.** $d^6/(2^2c^4)$ **33.** 1/6 **34.** $-17/9$ **35.** $-13/66$ **36.** 81/26 **37.** 35/18 **38.** 213/200 **39.** 9 **40.** 3 **41.** 4 **42.** 100 **43.** 4 **44.** -25 **45.** 2/3 **46.** 4/3 **47.** 1/32 **48.** 1/5 **49.** 4/3 **50.** 1000/1331 **51.** 2^2 **52.** $27^{1/3}$ **53.** 4^2 **54.** 1 **55.** r **56.** $12^3/y^8$

57. $1/(2^2 \cdot 3k^{5/2})$ or $1/(12k^{5/2})$ **58.** $1/(2p^2)$ **59.** $a^{2/3}b^2$ **60.** $y/(x^{4/3}z^{1/2})$ **61.** $h^{1/3}t^{1/5}/k^{2/5}$ **62.** m^3p/n
63. $3x^3(x^2 - 1)^{-1/2}$ **64.** $5(5x + 2)^{-1/2}(45x^2 + 3x - 5)$ **65.** $(2x + 5)(x^2 - 4)^{-1/2}(4x^2 + 5x - 8)$
66. $(4x^2 + 1)(2x - 1)^{-1/2}(4x^2 + 4x - 1)$

Section R.7 (page xlv)
1. 5 **2.** 6 **3.** -5 **4.** $5\sqrt{2}$ **5.** $20\sqrt{5}$ **6.** $4y^2\sqrt{2y}$ **7.** $7\sqrt{2}$ **8.** $9\sqrt{3}$ **9.** $2\sqrt{5}$ **10.** $-2\sqrt{7}$
11. $5\sqrt[3]{2}$ **12.** $7\sqrt[3]{3}$ **13.** $3\sqrt[3]{4}$ **14.** $xyz^2\sqrt{2x}$ **15.** $7rs^2t^5\sqrt{2r}$ **16.** $2zx^2y\sqrt[3]{2z^2x^2y}$ **17.** $x^2yz^2\sqrt[4]{y^3z^3}$
18. $ab\sqrt{ab}(b - 2a^2 + b^3)$ **19.** $p^2\sqrt{pq}(pq - q^4 + p^2)$ **20.** $5\sqrt{7}/7$ **21.** $-2\sqrt{3}/3$ **22.** $-\sqrt{3}/2$ **23.** $\sqrt{2}$
24. $-3(1 + \sqrt{5})/4$ **25.** $-5(2 + \sqrt{6})/2$ **26.** $-2(\sqrt{3} + \sqrt{2})$ **27.** $(\sqrt{10} - \sqrt{3})/7$ **28.** $(\sqrt{r} + \sqrt{3})/(r - 3)$
29. $5(\sqrt{m} + \sqrt{5})/(m - 5)$ **30.** $\sqrt{y} + \sqrt{5}$ **31.** $\sqrt{z} + \sqrt{11}$ **32.** $-2x - 2\sqrt{x(x + 1)} - 1$
33. $(p^2 + p + 2\sqrt{p(p^2 - 1)} - 1)/(-p^2 + p + 1)$ **34.** $-1/[2(1 - \sqrt{2})]$ **35.** $-2/[3(1 + \sqrt{3})]$
36. $-1/(2x - 2\sqrt{x(x + 1)} + 1)$ **37.** $(-p^2 + p + 1)/(p^2 + p - 2\sqrt{p(p^2 - 1)} - 1)$ **38.** $|4 - x|$
39. $|2y + 1|$ **40.** cannot be simplified **41.** cannot be simplified

CHAPTER 1

Section 1.1 (page 12)
1. not a function **3.** function **5.** function **7.** not a function

9. $(-2, -3), (-1, -2), (0, -1), (1, 0), (2, 1), (3, 2)$; range: $\{-3, -2, -1, 0, 1, 2\}$

11. $(-2, 17), (-1, 13), (0, 9)$ $(1, 5), (2, 1), (3, -3)$; range: $\{-3, 1, 5, 9, 13, 17\}$

13. $(-2, 13), (-1, 11), (0, 9), (1, 7), (2, 5), (3, 3)$; range: $\{3, 5, 7, 9, 11, 13\}$

15. $(-2, 3/2), (-1, 2), (0, 5/2), (1, 3), (2, 7/2), (3, 4)$; range $\{3/2, 2, 5/2, 3, 7/2, 4\}$

17. $(-2, 2), (-1, 0), (0, 0), (1, 2), (2, 6), (3, 12)$; range: $\{0, 2, 6, 12\}$

19. $(-2, 4), (-1, 1), (0, 0), (1, 1), (2, 4), (3, 9)$; range: $\{0, 1, 4, 9\}$

21. $(-2, 1), (-1, 1/2), (0, 1/3), (1, 1/4), (2, 1/5), (3, 1/6)$; range: $\{1, 1/2, 1/3, 1/4, 1/5, 1/6\}$

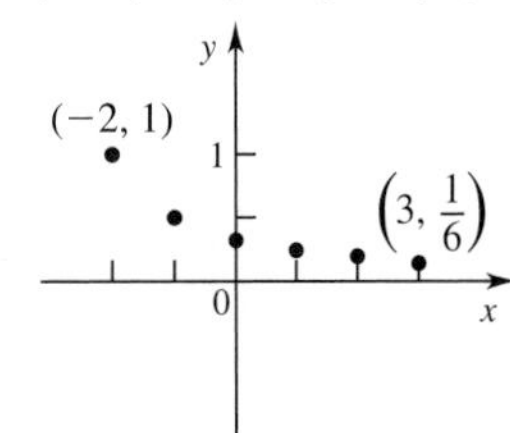

23. $(-2, -3), (-1, -3/2), (0, -3/5), (1, 0), (2, 3/7), (3, 3/4)$; range: $\{-3, -3/2, -3/5, 0, 3/7, 3/4\}$

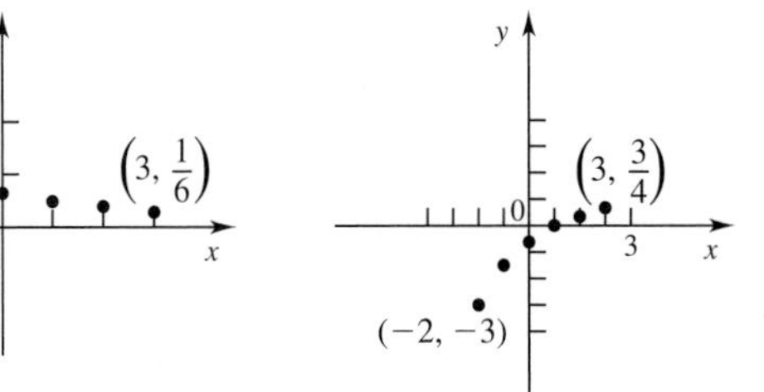

25. $(-\infty, 0)$

27. $[1, 2)$

29. $(-\infty, -9)$

31. $-4 < x < 3$ **33.** $x \le -1$ **35.** $-2 \le x < 6$ **37.** $x \le -4$ or $x \ge 4$ **39.** $(-\infty, \infty)$ **41.** $(-\infty, \infty)$ **43.** $[-4, 4]$ **45.** $[3, \infty)$ **47.** $(-\infty, -2) \cup (-2, 2) \cup (2, \infty)$ **49.** $(-\infty, \infty)$ **51.** $(-\infty, -1] \cup [5, \infty)$ **53.** $(-\infty, 2) \cup (4, \infty)$ **55.** domain: $[-5, 4]$; range: $[-2, 6]$ **57.** domain: $(-\infty, \infty)$; range: $(-\infty, 12]$ **59.** **(a)** 14 **(b)** -7 **(c)** $1/2$ **(d)** $3a + 2$ **(e)** $6/m + 2$ **61.** **(a)** 5 **(b)** -23 **(c)** $-7/4$ **(d)** $-a^2 + 5a + 1$ **(e)** $-4/m^2 + 10/m + 1$ or $(-4 + 10m + m^2)/m^2$ **63.** **(a)** $9/2$ **(b)** 1 **(c)** 0 **(d)** $(2a + 1)/(a - 2)$ **(e)** $(4 + m)/(2 - 2m)$ **65.** **(a)** 0 **(b)** 4 **(c)** 3 **(d)** 4 **67.** **(a)** -3 **(b)** -2 **(c)** -1 **(d)** 2 **69.** $6m - 20$ **71.** $r^2 + 2rh + h^2 - 2r - 2h + 5$ **73.** $9/q^2 - 6/q + 5$ or $(9 - 6q + 5q^2)/q^2$ **75.** function **77.** not a function **79.** function **81.** **(a)** $x^2 + 2xh + h^2 - 4$ **(b)** $2xh + h^2$ **(c)** $2x + h$ **83.** **(a)** $6x + 6h + 2$ **(b)** $6h$ **(c)** 6 **85.** **(a)** $1/(x + h)$ **(b)** $-h/[x(x + h)]$ **(c)** $-1/[x(x + h)]$

87. **(a)** 26,300 BTU per dollar **(b)** \$120 billion
(c) 23,500 BTU per dollar, \$85 billion, 1980
(d) none

89. **(a)** 29¢ **(b)** 52¢ **(c)** 52¢ **(d)** 98¢ **(e)** 75¢
(g)

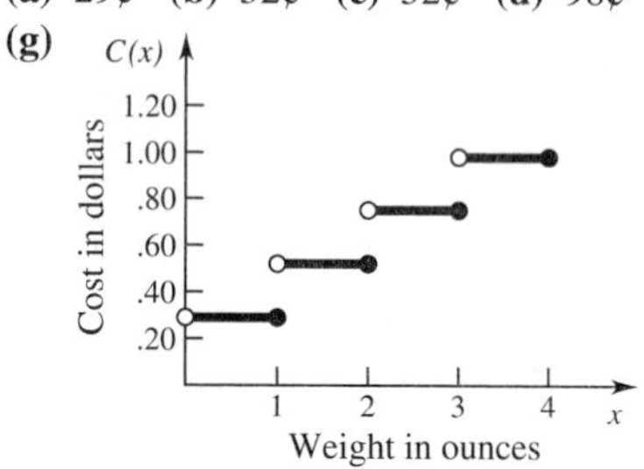

(h) x, the weight **(i)** $C(x)$, the cost to mail the letter

Section 1.2 (page 25)

1. 3/5 **3.** not defined **5.** 2 **7.** 5/9 **9.** not defined **11.** 2 **13.** .5785 **15.** 2/5 **17.** $-1/4$ **19.** $2x + y = 5$ **21.** $y = 1$ **23.** $x + 3y = 10$ **25.** $18x + 30y = 59$ **27.** $2x - 3y = 6$ **29.** $x = -6$ **31.** $5.081x + y = -4.69$

33.

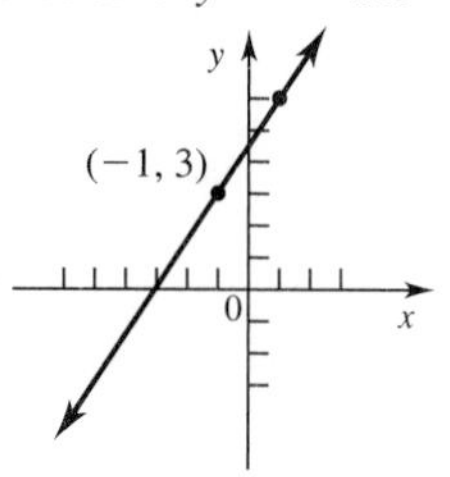

35.

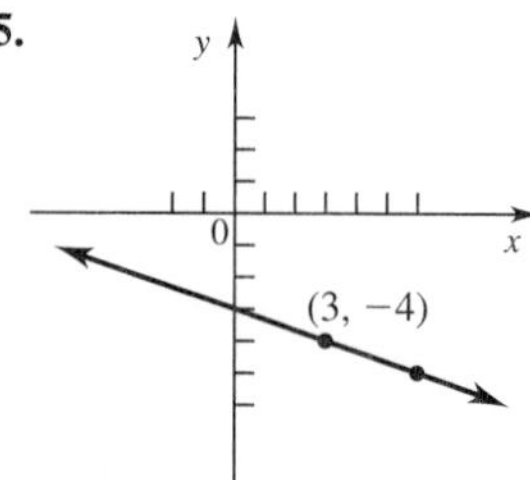

37.

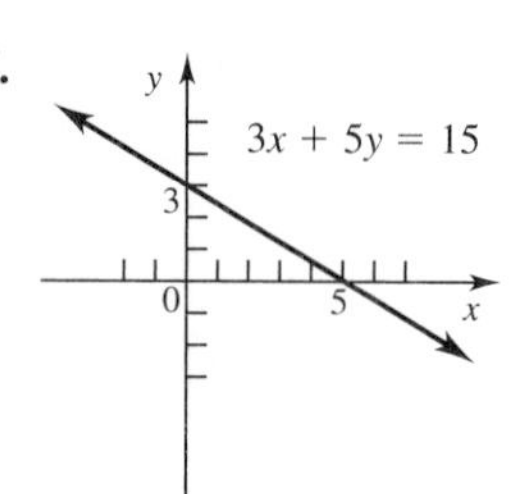

39.

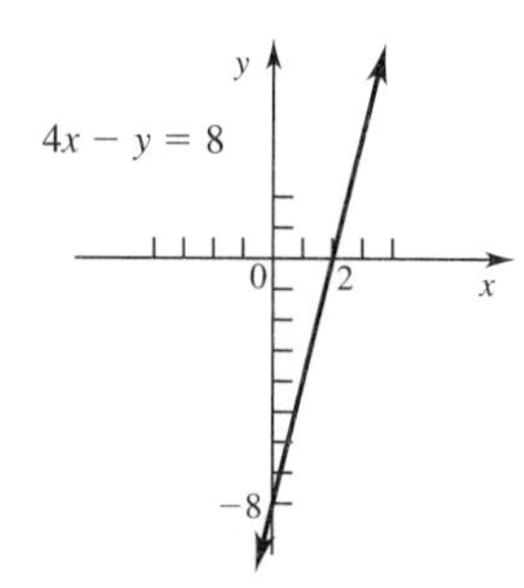

41.

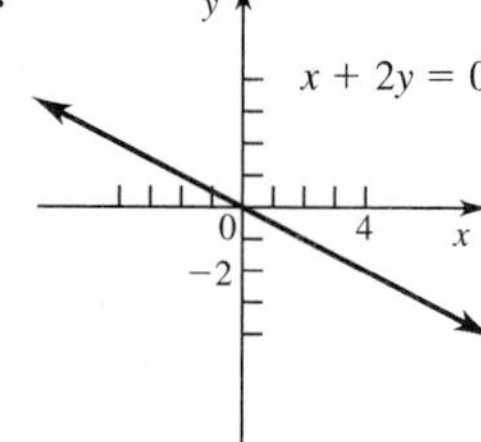

43.

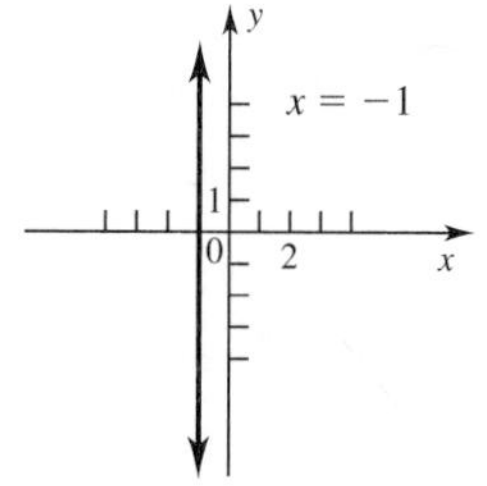

45.

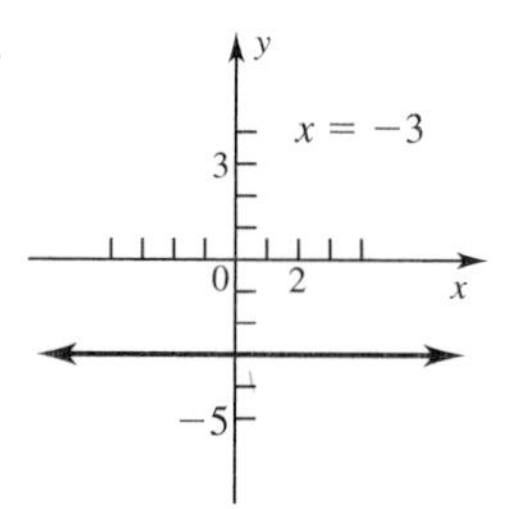

47. $x + 3y = 11$ **49.** $x - y = 7$ **51.** $-2x + y = 4$ **53.** $2x - 3y = -6$ **55.** no

63. **(a)**

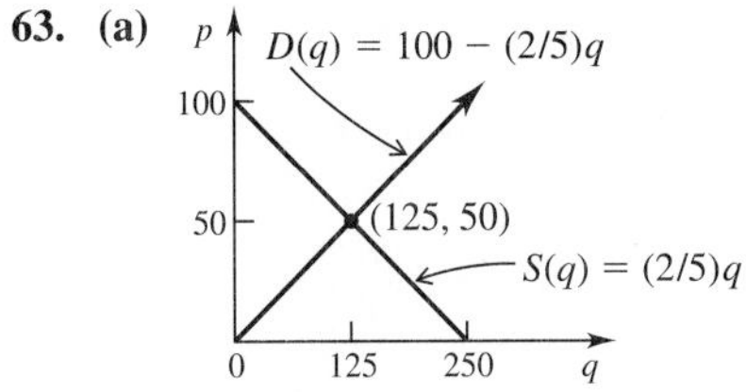

(b) 125 units **(c)** \$50

65. **(a)** \$52 **(b)** \$52 **(c)** \$52 **(d)** \$79 **(e)** \$106 **(f)**

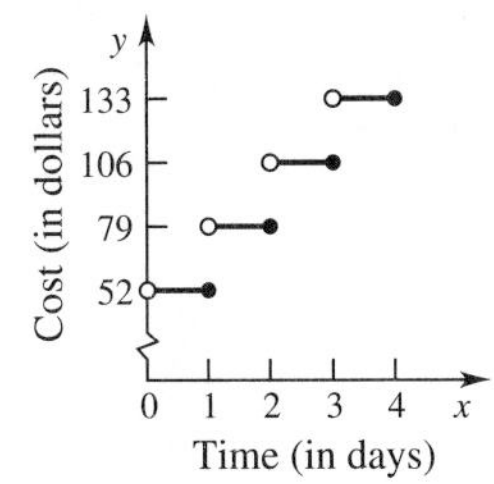

(g) yes **(h)** no

67. **(a)** $h = (8/3)t + 211/3$ **(b)** about 172 cm to 190 cm **(c)** about 45 cm **69.** **(a)** $m = 2.5$; $y = 2.5x - 70$ **(b)** 52% **71.** **(a)** $T = .03t + 15$ **(b)** approximately 2103

Section 1.3 (page 35)

1. If $C(x)$ is the cost of renting a saw for x hr, then $C(x) = 12 + x$. **3.** If $C(x)$ is the cost of parking for x half-hours, then $C(x) = 35x + 50$. **5.** $C(x) = 30x + 100$ **7.** $C(x) = 25x + 1000$ **9.** $C(x) = 50x + 500$ **11.** $C(x) = 90x + 2500$ **13.** **(a)** 2000 **(b)** 2900 **(c)** 3200 **(d)** yes **(e)** 300 **15.** **(a)** $y = (800{,}000/7)x + 200{,}000$ **(b)** about \$1,457,000 **(c)** 1997 **17.** **(a)** \$97 **(b)** \$97.097 **(c)** \$.097 or 9.7¢ **(d)** \$.097 or 9.7¢ **19.** **(a)** \$100 **(b)** \$36 **(c)** \$24 **21.** 500 units; \$30,000 **23.** Break-even point is 45 units; don't produce. **25.** Break-even point is -50 units; impossible to make a profit here. **27.** about \$140 billion in mid-1981 **29.** 81 yr **31.** approximately 4.3 m/sec **33.** **(a)** 32.5 min **(b)** 70 min **(c)** 145 min **(d)** 220 min **35.** **(a)** 14.4°C **(b)** 122°F **37.** $C = (5/9)(F - 32)$ **39.** **(a)** 240 **(b)** 200 **(c)** 160 **(d)** -8 students per hour of study; as required study increased, fewer students enrolled **(e)** 28 hr

Section 1.4 (page 44)

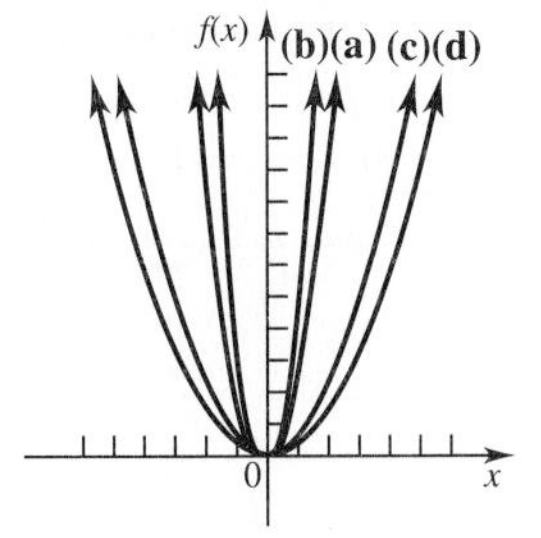

(e) As the coefficient increases in absolute value, the parabola becomes narrower.

3.

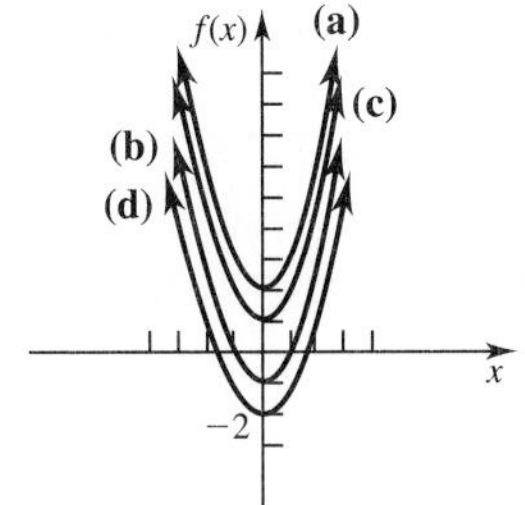

(e) The graphs are shifted upward or downward.

5. Vertex is $(-3, -4)$; axis is $x = -3$; x-intercepts are -1 and -5; y-intercept is 5.

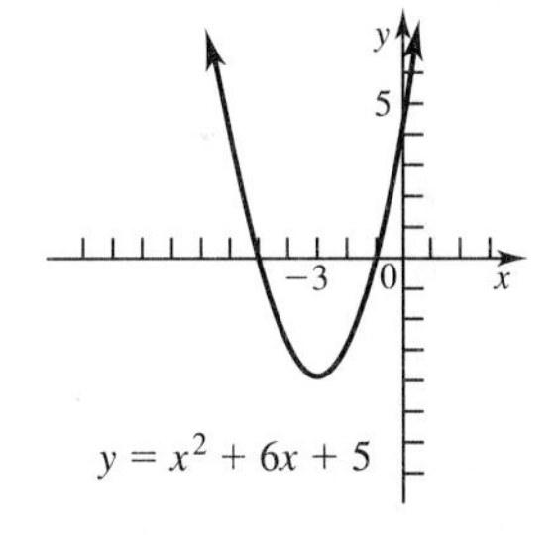

7. Vertex is $(2, 0)$; axis is $x = 2$; x-intercept is 2; y-intercept is 4.

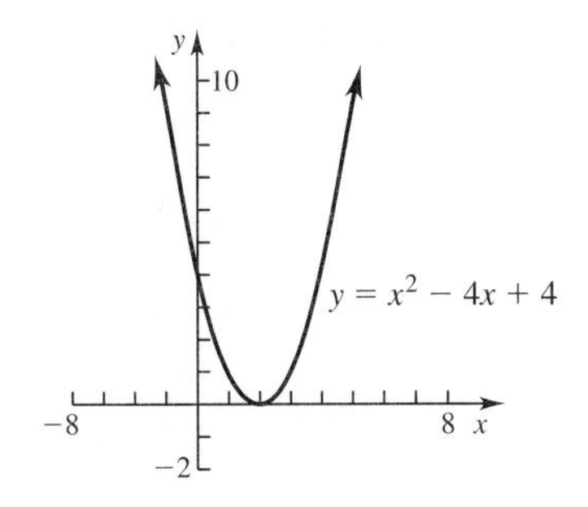

9. Vertex is $(-3, 2)$; axis is $x = -3$; x-intercepts are -2 and -4; y-intercept is -16.

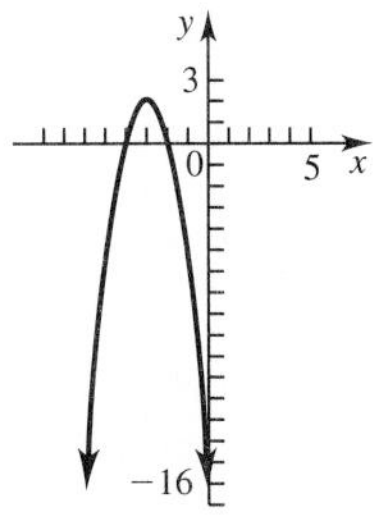

11. Vertex is $(-3, -34)$; axis is $x = -3$; x-intercepts are $-3 \pm \sqrt{17} \approx 1.12$ or -7.12;* y-intercept is -16.

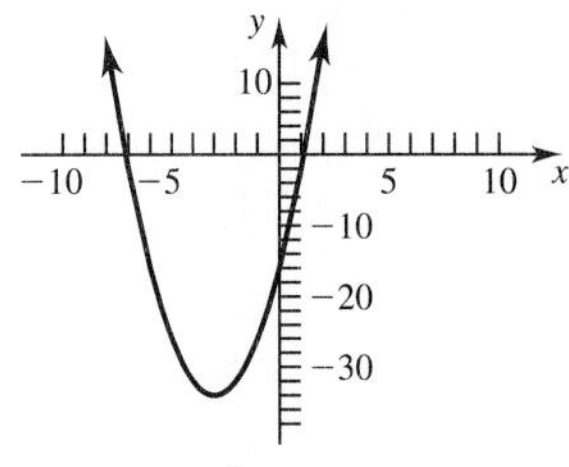

*The symbol ≈ means *approximately equal to.*

13. Vertex is $(-1, -1)$; axis is $x = -1$; x-intercepts are $-1 \pm \sqrt{3}/3 \approx -.42$ or -1.58; y-intercept is 2.

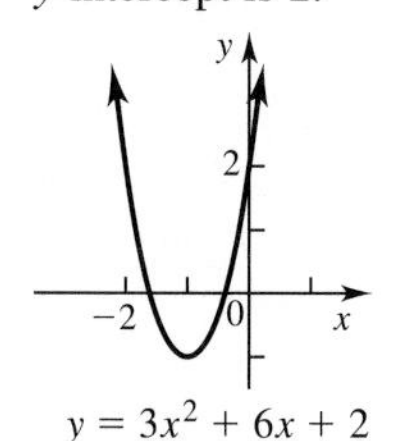

15. Vertex is $(3, 3)$; axis is $x = 3$; x-intercepts are $3 \pm \sqrt{3} \approx 4.73$ or 1.27; y-intercept is -6.

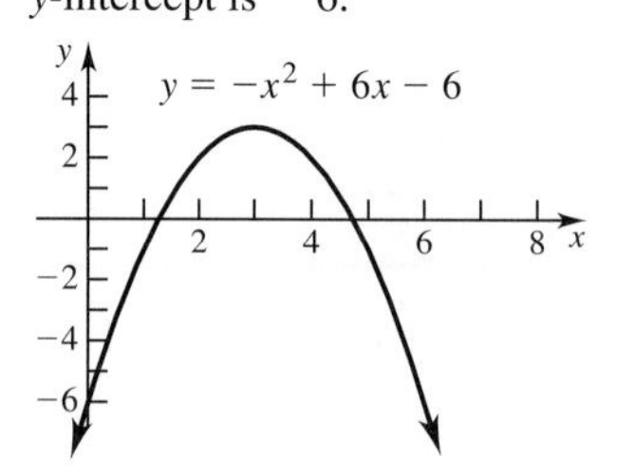

17. Vertex is $(4, 12)$; axis is $x = 4$; x-intercepts are 6 and 2; y-intercept is -36.

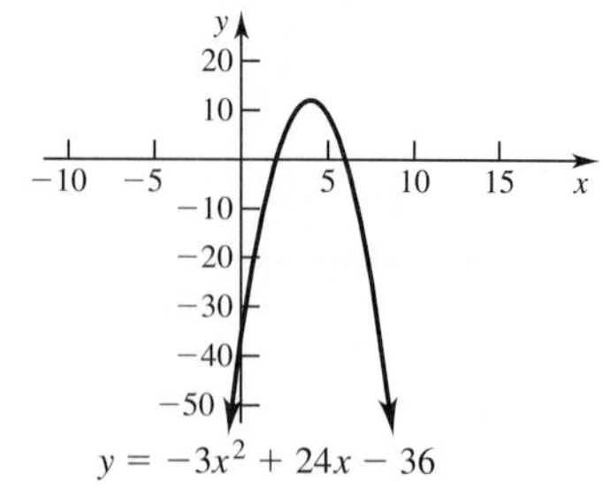

19. Vertex is $(-2, -2)$; axis is $x = -2$; x-intercepts are $-2 \pm 2\sqrt{5}/5 \approx -1.11$ or -2.89; y-intercept is 8.

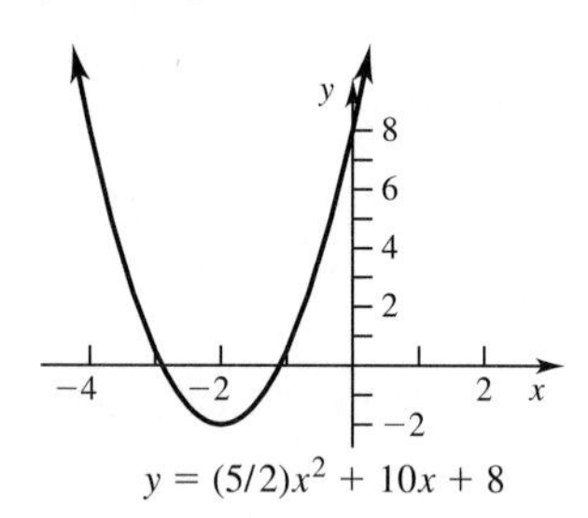

21. Vertex is $(2, -1)$; axis is $x = 2$; x-intercepts are $2 \pm \sqrt{6}/2 \approx 3.22$ or $.78$; y-intercept is $5/3$.

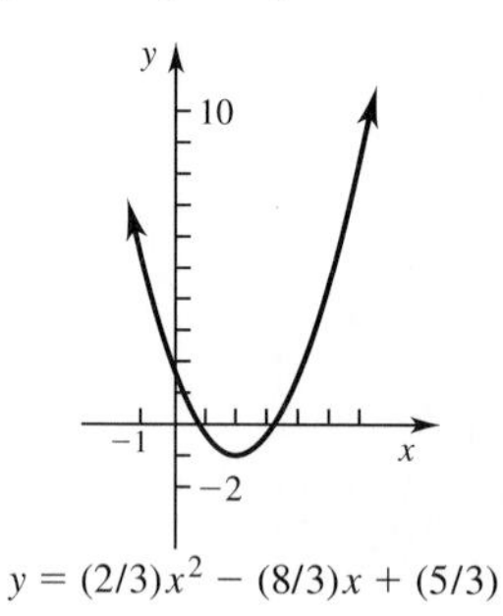

23. $x = 1/2$

25. Maximum revenue is \$5625; 25 seats are unsold.

27. **(a)** $R(x) = x(500 - x) = 500x - x^2$ **(b)**

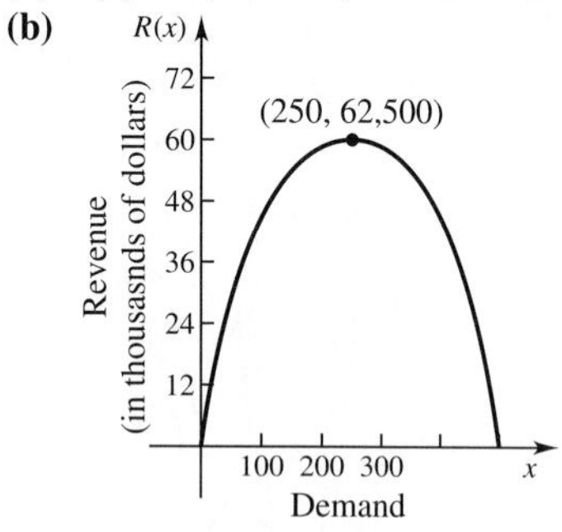

(c) \$250 **(d)** \$62,500

29. **(a)** $200 + 20x$ **(b)** $80 - x$ **(c)** $R(x) = 16{,}000 + 1400x - 20x^2$ **(d)** 35 **(e)** \$40,500 **31.** **(a)** 60 **(b)** 70 **(c)** 90 **(d)** 100 **(e)** 80 **(f)** 20 **33.** 16 ft; 2 sec **35.** 80 ft by 160 ft **37.** 10, 10 **39.** $10\sqrt{3}$ m ≈ 17.32 m

Section 1.5 (page 53)

1. 4, 6, etc. (true degree = 4); +

3. 5, 7, etc. (true degree = 5); +

5. 5, 7, etc. (true degree = 7); +

7. 7, 9, etc. (true degree = 7); −

9. $y = 0$; $x = 3$; no x-intercept, y-intercept $= 4/3$

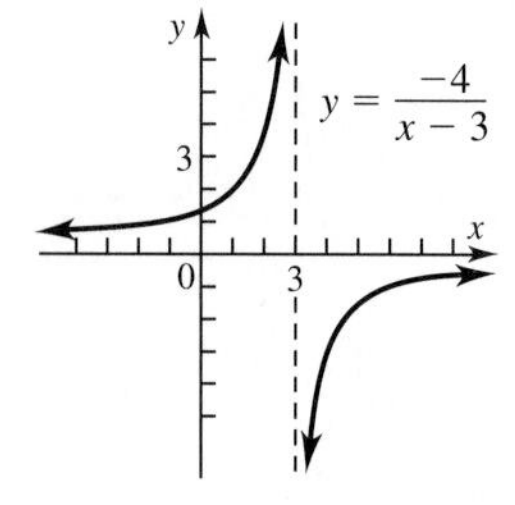

11. $y = 0$; $x = -3/2$; no x-intercept, y-intercept $= 2/3$

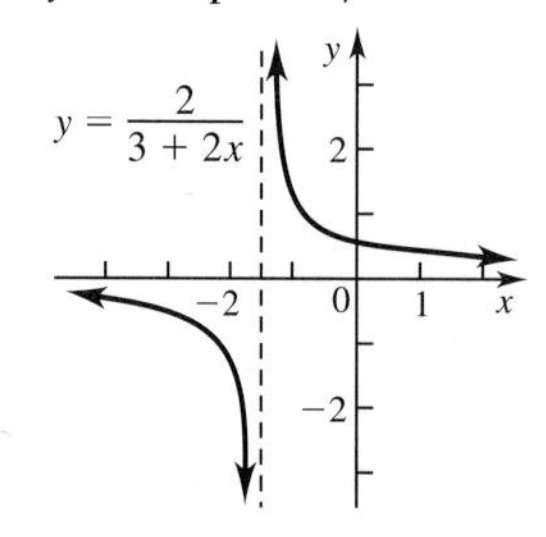

13. $y = 3$; $x = 1$; x-intercept $= 0$, y-intercept $= 0$

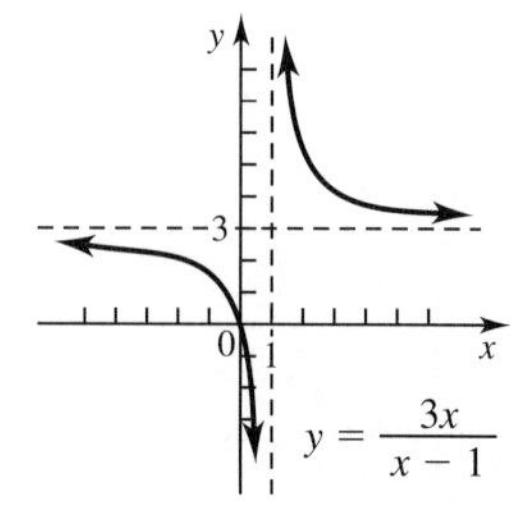

15. $y = 1$; $x = 4$; x-intercept $= -1$, y-intercept $= -1/4$

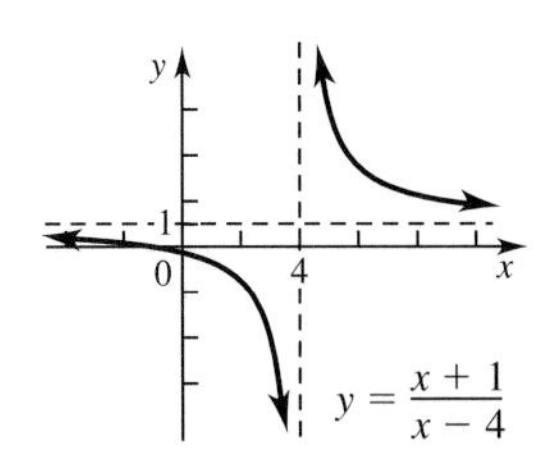

17. $y = -2/5$; $x = -4$; x-intercept $= 1/2$, y-intercept $= 1/20$

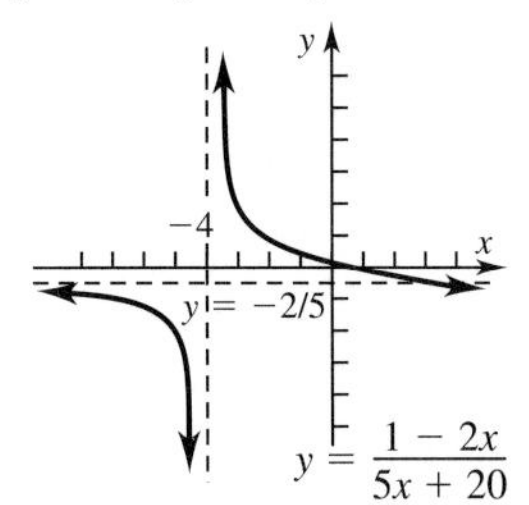

19. $y = -1/3$; $x = -2$; x-intercept $= -4$, y-intercept $= -2/3$

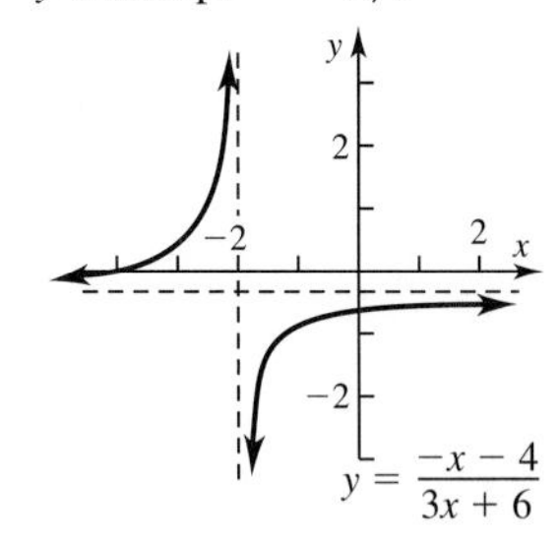

21. **(a)** \$12.50; \$10; \$6.25; \$4.76; \$3.85 **(b)** Probably $(0, \infty)$; it doesn't seem reasonable to discuss the average cost per unit of zero units.
(c)

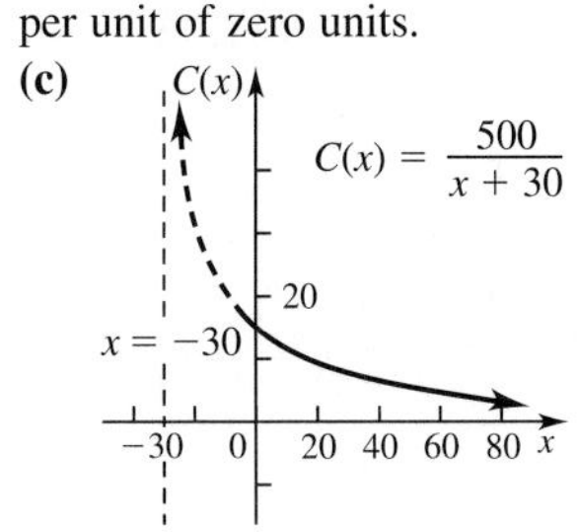

23. **(a)** \$54 billion **(b)** \$504 billion **(c)** \$750 billion **(d)** \$1104 billion
(e) $y = x(100 - x)(x^2 + 500)$; Revenue (in millions); Tax rate (in percent)

25. **(a)** \$42.9 million **(b)** \$40 million **(c)** \$30 million **(d)** \$0
(e) $y = \frac{60x - 6000}{x - 120}$; $x = 120$; Revenue (in millions of dollars); Tax rate (in percent)

27. **(a)** \$6700 **(b)** \$15,600 **(c)** \$26,800 **(d)** \$60,300 **(e)** \$127,300 **(f)** \$328,300 **(g)** \$663,300 **(h)** no
(i)

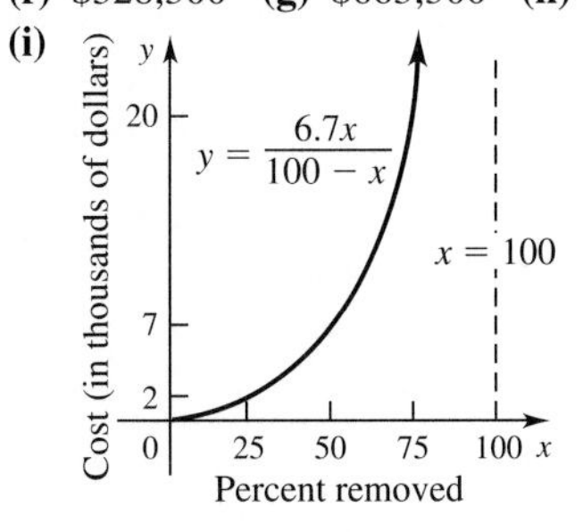

29. **(a)** $g(x) = -.006x^4 + .140x^3 - .053x^2 + 1.79x$
(b) no

31. **(a)** $A(x) = -.015x^3 + 1.058x$; Concentration (in tenths of percent); Time (in hours)
(b) between 4 and 5 hr, but closer to 5 hr **(c)** from less than 1 hr to about 8.4 hr

33. **(a)** $[0, \infty)$
(b)

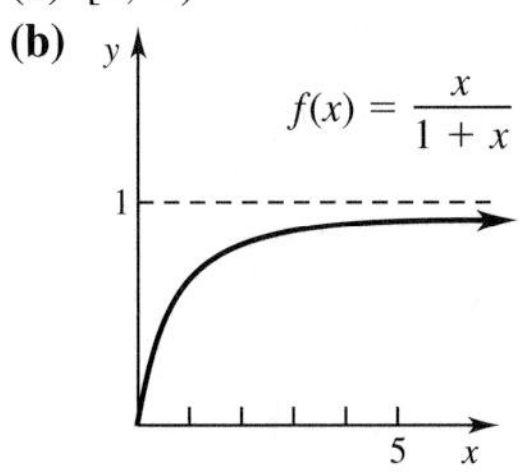

(c)

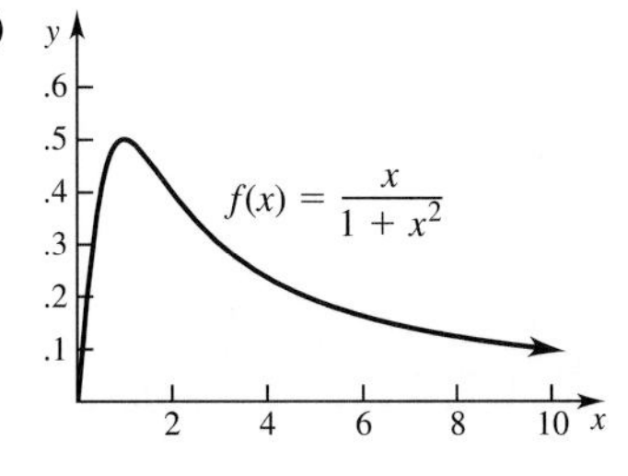

(d) Increasing b makes the next generation smaller when this generation is larger.

35. **(a)** $R = -1000$ **(b)** $G(R) = 1$
(c)

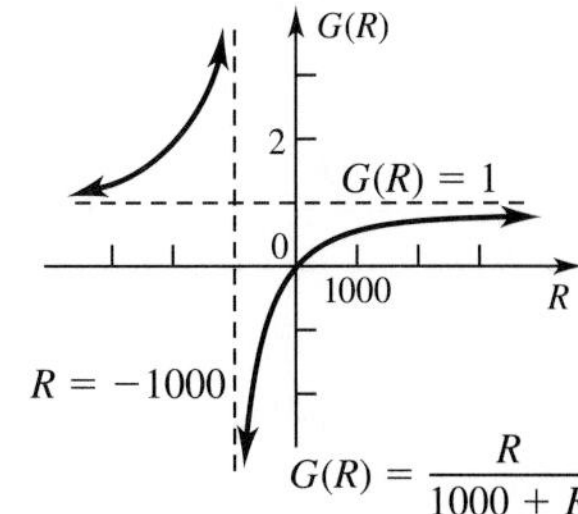

37. (a)

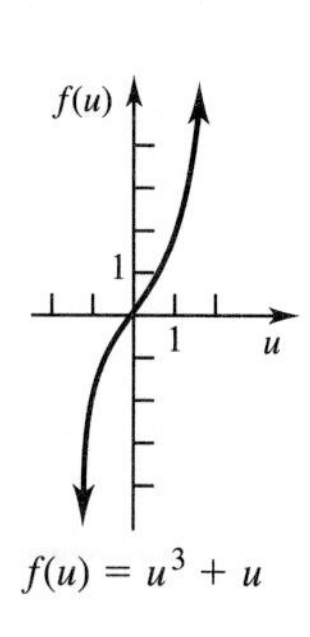

$f(u) = u^3 + u$

(b)

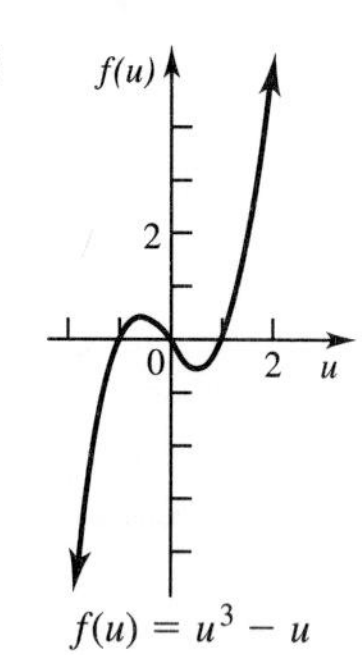

$f(u) = u^3 - u$

39. (a) about \$10,000
(b) about \$20,000

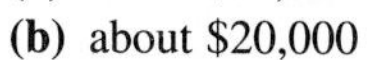

41.

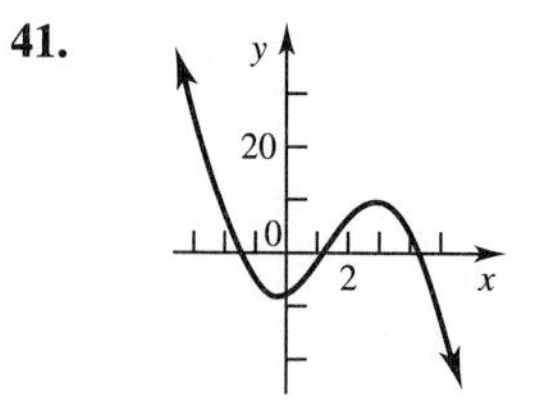

$f(x) = -x^3 + 4x^2 + 3x - 8$

43.

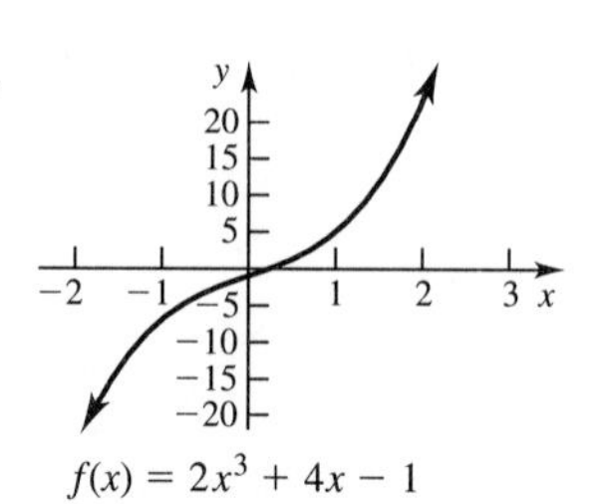

$f(x) = 2x^3 + 4x - 1$

45.

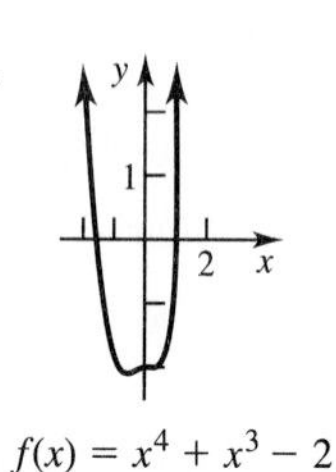

$f(x) = x^4 + x^3 - 2$

47.

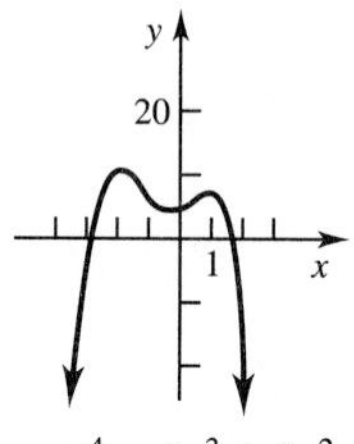

$f(x) = -x^4 - 2x^3 + 3x^2 + 3x + 5$

49.

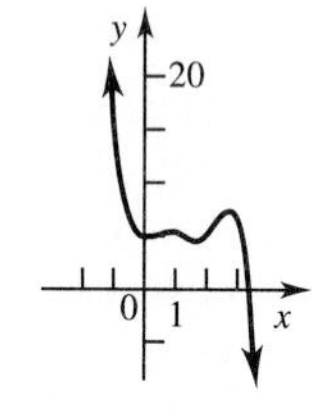

$f(x) = -x^5 + 6x^4 - 11x^3 + 6x^2 + 5$

51. no horizontal asymptote; vertical asymptote: $x = -3/2$

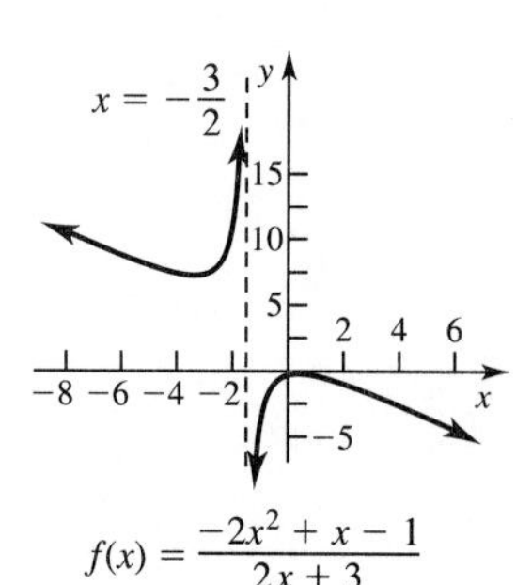

$f(x) = \dfrac{-2x^2 + x - 1}{2x + 3}$

53. horizontal asymptote: $y = 2$; vertical asymptotes: $x = 1$, $x = -1$

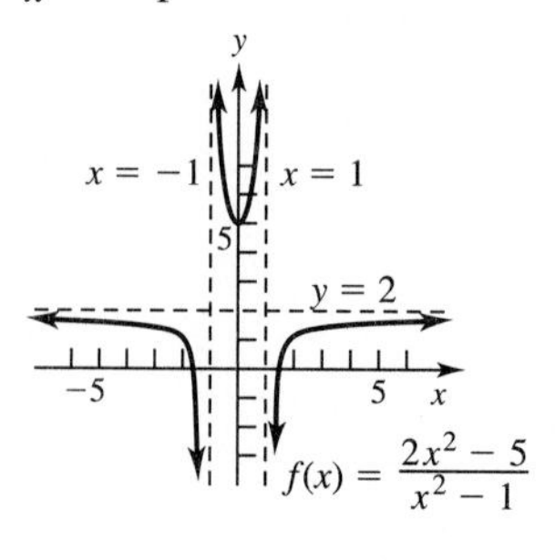

$f(x) = \dfrac{2x^2 - 5}{x^2 - 1}$

55. horizontal asymptote: $y = -2$; vertical asymptotes: $x = \sqrt{10}$, $x = -\sqrt{10}$

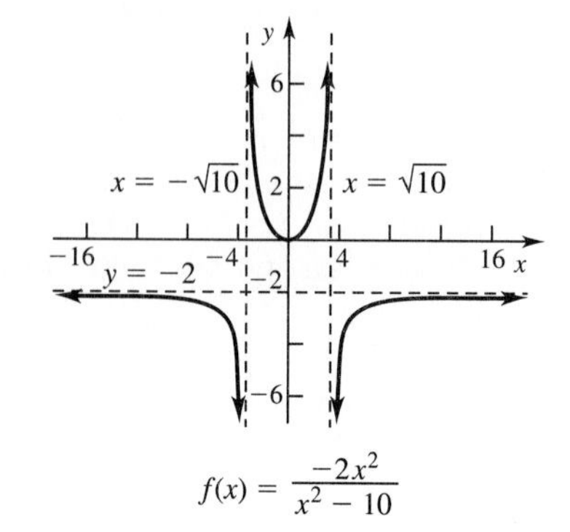

$f(x) = \dfrac{-2x^2}{x^2 - 10}$

Section 1.6 (page 61)

1.

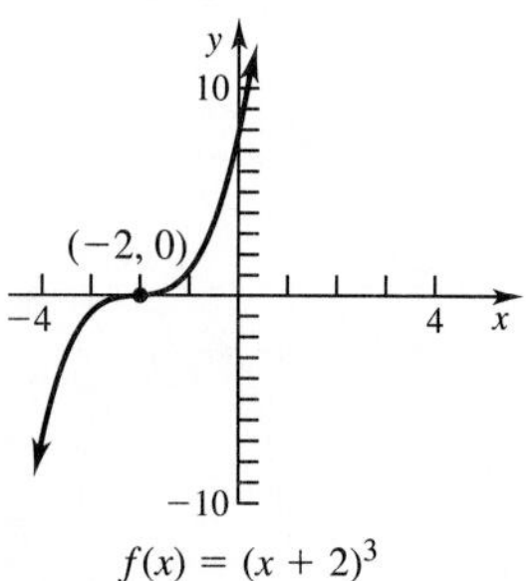

$f(x) = (x + 2)^3$

3.

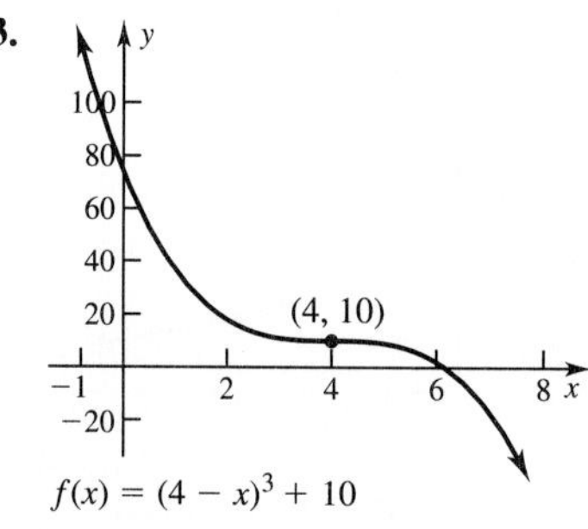

$f(x) = (4 - x)^3 + 10$

5.

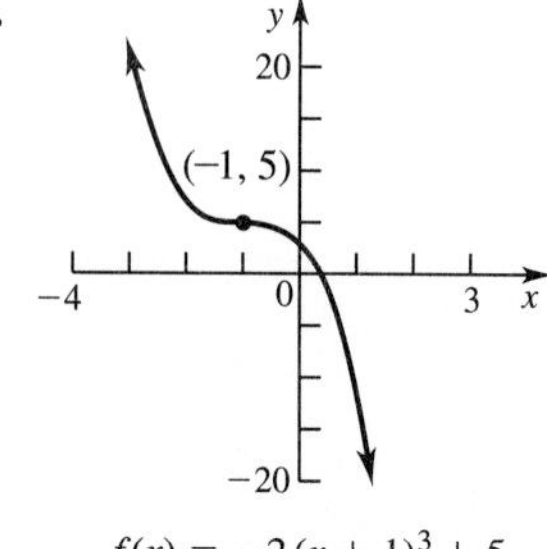

$f(x) = -2(x + 1)^3 + 5$

7.

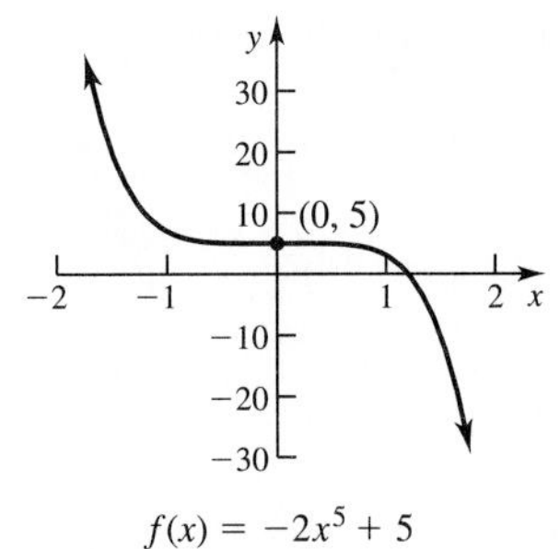

9.

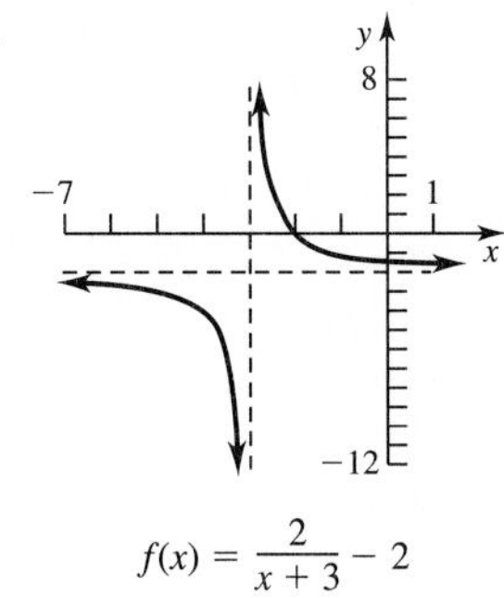

11.

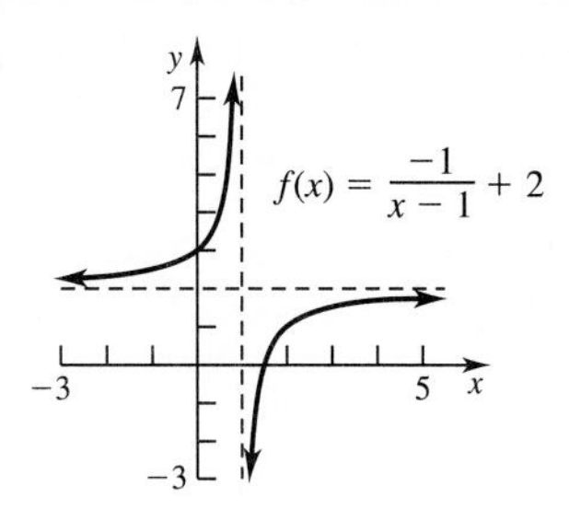

13.

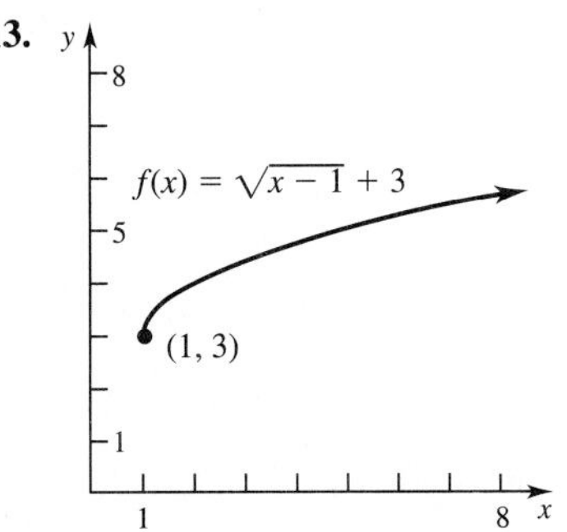

15.

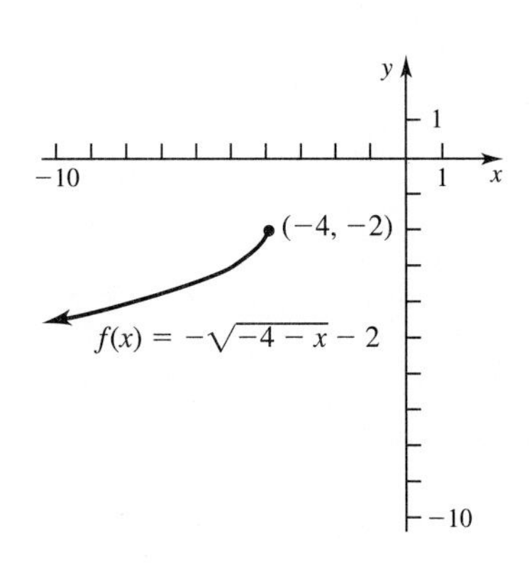

17.

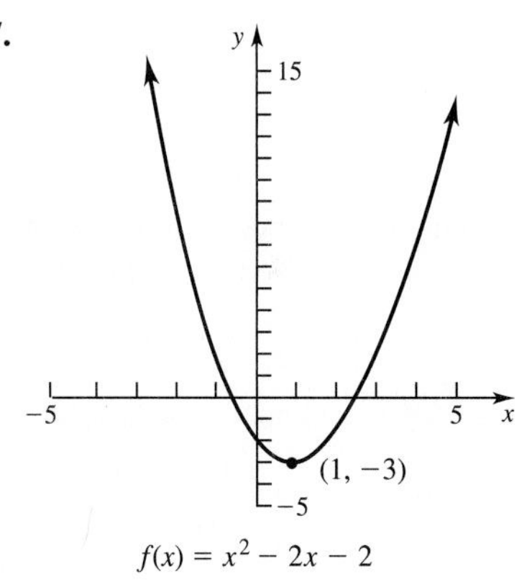

19.

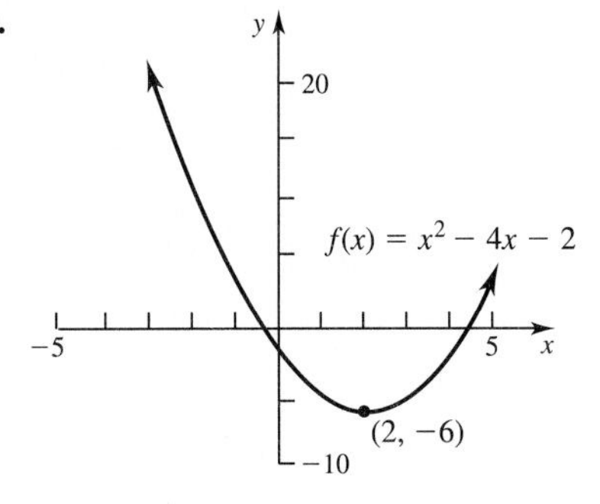

21.

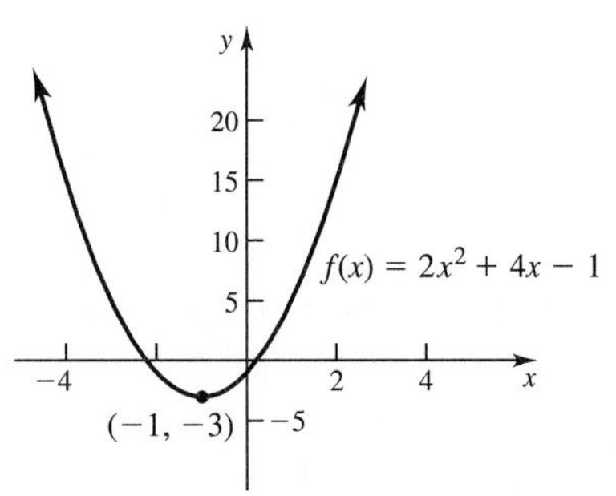

23.

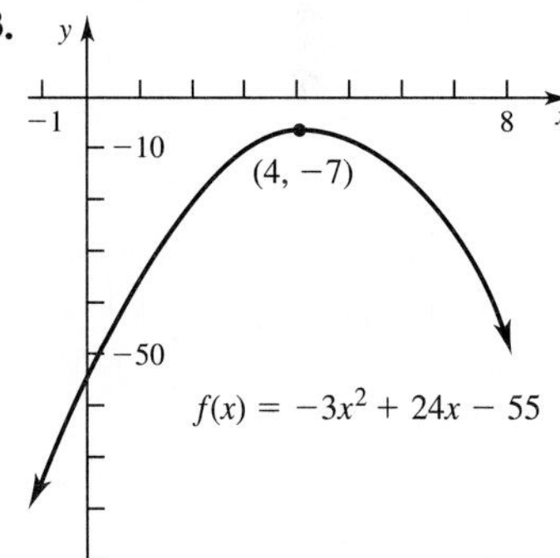

25.

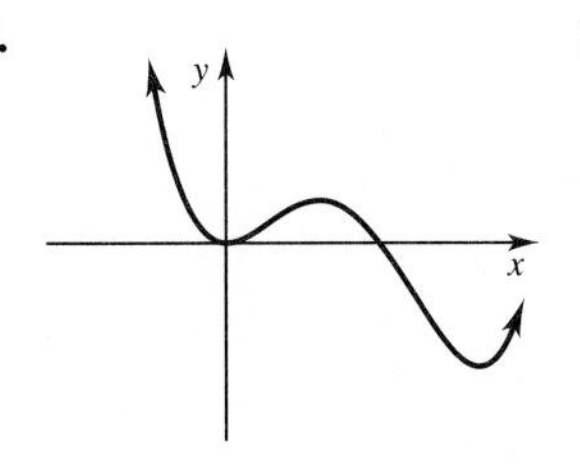

27.

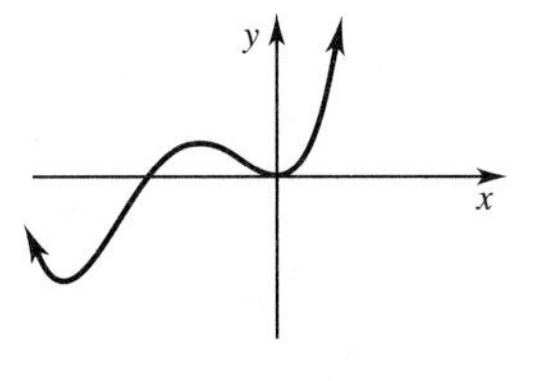

29.

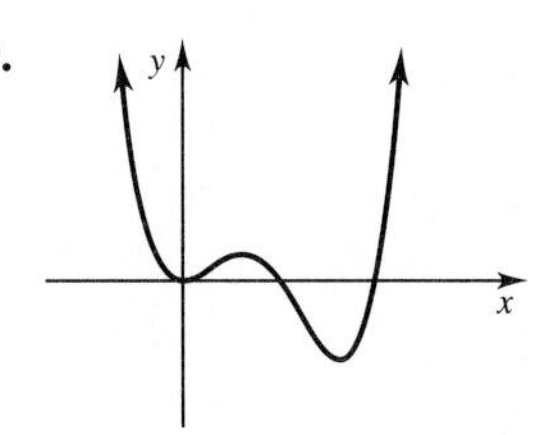

31.

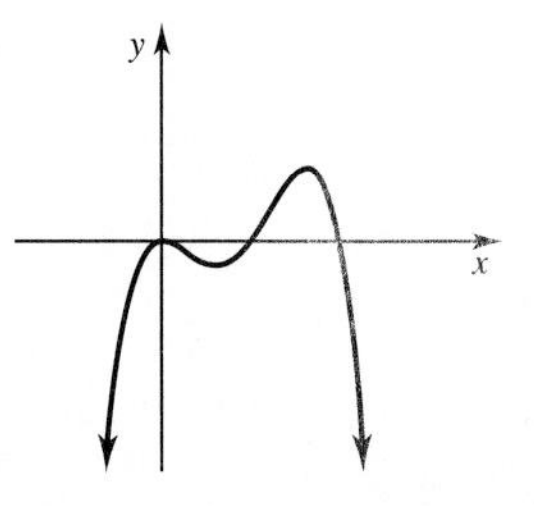

33. **(c)**

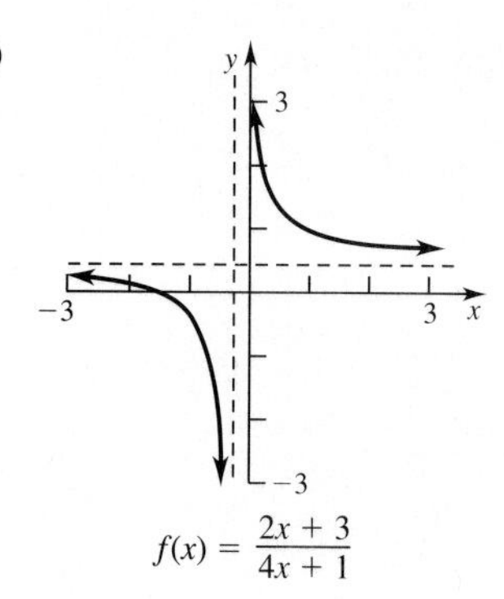

(d)

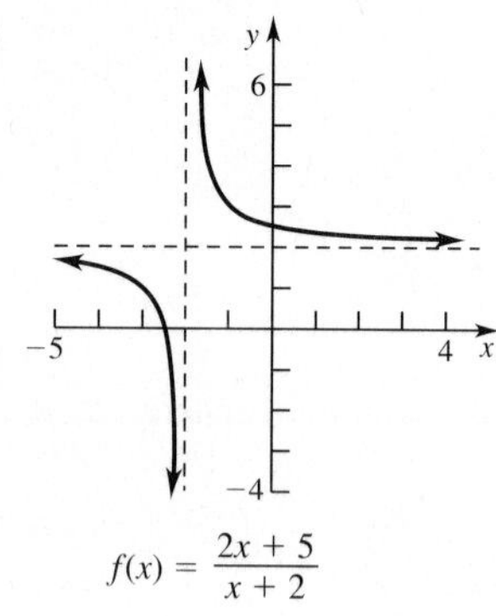

35. **(a)** \$0 **(b)** \$40,000
(c)

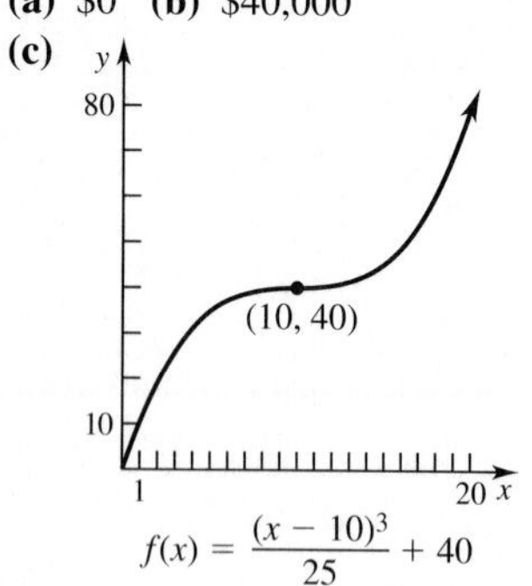

(d) $A(x) = [(x - 10)^3/25 + 40]/x$
(e) \$4 **(f)** \$52

Chapter 1 Review Exercises (page 63)

5. $(-3, -16/5), (-2, -14/5), (-1, -12/5), (0, -2), (1, -8/5), (2, -6/5), (3, -4/5)$;
range: $\{-16/5, -14/5, -12/5, -2, -8/5, -6/5, -4/5\}$

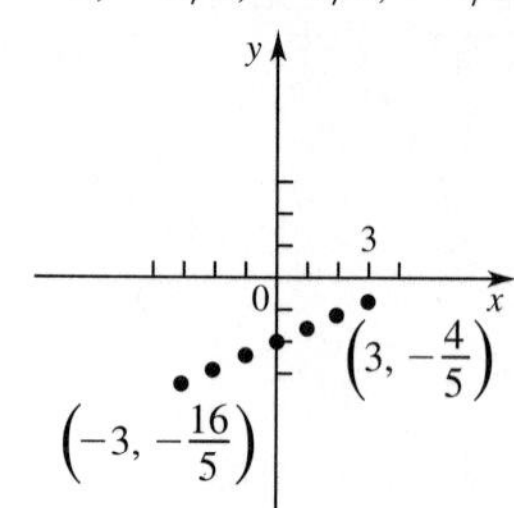

7. $(-3, 20), (-2, 9), (-1, 2), (0, -1), (1, 0), (2, 5), (3, 14)$;
range: $\{-1, 0, 2, 5, 9, 14, 20\}$

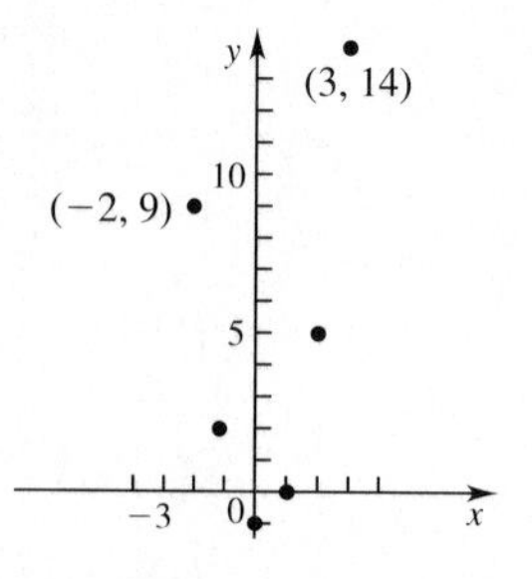

9. $(-3, 7), (-2, 2), (-1, -1), (0, -2), (1, -1), (2, 2), (3, 7)$;
range: $\{-2, -1, 2, 7\}$

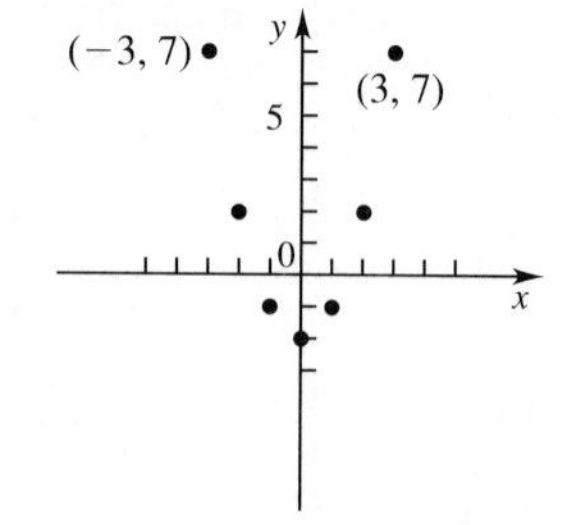

11. $(-3, 1/5), (-2, 2/5), (-1, 1), (0, 2), (1, 1), (2, 2/5), (3, 1/5)$;
range: $\{1/5, 2/5, 1, 2\}$

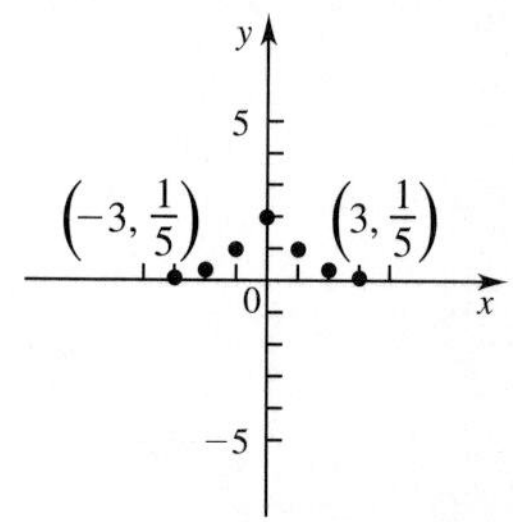

13. $(-3, -1), (-2, -1), (-1, -1), (0, -1), (1, -1), (2, -1), (3, -1)$; range: $\{-1\}$

15. **(a)** 23 **(b)** -9 **(c)** -17 **(d)** $4r + 3$ **17.** **(a)** -28 **(b)** -12 **(c)** -28 **(d)** $-r^2 - 3$ **19.** **(a)** -13 **(b)** 3
(c) -32 **(d)** 22 **(e)** $-k^2 - 4k$ **(f)** $-9m^2 + 12m$ **(g)** $-k^2 + 14k - 45$ **(h)** $12 - 5p$

21.

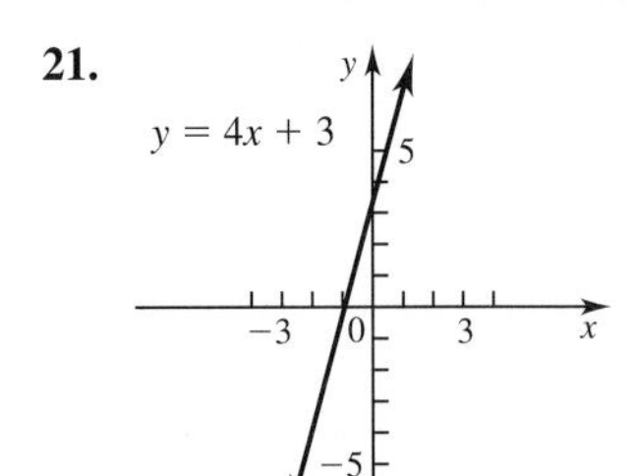

23.

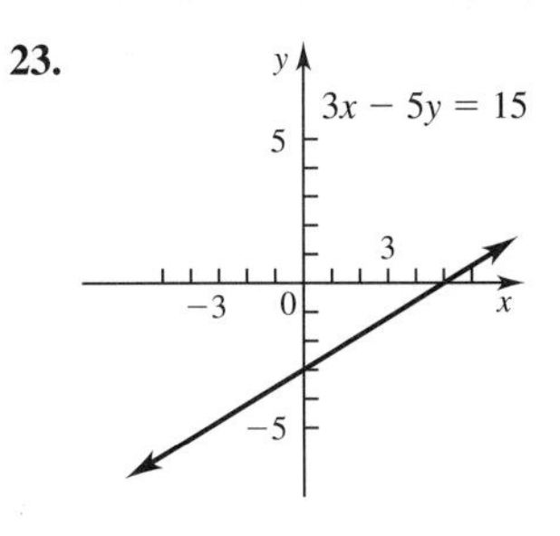

25.

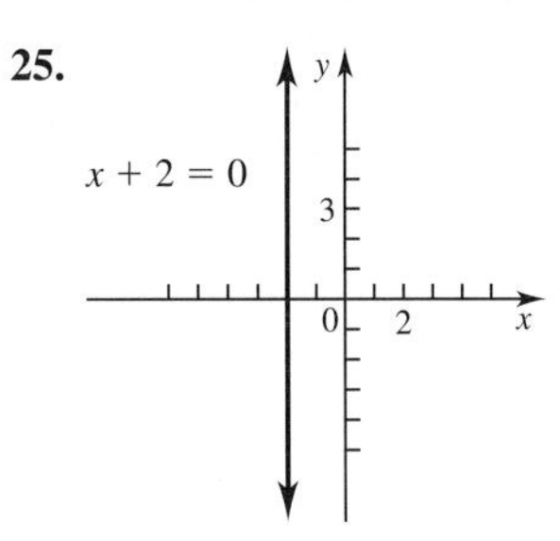

27.

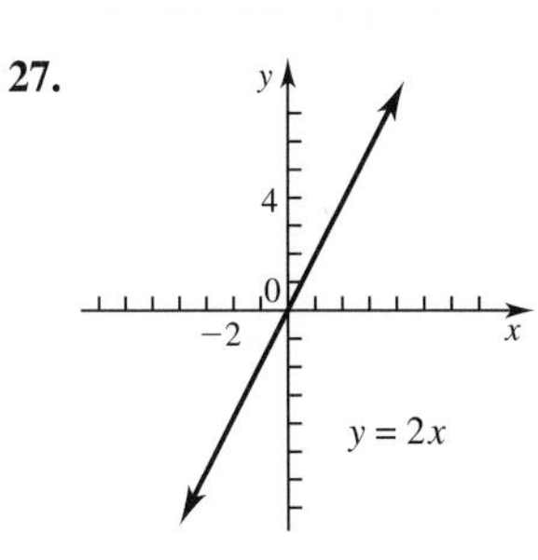

29. $1/3$ **31.** $-2/11$ **33.** $-2/3$ **35.** not defined **37.** $2x - 3y = 13$ **39.** $5x + 4y = 17$ **41.** $x = -1$
43. $5x - 8y = -40$ **45.** $y = -5$

47.

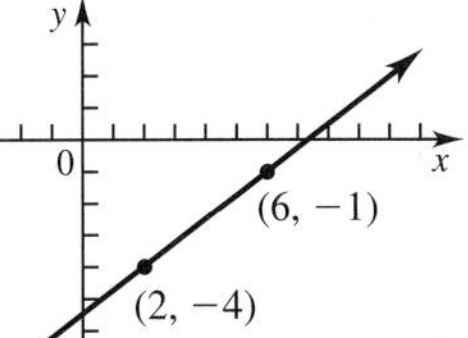

49.

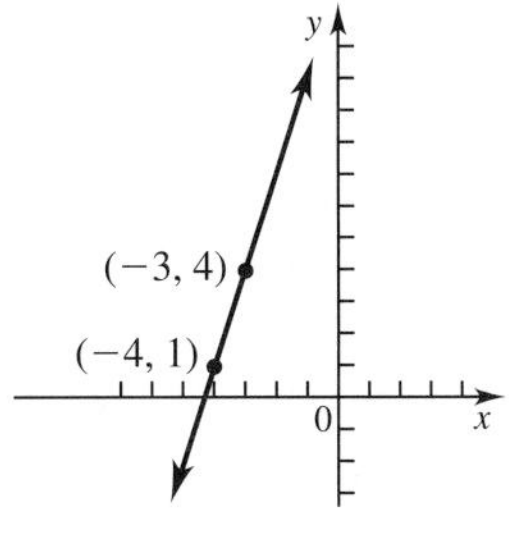

51.

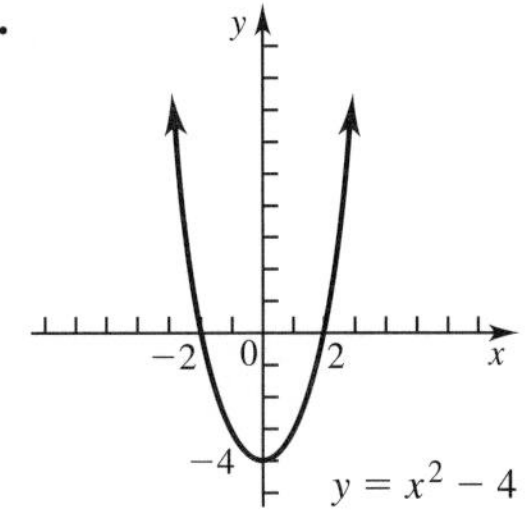

53.

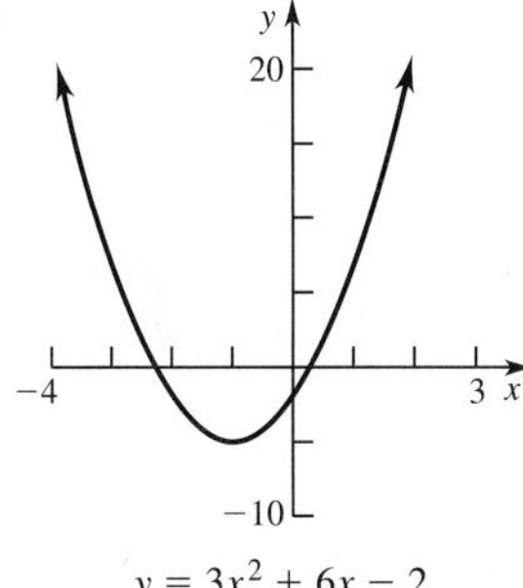

55.

y = x² − 4x + 2

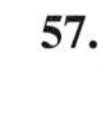

57.

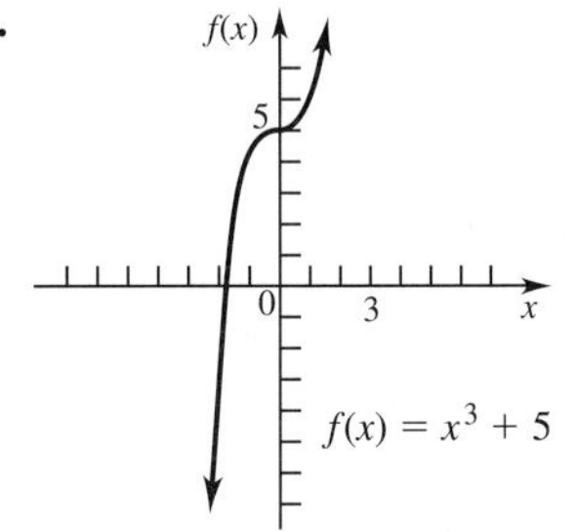

59.

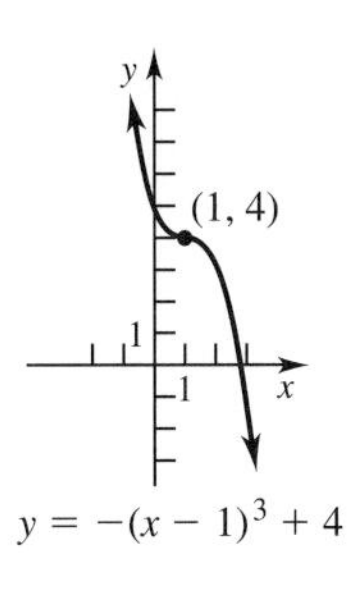

61.

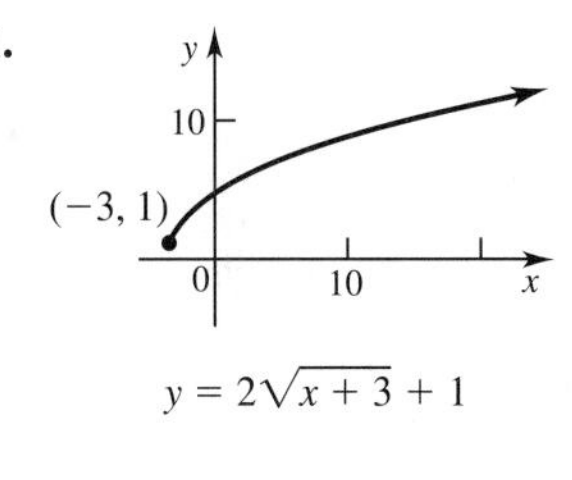

63.

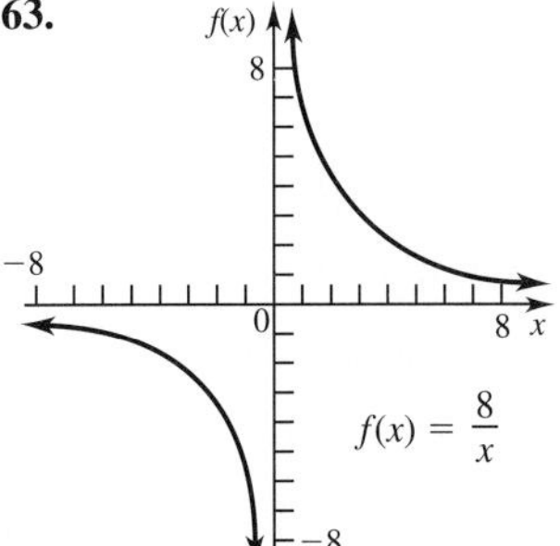

65.

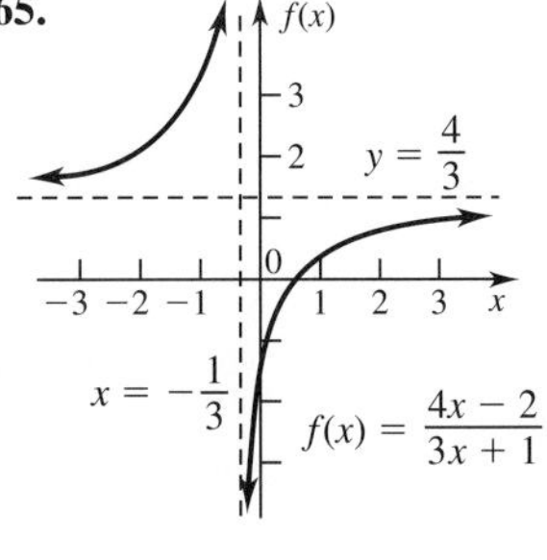

67. $y = 3(x + 1)^2 - 5$

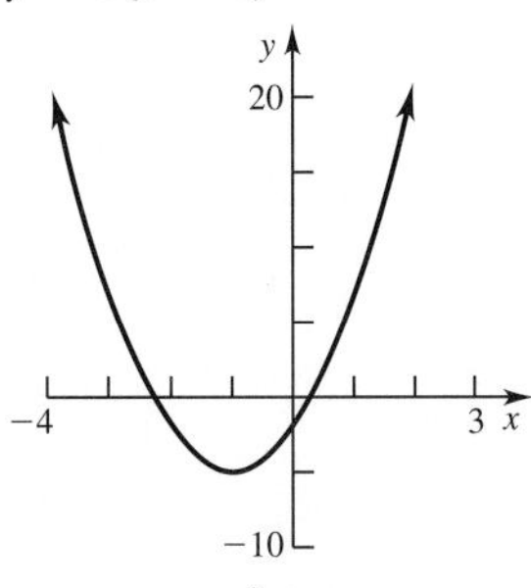

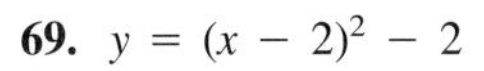

69. $y = (x - 2)^2 - 2$

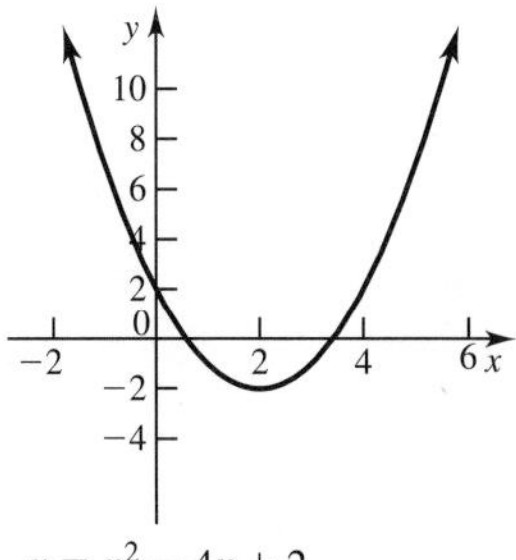

71. **(a)** 7/6; 9/2 **(b)** 2; 2 **(c)** 5/2; 1/2 **(d)**

(e) 15 **(f)** 2

73. $C(x) = 30x + 60$; $A(x) = 30 + 60/x$

75. $C(x) = 46x + 120$; $A(x) = 46 + 120/x$

77. **(a)** \$80 **(b)** \$80 **(c)** \$80 **(d)** \$120 **(e)** \$160 **(f)**

(g) x, the number of days **(h)** $C(x)$, the cost

79. **(a)** halfway through 1984 **(b)** 160,000 annually **(c)** [0, 160,000]

81. **(a)** approximately 1.2 yr and 9.8 yr **(b)** $f(A) > g(A)$ for $2 < A < 9.8$; $g(A) > f(A)$ for $A < 1.2$ or $A > 9.8$

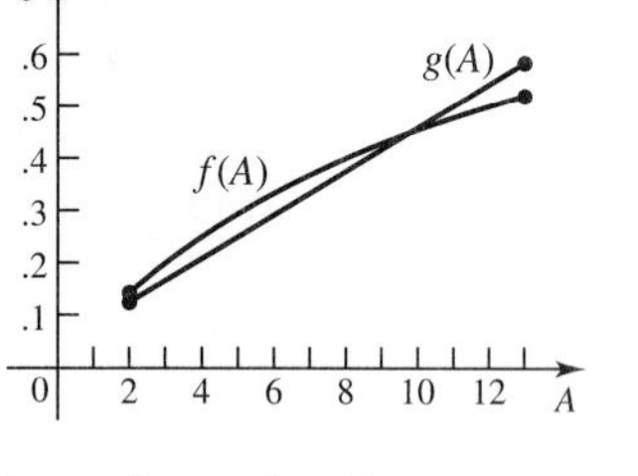

(c) at 5 yr and at 13 yr

83. **(a)** approximately 5750 rpm **(b)** approximately 310 horsepower **(c)** approximately 280 ft. lbs.

85. **(a)**

(b) $\dfrac{2}{(x+1)(x+2)}$

(c) $\dfrac{5x+3}{x(x+1)}$

(d) $\dfrac{-5x-6}{x(x+1)(x+2)}$

Extended Application (page 67)

1. 4.8 million units **2.**

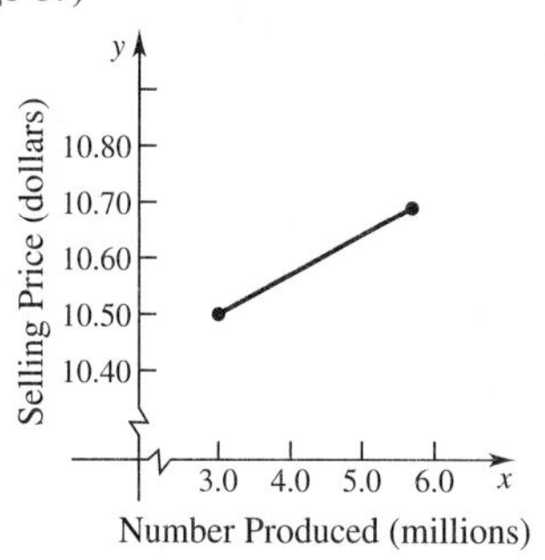

3. In the interval under discussion (3.1 million to 5.7 million units), the marginal cost always exceeds the selling price.

4. **(a)** 9.87; 10.22 **(b)**

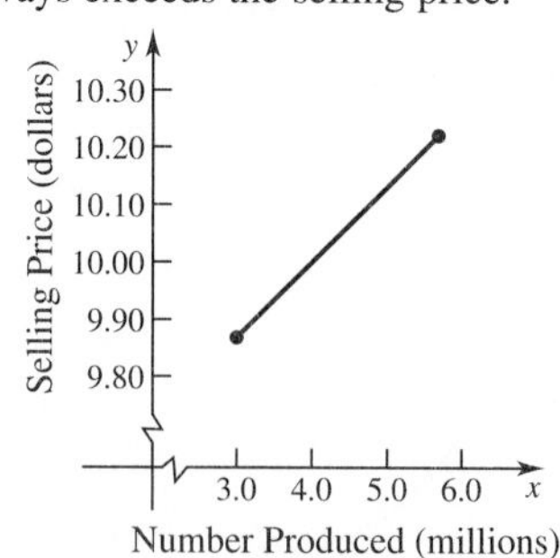

(c) .83 million units, which is not in the interval under discussion

CHAPTER 2

Section 2.1 (page 79)

1. 3 **3.** does not exist **5.** 1 **9.** 2 **11.** 10 **13.** does not exist **15.** 8 **17.** 2 **19.** 4 **21.** 512 **23.** 3/2 **25.** 6 **27.** -5 **29.** 4 **31.** $-1/9$ **33.** 1/10 **35.** $2x$ **37. (a)** does not exist **(b)** $x = -2$ **(c)** If $x = a$ is an asymptote for the graph of $f(x)$, then $\lim_{x\to a} f(x)$ does not exist. **39.** discontinuous at -1; $f(-1)$ does not exist; $\lim_{x\to -1} f(x) = 1/2$ **41.** discontinuous at 1; $f(1) = 2$; $\lim_{x\to 1} f(x) = -2$ **43.** discontinuous at -5 and 0; $f(-5)$ and $f(0)$ do not exist; $\lim_{x\to -5} f(x)$ does not exist; $\lim_{x\to 0} f(x) = 0$ **45.** no; no; yes **47.** yes; no; yes **49.** yes; no; yes **51.** yes; yes; yes **53.** no; yes; yes **55. (a)** 3 **(b)** does not exist **(c)** 2 **(d)** 16 months **57. (a)** \$520 **(b)** \$600 **(c)** \$630 **(d)** \$1200 **(e)** \$1250 **(f)** at $x = 150$ and $x = 400$ **59. (a)** \$120 **(b)** \$150 **(c)** \$150 **(d)** \$150 **(e)** \$180 **(f)** \$150 **(g)** \$150 **(h)** at $t = 1, 2, 3, 4, 7, 8, 9, 10, 11$ **61.** at $t = m$

Section 2.2 (page 90)

1. 5 **3.** 8 **5.** 1/3 **7.** $-1/3$ **9.** 17 **11.** 25 **13.** 50 **15.** -16 **17.** 0 **19.** increasing **21. (a)** 3 **(b)** 0 **(c)** $-9/5$ **(d)** Sales increase in years 0–4, stay constant until year 7, and then decrease. **(e)** Many answers are possible; one example might be Walkman radios. **23. (a)** $-3/2$ **(b)** -1 **(c)** 1 **(d)** $-1/2$ **(e)** 1988–1989 **(f)** stabilizing after a decline from 1985 to 1987 **25.** 11 **27. (a)** -25 boxes per dollar **(b)** -20 boxes per dollar **(c)** -30 boxes per dollar **(d)** Demand is decreasing. Yes, a higher price usually reduces demand. **29. (a)** \$11 million, $-$\$1 million **(b)** -1 million, 9 million **(c)** Civil penalities were increasing from 1987 to 1988 and decreasing from 1988 to 1989. Criminal penalties decreased slightly from 1987 to 1988, then increased from 1988 to 1989. This indicates that criminal penalties began to replace civil penalties in 1988. **(d)** About a \$3.5 million increase. The general trend was upward. More is being done to impose penalties for polluting. **31. (a)** $-$\$.5 billion, $-$\$5 billion **(b)** $-$\$.5 billion, $-$\$3.2 billion **(c)** $-$\$1.2 billion, $-$\$1.2 billion **(d)** cocaine in 1989–1990 **33. (a)** 5 ft/sec **(b)** 1 ft/sec **(c)** 1 ft/sec **(d)** 6 ft/sec **(e)** 1 ft/sec **(f)** The car is speeding up from 0 to 2 sec, then slowing down from 2 to 6 sec, maintaining constant velocity from 6 to 10 sec, then speeding up from 10 to 12 sec, and finally, slowing down again from 12 to 18 sec.

Section 2.3 (page 105)

1. 27 **3.** 1/8 **5.** 1/8 **7.** $y = 8x - 9$ **9.** $5x + 4y = 20$ **11.** $3y = 2x + 18$ **13.** 2 **15.** 1/5 **17.** 0 **19. (a)** 0 **(b)** 1 **(c)** -1 **(d)** does not exist **(e)** m **21.** at $x = -2$ **23.** $f'(x) = 12x - 4$; 20; -4; -40 **25.** $f'(x) = -9$; -9; -9; -9 **27.** $f'(x) = -6/x^2$; $-3/2$; does not exist; $-2/3$ **29.** $f'(x) = -3/(2\sqrt{x})$; $-3/(2\sqrt{2})$; does not exist; does not exist **31.** -6 **33.** has derivatives for all values of x **35.** -5; -3; 0; 2; 4 **37. (a)** $(a, 0)$ and (b, c) **(b)** $(0, b)$ **(c)** $x = 0$ and $x = b$ **39. (a)** distance **(b)** velocity **41. (a)** -1; the debt is decreasing at the rate of about 1% per month. 2; the debt is increasing at the rate of about 2% per month. 0; the debt is not changing at this point. **(b)** February to March, April to May, and August to September **(c)** It is increasing. **43. (a)** \$16 per table **(b)** \$16 **(c)** \$15.998 (or \$16) **(d)** The marginal revenue gives a good approximation of the actual revenue from the sale of the 1001st table. **45. (a)** 20 **(b)** 0 **(c)** -10 **(d)** at 5 hr **47. (a)** approximately 0; the power expended is not changing at that point **(b)** about .1; the power expended is increasing .1 unit for each 1 unit increase in speed **(c)** about .12; the power expended increases .12 units for each 1 unit increase in speed **(d)** The power level decreases to v_{mp}, which minimizes energy costs; then it increases at an increasing rate.

49. (a)

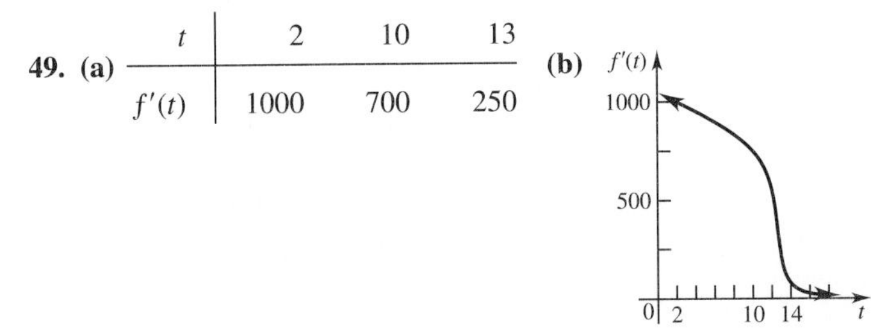

t	2	10	13
$f'(t)$	1000	700	250

(b)

Section 2.4 (page 119)

1. $y' = 30x^2 - 18x + 6$ **3.** $y' = 4x^3 - 15x^2 + (2/9)x$ **5.** $f'(x) = 9x^{.5} - 2x^{-.5}$ or $9x^{.5} - 2/x^{.5}$ **7.** $y' = -48x^{2.2} + 3.8x^{.9}$ **9.** $y' = 4x^{-1/2} + (9/2)x^{-1/4}$ or $4/x^{1/2} + 9/(2x^{1/4})$ **11.** $g'(x) = -30x^{-6} + x^{-2}$ or $-30/x^6 + 1/x^2$

13. $y' = -5x^{-6} + 2x^{-3} - 5x^{-2}$ or $-5/x^6 + 2/x^3 - 5/x^2$ **15.** $f'(t) = -4t^{-2} - 6t^{-4}$ or $-4/t^2 - 6/t^4$ **17.** $y' = -18x^{-7} - 5x^{-6} + 14x^{-3}$ or $-18/x^7 - 5/x^6 + 14/x^3$ **19.** $h'(x) = -x^{-3/2}/2 + 21x^{-5/2}$ or $-1/(2x^{3/2}) + 21/x^{5/2}$ **21.** $y' = 2x^{-4/3}/3$ or $2/(3x^{4/3})$ **23.** $-40x^{-6} + 36x^{-5}$ or $-40/x^6 + 36/x^5$ **25.** $(-9/2)x^{-3/2} - 3x^{-5/2}$ or $-9/(2x^{3/2}) - 3/x^{5/2}$ **27.** $-14/3$ **29.** (c) **33.** -15; $15x + y = 14$ **35.** 1 **37.** $(-1/2, -19/2)$ **41.** **(a)** 30 **(b)** 4.8 **(c)** -10 **43.** **(a)** 100 **(b)** 1 **45.** **(a)** $C'(x) = 2$ **(b)** $R'(x) = 6 - x/500$ **(c)** $P'(x) = 4 - x/500$ **(d)** $x = 2000$ **(e)** \$4000 **47.** **(a)** $V'(r) = 160\pi r$ **(b)** 640π cu mm **(c)** 960π cu mm **(d)** 1280π cu mm **(e)** The volume increases. **49.** **(a)** 100 **(b)** 1 **(c)** $-.01$; the percent of acid is decreasing at the rate of .01 per day after 100 days. **51.** **(a)** $v(t) = 50t - 9$ **(b)** -9; 241; 491 **53.** **(a)** $v(t) = -6t^2 + 8t$ **(b)** 0; -110; -520 **55.** **(a)** 0 ft/sec; -32 ft/sec **(b)** 2 sec **(c)** 64 ft

Section 2.5 (page 127)

1. $y' = 4x + 3$ **3.** $y' = 18x^2 - 6x + 4$ **5.** $y' = 6t^2 - 4t - 12$ **7.** $y' = 40x^3 - 60x^2 + 16x - 16$ **9.** $y' = 98x - 84$ **11.** $g'(t) = 36t^3 + 24t$ **13.** $y' = 3x^{1/2} - 3x^{-1/2}/2 - 2$ or $3x^{1/2} - 3/(2x^{1/2}) - 2$ **15.** $g'(x) = -12 + 15x^{-1/2}$ or $-12 + 15/x^{1/2}$ **17.** $f'(x) = 94/(8x + 1)^2$ **19.** $y' = 8/(2x - 11)^2$ **21.** $y' = 2/(1 - t)^2$ **23.** $y' = (x^2 + 6x - 12)/(x + 3)^2$ **25.** $y' = (-24x^2 - 2x + 6)/(4x^2 + 1)^2$ **27.** $k'(x) = (x^2 - 4x - 12)/(x - 2)^2$ **29.** $r'(t) = [-\sqrt{t} + 3/(2\sqrt{t})]/(2t + 3)^2$ or $(-2t + 3)/[2\sqrt{t}(2t + 3)^2]$ **31.** In the first step, the numerator should be $(x^2 - 1)2 - (2x + 5)(2x)$. **33.** $y = -2x + 9$ **35.** **(a)** \$22.86 per unit **(b)** \$12.92 per unit **(c)** $(3x + 2)/(x^2 + 4x)$ per unit **(d)** $\overline{C}'(x) = (-3x^2 - 4x - 8)/(x^2 + 4x)^2$ **37.** **(a)** $G'(20) = -1/200$; go faster **(b)** $G'(40) = 1/400$; go slower **39.** **(a)** $s'(x) = m/(m + nx)^2$ **(b)** $1/2560 \approx .000391$ mm per ml **41.** **(a)** -100 **(b)** $-1/100$ or $-.01$

Section 2.6 (page 136)

1. 1122 **3.** 97 **5.** $256k^2 + 48k + 2$ **7.** $(3x + 95)/8$; $(3x + 280)/8$ **9.** $1/x^2$; $1/x^2$ **11.** $\sqrt{8x^2 - 4}$; $8x + 10$ **13.** $\sqrt{(x - 1)/x}$; $-1/\sqrt{x + 1}$ **17.** If $f(x) = x^{2/5}$ and $g(x) = 5 - x$, then $y = f[g(x)]$. **19.** If $f(x) = -\sqrt{x}$ and $g(x) = 13 + 7x$, then $y = f[g(x)]$. **21.** If $f(x) = x^{1/3} - 2x^{2/3} + 7$ and $g(x) = x^2 + 5x$, then $y = f[g(x)]$. **23.** $y' = 5(2x^3 + 9x)^4(6x^2 + 9)$ **25.** $f'(x) = -288x^3(3x^4 + 2)^2$ **27.** $s'(t) = 144t^3(2t^4 + 5)^{1/2}$ **29.** $f'(t) = 32t/\sqrt{4t^2 + 7}$ **31.** $r'(t) = 4(2t^5 + 3)(22t^5 + 3)$ **33.** $y' = (x^2 - 1)(7x^4 - 3x^2 + 8x)$ **35.** $y' = [(5x^6 + x)(125x^6 + 5x)]/\sqrt{2x}$ **37.** $y' = -30x/(3x^2 - 4)^6$ **39.** $p'(t) = [2(2t + 3)^2(4t^2 - 12t - 3)]/(4t^2 - 1)^2$ **41.** $y' = (-30x^4 - 132x^3 + 4x + 8)/(3x^3 + 2)^5$ **43.** **(a)** -2 **(b)** $-24/7$ **45.** $D(c) = (-c^2 + 10c + 12{,}475)/25$ **47.** **(a)** \$101.22 **(b)** \$111.86 **(c)** \$117.59 **49.** **(a)** $-\$1050$ **(b)** $-\$457.06$ **51.** \$400 per additional worker **53.** $P[f(a)] = 18a^2 + 24a + 9$ **55.** $A[r(t)] = A(t) = 4\pi t^2$; this function gives the area of the pollution in terms of the time since the pollutants were first emitted. **57.** **(a)** $-.5$ **(b)** $-1/54 \approx -.02$ **(c)** $-1/128 \approx -.008$ **(d)** always decreasing; the derivative is negative for all $t \geq 0$

Chapter 2 Review Exercises (page 140)

3. 4 **5.** does not exist **7.** 4 **9.** 17/3 **11.** 8 **13.** -13 **15.** 1/6 **17.** discontinuous at x_2 and x_4 **19.** yes; no; yes; no **21.** yes; no; yes **23.** yes; yes; yes **25.** **(a)** 1/4 **(b)** 0 **(c)** $-3/2$ **27.** -60; -20 **29.** $-5/4$; -5 **31.** $y' = 10x + 6$ **33.** -2; $y + 2x = -4$ **35.** $-3/4$; $3x + 4y = -9$ **37.** $-4/3$; $4x + 3y = 11$ **39.** $-4/5$, $4x + 5y = -13$ **43.** $y' = 3x^2 - 8x$ **45.** $y' = 6x^{-3}$ or $6/x^3$ **47.** $f'(x) = -6x^{-2} - x^{-1/2}$ or $-6/x^2 - 1/x^{1/2}$ **49.** $y' = -20t^3 + 42t^2 - 16t$ **51.** $p'(t) = 98t^{3/4} - 12t^{-1/4}$ or $98t^{3/4} - 12/t^{1/4}$ **53.** $y' = -2x^{-3/5} - 54x^{-8/5}$ or $-2/x^{3/5} - 54/x^{8/5}$ **55.** $r'(x) = 16/(2x + 1)^2$ **57.** $y' = (4x^3 + 7x^2 - 20x)/(x + 2)^2$ **59.** $k'(x) = 30(5x - 1)^5$ **61.** $y' = -12(8t - 1)^{-1/2}$ or $-12/(8t - 1)^{1/2}$ **63.** $y' = 4x(3x - 2)^4(21x - 4)$ **65.** $s'(t) = (-4t^3 - 9t^2 + 24t + 6)/(4t - 3)^5$ **67.** $(4x^{1/2} + x + 1)/[2x^{1/2}(1 - x)^2]$ **69.** $(2 - x)/[2x^2(x - 1)^{1/2}]$ **71.** does not exist at $t = -2$ **73.** (d) **75.** $\overline{C}'(x) = (-3x - 4)/[2x^2(3x + 2)^{1/2}]$ **77.** $\overline{C}'(x) = [(4x + 3)^3(12x - 3)]/x^2$ **79.** **(a)** \$88.89 is the approximate increase in profit from selling the fifth unit. **(b)** \$99.17 is the approximate increase in profit from selling the thirteenth unit. **(c)** \$99.72 is the approximate increase in profit from selling the twenty-first unit. **(d)** The marginal profit is increasing as the number sold increases. **81.** **(a)** $R'(x) = 16 - 6x$ **(b)** -44; an increase of \$100 spent on advertising when advertising expenditures are \$1000 will result in revenue decreasing by \$44 **83.** **(a)** \$3.40 **(b)** \$3.28 **(c)** \$3.18 **(d)** \$3.15 **(e)** \$10.15 **(f)** \$15.15 **(g)** $[0, \infty)$ **(h)** no **(i)** $\overline{P}(x) = 15 + 25x$ **(j)** $\overline{P}'(x) = 25$ **(k)** no; the profit per pound never changes, no matter how many pounds are sold **85.** **(a)** none **(b)** none **(c)** $(-\infty, \infty)$ **(d)** Since the derivative is always negative, the graph of $g(x)$ is always decreasing. **87.** **(a)** $(-1, 1)$ **(b)** $x = -1$ **(c)** $(-\infty, -1)$, $(1, \infty)$ **(d)** The derivative is 0 when the graph of $G(x)$ is at a low point. It is positive where $G(x)$ is increasing and negative where $G(x)$ is decreasing.

CHAPTER 3

Section 3.1 (page 153)

1. (a) $(1, \infty)$ **(b)** $(-\infty, 1)$ **3. (a)** $(-\infty, -2)$ **(b)** $(-2, \infty)$ **5. (a)** $(-\infty, -4), (-2, \infty)$ **(b)** $(-4, -2)$ **7. (a)** $(-7, -4), (-2, \infty)$ **(b)** $(-\infty, -7), (-4, -2)$ **9. (a)** $(-6, \infty)$ **(b)** $(-\infty, -6)$ **11. (a)** $(-\infty, 3/2)$ **(b)** $(3/2, \infty)$ **13. (a)** $(-\infty, -3), (4, \infty)$ **(b)** $(-3, 4)$ **15. (a)** $(-\infty, -3/2), (4, \infty)$ **(b)** $(-3/2, 4)$ **17. (a)** none **(b)** $(-\infty, \infty)$ **19. (a)** none **(b)** $(-\infty, -1), (-1, \infty)$ **21. (a)** $(-4, \infty)$ **(b)** $(-\infty, -4)$ **23. (a)** none **(b)** $(1, \infty)$ **25. (a)** $(0, \infty)$ **(b)** $(-\infty, 0)$ **27. (a)** $(0, \infty)$ **(b)** $(-\infty, 0)$ **29.** vertex: $(-b/(2a), (4ac - b^2)/(4a))$; increasing on $(-b/(2a), \infty)$, decreasing on $(-\infty, -b/(2a))$ **31. (a)** nowhere **(b)** everywhere **33.** $[0, 1125)$ **35.** after 10 days **37. (a)** $(0, 3)$ **(b)** $(3, \infty)$ (Remember: x must be at least 0.) **39. (a)** $(1000, 6100)$ **(b)** $(6100, 6500)$ **(c)** $(1000, 3000)$, $(3600, 4200)$ **(d)** $(3000, 3600)$, $(4200, 6500)$

Section 3.2 (page 165)

1. relative minimum of -4 at 1 **3.** relative maximum of 3 at -2 **5.** relative maximum of 3 at -4; relative minimum of 1 at -2 **7.** relative maximum of 3 at -4; relative minimum of -2 at -7 and -2 **9.** relative minimum of -44 at -6 **11.** relative maximum of 8.5 at -3 **13.** relative maximum of -8 at -3; relative minimum of -12 at -1 **15.** relative maximum of $827/96$ at $-1/4$; relative minimum of $-377/6$ at -5 **17.** relative maximum of 57 at 2; relative minimum of 30 at 5 **19.** relative maximum of -4 at 0; relative minimum of -85 at 3 and -3 **21.** relative maximum of 0 at $8/5$ **23.** relative maximum of 1 at -1; relative minimum of 0 at 0 **25.** no relative extrema **27.** relative minimum of 0 at 0 **29.** relative maximum of 0 at 1; relative minimum of 8 at 5 **31.** $(2, 7)$ **33.** $(5/4, -9/8)$ **35. (a)** 40 **(b)** 15 **(c)** 375 **37. (a)** 250 **(b)** 10 **(c)** 800 **39.** $q = 5$; $p = 275/6$ **41.** 4:44 P.M.; 5:46 A.M. **43.** 10 min

Note for Exercises 45 and 47: Your answers may be slightly different from the given answers because of different calculators or computer programs. **45.** relative maximum of 6.2 at .085, relative minimum of -57.7 at 2.2 **47.** relative maximum of 280 at -5.1 and of -18.96 at .89, relative minimum of -19.08 at .56

Section 3.3 (page 172)

1. absolute maximum at x_3; no absolute minimum **3.** no absolute extrema **5.** absolute minimum at x_1; no absolute maximum **7.** absolute maximum at x_1; absolute minimum at x_2 **9.** absolute maximum at 0; absolute minimum at -3 **11.** absolute maximum at -1; absolute minimum at -5 **13.** absolute maximum at -2; absolute minimum at 4 **15.** absolute maximum at -2; absolute minimum at 3 **17.** absolute maximum at 6; absolute minimum at -4 and 4 **19.** absolute maximum at 0; absolute minimum at 2 **21.** absolute maximum at 6; absolute minimum at 4 **23.** absolute maximum at $\sqrt{2}$; absolute minimum at 0 **25.** absolute maximum at -3 and 3; absolute minimum at 0 **27.** absolute maximum at 0; absolute minimum at -4 **29.** absolute maximum at 0; absolute minimum at -1 and 1 **31.** 1000 manuals; more than \$2.20 **33. (a)** 341 **(b)** 859.4 **35.** 6 months; 6% **37.** 25; 16.1 **39.** The piece formed into a circle should have length $12\pi/(4 + \pi)$ ft, or about 5.28 ft. *Note for Exercises 41 and 43:* Your answers may be slightly different from the given answers because of different calculators or computer programs. **41.** absolute maximum at 0; absolute minimum at -2.4 **43.** absolute maximum at 0; absolute minimum at .74

Section 3.4 (page 184)

1. $f''(x) = 18x$; 0; 36; -54 **3.** $f''(x) = 36x^2 - 30x + 4$; 4; 88; 418 **5.** $f''(x) = 6$; 6; 6; 6 **7.** $f''(x) = 6(x + 4)$; 24; 36; 6 **9.** $f''(x) = 10/(x - 2)^3$; $-5/4$; $f''(2)$ does not exist; $-2/25$ **11.** $f''(x) = 2/(1 + x)^3$; 2; $2/27$; $-1/4$ **13.** $f''(x) = -1/[4(x + 4)^{3/2}]$; $-1/32$; $-1/[4(6^{3/2})] \approx -.0170$; $-1/4$ **15.** $f''(x) = (-6/5)x^{-7/5}$ or $-6/(5x^{7/5})$; $f''(0)$ does not exist; $-6/[5(2^{7/5})] \approx -.4547$; $-6/[5(-3)^{7/5}] \approx .2578$ **17.** $f'''(x) = -24x$; $f^{(4)}(x) = -24$ **19.** $f'''(x) = 240x^2 + 144x$; $f^{(4)}(x) = 480x + 144$ **21.** $f'''(x) = 18(x + 2)^{-4}$ or $18/(x + 2)^4$; $f^{(4)}(x) = -72(x + 2)^{-5}$ or $-72/(x + 2)^5$ **23.** $f'''(x) = -36(x - 2)^{-4}$ or $-36/(x - 2)^4$; $f^{(4)}(x) = 144(x - 2)^{-5}$ or $144/(x - 2)^5$ **25.** concave upward on $(2, \infty)$; concave downward on $(-\infty, 2)$; point of inflection at $(2, 3)$ **27.** concave upward on $(-\infty, -1)$ and $(8, \infty)$; concave downward on $(-1, 8)$; points of inflection at $(-1, 7)$ and $(8, 6)$ **29.** concave upward on $(2, \infty)$; concave downward on $(-\infty, 2)$; no points of inflection **31.** always concave upward; no points of inflection **33.** concave upward on $(-1, \infty)$; concave downward on $(-\infty, -1)$; point of inflection at $(-1, 44)$ **35.** concave upward on $(-\infty, 3/2)$; concave downward on $(3/2, \infty)$; point of inflection at $(3/2, 525/2)$ **37.** concave upward on $(5, \infty)$; concave downward on $(-\infty, 5)$; no points of inflection **39.** concave upward on $(-10/3, \infty)$; concave downward on $(-\infty, -10/3)$; point of inflection at $(-10/3, -250/27)$ **41.** relative maximum at -5 **43.** relative maximum at 0; relative minimum at $2/3$ **45.** relative minimum at -3 **47. (a)** car phones and CD players; the rate of growth of sales will now decline **(b)** food processors; the rate of decline of sales is starting to slow **49.** $(22, 6517.9)$ **53. (a)** at 4 hr **(b)** 1160 million **55. (a)** after 3 hr **(b)** $2/9\%$

57. $v(t) = 16t + 4$; $a(t) = 16$; $v(0) = 4$ cm/sec; $v(4) = 68$ cm/sec; $a(0) = 16$ cm/sec^2; $a(4) = 16$ cm/sec^2 **59.** $v(t) = -15t^2 - 16t + 6$; $a(t) = -30t - 16$; $v(0) = 6$ cm/sec; $v(4) = -298$ cm/sec; $a(0) = -16$ cm/sec^2; $a(4) = -136$ cm/sec^2 **61.** $v(t) = 6(3t + 4)^{-2}$ or $6/(3t + 4)^2$; $a(t) = -36(3t + 4)^{-3}$ or $-36/(3t + 4)^3$; $v(0) = 3/8$ cm/sec; $v(4) = 3/128$ cm/sec; $a(0) = -9/16$ cm/sec^2; $a(4) = -9/1024$ cm/sec^2 **63. (a)** -96 ft/sec **(b)** -160 ft/sec **(c)** -256 ft/sec **(d)** -32 ft/sec^2 *Note for Exercises 65 and 67:* Your answers may be slightly different from the given answers because of different calculators or computer programs. **65. (a)** increasing on $(0, 2)$ and $(4, 5)$; decreasing on $(-5, -.5)$ and $(2.5, 3.5)$ **(b)** minima between $-.5$ and 0, and between 3.5 and 4; a maximum between 2 and 2.5 **(c)** concave upward on $(-5, .5)$ and $(3.5, 5)$; concave downward on $(.5, 3.5)$ **(d)** inflection points between .5 and 1, and between 3 and 3.5 **67. (a)** decreasing on $(-1, 1)$; increasing on $(1.2, 2)$ **(b)** minimum between 1 and 1.2 **(c)** concave upward on $(-1, 0)$ and $(.8, 2)$; concave downward on $(0, .8)$ **(d)** inflection points between .6 and .8, and at 0

Section 3.5 (page 197)

1. 3 **3.** 3/5 **5.** 1/2 **7.** 1/2 **9.** 0 **11.** 0

13.

$f(x) = -2x^3 - 9x^2 + 108x - 10$

15.

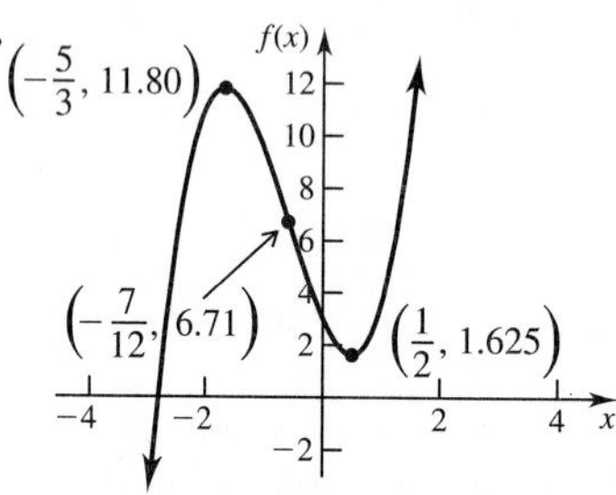

17.

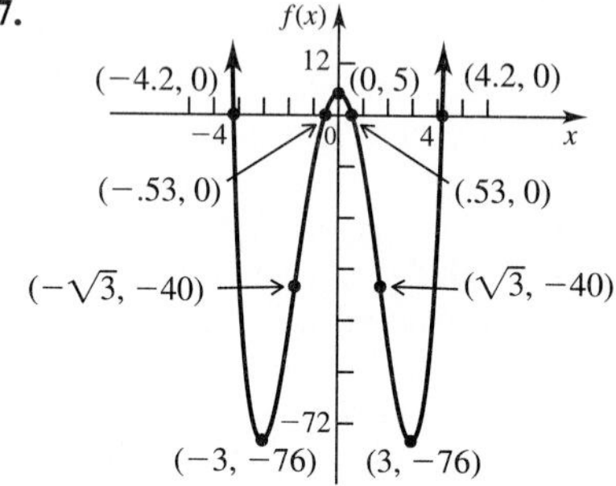

$f(x) = x^4 - 18x^2 + 5$

19.

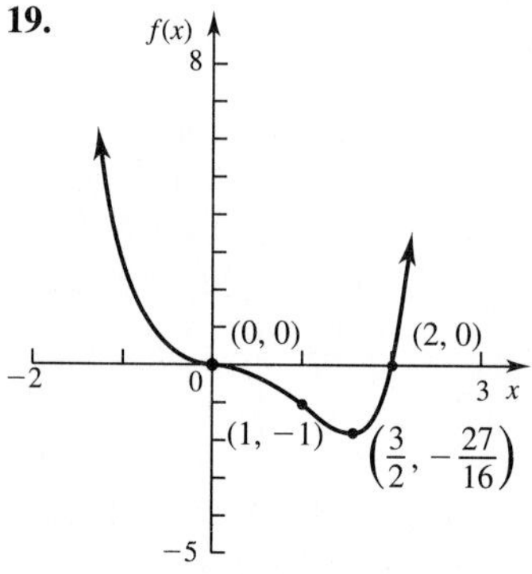

$f(x) = x^4 - 2x^3$

21.

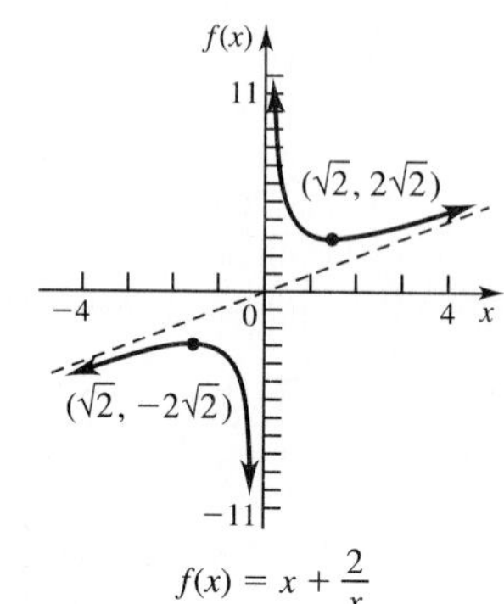

23.

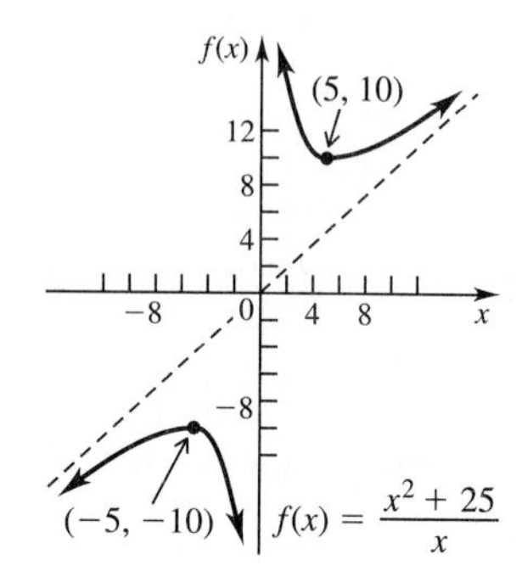

25.

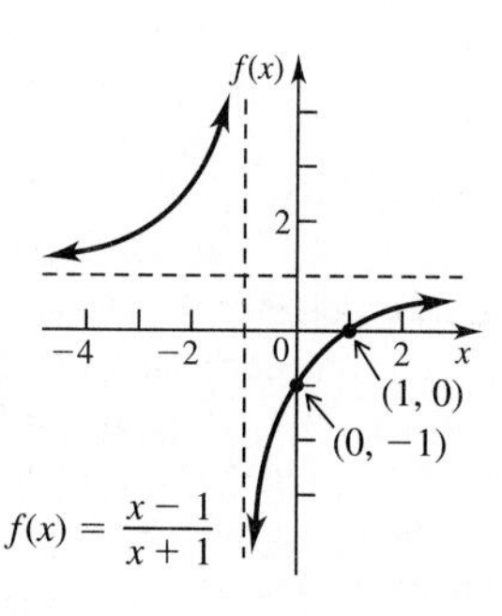

27.

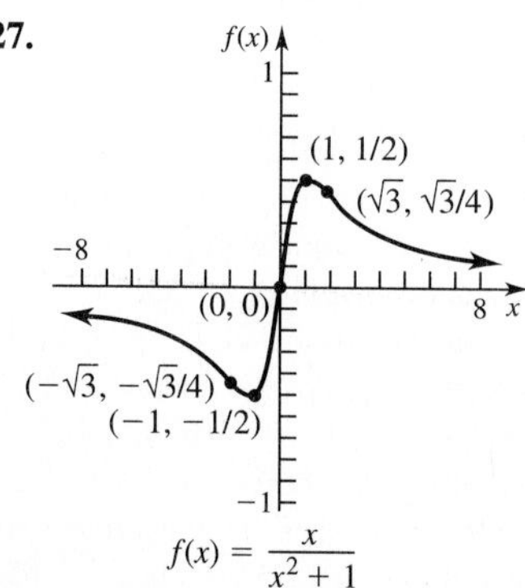

29.

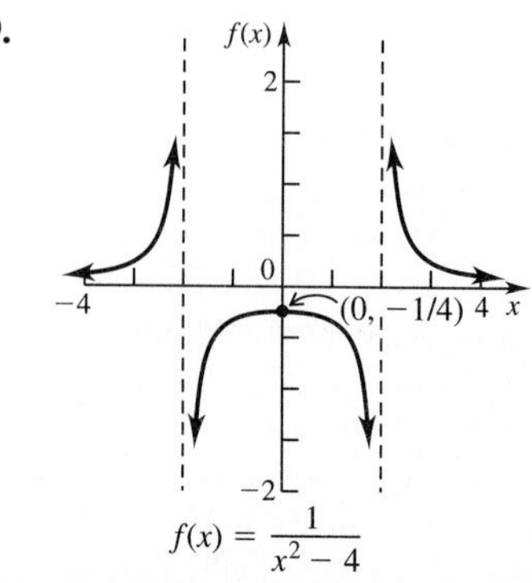

31.

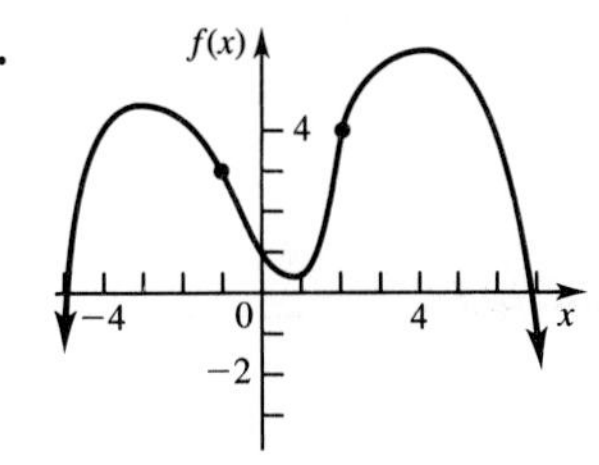

33.

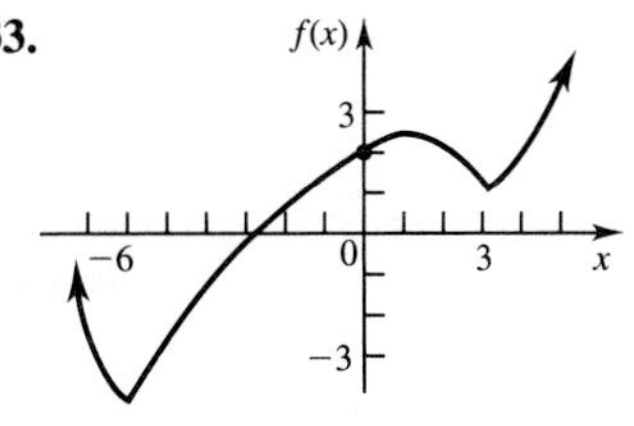

35. 6; the average cost approaches 6 as the number of tapes produced becomes very large. **37.** 0; the concentration of the drug in the bloodstream approaches 0 as the number of hours after injection increases. **39. (a)** 1.5 **41. (a)** -2 **43. (a)** 8

Chapter 3 Review Exercises (page 199)

5. increasing on $(5/2, \infty)$; decreasing on $(-\infty, 5/2)$ **7.** increasing on $(-4, 2/3)$; decreasing on $(-\infty, -4)$ and $(2/3, \infty)$ **9.** never increasing; decreasing on $(-\infty, 4)$ and $(4, \infty)$ **11.** relative maximum of -4 at 2 **13.** relative minimum of -7 at 2 **15.** relative maximum of 101 at -3; relative minimum of -24 at 2 **17.** $f''(x) = 36x^2 - 10$; 26; 314 **19.** $f''(x) = -68(2x + 3)^{-3}$ or $-68/(2x + 3)^3$; $-68/125$; $68/27$ **21.** $f''(t) = (t^2 + 1)^{-3/2}$ or $1/(t^2 + 1)^{3/2}$; $1/2^{3/2} \approx .354$; $1/10^{3/2} \approx .032$ **23.** absolute maximum of $29/4$ at $5/2$; absolute minimum of 5 at 1 and 4 **25.** absolute maximum of 39 at -3; absolute minimum of $-319/27$ at $5/3$ **27.** does not exist **29.** $1/5$ **31.** $3/4$

33.

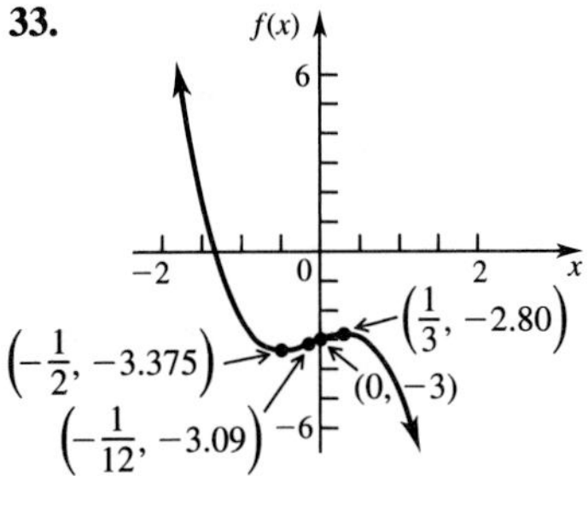

35.

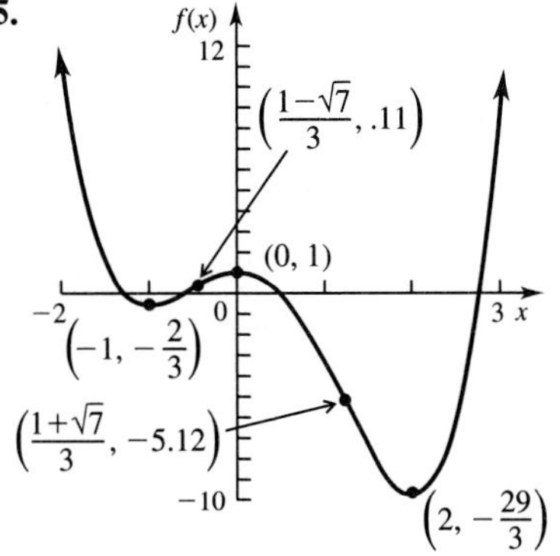

37.

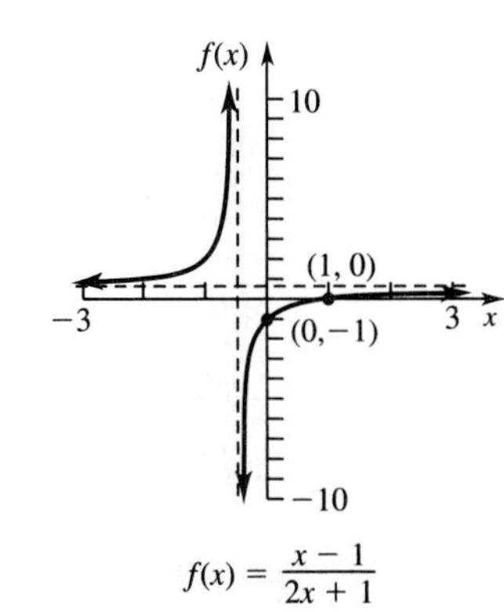

39.

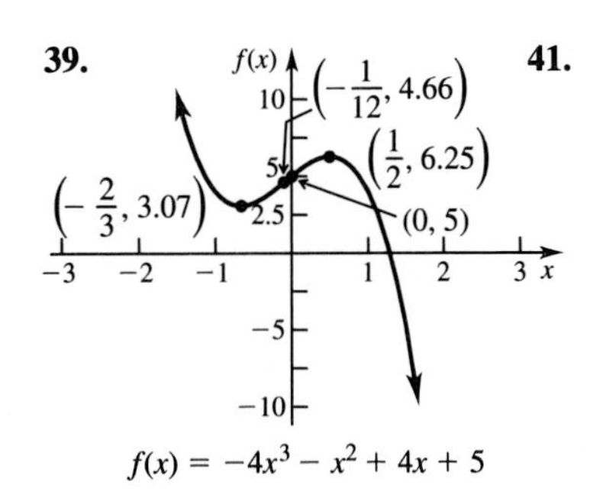

41.

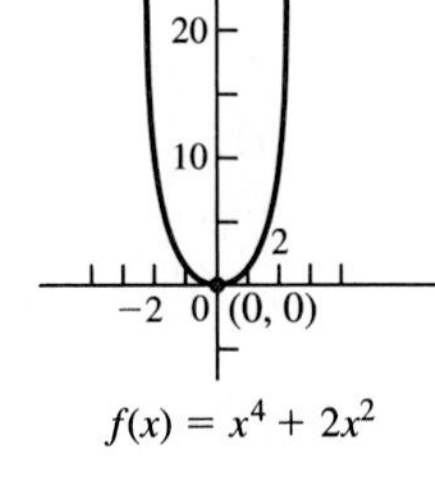

43.

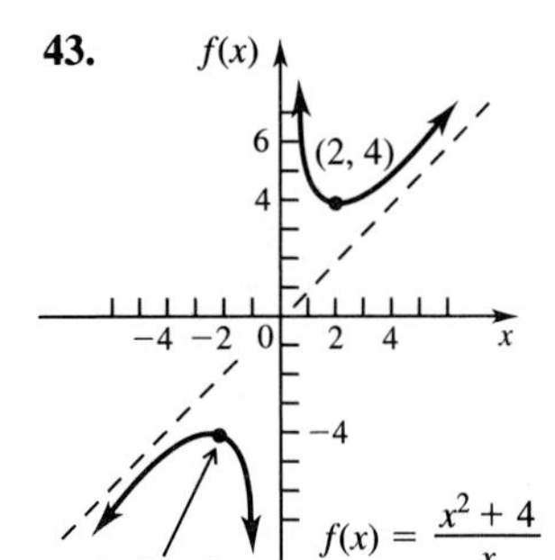

45.

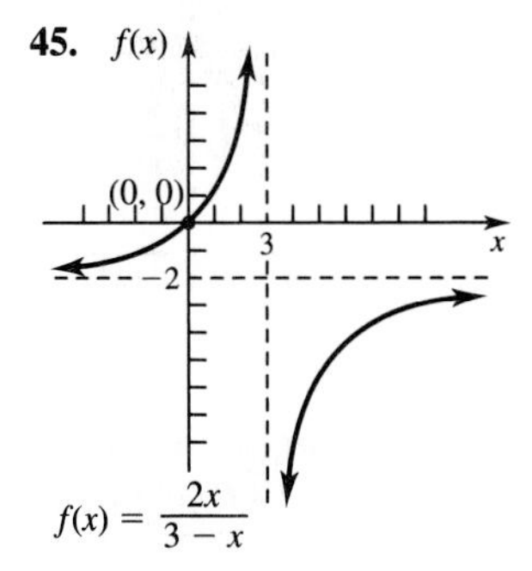

CHAPTER 4

Section 4.1 (page 209)

1. (a) $y = 100 - x$ **(b)** $P = x(100 - x)$ **(c)** $[0, 100]$ **(d)** $P' = 100 - 2x$; $x = 50$ **(e)** $P(0) = 0$; $P(100) = 0$; $P(50) = 2500$ **(f)** 2500; 50 and 50 **3.** 100; 100; 20,000 **5.** 100; 50; 500,000 **7.** 10; 0; 0 **9. (a)** $1200 - 2x$ **(b)** $A(x) = 1200x - 2x^2$ **(c)** 300 m **(d)** 180,000 m^2 **11.** 405,000 m^2 **13.** 0 mi **15. (a)** $R(x) = 100{,}000x - 100x^2$ **(b)** 500 **(c)** 25,000,000 cents **17. (a)** $\sqrt{3200} \approx 56.6$ mph **(b)** \$45.24 **19.** \$2400 **21. (a)** 90 **(b)** \$405 **23.** 4 in. by 4 in. by 2 in. **25.** 3 ft by 6 ft by 2 ft **27.** 10 cm and 10 cm **29.** Radius is 1.08 ft, height is 4.34 ft, cost is \$44.11 using the rounded values for the height and radius. **31.** $3\sqrt{6} + 3$ by $2\sqrt{6} + 2$ **33.** point A **35.** 250 thousand **37.** 56.25 thousand **39.** 12.86 thousand **41.** Point P is $3\sqrt{7}/7 \approx 1.134$ mi from point A. **43. (d)** $\alpha = 5$; the current stays constant and the salmon swim at a constant velocity. **45.** radius = 5.206 cm; height = 11.75 cm **47.** radius = 5.242 cm;

height = 11.58 cm **49.** $A(x) = .01x^2 + .05x + .2 + 28/x$; $x = 10.41$ **51.** $A(x) = 30/x + 42x^{-1/2} + .2x^{1/2} + .03x^{3/2}$; $x = 23.49$

Section 4.2 (page 222)
3. 10 **5.** 10,000 **7.** 5000 **13.** **(a)** $E = p/(500 - p)$ **(b)** 12,500 **15.** **(a)** $E = 1$ **(b)** none **17.** **(a)** $E = 2$; elastic; a percentage increase in price will result in a greater percentage decrease in demand **(b)** $E = 1/2$; inelastic; a percentage change in price will result in a smaller percentage change in demand **19.** **(a)** $9/8$ **(b)** $q = 50/3$; $p = 5\sqrt{3}/3$ **(c)** 1
21. The demand function has a horizontal tangent line at the value of P where $E = 0$.

Section 4.3 (page 228)
1. $dy/dx = -4x/(3y)$ **3.** $dy/dx = -y/(y + x)$ **5.** $dy/dx = -3y^2/(6xy - 4)$ **7.** $dy/dx = (-6x - 4y)/(4x + y)$
9. $dy/dx = 3x^2/(2y)$ **11.** $dy/dx = y^2/x^2$ **13.** $dy/dx = -3x(2 + y)^2/2$ **15.** $dy/dx = -2xy/(x^2 + 3y^2)$
17. $dy/dx = -y^{1/2}/x^{1/2}$ **19.** $dy/dx = -y^{1/2}x^{-1/2}/(x^{1/2}y^{-1/2} + 2)$ **21.** $dy/dx = (4x^3y^3 + 6x^{1/2})/(9y^{1/2} - 3x^4y^2)$
23. $dy/dx = (8x(x^2 + y^3)^3 - 1)/(2 - 12y^2(x^2 + y^3)^3)$
25. $4y = 3x + 25$ **27.** $y = x + 2$ **29.** $24x + y = 57$ **31.** $x + 4y = 12$
33. **(a)** $3x + 4y = 50$; $-3x + 4y = -50$
(b)

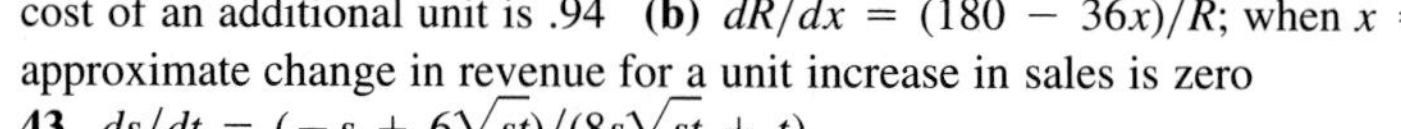

35. $2y = x + 1$ **37.** $dy/dx = -x/y$; there is no function $y = f(x)$ that satisfies $x^2 + y^2 + 1 = 0$. **39.** $dv/du = -(2v + 1)^{1/2}/(2u^{1/2})$
41. **(a)** $dC/dx = (x^{3/2} + 25)/(Cx^{1/2})$; when $x = 5$ the approximate increase in cost of an additional unit is .94 **(b)** $dR/dx = (180 - 36x)/R$; when $x = 5$ the approximate change in revenue for a unit increase in sales is zero
43. $ds/dt = (-s + 6\sqrt{st})/(8s\sqrt{st} + t)$

Section 4.4 (page 234)
1. 440 **3.** $-15/2$ **5.** $-5/7$ **7.** $1/5$ **9.** \$200 per month **11.** **(a)** Revenue is increasing at a rate of \$180 per day. **(b)** Cost is increasing at a rate of \$50 per day. **(c)** Profit is increasing at a rate of \$130 per day. **13.** Demand is decreasing at a rate of approximately 343 units per unit of time. **15.** $-.24$ mm/min **17.** .067 mm/min **19.** $-.370$
21. 7/6 ft/min **23.** 16π ft²/min **25.** 50π in³/min **27.** 1/16 ft/min **29.** 43.3 ft/min

Section 4.5 (page 242)
1. $dy = 12x\,dx$ **3.** $dy = (14x - 9)dx$ **5.** $dy = x^{-1/2}\,dx$ **7.** $dy = [-22/(x - 3)^2]dx$ **9.** $dy = (3x^2 - 1 + 4x)dx$
11. $dy = (x^{-2} + 6x^{-3})dx$ or $(1/x^2 + 6/x^3)dx$ **13.** -2.6 **15.** .1 **17.** .130 **19.** $-.023$ **21.** .24
23. $-.00444$ **25.** **(a)** -34 thousand pounds **(b)** -169.2 thousand pounds **27.** -5.625 housing starts **29.** about $-\$990{,}000$ **31.** $21{,}600\pi$ in³ **33.** **(a)** .347 million **(b)** $-.022$ million **35.** 1568π mm³ **37.** 80π mm²
39. -7.2π cm³ **41.** ± 1.224 in² **43.** $\pm .116$ in³

Chapter 4 Review Exercises (page 244)
5. $dy/dx = (-4y - 2xy^3)/(3x^2y^2 + 4x)$ **7.** $dy/dx = (-4 - 9x^{3/2})/(24x^2y^2)$ **9.** $dy/dx = (2y - 2y^{1/2})/(4y^{1/2} - x + 9y)$ (This form of the answer was obtained by multiplying both sides of the given function by $x - 3y$.) **11.** $dy/dx = [9(4y^2 - 3x)^{1/3} + 3]/(8y)$ **13.** $23x + 16y = 94$ **15.** 272 **17.** -2 **21.** $dy = 24x(x^2 - 1)^2dx$ **23.** $dy = (3x^2/2)(9 + x^3)^{-1/2}dx$ **25.** .00204 **29.** $x = 2$ and $y = 0$ **31.** **(a)** 600 boxes **(b)** \$720 **33.** radius = 1.684 in.; height = 4.490 in. **35.** 4434 **37.** 126 **39.** $E = k$; elastic when $k > 1$; inelastic when $k < 1$ **41.** 8/3 ft/min
43. $21/16 = 1.3125$ ft/min **45.** $\pm .736$ in² **47.** **(a)** $(2, -5)$, $(2, 4)$ **(b)** $(2, -5)$ is a relative maximum, and $(2, 4)$ is a relative minimum. **(c)** no

Extended Application (page 247)
1. $-C_1/m^2 + DC_3/2$ **2.** $m = \sqrt{2C_1/(DC_3)}$ **3.** about 3.33 **4.** $m^+ = 4$ and $m^- = 3$ **5.** $Z(m^+) = Z(4) = \$11{,}400$; $Z(m^-) = Z(3) = \$11{,}300$ **6.** 3 months; 9 trainees per batch

CHAPTER 5

Section 5.1 (page 256)

1.

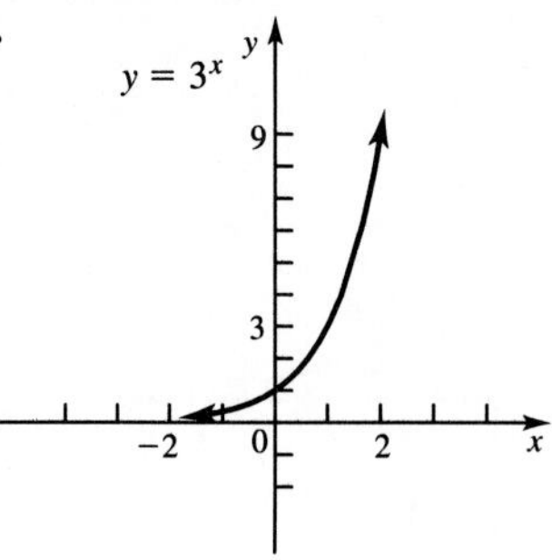

3.

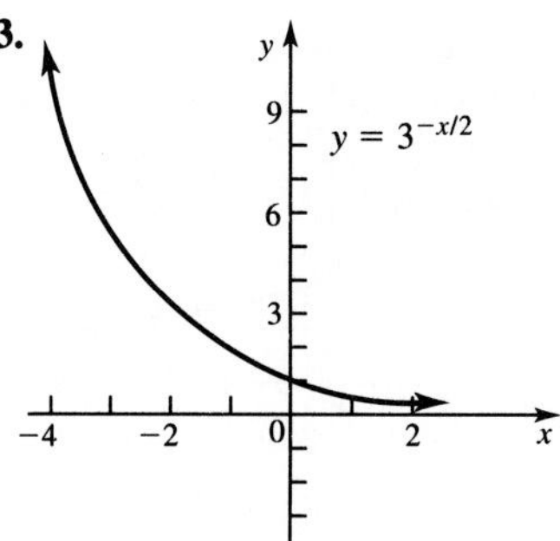

5.

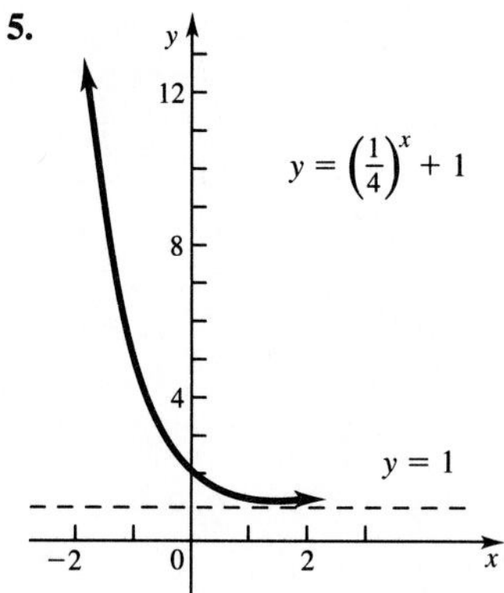

7.

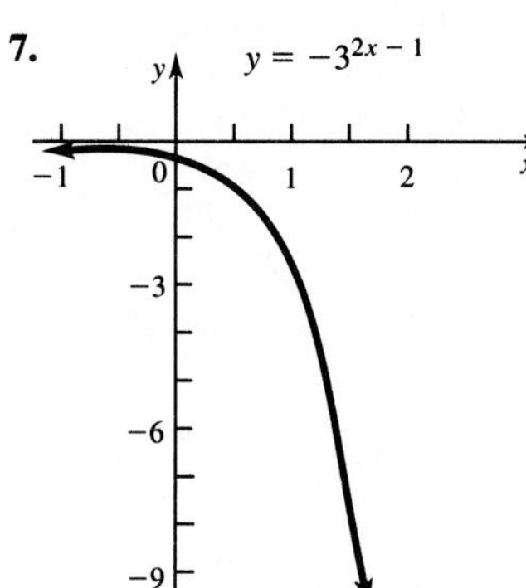

9.

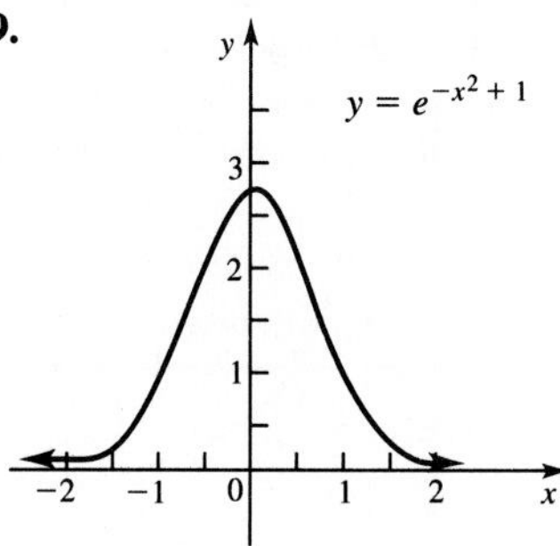

11.

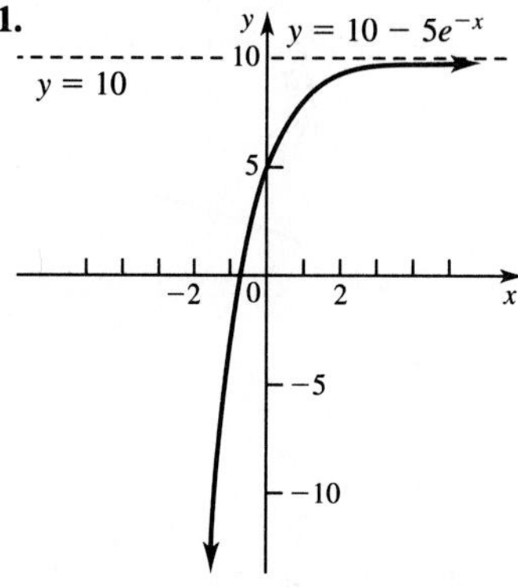

13.

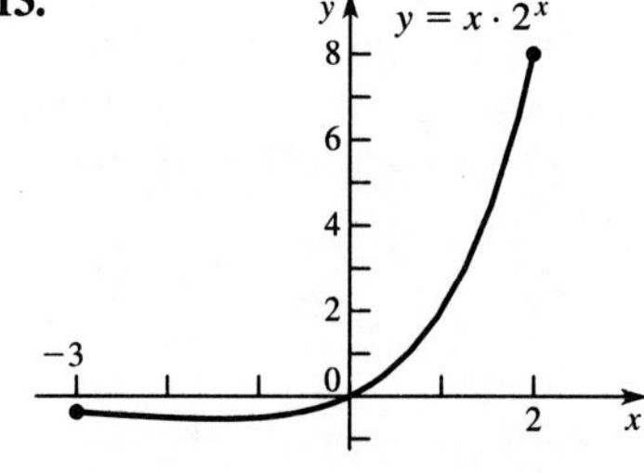

15.

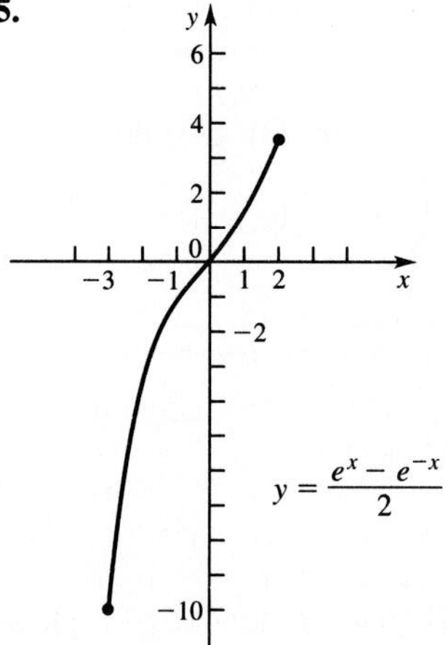

17. because 4 and 6 cannot easily be written as powers of the same base **19.** 3 **21.** -2 **23.** 6 **25.** 7/4 **27.** -2 **29.** 2, -2 **31.** 0, -1 **33.** 4, -2 **35.** **(a)** 1.718 **(b)** 1.052 **(c)** 1.005 **(d)** 1.001 **(e)** 1 **39.** **(a)** \$10,528.13 **(b)** \$10,881.50 **(c)** \$11,069.78 **(d)** \$11,199.99 **41.** \$31,427.49 **43.** **(a)** 12.5% **(b)** 11.9%

45. **(a)** about 207 **(b)** about 235 **(c)** about 249 **(d)**

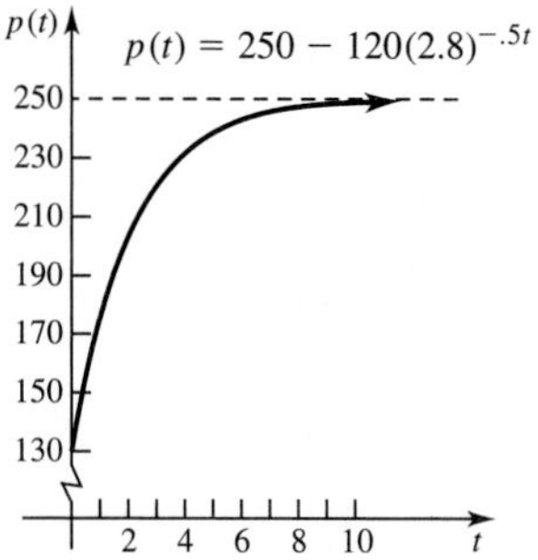

(e) It gets very close to 250. **(f)** 250

47. **(a)** The function gives 3727 million, which is very close. **(b)** 5341 million **(c)** 6395 million **49.** **(a)** 6 **(b)** 2 yr **(c)** 6 yr **51.** **(a)** 55 g **(b)** 10 mo **53.**

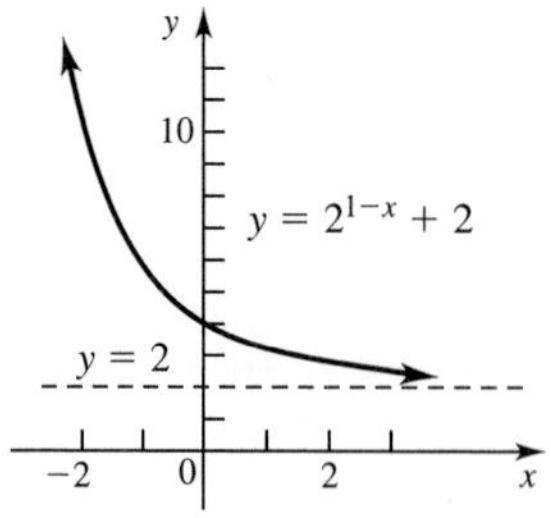

55.

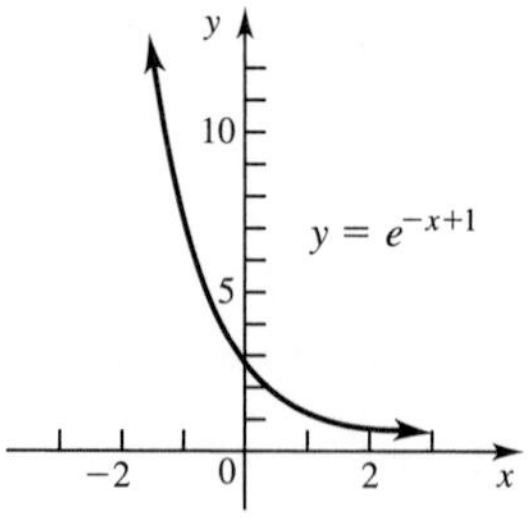

Section 5.2 (page 269)

1. $\log_2 8 = 3$ **3.** $\log_3 81 = 4$ **5.** $\log_{1/3} 9 = -2$ **7.** $2^7 = 128$ **9.** $25^{-1} = 1/25$ **11.** $10^4 = 10{,}000$ **13.** 2 **15.** 3 **17.** −2 **19.** −2/3 **21.** 1 **23.** 5/3 **25.** $\log_3 4$ **27.**

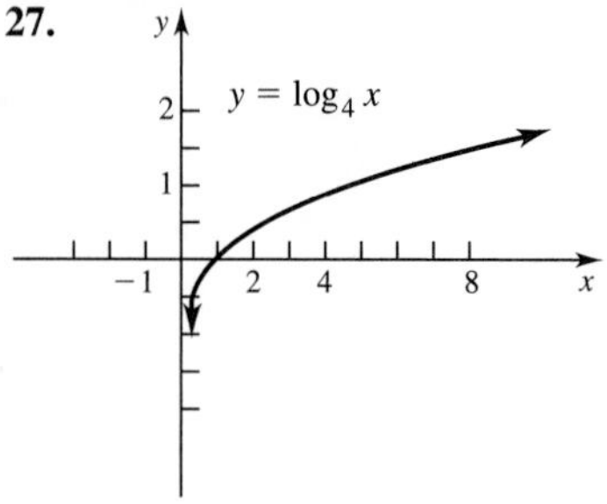

29.

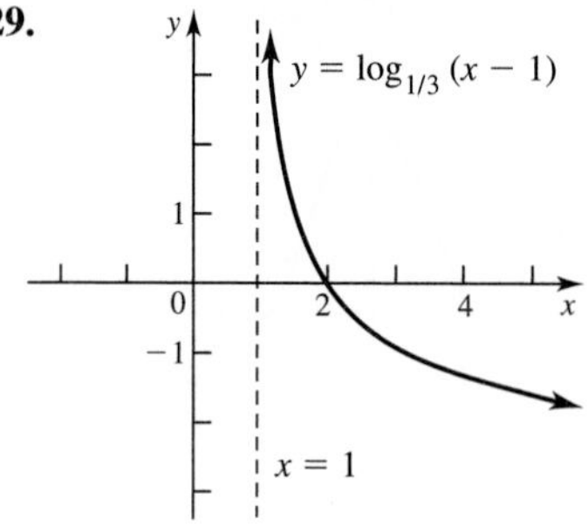

31.

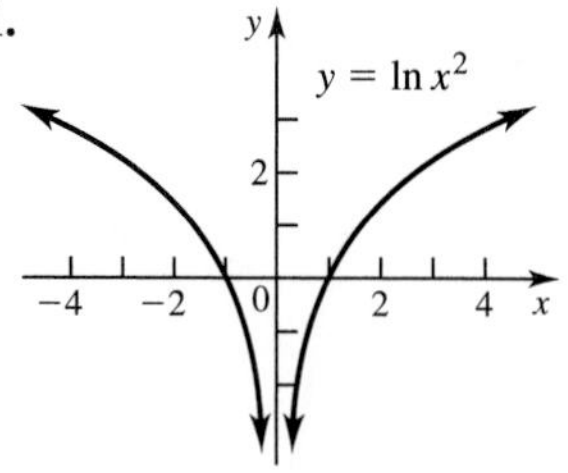

33. $\log_9 7 + \log_9 m$ **35.** $1 + \log_3 p - \log_3 5 - \log_3 k$ **37.** $\log_3 5 + (1/2)\log_3 2 - (1/4)\log_3 7$ **39.** $3a$ **41.** $2c + 3a + 1$ **43.** 1.86 **45.** 9.35 **47.** −.21 **49.** $x = 1/5$ **51.** $z = 2/3$ **53.** $r = 49$ **55.** $x = 1$ **57.** no solution **59.** $x = 1.79$ **61.** $y = 1.24$ **63.** $z = 2.10$ **67.** **(a)** 11.7 yr **(b)** 18.6 yr **(c)** 12 yr **69.** 1.589 **71.** 4.3 ml/min; 7.8 ml/min **73.** **(a)** 21 **(b)** 70 **(c)** 91 **(d)** 120 **(e)** 140 **(f)** about 2,300,000,000 I_0 **75.** **(a)** 1000 times greater **(b)** 1,000,000 times greater **77.**

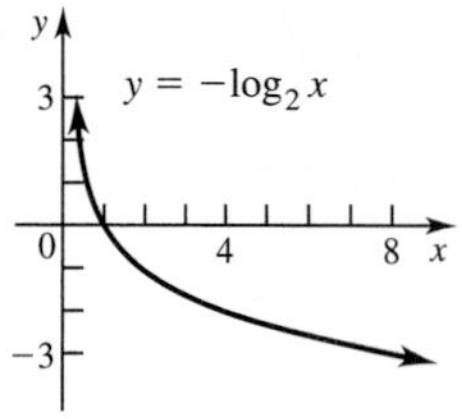

79.

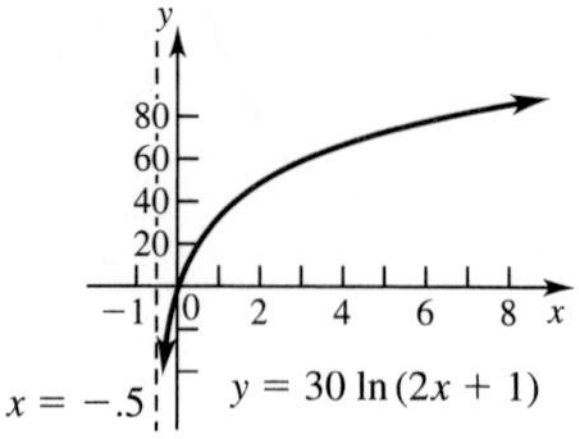

Section 5.3 (page 279)

1. 5.12% **3.** 10.25% **5.** 11.63% **7.** \$1043.79 **9.** \$4537.71 **11.** \$5248.14 **19.** **(a)** \$10.94 **(b)** \$11.27 **(c)** \$11.62 **21.** **(a)** the 10% investment compounded quarterly **(b)** \$622.56 **(c)** 10.38% and 10.24% **(d)** 2.95 yr **23.** 7.40% **25.** \$14,700.60 **27.** **(a)** \$13,459.43 **(b)** \$6540.57 **(c)** \$5140.53 **29.** **(a)** 200 **(b)** about 1/2 yr **(c)** no **(d)** yes; 1000 **31.** **(a)** $y = 100e^{.11t}$ **(b)** about 15 mo **33.** **(a)** $y = 50{,}000e^{-.102t}$ **(b)** about 6.8 hr **35.** **(a)** about 1100 **(b)** about 1600 **(c)** about 2300 **(d)** at about 1.8 decades **39.** **(a)** 0 **(b)** about 432 **(c)** about 497

(d) about 1.6 days **(e)** 500 **(f)**

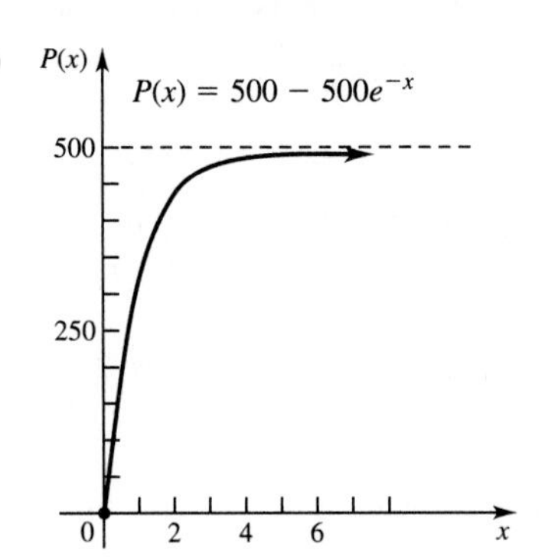

41. about 4100 yr old **43.** about 1600 yr
45. (a) 19.5 watts **(b)** about 173 days **(c)** No. It will approach 0 watts, but never be exactly 0.
47. (a) $y = 10e^{.0095t}$ **(b)** 42.7° C
49. about 30 min

Section 5.4 (page 288)

1. $y' = 1/x$ **3.** $y' = -1/(3 - x)$ or $1/(x - 3)$ **5.** $y' = (4x - 7)/(2x^2 - 7x)$ **7.** $y' = 1/[2(x + 5)]$
9. $y' = 3(2x^2 + 5)/[x(x^2 + 5)]$ **11.** $y' = -3x/(x + 2) - 3 \ln (x + 2)$ **13.** $y' = x + 2x \ln |x|$
15. $y' = [2x - 4(x + 3) \ln (x + 3)]/[x^3(x + 3)]$ **17.** $y' = (4x + 7 - 4x \ln x)/[x(4x + 7)^2]$
19. $y' = (6x \ln x - 3x)/(\ln x)^2$ **21.** $y' = 4(\ln |x + 1|)^3/(x + 1)$ **23.** $y' = 1/(x \ln x)$ **27.** $y' = 1/(x \ln 10)$
29. $y' = -1/[(\ln 10)(1 - x)]$ or $1/[(\ln 10)(x - 1)]$ **31.** $y' = 5/[2(\ln 5)(5x + 2)]$ **33.** $y' = 3(x + 1)/[(\ln 3)(x^2 + 2x)]$
35. minimum of $-1/e \approx -.3679$ at $x = 1/e \approx .3679$ **37.** minimum of $-1/e \approx -.3679$ at $x = 1/e \approx .3679$; maximum of $1/e$ at $-1/e$ **39.** maximum of $1/e \approx .3679$ at $x = e \approx 2.718$

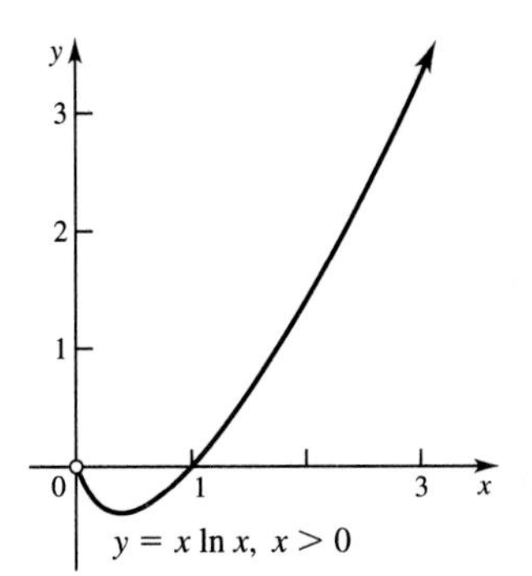

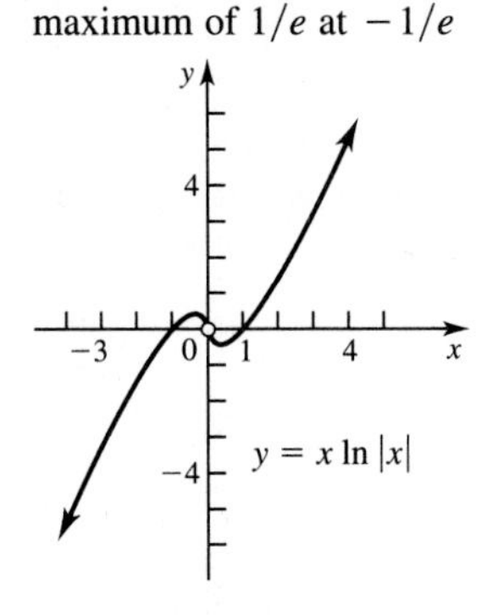

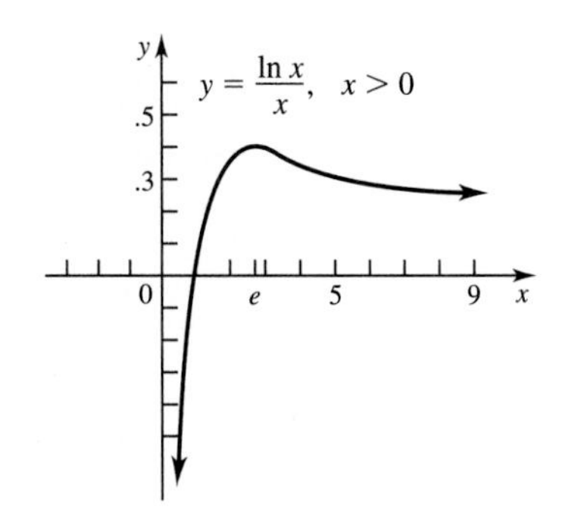

41. As $x \to \infty$, the slope approaches 0; as $x \to 0$, the slope becomes infinitely large. **45. (a)** $\frac{dR}{dx} = 100 + \frac{50(\ln x - 1)}{(\ln x)^2}$
(b) $112.48 **(c)** to decide whether it is reasonable to sell additional items **47. (a)** $R(x) = 100x - 10x \ln x$, $1 < x < 20{,}000$
(b) $dR/dn = 60(9 - \ln x) = 60[9 - \ln (6n)]$ **(c)** about 360; hiring an additional worker will produce an increase in revenue of about $360 **49.** 26.9; 13.1 **51.** minimum of $\approx .85$ at $x \approx .7$

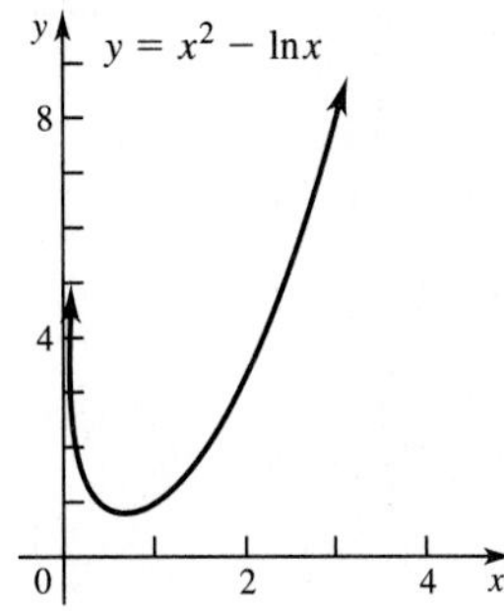

Section 5.5 (page 297)

1. $y' = 4e^{4x}$ **3.** $y' = -16e^{2x}$ **5.** $y' = -16e^{x+1}$ **7.** $y' = 2xe^{x^2}$ **9.** $y' = 12xe^{2x^2}$ **11.** $y' = 16xe^{2x^2-4}$ **13.** $y' = xe^x + e^x = e^x(x + 1)$ **15.** $y' = 2(x - 3)(x - 2)e^{2x}$ **17.** $y' = e^{x^2}/x + 2xe^{x^2} \ln x$ **19.** $y' = (xe^x \ln x - e^x)/[x(\ln x)^2]$ **21.** $y' = (2xe^x - x^2e^x)/e^{2x} = x(2 - x)/e^x$ **23.** $y' = [x(e^x - e^{-x}) - (e^x + e^{-x})]/x^2$ **25.** $y' = -20{,}000e^{.4x}/(1 + 10e^{.4x})^2$ **27.** $y' = 8000e^{-.2x}/(9 + 4e^{-.2x})^2$ **29.** $y' = 2(2x + e^{-x^2})(2 - 2xe^{-x^2})$ **31.** $y' = 5 \ln 8e^{5x \ln 8}$ or $y' = 5(\ln 8)(8^{5x})$ **33.** $y' = 6x(\ln 4)e^{(x^2+2)(\ln 4)}$ or $y' = 6x(\ln 4)4^{x^2+2}$ **35.** $y' = [(\ln 3)e^{\sqrt{x} \ln 3}]/\sqrt{x}$ or $y' = [(\ln 3)3^{\sqrt{x}}]/\sqrt{x}$

39. maximum of $1/e$ at $x = -1$; inflection point at $(-2, 2e^{-2})$

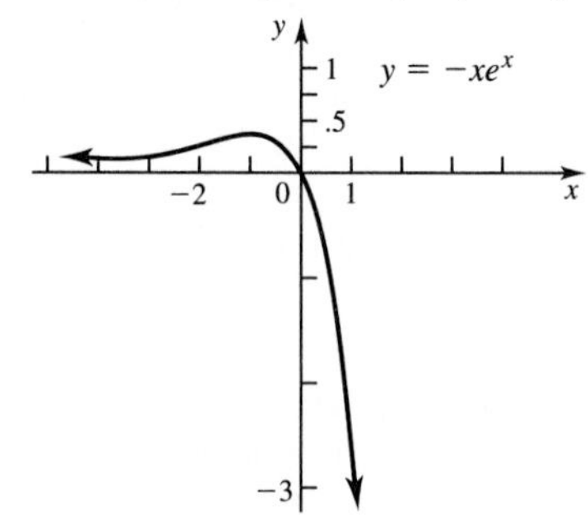

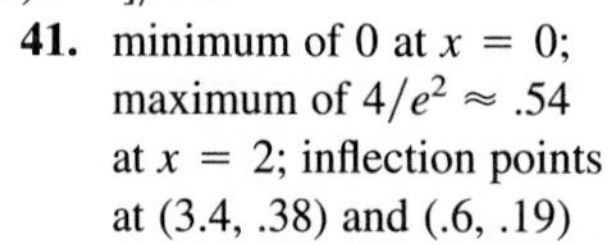

41. minimum of 0 at $x = 0$; maximum of $4/e^2 \approx .54$ at $x = 2$; inflection points at (3.4, .38) and (.6, .19)

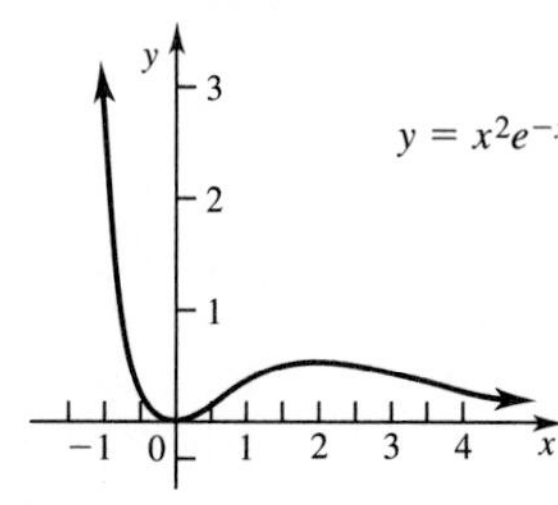

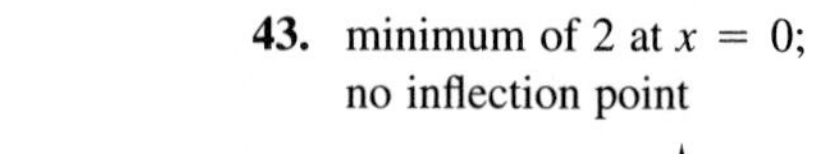

43. minimum of 2 at $x = 0$; no inflection point

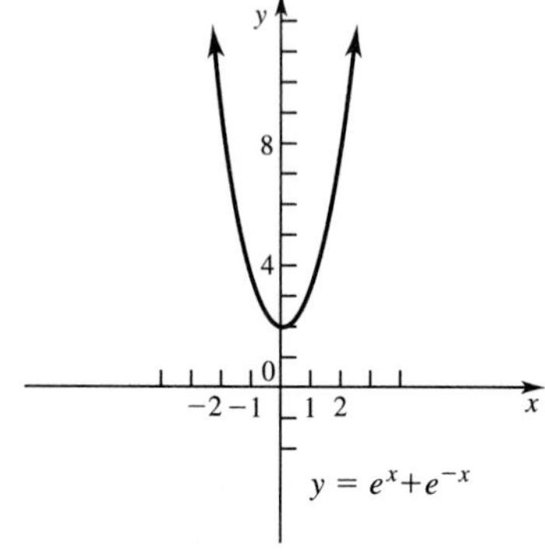

45. $f''(x) = e^x$, $f'''(x) = e^x$; $f^{(n)} = e^x$ **49.** **(a)** .98 **(b)** .82 **(c)** .14 **(d)** $-.0027$; the rate of change in the proportion wearable when $x = 100$ **(e)** It decreases. Yes, as time increases, the shoes wear out. **51.** **(a)** $E = 5/x$ **(b)** 5 **53.** **(a)** -46.0 **(b)** -27.9 **(c)** -10.3 **(d)** It is approaching 0. **(e)** no

55. no relative extrema; inflection point at (1, 0)

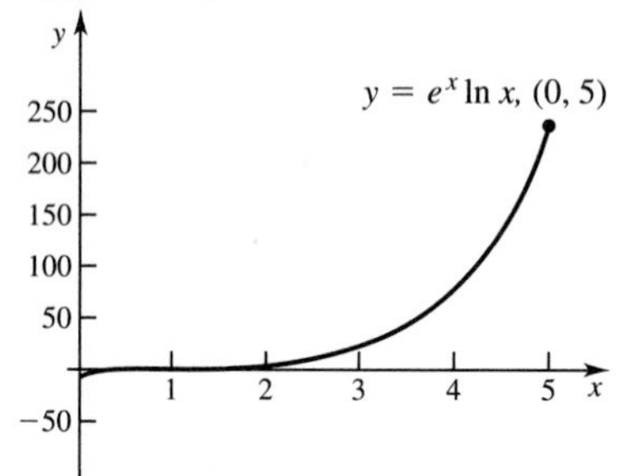

Chapter 5 Review Exercises (page 300)

3. -1 **5.** 2 **7.**

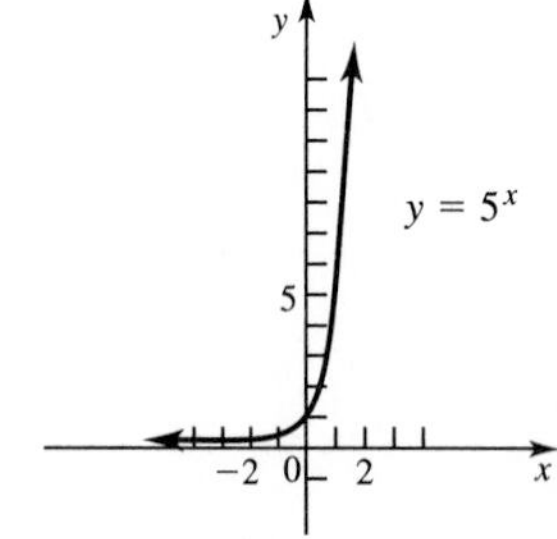

9.

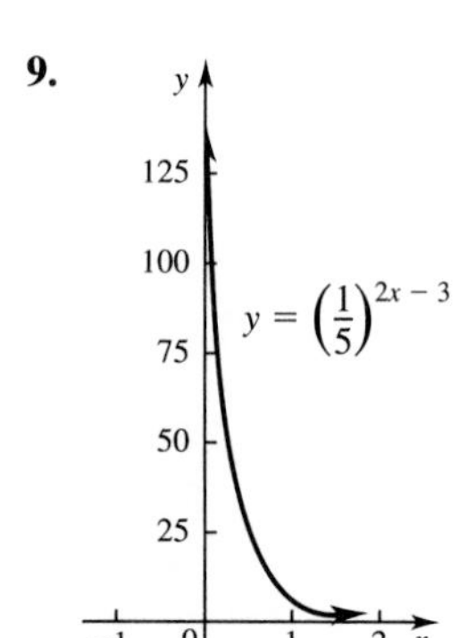

11.

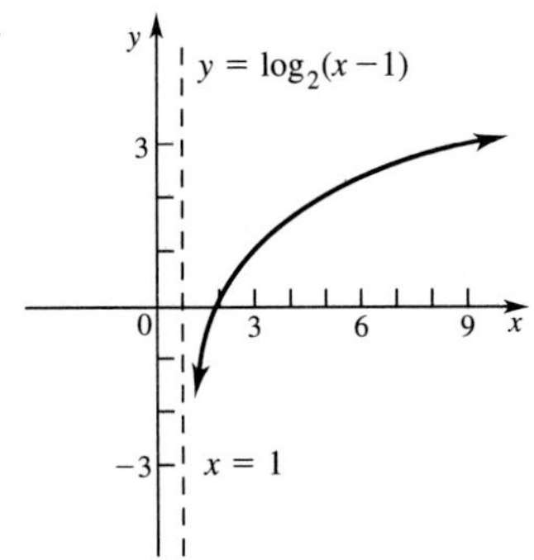

13.

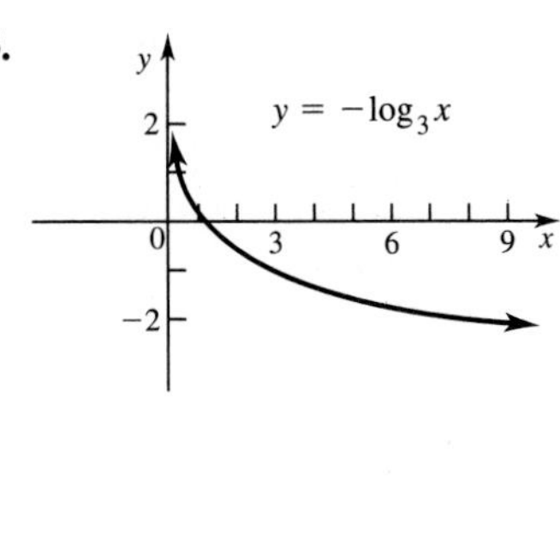

15. $\log_2 64 = 6$ **17.** $\ln 1.09417 = .09$ **19.** $2^5 = 32$ **21.** $e^{4.41763} = 82.9$ **23.** 4 **25.** $4/5$ **27.** $3/2$
29. $\log_5 (21k^4)$ **31.** $\log_2 (x^2/m^3)$ **33.** $p = 1.416$ **35.** $m = -1.807$ **37.** $x = -3.305$ **39.** $m = 1.7547$
41. $y' = -12e^{2x}$ **43.** $y' = -6x^2e^{-2x^3}$ **45.** $y' = 10xe^{2x} + 5e^{2x} = 5e^{2x}(2x + 1)$ **47.** $y' = 2x/(2 + x^2)$
49. $y' = (x - 3 - x \ln |3x|)/[x(x - 3)^2]$ **51.** $y' = [e^x(x + 1)(x^2 - 1) \ln (x^2 - 1) - 2x^2e^x]/[(x^2 - 1)(\ln (x^2 - 1))^2]$
53. $y' = 2(x^2 + e^x)(2x + e^x)$
55. relative minimum of $-e^{-1} = -.368$ at $x = -1$; inflection point at $(-2, -.27)$

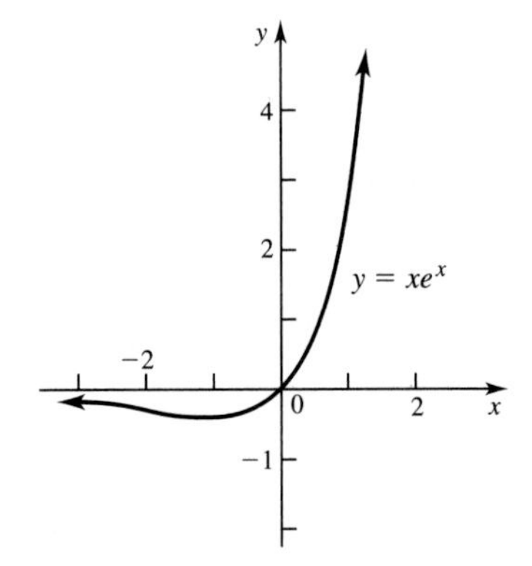

57. relative minimum of e^2 at $x = 2$; no inflection point

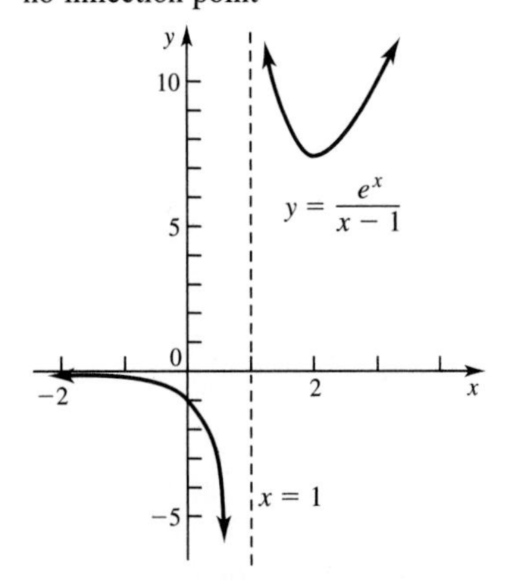

59. \$10,631.51 **61.** \$14,204.18 **63.** \$21,190.14 **65.** \$17,901.90 **67.** 9.38% **69.** 9.42% **71.** \$1494.52
73. \$31,468.86 **75.** **(a)** 100 **(b)** 100 **(c)** 1 day **77.** about 13.7 yr **79.** \$20,891.12 **81.** about 7.7 m
83. **(a)** when it is first injected; .08 **(b)** never **(c)** It approaches $c/a = .0769$ g **85.** **(a)** 0 yr **(b)** 1.85×10^9 yr
(c) $\dfrac{dt}{dr} = \dfrac{10.4958 \times 10^9}{\ln 2\,(1 + 8.33r)}$ **(d)** As r increases, t increases, but at a slower and slower rate. As r decreases, t decreases at a faster and faster rate. **87.** **(a)** 11 **(b)** 12.6 **(c)** 18.0 **(d)**

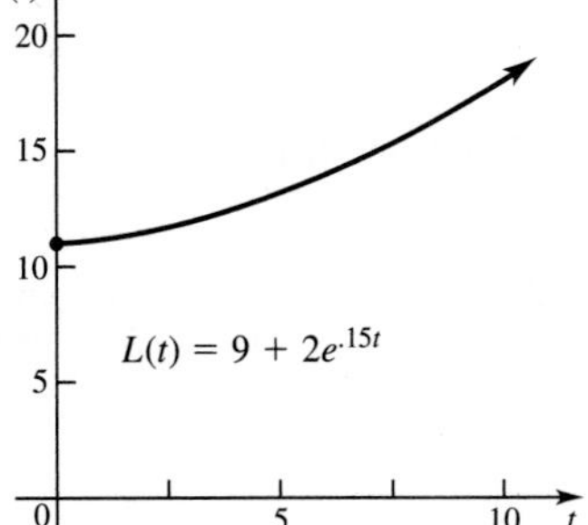

89. **(a)** $(-\infty, \infty)$ **(b)** $(0, \infty)$ **(c)** 1 **(d)** none **(e)** $y = 0$ **(f)** greater than 1 **(g)** between 0 and 1

Extended Application (page 304)

1. **(a)** $9216.65; 4.6 **(b)** $6787.27; 3.4 **(c)** 1.35; investment (a) will yield approximately 35% more after-tax dollars than investment (b) **2.** **(a)** $4008.12; 2.0 **(b)** $3684.37; 1.8 **(c)** 1.11; investment (a) will yield approximately 11% more after-tax dollars than investment (b) **3.** M is an increasing function of t. **4.** m is an increasing function of n. **5.** The conclusions are that the multiplier function is an increasing function of i. The advantage of the IRA over a regular account widens as the interest rate i increases and is particularly dramatic for high income tax rates.

CHAPTER 6

Section 6.1 (page 315)

1. $2x^2 + C$ **3.** $5t^3/3 + C$ **5.** $6k + C$ **7.** $z^2 + 3z + C$ **9.** $x^3/3 + 3x^2 + C$ **11.** $t^3/3 - 2t^2 + 5t + C$ **13.** $z^4 + z^3 + z^2 - 6z + C$ **15.** $10z^{3/2}/3 + C$ **17.** $2u^{3/2}/3 + 2u^{5/2}/5 + C$ **19.** $6x^{5/2} + 4x^{3/2}/3 + C$ **21.** $4u^{5/2} - 4u^{7/2} + C$ **23.** $-1/z + C$ **25.** $-1/(2y^2) - 2y^{1/2} + C$ **27.** $9/t - 2 \ln |t| + C$ **29.** $e^{2t}/2 + C$ **31.** $-15e^{-.2x} + C$ **33.** $3 \ln |x| - 8e^{-.5x} + C$ **35.** $\ln |t| + 2t^3/3 + C$ **37.** $e^{2u}/2 + 2u^2 + C$ **39.** $x^3/3 + x^2 + x + C$ **41.** $6x^{7/6}/7 + 3x^{2/3}/2 + C$ **43.** $f(x) = 3x^{5/3}/5$ **45.** $C(x) = 2x^2 - 5x + 8$ **47.** $C(x) = .2x^3/3 + 5x^2/2 + 10$ **49.** $C(x) = 3e^{.01x} + 5$ **51.** $C(x) = 3x^{5/3}/5 + 2x + 114/5$ **53.** $C(x) = x^2/2 - 1/x + 4$ **55.** $C(x) = 5x^2/2 - \ln x - 153.50$ **57.** $P(x) = x^2 + 20x - 50$ **59.** **(a)** $f(t) = -e^{-.01t} + k$ **(b)** .095 unit **61.** $v(t) = t^3/3 + t + 6$ **63.** $s(t) = -16t^2 + 6400$; 20 sec **65.** $s(t) = 2t^{5/2} + 3e^{-t} + 1$

Section 6.2 (page 323)

1. $2(2x + 3)^5/5 + C$ **3.** $-(2m + 1)^{-2}/2 + C$ **5.** $-(x^2 + 2x - 4)^{-3}/3 + C$ **7.** $(z^2 - 5)^{3/2}/3 + C$ **9.** $-2e^{2p} + C$ **11.** $e^{2x^3}/2 + C$ **13.** $e^{2t-t^2}/2 + C$ **15.** $-e^{1/z} + C$ **17.** $-8 \ln |1 + 3x|/3 + C$ **19.** $(\ln |2t + 1|)/2 + C$ **21.** $-(3v^2 + 2)^{-3}/18 + C$ **23.** $-(2x^2 - 4x)^{-1}/4 + C$ **25.** $[(1/r) + r]^2/2 + C$ **27.** $(x^3 + 3x)^{1/3} + C$ **29.** $(p + 1)^7/7 - (p + 1)^6/6 + C$ **31.** $2(5t - 1)^{5/2}/125 + 2(5t - 1)^{3/2}/75 + C$ **33.** $2(u - 1)^{3/2}/3 + 2(u - 1)^{1/2} + C$ **35.** $(x^2 + 12x)^{3/2}/3 + C$ **37.** $[\ln (t^2 + 2)]/2 + C$ **39.** $e^{2z^2}/4 + C$ **41.** $(1 + \ln x)^3/3 + C$ **43.** $(4/15)(x^{5/2} + 4)^{3/2} + C$ **45.** **(a)** $R(x) = (x^2 + 50)^3/3 + 137{,}919.33$ **(b)** 7 **47.** **(a)** $p(x) = -e^{-x^2}/2 + .01$ **(b)** It approaches $10,000. **49.** **(a)** $D(x) = 2 \ln |x + 9| - 2.11$ **(b)** 3.9 mg

Section 6.3 (page 333)

1. 18 **3.** 65 **5.** 20 **7.** 8 **9.** **(a)** 56 **(b)** $\int_0^8 (2x + 1)\,dx$ **11.** 32; 38 **13.** 15; 31/2 **15.** 20; 30 **17.** 16; 14 **19.** 12.8; 27.2 **21.** 2.67; 2.08 **23.** **(a)** 3 **(b)** 3.5 **(c)** 4 **25.** 9 **27.** $9\pi/2$ **29.** 6 **31.** **(a)** about 52 billion barrels **(b)** about 64 billion barrels **33.** **(a)** $83,000 **(b)** $89,000 **35.** about 33.2 liters **37.** about 2600 ft **39.** **(a)** about 690 BTUs **(b)** about 180 BTUs **41.** 1.91837 **43.** 25.7659

Section 6.4 (page 344)

1. -6 **3.** 3/2 **5.** 28/3 **7.** 13 **9.** 1/3 **11.** 76 **13.** 4/3 **15.** 112/25 **17.** $20e^{-.2} - 20e^{-.3} + 3 \ln 3 - 3 \ln 2 \approx 2.775$ **19.** $e^{10}/5 - e^5/5 - 1/2 \approx 4375.1$ **21.** 91/3 **23.** $447/7 \approx 63.857$ **25.** $(\ln 2)^2/2 \approx .24023$ **27.** 49 **29.** $1/4 - 1/(3 + e) \approx .075122$ **31.** $(6/7)(128 - 2^{7/3}) \approx 105.39$ **33.** 42 **35.** 76 **37.** 54 **39.** 41/2 **41.** $e^2 - 3 + 1/e \approx 4.757$ **43.** 1 **45.** 23/3 **47.** $e^3 - 2e^2 + e \approx 8.026$ **53.** **(a)** $22,000 **(b)** $62,000 **(c)** 4.5 days **55.** **(a)** $(9000/8)(17^{4/3} - 2^{4/3}) \approx \$46{,}341$ **(b)** $(9000/8)(26^{4/3} - 17^{4/3}) \approx \$37{,}477$ **(c)** It is slowly increasing without bound. **57.** no **59.** **(a)** 1.37 ft **(b)** .32 ft **61.** **(a)** 14.26 **(b)** 3.55 **63.** **(a)** $c'(t) = 1.2e^{.04t}$ **(b)** $\int_0^{10} 1.2e^{.04t}\,dt$ **(c)** $30e^{.4} - 30 \approx 14.75$ billion **(d)** about 12.8 yr **(e)** about 14.4 yr **65.** $5142.9(e^{.014T} - 1)$

Section 6.5 (page 355)

1. 15 **3.** 4 **5.** 23/3 **7.** 366.1667 **9.** 4/3 **11.** $5 + \ln 6 \approx 6.792$ **13.** $6 \ln (3/2) - 6 + 2e^{-1} + 2e \approx 2.6051$ **15.** $e^2 - e - \ln 2 \approx 3.978$ **17.** 1/2 **19.** 1/20 **21.** $3(2^{4/3})/2 - 3(2^{7/3})/7 \approx 1.6199$ **23.** **(a)** 8 yr

(b) about \$148 **(c)** about \$771 **25. (a)** 39 days **(b)** \$3369.18 **(c)** \$484.02 **(d)** \$2885.16 **27.** 12,931.66 **29.** 27

31. (a)

(b) (15, 375) **(c)** 4500 **(d)** 3375

33. (a) .019; the lower 10% of the income producers earn 1.9% of the total income of the population **(b)** .184; the lower 40% of the income producers earn 18.4% of the total income of the population **(c)** .384; the lower 60% of the income producers earn 38.4% of the total income of the population **(d)** .819; the lower 90% of the income producers earn 81.9% of the total income of the population **(e)**

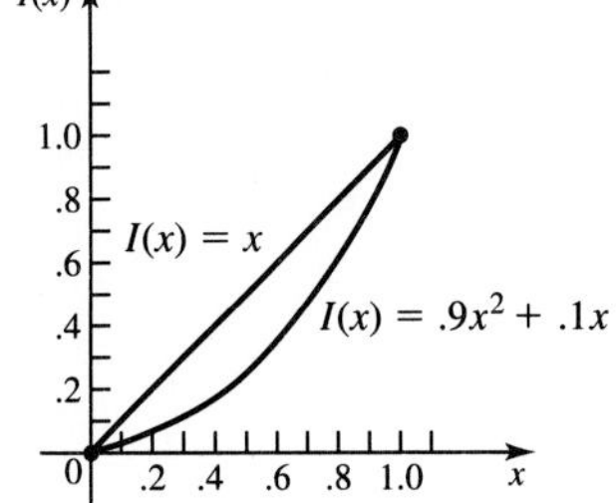

(f) .15

Note: Answers for Exercises 35 and 37 may vary slightly depending on the computer and the software used.
35. 161.2 **37.** 5.516

Chapter 6 Review Exercises (page 358)

5. $6x + C$ **7.** $x^2 + 3x + C$ **9.** $x^3/3 - 3x^2/2 + 2x + C$ **11.** $2x^{3/2} + C$ **13.** $2x^{3/2}/3 + 9x^{1/3} + C$ **15.** $2x^{-2} + C$ **17.** $-3e^{2x}/2 + C$ **19.** $2 \ln |x - 1| + C$ **21.** $e^{3x^2}/6 + C$ **23.** $(3 \ln |x^2 - 1|)/2 + C$ **25.** $-(x^3 + 5)^{-3}/9 + C$ **27.** $\ln |2x^2 - 5x| + C$ **29.** $-e^{-3x^4}/12 + C$ **31.** $2e^{-5x}/5 + C$ **33.** 20 **35.** 24 **39.** $965/6 \approx 160.83$ **41.** $559/648 \approx .863$ **43.** $8 \ln 6 \approx 14.334$ **45.** $5(e^{24} - e^4)/8 \approx 1.656 \times 10^{10}$ **47.** 19/15 **49.** 32/7 **51.** $5 - e^{-4} \approx 4.982$ **53.** 1/6 **55.** 32 **57.** $C(x) = 10x - x^2 + 4$ **59.** $C(x) = (2x - 1)^{3/2} + 145$ **61.** about \$96,000 **63.** 36,000 **65.** 2.5 yr; about \$99,000 **67.** $50 \ln 17 \approx 141.66$ **69.** approximately 4500 degree-days (using rectangles); actual value according to the National Weather Service: 4868 degree-days

Extended Application (page 362)

1. about 102 yr **2.** about 55.6 yr **3.** about 45.4 yr **4.** about 90 yr

CHAPTER 7

Section 7.1 (page 370)

1. $xe^x - e^x + C$ **3.** $(-5xe^{-3x})/3 - (5e^{-3x})/9 + 3e^{-3x} + C$ or $(-5xe^{-3x})/3 + (22e^{-3x})/9 + C$ **5.** $-5e^{-1} + 3 \approx 1.1606$ **7.** $11 \ln 2 - 3 \approx 4.6246$ **9.** $(x^2 \ln x)/2 - x^2/4 + C$ **11.** $e^4 + e^2 \approx 61.9872$ **13.** $(x^2e^{2x})/2 - (xe^{2x})/2 + (e^{2x})/4 + C$ **15.** $243/8 - (3\sqrt[3]{2})/4 \approx 29.4301$ **17.** $4x^2 \ln (5x) + 7x \ln (5x) - 2x^2 - 7x + C$ **19.** $[2x^2(x + 2)^{3/2}]/3 - [8x(x + 2)^{5/2}]/15 + [16(x + 2)^{7/2}]/105 + C$ or $(2/7)(x + 2)^{7/2} - (8/5)(x + 2)^{5/2} + (8/3)(x + 2)^{3/2} + C$ **21.** .13077 **23.** $-4 \ln |x + \sqrt{x^2 + 36}| + C$ **25.** $\ln |(x - 3)/(x + 3)| + C$ **27.** $(4/3) \ln |(3 + \sqrt{9 - x^2})/x| + C$ **29.** $-2x/3 + 2 \ln |3x + 1|/9 + C$ **31.** $(-2/15) \ln |x/(3x - 5)| + C$ **33.** $\ln |(2x - 1)/(2x + 1)| + C$ **35.** $-3 \ln |(1 + \sqrt{1 - 9x^2})/(3x)| + C$ **37.** $2x - 3 \ln |2x + 3| + C$ **39.** $1/[25(5x - 1)] - (\ln |5x - 1|)/25 + C$ **43. (a)** $(2/3)x(x + 1)^{3/2} - (4/15)(x + 1)^{5/2} + C$ **(b)** $(2/5)(x + 1)^{5/2} - (2/3)(x + 1)^{3/2} + C$ **45.** $-100e^{-6} + 10e^{-1} \approx 3.431$ **47.** $7\sqrt{65}/2 + 8 \ln (7 + \sqrt{65}) - 8 \ln 4 \approx 38.8$

Section 7.2 (page 379)

1. (a) 2.7500 **(b)** 2.6667 **(c)** $8/3 \approx 2.6667$ **3. (a)** 1.6833 **(b)** 1.6222 **(c)** $\ln 5 \approx 1.6094$ **5. (a)** 16 **(b)** 14.6667 **(c)** $44/3 \approx 14.6667$ **7. (a)** .9436 **(b)** .8374 **(c)** $4/5 = .8$ **9. (a)** 9.3741 **(b)** 9.3004 **(c)** $2\sqrt{17} + (1/2)\ln(4 + \sqrt{17}) \approx 9.2936$ **11. (a)** 5.9914 **(b)** 6.1672 **(c)** 6.2832; Simpson's rule

13. (a)

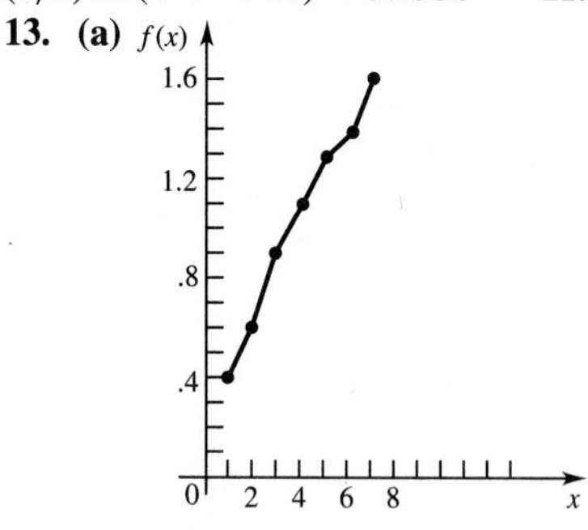

(b) 6.3 **(c)** 6.27 **15. (a)** 2.4759 **(b)** 2.3572 **17.** about 30 mcg/ml; this represents the total amount of drug available to the patient **19.** about 9 mcg/ml; this represents the total effective amount of the drug available to the patient

21. (a)

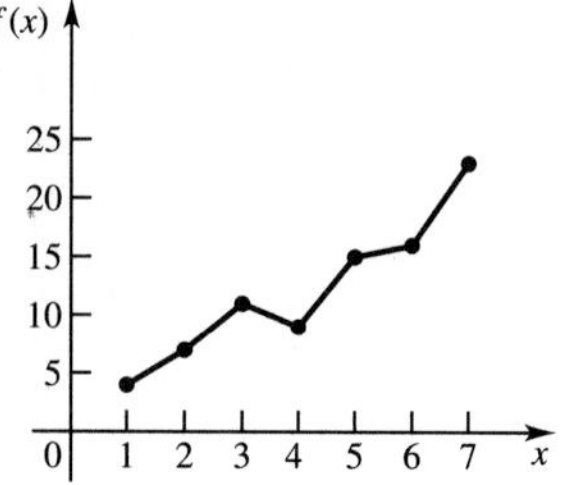

(b) 71.5 **(c)** 69.0

(*Note*: the answers for Exercises 23–30 may vary depending on the computer and the software used.)
23. trapezoidal: 12.6027; Simpson's: 12.6029 **25.** trapezoidal: 9.83271; Simpson's: 9.83377 **27.** trapezoidal: 14.5192; Simpson's: 14.5193 **29.** trapezoidal: 3979.24; Simpson's: 3979.24 **31. (a)** trapezoidal: .682673; Simpson's: .682689 **(b)** trapezoidal: .954471; Simpson's: .954500 **(c)** trapezoidal: .997292; Simpson's: .997300 **33.** The error is multiplied by 1/4. **35.** The error is multiplied by 1/16.

Section 7.3 (page 387)

1. $8\pi/3$ **3.** $364\pi/3$ **5.** $386\pi/27$ **7.** $3\pi/2$ **9.** 18π **11.** $\pi(e^4 - 1)/2 \approx 84.19$ **13.** $\pi \ln 4 \approx 4.36$ **15.** $3124\pi/5$ **17.** $16\pi/15$ **19.** $4\pi/3$ **21.** $4\pi r^3/3$ **23.** $\pi r^2 h$ **25.** 19/3 **27.** 38/15 **29.** $e - 1 \approx 1.718$ **31.** $(5e^4 - 1)/8 \approx 33.999$ **33. (a)** $110e^{-.1} - 120e^{-.2} \approx 1.2844$ **(b)** $210e^{-1.1} - 220e^{-1.2} \approx 3.6402$ **(c)** $330e^{-2.3} - 340e^{-2.4} \approx 2.2413$ **35. (a)** 80 **(b)** $505/6 \approx 84$ words per minute when $t = 5/6$ **(c)** 55 words per minute

Section 7.4 (page 396)

1. (a) \$5823.38 **(b)** \$19,334.31 **3. (a)** \$2911.69 **(b)** \$9667.16 **5. (a)** \$2637.47 **(b)** \$8756.70 **7. (a)** \$27,979.55 **(b)** \$92,895.37 **9. (a)** \$2.34 **(b)** \$7.78 **11. (a)** \$582.57 **(b)** \$1934.20 **13. (a)** \$9480.41 **(b)** \$31,476.07 **15.** \$74,565.94 **17.** \$28,513.76; \$54,075.81 **19.** \$4175.52

Section 7.5 (page 401)

1. 1/2 **3.** divergent **5.** −1 **7.** 1000 **9.** 1 **11.** 3/5 **13.** 1 **15.** 4 **17.** divergent **19.** 1 **21.** divergent **23.** divergent **25.** divergent **27.** $(2 \ln 4.5)/21 \approx .143$ **29.** $4(\ln 2 - 1/2)/9 \approx .086$ **31.** divergent **33.** divergent **35.** 1 **37.** 0 **41.** \$750,000 **43. (a)** \$75,000 **(b)** \$60,000 **45.** \$30,000 **47.** 833.33

Chapter 7 Review Exercises (page 402)

5. $[-2x(8 - x)^{5/2}]/5 - [4(8 - x)^{7/2}]/35 + C$ **7.** $xe^x - e^x + C$ **9.** $[(2x + 3)(\ln|2x + 3| - 1)]/2 + C$ **11.** $-(1/8)\ln|9 - 4x^2| + C$ **13.** $(1/12)\ln|(3 + 2x)/(3 - 2x)| + C$ **15.** $(1/3)\ln(6 + 4\sqrt{2})/(3 + \sqrt{5}) \approx .26677$ **17.** $(3e^4 + 1)/16 \approx 10.300$ **19.** $7(e^2 - 1)/4 \approx 11.181$ **21.** .4143; .3811 **23.** 3.983; 4.047 **25.** 10.28 **27.** 1.459 **29.** $125\pi/6 \approx 65.45$ **31.** $\pi(e^4 - e^{-2})/2 \approx 85.55$ **33.** $406\pi/15 \approx 85.03$ **35.** $7\pi r^2 h/12$ **37.** 13/6 **39.** divergent **41.** 1/10 **43.** divergent **45.** 3 **49.** \$10.55 million **51.** \$28,513.76 **53.** \$402.64 **55.** \$10,254.22 **57.** \$464.49 **59.** \$5715.89 **61.** \$555,555.56 **63.** 2000 gal **65. (a)** .68270 **(b)** .95450 **(c)** .99730 **(d)** .99994 **(e)** 1

Extended Application (page 406)

2. 10.65 **3.** 175.8 **4.** 441.46

CHAPTER 8

Section 8.1 (page 416)

1. **(a)** 6 **(b)** −8 **(c)** −20 **(d)** 43 **3.** **(a)** $\sqrt{43}$ **(b)** 6 **(c)** $\sqrt{19}$ **(d)** $\sqrt{11}$

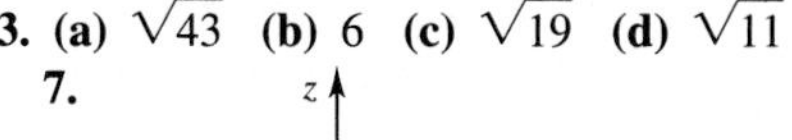

5.

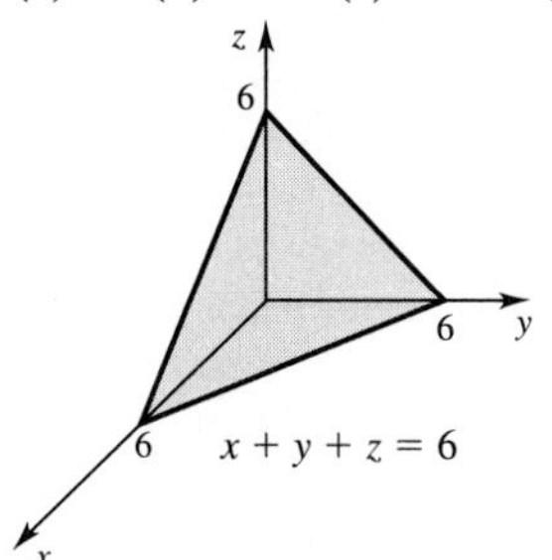

7.

$2x + 3y + 4z = 12$

9.

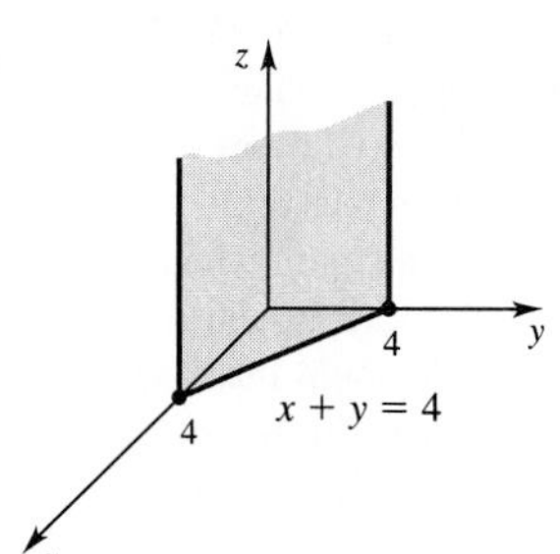

11.

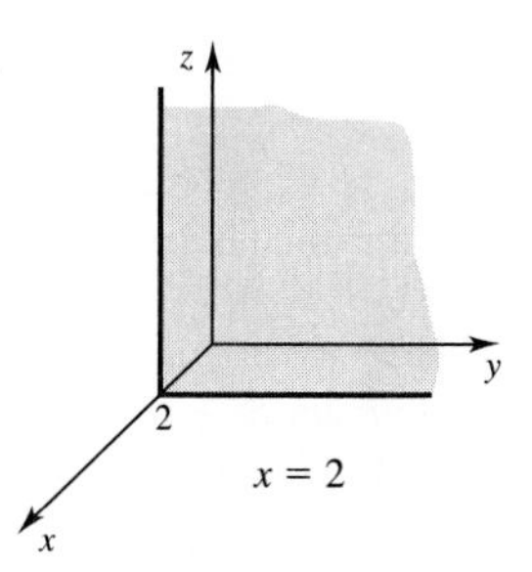

13.

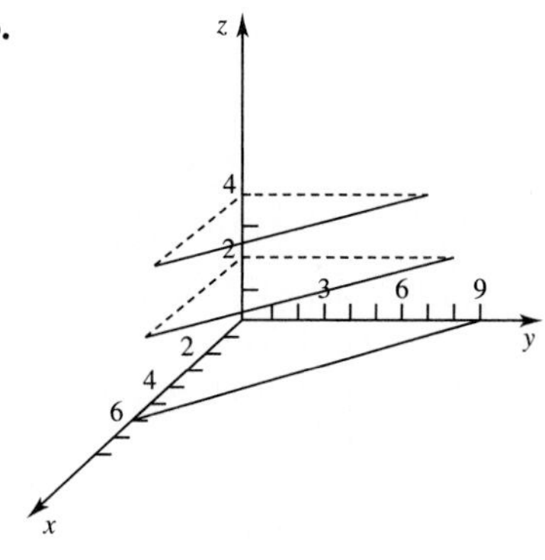

15.

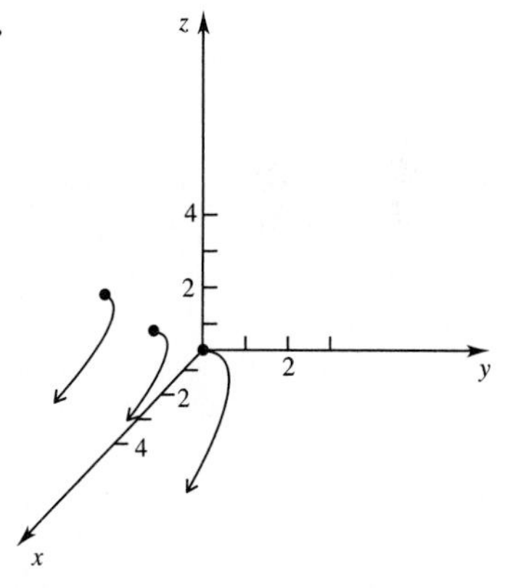

17. **(a)** 1987 (rounded) **(b)** 595 (rounded) **(c)** 359,768 (rounded) **19.** 1.197 (rounded); the IRA account grows faster

21. $y = (500^{5/3})/x^{2/3} \approx 31{,}498/x^{2/3}$

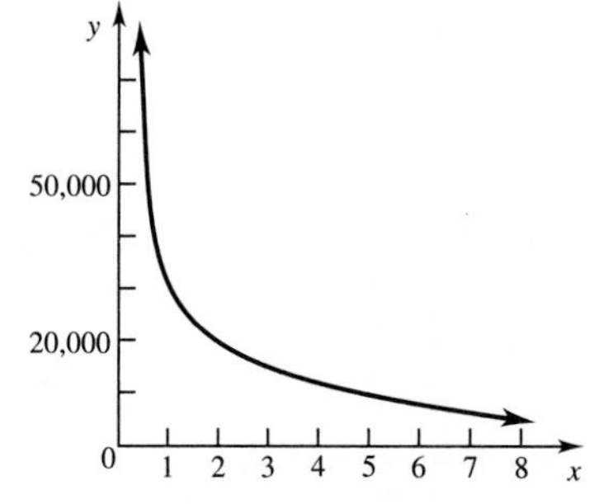

23. $C(x, y, z) = 200x + 100y + 50z$ **25.** **(a)** 1.89 sq m **(b)** 1.52 sq m **(c)** 1.73 sq m **(d)** Answers vary.

27. $g(L, W, H) = 2LW + 2WH + 2LH$ **29.** (c) **31.** (e) **33.** (b) **35.** **(a)** $18x + 9h$ **(b)** $-6y - 3h$

Section 8.2 (page 425)

1. **(a)** $24x - 8y$ **(b)** $-8x + 6y$ **(c)** 24 **(d)** −20 **3.** $f_x = -2y$; $f_y = -2x + 18y^2$; 2; 170 **5.** $f_x = 9x^2y^2$; $f_y = 6x^3y$; 36; −1152 **7.** $f_x = e^{x+y}$; $f_y = e^{x+y}$; e^1 or e; e^{-1} or $1/e$ **9.** $f_x = -15e^{3x-4y}$; $f_y = 20e^{3x-4y}$; $-15e^{10}$; $20e^{-24}$ **11.** $f_x = (-x^4 - 2xy^2 - 3x^2y^3)/(x^3 - y^2)^2$; $f_y = (3x^3y^2 - y^4 + 2x^2y)/(x^3 - y^2)^2$; $-8/49$; $-1713/5329$ **13.** $f_x = 6xy^3/(1 + 3x^2y^3)$; $f_y = 9x^2y^2/(1 + 3x^2y^3)$; 12/11; 1296/1297 **15.** $f_x = e^{x^2y}(2x^2y + 1)$; $f_y = x^3e^{x^2y}$; $-7e^{-4}$; $-64e^{48}$ **17.** $f_{xx} = 36xy$; $f_{yy} = -18$; $f_{xy} = f_{yx} = 18x^2$ **19.** $R_{xx} = 8 + 24y^2$; $R_{yy} = -30xy + 24x^2$; $R_{xy} = R_{yx} = -15y^2 + 48yx$ **21.** $r_{xx} = -8y/(x + y)^3$; $r_{yy} = 8x/(x + y)^3$; $r_{xy} = r_{yx} = (4x - 4y)/(x + y)^3$

23. $z_{xx} = 0$; $z_{yy} = 4xe^y$; $z_{xy} = z_{yx} = 4e^y$ **25.** $r_{xx} = -1/(x + y)^2$; $r_{yy} = -1/(x + y)^2$; $r_{xy} = r_{yx} = -1/(x + y)^2$ **27.** $z_{xx} = 1/x$; $z_{yy} = -x/y^2$; $z_{xy} = z_{yx} = 1/y$ **29.** $x = -4$; $y = 2$ **31.** $x = 0, y = 0$ or $x = 3, y = 3$ **33.** $f_x = 2x$; $f_y = z$; $f_z = y + 4z^3$; $f_{yz} = 1$ **35.** $f_x = 6/(4z + 5)$; $f_y = -5/(4z + 5)$; $f_z = -4(6x - 5y)/(4z + 5)^2$; $f_{yz} = 20/(4z + 5)^2$ **37.** $f_x = (2x - 5z^2)/(x^2 - 5xz^2 + y^4)$; $f_y = 4y^3/(x^2 - 5xz^2 + y^4)$; $f_z = -10xz/(x^2 - 5xz^2 + y^4)$; $f_{yz} = 40xy^3z/(x^2 - 5xz^2 + y^4)^2$ **39.** **(a)** 80 **(b)** 180 **(c)** 110 **(d)** 360 **41.** **(a)** \$206,800 **(b)** $f_p = 132 - 2i - .02p$; $f_i = -2p$; the rate at which weekly sales are changing per unit of change in price (f_p) or interest rate (f_i) **(c)** a weekly sales decrease of \$18,800 **43.** **(a)** 46.656 hundred units **(b)** $f_x(27, 64) = .6912$ hundred units and is the rate at which production is changing when labor changes by 1 unit (from 27 to 28) and capital remains constant; $f_y(27, 64) = .4374$ hundred units and is the rate at which production is changing when capital changes by 1 unit (from 64 to 65) and labor remains constant **(c)** Production would increase at a rate of $f_x(x, y) = (1/3)x^{-4/3}[(1/3)x^{-1/3} + (2/3)y^{-1/3}]^{-4}$. **45.** $.4x^{-.6}y^{.6}$; $.6x^{.4}y^{-.4}$ **47.** **(a)** 168 **(b)** 5448 **(c)** an increase in days since rain **49.** **(a)** .0112 **(b)** .783 **51.** **(a)** 4.125 lb **(b)** $\partial f/\partial n = (1/4)n$; the rate of change of weight loss per unit change in workouts **(c)** an additional loss of 3/4 lb **53.** **(a)** 2.04% **(b)** 1.26% **(c)** .05% is the rate of change of the probability for an additional semester of high school math; .003% is the rate of change of the probability per unit of change in the SAT score.

Section 8.3 (page 434)

1. saddle point at $(1, -1)$ **3.** relative minimum at $(-1, -1/2)$ **5.** relative minimum at $(-2, -2)$ **7.** relative minimum at $(15, -8)$ **9.** relative maximum at $(2/3, 4/3)$ **11.** saddle point at $(2, -2)$ **13.** saddle point at $(0, 0)$; relative minimum at $(4, 8)$ **15.** saddle point at $(0, 0)$; relative minimum at $(9/2, 3/2)$ **17.** saddle point at $(0, 0)$ **21.** relative maximum of $1\ 1/8$ at $(-1, 1)$; saddle point at $(0, 0)$; (a) **23.** relative minimum of $-2\ 1/16$ at $(0, 1)$ and at $(0, -1)$; saddle point at $(0, 0)$; (b) **25.** relative maximum of $1\ 1/16$ at $(1, 0)$ and $(-1, 0)$; relative minimum of $-15/16$ at $(0, 1)$ and $(0, -1)$; saddle points at $(0, 0)$, $(-1, 1)$, $(1, -1)$, $(1, 1)$, and $(-1, -1)$; (e) **29.** \$12 per unit of labor and \$40 per unit of goods produce a maximum profit of \$274,400 **31.** 12 units of electrical tape and 25 units of packing tape give the minimum cost of \$2237.

Section 8.4 (page 442)

1. $f(6, 6) = 72$ **3.** $f(4/3, 4/3) = 64/27 \approx 2.4$ **5.** $f(5, 3) = 28$ **7.** $f(20, 2) = 360$ **9.** $f(3/2, 3/2, 3) = 81/4 = 20.25$ **11.** $x = 6$, $y = 12$ **13.** 30, 30, 30 **17.** 60 ft by 60 ft **19.** make 10 large, no small **21.** 167 units of labor and 178 units of capital **23.** 50 m by 50 m **25.** radius 5 in and height 10 in **27.** 12.91 m by 12.91 m by 6.45 m **29.** 3 m by 3 m by 3 m

Section 8.5 (page 449)

3. $y' = 3.125x - 8.875$; 6.68; 13.08

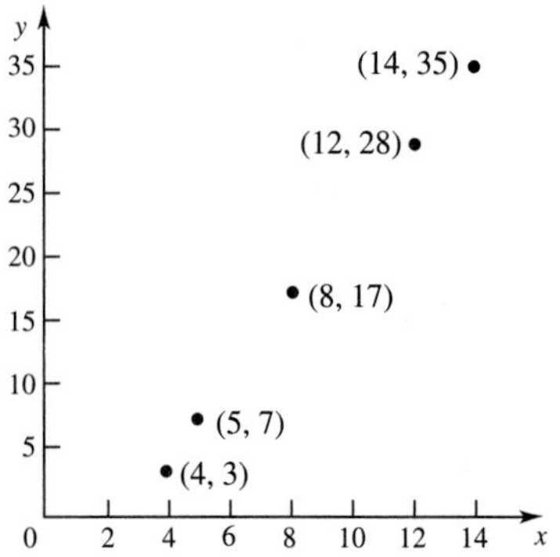

5. (a), (b)

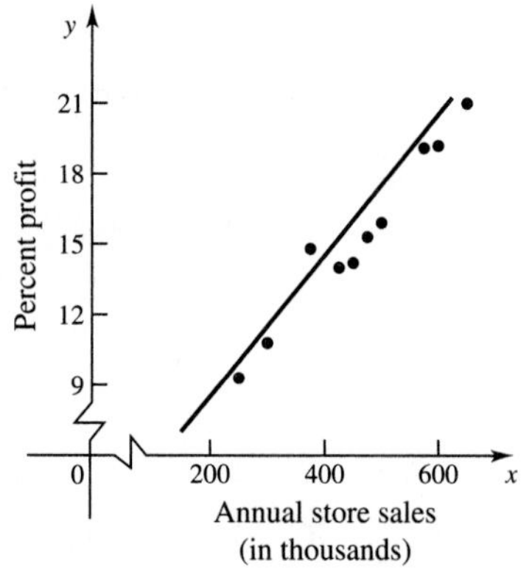

(b) $y' = .03x + 2.49$ **(c)** 23.5; 25.0

7. **(a)** $y' = 1.02x - 135$ **(b)** \$375,000 **(c)** There appears to be an approximately linear relationship. **9.** **(a)** \$26,920; \$23,340; \$19,770 **(b)** \$29,370; \$25,790; \$22,210 **(c)** \$42

11. **(a)**

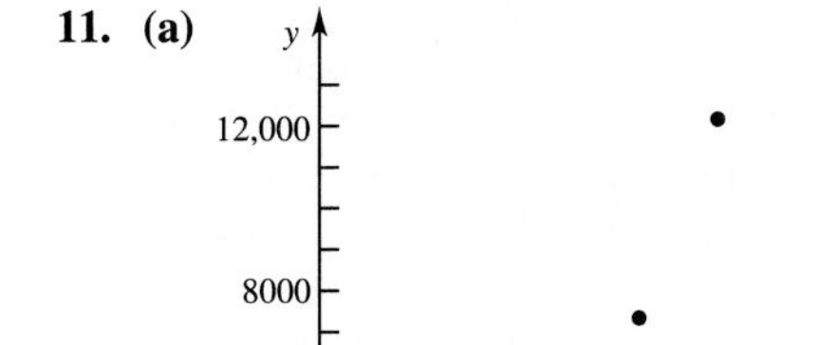

(b)

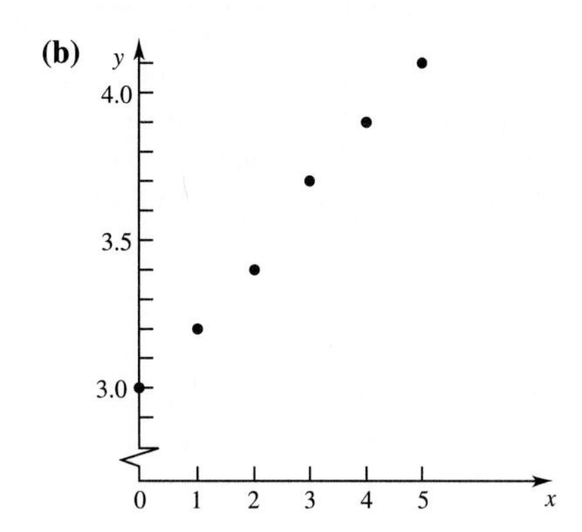

(c) $y' = .22x + 3.00$
(d) 4.54; about 35,000

13. **(a)** $y' = -.00041x + 265$ **(b)** 264; 262 **(c)** As the expenditure per pupil goes up, the mathematics proficiency score goes down. **15.** **(a)** $y' = 14.75 - .0067x$ **(b)** 12 **(c)** 11 **17.** $y' = 14.9x + 2820$ **(a)** 5060, 6990, 9080; the largest discrepancy is 80 BTUs, a good agreement. **(b)** 6250; 6500

Section 8.6 (page 457)

1. $dz = 36x^3\,dx - 15y^2\,dy$ **3.** $dz = [-2y/(x-y)^2]\,dx + [2x/(x-y)^2]\,dy$ **5.** $dz = (y^{1/2}/x^{1/2} - 1/[2(x+y)^{1/2}])\,dx + (x^{1/2}/y^{1/2} - 1/[2(x+y)^{1/2}])\,dy$ **7.** $dz = (3\sqrt{1-2y}\,)\,dx + [-(3x+2)/(1-2y)^{1/2}]dy$ **9.** $dz = [2x/(x^2+2y^4)]\,dx + [8y^3/(x^2+2y^4)]\,dy$ **11.** $dz = [y^2e^{x+y}(x+1)]\,dx + [xye^{x+y}(y+2)]\,dy$ **13.** $dz = (2x - y/x)\,dx + (-\ln x)\,dy$ **15.** $dw = (4x^3yz^3)\,dx + (x^4z^3)\,dy + (3x^4yz^2)\,dz$ **17.** $-.2$ **19.** $-.009$ **21.** $-.00769$ **23.** $-.335$ **25.** 20.73 cm^3 **27.** 142.4 in^3 **29.** .0769 units **31.** 6.65 cm^3 **33.** 2.98 liters **35.** 33.2 cm^2

Section 8.7 (page 468)

1. $93y/4$ **3.** $(1/9)[(48+y)^{3/2} - (24+y)^{3/2}]$ **5.** $(2x/9)[(x^2+15)^{3/2} - (x^2+12)^{3/2}]$ **7.** $6 + 10y$ **9.** $(1/4)e^{x+4} - (1/4)e^{x-4}$ **11.** $(1/2)e^{25+9y} - (1/2)e^{9y}$ **13.** 279/8 **15.** $(2/45)(39^{5/2} - 12^{5/2} - 7533)$ **17.** 21 **19.** $(\ln 2)^2$ **21.** $8\ln 2 + 4$ **23.** 256 **25.** $(4/15)(33 - 2^{5/2} - 3^{5/2})$ **27.** $-2\ln(6/7)$ or $2\ln(7/6)$ **29.** $(1/2)(e^7 - e^6 - e^3 + e^2)$ **31.** 48 **33.** 4/3 **35.** $(2/15)(2^{5/2} - 2)$ **37.** $(1/4)\ln(17/8)$ **39.** $e^2 - 3$ **41.** $97{,}632/105 \approx 929.83$ **43.** 128/9 **45.** ln 3 **47.** 64/3 **49.** 116 **51.** 10/3 **53.** $7(e-1)/3$ **55.** 16/3 **57.** $4\ln 2 - 2$ **59.** 13/3 **61.** $(e^7 - e^6 - e^5 + e^4)/2$ **63.** \$2583 **65.** \$933.33

Chapter 8 Review Exercises (page 471)

1. -19; -255 **3.** -1; $-5/2$

5.

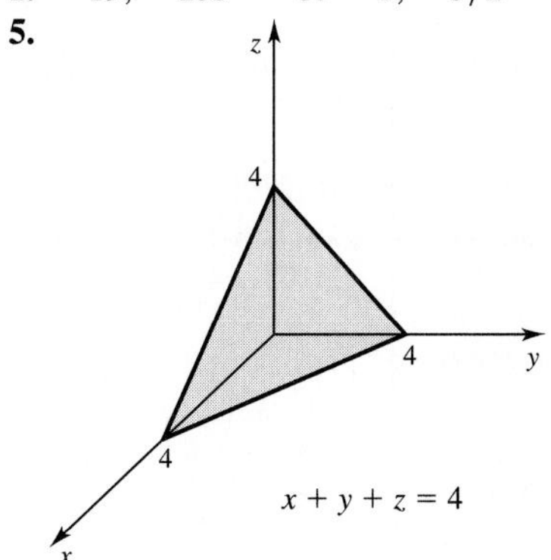

7.

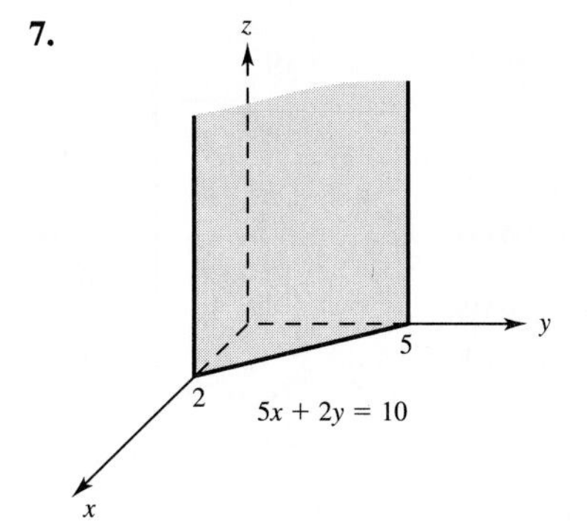

9.

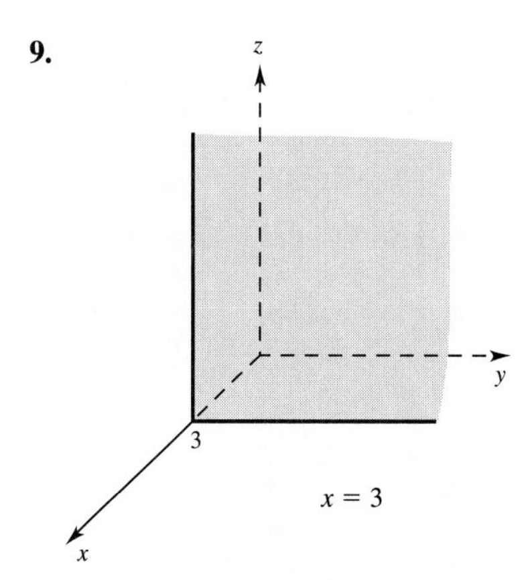

11. **(a)** $-10x + 7y$ **(b)** -15 **(c)** 7 **13.** $f_x = 27x^2y^2 - 5$; $f_y = 18x^3y$ **15.** $f_x = 4x/(4x^2+y^2)^{1/2}$; $f_y = y/(4x^2+y^2)^{1/2}$ **17.** $f_x = 2xe^{2y}$; $f_y = 2x^2e^{2y}$ **19.** $f_x = 4x/(2x^2+y^2)$; $f_y = 2y/(2x^2+y^2)$ **21.** $f_{xx} = 24xy^2$;

$f_{xy} = 24x^2y - 8$ **23.** $f_{xx} = 8y/(x - 2y)^3$; $f_{xy} = (-4x - 8y)/(x - 2y)^3$ **25.** $f_{xx} = 2e^y$; $f_{xy} = 2xe^y$ **27.** $f_{xx} = (-2x^2y^2 - 4y)/(2 - x^2y)^2$; $f_{xy} = -4x/(2 - x^2y)^2$ **29.** relative minimum at (0, 1) **31.** saddle point at (0, 2) **33.** saddle point at (3, 1) **35.** relative minimum at (1, 1/2); saddle point at (−1/3, 11/6) **37.** minimum of 0 at (0, 4); maximum of 256/27 at (8/3, 4/3) **39.** $x = 160/3$; $y = 80/3$ **41.** 5 in. by 5 in. by 5 in. **43.** $dz = (21x^2y)\,dx + (7x^3 - 12y^2)\,dy$ **45.** $dz = [xye^{x-y}(x + 2)]\,dx + [x^2e^{x-y}(1 - y)]\,dy$ **47.** $dw = 5x^4\,dx + 4y^3\,dy - 3z^2\,dz$ **49.** 1.7 **51.** $64y^2/3 + 40$ **53.** $(1/9)[(30 + 3y)^{3/2} - (12 + 3y)^{3/2}]$ **55.** $12y - 16$ **57.** $(3/2)[(100 + 2y^2)^{1/2} - (2y^2)^{1/2}]$ **59.** 1232/9 **61.** $2[(42)^{5/2} - (24)^{5/2} - (39)^{5/2} + (21)^{5/2}]/135$ **63.** 2 ln 2 or ln 4 **65.** 26 **67.** $(4/15)(782 - 8^{5/2})$ **69.** 1900 **71.** 1/2 **73.** 1/48 **75.** 3 **77. (a)** $\$(150 + \sqrt{10})$ **(b)** $\$(400 + \sqrt{15})$ **(c)** $\$(1200 + 2\sqrt{5})$ **79. (a)** $.6x^{-.4}y^{.4}$ or $.6y^{.4}/x^{.4}$ **(b)** $.4x^{.6}y^{-.6}$ or $.4x^{.6}/y^{.6}$ **81. (a)** $y' = 3.90x - 7.92$ **(b)** $154,800 **83.** a decrease of $13.42 **85.** 4.19 cu ft **87. (a)** $y' = .97x + 31.5$ **(b)** about 216 **(c)** 158, 169, 226; the predicted values are in the vicinity of the actual values, but not "close" **89.** 1.3 cm^3 **91. (a)** $200 spent on fertilizer and $80 spent on seed will produce maximum profit of $266 per acre. **(b)** Spend $200 on fertilizer and $80 on seed. **(c)** Spend $200 on fertilizer and $80 on seed.

Extended Application (page 476)

3. .48 in location 1, .10 in location 2; .41 on nonfeeding activities

CHAPTER 9

Section 9.1 (page 487)

1. $y = -x^2 + x^3 + C$ **3.** $y = 3x^4/8 + C$ **5.** $y^2 = x^2 + C$ **7.** $y = ke^{x^2}$ **9.** $y = ke^{(x^3 - x^2)}$ **11.** $y = Mx$ **13.** $y = Me^x + 5$ **15.** $y = -1/(e^x + C)$ **17.** $y = x^3 - x^2 + 2$ **19.** $y = -2xe^{-x} - 2e^{-x} + 44$ **21.** $y^2 = 2x^3/3 + 9$ **23.** $y = e^{x^2+3x}$ **25.** $y = -5/(5 \ln |x| - 6)$ **27.** $y^2/2 - 3y = x^2 + x - 4$ **29.** $y = (e^x - 3)/(e^x - 2)$ **35. (a)** $1011.75 **(b)** $1024.52 **(c)** No; if $x = 8$, the denominator becomes 0. **37.** about 11.6 yr **39.** about 260 **41.** $q = \sqrt{-4p^2 + C}$ **43.** about 4.4 cc **45.** about 387 **47. (a)** $k \approx .8$ **(b)** 11 **(c)** 55 **(d)** 2981 **49. (a)** $dy/dt = -.05y$ **(b)** $y = Me^{-.05t}$ **(c)** $y = 90e^{-.05t}$ **(d)** 55 g

Section 9.2 (page 494)

1. $y = 5/2 + Ce^{-2x}$ **3.** $y = 3 + Ce^{-x^2/2}$ **5.** $y = x \ln x + Cx$ **7.** $y = -1/2 + Ce^{x^2/2}$ **9.** $y = x^2/4 + x + Cx^{-2}$ **11.** $y = -x^3/2 + Cx$ **13.** $y = e^x + 99e^{-x}$ **15.** $y = -1 + 11e^{(x^2-1)/2}$ **17.** $y = x^2/7 + 2560/(7x^5)$ **19.** $y = (3 + 197e^{4-x})/x$ **21. (a)** $y = c/(p + kce^{-cx})$ **(b)** $y = (cy_0)/[py_0 + (c - py_0)e^{-cx}]$ **(c)** c/p **23.** $y = 1.02e^t + 9999e^{.02t}$ (rounded) **25.** $y = 50t + 2500 + 7500e^{.02t}$ **29.** The temperature approaches T_F, the temperature of the surrounding medium. **31. (a)** $T = 28e^{-.767t} + 22$ **(b)** 28.04° C

Section 9.3 (page 500)

1. 1.837 **3.** 4.315 **5.** 1.491 **7.** .087 **9.** −.540; −.520 **11.** 2.030; 2.042 **13.** 3.806; 4.759 **15.** −.025; −.036

17.

x_i	y_i	$f(x_i)$	$y_i - f(x_i)$
0	0	0	0
.2	0	.08772053	−.08772053
.4	.11696071	.22104189	−.10408118
.6	.26432197	.37954470	−.11522273
.8	.43300850	.55699066	−.12398216
1.0	.61867206	.75000000	−.13132794

19.

x_i	y_i	$f(x_i)$	$y_i - f(x_i)$
0	0	0	0
.2	.2	.1812692	.0187308
.4	.36	.32967995	.03032005
.6	.488	.45118836	.03681164
.8	.5904	.55067104	.03972896
1.0	.67232	.63212056	.04019944

21.

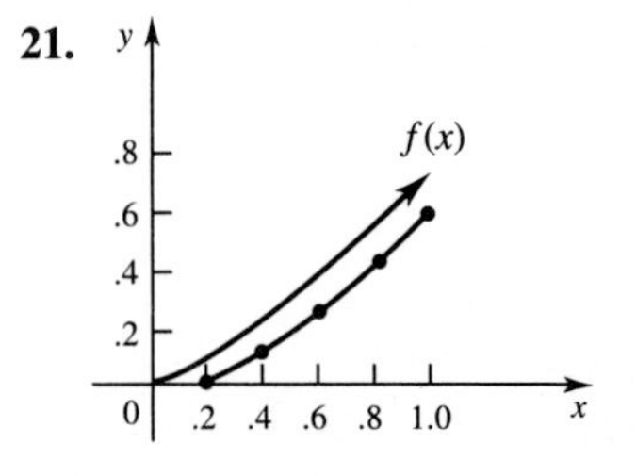

23.

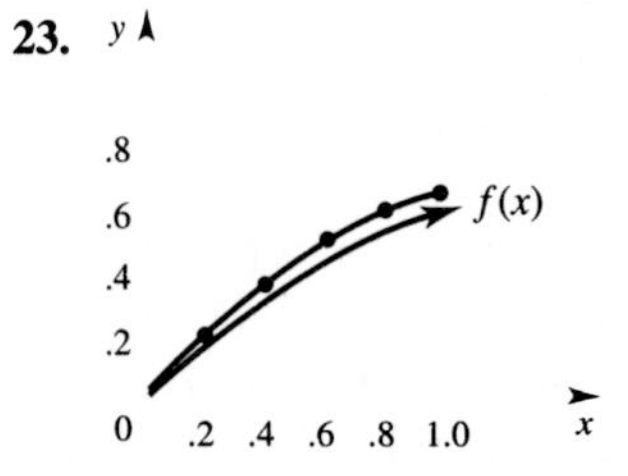

25. **(a)** 4.109 **(b)** $y = 1/(1 - x)$; y approaches ∞ **27.** **(a)** $dy/dt = .01y(100 - y) = y - .01y^2$ **(b)** about 14,000 **29.** about 75 **31.** about 360 people

Note: The answers for Exercises 33 and 35 may vary depending on the computer and the software used.

33. 2.334 **35.** 4.091

Section 9.4 (page 510)

1. \$31,048.20 **3.** about 5.7 yr **5.** **(a)** $3 \ln x - 2x + 2 \ln y - 3y = \ln 4 - 8$ **(b)** $x = 3/2$; $y = 2/3$ **7.** **(a)** $y = (24{,}995{,}000)/(4999 + e^{.25t})$ **(b)** 3672 **(c)** 91 **(d)** on the thirty-fourth day **9.** **(a)** $y = (50e^{.18t})/(1 + .005e^{.18t})$ or $y = 10{,}000e^{.18t}/(e^{.18t} + 199)$ **(b)** in about 29 days **11.** **(a)** $y = .005 + .015e^{-1.010t}$ **(b)** $Y = .00273 + .00727e^{-1.1t}$ **13.** **(a)** $y = (5.55e^{.45t})/(1 + .111e^{.45t})$ or $y = 50/(1 + 9e^{-.45t})$ **(b)** in about 6 days **15.** **(a)** $y = 347e^{-4.24e^{-.1t}}$ **(b)** in about 5.5 days **17.** **(a)** $y = [2(t + 100)^3 - 1{,}800{,}000]/(t + 100)^2$ **(b)** about 250 lb **(c)** It continues to increase. **19.** **(a)** $y = 20e^{-.02t}$ **(b)** about 6 lb **(c)** It continues to decrease. **21.** **(a)** $y = [.1(t + 100)^2 - 500]/(t + 100)$ **(b)** about 9.2 g

Chapter 9 Review Exercises (page 512)

5. $y = x^4/2 + 3x^2 + C$ **7.** $y = 4e^x + C$ **9.** $y^2 = 3x^2 + 2x + C$ **11.** $y = (Cx^2 - 1)/2$ **13.** $y = 12/5 + Ce^{-5x}$ **15.** $y = 1 + Ce^{-x^2/6}$ **17.** $y = x^3/3 - 5x^2/2 + 1$ **19.** $y = -5e^{-x} - 5x + 22$ **21.** $y = 2e^{5x-x^2}$ **23.** $y^2 + 6y = 2x - 2x^2 + 352$ **25.** $y = 1 + 7e^{-x^3/3}$ **27.** $y = 3/2 + (27e^{x^2})/2$ **31.** 2.343

33.

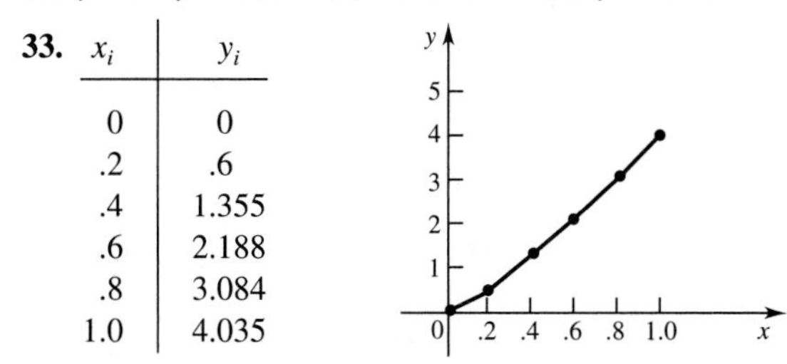

x_i	y_i
0	0
.2	.6
.4	1.355
.6	2.188
.8	3.084
1.0	4.035

35. **(a)** \$5800 **(b)** \$25,100 **37.** about 13.9 yr **39.** about 219 **41.** 17.3 min **43.** **(a)** $y = 489{,}300/(699 + e^{1.140t})$ **(b)** about 135 **(c)** after 5.7 wk **45.** **(a)** $y = 100/(1 + 24e^{-.481x})$ **(b)** about 32 **47.** **(a)** about 170 million **(b)** about 207 million **(c)** about 329 million **49.** 213° **51.** **(a)** $v = (B/K)(e^{2BKt} - 1)/(e^{2BKt} + 1)$ **(b)** B/K

Extended Application (page 516)

1. **(a)** 2.4 yr **(b)** 3.1 yr **2.** **(a)** 28.2 yr **(b)** 37.1 yr **3.** **(a)** 173 yr **(b)** 228 yr

CHAPTER 10

Section 10.1 (page 524)

1. yes **3.** yes **5.** no; $\int_0^3 4x^3\,dx \neq 1$ **7.** no; $\int_{-2}^2 (x^2/16)\,dx \neq 1$ **9.** $k = 3/14$ **11.** $k = 3/125$ **13.** $k = 2/9$ **15.** $k = 1/12$ **17.** 1 **21.** **(a)** .4226 **(b)** .2071 **(c)** .4082 **23.** **(a)** .3935 **(b)** .3834 **(c)** .3679 **25.** **(a)** .9975 **(b)** .0024 **27.** **(a)** .2679 **(b)** .4142 **(c)** .3178 **29.** **(a)** $4/7 \approx .5714$ **(b)** $5/21 \approx .2381$ **31.** **(a)** .7 **(b)** .5 **(c)** .2

Section 10.2 (page 530)

1. $\mu = 5$; $\text{Var}(x) \approx 1.33$; $\sigma \approx 1.15$ **3.** $\mu = 14/3 \approx 4.67$; $\text{Var}(x) \approx .89$; $\sigma \approx .94$ **5.** $\mu \approx 2.83$; $\text{Var}(x) \approx .57$; $\sigma \approx .76$ **7.** $\mu = 4/3 \approx 1.33$; $\text{Var}(x) = 2/9 \approx .22$; $\sigma \approx .47$ **11.** **(a)** 5.40 **(b)** 5.55 **(c)** 2.36 **(d)** .54 **(e)** .60 **13.** **(a)** $4/3 \approx 1.33$ **(b)** .22 **(c)** .47 **(d)** .56 **(e)** .63 **15.** **(a)** 5 **(b)** 0 **17.** **(a)** 4.83 **(b)** .055 **19.** **(a)** $\sqrt[4]{2} \approx 1.19$ **(b)** .18 **21.** **(a)** 310.3 hr **(b)** 267 hr **(c)** .206 **23.** **(a)** 2 mo **(b)** 2 mo **(c)** .632 **25.** **(a)** 2 **(b)** .89 **(c)** .62 **27.** **(a)** 1.806 **(b)** 1.265 **(c)** .1886

Section 10.3 (page 541)

1. **(a)** 4.4 cm **(b)** .23 cm **(c)** .29 **3.** **(a)** 33.33 yr **(b)** 33.33 yr **(c)** .23 **5.** **(a)** 1 day **(b)** 1 day **(c)** .23 **7.** 49.98% **9.** 8.01% **11.** -1.28 **13.** .92 **19.** $m = (-\ln .5)/a$ or $m = (\ln 2)/a$ **21.** **(a)** \$47,500 **(b)** .47 **23.** **(a)** $f(x) = .235e^{-.235x}$ on $[0, \infty)$ **(b)** .095 **25.** **(a)** .1587 **(b)** .7698 **27.** **(a)** 28 days **(b)** .375 **29.** **(a)** 1 hr **(b)** .39 **31.** **(a)** about 58 min. **(b)** .09 **33.** 1.00002 **35.** 8.000506 **37.** **(a)** $\mu = 2.7416 \times 10^{-8} \approx 0$ **(b)** $\sigma = .999433 \approx 1$

Chapter 10 Review Exercises (page 543)

1. probabilities **3.** $\int_a^b f(x)\,dx = 1$, and $f(x) \ge 0$ for all x in $[a, b]$. **5.** not a probability density function **7.** probability density function **9.** $k = 1/9$ **11. (a)** $1/5 = .2$ **(b)** $9/20 = .45$ **(c)** .54 **13.** (b) **15.** $\mu = 4$; $\text{Var}(x) = .5$; $\sigma \approx .71$ **17.** $\mu = 5/4$; $\text{Var}(x) = 5/48 \approx .10$; $\sigma \approx .32$ **19.** $m = .60$; .02 **21. (a)** 100 **(b)** 100 **(c)** .86 **23.** 31.21% **25.** 27.69% **27.** 11.51% **29.** $-.05$ **31.** .911 **33. (a)** 6 **(b)** 6 **(c)** .37 **35.** .63 **37. (a)** 22.68° C **(b)** .48 **39.** .1056 **41.** .2206 **43. (a)** exponential **(b)** domain: $[0, \infty)$; range: $(0, 1]$ **(c)**

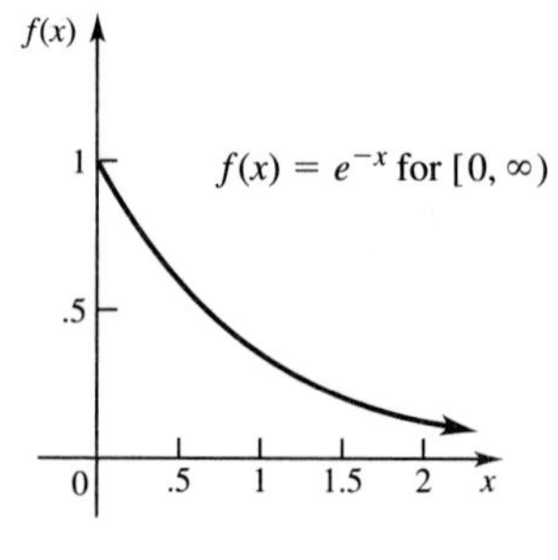

(d) $\mu = 1$; $\sigma = 1$ **(e)** .86

Extended Application (page 547)

1. Expected profit $= -.04A^2 + 640A + 640{,}000$ **2.** Planting 8000 acres will give maximum profit.

CHAPTER 11

Section 11.1 (page 559)

1. $\pi/3$ **3.** $5\pi/6$ **5.** $7\pi/6$ **7.** $13\pi/6$ **9.** 315° **11.** 330° **13.** 288° **15.** 48° **17.** $4/5$; $-3/5$; $-4/3$; $-3/4$; $-5/3$; $5/4$ **19.** $4/5$; $3/5$; $4/3$; $3/4$; $5/3$; $5/4$ **21.** + + + + + + **23.** − − + + − − **25.** $\sqrt{3}/3$; $\sqrt{3}$; 2 **27.** $\sqrt{3}/2$; $\sqrt{3}/3$; $2\sqrt{3}/3$ **29.** -1; -1 **31.** $-\sqrt{3}/2$; $-2\sqrt{3}/3$ **33.** $\sqrt{3}/2$ **35.** 1 **37.** $2\sqrt{3}/3$ **39.** -1 **41.** $-\sqrt{3}/2$ **43.** $-\sqrt{2}$ **45.** $-\sqrt{3}$ **47.** 1/2 **49.** .6293 **51.** 7.1154 **53.** .3907 **55.** .1564 **57.**

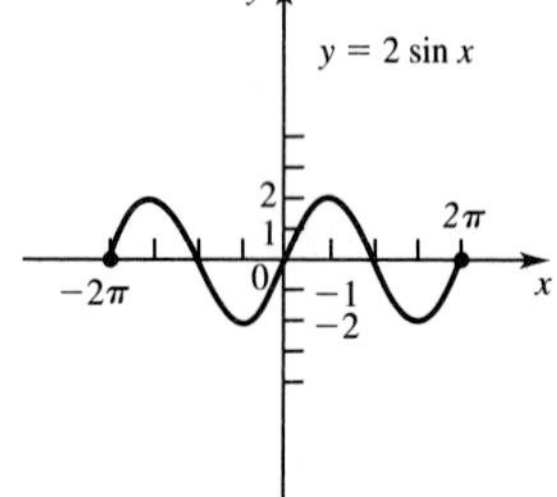

59.

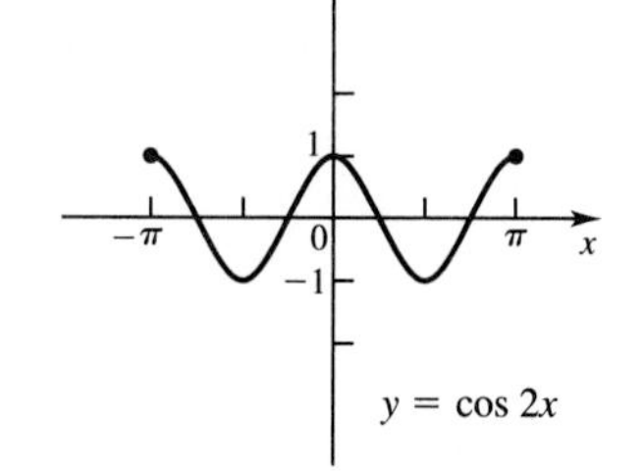

61.

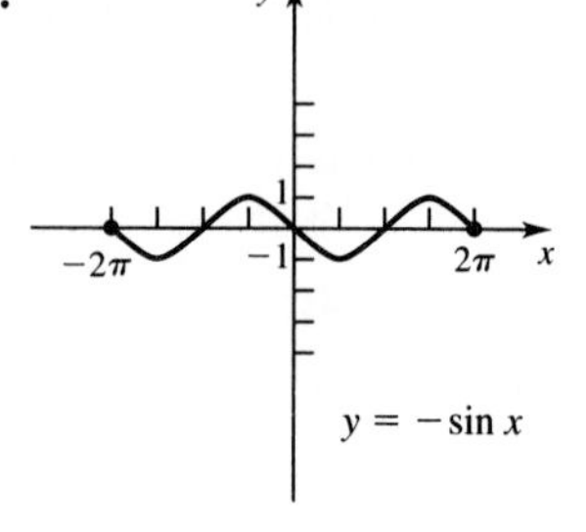

63.

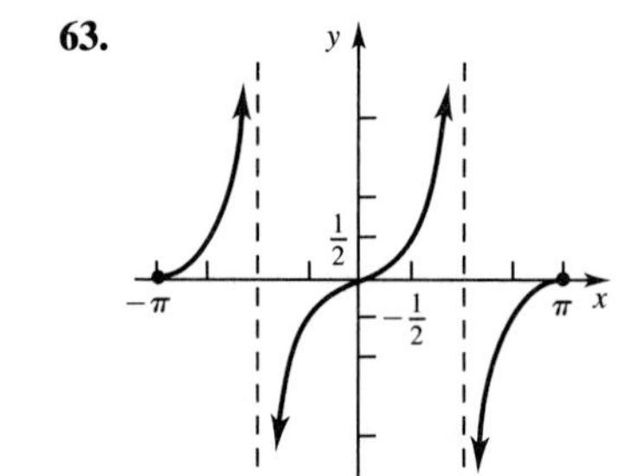

65. (a) 1000 **(b)** 750 **(c)** 500 **(d)** 0 **(e)** 500 **(f)**

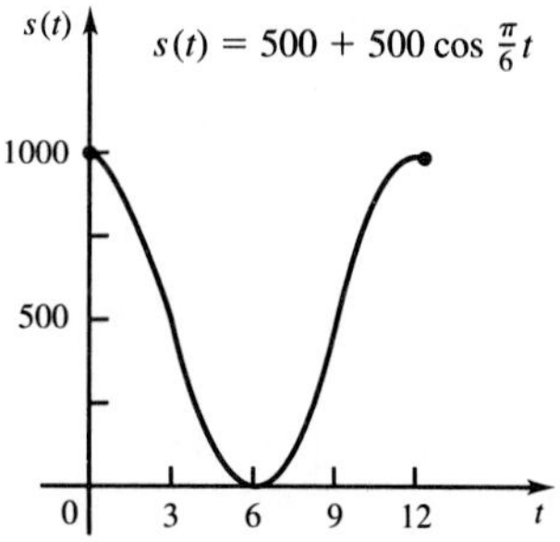

67. 2×10^8 m/sec **69.** 120° **71. (a)** 30° F **(b)** 44° F **(c)** 66° F **(d)** 84° F **(e)** 88° F **73.** 2.2 m, 7.6 m **75.** 180 ft

Section 11.2 (page 566)
1. $y' = 12 \cos 6x$ **3.** $y' = 108 \sec^2 (9x + 1)$ **5.** $y' = -4 \cos^3 x \sin x$ **7.** $y' = 5 \tan^4 x \sec^2 x$
9. $y' = -5(4x \cos 4x + \sin 4x)$ **11.** $y' = (x \cos x - \sin x)/x^2$ **13.** $y' = 5e^{5x} \cos e^{5x}$ **15.** $y' = (\cos x)e^{\sin x}$
17. $y' = (2/x) \cos (\ln 4x^2)$ **19.** $y' = (2x \cos x^2)/\sin x^2$ or $2x \cot x^2$ **21.** $y' = 6 \cos x/(3 - 2 \sin x)^2$
23. $y' = \sqrt{\sin 3x}\,(\sin 3x \cos x - 3 \sin x \cos 3x)/[2\sqrt{\sin x}\,(\sin^2 3x)]$ **25.** $-\csc^2 x$ **27.** $-\csc x \cot x$
29. (a) 1 **(b)** -2 **(c)** -1 **(d)** -2 **(e)** -1 **(f)** 2

31. (a)

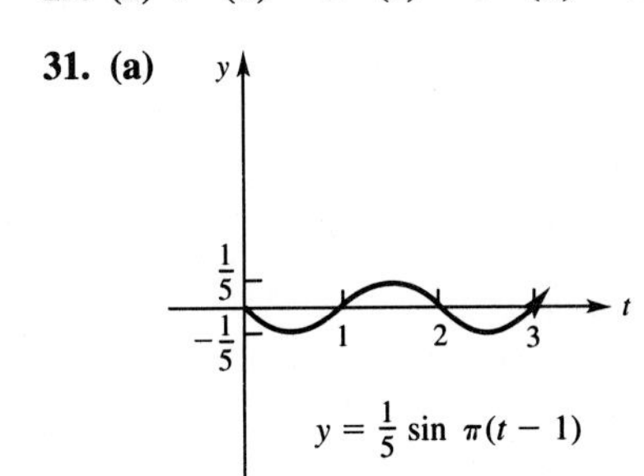

(b) $v = dy/dt = (\pi/5) \cos \pi(t - 1)$; $a = d^2y/dt^2 = \left(-\frac{\pi^2}{5}\right) \sin \pi(t - 1)$

(d) At $t = 1.5$, acceleration is negative, arm is moving clockwise and is at an angle of 1/5 from vertical; at $t = 2.5$, acceleration is positive, arm is moving counterclockwise and is at an angle of $-1/5$ from vertical; at $t = 3.5$, acceleration is negative, arm is moving clockwise and is at an angle of 1/5 from vertical.

33. (a) $5/\pi$ rev/min **(b)** $5/(2\pi)$ rev/min **35.** 20.81 ft

Section 11.3 (page 571)
1. $(1/5) \sin 5x + C$ **3.** $5 \sin x - 2 \cos x + C$ **5.** $(-\cos x^2)/2 + C$ **7.** $-3 \tan 2x + C$ **9.** $(1/8) \sin^8 x + C$
11. $(2/3) \sin^{3/2} x + C$ **13.** $-\ln |1 + \cos x| + C$ **15.** $(1/6) \sin x^6 + C$ **17.** $-4 \ln |\cos (x/4)| + C$
19. $(1/3) \ln |\sin x^3| + C$ **21.** $-\cos e^x + C$ **23.** $(-6/5)x \sin 5x - (6/25) \cos 5x + C$ **25.** $-8x \cos x + 8 \sin x + C$ **27.** $-(3/4)x^2 \sin 8x - (3/16)x \cos 8x + (3/128) \sin 8x + C$ **29.** $1 - \sqrt{2}/2$ **31.** $-\ln (1/2)$ or $\ln 2$
33. $\sqrt{3}/2 - 1$ **35.** 1/2 **37.** 6000

Section 11.4 (page 577)
1. $-\pi/3$ **3.** $\pi/4$ **5.** $-\pi/2$ **7.** $\pi/3$ **9.** $3\pi/4$ **11.** $-\pi/3$ **13.** $-8°$ **15.** 154° **17.** 22° **19.** 48°
21. $-42°$ **23.** 68° **25.** $y' = 12/\sqrt{1 - 144x^2}$ **27.** $y' = 3/(1 + 9x^2)$ **29.** $y' = -2/(|x|\sqrt{x^2 - 4})$
31. $y' = 1/[x(1 + [\ln |7x|]^2)]$ **33.** $y' = 1/[(1 + (x + 1)^2) \tan^{-1} (x + 1)]$ **37.** $(1/4) \tan^{-1} x^4 + C$
39. $\sin^{-1} e^x + C$ **41.** $(4/3) \sin^{-1} 3x + C$ **43.** $\tan^{-1} \ln |x| + C$ **45.** $\pi/4$ **47.** $\pi/6$ **49.** $\pi/2$
51. (a) 31° **(b)** 37° **(c)** 35° **(d)** 31° **53.** $2\sqrt{3}$ ft $\approx$ 3.46 ft

Section 11.5 (page 581)
1. 1 **3.** -1 **5.** 1 **7.** about 1490 **9.** about 2610 **11.** about 5380 **13.** about 7010 **15.** about 1330
17.

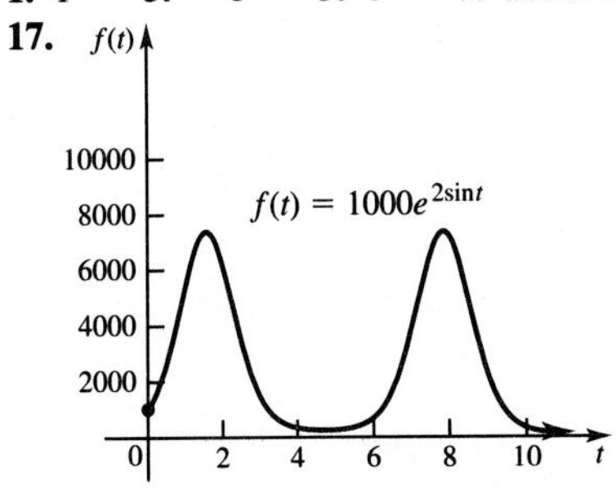

21. about 241 cm **25.** $L_2 = s/(\sin \theta)$ **27.** $L_1 = L_0 - s \cot \theta$
29. $R_2 = k \cdot L_2/r_2^4$ **31.** $R = k(L_0 - s \cot \theta)/r_1^4 + k(s/\sin \theta)/r_2^4$
33. $0 = k(s \csc^2 \theta)/r_1^4 + k(-s \cos \theta/\sin^2 \theta)/r_2^4$ **35.** $\cos \theta = r_2^4/r_1^4$
37. 84° to the nearest degree

Chapter 11 Review Exercises (page 583)
5. $\pi/2$ **7.** $7\pi/6$ **9.** 2π **11.** 1260° **13.** 81° **15.** 117° **17.** $\sqrt{3}/2$ **19.** $\sqrt{3}/2$ **21.** $2\sqrt{3}/3$ **23.** 1/2
25. $-\sqrt{2}$ **27.** .7314 **29.** 6.314 **31.** .9945 **33.** .8290 **37.** $-\pi/4$ **39.** $\pi/3$ **41.** $\pi/3$

43.

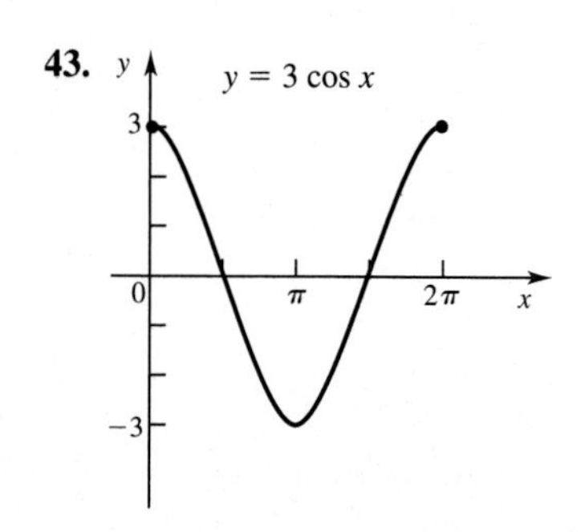

45.

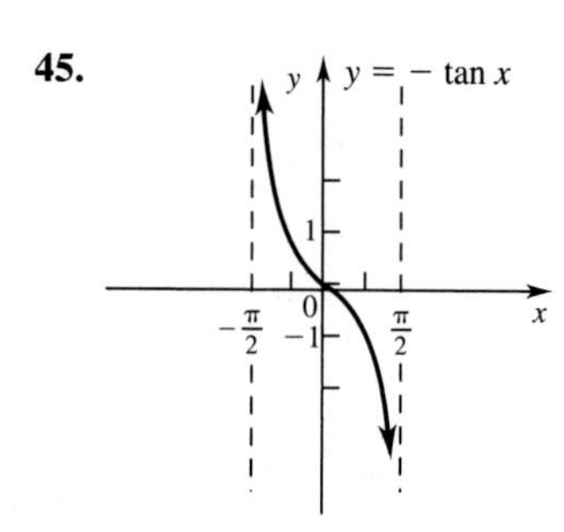

47. $y' = -28 \cos 7x$ **49.** $y' = 8x \sec^2 (4x^2 + 3)$ **51.** $y' = -18 \cos^5 x \sin x$ **53.** $y' = -20x^4 \csc^2 (4x^5)$
55. $y' = -x^2 \sin x + 2x \cos x$ **57.** $y' = 2 \cos x/(\sin x + 1)^2$ **59.** $y' = (\sin x - x \cos x + 2 \cos x)/\sin^2 x$
61. $y' = -\tan x$ **63.** $y' = -3/\sqrt{1 - 9x^2}$ **65.** $y' = 10x/(1 + 25x^4)$ **67.** $-(1/2) \cos 2x + C$
69. $-(1/9) \ln |\cos 9x| + C$ **71.** $5 \tan x + C$ **73.** $-(1/6) \cos 3x^2 + C$ **75.** $-(2/3) (\cos x)^{3/2} + C$
77. $-(1/22) \ln |\cos 11x^2| + C$ **79.** $2 (\sin x)^{7/2}/7 + C$ **81.** $3 \sin^{-1} x + C$ **83.** $-(6/5) \tan^{-1} x^5 + C$ **85.** 1
87. 20π **93.** $\theta = \pi/4$ or 45° **95.** **(d)** 8.63 inches **(e)** 5.15 inches

INDEX

5.2 SUBSTITUTION METHOD

	Form of the Integral	*Form of the Antiderivative*
1.	$\int [u(x)]^n \cdot u'(x)\,dx, \quad n \neq -1$	$\dfrac{[u(x)]^{n+1}}{n+1} + C$
2.	$\int e^{u(x)} \cdot u'(x)\,dx$	$e^{u(x)} + C$
3.	$\int \dfrac{u'(x)\,dx}{u(x)}$	$\ln \lvert u(x) \rvert + C$

5.4 FUNDAMENTAL THEOREM OF CALCULUS

Let f be continuous on the interval $[a, b]$, and let F be *any* antiderivative of f. Then

$$\int_a^b f(x)\,dx = F(b) - F(a) = F(x)\Big|_a^b.$$

6.1 INTEGRATION BY PARTS

If u and v are differentiable functions, then

$$\int u\,dv = uv - \int v\,du.$$

6.2 TRAPEZOIDAL RULE

Let f be a continuous function on $[a, b]$ and let $[a, b]$ be divided into n equal subintervals by the points $a = x_0, x_1, x_2, \ldots, x_n = b$. Then, by the trapezoidal rule,

$$\int_a^b f(x)\,dx \approx \left(\frac{b-a}{n}\right)\left[\frac{1}{2}f(x_0) + f(x_1) + \cdots + f(x_{n-1}) + \frac{1}{2}f(x_n)\right].$$

6.2 SIMPSON'S RULE

Let f be a continuous function on $[a, b]$ and let $[a, b]$ be divided into an even number of n of equal subintervals by the points $a = x_0, x_1, x_2, \ldots, x_n = b$. Then, by Simpson's rule,

$$\int_a^b f(x)\,dx \approx \frac{b-a}{3n}[f(x_0) + 4f(x_1) + 2f(x_2) + 4f(x_3) + \cdots + 2f(x_{n-2}) + 4f(x_{n-1}) + f(x_n)].$$

6.5 IMPROPER INTEGRALS

If f is continuous on the indicated interval and if the indicated limits exist, then

$$\int_a^{\infty} f(x)\,dx = \lim_{b\to\infty} \int_a^b f(x)\,dx,$$

$$\int_{-\infty}^{b} f(x)\,dx = \lim_{a\to-\infty} \int_a^b f(x)\,dx,$$

$$\int_{-\infty}^{\infty} f(x)\,dx = \int_{-\infty}^{c} f(x)\,dx + \int_c^{\infty} f(x)\,dx,$$

for real numbers a, b, and c.

7.2 PARTIAL DERIVATIVES (INFORMAL DEFINITION)

The partial derivative of f with respect to x is the derivative of f obtained by treating x as a variable and y as a constant.

The partial derivative of f with respect to y is the derivative of f obtained by treating y as a variable and x as a constant.